METAL IONS IN BIOLOGICAL SYSTEMS

VOLUME 31

Vanadium and Its Role in Life

METAL IONS IN BIOLOGICAL SYSTEMS

Edited by

Helmut Sigel
and Astrid Sigel

Institute of Inorganic Chemistry
University of Basel
CH-4056 Basel, Switzerland

VOLUME 31

Vanadium and Its Role in Life

Marcel Dekker, Inc. New York • Basel • Hong Kong

ISBN: 0–8247–9383–8

This book is printed on acid-free paper.

MARCEL DEKKER, INC.
270 Madison Avenue, New York, New York 10016

Current printing (last digit):
10 9 8 7 6 5 4 3 2 1

PRINTED IN THE UNITED STATES OF AMERICA

Preface to the Series

Recently, the importance of metal ions to the vital functions of living organisms, hence their health and well-being, has become increasingly apparent. As a result, the long-neglected field of "bioinorganic chemistry" is now developing at a rapid pace. The research centers on the synthesis, stability, formation, structure, and reactivity of biological metal ion-containing compounds of low and high molecular weight. The metabolism and transport of metal ions and their complexes is being studied, and new models for complicated natural structures and processes are being devised and tested. The focal point of our attention is the connection between the chemistry of metal ions and their role for life.

No doubt, we are only at the brink of this process. Thus, it is with the intention of linking coordination chemistry and biochemistry in their widest sense that the Metal Ions in Biological Systems series reflects the growing field of "bioinorganic chemistry." We hope, also, that this series will help to break down the barriers between the historically separate spheres of chemistry, biochemistry, biology, medicine, and physics, with the expectation that a good deal of future outstanding discoveries will be made in the interdisciplinary areas of science.

Should this series prove a stimulus for new activities in this fascinating "field," it would well serve its purpose and would be a satisfactory result for the efforts spent by the authors.

Fall 1973

Helmut Sigel
Institute of Inorganic Chemistry
University of Basel
CH-4056 Basel, Switzerland

Preface to Volume 31

Vanadium was discovered in the first part of the nineteenth century. The earth's crust contains 0.014% of this element, which corresponds approximately to the abundance of zinc. With a concentration of 20–35 nM in seawater, vanadium is the most abundant transition metal ion in the aquasphere.

The physiological effects of vanadium are in many instances a consequence of the excellent complexation properties of the oxovanadium(IV) or vanadyl ion (VO^{2+}), which is one of the most stable diatomic ions known, as well as of the chemical similarity between phosphate and vanadate(V). In addition, there are organisms such as ascidians in which the principal oxidation state of vanadium is +III.

An aqueous solution of vanadate contains a series of ionic oxovanadium(V) species, whose relative abundances depend on concentration, pH, ionic strength, temperature, and the presence of other substances. The same is true for vanadium(III) and vanadium(IV) species. In this connection the high flexibility of the coordination sphere of VO^{2+} also needs to be emphasized, as this property undoubtedly accounts for the large number of VO^{2+}–protein complexes. These and further aspects of the various oxidation states of vanadium and their binding properties to biogenic ligands including nucleotides and proteins are covered in the first seven chapters, demonstrating that the development of the chemistry of vanadium and its compounds is still in full bloom, although much has been achieved during the past decade. Similarly, exciting discoveries in nature are still being made, such as the detection of high levels of vanadium in the fan worm *Pseudopotamilla occelata* in 1993.

The next seven chapters deal with topics such as the inhibition and stimulation of enzyme activity by vanadium species, the role of vanadium in haloperoxidases and nitrogenases (discovered about a decade ago), and the occurrence of vanadium in the mushroom *Amanita muscaria*, in ascidians, and in the above-mentioned polychaete worm.

Despite all the accumulated knowledge, vanadium still lacks a clearly defined role in animal systems. It has been touted as an essential trace element, but it is not yet unequivocally accepted as an essential nutrient for higher animals. To date, the best evidence for a biological role of vanadium does not come from the animal kingdom, but from bacteria (nitrogen fixation in *Azotobacter* species) and from plants (haloperoxidases in marine algae). If vanadium has a functional role in higher animals, it is clear that only ultratrace levels are required to fulfill it. These issues as well as the transport and metabolism of vanadium, its toxicity, and dietary intake are considered in Chapters 15 and 16, while the insulin-mimetic properties and antitumor activities of certain vanadium compounds are treated in Chapters 17 and 18.

The volume closes with a chapter discussing the spectroscopic methods used successfully to prove the nature of vanadium centers in a variety of biological systems and a final chapter summarizing analytical methods for the determination of vanadium at trace levels in biological and environmental materials, a topic still controversial with respect to the validity and applicability of certain methods.

Helmut Sigel
Astrid Sigel

Contents

Contributors

Numbers in parentheses indicate the pages on which the authors' contributions begin.

Enrique J. Baran Quimica Inorgánica (QUINOR), Facultad de Ciencias Exactas, Universidad Nacional de la Plata, C.C. 962, 1900-La Plata, Argentina (FAX: 54-21-259485) (129)

Mary Battell Faculty of Pharmaceutical Sciences, University of British Columbia, 2146 East Mall, Vancouver, BC, V6T 1Z3, Canada (FAX: 1-604-822-3035) (575)

Ernst Bayer Institut für Organische Chemie der Universität, Auf der Morgenstelle 1, D-72076 Tübingen, Germany (FAX: 49-7071-295246) (407)

Martin Boddin Hochschule für Technik, Wirtschaft und Kultur (FH) Leipzig, Fachbereich Informatik, Mathematik und Naturwissenschaften, FG Chemie, Postfach 66, D-04251 Leipzig, Germany (FAX: 49-341-5629-316) (45)

N. Dennis Chasteen Department of Chemistry, University of New Hampshire, Parsons Hall, Durham, NH 03824-3598, USA (FAX: 1-603-862-4278) (231)

David Collison Department of Chemistry, University of Manchester, Oxford Road, Manchester, M13 9PL, UK (FAX: 44-61-2754616) (617)

Debbie C. Crans Department of Chemistry and Cell and Molecular Biology Program, Colorado State University, Fort Collins, CO 80523, USA (FAX: 1-303-491-1801) (147/287)

Cirila Djordjevic Department of Chemistry, The College of William and Mary, P.O. Box 8795, Williamsburg, VA 23185-8795, USA (FAX: 1-804-221-2715) (595)

Robert R. Eady AFRC-IPSR Nitrogen Fixation Laboratory, The University of Sussex, Falmer, Brighton, BN1 9RQ, UK (FAX: 44-273-678-133) (363)

Patrick Frank Department of Chemistry, Stanford University, Stanford, CA 94305-5080, USA (FAX: 1-415-723-4817) (423)

C. David Garner Department of Chemistry, University of Manchester, Oxford Road, Manchester, M13 9PL, UK (FAX: 44-61-2754616) (617)

Keith O. Hodgson Department of Chemistry, Stanford University, Stanford, CA 94305-5080, USA (FAX: 1-415-723-4817) (423)

Toshiaki Ishii Division of Marine Radioecology, National Institute of Radiological Sciences, 3609 Isozaki, Nakaminato, Ibaraki 311-12, Japan (FAX: 81-292-65-9214) (491)

Kan Kanamori Department of Chemistry, Faculty of Science, Toyama University, Gofuku 3190, Toyama 930, Japan (45)

Kenneth Kustin Department of Chemistry, Brandeis University, Waltham, MA 02254-9110, USA (FAX: 1-617-736-2516) (511)

Frank E. Mabbs Department of Chemistry, University of Manchester, Oxford Road, Manchester, M13 9PL, UK (FAX: 44-61-2754616) (617)

Marvin W. Makinen Department of Biochemistry and Molecular Biology, The University of Chicago, 920 East 58th Street, Chicago, IL 60637, USA (FAX: 1-312-702-0439) (89)

John H. McNeill Faculty of Pharmaceutical Sciences, University of British Columbia, 2146 East Mall, Vancouver, BC, V6T 1Z3, Canada (FAX: 1-604-822-3035) (575)

Roland Meier Hochschule für Technik, Wirtschaft und Kultur (FH) Leipzig, Fachbereich Informatik, Mathematik und Naturwissenschaften, FG Chemie, Postfach 66, D-04251 Leipzig, Germany (FAX: 49-341-5629-316) (45)

Steffi Mitzenheim Hochschule für Technik, Wirtschaft und Kultur (FH) Leipzig, Fachbereich Informatik, Mathematik und Naturwissenschaften, FG Chemie, Postfach 66, D-04251 Leipzig, Germany (FAX: 49-341-5629-316) (45)

Andras Muhlrad Department of Oral Biology, The Hebrew University–Hadassah School of Dental Medicine, 91010 Jerusalem, Israel (211)

Devkumar Mustafi Department of Biochemistry and Molecular Biology, The University of Chicago, 920 East 58th Street, Chicago, IL 60637, USA (FAX: 1-312-702-0439) (89)

Izumi Nakai Department of Chemistry, University of Tsukuba, Tsukuba 305, Japan (491)

Koji Nakanishi Department of Chemistry, Columbia University, New York, NY 10027, USA (FAX: 1-212-932-1289) (423)

Forrest H. Nielsen USDA, ARS, Grand Forks Human Nutrition Research Center, 2420 2nd Avenue North, P.O. Box 9034, University Station, Grand Forks, ND 58202-9034, USA (FAX: 1-701-795-8395) (543)

Kenji Okoshi Department of Biotechnology, Senshu University of Ishinomaki, Ishinomaki 986, Japan (491)

Chris Orvig Medicinal Inorganic Chemistry Group, Department of Chemistry, University of British Columbia, 2036 Main Mall, Vancouver, BC, V6T 1Z1, Canada (FAX: 1-604-822-2847) (575)

Dieter Rehder Institut für Anorganische und Angewandte Chemie der Universität, Martin-Luther-King-Platz 6, D-20146 Hamburg, Germany (FAX: 49-40-4123-2893) (1)

Israel Ringel Department of Pharmacology, The Hebrew University–Hadassah Medical School, P.O.B. 1172, 91010 Jerusalem, Israel (FAX: 972-2-431094) (211)

William E. Robinson Environmental Sciences Program, University of Massachusetts—Boston, 100 Morrissey Blvd., Boston, MA 02125-3393, USA (511)

Daniel E. Ryan Department of Chemistry, Columbia University, New York, NY 10027, USA (FAX: 1-212-932-1289) (423)

Hans G. Seiler Institut für Anorganische Chemie, Universität Basel, Spitalstrasse 51, CH-4056 Basel, Switzerland (FAX: 41-61-3221554) (671)

Mitchell J. Smith Natural Products Division, CFSAN, Food and Drug Administration (HFS-345), 200 C Street, SW, Washington, DC 20204, USA (FAX: 1-202-205-4758 or 4422) (423)

Paul J. Stankiewicz Department of Chemistry, Simon Fraser University, Burnaby, British Columbia, V5A 1S6, Canada (FAX: 1-604-291-3765) (249/287)

Katherine H. Thompson Faculty of Pharmaceutical Sciences, University of British Columbia, 2146 East Mall, Vancouver, BC, V6T 1Z3, Canada (FAX: 1-604-822-3035) (575)

Alan S. Tracey Department of Chemistry, Simon Fraser University, Burnaby, British Columbia, V5A 1S6, Canada (FAX: 1-604-291-3765) (249/287)

Hans Vilter Gesellschaft für Biotechnologische Forschung mbH, Mascheroder Weg 1, D-38124 Braunschweig, Germany. Present address: Zurmaienerstrasse 16, D-54292 Trier, Germany (Tel: 49-651-27494) (325)

Contents of Previous Volumes

*Out of print.

Volume 11. Metal Complexes as Anticancer Agents

Volume 12. Properties of Copper

---

*Out of print.

Volume 16. Methods Involving Metal Ions and Complexes in Clinical Chemistry

Volume 17. Calcium and Its Role in Biology

Volume 18. Circulation of Metals in the Environment

Volume 21. Applications of Nuclear Magnetic Resonance to Paramagnetic Species

Volume 24. Aluminum and Its Role in Biology

Volume 25. Interrelations Among Metal Ions, Enzymes, and Gene Expression

Volume 26. Compendium on Magnesium and Its Role in Biology, Nutrition, and Physiology

Volume 27. Electron Transfer Reactions in Metalloproteins

Volume 28. Degradation of Environmental Pollutants by Microorganisms and Their Metalloenzymes

Volume 29. Biological Properties of Metal Alkyl Derivatives

Volume 33: Probing of Nucleic Acids by Metal Ion Complexes of Small Molecules (tentative)

Other volumes are in preparation.

Comments and suggestions with regard to contents, topics, and the like for future volumes of the series are welcome.

The following Marcel Dekker, Inc. books are also of interest for any reader dealing with metals or other inorganic compounds:

Handbook on Toxicity of Inorganic Compounds
edited by Hans G. Seiler and Helmut Sigel, with Astrid Sigel
In 74 chapters, written by 84 international authorities, this book covers the physiology, toxicity, and levels of tolerance, including prescriptions for detoxification, for all elements of the Periodic Table (up to atomic number 103). The book also contains short summary sections for each element, dealing with the distribution of the elements, their chemistry, technological uses, and ecotoxicity as well as their analytical chemistry.

Handbook on Metals in Clinical and Analytical Chemistry
edited by Hans G. Seiler, Astrid Sigel, and Helmut Sigel
This book is written by 80 international authorities and covers over 3500 references. The first part (15 chapters) focuses on sample treatment, quality control, etc., and on the detailed description of the analytical procedures relevant for clinical chemistry. The second part (43 chapters) is devoted to a total of 61 metals and metalloids; all these contributions are identically organized covering the clinical relevance and analytical determination of each element as well as, in short summary sections, its chemistry, distribution, and technical uses.

METAL IONS IN BIOLOGICAL SYSTEMS

VOLUME 31

Vanadium and Its Role in Life

1
Inorganic Considerations on the Function of Vanadium in Biological Systems

Dieter Rehder

Institut für Anorganische und
Angewandte Chemie der Universität
Martin-Luther-King-Platz 6
D-20146 Hamburg, Germany

1. INTRODUCTION

1.1. Discovery, Occurrence, and Uses of Vanadium

Vanadium was discovered in 1802/1803 by the Spanish mineralogist A. M. Del Rio in Mexico in brown lead ore and named "panchromo," a name referring to the "panchromism" of vanadium when passing through its various oxidation states. It was renamed "eritrono" (erythronium) by Del Rio because "salts of the new element turned red when treated with fire and acids". Shortly thereafter, becoming convinced that he had merely discovered a new occurrence for chromium, he withdrew his discovery. The element was rediscovered in 1831 by the Swedish chemist N. G. Sefström in smelting products from iron ores of southern Sweden and named "vanadin" (vanadium) in honor of Vanadis, the northern Germanic goddess of beauty, more commonly known as Freya. In the same year, Wöhler established the identity of erythronium and vanadium [1].

The vanadium content of seawater, where the ions $H_2VO_4^-$ and, to a lesser extent, HVO_4^{2-} and $NaHVO_4^-$ are present, amounts to 20–35 nM. Vanadium is thus the most abundant transition metal in the aquasphere. The earth's crust contains 0.014% of vanadium, which is about the abundance of zinc. Vanadium ores is rare. Examples are patronite (VS_4) and vanadinite ($Pb_5(VO_4)_3Cl$). Vanadinite is isomorphous with chloroapatite [$Ca_5(PO_4)_3Cl$], and was a major component in the brown lead ore studied by Del Rio. Vanadium is widely dispersed in iron ores, clays, and basalts. The main source for vanadium smelting is titanomagnetite, which contains up to 0.8% vanadium. Another source for vanadium, with increasing importance for environmental problems, is fossilized organic matter, mainly crude oil, oil shales, and bituminous limestone. Crude oil may contain up to 4% vanadium mainly in the form of VO^{2+} porphyrins [2]. These are transformed to V_2O_5 on burning which, when emitted, may cause health hazards and catalyzes the conversion of SO_2 to SO_3 (and hence to sulfuric acid). Elemental vanadium is rarely used. Its main source of application is as an additive in steel (ferrovanadium). V_2O_5 is used as a catalyst in the industrial production of sulfuric acid, and vanadium oxides, vanadates, oxovanadium halides and alkoxides are employed in various catalytically conducted oxidation, condensation, and polymerization reactions (see Sec. 5). The system V_2O_5-GeO_2 is used in semiconductor techniques. A few vanadium compounds seem to have therapeutic effects. Among these are insulin-mimetic effects of vanadates, peroxovanadates [3], and various coordination compounds of V^V and V^{IV} [4], and antitumor effects of vanadocene dichloride [5] and polyoxovanadates [6]. VO^{2+} complexes of hydrazones have been shown to be active against mycobacteria causing tuberculosis [7a]. More generally, the pharmacological effects of vanadium, important also for the assessment of its toxicity, include a diuretic and natriuretic potential, lowering of cholesterol and glucose levels, and the contraction of blood vessels, resulting in cardiac effects [7b].

1.2. Availability and Physiological Effects

Physiological effects are in many cases a consequence of the good complexation behavior of the vanadyl cation (VO^{2+}), on the one hand, and the chemical similarity between phosphate and vanadate, on the other hand. Vanadyl ($[VO(H_2O)_4OH]^+$) and vanadate ($H_2VO_4^-$) are the main species present in solution under physiological conditions (pH 7, c(V) in human tissues = 0.1–1 μM). Other vanadium species, like the cations VO_2^+, VO^{3+}, VO^{2+}, V^{4+}, and V^{3+}, are available in neutral media as coordination compounds only. The free ions $[VO(H_2O)_5]^{2+}$ and

$[V(H_2O)_6]^{3+}$ undergo hydrolysis to mono- and dinuclear hydroxy species (cf. the stability constants for the protonation/deprotonation equilibria in Table 1), and eventually hydroxides precipitate. $[VO_2(H_2O)_4]^+$, the species present in strongly acidic solution, converts to vanadates beyond pH 2 (see Sec. 2).

Vanadate and vanadyl are joint by the redox couple in Eq. (1):

$$H_2VO_4^- + 4H^+ + e^- \rightleftharpoons VO^{2+} + 3H_2O;\ E^0 = 1.31\ \text{V} \qquad \qquad E_{\text{pH}=7} = -0.34\ \text{V} \tag{1}$$

Comparison with the redox potentials of the couples $H_2O/(1/2O_2 + 2H^+)$ ($E = +1.24$ V at pH 7) and $(NADH + H^+)/(NAD^+ + 2H^+)$ ($E = -0.32$ V at pH 7) demonstrates the ease of physiological interconversion between V^V and V^{IV}. Vanadate appears to be the more toxic species, and its intracellular reduction hence contributes to its detoxification. As a competitor for phosphate, vanadate inhibits and stimulates many phosphate-metabolizing enzymes. A well-documented example is the inhibition of the sodium-potassium pump (Na,K-ATPase) [9a]. While there are many parallels between vanadate and phosphate in their chemistry and biological action, there are also notable differences. Thus, vanadate ($H_2VO_4^-$) carries one negative charge, phosphate two (HPO_4^{2-}); vanadate tends to oligomerize at concentrations above 0.1 mM, phosphate does not; vanadium, as a transition metal, has available

TABLE 1

Equilibrium Constants for the Hydrolysis of Selected Vanadium Species [8]

Deprotonated form[a]	Protonated form[a]	$-\log K$
$[V(OH)]^{2+}$	V^{3+}	2.85
$[V(OH)_2]^+$	$[V(OH)]^{2+}$	3.85
$[V_2(OH)_2]^{4+}$	$2V^{3+}$	3.89
$[VO(OH)]^+$	VO^{2+}	6.0
$[(VO)_2(OH)_2]^{2+}$	$2VO^{2+}$	6.88
$H_2VO_4^-$	HVO_4^{2-}	7.91

[a]Water of hydration omitted.

low-lying d orbitals which can be used to stabilize the coordination numbers 5, 6, 7, and 8, while the biologically relevant phosphorus compounds exclusively are tetrahedral (phosphorus may attain the coordination number 5 in transition states, e.g., in the enzymatic hydrolysis of the phosphoester bond [9b]); further, vanadate, but not phosphate, is easily reducible under physiological conditions. Finally, the common view of a phosphate analog four-coordinate, tetrahedral vanadate has been challenged recently (see next section). Inhibitory effects have also been reported for V^{IV} [10], and both vanadate and VO^{2+} exhibit an insulin-mimetic effect [11].

1.3. Scope

In the present article, I will emphasize the aquatic chemistry of vanadate, on the one hand, and the complexation behavior of V^V (VO^{3+}, VO_2^+) and V^{IV} (V^{4+}, VO^{2+}), on the other hand. Many of these compounds attain model character for biogenic vanadium species such as vanadate-dependent haloperoxidases (V^V) from marine brown algae [12] and amavadin (V^{IV}) from the fly agaric toadstool [13,14]. Other oxidation states besides +V and +IV have been found in vanadium-accumulating organisms (ascidians [15,16] and fan worms [17]), where part of the vanadium is in the +III state), and proposed for the reduced form of vanadium nitrogenases (V^{II}) from *Azotobacter* [18]. Model compounds for these systems will be surveyed. The change of vanadium between different oxidation states, and processes employing vanadium catalysts, an area of interest in the context of enzymes containing vanadium in the active center (haloperoxidase and nitrogenase), will also be addressed in separate sections. Recent reviews covering various aspects of the role of vanadium in living organisms are referenced under [12] and [19–23].

2. VANADATES

The simplest forms of vanadate, orthovanadate VO_4^{3-} and metavanadate VO_3^-, exist in the solid state only or, in the case of orthovanadate, in strongly alkaline solutions. What is commonly termed "metavanadate" for solutions of $NaVO_3$ in water is actually a mixture of mono- and oligovanadates, the composition of which depends on the pH, the overall vanadium concentration, and the ionic strength of the medium. Mono-, di-, linear tetra- and decavanadate may be protonated, with the degree of protonation again depending on the pH. An overview of the species which have been detected in water by potentiometric and nuclear magnetic resonance (NMR) (mainly ^{17}O and ^{51}V [24]) techniques is given, together with the formation constants and δ (^{51}V) shift values, in Table 2. The main species present in solution, monovanadate, divanadate, cyclic tetra- and pentavanadate, interconvert on a millisecond scale; equilibration between these vanadates and decavanadate takes hours to days [26], i.e., decavanadate, once it has formed, decays only slowly even in the pH range (>6.3), where it is thermodynamically unstable. Not included in Table 2 are trivanadate ($V_3O_{10}^{5-}$), which appears to be present as a transient species [26] or at high ionic strengths only [24], and hexavanadate ($V_6O_{18}^{6-}$, $\delta = -586$ ppm), which is observed at high ionic strength [27]. A linear and a cyclic trivanadate ($V_3O_{10}^{5-}$ and $V_3O_9^{3-}$, respectively) have been assigned on the basis of Raman spectra in alkaline solutions [28a]. Tridecavanadate, a short-lived species in freshly acidified solutions of "*meta*-vanadate", has recently been described [28b]. In strongly acidic solutions, VO_2^+ is formed. In alkaline solutions (pH >9), where HVO_4^{2-} and $V_2O_7^{4-}$ coexist, one has to take into consideration effective ion pairing, i.e., the species $NaHVO_4^-$ and $NaV_2O_7^{3-}$ at $c(Na^+) > 0.1$ M [29].

^{51}V NMR spectroscopy is a powerful tool for the examination of vanadium speciation [24,30,31]. $\delta(^{51}V)$ values are sensitive to the nature of the ligand functions directly coordinated to vanadium and also, within a group of alike functions, to the coordination number and coordination geometry. Linewidth may be used for qualitative evaluation of the geometry, but they are also sensitive to steric factors and exchange phenomena. Figure 1 shows a set of typical ^{51}V NMR spectra, and Figure 2 contains a distribution and a predominance diagram, obtained from a quantitative evaluation by ^{51}V NMR and electromotoric force (EMF) measurements.

In the solid state, monovanadate (VO_4^{3-}, HVO_4^{2-}) attains the expected tetrahedral geometry (V_1' in Fig. 3), as has been shown by X-ray absorption [33] and X-ray diffraction spectroscopy [34], and this has

TABLE 2

Vanadate Speciation in Aqueous Solution[a]

Species	Abbrev.	p,q	$\log \beta_{pq}$[b]	pK_a	$\delta(^{51}V)$[c]
VO_4^{3-}	V_1	−2,1	−21.31		−541.2
HVO_4^{2-}	V_1'	−1,1	−7.91	13.4	−538.8[d]
$H_2VO_4^-$	V_1''	0,1		7.91	−560.4[d]
$V_2O_7^{4-}$	V_2	−2,2	−15.13		−561.0
$HV_2O_7^{3-}$	V_2'	1,2	−5.39	9.74	−563.5
$H_2V_2O_7^{2-}$	V_2''	0,2	2.90	8.29	−572.7
$V_4O_{13}^{6-}$		−2,4	−8.5		[e]
$HV_4O_{13}^{5-}$		−1,4	0.4	8.9	[e]
$V_4O_{12}^{4-}$	V_4	0,4	10.04		−577.6
$V_5O_{15}^{5-}$	V_5	0,5	12.43		−586.0
$V_{10}O_{28}^{6-}$	V_{10}	4,10	51.98		−422, −496, −513
$HV_{10}O_{28}^{5-}$	V_{10}'	5,10	58.12	6.14	−524, −500, −516
$H_2V_{10}O_{28}^{4-}$	V_{10}''	6,10	61.80	3.68	−425, −506, −524
$H_3V_{10}O_{28}^{3-}$	V_{10}'''	7,10	63.37	1.57	−427, −515, −534
VO_2^+		2,1	6.97		−545

[a]Ionic strength = 0.6 M Na(Cl). Data taken from [25], δ values from [24] and [30].
[b]$\log \beta$ for the equilibria $p\mathrm{H}^+ + q(\mathrm{H_2VO_4^-}) \rightleftharpoons (\mathrm{H^+})_p(\mathrm{H_2VO_4^-})_q$.
[c]Relative $VOCl_3$.
[d]−538.8 (δ_1) and −560.4 (δ_2) are limiting shift values. Observed shifts δ_{obs} corresponding to the position of the equilibrium $HVO_4^{2-} + H^+ \rightleftharpoons H_2VO_4^-$ can be calculated from $\mathrm{pH} = \mathrm{p}K_a + \log(\delta_2 - \delta_{obs})/(\delta_{obs} - \delta_1)$.
[e]Values between −566 and −585 ppm have been reported.

also been the assumed structure in solution, where monovanadate shows a sharp ^{51}V NMR signal, compatible with a vanishing electric field gradient at the ^{51}V nucleus under local T_d symmetry. Comparative electron absorption spectroscopy (UV-VIS) and magnetic circular dichroism (MCD) spectra, on the other hand, favor a pentacoordinate, trigonal-bipyramidal structure (V_1* in Fig. 3) [35]. The two alternatives point to an analogy between vanadate and phosphate, on the one hand,

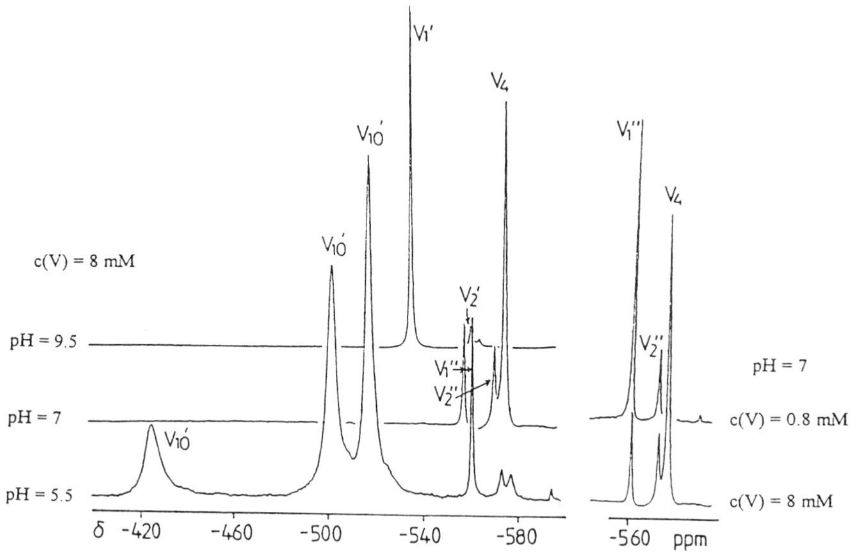

FIG. 1. 94.7-MHz ^{51}V NMR spectra of aqueous vanadate solutions at an ionic strength of 0.6 M Na(Cl), demonstrating the varying compositions with varying $c(V)$ and pH. See Table 2 for abbreviations.

and vanadate and the (trigonal-bipyramidal) transition state of the enzymatic splitting of a phospho-ester bond on the other hand. Vanadate may hence be considered a ground state analog *and* a transition state analog of phosphate [36]. The structures of di- (V_2), tetra- (V_4, V_4', and V_4*) and pentavanadate (V_5) (Fig. 3) are based on corner-linked tetrahedral VO_4 units, while the vanadium ions are centers of octahedra in the decavanadates (V_{10}'' in Fig. 3).

Although oligovanadates cannot be present at common physiological concentrations, the species V_n with $n > 1$ may be important at toxic levels, or in special cell compartments, where accumulation of vanadium occurs. Inhibitory actions of V_n have been noted in several cases. Examples are the inhibition of seminal fluid phosphatase by V_2 [41], of dehydrogenases and aldolases by V_2 and V_4 [42,43], of Cu-Zn-superoxide dismutase by V_4 [44], and of kinases and phosphorylases by V_{10} [45,46]. V_{10} has also been reported to inhibit the Ca^{2+} release induced by inositol triphosphate [47].

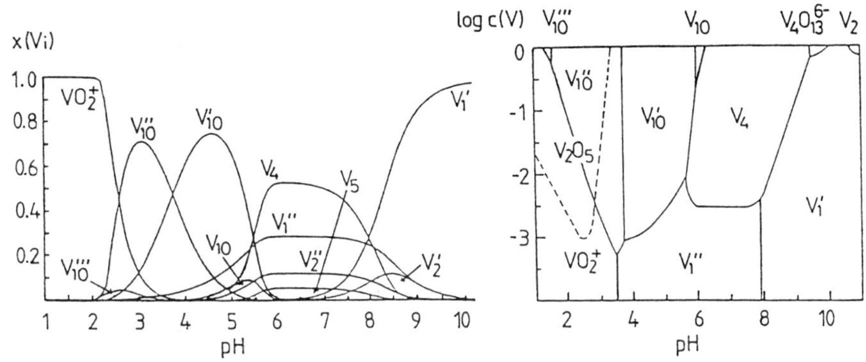

FIG. 2. Distribution diagram (left) and predominance diagram (right) for vanadium species in aqueous solution, $I = 0.6$ M Na(Cl). c(V) in (left) is 1.25 mM. See Table 2 for abbreviations. (Redrawn from Refs. [29] and [32].)

3. VANADIUM COMPOUNDS OF SOME INORGANIC AND RELATED MOLECULES

3.1. General

In this section, the main objective is to introduce vanadate- and vanadyl-phosphate complexes including vanadate analogs of ADP/ATP on the one hand, and peroxovanadates on the other hand. The latter are of considerable interest in the context of their insulin-mimetic effect [3], their possible anticancer effect [48], and as intermediates in vanadium redox chemistry (Sec. 4) and haloperoxidase activity [49–51]. The related hydroxylamido complexes, an organic derivative of which is present in amavadin from the fly agaric toadstool [13,14], will also be discussed here. Vanadate/vanadyl phosphate systems appear to be involved in enzyme activity. Thus, the 2,3-diphosphoglycerate phosphatase activity of phosphoglycerate mutase is stimulated by a mixed phosphate-vanadate anhydride more effectively than by simple vanadate [52], and vanadium analogs of NADPH have been found to be recognized as cofactors by alcohol dehydrogenase [53]. Other anions besides phosphate may interact with vanadate, as has been shown by the structural characterization of a tetranuclear cluster containing the V_4O_4 core, bridged by two bidentate sulfate groups [54]. Fe^{3+} may form

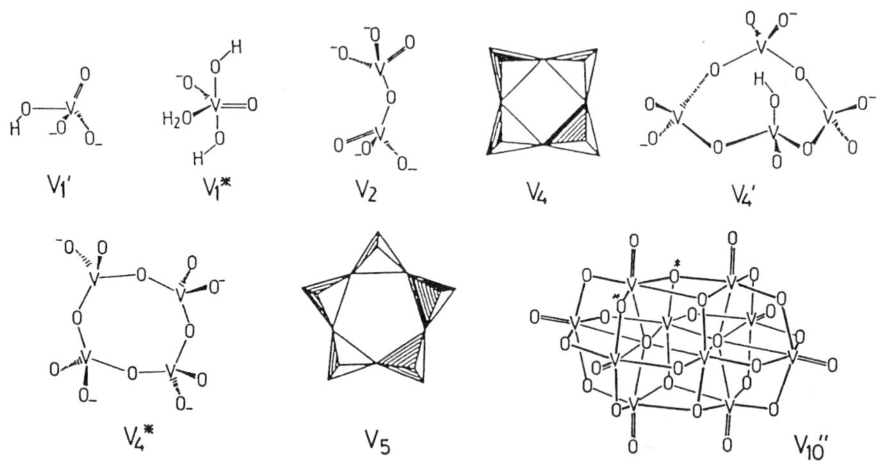

FIG. 3. Structures of vanadates. V_1' and V_2 are based on an X-ray structure analysis of $[Co(en)_3]_2[V_2O_7][HVO_4]\cdot 6H_2O$ [34], V_1^* $(H_2VO_4(H_2O)^-)$ has been proposed on the basis of MCD and electronic absorption spectra [35]. For the puckered (D_{2d}) V_4 and V_4', and the planar (D_{4h}) V_4^*, see [37–39], for [4-Et-py]$_4V_{10}''$ [40]; protonation sites in the latter are indicated by an asterisk.

iron vanadates and vanadatoferrates with V^V and V^{IV} [55], but the relevance of these compounds on a physiological level is debatable. Thiovanadates and other vanadium compounds containing the S^{2-} and S_2^{2-} ligands will be discussed in Sec. 6.1.1. (model compounds for the cofactor in vanadium nitrogenase). The interaction of vanadate with biogenic ligands in aqueous solution will be briefed in the appropriate subsections of Sec. 6. They are addressed in more detail in Chap. 5 of this book.

3.2. The Vanadium-Phosphorus System

Below pH 7, the main equilibrium species present in the system H^+/vanadate/phosphate is $H_4PV_{14}O_{42}^{5-}$ [32]. In the physiological pH range, mixed anhydrides between vanadate and phosphate of composition $HVPO_7^{3-}/H_2VPO_7^{2-}$ (VP), $HVP_2O_{10}^{4-}$ (VP_2), and $H_2VO_2(P_2O_7)_2^{5-}$ (VP_4) are formed, in rapid exchange with their hydrolysis products [56]. The formation constant K for the formation of VP [Eq. (2)] at pH 7.5

indicates that VP is less stable toward hydrolysis than divanadate by two orders of magnitude (cf. Table 2), but more stable than pyrophosphate by six orders of magnitude. A mixed vanadate-arsenate, analogous to VP, and with about the same formation constant has also been described [56].

$$V_1 + HPO_4^{2-} \rightleftharpoons VP; \qquad K = 11.7\ M^{-1} \tag{2}$$

While VP and VP_2 may be considered structural analogs to divanadate/diphosphate and trivanadate/triphosphate, the structure proposed for VP_4 (**1** in Fig. 4), an octahedron with two chelating pyrophosphates in the coordination sphere of vanadium [56], has no similarities with that of pentavanadate (V_5, Fig. 3). There is no direct structural information available on discrete mixed vanadate-phosphates. The structure reported for the vanadium-organophosphonate complex $V_2O_4(RPO_3)_2$ (R = 2-$MeC_6H_4CH_2$) (**2** in Fig. 4) consists of corner-sharing VO_4 and PRO_3 tetrahedra [57] and hence resembles cyclic tetravanadate (V_4, Fig. 3) and other cyclic V_4O_4 cores (vide infra).

The analogy between vanadate-phosphate anhydrides and di-/triphosphate extends to sugar phosphates [58,59] and nucleotides

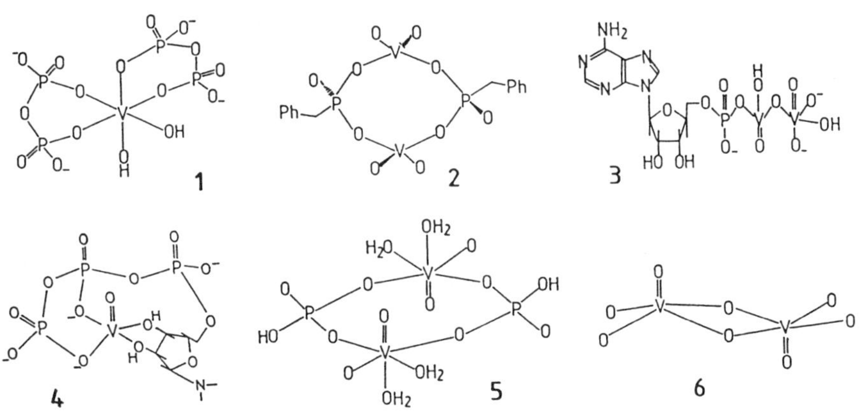

FIG. 4. Structures of vanadate and vanadyl phosphates. **1**, **3**, and **4** are assumed structures. The structures of **2** (a discrete cluster [57]) and **5** (the basic structural element of the polymer $VO(H_2O)_2(HPO_4)\cdot 2H_2O$ [65a]) have been determined by X-ray crystallography. **6** [65b] represents a structure element of the (1,0,0) surface of $(VO)_2P_2O_7$ (see Sec. 5 for discussion).

[59,60]. Adenosine monophosphate (AMP), e.g., in addition to forming cyclic esters with the ribose moiety (Sec. 6.2), picks up mono- and divanadate to yield AVP and AV_2P, which are analogous to ADP and ATP [Eqs. (3) and (4); **3** in Fig. 4]. The formation constants K at pH 7 compare to that for the formation of VP [Eq. (2)] [60].

$$V_1 + AMP \rightleftharpoons AVP; \qquad K = 9.2(7)\ M^{-1} \tag{3}$$

$$V_2 + AMP \rightleftharpoons AV_2P; \qquad K = 12(2)\ M^{-1} \tag{4}$$

Vanadyl ions VO^{2+} are more strongly bonded to ATP than vanadate(V). In the acidic to neutral region, a blue complex has been isolated, where the vanadium center is probably bonded to the β- and γ-phosphate *and* the ribose [61] as shown for **4** in Fig. 4. Bidentate coordination of the α- and β-phosphate of ATP, and bridging coordination of the γ-phosphate, have also been inferred from spectral evidence for the (presumably polymeric) complexes $VO(H_2ATP)(NN)$ (NN = 2,2′-bipyridyl or 1,10-phenanthroline) [62]. Evidence for the formation of vanadyl-phosphate complexes in aqueous solution has been obtained from electron paramagnetic resonance (EPR) and electron absorption spectroscopy [63]. Below pH 4.5, the complexes $VO(HPO_4^{2-}/H_2PO_4^-)$ are present, with the phosphate acting as a monodentate ligand. Above pH 4.5, precipitation and almost complete depletion of vanadium occurs. In the pH range of 8–11, $VO(PO_4^{3-})_2$ and $(OH)_2VO(PO_4^{3-})$ may exist, both with the phosphate in the bidentate coordination mode. Again, structure information on discrete molecules is not available. In the coordination polymers underlying the vanadyl phosphates $VOPO_4$ [64] and $VO(H_2O)_2HPO_4{\cdot}2H_2O$ [65a] (**5** in Fig. 4), highly distorted VO_6 octahedra are bridged by PO_4 tetrahedra. This is also the building principle in $(VO)_2P_2O_7$ [65b], a key compound in alkane oxidation catalyzed by VPO catalysts (see Sec. 5, and structure **6** in Fig. 4), and in the mixed valence compound $[(V^{IV}O)_2V^{III}(PO_4)_2(HPO_4)(H_2PO_4)(H_2O)_2]^{2-}$ [66].

3.3. Peroxovanadates and Hydroxylamido Complexes

Peroxide (O_2^{2-}) and hydroxylamide (H_2NO^-) including its organic derivatives are isoelectronic and exhibit a like complexation behavior insofar as either ligand may coordinate to a metal ion in the side-on fashion depicted schematically below. They are "electronically similar", meaning a similar effective ligand electronegativity and polarizability, as has been inferred from, inter alia, the ^{51}V NMR characteristics of comparable peroxo and hydroxylamido complexes (both ligands usually induce rather high ^{51}V shielding [67]; Table 3). While peroxide coordinates to V^V with and without additional ligand systems, the $V(H_2NO)$ moiety needs to be stabilized by supporting ligands.

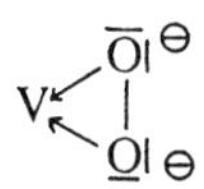

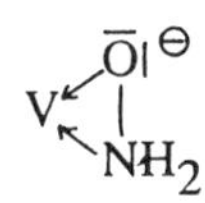

The composition of peroxo complexes is mainly influenced by the ratio vanadate/peroxide and by the pH. Of the approximately 15 peroxovanadates that have been found in solution, only three or four, listed in Table 3, exist at about physiological pH [24,68–70]. The main species

TABLE 3

$\delta(^{51}V)$ Chemical Shifts of Selected Oxoperoxo- and Oxohydroxylamidovanadium Complexes

Compound[a]	$\delta(^{51}V)$[b] (ppm)	Ref.
$[VO(O_2)H_2O]^-$	−694	69
$[VO(O_2)(H_2O)dipic]^-$	−595	67
$[VO(ONH_2)(H_2O)dipic]$	−672	67
$[VO(O_2)_2(OH)H_2O]^{2-}$	−739	69
$[VO(O_2)_2pic]^{2-}$	−573	67
$[VO(ONH_2)_2pic]$	−883	67
$[VO_2(O_2)_2]^{3-}$	−769	24
$[\{VO(O_2)_2\}_2\mu\text{-}OH]^{3-}$	−767	69
$[\{VO(ONEt_2)_2\}_2\mu\text{-}O]$	−711[c]	67

[a]In several cases, formulations and assignments are under debate.
[b]In water, if not indicated otherwise.
[c]In $CHCl_3$.

present at pH 7 and with excess peroxide carry two peroxo ligands per vanadium, i.e., $[VO(O_2)_2(OH)H_2O]^{2-}$ and $[\{VO(O_2)_2\}_2\mu\text{-}OH]^{3-}$ [69]. A large number of peroxo complexes containing, in addition to the peroxo group, other ligands such as F^-, oxalate, 2,2′-bipyridyl, picolinate (Table 3), and polyaminocarboxylates have been described; see ref. [68] for a summary. A solution of V_2O_5 in phosphoric acid, when treated with H_2O_2 and adjusted to pH 6, contains the anion $[VO(O_2)_2HPO_4]^{3-}$ [71], again demonstrating the preferential formation of bis(peroxo) complexes at ambient pH.

4. REDOX CHEMISTRY OF VANADIUM

Biologically relevant oxidation states of vanadium are V^{II}/V^{III} (in nitrogenase), V^{III}/V^{IV} (in sea squirts and fan worms), V^{IV} (in intracellular media and in amavadin), and V^{V} (in seawater, in extracellular media, and in the active center of vanadate-dependent haloperoxidases). The ease of the intercoversion between vanadate and vanadyl (cf. also Introduction) suggests the presence of both V^{V} and V^{IV} in the intra- *and* extracellular fluids. Half-lives for the air oxidation of V^{IV}-albumin and V^{IV}-transferrin complexes and for the reduction of V^{V} to V^{IV} by endogenous reducing agents in serum fall in the range 5–30 min at pH 7.5 [72]. Whether V^{III} may also be present inside cells of mammalian tissues does not seem to have been investigated systematically (V^{III} is EPR-silent and usually undetectable by ^{51}V NMR), but there is evidence for the possibility of intracellular reduction of V^{V} and V^{IV} to V^{III} [73]. Undoubtedly, V^{III} plays a role in invertebrates, indicated by, e.g., the existence of V^{III} in fan worms and the vanadocytes of sea squirts, the in vitro reduction of V^{V} to V^{IV} and V^{III} by tunichromes (pigments in the tunique of vanadium accumulating sea squirts) [74], and the reduction of vanadate to V^{IV} and V^{III} by *Pseudomonas vanadiumreductans* and *P. isachenkovii* in the presence of H_2, CO, or organic electron donors [75]. A point of interest in the context of biomineralization is the detection of a mineral similar to sherwoodite [essentially a calcium oxovanadate (V,IV)] in cultures of these vanadate-reducing bacteria [75b].

In Table 4, redox potentials are listed for simple redox couples in aqueous solution. A few more complex couples, involving strongly

TABLE 4

Redox Potentials E^0 (V) in Aqueous Media, Relative to the Hydrogen Electrode

Oxidized form[a]	Reduced form	E^0	Ref.
V^{3+}(ox)	V^{2+}(ox)	−0.89	76a
V^{3+}	V^{2+}	−0.242(2)[b]	76b
$VO^{2+} + 2H^+$	V^{3+}	0.309(2)[b]	76b
$[VO(edta)]^{2-} + 2H^+$	$[V(edta)]^-$	0.80	76a
$VO_2^+ + 2H^+$	VO^{2+}	1.016(2)[b]	76b
$H_2VO_4^- + 4H^+$	VO^{2+}	1.314	76a
VO(OH)rhod	VO(rhod)	0.51	20
$[VO_2(nta)]^{2-} + 2H^+$	$[VO(nta)H_2O]^-$	0.105	20
VO(ehgs)	$[VO(ehgs)]^-$	0.531	20

[a]For abbreviations of the ligands, see Abbreviations.
[b]In 1 M NaCl.

coordinating ligands, have been included in order to demonstrate the influence of ligation of biogenic or related ligands [73], a fact which is sometimes not adequately taken into account when discussing the redox behavior of the couples V^V/V^{IV} or V^{IV}/V^{III}. The aqueous redox chemistry of V^{III} will be addressed in detail in Chapter 2 of this book. I will therefore concentrate on selected examples related to the pair vanadate(V^V)/vanadyl(V^{IV}).

Despite what is predicted thermodynamically, V^{IV} is more easily oxidized by O_2 than V^{III}, the former hence being kinetically superior to the latter. In both cases, oxidation is inhibited by ATP and albumin, and accelerated by the siderophore desferrioxamine and by histidine [73].

NADH and NADPH are oxidized by VO^{2+} in the presence of oxygen, and a mechanism involving vanadium-bound superoxide and hydroperoxide intermediates [Eq. (5)] has been proposed [77]. The intermediate formation of O_2^-, HO_2^-, and H_2O_2 in O_2 oxidation of V^{IV} in alkaline solution was in fact suggested three decades ago [78]. H_2O_2 reacts with VO^{2+} according to Eq. (6) and hence analogously to organic hydroperoxides [Eq. (7)]. The rate constants for the formation of the V^{IV}-

alkyl and V^{IV}-hydroperoxide intermediates are $k = 4.2$ [Eq. (6), R = *t*Bu] and $k = 52\ M^{-1}$ [Eq. (7)], respectively [79].

$$V^{IV} + O_2 \rightarrow V^{V}\text{–}O_2^- \xrightarrow{O_2,\ NADH} V^{IV}\text{–OOH} + NAD^+ + O_2^- \tag{5}$$

$$VO^{2+} + H_2O_2 \rightarrow O{=}V\text{–OOH} \rightarrow VO_2^+ + HO \tag{6}$$

$$VO^{2+} + ROOH \rightarrow O{=}V\text{–OOR} \rightarrow VO_2^+ + RO \tag{7}$$

Whether vanadate(V) plays a role as a direct oxidizing agent or as a mediator in NAD(P)H oxidation is under debate. The existence of a vanadate-dependent NADH oxidase in cardiac cell membranes [80a] and the discovery that flavoenzymes can act as NAD(P)H-dependent vanadate reductases [80b] appear to support a functional role of vanadate [80]. Further support comes from the observation that vanadate is complexed to and reduced by the cofactor NADPH of yeast glucose-6-phosphate dehydrogenase, an enzyme which is inhibited by both V^V and V^{IV} [81a]. On the other hand, glycerol-3-phosphate dehydrogenase, which carries a thiol group in the cofactor (NAD^+) binding site, is inhibited by divanadate and tetravanadate by nonreductive interaction [42].

The microsomal reduction of vanadate by NADH is accompanied by the generation of free hydroxyl radicals without exogenous H_2O_2 [81b], as is the reaction of RNA-bound VO^{2+} with H_2O_2 [Eq. (8)] [81c].

$$VO^{2+}\text{–RNA} + H_2O_2 \rightarrow HO^{\bullet}\ (\rightarrow \text{uracil C6-centered radical}) \tag{8}$$

Cysteine residues in proteins are a main target for oxidation also by peroxide and/or V^V. The neat oxidation of cysteine to cystine by $[VO(O_2)_2pic]^{2-}$ [82] may be considered a model reaction for this oxidation. Several other redox reactions involving vanadium coordination compounds as catalysts are introduced in Sec. 5.2.

5. VANADIUM IN CATALYSIS

5.1. General

As has been shown in the preceding section, vanadate and vanadyl may act as mediators in NAD(P)-driven redox processes, a reaction type resembling in vitro hydrogenation and dehydrogenation/oxidation catalyzed by systems such as vanadium oxides, (hetero)polyoxovanadates, vanadium alkoxides, $VO(acac)_3$, and other coordination compounds. The substrates may be organic or inorganic, the best known example of the latter category being the catalytic conversion of SO_2 to SO_3 by oxygen in the presence of V_2O_5. Many of these redox-active systems are also effective in olefin polymerization.

Of the two enzymes with vanadium in the active site (a nitrogenase and a peroxidase), the vanadium center in vanadium nitrogenase is likely to change its oxidation state between +II and +III during turnover, while in vanadate-dependent peroxidase the +V state seems to be retained (the reduced V^{IV} enzyme is inactive). Nitrogenase catalyzes the reductive protonation of N_2 to NH_4^+ (plus small amounts of N_2H_4), of acetylene to ethylene and further to ethane, and also exhibits a hydrogenase activity. Vanadate-dependent haloperoxidase catalyzes the halogenation of organic molecules via the oxidation of inorganic halide by peroxide. In the absence of a suitable substrate, singlet oxygen evolves. Both enzymes hence belong to the category of oxidoreductases. The main objective of this section therefore is the treatment of model redox reactions catalyzed by V catalysts, but a brief survey of other classes of reactions will also be given.

5.2. Vanadium in Redox Catalysis

V^{IV} complexes (e.g., vanadyl oxalate) and the system VO_2^+/H^+ have been shown to exhibit "bromoperoxidase activity" in that they catalyze the bromination of organic substrates such as trimethoxybenzene in the presence of H_2O_2 and bromide [50,83]. Various reactions are known, where alkanes, cycloalkanes or aromates are oxidized to alcohols, aldehydes and ketones by H_2O_2 under catalytic action of vanadium compounds [84,85] (see Eq. (9a) for an example). The active catalyst in these cases probably is a peroxovanadium species. This is suggested by the fact that peroxo complexes such as $VO(O_2)pic(H_2O)_2$ [85] and $[VO(O_2)(ON)_2]^-$ (ON = pyrazine-2-carboxylate) [84b] are effective catalysts. Further, $VO(O_2)pic(H_2O)_2$ has been used in stoichiometric oxidation of alkenes [Eq. (9b)] [85b]. Based on mechanistic considerations and structure evidence, an open peroxo or hydroperoxo intermediate (see **I** to **III** in the schematic representation below) has been postulated as the active form, with a preference given to the biradical **II** [85b].

Epoxidation may also be carried out by organic hydroperoxides, as shown in Eq. (10), which is catalyzed by $VO(acac)_2$ and leads to the erythro isomer as the main product [86].

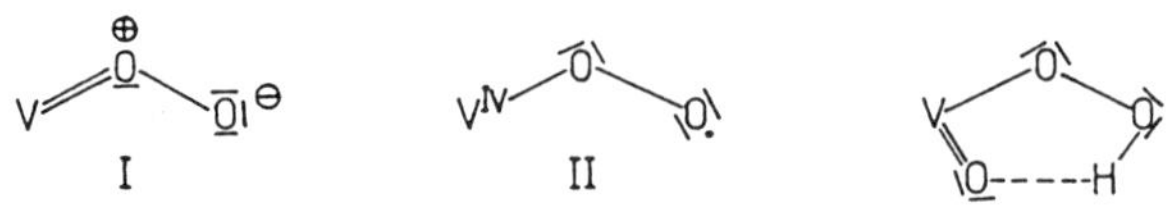

I II III

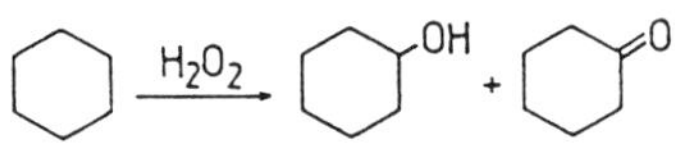

(9a)

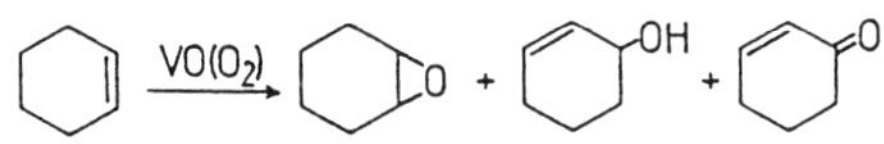

(9b)

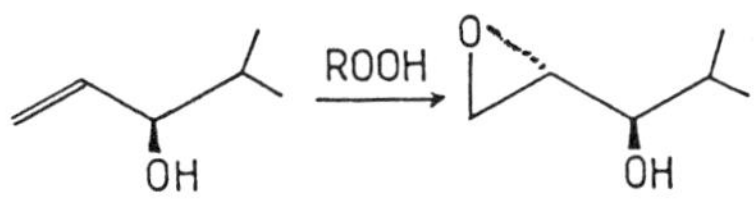

(10)

Oxidation of various organic substrates by oxygen can be carried out in the presence of vanadium oxides, oxides supported on silica gel [87a], divanadate [87b], phosphovanadates [65b,87b,88a,b], and other heteropolyvanadates [88c]. Vanadyl phosphate $[(VO)_2P_2O_7]$ has attained a key role in studies of the mechanism of catalysis, and the 14-electron oxidation of butane to maleic anhydride [Eq. (11)] has been employed as *the* model reaction [65b]. A certain number of V^V centers alongside the V^{IV} sites imposed by the composition of the catalyst are necessary for optimal activity. In the surface important for the catalytic oxidation, the (1,0,0) surface, the vanadium sites are essentially penta-coordinate (cf. **6** in Fig. 4). A furane intermediate as shown in Eq. (11)

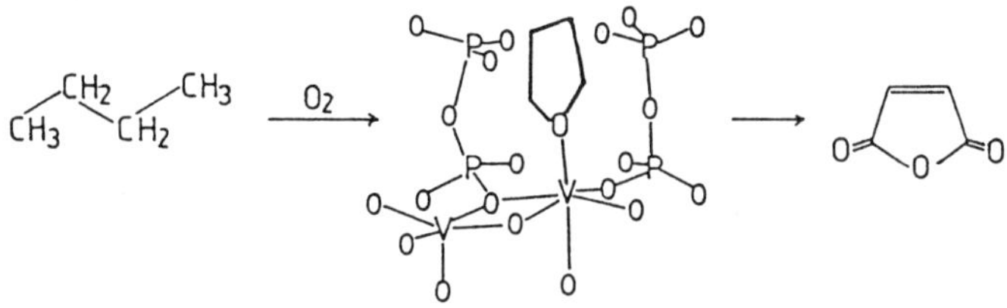

(11)

might form, oriented with surface-pendant pyrophosphate groups, which act as the sites that assist hydrogen abstraction from the furane molecule [65b].

Optically active complexes of composition [$V^{IV}/V^{V}O$(*ONNO*)] (where *ONNO* is a Schiff base of the salen type) catalyze the asymmetric oxidation of organic sulfides to sulfoxides with organic hydroperoxide [89] [Eq. (12)].

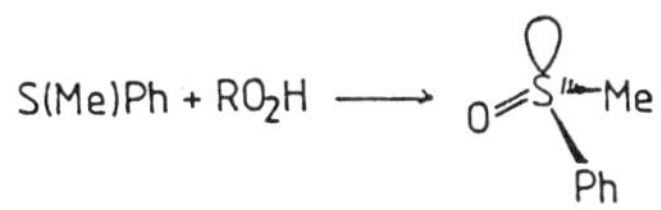
(12)

5.3. Vanadium Catalysts in Other Than Redox Reactions

The reactions shown in Eqs. (13) [90] and (14) [91] are catalyzed by the system Zn/VCl_3. The reducing agent is zinc. The species present in this system, [$\{VCl(THF)_3\}_2(\mu\text{-}Cl)_2$][$Zn_2Cl_6$] (a V^{II} compound), may be considered the catalyst, but an even lower-valent vanadium intermediate possibly is the active species, as suggested by the improvement of activity, if a phosphite (which stabilizes low-valent vanadium) is present as a cocatalyst in the case of the stereospecific reduction of dibromocyclopropane [Eq. (13)], or by employing $CpV(CO)_4$ (with V^I) as an alternative catalyst [90]. Equation (14) represents a coupling reaction leading to pinacol derivatives. In another example for reductive coupling, Eq. (15), the catalyst is the system Li[AlH_4]/VCl_3 [92]. Again, a low-valent vanadium complex is likely to be the catalytically active form. Equations (16)–(19) are selected examples for nonreductive reactions catalyzed by $V^{IV}O$(camphorate) (Eq. (16), [93]), $VOCl_2$(OR) (Eq. (17), [94]), V_2O_5 (Eq. (18); the addition of trifluoroacetic acid is selectively obeying Markovnikov's rule [95]), or $(Me_3SiCH_2)_2VO$(silsesquioxane) (Eq. (19), [96]).

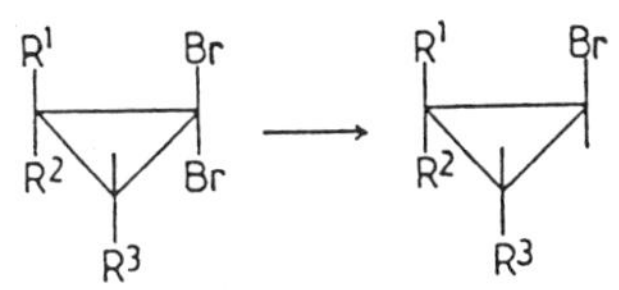
(13)

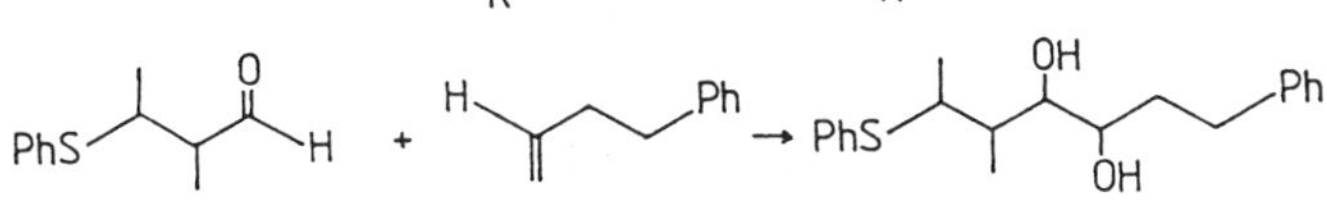
(14)

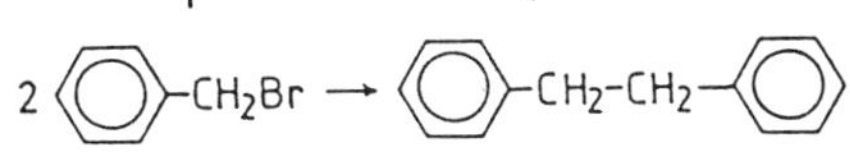
(15)

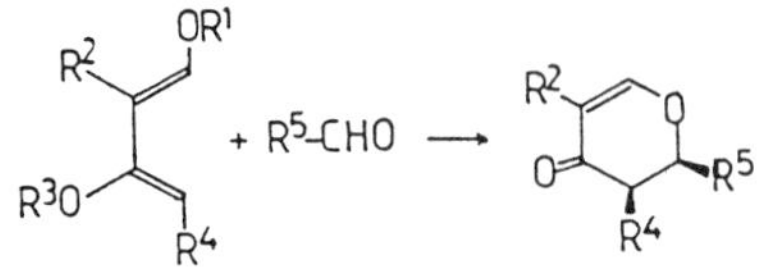

(16)

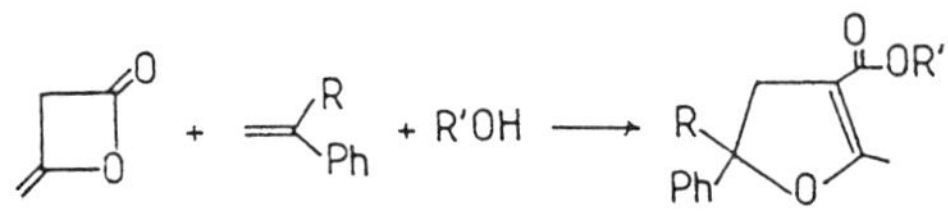

(17)

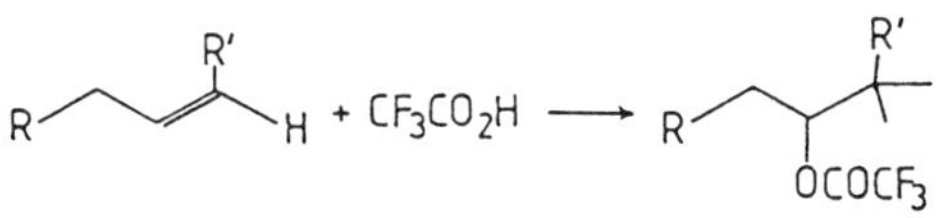

(18)

$$H_2C{=}CH_2 \longrightarrow -\!\!\left[H_2C{-}CH_2\right]\!\!-_n \tag{19}$$

6. COORDINATION COMPOUNDS AS MODELS FOR BIOGENIC VANADIUM SYSTEMS

With the increasing acceptance of a biological role of vanadium, the number of model compounds, or compounds which are being introduced as biomimetic complexes, has increased dramatically during the past few years. For this report, I have selected a small number of complexes exhibiting structural or functional similarities to known biogenic vanadium compounds, or to systems discussed as possible reactants between vanadate or vanadyl and physiologically available ligands.

6.1. Models for Vanadium Nitrogenase

6.1.1. Structural Models (Thio Ligation)

Thio analogs of vanadate and divanadate, i.e., HVS_4^{2-} and $V_2S_7^{4-}$, and a number of mixed oxothiovanadates have been known for some time and are often used as precursor compounds in self-assembly reactions leading to cuboidal clusters such as **7** [97a] and **8** [97b] in Fig. 5, which represent close structural models for the $FeVS_4$ cluster portions of the FeV cofactor in vanadium nitrogenase. Cluster **8** can be reversibly reduced by two one-electron redox steps, with the changes in electron density largely occurring in the Fe_3S_4 fragment. Comparison of the core fragments $[VFe_3S_4]^{2+}$ and $[MoFe_3S_4]^{3+}$ illustrates that the vanadium-

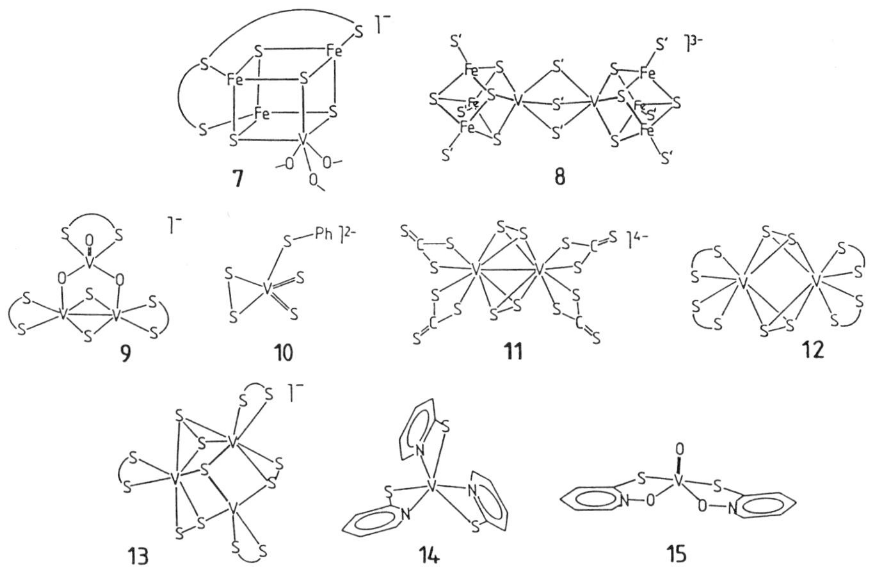

FIG. 5. Sulfido and thiolate complexes as structural models for the cofactor in vanadium nitrogenase (**7** and **8**) or the coordination environment of vanadium in this enzyme (**9–15**). *SSS* in **7** is a tridentate mercaptophenolate; the three *O* ligands on vanadium are dmf or dmso, S′ in **8** is SEt. The *SS* ligands in **9**, **12**, and **13** are *N,N*-dialkyldithiocarbamate ($R_2NCS_2^{2-}$).

containing cluster sustains the more reduced Fe_3S_4 fragment, a fact which might be of relevance when comparing functional differences between vanadium and molybdenum nitrogenases, imparted by the fine tuning of the hetero metal.

The clusters **7** and **8** contain the ligand μ_3-S^{2-}. μ_2-S^{2-} or terminal S^{2-} (as in simple thiovanadates) are constituents in the V^{IV} complex **9** [98] and the V^{V} complex **10** [99] (Fig. 5). An additional interesting feature in **10** is the dithioperoxo (disulfide) group, which becomes bridging, as **10** is reacted with CS_2 to form the dinuclear trithiocarbonato V^{IV} complex **11**. The two d^1 centers in **11** are coupled to the extent where the population of the triplet state is negligibly small [99]. μ-S_2^{2-} is also found in compounds **12** [100a] and **13** [100b]. **13** also contains a trebly bridging S^{2-}. The coordination compounds **14** [101] and **15** [102], with 2-mercaptopyridine and 2-mercaptopyridine-*N*-oxide, respectively, are examples for the hybrid coordination of ligands containing, alongside a thiolate group, an *N* or *O* function.

6.1.2. *Functional Models*

There are many batch reactions with systems containing low-valent metal ions which are able to reduce nitrogen. In order to achieve the status of a functional model for nitrogenase, a few conditions are to be fulfilled: (1) The compound should be defined; (2) at least one of the products (NH_3 and some N_2H_4 in the case of vanadium nitrogenase) produced by enzymatic action should be formed in appreciable yields; (3) the electron flow should involve the metal center; and (4) intermediates should be backed up by (related and) well-characterized complexes.

V^{II} in the presence of dicarboxylic acids and in slightly alkaline media reduces N_2 to hydrazine [Eq. (20)]. A model has been proposed (**16** in Fig. 6), according to which activation of the nitrogen occurs via dinuclear, N_2-bridged intermediates [103]. μ-N_2 complexes of vanadium in low-oxidation states have in fact been synthesized (**17–19**) [104–106],

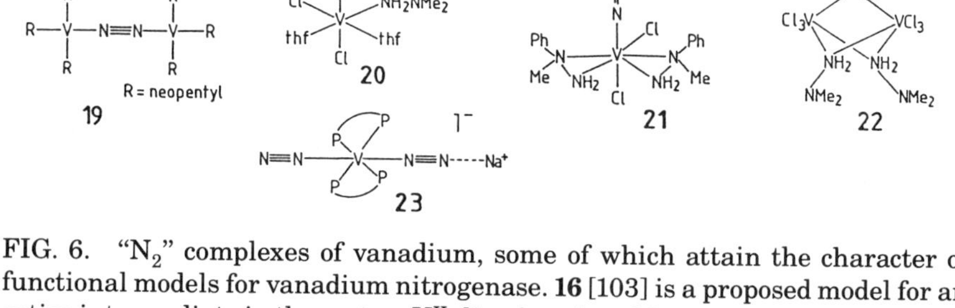

FIG. 6. "N_2" complexes of vanadium, some of which attain the character of functional models for vanadium nitrogenase. **16** [103] is a proposed model for an active intermediate in the system V^{II}-dicarboxylate-N_2. **17** [104], **18** [105], and **19** [106] are formulated here as μ-N_2 complexes; alternatively, they might be formulated as diazenido(2-) complexes. For **20** and related complexes, see [107]. **21** [108] contains two side-on bonded hydrazines and one end-on hydazido(2-) (V^V) or isodiazene (V^{III}), **22** [109] three end-on bridging hydrazines. The *PP* ligand in **23** [110] is bis(diphenylphosphino)ethane (dppe).

although with other ligand systems, i.e., those forming vanadium-carbon σ bonds. **17** reacts with protic solvents to form ammonia [104b] [Eq. (21)], while with **18**, about half of the complexed N_2 is converted to ammonia and hydrazine. No NH_3 or N_2H_4 is formed on treatment of **19** with acids. This compound, which has been added here for completeness, therefore does not attain the status of a functional model.

$$4\,V^{II} + N_2 + 5\,H^+ \rightarrow 4\,V^{III} + N_2H_5^+ \tag{20}$$

$$2\,V^{II} + N_2 + 2\,H^+ \rightarrow 2\,V^{III} + 2/3\,NH_3 + 2/3\,N_2 \tag{21}$$

The bond distances d(NN) in **17–19**, 1.23–1.28 Å, can be interpreted in terms of a diazenido(2-) rather than a dinitrogen ligand [d(NN) in free N_2 is 1.10 Å]. Diazene or hydrazine/hydrazide are likely intermediates in N_2 fixation. Compounds **20–22** [107–109] in Fig. 6, containing organic derivatives of these ligands, may be considered to model such intermediates.

The intermediate formation of hydrazide or hydrazinyl has also been proposed for the formation of ammonia (plus small amounts of hydrazine) on treatment of **23** with H^+ [Eq. (22)] [110]. As evidenced by $d(NN) = 1.13$ Å, **23** is a true bis(dinitrogen) complex, containing N_2 in the end-on, nonbridging coordination mode (although close ion pair contacts with Na^+ are essential for its stability). In Eqs. (20)–(22), the electrons for the reduction of N_2 are delivered by the metal ion, hence a metal-mediated reductive protonation of N_2.

$$V^{-I}(N_2)_2 + 3\,H^+ \rightarrow V^{II}(NHNH_2) + N_2 \xrightarrow{+\,H^+} V^{II} + NH_4^+\ (+\ N_2H_5^+) + \tfrac{1}{2}\,N_2 \tag{22}$$

Reduction of N_2 to NH_3 or N_2H_4 (depending on the reaction conditions) by the system V^{II}-Mg^{2+}-OH^- has long been known by the work of, inter alia, Shilov et al. [111], who recently succeeded in isolating a complex V^{II}-V^{III} compound of composition $Na_2V_4(OMe)_2(cat)_4$-$(Hcat)_2(MeOH)_3$ [cat = 3,5-bis(*tert*-butyl)catecholate(2-)], which, in methanolic solution, evolves H_2 and NH_3 according to Eqs. (23) and (24) [112] and hence may be viewed at as functional models for both the nitrogenase and hydrogenase activity of vanadium nitrogenase.

$$2\,V^{II} + 2\,MeOH \rightarrow 2\,V^{III} + H_2 + 2\,MeO^- \tag{23}$$

$$6\,V^{II} + N_2 + 6\,MeOH \rightarrow 6\,V^{III} + 2\,NH_3 + 6\,MeO^- \tag{24}$$

6.2. Carboxylato, Alkoxo, and Aryloxo Complexes

The functionalities RCO_2^- and RO^- exhibit an important binding capacity to oxophilic V^V (and V^{IV}). Model ligands usually contain one or more of these groups. As far as vanadium-protein interaction is concerned, they are provided by the side chain functions of aspartate and glutamate, serinate and tyrosinate. Under this subsection, I will review homoleptic systems (formally) derived from vanadate(V), i.e., complexes formed with alcohols and phenols (vanadate esters), carboxylates (mixed anhydrides of orthovanadic and carboxylic acids), and hydroxycarboxylates. A few structurally characterized complexes of V^{IV} and V^{III} will be included. Amino acid complexes will be treated, together with peptide complexes, in Sec. 6.3.2.

In aqueous media, which will be dealt with in more detail in Chaps. 5 and 8 of this book, complex formation between vanadate and these ligands competes with the (usually much stronger) tendency of monovanadate to oligomerize (Sec. 2). Simple monofunctional alcohols and phenols form tetrahedral mono- and diesters, typically of composition $HVO_3(OR)^-$ (**24** in Fig. 7) and $VO_2(OR)_2^{2-}$, and complex formation constants K around 0.1 (alcohols) and 1 M^{-1} (phenols) [113]. The monoesters underly protonation/deprotonation equilibria. Esters derived from divanadate have also been reported. Glycol and other bi- and

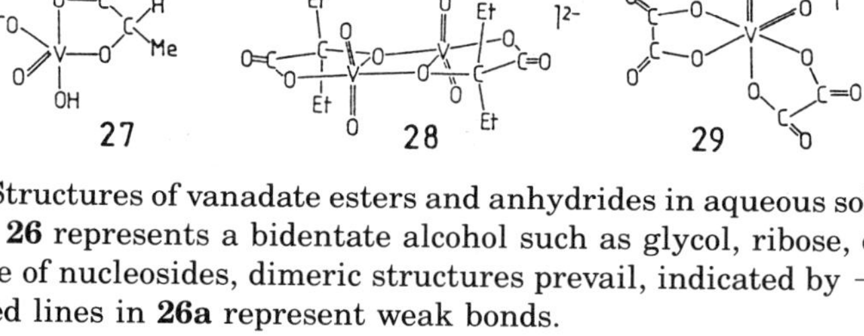

FIG. 7. Structures of vanadate esters and anhydrides in aqueous solution. *OO* in **25** and **26** represents a bidentate alcohol such as glycol, ribose, or inosine. In the case of nucleosides, dimeric structures prevail, indicated by −(V) in **25**. The dashed lines in **26a** represent weak bonds.

polyfunctional alcohols yield pentacoordinate cyclic in addition to the tetrahedral esters [113a,114]. The preferred positions for coordination of sugar derivatives like nucleosides are the oxygens on the 2′ and 3′ carbons of ribose (although involvement of the 5′ position has also been reported), and the complexes are usually dimeric with a 2:2 stoichiometry and K values between 10^5 and 10^8 M^{-3} [114]. Pentacoordination (**25**) [114,115a], hexacoordination (**26b**) [115b], and a coordination mode intermediate between the coordination numbers 5 and 6 (**26a** in Fig. 7) [116a] have been proposed. Although the discussion on the correct structure is not yet settled, evidence has been provided which favors a cyclic ester with vanadium in a trigonal-bipyramidal array [115a].

Lactate forms several complexes with vanadate, among these a tetrahedral ester (**24**), a pentacoordinate complex (**25**) with mixed alkoxide/carboxylate coordination, and a presumably octahedral complex (**27**) [116b]. For a first orientation on the geometry of the complexes in water, $\delta(^{51}V)$ values may be employed: Typical $\delta(^{51}V)$ values for tetrahedral, trigonal-bipyramidal, and octahedral complexes are approx. −556, −523, and −536 ppm, respectively. A structurally characterized dimeric complex has been obtained with 2-hydroxy-2-ethylbutanoic acid [117] (**28** in Fig. 7), with the alkoxides in bridging positions. The complexation behavior of monofunctional, monocarboxylic acids in aqueous solution has not been investigated. Oxalate forms two complexes: $[VO_2(ox)H_2O]^-$ and $[VO_2(ox)_2]^{3-}$ [118a]. At least in the latter, the coordination geometry is octahedral (**29**) and hence corresponds to that found in the solid state [118b].

In other solvents than water, the triesters $[VO(OR)_3]_n$ are obtained. In the solid state, the esters are polymers (R = Me) or dimers (other R; **30** in Fig. 8) [119]. In solution, concentration and temperature-dependent equilibria between monomers, dimers, and, in some cases, higher aggregates have been noted [120]; for bulky R such as *i*Pr and *t*Bu, only monomers are present. In the dimers, vanadium is in the center of a distorted trigonal bipyramid, and hence they may be considered transition state analogs for the pentacoordinate phosphorus intermediate in the enzymatic cleavage of the phosphoester bond in, e.g., polynucleotides. In the presence of water, the triesters hydrolyze to form diesters. The complete hydrolysis to vanadates runs through a decavanadate ester [Eq. (25)] [121]. In the case of bulky substituents R, hydrolysis is a slow process: For RO^- = 1-adamantoxy, the hydrolysis constant $[VO(OR)_2OH][ROH]/$

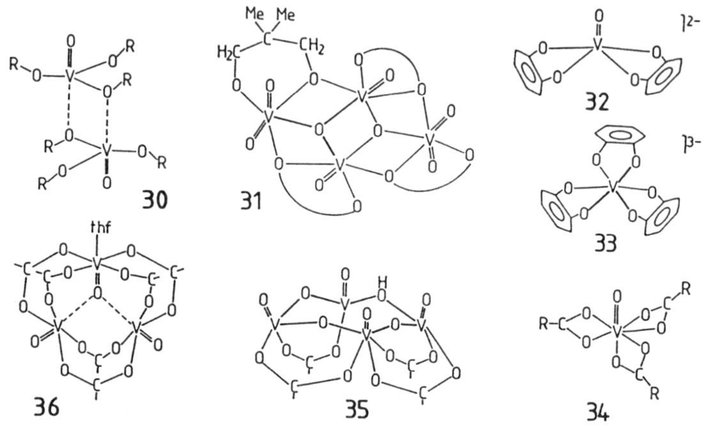

FIG. 8. Homoleptic alkoxo, aryloxo, and carboxylato complexes of vanadium which have been characterized by single-crystal X-ray diffraction. R in **30** is cyclopentyl [119], *OO* in **31** represents 2,2-dimethylpropanediolate(2-) [125]. For **32** and **33** see [127]. The carboxylates in **34–36** are *t*Bu-formiate (**34**, [128]), *t*Bu-acetate (**35** [124]), and acetate or benzoate (**36**, [20]). The cluster **35** has been crystallized with K^+ accommodated in the bole-like cavity.

$[VO(OR)_3][H_2O]$ is 0.4 [122]. Vanadate esters with bulky groups R are also rather stable against reduction. However, reduction to V^{IV} is induced by ascorbic acid and β-mercaptoethanol [122]. Alkoxo compounds of V^{IV}, i.e., $VO(OR)_2$ and $V(OR)_4$, have been characterized [123].

$$VO(OR)_3 + H_2O \rightarrow VO(OR)_2OH \rightarrow VO_2(OR)_2^-$$
$$\downarrow$$
$$H_xV_{10}O_{28}R_y^{(6-x-y)-} \rightarrow \text{Vanadates} \qquad (25)$$

A number of polynuclear oxovanadates with alkoxo groups in the periphery have been described [124]. Among these are a tetranuclear V^V complex of 2,2-dimethylpropanediol(1,3) (**31** in Fig. 8) [125], and hexa- and decanuclear mixed valence (V^V/V^{IV}) complexes of tridentate tripod alcohols [126]. The clusters containing the decameric core are directly related to decavanadate $H_xV_{10}O_{28}^{(6-x)-}$. Homoleptic vanadate esters with phenols have not been structurally characterized, but catecholate complexes of V^{IV} (**32** in Fig. 8) and V^{III} (**33**) are known [127].

Monocarboxylate may form mononuclear, pentagonal-bipyramidal V^V complexes (**34** in Fig. 8) [128], trimeric $V^{IV}/V^{III}/V^{II}$ complexes [124] (see compound **35** in Fig. 8 for a VO^{2+} species, which is usually formed if V^V carboxylates are in contact with water), and mixed valence V^V-V^{IV} complexes with tetrameric (**36**) [124], pentameric [129], and hexameric cores [128]. Except in the monomeric complex, where the carboxylate acts as a chelating ligand, the bidentate bridging coordination mode is found exclusively.

6.3. Models for the Vanadium-Protein Interaction

6.3.1. Complexes Modeling Histidine and Amide Binding

In the previous section, I reviewed recent results on complexes which model the binding of acidic and hydroxylic side chain functions of proteins. An additional side chain group, the imidazole moiety of histidine, has also been shown to be involved in vanadium binding, and the peptide linkages of the protein backbone, either through the carbonyl oxygen or through the deprotonated amide nitrogen, should also be considered potential binding sites for vanadium, as suggested by the mode of interaction between vanadate and small peptides (Sec. 6.3.2).

The vanadyl (VO^{2+}) ion has been fruitfully employed as a spin probe in the study of metal binding to proteins. In a recent ESEEM study of $VO(im)_4$ (im = imidazole), it has been shown that the single electron of the VO^{2+} group is not only coupled to the directly bonded, aromatic N, but also weakly to the remote N of the imidazole ligand [130]. ESEEM spectroscopy has also provided evidence for N coordination (presumably histidine) to vanadium in reduced (V^{IV}) vanadate-dependent peroxidase. **37** [131] and **38** [132] in Fig. 9 are examples for structurally characterized vanadyl-imidazole complexes, and **39** is an imidazole complex of V^V [133]. Compounds **40–42** [134,135] represent examples for the coordination of deprotonated amide. The V–N bond lengths range from 2.045 to 2.190 Å in the case of the imidazole complexes and from 1.970 to 2.026 Å in the amide complexes.

6.3.2. Amino Acid and Peptide Complexes

Amide binding has not yet been established by structure analyses in vanadium-peptide interaction, but there is strong evidence for amide-N binding in the aqueous vanadate-peptide systems (vide infra), and

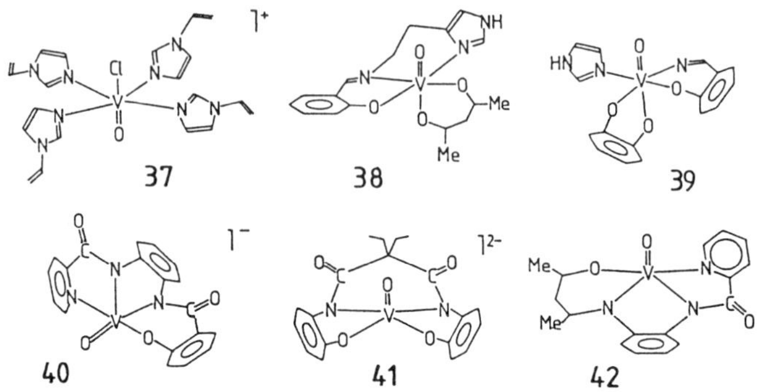

FIG. 9. Complexes modeling binding of the imidazole moiety of histidine (**37–39**) and the binding of deprotonated amide-N of the peptide linkage (**40–42**).

chelation of saccharine to VO^{2+} through the carbonyl group and the deprotonated amide has been proposed for the complex [VO(OH)$^+$-saccharinate(1-)] [136]. There are a few examples where direct coordination of vanadium to proteins has been detected by X-ray crystallography. Among these are the coordination of vanadyl to xylose isomerase, where histidine binding through the aromatic N is involved [137], and vanadate coordination in the ternary vanadate/uridine/ribonuclease A complex, where binding occurs through the 2′ and 3′ positions of ribose [138a]. In the binary vanadate/ribonuclease-T_1 complex, $H_2VO_4^-$ is linked to the amino acids in the active center via hydrogen bonds [138b].

No structural reports are available on complexes formed between vanadium and small peptides, but several structures on amino acid complexes and complexes with ligand systems containing amino acids have been reported and are collated in Fig. 10. Thus, histidine, proline, and cysteine form homoleptic, hexacoordinate complexes with V^{III}, in which the amino acids act as bifunctional (proline, **43** [139]) and trifunctional (cysteine, **44** [140]) or bi- plus trifunctional ligands (histidine, **45** [141]). Schiff base ligands containing amino acids as the amine constituent coordinate through one of the carboxylic oxygens. The vanadium center may be hexacoordinate, as in the V^{IV} complex formed with sal-gly (the Schiff base from salicyl aldehyde and glycine; **46** in Fig. 10 [142]), or pentacoordinate as in the binuclear V^{IV}/V^V complex **47**, containing serine [143], and in the V^V dioxo complex **48** with histidine [144].

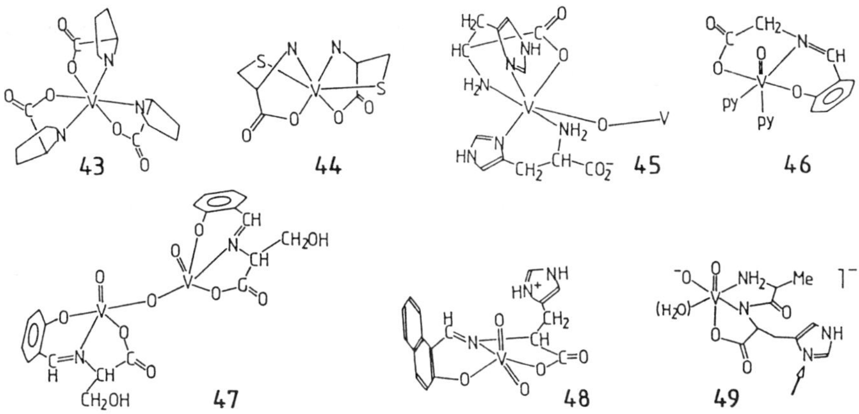

FIG. 10. Structures of amino acid and peptide complexes. **43–48** are based on X-ray structure analyses. Only one half of the molecule is shown in the case of **45** (the angle at the bridging oxygen is 153.9°). **49** (the arrow in **49** indicates the protonation site) is based on indirect evidence (see text).

Neither in **47** nor in **48** is the side chain (CH_2-OH, CH_2-imidazole) of the amino acid employed in coordination. In the histidine complex **48**, this is due to protonation at the imino-N. In compound **48**, there is a complex intermolecular hydrogen bonding network involving one of the oxo groups, the carboxylate oxygens, and the imidazole nitrogens [144]. The implications of such an H bond interaction as a model for hydrogen bonding networks in proteins containing His and Glu/Asp is apparent.

In contrast to **45**, but in analogy to **48**, histidine also "avoids" complexation through the imidazole moiety in the aqueous vanadate/Ala-His system [25], where the following two complexes are formed: a neutral complex of composition $[H_2VO_4^-/\text{Ala-His}(H^+)]^0$ with a formation optimum at pH 5.8 and log $\beta = 9.44(5)$ (for the formation from $H_2VO_4^-$, Ala-His, and H^+), and an anionic complex of composition $[H_2VO_4^-/\text{Ala-His}]^-$ (**49** in Fig. 10) with a formation optimum at pH 7.5 and a log $\beta = 2.55(4)$. As in **48**, protonation in **49** occurs at the imino-N. The requirement of the N and C termini and the deprotonated amide-N as shown in **49** has been inferred from multinuclear NMR spectroscopy and the fact that neither N- nor O-protected peptides nor peptides with the alkylated amide (sarcosine, Gly-Pro) form stable complexes [145]. On the other hand, peroxovanadates coordinate peptides in a monodentate fashion through the amino or the carboxylate group (an exception is Gly-Gly, which forms an additional complex, where this dipeptide acts in the tridentate fashion) [146a]. Amino acids themselves form weak complexes with vanadate [145b] and peroxovanadate [146b], with complex formation constants comparable to those found for lactic acid (Sec. 6.2), i.e., around 0.5 M^{-1} [145b]. The participation of deprotonated amide in coordination to vanadate is also evidenced by a kinetic investigation of the vanadate/Gly-Tyr system [147]; the complex formation in water (the formation constant amounts to 120 M^{-1}) follows two main pathways, one with and one without acid catalysis. The rate constants k are 0.017(1) $s^{-1}\ M^{-2}$ (acid catalysis) and 0.025(3) $s^{-1}\ M^{-1}$, i.e., complex formation is slower by orders of magnitude than for other V^{V} complexes ($k \sim 10^4\ s^{-1}\ M^{-1}$) [147]. In some cases, the vanadate/amino acid systems are redox-labile. Histidine, e.g., although forming weak complexes with vanadate, slowly reduces V^{V} to V^{IV} [25].

Amino acid complexes of VO^{2+} are much stabler than those of vanadate. Various of these systems, containing the amino acids glutamic acid, alanine, glycine, serine, and cysteine, have been investigated by combined potentiometric, EPR, and UV-VIS methods [148]. As an example, the constant K for the complex formation of VO^{2+}-Ala^{-} [see Eq. (26)] amounts to $7 \times 10^6\ M^{-1}$; compare 0.49 and 0.19 M^{-1} for the two complexes formed between vanadate and alanine [145b].

$$VO^{2+} + Ala^{-} \rightleftharpoons [VO\text{-}Ala]^{+} \qquad (26)$$

7. MISCELLANEOUS

Recently, non-oxo ("naked") V^{IV} and V^{V} complexes have become of considerable interest because of their potential as models for naturally occurring vanadium compounds such as amavadin or vanadium compounds of naturally occurring ligands, such as $[V(\text{enterobactin})]^{2-}$. Two examples for structurally characterized non-oxo complexes of V^{IV}, **50** [149] and **51** [150], are depicted in Fig. 11. Both complexes can be reversibly oxidized and reduced electrochemically [Eqs. (27) and (28)]. **51** can also be oxidized by Ag^{+} to yield the cationic V^{V} complex. Amavadin, the vanadium compound from mushrooms of the genus *Amanita*, is a non-oxo V^{IV} complex. It has been suggested that ama-

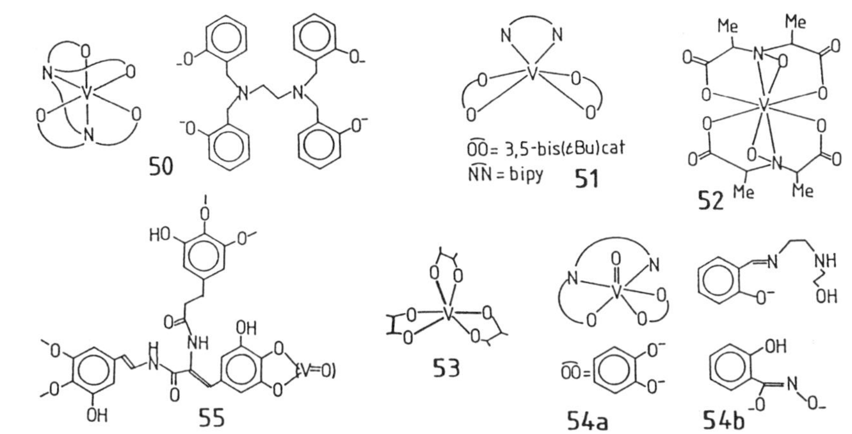

FIG. 11. Miscellaneous coordination compounds of vanadium with biological relevance. **50–53** are non-oxo complexes. **50** and **51** are model compounds, which undergo reversible redox reactions [cf. Eqs. (27) and (28)]. *NN* and *OO* in **51** are bipyridyl or phenanthroline, and 3,5-bis(*tert*-butyl)catecholate(2-), respectively. **52** is the oxidized form of amavadin, and **53** the V^{IV} analog of Fe^{III}-enterobactin. **54a** and **54b** are examples of complexes with noninnocent ligands [catecholate(2-) in the case of **54a**, salicylhydroxamate(2-) in the case of **54b**]. **55** models the supposed template effect of VO^{2+} in tunichrome synthesis in vanadium-accumulating sea squirts.

vadin plays a role as a mediator for electron transfer to an oxidase [13]. In this context, the synthesis and structural characterization of oxidized, V^V-amavadin [14] (**52** in Fig. 11) is of considerable interest [14]. The ligand *N,N'*-bis(propionato)hydroxylamide coordinates through the carboxylate and the R_2NO^- groups, the latter in the side-on fashion, i.e., analogous to the coordination mode usually encountered for inorganic hydroxylamide complexes (cf. Sec. 3.3). **53** in Fig. 11 shows the immediate coordination sphere of V^{IV} in $[V(\text{enterobactin})]^{2-}$ (the approximate point symmetry is C_3). This structural motif is reminiscent of the V^{III}-tris(catecholate) complex **33** in Fig. 8. The distortion from octahedral symmetry in **53**, as denoted by the twist angle (0° for the trigonal prism, 60° for the octahedron), is 28° [151].

$$[V^VL]^+ \overset{E_1}{\rightleftharpoons} [V^{IV}L] \overset{E_2}{\rightleftharpoons} [V^{III}L]^- \tag{27}$$

$$E_1 = +0.16, E_2 = -0.84 \text{ V vs. Fc in DMF}$$

$$[V^{V}(OO)_2(NN)]^+ \overset{E_1}{\rightleftharpoons} [V^{IV}(OO)_2(NN)] \overset{E_2}{\rightleftharpoons} [V^{III}(OO)_2(NN)]^- \quad (28)$$

$$E_1 = +0.52, E_2 = -0.79 \text{ V vs. NHE in MeCN}$$

Another point of interest, related to the necessity of creating compounds that model uncommon ligation spheres of vanadium, are complexes with so-called innocent ligands, i.e., ligands affecting low-energy charge-transfer transitions and hence unusual electronic conditions particularly in V^V (d^0) systems. **54a** and **54b** [133] in Fig. 11 are examples, containing a tridentate "innocent" Schiff base ligand and a bidentate noninnocent [catecholate(2-); salicylhydroxamate(2-)] ligand. Among the surprising features of this class of complexes is the exceptionally low ^{51}V shielding, manifest in $\delta(^{51}V)$ values of +221 ppm (**54a**) and −132 ppm (**54b**) as compared to $\delta(^{51}V)$ around approx. −530 ppm in hexacoordinate complexes with an *ON* ligand set [152].

The excellent coordination behavior of catecholates toward V^{IV} (and V^V) (**33** in Fig. 8, **39** in Fig. 9, **51**, **53**, and **54a** in Fig. 11) has led to the suggestion that VO^{2+} acts as a template in the synthesis of tunichromes, the dark pigments in the tunique of the those tunicates that accumulate vanadium. These pigments contain oligopeptides based on hydroxy-dopa units, which have been shown to form very stable complexes ($K \sim 10^{23}$) with VO^{2+} in vitro. A proposed structure of such a complex, compound **55** [153], is shown in Fig. 11.

8. CONCLUSION

A profound knowledge of the classic, modern, and future-oriented inorganic chemistry of vanadium is essential for a competent understanding of the function of this metal in living organisms. In this chapter, I have shown that even simple inorganic compounds, like vanadate, can act on a physiological level simply by competition with the chemically similar phosphate, a similarity which becomes evident on inspection of the short version of the periodic table, where the elements vanadium and phosphorus "coexist" in the same box. An essential part in dealing with inorganic aspects of bioinorganic chemistry is the coordination chemistry of the metal ions and its relation to the structure/function synergism in active sites of metalloproteins. Studies in this field range from classic to modern chemistry and provide many future aspects, one of which is the appropriate ligand design to model the correct properties of a biologically relevant compound. I have highlighted several of the future-oriented aspects of vanadium coordination chemistry in the previous section. Other innovative fields in inorganic chemistry will attain increasing importance, such as (1) solid state chemistry in the context of biomineralization (cf. the production of mineralic vanadium compounds by vanadate-reducing bacteria mentioned in Sec. 4), and (2) organovanadium chemistry, e.g., in the context of modeling the alkyne reductase activity of vanadium nitrogenase by generating alkynevanadium complexes as possible active intermediates [154].

ABBREVIATIONS

acac	acetylacetonate(1-)
ADP	adenosine 5′-diphosphate
AMP	adenosine 5′-monophosphate
ATP	adenosine 5′-triphosphate
bipy	2,2′-bipyridyl
c	concentration
cat	catecholate(2-)
Cp	cyclopentadienyl(1-)
dipic	dipicolinate(2-)
DMF	dimethylformamide
dmso	dimethylsulfoxide
dopa	dihydroxyphenylalanine
dppe	diphenylphosphinoethane
edta	ethylenediammine-*N*,*N*,*N*′,*N*′-tetraacetate(4-)
ehgs	*N*-[2-(*o*-salicylideneamino)ethyl](*o*-hydroxyphenyl)-glycinate(3-)
EMF	electromotoric force (potentiometry)
EPR	electron (spin) paramagnetic resonance spectroscopy
ESEEM	electron spin echo envelope modulation
Et	ethyl
Fc	ferrocene
im	imidazole

*i*Pr	isopropyl
MCD	magnetic circular dichroism
Me	methyl
mes	mesitylene
NAD	nicotine adenine dinucleotide, oxidized form
NADH	nicotine adenine dinucleotide, reduced form
NHE	normal (standard) hydrogen electrode
NMR	nuclear magnetic resonance spectroscopy
nta	nitrilotriacetate(3-)
ox	oxalate(2-)
Ph	phenyl
pic	picolinate(1-)
py	pyridine
rhod	rhodotorulate(2-)
salen	double Schiff base from salicyl aldehyde and ethylenediamine
*t*Bu	tertiary butyl
UV-VIS	electron absorption spectroscopy
x	mole fraction

REFERENCES

1. G. Hoppe, J. Siemroth, and F. Damaschun, *Chem. Erde*, *50*, 81 (1990).
2. E. W. Baker and S. E. Palmer, in *The Porphyrins* I (D. Dolphin, ed.), Academic Press, New York, 1978.
3. B. Leighton, G. J. S. Cooper, C. DaCosta, and E. A. Foot, *Biochem. J.*, *276*, 289 (1991).
4. H. Sakurai, K. Tsuchiya, M. Nakatsuka, J. Kawada, S. Ishikawa, H. Yoshida, and M. Komatsu, *J. Clin. Biochem. Nutr.*, *8*, 193 (1990).
5. P. Köpf-Maier and H. Köpf, *Z. Naturforsch.*, *B34*, 805 (1979).
6. T. Yamase, in *Polyoxometalates: From Platonic Solids to Anti-Retroviral Activity* (M. T. Pope and A. Müller, eds.), Kluwer, Dordrecht, 1994, p. 337 ff.
7. (a) A. Maiti and S. Ghosh, *J. Inorg. Biochem.*, *36*, 131 (1989). (b) K. Greef, A. Kleinschmidt, J. A. Köhler, K. E. Loose, B.

Lüderitz, E. Lüthy, and H. Spitzbarth, *Herz Kreislauf*, *11*, 501 ff. (1979).

8. *Gmelins Handbuch der Anorganischen Chemie: Vanadium*, Part A, Verlag Chemie, Weinheim, 1968, p. 529 ff.
9. (a) L. Josephson and L. C. Cantley, Jr., *Biochemistry*, *16*, 4572 (1977). (b) C. N. Pace, U. Heinemann, U. Hahn, and W. Saenger, *Angew. Chem., Int. Ed. Engl.*, *30*, 343 (1991).
10. M. D. Cohen, A. C. Sen, and C. I. Wei, *Inorg. Chim. Acta*, *138*, (1987) 179.
11. (a) G. Vandorpe, M. Bollen, E. van Herck, R. Boullion, and W. Stalmans, *FEBS Lett.*, *309*, 209 (1992). (b) H. Sakurai, K. Tsuchiya, M. Nakatsuka, M. Sofue, and J. Kawada, *J. Endocrinol.*, *126*, 451 (1990).
12. J. Hormes, U. Kuetgens, R. Chauvistre, W. Schreiber, N. Anders, H. Vilter, D. Rehder, and C. Weidemann, *Biochim. Biophys. Acta*, *956*, 293 (1988).
13. J. J. R. Fraústo da Silva, *Chem. Spec. Bioavailab.*, *1*, 139 (1989).
14. E. M. Armstrong, R. L. Beddoes, L. J. Calviou, J. M. Charnock, D. Collison, N. Ertok, J. H. Naismith, and C. D. Garner, *J. Am. Chem. Soc.*, *115*, 807 (1993).
15. S. Lee, K. Kustin, W. E. Robinson, R. B. Frankel, and K. Spartalian, *J. Inorg. Biochem.*, *33*, 183 (1988).
16. S. G. Brand, C. J. Hawkins, A. T. Marshall, G. W. Nette, and D. L. Parry, *Compt. Biochem. Physiol.*, *B93*, 425 (1989).
17. T. Ishii, I. Nakai, C. Numaco, K. Okishi, and T. Otake, *Naturwissenschaften*, *80*, 268 (1993).
18. B. E. Smith, R. R. Eady, D. J. Lowe and C. Gormal, *Biochem. J.*, *250*, 299 (1988).
19. N. D. Chasteen (ed.), *Vanadium in Biological Systems*, Kluwer, Dordrecht, 1990.
20. D. Rehder, *Angew. Chem., Int. Ed. Engl.*, *30*, 148 (1991).
21. D. Rehder, *BioMetals*, *5*, 3 (1992).
22. A. Butler and C. J. Carrano, *Coord. Chem. Rev.*, *109*, 61 (1991).
23. R. Wever and K. Kustin, *Adv. Inorg. Chem.*, *35*, 81 (1990).
24. O. W. Howarth, *Progr. Nucl. Magn. Reson. Spectr.*, *22*, 453 (1990).
25. K. Elvingson, M. Fritzsche, D. Rehder, and L. Pettersson, *Acta Chem. Scand.*, in press.

26. D. C. Crans, C. D. Rithner, and L. A. Theisen, *J. Am. Chem. Soc.*, *112*, 2901 (1990).
27. L. Petterson, B. Hedman, A. M. Nenner, and I. Andersson, *Acta Chem. Scand.*, *A39*, 499 (1985).
28. (a) A. M. Amado, M. Aureliano, P. J. A. Ribeiro-Claro, and J. J. C. Teixeira-Dias, *J. Raman Spectrosc.*, *24*, 699 (1993). (b) L. Pettersson, I. Andersson, and O. W. Howarth, *Inorg. Chem.*, *31*, 4032 (1992).
29. L. Pettersson, B. Hedman, I. Andersson, and N. Ingri, *Chem. Script.*, *22*, 254 (1983).
30. D. Rehder, in *Transition Metal NMR* (P. S. Pregosin, ed.), Elsevier, Amsterdam, 1991, p. 1 ff.
31. D. Rehder, in *Vanadium in Biological Systems* (N. D. Chasteen, ed.), Kluwer, Amsterdam, 1990, ch. X.
32. L. Pettersson in *Polyoxometalates: From Platonic Solids to Anti-Retroviral Activity* (M. T. Pope and A. Müller, eds.), Kluwer, Dordrecht, 1994, p. 27 ff.
33. M. Nabavi, F. Taulelle, C. Sanchez, and M. Verdaguer, *J. Phys. Chem. Solids*, *51*, 1375 (1990).
34. S. Aschwanden, H. W. Schmalle, A. Reller, and H. R. Oswald, *Mat. Res. Bull.*, *6*, 575 (1993).
35. S. E. Harnung, E. Larsen, and J. Pedersen, *Acta Chem. Scand.*, *47*, 674 (1993).
36. D. C. Crans, *Comments Inorg. Chem.*, in press.
37. V. W. Day, W. G. Klemperer, and A. Yagasaki, *Chem. Lett.*, 1267 (1990).
38. J. Fuchs, S. Mahjour, and J. Pickardt, *Angew. Chem., Int. Ed. Engl.*, *15*, 374 (1976).
39. P. Román, A. San José, A. Luque, and J. M. Gutierrez-Zorilla, *Inorg. Chem.*, *32*, 775 (1993).
40. T. Debaerdemaeker, J. M. Arrieta, and J. M. Amigo, *Acta Cryst.*, *B38*, 2465 (1982).
41. D. C. Crans, C. M. Simone, A. K. Saha, and R. H. Glew, *Biochem. Biophys. Res. Commun.*, *165*, 264 (1989).
42. D. C. Crans and C. M. Simone, *Biochemistry*, *30*, 6734 (1991).
43. D. C. Crans, K. Sudhakar, and T. Zamborelli, *Biochemistry*, *31*, 6812 (1992).

44. L. Wittenkeller, A. Abraha, R. Ramasamy, D. M. Defreitas, L. A. Theisen, and D. C. Crans, *J. Am. Chem. Soc.*, *113*, 7872 (1991).
45. D. W. Boyd, K. Kustin, and M. Niwa, *Biochem. Biophys. Acta*, *827*, 472 (1985).
46. G. Soman, Y. C. Chang, and D. J. Graves, *J. Biochem.*, *22*, 4994 (1983).
47. K. J. Föhr, Y. Wahl, R. Engling, T. P. Kemmer, and M. Gratzl, *Cell Calcium*, *12*, 735 (1991).
48. C. Djordjevic and G. L. Wampler, *J. Inorg. Biochem.*, *25*, 51 (1985).
49. H. Sakurai and K. Tsuchiya, *FEBS Lett.*, *260*, 109 (1990).
50. R. I. de la Rosa, M. J. Clague, and A. Butler, *J. Am. Chem. Soc.*, *114*, 760 (1992).
51. A. Butler and J. V. Walker, *Chem. Rev.*, *93*, 1937 (1993).
52. S. Liu, M. J. Gresser, and A. S. Tracey, *Biochemistry*, *31*, 2677 (1992).
53. D. C. Crans, R. W. Marshman, R. Nielsen, and I. Felty, *J. Org. Chem.*, *58*, 2244 (1993).
54. Y. Chang, Q. Chen, M. I. Khan, J. Salta, and J. Zubieta, *J. Chem. Soc., Chem. Commun.*, 1872 (1993).
55. A. A. Ivakin, M. V. Kruchinina, and O. V. Koryakova, *Russ. J. Inorg. Chem.*, *35*, 1523 (1990).
56. M. J. Gresser, A. S. Tracey, and K. M. Parkinson, *J. Am. Chem. Soc.*, *108*, 6229 (1986).
57. Q. Chen, J. Salta, and J. Zubieta, *Inorg. Chem.*, *32*, 4485 (1993).
58. A. S. Tracey and M. J. Gresser, *Inorg. Chem.*, *27*, 2695 (1988).
59. C. F. G. C. Geraldes and M. M. C. A. Castro, *J. Inorg. Biochem.*, *35*, 79 (1989).
60. A. S. Tracey, M. J. Gresser, and S. Liu, *J. Am. Chem. Soc.*, *110*, 5869 (1988).
61. H. Sakurai, T. Goda, and S. Shimomura, *Biochem. Biophys. Res. Commun.*, *108*, 1421 (1982)
62. E. G. Ferrer and E. J. Baran, *Trans. Met. Chem.*, *16*, 599 (1991).
63. E. Alberico and G. Micera, *Inorg. Chim. Acta*, *215*, 225 (1994).
64. B. Jordan and C. Calvo, *Can. J. Chem.*, *51*, 2621 (1973).
65. (a) J. T. Wrobleski, *Inorg. Chem.*, *27*, 946 (1988). (b) J. R. Ebner and M. R. Thompson, *Catalysis Today*, *16*, 51 (1993).

66. R. C. Haushalter, Z. Wang, M. E. Thompson, J. Zubieta, and C. J. O'Connor, *Inorg. Chem.*, *32*, 3966 (1993).
67. D. Rehder and K. Wieghardt, *Z. Naturforsch*, *B36*, 1251 (1981).
68. P. Schwendt, Composition and Structure of Vanadium(V) Peroxo Complexes in Aqueous Solutions and in the Solid State, Proc. 13. Conf. Coord. Chem., Bratislava, CSFR, 1991, pp. 243–246.
69. N. J. Campbell, A. C. Dengel, and W. P. Griffith, *Polyhedron*, *8*, 1379 (1989).
70. A. T. Harrison and O. W. Howarth, *J. Chem. Soc., Dalton Trans.*, 1173 (1985).
71. M. Bhattacharjee, M. K. Chaudhuri, and P. C. Paul, *Can. J. Chem.*, *70*, 2245 (1972).
72. N. D. Chasteen, J. K. Grady, and C. E. Hoolway, *Inorg. Chem.*, *25*, 2754 (1986).
73. A. Stern, A. J. Davison, Q. Wu, and J. Moon, *Arch. Biochem. Biophys.*, *299*, 125 (1992).
74. D. E. Ryan, N. D. Ghatlia, A. E. McDermott, N. J. Turro, K. Nakanishi, and K. Kustin, *J. Am. Chem. Soc.*, *114*, 6959 (1992).
75. (a) N. A. Yukova, N. D. Saveljeva, and N. N. Lyalikova, *Mikrobiologiya*, *62*, 597 (1993). (b) N. N. Lyalikova and N. A. Yukova, *Geomicrobiol. J.*, *10*, 15 (1992).
76. (a) *Gmelins Handbuch der Anorganischen Chemie, Vanadium*, Part A, Verlag Chemie, Weinheim, 1968, p. 466 ff. (b) M. Fiore, S. Orecchio, V. Romano, and R. Zingales, *J. Chem. Soc., Dalton Trans.*, 799 (1993).
77. T. Ozawa, A. Hanaki, and F. Takazawa, *Inorg. Chim. Acta*, *188*, 163 (1991).
78. G. A. Dean and J. F. Herringshaw, *Talanta*, *10*, 793 (1963).
79. R. Ma, A. Bacac, and J. H. Espenson, *Inorg. Chem.*, *31*, 1925 (1992).
80. (a) E. Erdman, W. Krawietz, G. Philipp, I. Hackbarth, W. Schmitz, H. Scholz, and F. L. Crane, *Nature*, *282*, 335 (1979). (b) X. Shi and N. S. Dalal, *Arch. Biochem. Biophys.*, *302*, 300 (1993).
81. (a) F. Vyskočil, J. Teisinger, and H. Dlouhá, *Nature*, *286*, 516 (1980). (b) X. Shi and N. S. Dalal, *Free Radical Res. Commun.*, *17*, 369 (1992). (c) A. J. Carmichael, *FEBS Lett.*, *261*, 165 (1990).

82. A. Shaver, J. B. Ng, D. A. Hall, B. Soo Lum, and B. I. Posner, *Inorg. Chem.*, *32*, 3109 (1993).

83. H. Sakurai and K. Tsuchiya, *FEBS Lett.*, *260*, 109 (1990).

84. (a) P. R. Hari Prasad Rao, A. V. Ramaswamy, and P. Ratnasamy, *J. Catal.*, *141*, 604 (1993). (b) G. B. Shul'pin, D. Attanasio, and L. Suber, *J. Catal.*, *142*, 147 (1983).

85. (a) M. Bianchi, M. Bonchio, V. Conte, F. Coppa, F. DiFuria, G. Modena, S. Moro, and S. Standen, *J. Mol. Catal.*, *83*, 107 (1993). (b) H. Mimoun, L. Saussine, E. Daire, M. Postel, J. Fischer, and R. Weiss, *J. Am. Chem. Soc.*, *105*, 3101 (1993). (c) P. R. Hari Prasad Rao and A. V. Ramaswamy, *J. Chem. Soc., Chem. Commun.*, 1245 (1992).

86. B. E. Rossiter, T. R. Verhoeven, and K. P. Sharpless, *Tetrahedron Lett.*, 4733 (1979).

87. (a) C. M. Sorensen and R. S. Weber, *J. Catal.*, *142*, 1 (1993). (b) P. M. Michalakos, M. C. Kung, and H. H. Kung, *J. Catal.*, *140*, 226 (1993).

88. (a) G. Centi, *Catal. Today*, *16*, 5 (1993). (b) I. Matsuura, *Catal. Today*, *16*, 123 (1993). (c) B. El Ali, J. M. Brégeault, J. Mercier, J. Martin, C. Martin, and O. Convert, *J. Chem. Soc., Chem. Commun.*, 825 (1989).

89. K. Nakajima, K. Kojima, M. Kojima, and J. Fujita, *Bull. Chem. Soc. Jpn.*, *63*, 2620 (1990).

90. T. Hirao, K. Hirano, T. Hasegawa, Y. Ohshiro, and I. Ikeda, *J. Org. Chem.*, *58*, 6529 (1993).

91. E. A. Kraynack and S. F. Pedersen, *J. Org. Chem.*, *58*, 6114 (1993).

92. T. L. Ho and G. A. Olah, *Synthesis*, 170 (1977).

93. A. Togni, G. Rist, G. Rihs, and A. Schweiger, *J. Am. Chem. Soc.*, *115*, 1908 (1993).

94. T. Hirao, T. Fuji, and Y. Ohshiro, *J. Organomet. Chem.*, *407*, C1 (1991).

95. B. M. Choudary and P. N. Reddy, *J. Chem. Soc., Chem. Commun.*, 405 (1993).

96. F. J. Feher and R. L. Blanski, *J. Am. Chem. Soc.*, *114*, 5886 (1992).

97. (a) S. Ciurli and R. H. Holm, *Inorg. Chem.*, *28*, 1685 (1989). (b) W. Cen, S. C. Lee, J. Li, F. M. MacDonnell, and R. H. Holm, *J. Am. Chem. Soc.*, *115*, 9515 (1993).

98. F. Lin, R. L. Beddoes, D. Collison, C. D. Garner, and F. E. Mabbs, *J. Chem. Soc., Chem. Commun.*, 496 (1993).

99. S. C. Sendlinger, J. R. Nicholson, E. B. Lobkovsky, J. C. Huffman, D. Rehder, and G. Christou, *Inorg. Chem.*, *32*, 204 (1993).

100. (a) E. R. T. Tiekink, X. F. Yan, and C. G. Young, *Aust. J. Chem.*, *45*, 897 (1992). (b). Y. Yang, Q. Liu, and D. Wu, *Inorg. Chim. Acta*, *208*, 85 (1993).

101. G. Henkel, B. Krebs, and W. Schmidt, *Angew. Chem., Int. Ed. Engl.*, *31*, 1366 (1992).

102. W. Tsagkalidis, D. Rodewald, D. Rehder, and V. Vergopoulos, *Inorg. Chim. Acta*, *219*, 213 (1994).

103. B. Folkesson and R. Larsson, *Acta Chem Scand.*, *A33*, 347 (1979).

104. (a) J. J. H. Ederna, A. Meetsma, and S. Gambarotta, *J. Am. Chem. Soc.*, *111*, 6878 (1989). (b) G. J. Leigh, A. Prieto-Alcón, and J. Sanders, *J. Chem. Soc., Chem. Commun.*, 921 (1991).

105. V. R. Ferguson, E. Solari, C. Floriani, A. Chiesi-Villa, and C. Rizzoli, *Angew. Chem., Int. Ed. Engl.*, *32*, 396 (1993).

106. J. K. F. Buijink, A. Meetsma, and J. H. Teuben, *Organometallics*, *12*, 2004 (1993).

107. W. Tsagkalidis, C. Woitha, and D. Rehder, *Inorg. Chim. Acta.*, *205*, 239 (1993).

108. J. Bultitute, L. F. Larkeorthy, D. C. Povey, G. W. Smith, J. R. Dilworth, and G. J. Leigh, *J. Chem. Soc., Chem. Commun.*, 1748 (1986).

109. C. Le Floc'h, R. A. Henderson, D. L. Hughes, and R. L. Richards, *J. Chem. Soc., Chem. Commun.*, 175 (1993).

110. D. Rehder, C. Woitha, W. Priebsch, and H. Gailus, *J. Chem. Soc., Chem. Commun.*, 364 (1992); H. Gailus, C. Woitha, and D. Rehder, *J. Chem. Soc., Dalton Trans.*, in press.

111. A. Shilov, N. Denisov, O. Efimov, N. Shuvalov, N. Shuvalova, and A. Shilova, *Nature*, *231*, 460 (1971).

112. N. P. Luneva, S. A. Mironova, A. E. Shilov, M. Yu. Antipin, and Y. T. Struchkov, *Angew. Chem.*, *105*, 1240 (1993).

113. (a) A. S. Tracey and M. J. Gresser, *Can. J. Chem.*, *66*, 2570 (1988). (b) B. Galeffi and A. S. Tracey, *Can. J. Chem.*, *67*, 2565 (1988).

114. A. S. Tracey, J. S. Jaswal, M. J. Gresser, and D. Rehder, *Inorg. Chem.*, *29*, 4283 (1990).

115. (a) X. Zhang and A. S. Tracey, *Acta Chem. Scand.*, *46*, 1170 (1992). (b) D. C. Crans, S. E. Harnung, E. Larsen, P. K. Shin, L. A. Theisen, and I. Trabjerg, *Acta Chem. Scand.*, *45*, 456 (1991).

116. (a) J. Richter and D. Rehder, *Z. Naturforsch.*, *B46*, 1613 (1991). (b) A. S. Tracey, M. J. Gresser, and K. M. Parkinson, *Inorg. Chem.*, *26*, 629 (1987).

117. T. W. Hambley, R. J. Judd, and P. A. Lay, *Inorg. Chem.*, *31*, 345 (1992).

118. (a) P. M. Ehde, L. Pettersson, and J. Glaser, *Acta Chem. Scand.*, *45*, 998 (1991). (b) W. R. Scheidt, C. Tsai, and J. L. Hoard, *J. Am. Chem. Soc.*, *93*, 3867 (1971).

119. (a) W. Priebsch and D. Rehder, *Inorg. Chem.*, *29*, 3013 (1990). (b) F. Hillerns, F. Olbrich, U. Behrens, and D. Rehder, *Angew. Chem., Int. Ed. Engl.*, *31*, 447 (1992).

120. K. Paulsen, D. Rehder, and D. Thoennes, *Z. Naturforsch.*, *A33*, 834 (1978).

121. H. Langbein, A. Polte, R. Lang, and G. Großmann, *Z. Naturforsch.*, *B46*, 1509 (1991)

122. D. C. Crans, H. Chen, and R. A. Felty, *J. Am. Chem. Soc.*, *114*, 4543 (1992).

123. (a) E. P. Tureskova and N. Ya. Turova, *Koord. Khim.*, *15*, 373 (1989). (b) A. Haaland, K. Rypdal, H. V. Volden, and R. A. Andersen, *J. Chem. Soc., Dalton Trans.*, 891 (1992).

124. D. Rehder, in *Polyoxometalates: From Platonic Solids to Anti-Retroviral Activity* (M. T. Pope and A. Müller, eds.), Kluwer, Dordrecht, 1994, p. 162.

125. D. C. Crans, R. W. Marshman, M. S. Gottlieb, O. P. Anderson, and M. M. Miller, *Inorg. Chem.*, *31*, 4939 (1992).

126. (a) Q. Chen, D. P. Goshorn, C. P. Scholes, X. L. Tan, and J. Zubieta, *J. Am. Chem. Soc.*, *114*, 4667 (1992). (b) M. I. Khan, Q. Chen, D. P. Goshorn, and J. Zubieta, *Inorg. Chem.*, *32*, 672 (1993).

127. S. R. Cooper, Y. B. Koh, and K. N. Raymond, *J. Am. Chem. Soc.*, *104*, 5092 (1982).

128. D. Rehder, W. Priebsch, and M. von Oeynhausen, *Angew. Chem., Int. Ed. Engl.*, *28*, 1221 (1989).

129. D. D. Heinrich, K. Folting, W. E. Streib, J. C. Huffman, and G. Christou, *J. Chem. Soc., Chem. Commun.*, 1411 (1989).

130. S. A. Dikanov, C. Burgard, and J. Hüttermann, *Chem. Phys. Lett.*, *212*, 493 (1993).

131. L. J. Calviou, J. M. Arber, D. Collison, C. D. Garner, and W. Clegg, *J. Chem. Soc., Chem. Commun.*, 654 (1992).

132. C. R. Cornman, J. Kampf, M. S. Lah, and V. L. Pecoraro, *Inorg. Chem.*, *31*, 2035 (1992).

133. C. R. Cornman, G. J. Colpas, J. D. Hoeschele, J. Kampf, and V. L. Pecoraro, *J. Am. Chem. Soc.*, *114*, 9925 (1992).

134. T. A. Kabanos, A. D. Keramidas, A. P. Papaioannou, and A. Terzis, *J. Chem. Soc., Chem. Commun.*, 643 (1993).

135. (a) A. S. Borovik, T. M. Dewey, and K. N. Raymond, *Inorg. Chem.*, *32*, 413 (1993). (b) G. R. Hanson, T. A. Kabanos, A. D. Keramidas, D. Mentzafos, and A. Terzis, *Inorg. Chem.*, *31*, 2587 (1992).

136. E. G. Ferrer, S. B. Etcheverry, and E. J. Baran, *Monatsh. Chem.*, *124*, 355 (1993).

137. R. Bogumil, J. Hüttermann, R. Kappl, R. Stabler, C. Dudfeldt, and H. Witzel, *Eur. J. Biochem.*, *196*, 305 (1991).

138. (a) B. Borah, C. W. Chen, W. Egan, M. Miller, A. Wlodawer, and J. S. Cohen, *Biochemistry*, *24*, 2058 (1985). (b) D. Kostrewa, H. W. Choe, U. Heinemann, and W. Saenger, *Biochemistry*, *28*, 7592 (1989).

139. C. P. Magill, C. Floriani, A. Chiesi-Villa, and C. Rizzoli, *Inorg. Chem.*, *32*, 2729 (1993).

140. H. Maeda, K. Kanamori, H. Michibata, T. Konno, K. Okamoto, and J. Hidaka, *Bull Chem. Soc. Jpn.*, *66*, 790 (1993).

141. K. Kanamori, M. Teraoka, H. Maeda, and K. Okamoto, *Chem. Lett.*, 1731 (1993).

142. I. Cavaco, J. C. Pessoa, D. Costa, M. T. Duarte, R. D. Gillard, and P. Matias, *J. Chem. Soc., Dalton Trans.*, 149 (1994).

143. J. C. Pessoa, A. L. Silva, A. L. Vieira, L. F. Vilas-Boas, P. O'Brien, and P. Thornton, *J. Chem. Soc., Dalton Trans.*, 1745 (1992).

144. V. Vergopoulos, W. Priebsch, M. Fritzsche, and D. Rehder, *Inorg. Chem.*, *32*, 1844 (1993).

145. (a) D. Rehder, *Inorg. Chem.*, *27*, 4312 (1988). (b) J. S. Jaswal and A. S. Tracey, *Can. J. Chem.*, *69*, 1600 (1991).

146. (a) A. S. Tracey and S. J. Jaswal, *J. Am. Chem. Soc.*, *114*, 3835 (1992). (b) A. S. Tracey and J. S. Jaswal, *Inorg. Chem.*, *32*, 4235 (1993).

147. D. C. Crans, H. Holst, and D. Rehder, *J. Am. Chem. Soc.*, submitted.

148. (a) J. C. Pessoa, J. L. Antunes, L. F. Vilas-Boas, and R. D. Gillard, *Polyhedron*, *11*, 1449 (1992). (b) J. C. Pessoa, L. F. Vilas-Boas, R. D. Gillard, and R. J. Lancashire, *Polyhedron*, *7*, 1245 (1988).

149. A. Neves, A. S. Ceccato, I. Vencato, Y. P. Mascarenhas, and C. Erasmus-Buhr, *J. Chem. Soc., Dalton Trans.*, 652 (1992).

150. T. A. Kabanos, A. M. Z. Slawin, D. J. Williams, and J. D. Woollins, *J. Chem. Soc., Dalton Trans.*, 1423 (1992).

151. T. B. Karpishin, T. M. Dewey, and K. N. Raymond, *J. Am. Chem. Soc.*, *115*, 1842 (1993).

152. D. Rehder, C. Weidemann, A. Duch, and W. Priebsch, *Inorg. Chem.*, *27*, 584 (1988).

153. E. M. Oltz, R. C. Bruening, M. J. Smith, K. Kustin, and K. Nakanishi, *J. Am. Chem. Soc.*, *110*, 6162 (1988).

154. H. Gailus, H. Maelger, and D. Rehder, *J. Organomet. Chem.*, *465*, 181 (1994).

2
Solution Properties of Vanadium(III) with Regard to Biological Systems

Roland Meier,[1] Martin Boddin,[1]
Steffi Mitzenheim,[1] and Kan Kanamori[2]

[1]Hochschule für Technik, Wirtschaft und Kultur (FH) Leipzig, Fachbereich Informatik, Mathematik und Naturwissenschaften, FG Chemie, P.O. Box 66, D-04251 Leipzig, Germany, and

[2]Department of Chemistry, Faculty of Science, Toyama University, Gofuku 3190, Toyama 930, Japan

1. INTRODUCTION

1.1. Vanadium(III) in Biological Systems

This chapter is devoted to the biological relevance of vanadium in its oxidation state +3. From a strictly inorganic point of view it is rather unexpected that vanadium, which can adopt oxidation numbers from +5 down to +2 in aqueous solution, should occur in its trivalent state in a living system. In an oxygen-containing environment, vanadium(III) is usually oxidized to vanadium(IV) and vanadium(V). The rate of oxidation is dependent on such features as acidity and the coordination environment of vanadium. Therefore, it seems necessary to present a brief historical excursion relating the chemistry of vanadium(III) to biological systems.

Significant amounts of vanadium in a biological system were first discovered in 1911 by the German physiologist M. Henze in the blood of the ascidian *Phallusia mammillata* [1]. This discovery has attracted large interest and it was expected that vanadium has a distinct physiological function in these primitive sea animals [2,3].

The ascidians constitute a class of marine animals (Ascidiacea) distantly related to the vertebrates. They are commonly called "sea squirts" because of their tendency to expel jets of water from their body cavity and through their siphons when agitated. All species have basically the same simple structure consisting of two apertures atop a saclike body, from which they get their name (the word *ascidian* comes from the Greek word *askidion*, meaning "small wine skin"). The Ascidiacea form the major class in the subphylum Tunicata, which also includes the free swimming thaliceans (salps) and larvaceans. Tunicates get their name from their tough, often translucent, tunic-like body covering called a test.

In their larval stage ascidians possess a dorsal, hollow nerve cord and notochord-like cells in their larval tail which are lost on metamorphosis to the sedentary adult form. These characteristics, together with the pharyngeal gill slits which are retained in the branchial basket of the adult animal, place ascidians and tunicates in general in the phylum Chordata to which all higher vertebrates belong. While in the adult form these animals lack a backbone and are invertebrates, they occupy a unique niche in the animal kingdom by simultaneously being classified as chordates. Ascidians, and tunicates in general, can indeed be considered man's closest living invertebrate relatives and represent an important juncture in the evolutionary tree between invertebrates and vertebrates [4,5].

After a period of speculation in terms of the vanadium oxidation number in the ascidian blood [1,6–9], the first unambiguous proof for +3 as the in vivo occurring vanadium oxidation state was delivered by Lybing [10]. Based on a detailed comparison of the optical spectra of blood cell lysates from *Ascidia obliqua* with those of solutions containing $[V(H_2O)_6]^{3+}$ and its hydrolysis products, the close similarity became evident.

Soon after the appearance of Lybing's report, Boeri and Ehrenberg [11] provided further evidence for V(III) as the in vivo oxidation state in a comparative spectrophotometric and magnetochemical study of blood cells from *A. obliqua* and *P. mammillata*. In the same year, Bielig and coworkers [12] published a detailed investigation of the chemistry occurring in cytosols of *P. mammillata*. The title of their publication, "Haemovanadin—A Sulfato Complex of Vanadium(III)", shows that "haemovanadin" is not used an as expression for porphyrin-bound V(III) but rather to indicate the presence of one or more sulfato ligands. (An attempt to clarify the confusion around "haemovanadin" can be found in Sec. 3.2.3 of [23].)

The conclusions drawn in these early reports about the status of V(III) in blood cell lysates experienced new attention after studies of V(III) in intact blood cells with more sophisticated analytic techniques [2,3,27]. Investigations of intact *Ascidia ceratodes* blood cells by ^{1}H NMR measurements [13] and sulfur K-edge X-ray absorption spectroscopy [14–16] have provided evidence for the coordination of at least one sulfato ligand to V(III). In addition, Tullius et al. [17] observed no changes in the coordination environment around the V(III) center upon cell lysis. This was concluded because the V(III) X-ray absorption edge features (Fig. 2 in [17]) and the appropriate EXAFS data (Figs. 5–7 in [17]) do not differ between intact and lysed cells. Thus, the suggestions of Lybing [10], Boeri and Ehrenberg [11], and Bielig et al. [12] that V(III) sulfato complexes should be present in lysate solutions from blood cells of *A. oliqua* and *P. mammillata* could be valid also for the status of V(III) in the vanadocytes of both these species.

A striking physiological feature of the ascidians is their efficiency to accumulate vanadium. From the surrounding seawater, where vanadium is present as V(V) in nanomolar quantities, it is accumulated via a V(V)/V(IV)/V(III) reduction process by the blood cells against an at least 10^6-fold concentration gradient and stored there in the oxidation state +3 [18]. Evidence was presented that vanadium uptake from the seawater occurs in the branchial basket [18,19]. Chelate-bound V(IV) was found using electron paramagnetic resonance (EPR) to characterize vanadium binding in the branchial net [20]. Thus, a V(IV) species should be the initial reduction product. In a second step the reduction of V(IV) to V(III) should occur either extra- or intracellularly [27]. Current experimental evidence clearly favors a two-step reduction process during vanadium accumulation by the ascidians, in contrast to suggestions that a direct intracellular reduction of V(V) to V(III) takes place [27].

Further results related to various aspects of the vanadium biochemistry in the ascidians that appeared since Henze's work have been collected in a number of reviews [2,3,21–27]. However, despite all the tremendous efforts over the last 80 years, the physiological role of vanadium in these animals is far from being understood and we are still at a level that has already been characterized in [13]: "The number of unanswered questions greatly exceeds the answered ones".

Until recently, the ascidians were believed to be the only sea animals which accumulate vanadium. However, the fan worm *Pseudopotamilla occelata* was recently shown to be another sea animal containing large amounts of vanadium [28] (see also Chapter 14 of this book). Using X-ray absorption spectroscopy with synchrotron excitation (XANES/EXAFS) the vanadium oxidation state was estimated to be +3. Also, the X-ray features of the V(III) form in *P. occelata* are quite similar to that of the ascidians [17]. This led Ishii et al. to the conclusion [28] that the coordination modes of V(III) in both animals are comparable. Application of electron microprobe analysis has revealed that most of the vanadium is concentrated in the epidermis of the bipinnate radiole, a localization similar to the membrane-bound vanadium fraction that will be discussed below.

The biological effects of vanadium in the mammalian biochemistry have been studied and discussed mainly in relation to V(V) and V(IV), and appropriate effects of V(III) were not considered because of its susceptibility to oxidation. It has to be mentioned, however, that the in vitro treatment of the serum iron transport protein transferrin with V(III) revealed the formation of a very stable complex [29]. A later study has shown that the V(III)-transferrin derivative is even more resistant against oxidation than its V(IV) analog [30].

A detailed literature search and numerous discussions with colleagues revealed the need to focus our presentation onto the hydrolytic reactions of V(III). This area of V(III) chemistry has been poorly explored [31]. From a theoretical point of view [33] and based on our own experience with the chemistry of V(III), many features of V(III) are similar to Fe(III). Thus, there should be similarities between the hydrolysis reactions of V(III) as compared to analogous processes taking place with Fe(III). This is important because the hydrolysis of iron(III) is a field that has been studied more frequently than that of V(III) [32–35].

1.2. Chemical Aspects of the Vanadium Biochemistry in Ascidians

Two opposing views in terms of the questions for the intracellular acidity in vanadium-accumulating blood cells of ascidians and the coordination mode of V(III) became bones of contention in the last 20 years:

1. The intracellular pH is low and nonchelated aqua-vanadium(III) is present [13,17,27,36–39], or
2. The intracellular pH is near neutral and chelated V(III) is the prevailing in vivo form [40–45].

We will show below that this conflict is unnecessary and has a retarding influence on the progress in this field. It is also unfortunate that results from papers published in the 1950s and 1960s have been neglected or misinterpreted. A crucial example is the question of whether the in vivo vanadium is involved in oxygen transport or not. This was already answered in Henze's 1911 paper [1]. There it was explained that the blood of *P. mammillata* has no specific oxygen-binding capacity. The original experiments on this topic were done by Winterstein in 1909, even earlier than the discovery of vanadium in ascidians [46]. Nevertheless, this problem was addressed again in the 1970s leading to the same answer as that given by Henze [47].

In the course of this chapter we would like to emphasize the possibility that more than one chemical form of V(III) exists in the blood cells. This is connected to a publication dealing with the histogenesis of the vanadocytes by Kalk in 1963 [48]. In this work, the histogenesis of blood cells from *Ascidia pygmaea* was followed by electron microscopy. It was found that in an initial stage of development, vanadium is seen to accumulate on the vacuolar membranes of amoebocytes. The accumulation of this membrane-bound vanadium becomes so tremendous that at a later stage vanadium-containing granules are separated from the membrane into the intravacuolar region, which should take place in the signet ring cells. A ferritin-like storage of vanadium was suggested to occur in the compartment cells. Different pH values are believed to be connected with the different forms of vanadium. There seemed to be zones of near-neutral pH at the vacuolar membrane of the vanadophores in an earlier stage of development. The insoluble vanadium was suggested to become soluble again accompanied by a drop in pH in the morula cells.

Some of the suggestions by Kalk were later criticized. The criticism concerned her ideas related to cell differentiation that were based on an earlier model by Endean [102]. Scippa et al. presented a different view on the cell types that are passed during development from a lymphocyte cell to a mature morula cell [103]. Also, her OsO_4-staining approach for localization of vanadium may indicate the presence of other reducing agents instead of V(III) [16]. Nevertheless, the presence of different chemical forms of V(III) in various blood cell types was suggested for the first time by Kalk [48]. Her vision-like ideas found support in several related papers, as discussed below.

Bielig et al. correlated the vanadium accumulation efficiency with animal body weight [19]. It was found that younger animals show a much faster vanadium uptake than adult ones. Based on this relationship, Bielig et al. drew the conclusion that a relation exists between the chemistry during accumulation and the evolutionary position of the animal [18].

In a study using different blood cell types from *P. mammillata* that were cryofixed and investigated by X-ray microanalysis, the following locations of intracellular vanadium were found [49,50]:

—A high vandium content in a zone of the vacuolar membrane in vacuolated amoebocytes
—Large amounts of vanadium in electron-dense granules of granular amoebocytes
—High contents of vanadium in the vacuolar membrane and in an intravacuolar zone of the signet ring cells
—Large amounts of vanadium in electron-dense granules of a new type of compartment cells [49,50]

These cell types in which membrane-bound vanadium was found are identical with those in which Kalk observed the same phenomenon [48]. It has to be mentioned in this connection that the detection limit of this technique is in the order of 1 mM, which makes it difficult to visualize any solvated V(III) present in such concentrations in intravacuolar fluids.

From cell separation studies of Michibata and coworkers with blood cells from *Ascidia ahodori* it emerged that in the course of fractionation (Ficoll layering of cell suspension and concurrent density centrifugation) 60% of the vanadium originally present gets lost [51]. We assume that the V(III) originally present in a soluble form was the fraction that was lost when the washed cells were analyzed by neutron activation analysis. Only the membrane-bound vanadium might have persisted the washing procedure.

In line with Michibata's results are those published by Roman et al. [52]. These workers found significant levels of vanadium in cell residues after lysis with blood cells from *Ascidia dispar*. It was suspected that nonlysed metal ions should be associated with membranes.

These facts show that the conclusions by Kalk [48] need more attention in future work. If her ideas are correct, the intracellular pH as well as the chemical form of intracellular V(III) has to be considered as an evolutionary phenomenon. If one could look into a vanadocyte vacuole of a mature ascidian, it seems even very feasible to see a higher pH value in the vanadium deposits around the membrane and in the globules whereas low pH values (as cited in the previous section) would become visible at places where the mobilized V(III) resides. Therefore, it is dangerous to take the intracellular pH of vanadocytes as a fixed value. The cells are in a developmental process [24] during which the status of vanadium might be changed. Therefore, the presence of vanadium is not only connected with one special cell type and if we speak about vanadocytes we have in mind cells which accumulate vanadium.

2. VANADIUM(III) IN AQUEOUS SOLUTION

2.1. Experimental Techniques

The study of solutions containing the V(III) aqua ion is difficult. One has to exclude impurities of V(IV) and the tendency of $[V(H_2O)_6]^{3+}$ to form complexes with anions of supporting electrolytes has to be considered. It became readily evident in the course of our study that commercial VCl_3 has a variable V(IV) content mainly dependent on time of storage and frequency of bottle opening. Thus, all sample solutions were carefully freed from V(IV) in the following way: Controlled potential electrochemical reduction was conducted at −1.00 V vs. Ag/AgCl over a mercury pool working electrode until the reduction current reached its background value. Because of the peculiarities of the $[VO(H_2O)_5]^{2+}$ reduction, one is only able to reduce this ion directly to $[V(H_2O)_6]^{2+}$. After complete reduction of the residual V(IV), the solution color is a clear violet. One obtains a solution that contains exclusively V(II) which was checked spectrophotometrically [68]. The next step of the procedure was a controlled potential oxidation of the V(II) solution back to V(III) at −0.2 V vs. Ag/AgCl. It was important to ensure that the starting solution for any kind of titration contains solely $[V(H_2O)_6]^{3+}$.

This was proved by recording UV-visible spectra in a thermostated sample holder and comparing the spectral features obtained with data from the literature (Table 1). A critical question was not only the removal of V(IV) residues but whether a V(III) complexation with chloride ion took place. Our solutions of pH 1.0 [I = 1.0 M (NaCl)] that were used for subsequent titrations had the following absorption maxima and molar absorbances: λ_1 = 397 nm, ϵ_1 = 9.5 M^{-1} cm^{-1}, λ_2 = 588 nm, ϵ_2 = 6.0 M^{-1} cm^{-1}. These data are, in terms of the band maxima, very similar to those from spectra recorded with single crystals containing $[V(H_2O)_6]^{3+}$ (λ_1 = 395 nm, λ_2 = 578 nm) [70]. Apparently, the presence of chloride ions does not lead to any appreciable formation of chloro complexes because our UV-VIS spectra (absorption maxima, extinction coefficients; see Table 1) are in agreement with previous data obtained in perchlorate media [68,69,71].

The evaluation of the pH dependence of redox potentials was studied using the program EFIT [72]. EFIT is a general computer program for fitting measured redox potentials to calculated $E^{0\prime}$ vs. pH

TABLE 1

UV-VIS Parameters for $[V(H_2O)_6]^{3+}$ in Acidic Aqueous Electrolytes Compared with Literature Data

Medium	Ionic strength	pH	$\lambda_{max\,1}$ (nm)	$\epsilon_{max\,1}$ ($M^{-1}\,cm^{-1}$)	$\lambda_{max\,2}$ (nm)	$\epsilon_{max\,2}$ ($M^{-1}\,cm^{-1}$)	Ref.
0.5 M $HClO_4$	0.5	—	400	—	590	—	98
0.5 M HCl	0.5	—	400	—	590	—	
5 M HCl	5	—	400	—	620	—	
0.5 M H_2SO_4	1.5	—	400	—	605	—	
5 M H_2SO_4	15	—	415	—	645	—	
0.1 M $HClO_4$	0.1	1	400	8.3	590	5.6	69
0.2 M $HClO_4$	0.2	0.69	400	8.7	590	5.8	
0.87 M $HClO_4$	0.9	0.07	400	9.4	590	5.9	
0.94 M $HClO_4$	1.2	0.1	400	8.3	~585	5.5	71
1 M $HClO_4$	1	0	400	8.3	585	5.8	67
0.5 M HCl/0.5 M NaCl	1	0.3	400	9.3	590	6.0	
0.1 M HCl/0.9 M NaCl	1	0.92	397	9.5	588	6.0	
0.1 M HCl/0.9 M NaCl/0.15 M Na_2SO_4	1.45	0.92	400	11.2	606	6.9	89
3 M $(NH_4)_2SO_4$	9	2.00	406	15.2	617	9.1	89

plots for cases where multiple equilibria occur. In each iteration step, the mass balance equations are solved for the free concentrations of the metal ions using the subroutine COGS taken from a BASIC version of COMICS [73]. The fitting function is the Nernst equation and the equilibrium constants (hydrolytic and complex formation constants) are determined by means of the Marquardt algorithm [74]. The setup of EFIT is very similar to an approach introduced by Gampp [75,76].

For evaluation of spectrophotometric titrations we applied the programs DIMFIT [67] and SPECFIT [77], respectively. The species distribution diagrams were computed with COMICS [73]. In the course of a DIMFIT run, the mass balance equations are solved using the subroutine COGS taken from the program COMICS [73]. The equilibrium constants are refined by any nonlinear least squares method. In the present case, the Marquardt approach was used [74]. The fitting function was Beer's law in the form $A_{430} = \epsilon_{\mathrm{dimer}} \cdot c_{\mathrm{dimer}} \cdot d$, where d is the optical pathlength in cm.

The equipment used for electrochemical and spectrophotometric measurements as well as the calibration is described elsewhere [67,72]. Calibration of the pH-measuring device was done using commercial U.S. National Bureau of Standards (NBS)-buffers of pH 1.68, 4.01, and 6.86. A description regarding the exclusion of influence by liquid junction potentials on the equilibrium data is given in [66].

The redox potentials of the V(III)/V(II) electrode reactions were measured in dependence on pH using cyclic voltammetry at a hanging mercury drop electrode. For the evaluations of equilibrium data regarding the interactions of V(III) with sulfate, we assumed that the mass balance $[SO_4]_t = [HSO_4^-] + [SO_4^{2-}]$ is operative because we never worked at very high H^+ concentrations where hydrogen sulfate is also protonated. The appropriate equilibrium constants for $K^{\mathrm{H}}_{\mathrm{HSO_4}} = [H^+]\,[SO_4^{2-}]/[HSO_4^-]$ were taken from the U.S. National Institute of Standards and Technology (NIST) data base for $I = 1.0$ M ($\log K^{\mathrm{H}}_{\mathrm{HSO_4}} = -1.1$) and for $I = 3.0$ M ($\log K^{\mathrm{H}}_{\mathrm{HSO_4}} = -0.9$) [95]. During all computations, the first protonation constant of sulfate was implemented in the input file of the appropriate program (EFIT, DIMFIT, as well as SPECFIT). Thus we never obtained and are not presenting now constants modified by consideration of side reaction coefficients, but equilibrium constants with SO_4^{2-} as the ion participating in the complex formation.

2.2. Hydrolysis of Aqua-Vanadium(III)

For solutions containing the V(III)-aqua system, the existence of the species $[V(H_2O)_6]^{3+}$, $[V(OH)(H_2O)_5]^{2+}$, $[V(OH)_2(H_2O)_4]^{+}$, and $[V_2(\mu_2\text{-}O)\text{-}(H_2O)_{10}]^{4+}$ is well established [53–57]. The V^{3+} ion has been shown to exist as the distorted octahedral $[V(H_2O)_6]^{3+}$ complex in $[V(H_2O)_6]\text{-}[H_5O_2](CF_3SO_3)_4$ [58] and in the alumn $Cs[V(H_2O)_6](SO_4)_2{\cdot}6H_2O$ [59]. The species $[V(OH)(H_2O)_5]^{2+}$ and $[V(OH)_2(H_2O)_4]^{+}$ derived by deprotonation of $[V(H_2O)_6]^{3+}$ are expected to possess structures as shown in Fig. 4 below. For convenience in the formulation of the expected stepwise oligomerization, the bishydroxo monomer complex $[V(OH)_2(H_2O)_4]^{+}$ was formulated as the cis isomer. However, we do not rule out the existence of the trans isomer in solution. The proposed structure of the dimer as singly μ-oxo-bridged species is in contrast to an earlier suggestion of Pajdowski and Jezowska-Trzebiatowska who expected a bis-μ-hydroxo-bridged complex [57] based on the effective magnetic moment of 2.7 μ_B, measured at room temperature in solutions containing the dinuclear species. Their assumption was based on the idea that μ-oxo bridging should result in a strong spin coupling through the bridge and disappearance of the paramagnetism. However, from the systematic investigation of μ-oxo-bridged V(III) dimers by Wieghardt and coworkers, it follows that the bridging type cannot be assigned by means of the room temperature magnetic moment alone [60–63]. They showed that μ-oxo bridging does not necessarily give rise to diamagnetic dimers of V(III). Important features in these dinuclear systems are the bridging angle and the V(III) · · · V(III) distance, which influence the extent of intramolecular spin exchange coupling and the positions of the V-O-V charge transfer band. Nevertheless, and despite bridging, the spins of both vanadium(III) ions (d^2) remain fully aligned [63].

Another interesting property of the dinuclear V(III) species is a resonance effect in the Raman spectra [64]. The Raman band due to the first overtone of the antisymmetric V-O-V stretching is strongly resonance-enhanced, which parallels the behavior of dinuclear μ-oxo molybdenum(V) complexes [65] delivering decisive support for the μ-oxo bridging mode in $[V_2(\mu_2\text{-}O)(H_2O)_{10}]^{4+}$. Thus, the dinuclear complex formed as a result of hydrolysis in the V(III)-aqua system is proposed possessing the structure shown below in Fig. 4.

It was clear for some time that the speciation of the V(III)-aqua system requires a thorough investigation [31]. In analogy to the Fe(III)-aqua system [34], we expected additional species at pH $\geqslant$ 3 and at $[V(III)]_t \geqslant 10^{-3}$ M. This led us to do a careful reinvestigation of the V(III) hydrolysis reactions [66,67].

2.2.1. *Speciation by Redox Measurements*

The strategy for obtaining reliable equilibrium constants for the different protolytic reactions followed two lines. The first was to measure the redox potential in dependence on pH at low $[V(III)]_t$. Initially, species distribution curves were calculated to establish the extent of dimer formation in dependence on $[V(III)]_t$. As input data for these computations the values for the equilibrium constants of reactions (1)–(3) [for definition of overall equilibrium constants, see Eq. (5)] were taken from Pajdowski's papers [54,56].

$$[V(H_2O)_6]^{3+} + OH^- \overset{\beta_{11}}{\rightleftharpoons} [V(H_2O)_5(OH)]^{2+} + H_2O \quad (1)$$

$$2\,[V(H_2O)_6]^{3+} + 2\,OH^- \overset{\beta_{22}}{\rightleftharpoons} [V_2O(H_2O)_4]^{4+} + 2\,H_2O \quad (2)$$

$$[V(H_2O)_6]^{3+} + 2\,OH^- \overset{\beta_{12}}{\rightleftharpoons} [V(H_2O)_4(OH)_2]^{+} + 2\,H_2O \quad (3)$$

It became clear that at $[V(III)]_t \leqslant 1$ mM the dimerization can be neglected. Therefore, one should be able to estimate the equilibrium constants for reactions (1) and (3) (i.e., β_{11} and β_{12}) without considering the dimerization. Thus, β_{11} and β_{22} were obtained by measuring the V(III)/V(II) redox potential in dependence on pH in the range $1.0 \leqslant \mathrm{pH} \leqslant 3.5$.

The basis for this approach is a reliable and accurate value of the V(III)/V(II) redox potential:

$$[V(H_2O)_6]^{3+} + e^- \overset{E^{0'}}{\rightleftharpoons} [V(H_2O)_6]^{2+} \quad (4)$$

It is well established that the redox couple $[V(H_2O)_6]^{3+}/[V(H_2O)_6]^{2+}$ gives rise to a reversible electrode reaction if cyclic voltammograms are recorded with sweep rates up to ~ 100 mV/sec [81]. This enabled us to estimate the formal potential of reaction (4) in a straightforward manner. Figure 1a shows the results of a series of cyclic voltammograms (CVs) recorded at pH 1.0 in dependence on scan rate. The separation

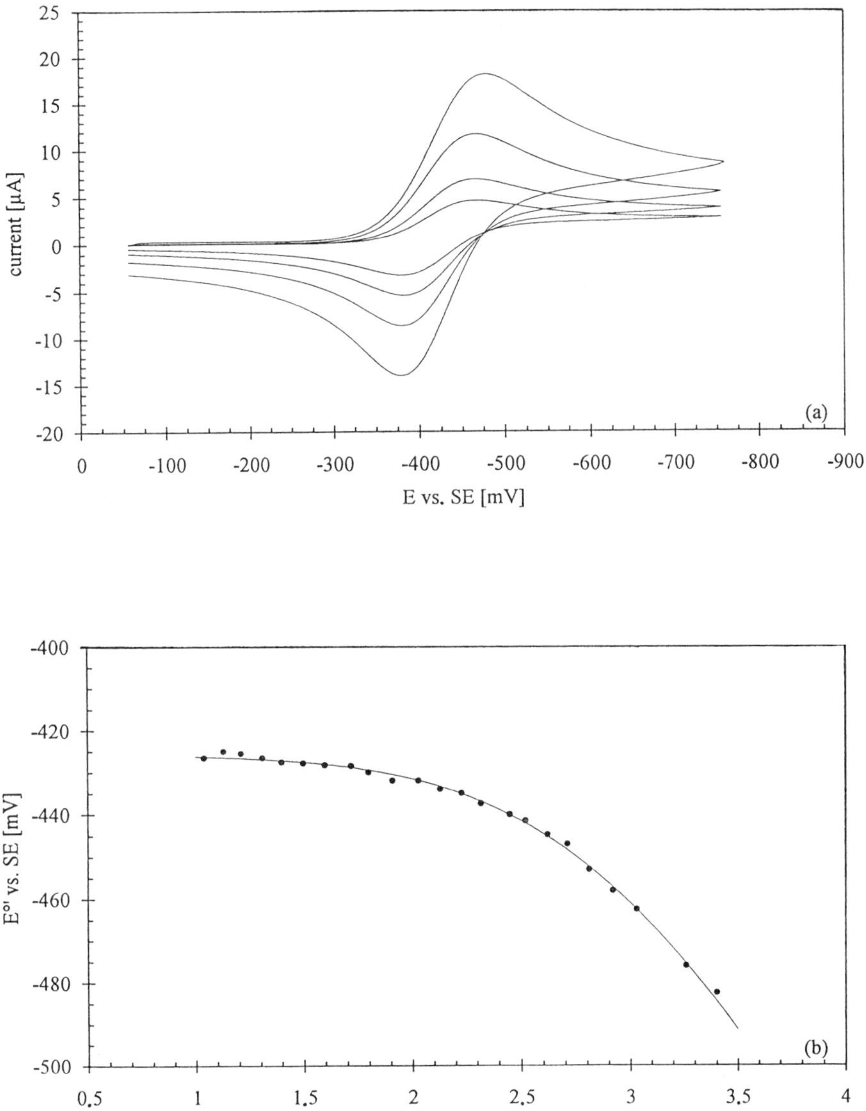

FIG. 1. (a) Cyclic voltammograms of 1 mM $[V(III)]_t$ in 0.5 M NaCl/0.5 M HCl (25°C) recorded at a hanging mercury drop electrode, scan rates applied (from top to bottom): 0.5, 0.2, 0.1, and 0.05 V s^{-1}. (b) $E^{0'}$ vs. pH plot for the V(III/II) couple at pH 1.0–3.5, 1 mM $[V(III)]_t$, $I = 1.0$ M (NaCl), 25°C, solid line: fit to the experimental points (EFIT).

between cathodic and anodic peak potentials remains close to 60 mV for all scan rates, the peak currents divided by the square root of scan rates yield a constant ratio, and the quotient i_{pa} (anodic peak current)/i_{pc} (cathodic peak current) is essentially unity. Thus, chemical and electrochemical reversibility is given and the potentials measured in this way possess thermodynamic relevance [66].

The formal potential measured for reaction (4) at $I = 1.0$ M (NaCl) with $E^{0\prime} = -426 \pm 0.5$ mV vs. Ag/AgCl/KCl$_{sa}$ (25°C; pH 1.0) is in excellent agreement with the values from previous studies [78–82]. The value found in 3M NaCl is $E^{0\prime} = -412 \pm 0.5$ mV vs. Ag/AgCl/KCl$_{sa}$ (25°C; pH 1.0) and has been interpreted as a result of a different variation of the activity coefficients of V(II) as compared to V(III) when going from 1.0 to 3.0 M ionic strength [66].

Based on the knowledge of the pH-independent formal potential, the pH dependence of $E^{0\prime}$ was evaluated. In Fig. 1b, $E^{0\prime}$ values resulting from measurements between pH 1.0 and 3.5 are shown. The curve passing through the experimental points is a computed fit [66,72].

The slope of the plot in Fig. 1b is slowly increasing to a value of -46 mV/pH between pH 3.2 and 3.4. A value of -119 mV/pH is reached at pH > 3.4 [85]. Because the hydrolysis of $[V(H_2O)_6]^{2+}$ starts only at pH > 7.0 it is clear that the shape of the curve in Fig. 1b is caused by the stepwise occurrence of the equilibria (1) and (3). The values for log β_{11} as well as log β_{12} resulting from the fit in Fig. 1b are shown in Table 2.

The dimerization equilibrium (2) leads to a negative shift of the electrode potential at constant pH and elevated values of $[V(III)]_t$ as predicted by the Nernst equation. This was used to estimate β_{22} at pH 2.5 and $[V(III)]_t \geqslant 5$ mM.

2.2.2. *Speciation by Spectrophotometric Titrations*

The first systematic approach to clarify the speciation of aqua-V(III) using photometric titrations was published by Furman and Garner [71]. It was noticed that complex formation occurs even at low pH if the solutions contained sulfuric acid (see absorption maxima at 5 M H_2SO_4 in Table 1). They observed the formation of a strong band with an absorption maximum at 430 nm as a result of the pH increase. The species that should give rise to this absorption was believed to be $[V(H_2O)_5(OH)^{2+}]$.

TABLE 2

Equilibrium Constants for Hydrolysis of $[V(H_2O)_6]^{3+}$ Obtained by Redox and Spectrophotometric Measurements

$[V(H_2O)_6]^{3+}$ Hydrolysis						
$-\log\beta_{11}$	$-\log\beta_{22}$	$-\log\beta_{12}$	$-\log\beta_{38}$	$-\log\beta_{412}$		Ref.
2.59 ± 0.01	3.53 ± 0.08	6.48 ± 0.07	—	—	a,1	67
2.60 ± 0.01	3.59 ± 0.01	6.83 ± 0.03	23.3 ± 0.4	33.80 ± 0.04	a,2	67
2.70	—	—	—	—	b	71
2.51	—	—	—	—	c	11
2.52	—	—	—	—	d	53
2.70	4.63	—	—	—	e	55
2.57	3.90	6.27	—	—	f	54
2.85	3.90	6.70	—	—	g	56
pK_{a1}	$\log K_d$	pK_{a2}	$-\log K_{tr}$	$-\log K_{te}$		
2.59 ± 0.01	1.65 ± 0.09	3.89 ± 0.08	—	—	a,1	67
2.60 ± 0.01	1.61 ± 0.02	4.23 ± 0.04	12.9 ± 0.5	3.67 ± 0.5	a,2	67
2.70	—	—	—	—	b	71
2.51	—	—	—	—	c	11
2.52	—	—	—	—	d	53
2.70	0.77	—	—	—	e	55
2.57	1.24	3.70	—	—	f	54
2.85	1.80	3.85	—	—	g	56

[1]Redox measurements.
[2]Spectrophotometric measurements.
[a]I = 1.0 M (NaCl), 25°C.
[b]I = 1.0 M ($NaClO_4$), 25°C.
[c]Conditions not mentioned.
[d]I = 1 M (1 M $NaClO_4$), 25°C.
[e]I = 1 M (1 M $LiClO_4$), 23°C.
[f]I = 1 M (1 M KCl), 22.5°C.
[g]I = 1 M (1 M NaCl), 20°C.
Note: For definition of the β values, see Eqs. (1)–(3) and (7) and (8); for the K values see Fig. 4.

Soon after the appearance of this paper, spectrophotometric studies of cell lysates from *Ascidia obliqua* [10,11] revealed that the spectra of the lysis solutions looked very similar to those obtained by Furman and Garner [71]. Correlations of the absorbance at 430 nm with pH at different excess concentrations of sulfate led Boeri and Ehrenberg to the conclusion that the hydrolysis of trivalent vanadium is sulfate ion–sensitive [11].

In a study by Gandeboeuf and Souchay [53], it was found that the 430-nm maximum is caused by the appearance of a V(III)-aqua dimer. Plots similar to those obtained by Boeri and Ehrenberg were presented revealing that the dimer absorbance has an unusual pH dependence. After a steady increase between pH 2 and 3, a sudden absorbance drop was noticed at pH > 3.0. It was again observed that the hydrolytic properties of aqua-V(III) in the presence of sulfate differ to solutions containing a noncomplexing supporting electrolyte.

Pajdowski's investigations of V(III) hydrolysis in the presence of chloride ions [54,56,57] gave similar results in terms of the pH dependence of the dimer absorbance ($\epsilon_{dimer} = 3300\ M^{-1}\ cm^{-1}$ [56]) as compared to those of Gandeboeuf and Souchay [53]. He determined the equilibrium constants of reactions (1)–(3) by pH potentiometry [54] as well as spectrophotometric titrations [56], and delivered the first quantitative data regarding the extent of dimer formation (see Table 2). The appearance of higher aggregated V(III) species was made responsible for the drastic decrease of the 430-nm band at pH >3.0.

The peculiarities occurring in the UV-VIS spectra of V(III)-aqua solutions as a result of hydrolysis are shown in Fig. 2a–c. The spectra were recorded at pH 1.5–4.5 with a solution containing 4 mM $[V(III)]_t$ [$I = 1.0$ M (NaCl)]. Because of the high molar absorbance of the dimer at 430 nm, the spectrum of $[V(H_2O)_5(OH)]^{2+}$ is not observed. Attempts to record this spectrum were even unsuccessful with $[V(III)]_t < 0.5$ mM because the spectroscopic contributions of this species are again completely superimposed by the dimer absorbance between pH 1.0 and 2.0. Thus, β_{11} [Eq. (1)] was obtained by application of DIMFIT as described above.

On the other hand, the intensity of the dimer spectrum is of importance since the pH dependence of the dimer absorbance can be used to evaluate the equilibrium constants belonging to the mass balance of the V(III) hydrolysis between pH 1 and 4. Formation and decay of $[V_2(\mu_2\text{-}O)(H_2O)_{10}]^{4+}$ is coupled with all other equilibria [see Eqs. (1)–(3), (7)–(13)]. The spectra in Fig. 2a indicate a continuous growing of the dimer absorbance as a consequence of a stepwise pH increase from 1.46 to 3.5. A further increase of the pH leads to a drop of the dimer peak (Fig. 2b). Over a small pH interval (between pH 3.71 and

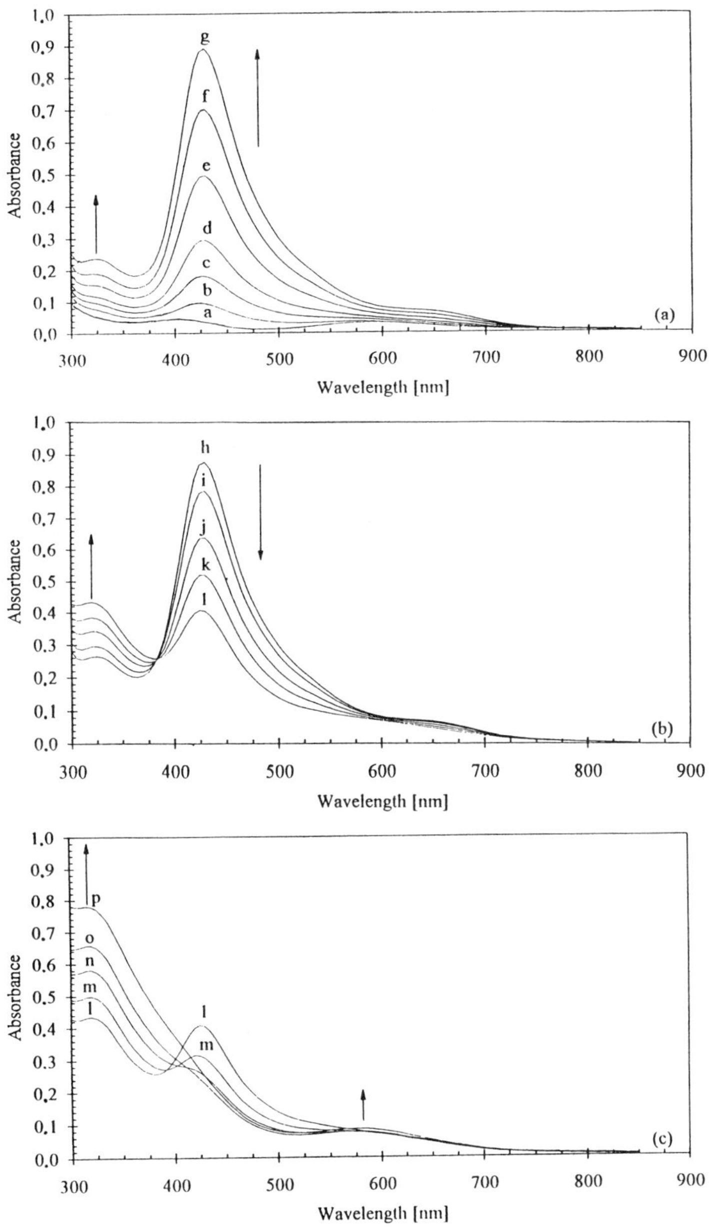

FIG. 2. UV-VIS spectra of 4 mM $[V(III)]_t$ in dependence on pH at I = 1.0 M (NaCl) and 25°C. (a) pH: a = 1.46, b = 2.00, c = 2.25, d = 2.46, e = 2.75, f = 3.04, g = 3.50. (b) pH: h = 3.71, i = 3.80, j = 3.86, k = 3.91, l = 3.96. (c) pH: l = 3.96, m = 4.05, n = 4.16, o = 4.28, p = 4.45.

3.91), an isosbestic point appears at 370 nm and a band with a maximum at 330 nm becomes more dominant. Finally, as shown in Fig. 2c, the spectra no longer pass through the isosbestic point, the dimer band disappears completely, and the absorption at 330 nm shows a larger increase.

Two plots of the absorbance at 430 nm vs. pH from spectra of solutions containing 5 mM and 10 mM $[V(III)]_t$ are shown in Fig. 3. It was already demonstrated by Pajdowski [56] that plots as those of Fig. 3 reflect the molar fraction of the dimer relative to $[V(III)]_t$. To be able to use the pH dependence of the dimer absorbance for the determination of equilibrium constants along the mass balance of $[V(III)]_t$, the molar absorbance of the species $[V_2(\mu_2\text{-}O)(H_2O)_{10}]^{4+}$ must be known. The value of the ϵ_{dimer} at 430 nm was evaluated graphically and further refined using DIMFIT [67]. We obtained $\epsilon_{dimer} = 3000 \pm 50\ M^{-1}\ cm^{-1}$.

The estimated ϵ_{dimer} was used as a fixed input parameter for computations with DIMFIT. Various equilibrium models were tested before a satisfying fit to the experimental data was obtained. The suitability of the equilibrium models was tested according to the magnitude of the following statistical criteria: sum of error squares, goodness of fit, Abbe criterion, and the standard deviation of the individual equilibrium constants that had to be refined [88].

Among the possibilities tested only one model gave a sufficiently good fit. This particular model is described by a mass balance distribution over equilibria (1)–(3) and (7)–(8). Thus, equilibria (1)–(3) were completed by the successive formation of trimeric [Eq. (7)] and tetrameric species [Eq. (8)] together with a common definition of the mass balance. According to the tests with DIMFIT, these reactions take place with the consumption of 8 and 12 mol of hydroxide, respectively.

$$[V(III)]_t = [V(H_2O)_6^{3+}] + \Sigma\ m[V_m(OH)_k] \tag{5}$$

$$\beta_{mk} = [V_m(OH)_k^{3m+(-k)}]/[V(H_2O)_6^{3+}]^m[OH^-]^k \tag{6}$$

$$3\,[V(H_2O)_6]^{3+} + 8\,OH^- \overset{\beta_{38}}{\rightleftharpoons} [V_3(OH)_8(H_2O)_{10}]^+ + 8\,H_2O \tag{7}$$

$$4\,[V(H_2O)_6]^{3+} + 12\,OH^- \overset{\beta_{4\,12}}{\rightleftharpoons} [V_4(OH)_{12}(H_2O)_{12}] + 12\,H_2O \tag{8}$$

The equilibrium constants obtained by a fit over both plots in Fig. 3 are presented in Table 2 together with the values that resulted from measurements of $E^{0\prime}$ vs. pH. A comparison is provided in Table 2 also

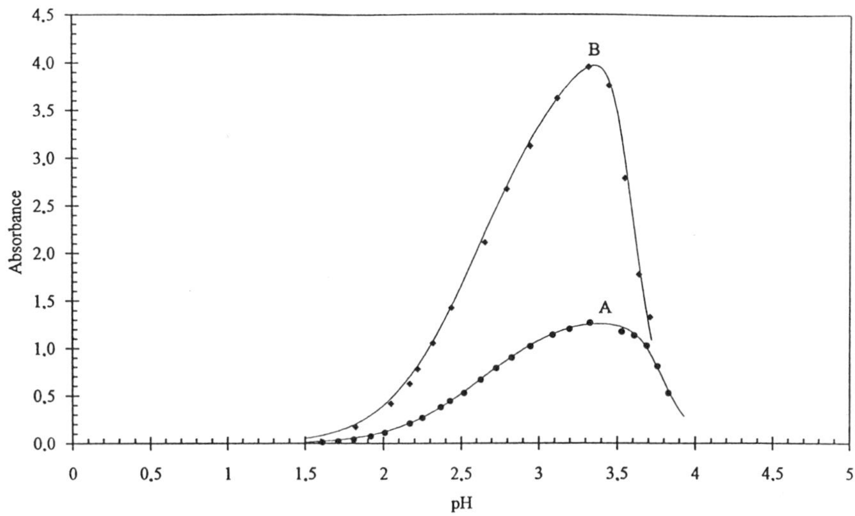

FIG. 3. Absorbances measured at 430 nm in dependence on pH. Curve A: 5 mM $[V(III)]_t$. Curve B: 10 mM $[V(III)]_t$ (I = 1.0 M (NaCl), 25°C). The solid lines are the fit to the experimental points (DIMFIT).

with previously published data. Input values in the form of overall equilibrium constants were unavoidable for the computations with DIMFIT [67]. However, the lucidity of such expressions in terms of chemical reactions taking place in solution is limited. Also, the formulas written for dimer, trimer, and tetramer are preliminary in character.

A better understanding of the V(III) hydrolysis as a step-by-step process becomes possible using successive equilibrium constants. These are introduced in Fig. 4 ([Eqs. (9)–(13)]; for magnitudes of the K values see Table 2). Figure 4 was drawn to show proposed solution structures for all hydrolytic species. The structures suggested for the trimer and the tetramer require some comment. The expressions for the composition of these species given in square brackets are deviated from proposed solution structures. The assignments of the solution structures of the trimer and tetramer have been made based on the number of protons released in the course of the overall formation reactions [this number is equal to the number of hydroxide ions consumed in the course of reactions (7) and (8)]. Both in terms of the overall equilibrium

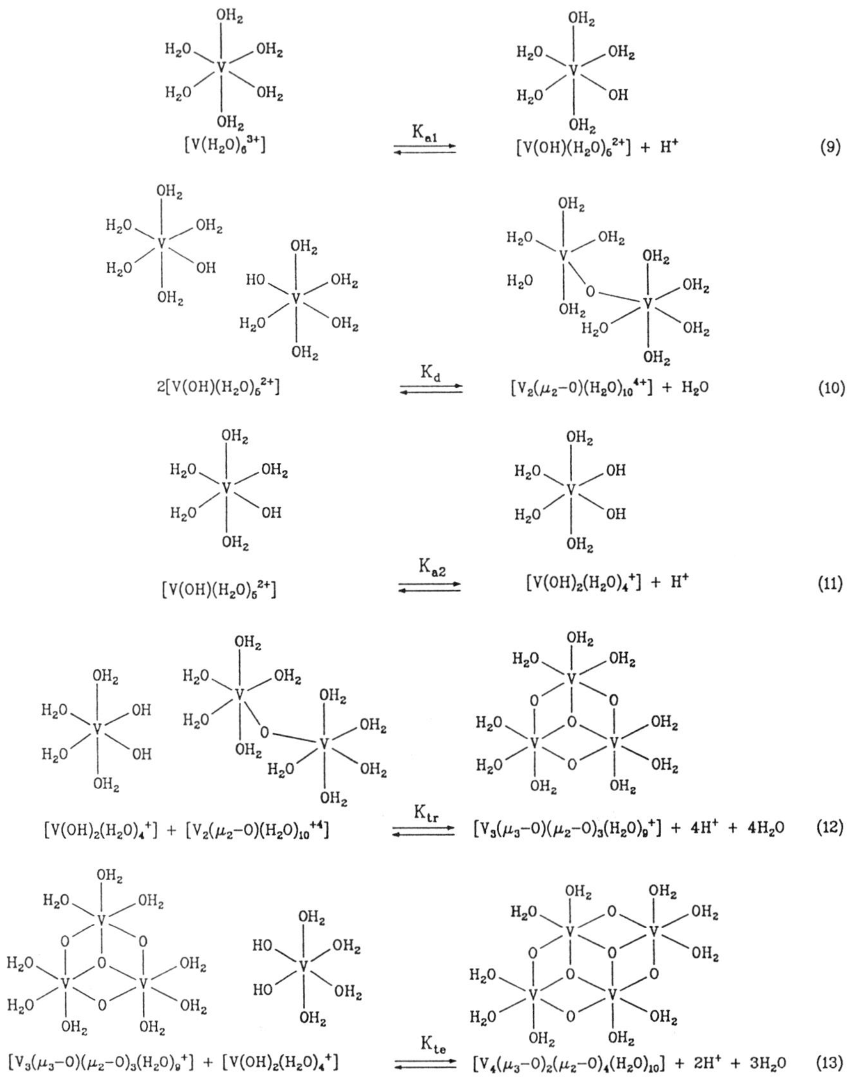

FIG. 4. Hydrolytic reactions occurring in the V(III)-aqua system in the framework of successively occurring equilibria (equilibria (9)–(13)): the equilibrium constants are defined as K values and proposed solution structures are drawn for each hydrolytic species.

model as well as regarding the solution structures proposed, the good fit of the curves through the experimental points in Fig. 3 can be taken as an indication for the quality of our approach.

The trimer structure is similar to that of trimeric V(III) complexes with carboxylato ligands. Cotton et al. showed that the $[V_3(\mu_3\text{-}O)]^{6+}$ core is a typical structural feature of these trimers [83]. A structure completely identical with that given for $[V_3(\mu_3\text{-}O)(\mu_2\text{-}O)_3(H_2O)_9]^+$ has been suggested for the isoelectronic (d^2) Mo(IV) trimer $[Mo_3(\mu_3\text{-}O)(\mu_2\text{-}O)_3\text{-}(H_2O)_9]^{3+}$ [84–86].

A number of tetrameric complexes with 3d M(III) ions have a central core identical with that proposed for $[V_4(\mu_3\text{-}O)_2(\mu_2\text{-}O)_4(H_2O)_{10}]$. The unit $[M_4(\mu_3\text{-}O)_2]^{8+}$ has been found in an Mn(III) tetramer [90], several synthetic Fe(III) tetramers [87,91,92], as well as in the mineral amarantite [93].

Based on the equilibrium constants given in Table 2, the species distribution diagrams shown in Fig. 5a (1 mM $[V(III)]_t$), 5b (5 mM $[V(III)]_t$), 5c (10 mM $[V(III)]_t$), and 5d (0.35 M $[V(III)]_t$) were calculated. Important trends related to the reaction pathways shown in Fig. 4 can be found. Dimer formation starts only after a certain amount of the monohydroxo complex has been formed. Formation of the tetramer (te) requires initial trimer (tr) formation while the latter starts only if substantial amounts of dimer and the species $[V(OH)_2(H_2O)_4]^+$ are present. Despite the fact that the formation of trimer and tetramer begins only at pH >3 at all $[V(III)]_t$ values, the starting pH for these processes is clearly concentration-dependent as is the case with the dimerization.

There is a large gap between the magnitudes of K_{tr} and K_{te} (Table 2) revealing that the tetramer is about nine orders of magnitudes more stable than the trimer. Therefore, the trimer seems to appear only as a transitional species in the course of tetramer formation. The fact that the trimer was present only as minor species under the experimental conditions relevant to Fig. 3 is the reason that $\log \beta_{38}$ has a distinctly larger standard deviation than all other constants in Table 2.

In connection with the buildup of the $[Fe_4(\mu_3\text{-}O)_2]^{8+}$ core [87], the suggestion was made that the tetramer might simply be formed by dimerization of two $[Fe_2(\mu_2\text{-}O)]^{4+}$ units. The distributions shown in Fig. 5 clearly indicate that such a pathway is impossible in the course of V(III)-aqua hydrolysis.

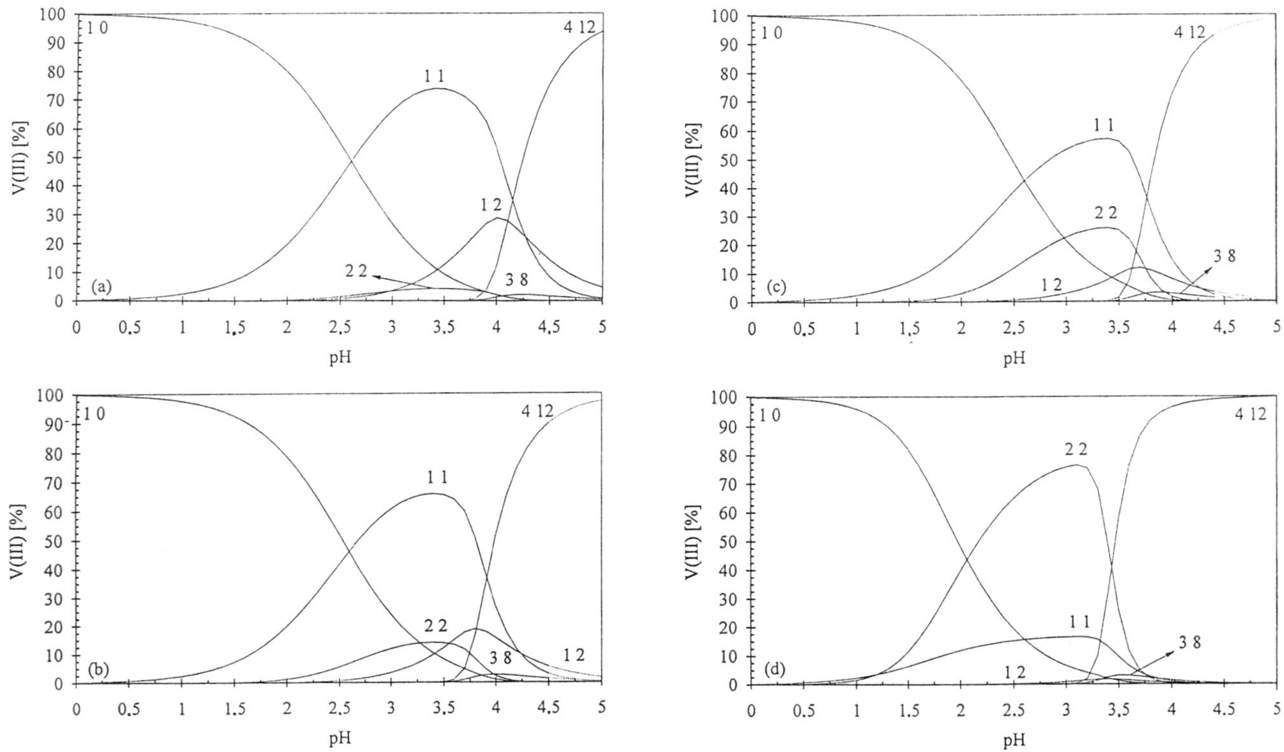

FIG. 5. Species distributions in the V(III)-aqua system in dependence on pH ($0 \leqslant pH \leqslant 5$) at different $[V(III)]_t$ values. (a) 1 mM $[V(III)]_t$, (b) 5 mM $[V(III)]_t$, (c) 10 mM $[V(III)]_t$, (d) 350 mM $[V(III)]_t$. Hydrolytic species are defined according to the index of the corresponding overall equilibrium constants [Eqs. (1)–(3) and (6)–(8)], overall equilibrium constants used for computation: see the second line in Table 2. Regarding the meaning of the dotted line portions, see final paragraph of Sec. 2.2.2.

Important for the assignment of the spectral features occurring at pH >3.5 is a comparison of Fig. 2b and 2c with Fig. 5b. This comparison clearly reveals that the isosbestic point in the spectra between pH 3.71 and 3.91 is not indicating the presence of only two species. Rather, it occurs by mere chance because the dimer concentration decreases while the tetramer concentration increases drastically and, at the same time, the concentration of $[V(OH)_2(H_2O)_4]^+$ has passed just a maximum and the trimer concentration approaches a maximum. By means of independent computations [67], we were able to show that the small increase of absorbance at 330 nm until pH 3.7 is caused by the dimer, followed by a region where both species deliver contributions, while at pH >4.2 this maximum is solely caused by the tetramer.

In the diagrams shown in Fig. 5, dotted lines appear for the relative concentration of species at pH ⩾4.5 because precipitation of a solid starts at this pH value.

2.3. Complex Formation of V(III) with Sulfate

2.3.1. Speciation by Redox Measurements

The presence of sulfate at places where intracellular V(III) occurs is a phenomenon as old as the discovery of vanadium in ascidians [1]. Application of modern analytical techniques for analysis of intracellular locations of vanadium in ascidians has shown in all studies also the presence of at least equimolar amounts of sulfate [27]. It seems reasonable to assume that a certain fraction of sulfate is coordinated monodentately to the V(III) center. In vitro proof for this suggestion was delivered in a solution Raman study [64]. The coordination of sulfate to V(III) in vivo was suggested based on broadening of the sulfate feature in sulfur-K XANES (x-ray absorption near edge structure) spectra [14–16]. It is thus highly desirable to have a detailed knowledge about the speciation of V(III) in the presence of sulfate.

Equilibrium constants published so far concerning hydrolysis of V(III) in the presence of sulfate as well as the formation of the complex $[V(SO_4)(H_2O)_5]^+$ are given in Table 3. In analogy to the Fe(III)-sulfate system, where the formation of $[Fe(SO_4)_2(H_2O)_4]^-$ is well established [94], we expected also the existence of $[V(SO_4)_2(H_2O)_4]^-$ in addition to the 1:1 complex reported so far. A proof for the occurrence of this species

TABLE 3

Equilibrium Constants for Hydrolysis and Complex Formation of $[V(H_2O)_6]^{3+}$ in the Presence of Sulfate

$[V(H_2O)_6]^{3+}$						
$\log \beta_{101}$	$\log \beta_{102}$	$-\log \beta_{111}$	$-\log \beta_{222}$	$-\log \beta_{121}$		Ref.
1.50 ± 0.05	1.9 ± 0.1	1.74 ± 0.05	[1.5]	[5.66]	1a	89
1.36 ± 0.02	1.95 ± 0.01	1.89 ± 0.03	[1.8]	[5.54]	1b	89
1.50 ± 0.04	[1.8]	1.85 ± 0.04	1.79 ± 0.01	6.40 ± 0.10	2	89
1.45	—	—	—	—	3	97
$\log K_1$	$\log K_2$	pK_{a1}	$\log K_d$	pK_{a2}		
1.50 ± 0.05	0.4 ± 0.2	3.24 ± 0.10	—	—	1a	89
1.36 ± 0.02	0.49 ± 0.03	3.23 ± 0.05	—	—	1b	89
1.50 ± 0.04	[0.3]	3.35 ± 0.08	1.91 ± 0.03	4.55 ± 0.10	2	89
1.45	—	—	—	—	3	97
—	—	2.92	—	6.44	4	96
—	—	—	—	5.24	4	10

[1a]Redox measurements, $I = 1.0$ M (NaCl, Na_2SO_4), 25°C.
[1b]Redox measurements, $I = 3.0$ M (NaCl, Na_2SO_4), 25°C.
[2]Spectrophotometric measurements, $I = 1.0$ M (NaCl, Na_2SO_4), 25°C.
[3]$I = 1$ M ($NaClO_4$), 25°C.
[4]Conditions not mentioned.

in solution was especially interesting because it was postulated to exist by Bielig et al. [12].

As was inferred in initial studies of the interactions of sulfate with Fe(III) [94], the shift of the V(III/II) redox potential in dependence on pH and increasing sulfate excess should be a good indication regarding the extent of complex formation with sulfate.

Figure 6 shows the $E^{0'}$ vs. pH dependence of the V(III/II) couple at v arious ratios $[V(III)]_t$:$[SO_4]_t$ [$I = 3.0$ M (NaCl); 25°C]. The solid lines result from an overall fit through all experimental points shown and the equilibrium constants obtained are given in Table 3.

To obtain a reasonable fit to the experimental points, an equilibrium model including formation of $[V(SO_4)(H_2O)_5]^+$, $[V(SO_4)_2(H_2O)_4]^-$, $[V(OH)(SO_4)(H_2O)_4]$, $[V_2(OH)_2(SO_4)_2(H_2O)_8]$, and $[V(OH)_2(SO_4)(H_2O)_3]^-$ had to be considered [89]. Appropriate overall equilibrium expressions

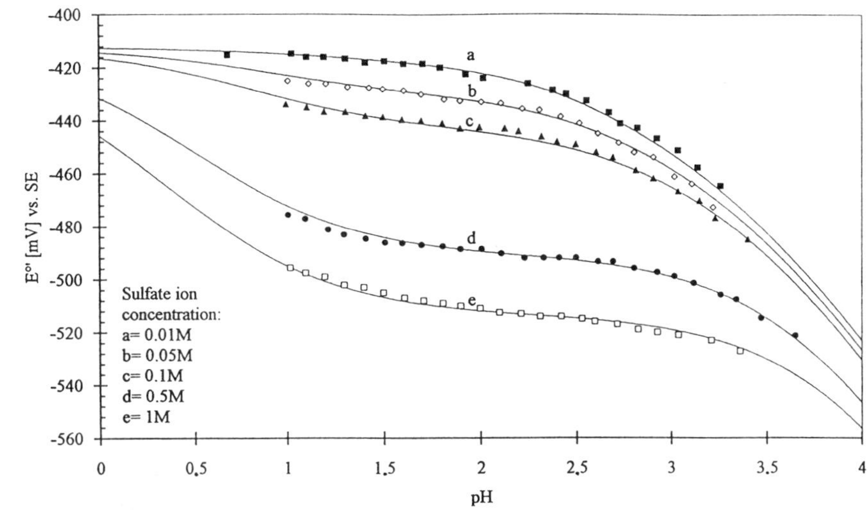

FIG. 6. $E^{0\prime}$ vs. pH plot for the V(III/II) couple at pH 1.0–3.5 in the presence of various sulfate concentrations but at a constant ionic strength (I = 3.0 M (NaCl, $NaSO_4$), 25°C): a = 0.01 M $[SO_4]_t$, b = 0.05 M $[SO_4]_t$, c = 0.1 M $[SO_4]_t$, d = 0.5 M $[SO_4]_t$, e = 1.0 M $[SO_4]_t$.

are formulated in Eqs. (16)–(20). EFIT computations were done to refine the constants β_{101}, β_{102}, and β_{111}. Fixed values for β_{222} and β_{121} (see Table 3, values in square brackets) were included in the computations. These were changed (as fixed input values) as long as statistical criteria for an optimal fit (see Sec. 2.2.2) were fulfilled.

Complementary measurements have been also done in solutions with I = 1.0 M (NaCl), 25°C. At this ionic strength, however, one is confined by the adjustment of high ratios $[SO_4]_t$:$[V(III)]_t$ to a limiting value of $[SO_4]_t$ = 0.333 M. All computations for the evaluation of equilibrium constants in the V(III)-sulfate system were done using a mass balance including the subsequent equilibria:

$$[V(III)]_t = [V(H_2O)_6^{3+}] + \Sigma m\,[V_m(OH)_k(SO_4)_n(H_2O)_{6-k-n}] \tag{14}$$

$$\beta_{mkn} = [V_m(OH)_k(SO_4)_n^{3m-(k+2n)}]/[V(H_2O)_6^{3+}]^m[OH]^k[SO_4]^n \tag{15}$$

$$[V(H_2O)_6]^{3+} + SO_4^{2-} \overset{\beta_{101}}{\rightleftharpoons} [V(SO_4)(H_2O)_5]^{+} + H_2O \tag{16}$$

$$[V(H_2O)_6]^{3+} + 2\,SO_4^{2-} \overset{\beta_{102}}{\rightleftharpoons} [V(SO_4)_2(H_2O)_4]^{-} + 2\,H_2O \tag{17}$$

$$[V(H_2O)_6]^{3+} + OH^- + SO_4^{2-} \overset{\beta_{111}}{\rightleftharpoons} [V(OH)(SO_4)(H_2O)_4] + H_2O \quad (18)$$

$$2\,[V(H_2O)_6]^{3+} + 2\,OH^- + 2\,SO_4{}^{2-} \overset{\beta_{222}}{\rightleftharpoons} [V_2(OH)_2(SO_4)_2(H_2O)_8] + 4\,H_2O \quad (19)$$

$$[V(H_2O)_6]^{3+} + 2\,OH^- + SO_4^{2-} \overset{\beta_{121}}{\rightleftharpoons} [V(OH)_2(SO_4)(H_2O)_4]^- + 2\,H_2O \quad (20)$$

The expressions for the appropriate successive equlibria are given in Fig. 7 together with the proposed solution structures for the individual species in the V(III)-sulfato system.

2.3.2. Speciation by Spectrophotometric Titrations

As described in the previous section, the interaction of V(III) with sulfate requires consideration of the successive formation of the 1:1 and 1:2 V(III)-sulfato complexes. It was also found by us that the coordinated water molecules of $[V(SO_4)(H_2O)_5]^+$ release protons at slightly higher and slightly lower pH values than is the case with the V(III)-hexaaqua ion (for comparison, see the pK_{a1} and pK_{a2} values in Tables 2 and 3). Because the limiting spectra of $[V(SO_4)(H_2O)_5]^+$ and $[V(SO_4)_2(H_2O)_4]^-$ should be helpful regarding the spectral V(III) speciation in intact ascidian blood cells and their lysates [13,99], it was one of our goals to obtain these spectra (it has been assumed in ref. [99] that the spectral parameters of V(III) in 5 M H_2SO_4 (Table 1) are only those of $[V(SO_4)_2(H_2O)_4]^-$. Furthermore, it was of interest if there is a general agreement with the equilibrium constants determined from $E^{0\prime}$ vs. pH correlations. It was decided to evaluate the dependence of the spectra on pH and the ratio $[V(III)]_t$:$[SO_4]_t$ along two ways: titrations at low pH (pH $<$2) should be evaluated using SPECFIT while DIMFIT should be applied for titrations over a broader pH range.

Figure 8a shows spectra recorded at the constant pH 0.92 [I = 1.0

FIG. 7. Hydrolytic and complex formation reactions occurring in the V(III)-sulfato system in the framework of successively occurring equilibria [Eqs. (21)–(27): the equilibrium constants are defined as K values (except for trimer (26) and tetramer (27) formation) and proposed solution structures are drawn for each hydrolytic species.

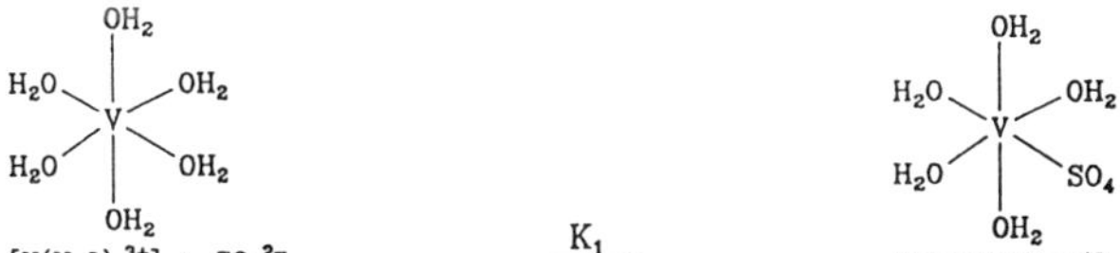

$$[V(H_2O)_6^{3+}] + SO_4^{2-} \overset{K_1}{\rightleftharpoons} [V(SO_4)(H_2O)_5^{+}] \tag{21}$$

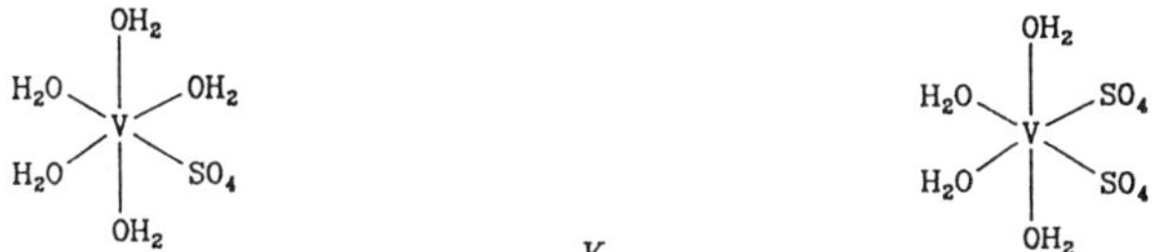

$$[V(SO_4)(H_2O)_5^{+}] + SO_4^{2-} \overset{K_2}{\rightleftharpoons} [V(SO_4)_2(H_2O)_4^{-}] \tag{22}$$

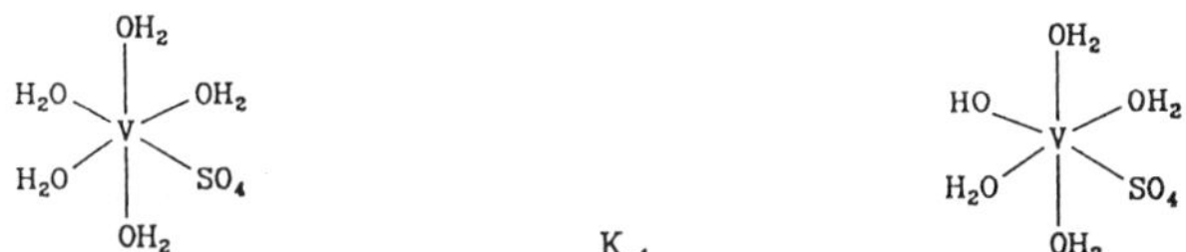

$$[V(SO_4)(H_2O)_5^{+}] \overset{K_{a1}}{\rightleftharpoons} [V(SO_4)(OH)(H_2O)_4] + H^+ \tag{23}$$

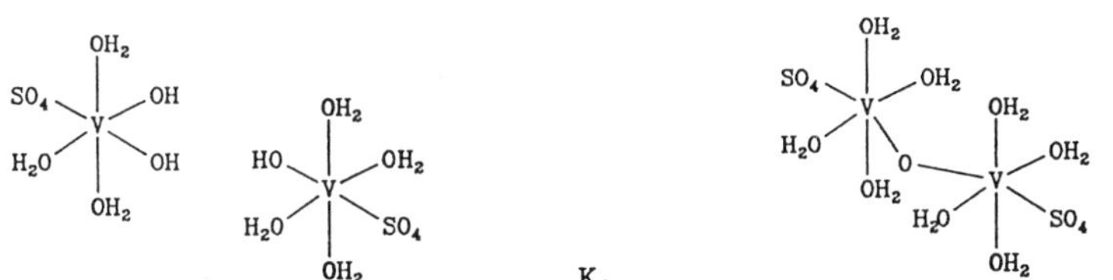

$$2[V(OH)(SO_4)(H_2O)_4] \overset{K_d}{\rightleftharpoons} [V_2(\mu_2\text{-}O)(SO_4)_2(H_2O)_8] + H_2O \tag{24}$$

$$[V(SO_4)(OH)(H_2O)_4] \overset{K_{a2}}{\rightleftharpoons} [V(SO_4)(OH)_2(H_2O)_3^{-}] + H^+ \tag{25}$$

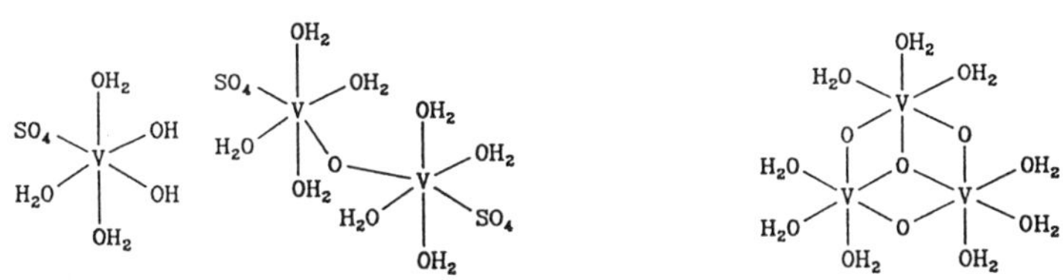

$$[V(OH)_2(SO_4)(H_2O)_3^{-}] + [V_2(\mu_2\text{-}O)(SO_4)_2(H_2O)_8] \rightleftharpoons [V_3(\mu_3\text{-}O)(\mu_2\text{-}O)_3(H_2O)_9^{+}] + 4H^+ + 3SO_4^{2-} + H_2O \tag{26}$$

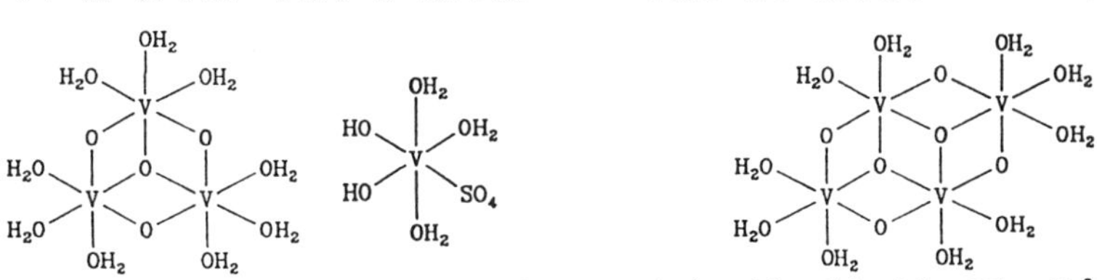

$$[V_3(\mu_3\text{-}O)(\mu_2\text{-}O)_3(H_2O)_9^{+}] + [V(OH)_2(SO_4)(H_2O)_3^{-}] \rightleftharpoons [V_4(\mu_3\text{-}O)_2(\mu_2\text{-}O)_4(H_2O)_8] + 2H^+ + SO_4^{2-} + 2H_2O \tag{27}$$

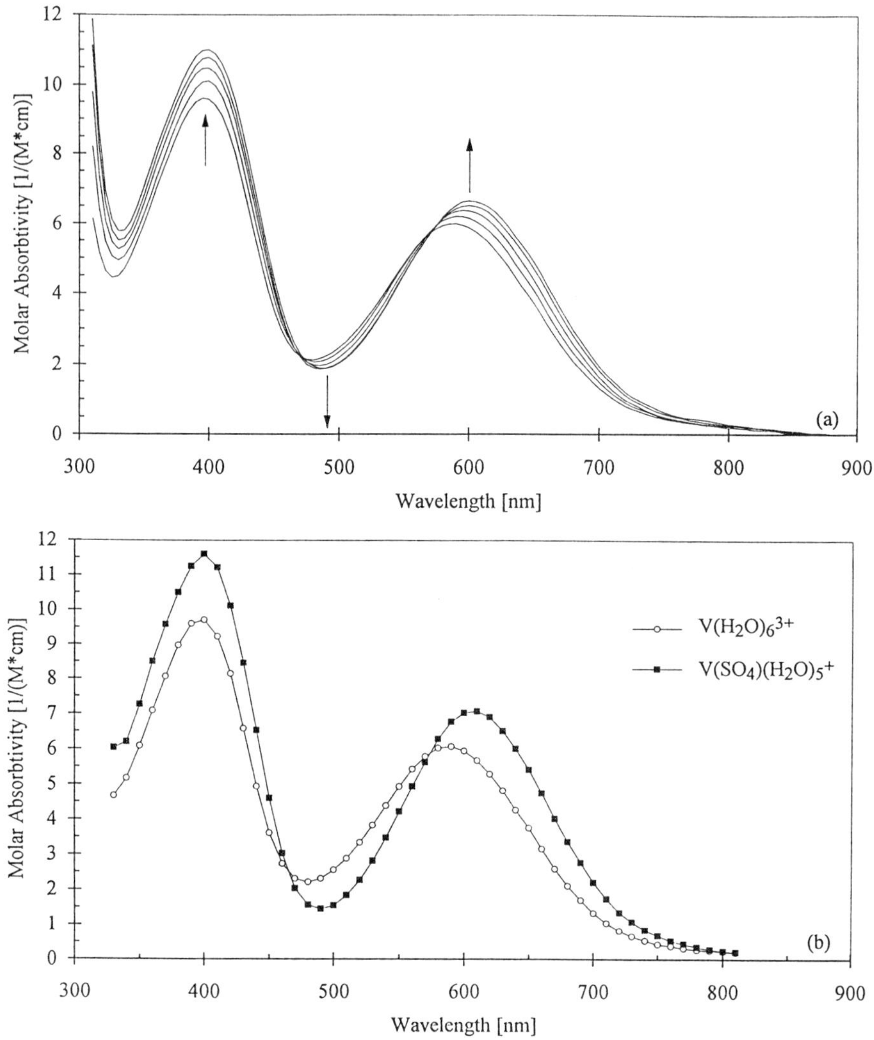

FIG. 8. (a) Spectral changes resulting from increasing values of $[SO_4]_t$ at constant $[(VIII)]_t$ = 10 mM and the constant pH of 0.92 at I = 1.0 M (NaCl, Na_2SO_4) and 25°C. $[SO_4]_t$ (from bottom to top at 400 and 600 nm): 0 M, 0.02 M, 0.05 M, 0.1 M, 0.15 M. (b) Limiting spectra for $[V(H_2O)_6]^{3+}$ and $[V(SO_4)(H_2O)_5]^+$ as computed by SPECFIT.

M (NaCl)] and $[V(III)]_t$ = 10 mM at various ratios $[V(III)]_t$:$[SO_4]_t$. Two isosbestic points are visible at 468 and 572 nm. Applying SPECFIT, the $\log \beta_{101}$ value given in Table 3 was obtained. The limiting spectrum obtained for $[V(SO_4)(H_2O)_5]^+$ is compared in Figure 8b with that of $[V(H_2O)_6]^{3+}$.

The spectrum in Fig. 8a recorded with the highest sulfate excess clearly has left both isosbestic points because the formation of $[V(SO_4)_2$-$(H_2O)_4]^-$ begins. It became clear that under these conditions $\log \beta_{102}$ could be obtained only as a rough estimate (see Table 3, value in square brackets). The magnitude of this formation constant is too low to be determined with good precision at I = 1.0 M. Calculations showed that even at $[SO_4]_t$ = 3.0 M (corresponding to $[V(III)]_t$:$[SO_4]_t$ = 1:300 under the present conditions) and the pH at which the $[V(SO_4)_2(H_2O)_4]^-$ concentration has its maximum, only 80% of $[V(III)]_t$ are present as the 1:2 sulfato complex. Therefore, no limiting spectrum for pure $[V(SO_4)_2$-$(H_2O)_4]^-$ could be obtained. The spectral data in Table 1 for V(III) in 5 M H_2SO_4 and 3 M $(NH_4)_2SO_4$ show that larger quantities of this species can be generated at high $[SO_4]_t$ and an exact limiting spectrum for $[V(SO_4)_2(H_2O)_4]^-$ might be obtainable in future work.

If spectra are recorded in the system V(III)-sulfate in dependence on pH over the pH interval $1.0 < \mathrm{pH} < 4.5$ and a comparison with spectra obtained with a sulfate-free solution having the same V(III) content, a picture is obtained as shown in Fig. 9. Quantitative evaluation of the plot resulting from the sulfate containing solution using DIMFIT provided evidence that a sulfate-containing dimer is formed under the conditions in Fig. 9 (for $\log \beta_{222}$ see Table 3). The existence of a binuclear V(III) complex containing coordinated sulfate was already inferred from results of a kinetic study [100]. It is of about the same stability as its aqua counterpart and has an $\epsilon_{dimer} = 3000 \pm 50\ M^{-1}\ cm^{-1}$ [89].

The shift of the whole plot toward higher pH values in the presence of sulfate is caused by the increased value of pK_{a1} = 3.35 [reaction (23) in Figure 7] of $[V(SO_4)(H_2O)_5]^+$ as compared to pK_{a1} = 2.60 for $[V(H_2O)_6]^{3+}$. It became clearly evident during our study that no sulfate-containing trimeric or tetrameric species are formed. Instead, it was concluded that the high-pH branch of plot B (dimer-consuming pathways) in Fig. 9 is caused by the occurrence of reactions (26) and (27) (see Fig. 7). Finally, a pH-dependent species distribution diagram is shown (Fig. 10) corresponding to the conditions given in Fig. 9 for the presence of sulfate.

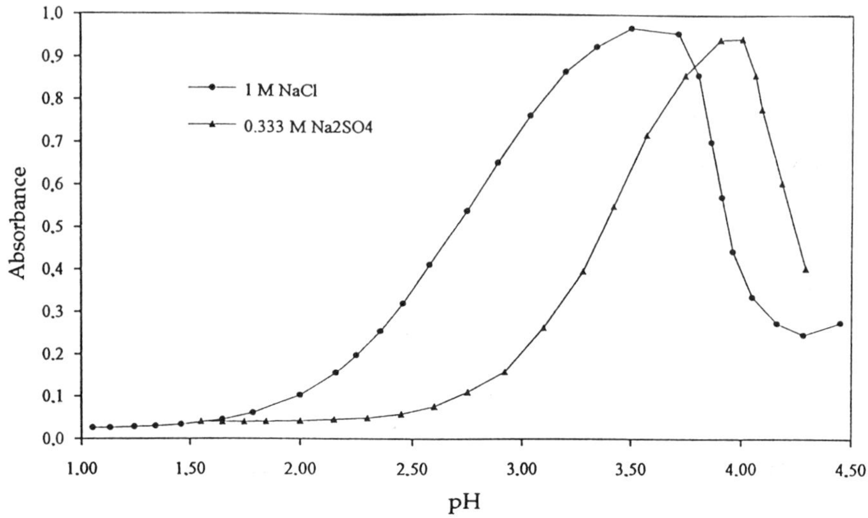

FIG. 9. Absorbances measured at 430 nm in dependence on pH in the absence and presence of sulfate; $[V(III)]_t$ = 5 mM; 25°C. Circles: no sulfate present, I = 1.0 M (NaCl). Triangles: 0.333 M $[SO_4]_t$, I = 1.0 M (Na_2SO_4).

3. VANADIUM(III) IN ASCIDIAN BLOOD CELLS

3.1. UV-VIS Spectra of Blood Cell Lysates

Lybing was able to confirm that V(III) is the prevailing intracellular vanadium oxidation state based on the UV-VIS spectra of *Ascidia obliqua* blood cell lysates [10]. Thus, it is not useless to also investigate cell lysates despite the fact that such an approach is halfway between in vivo and in vitro. The problems that one might have with spectra from blood cell lysates were described by Boeri and Ehrenberg in the following way [11]: "The properties of the vanadium in the vanadium hemolysates do not give any conclusive information about the properties of vanadium in the much more condensed system inside the vanadocyte. The vanadium in the cell may well to some extent be bound to the cell membrane or to some protein. The state of valence is probably the same as in the hemolysate".

Therefore, the spectral data from lysis solutions have to be analyzed with caution. The main problem in this regard is that the usual

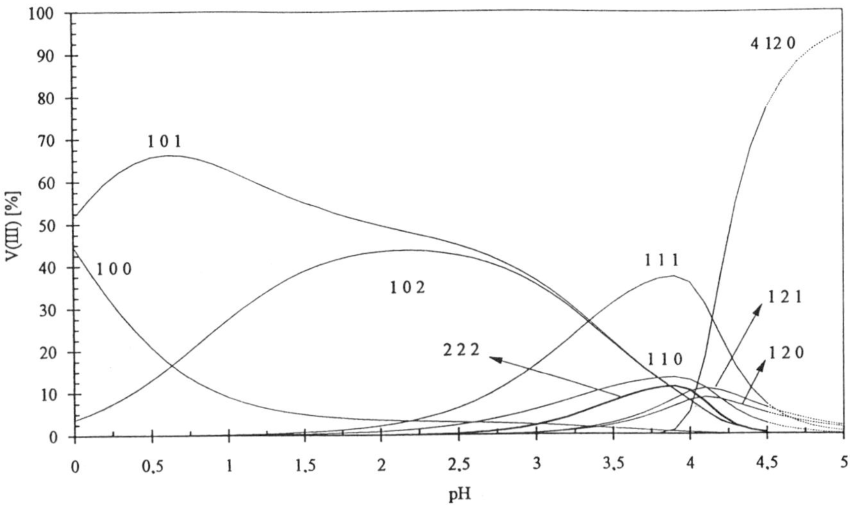

FIG. 10. Species distribution in the V(III)-sulfato system in dependence on pH ($0 \leq pH \leq 5$) for 5 mM $[V(III)]_t$ and 0.333 M $[SO_4]_t$ at 25°C. Hydrolytic species are defined according to the indices of the corresponding overall equilibrium constants [Eqs. (16)–(20)]; the overall equilibrium constants used for the computation are given in line 2 of Table 3.

cell lysis experiments are done with unfractioned blood such that intracellular compounds from non-vanadium-containing cells might appear in the lysates. An inspection of the spectra obtained with lysis solution from different ascidians shall illustrate the problems. Spectra resulting from lysates of *A. obliqua* and *P. mammillata* are compared in Fig. 11a and b [11]. These spectra are important because they were recorded under identical experimental conditions. The spectrum with number 7 in Fig. 11a resulted from a solution produced by lysis of *A. obliqua* blood cells and contains no exogenous chemicals except distilled water. The pH in the lysate was 1.75 and the solution was analyzed to have $[V(III)]_t = 7.79$ mM and $[SO_4]_t = 55$ mM. Because this spectrum is readily reproducible in vitro by preparing a solution with identical $[V(III)]_t$ and $[SO_4]_t$ contents and an adjusted pH of 1.75, one can calculate the V(III) speciation of this solution using the constants obtained from the spectrophotometric titrations in the presence of sulfate (Table 3):

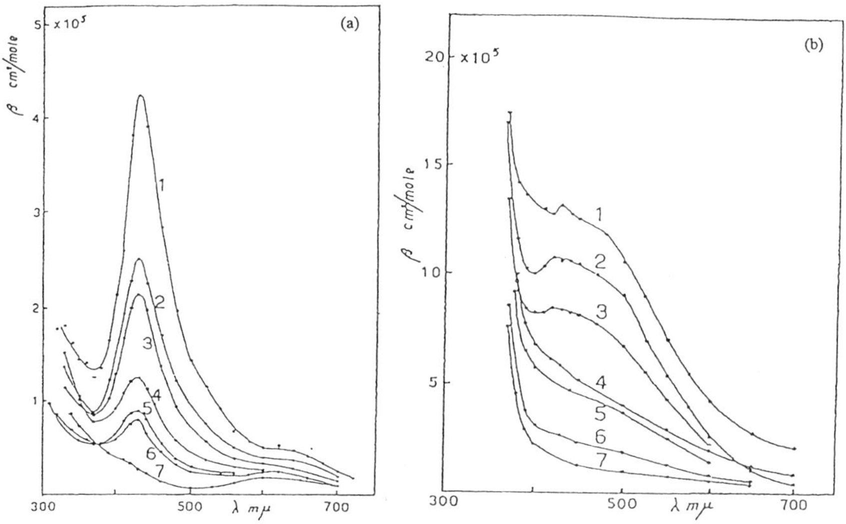

FIG. 11. (a) Spectra of a blood cell lysate from *Ascidia obliqua* $[V(III)]_t = 7.79$ mM, $[SO_4]_t = 55$ mM; initial solution (7) is lysate as obtained, pH subsequently changed by adding hydroxide, (No.;pH), (7: 1.75), (6: 2.06), (5: 2.16), (4: 2.23), (3: 2.44), (2: 2.59), (1: 2.98). (b) Spectra of a blood cell lysate from *Phallusia mammillata*, $[V(III)]_t$ not given $[SO_4]_t = 20$ mM; initial solution (7) is lysate as obtained, pH subsequently changed by adding hydroxide, (No.;pH), (7: 2.13), (6: 2.28), (5: 2.36), (4: 2.64), (3: 3.17), (2: 3.30), (1: 3.50). (Reproduced with permission from [11]. Copyright by Academic Press.)

50% $[V(SO_4)(H_2O)_5]^+$
40% $[V(H_2O)_6]^{3+}$
6% $[V(OH)(H_2O)_5]^+$
4% $[V(SO_4)_2(H_2O)_4]^-$

This species distribution is striking in connection with the following: Based on studies by a variety of techniques during investigations of intact blood cells from *Ascidia ceratodes*, Carlson et al. suggested the overall formula $[V(SO_4)(H_2O)_{4\text{–}5}]^+$ for the intracellular $[V(III)]_t$ [38]. It is also important that they concluded that the intracellular pH was 1.8 ± 0.1 in their case, based on a noninvasive approach for measurement. This pH value is quite close to that in solution 7 of Fig. 11a. It seems possible, therefore, that Boeri and Ehrenberg obtained a lysate in which the original distribution of $[V(III)]_t$ was not markedly changed during lysis. Boeri and Ehrenberg [11] added exogenous hydroxide to the *A. obliqua* lysate to increase the pH and obtained the spectra 6 − 1 (Fig. 11a). While doing so, the intensive absorbance of dinuclear V(III) is subsequently increased.

These spectral changes with pH are virtually identical with those shown in Fig. 2. The composition of the final solution 1 with a pH of 2.98 can be computed to be:

36% $[V(OH)(H_2O)_5]^{2+}$
24% $[V(SO_4)(H_2O)_5]^{+}$
15% $[V(H_2O)_6]^{3+}$
11% $[V(OH)(SO_4)(H_2O)_4]$
8% $[V(\mu_2\text{-}O)(H_2O)_{10}]^{4+}$
3% $[V(SO_4)_2(H_2O)_4]^{-}$
2% $[V(OH)_2(H_2O)_4]^{+}$
1% $[V(\mu_2\text{-}O)(SO_4)_2(H_2O)_8]$

The contents of the two different binuclear ions are such that the absorbance at 430 nm is almost exclusively governed by $[V(\mu_2\text{-}O)\text{-}(H_2O)_{10}]^{4+}$. Generation of $[V(\mu_2\text{-}O)(SO_4)_2(H_2O)_8]$ would require a much larger excess of $[SO_4]_t$.

The spectra in Fig. 11b were recorded with a lysate from *P. mammillata* blood cells and reflect also appropriate spectral changes with pH. However, the differences to the spectra in Fig. 11a become apparent. The initial solution 7 shows no spectral feature corresponding to V(III). The growing of the absorbance at wavelengths shorter than 400 nm is caused by a band with a maximum at 300 nm [99]. A gradual increase of pH leads to the spectra 6 − 1 with concurrent formation of new absorbance features between 400 and 500 nm. Boeri and Ehrenberg [11], as well as Bielig et al. (they did similar experiments) [99], assume that these spectral changes might be caused by complex formation with a chelating ligand. We are unable at present to clarify this question because the nature of the ligands responsible for this probable complex formation is not known with certainty (see discussion below). However, it is useful to extend our considerations also to the lysate spectra from *Ascidia ceratodes*.

In a compilation of lysate spectra from various ascidian species, Carlson noted the similarity of the basic spectral features between *A. ceratodes* and *P. mammillata* [110]. In both cases, the typical absorbances of aqua-V(III) at around 400 and 600 nm are superimposed by a very intensive band with a maximum at 335 nm in the lysate from *A. ceratodes* and at 300 nm in those from *P. mammillata*. The absorbance of the 335-nm peak is at least 50-fold greater in *A. ceratodes* lysates than the 300-nm absorbance in the case of *P. mammillata*. Carlson summarized the findings related to the intensive bands in the near-UV as follows [110]: "The 335 nm peak of high absorbance occurs in the hemolysate spectrum of *Ascidia ceratodes*, a 300 nm λ_{max} peak of lower absorbance occurs in the hemolysate spectrum of *Phallusia mammillata*, a peak of this absorbance is absent from the spectra of *Ascidia obliqua* and *Ascidiella aspersa*. Therefore, a component λ_{max} 300–335 nm does not appear to be essential to the hemovanadin system". With the current knowledge regarding the composition of blood cell lysates of ascidians it is clear that both peaks at 335 and 300 nm were caused by tunichromes [106,107]. These compounds were not yet discovered in 1977 when Carlson did his compilation [110]. Tunichromes belong to a class of oxygen-sensitive organic pigments which have been isolated from ascidian blood cells. These substances are oligopeptides derived from the amino acids 3,4-dihydroxyphenylalanine (DOPA) and 3,4,5-trihydroxyphenylalanine (TOPA) [104,105]. Tunichromes were considered as in vivo occurring compounds, located in the morula cells. However, this became uncertain recently. In the framework of a doctoral dissertation [106], a detailed search was started to find tunichrome in intact blood cells of the species *A. ceratodes* and *P. julinea*. According to the findings from Taylor's work that were published in the meantime [107], there is no evidence for tunichrome in intact cells of these two ascidian species.

At this moment the question remains regarding the origin and role of tunichrome. The data accumulated by Taylor et al. [107] suggest that both lysis of morula (low in vanadium and high in tunichrome precursor contents) as well as compartment and signet ring (high in vanadium content) cells are necessary for the formation of tunichrome. It was therefore concluded that tunichromes are degradation products of a protein [107]. This interpretation agrees with a recent suggestion of Waite [108] that evidence is needed to show that the tunichromes are not degradation products of a DOPA- or tyrosine-rich protein, such as ferreascidin [109]. A detailed model describing the formation of tunichrome as a result of cell lysis is presented in [45].

Thus we conclude that the pH-dependent spectral changes in *P.*

mammillata lysates (Fig. 11b) might be caused by successive complex formation with tunichrome formed upon cell lysis. These spectral features were not present in the initial solution. Thus given the case that complex formation is indicated by the overlapping bands between 400 and 500 nm, this was caused by addition of exogenous hydroxide. Also, no publication appeared as yet showing pH-dependent spectra of solutions prepared by mixing of isolated tunichrome and V(III).

The alterations in the spectra of blood cell lysates from different ascidians show that the composition of these solutions is not identical. This should then also be true for the components in the cells of different species. However, the chemical status of vanadium(III) in the blood cells could be identical in the different species because the tunichrome formed as a result of cell lysis has superimposed the spectral features of V(III) in the initial lysates. The comparison between the V(III) speciation in a blood cell lysate from *A. obliqua* with that of *A. ceratodes* resulting from noninvasive probing with different methods is a tantalizing hint in favor of this conclusion.

3.2. UV-VIS Spectra of Intact Blood Cells

We will confine the discussion under this heading to a comparison between a UV-VIS spectrum of a dried blood cell from *A. obliqua* as prepared by Lybing [10] (Fig. 12b) and a spectrum of the species $[V_4(\mu_3\text{-}O)_2(\mu_2\text{-}O)_4(H_2O)_{10}]$ (Fig. 12a) resulting from our ongoing research dealing with V(III) speciation [67,89,101].

It was demonstrated in the previous section that blood cell lysis from *A. obliqua* yields mainly V(III) and sulfate in the lysates. Thus, no soluble, intensively absorbing compounds have accompanied the V(III) on its way from intracellular locations into the lysis solution. It could then be that there are also no components in the cells interfering with the absorbances of V(III).

This assumption is verified by the close similarity of the spectra in Fig. 12a and b. The spectrum in Fig. 12a was obtained at pH 4.18 with a solution containing 10 mM $[V(III)]_t$. For this solution, one can calculate that 80% of $[V(III)]_t$ is present as $[V_4(\mu_3\text{-}O)_2(\mu_2\text{-}O)_4(H_2O)_{10}]$. Thus, what we see in Fig. 12a is essentially a spectrum of the tetrameric V(III) complex derived by hydrolysis.

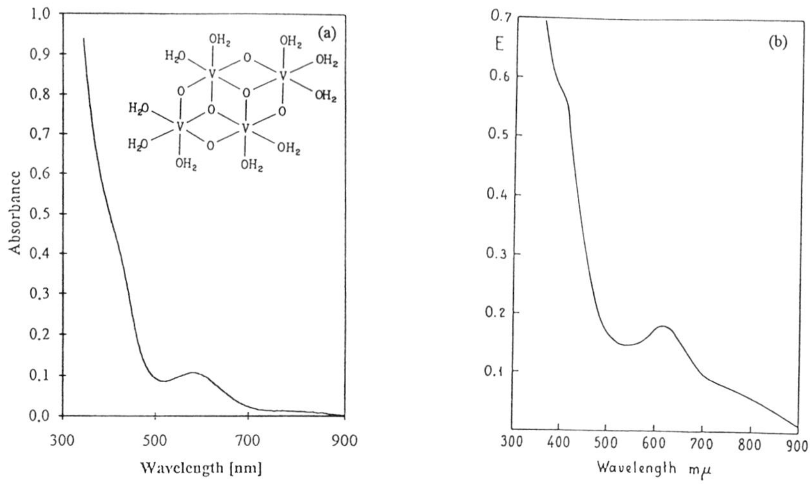

FIG. 12. Spectra of (a) a solution containing $[V(III)]_t = 10$ mM at $I = 1.0$ M (NaCl), 25°C, and pH 4.18 and (b) of a dried blood cell from *Ascidia obliqua*. (Fig. 12b was reproduced with permission from [10]. Copyright by the Royal Swedish Chemical Society.)

The spectrum of the dried *A. obliqua* blood cell is given in Fig. 12b. It is virtually identical with that in Fig. 12a except for a spectral component between 700 and 800 nm that is not present in the latter. This feature belongs apparently to some amount of V(IV) formed by aerial oxidation during the drying process. The similarity of both these spectra gives rise to the following conclusions: It seems certain that the spectrum of the dried blood cell shows a vanadium fraction that is mainly present in the form of the tetramer described above. It could be that this tetrameric V(III) complex represents the membrane-bound vanadium and that the mobile V(III) that can leave the cells during lysis was also leached out during drying. Drying of a cell should be connected with some leaching. However, these suggestions await experimental proof.

4. CONCLUSIONS

In the present chapter, an overview has been given regarding the hydrolytic reactions in which V(III) is involved in aqueous solution. The hydrolysis of V(III) in the presence of sulfate was also described. Future studies of the V(III) speciation in ascidian blood cells or related biological systems should make use of the equilibrium constants presented.

It is suggested that not only one chemical form of V(III) should exist in ascidian blood cells. The status of V(III) in the ascidians might be connected with the histogenesis of the blood cells. Furthermore, even in one and the same cell, mobile monomeric and membrane bound oligomeric V(III) might coexist.

The species distribution diagrams presented and discussed in this work should become a common tool for workers dealing with in vivo V(III) speciation. This is especially true in the light of the intravacuolar pH values that have recently been measured using noninvasive approaches: *Ascidia ceratodes* pH 1.8 ± 0.1 [37], *Ascidia gemmata* pH 1.88 [27], *Ascidia ahodori* pH 3.07 [27], and *Ascidia sydneiensis samea* pH 1.52 [27]. Together with the very large intravacuolar vanadium(III) concentrations that have been calculated (e.g., 0.35 M in *Ascidia gemmata* [27]), these pH values are just located in intervals where very small changes of pH cause tremendous alterations in the speciation (for illustration, see Fig. 5d).

There is also a very strong support for the suggestion that a certain part of the intracellular V(III) is membrane-bound: If all of 0.35 M $[V(III)]_t$ in the vanadocytes of *A. gemmata* would be present in a solvated form, the cells would look dark brown because of the presence of about 40% (0.07 M) of the species $[V_2(\mu_2\text{-}O)(H_2O)_{10}]^{4+}$ in the intravacuolar fluids. This would result in a local absorbance of $A = 21$ at 430 nm with ϵ_{dimer}=3000 M^{-1} cm^{-1} and a layer thickness of 0.1 mm. However, dark brown ascidian blood cells have never been reported. Thus, not all of the quantity $[V(III)]_t$ = 0.35 M in the blood cells of *A. gemmata* [27] can be present as soluble V(III). It seems to be certain that the soluble fraction of V(III) in these cell is distinctly smaller than 0.35 M. In other words, $[V(III)]_t$ might be divided into a soluble, mobile and a membrane-bound fraction.

It is now the time to start new investigations to prove these suggestions. A number of possible biological functions of the intracellularly occurring V(III) could at this point be discussed. We feel, however, that such speculations must await a better knowledge about the chemical forms in which V(III) occurs in the ascidians. Future studies must be done using as many analytical methods as possible for the in vivo speciation of V(III). These should be noninvasive methods that are as specific as possible for V(III). If the in vivo chemistry of V(III) in the ascidians is clarified on a better level, it will become much easier to elucidate its long suspected physiological role.

ABBREVIATIONS AND DEFINITIONS

CV	cyclic voltammetry
DOPA	3,4-dihydroxyphenylalanine
EPR	electron paramagnetic resonance
EXAFS	extended X-ray absorption fine structure
NBS	U.S. National Bureau of Standards (now NIST)
NIST	U.S. National Institute of Standards and Technology
te	tetramer
TOPA	3,4,5-trihydroxyphenylalanine
tr	trimer
XANES	X-ray absorption near edge structure
UV-VIS	ultraviolet-visible absorption spectrophotometry

Use has been made throughout the text of the abbreviation $[V(III)]_t$. This expression represents the total concentration of V(III) in any given system according to the mass balance. The general mass balance for the V(III)-aqua system is given in Eq. (5); the same expression for the V(III)-sulfato system is defined in Eq. (14). The particular concentration for a species of interest is computed using a program designed for species distributions. For workers unskilled with such calculations the present authors may be contacted for support.

The corresponding meaning of $[SO_4]_t$ is explained in Sec. 2.1.

ACKNOWLEDGMENT

R. M. thanks the Deutsche Forschungsgemeinschaft for generous and continuous support of the research work related to this chapter (grants Me 1148/2-1 and Me 1148/2-2). For the support with a stipend, R. M. is grateful to the Alfried-Krupp-von-Bohlen-und-Halbach-Stiftung (Essen) as well as to the Alexander-von-Humboldt-Stiftung (Bonn).

REFERENCES

1. M. Henze, *Hoppe-Seyler's Z. Physiol. Chem.*, *72*, 494 (1911).
2. H. Michibata, *Zool. Sci.*, *6*, 639 (1989) and references cited therein.
3. H. Michibata and H. Sakurai, in *Vanadium in Biological Systems* (N. D. Chasteen, ed.), Kluwer, Dordrecht, 1990, pp. 153–171.
4. R. Buchsbaum, *Animals without Backbones*, 2nd Ed., University of Chicago Press, Chicago, 1972, pp. 316–318.
5. A. J. Waterman, *Chordate Structure and Function*, Collier Macmillan Int. Edit., New York, 1971, pp. 42–90.
6. M. Henze, *Hoppe-Seyler's Z. Physiol. Chem.*, *79*, 215 (1912).
7. M. Henze, *Hoppe-Seyler's Z. Physiol. Chem.*, *86*, 341 (1913).
8. M. Henze, R. Stöhr, and R. Müller, *Hoppe-Seyler's Z. Physiol. Chem.*, *213*, 125 (1932).
9. E.Boeri, *Arch. Biochem. Biophys.*, *37*, 449 (1952).
10. S. Lybing, *Ark. Kem.*, *6*, 261 (1953).
11. E. Boeri and A. Ehrenberg, *Arch. Biochem. Biophys.*, *50*, 404 (1954).
12. H.-J. Bielig, E. Bayer, L. Califano, and L. Wirth, *Pubbl. Staz. Zool. Napoli*, *24*, 26 (1954).
13. R. M. K. Carlson, *Proc. Natl. Acad. Sci. USA*, *72*, 2217 (1975).
14. P. Frank, B. Hedman, R. M. K. Carlson, T. A. Tyson, A. L. Roe, and K. O. Hodgson, *Biochemistry*, *26*, 4975 (1987).
15. P. Frank, B. Hedman, R. M. K. Carlson, and K. O. Hodgson, *J. Inorg. Biochem.*, *43*, 408 (1991).
16. P. Frank, personal communication, 1994.
17. T. D. Tullius, W. O. Gillum, R. M. K. Carlson, and K. O. Hodgson, *J. Am. Chem. Soc.*, *102*, 5670 (1980).
18. W. Rummel, H.-J. Bielig, W. Forth, K. Pfleger, W. Rüdiger, and E. Seifen, *Protides Biol. Fluids*, *14*, 205 (1966).
19. H.-J. Bielig, E. Jost, K. Pfleger, W. Rummel, and E. Seifen, *Hoppe-Seyler's Z. Physiol. Chem.*, *325*, 122 (1961).
20. H. Sakurai, J. Hirata, and H. Michibata, *Inorg. Chim. Acta*, *152*, 177 (1988).

21. H.-J. Bielig, E. Bayer, H.-D. Dell, G. Rohns, H. Möllinger, and W. Rüdiger, *Protides Biol. Fluids*, *14*, 197 (1966).
22. D. B. Carlisle, *Proc. Roy. Soc. London, Ser. B*, *171*, 171 (1968).
23. W. R. Biggs and J. H. Swinehart, in *Metal Ions in Biological Systems: Biological Actions of Metal Ions*, Vol. 6 (H. Sigel, ed.), Marcel Dekker, New York, 1976, pp. 141–196.
24. R. K. Wright, in *Invertebrate Blood Cells*, Vol. 2 (N. A. Ratcliffe and A. F. Rowley, eds.), Academic Press, London, 1981, pp. 565–626.
25. K. Kustin, G. C. McLeod, T. R. Gilbert, and L. B. R. 4th Briggs, *Structure and Bonding*, *53*, 139 (1983).
26. A. Butler and C. J. Carrano, *Coord. Chem. Rev.*, *109*, 61 (1991).
27. H. Michibata, *Adv. Biophys.*, *29*, 105 (1993).
28. T. Ishii, I. Nakai, C. Numako, K. Okoshi, and T. Otake, *Naturwissenschaften*, *80*, 268 (1993).
29. I. Bertini, G. Canti, and C. Luchinat, *Inorg. Chim. Acta*, *67*, L21 (1982).
30. I. Bertini, C. Luchinat, and L. Messori, *J. Inorg. Biochem.*, *25*, 57 (1985).
31. C. F. Baes, Jr., and R. E. Mesmer, *The Hydrolysis of Cations*, John Wiley and Sons, New York, 1976, p. 199.
32. C. M. Flynn Jr., *Chem. Rev.*, *84*, 31 (1984) and references cited therein.
33. P. L. Brown, R. N. Sylva, and J. Ellis, *J. Chem. Soc., Dalton Trans.*, 723 (1985).
34. G. H. Khoe, P. L. Brown, R.N. Sylva, and R. G. Robins, *J. Chem. Soc., Dalton Trans.*, 1901 (1986).
35. G. H. Khoe and R. G. Robins, *J. Chem. Soc., Dalton Trans.*, 2015 (1988).
36. B. J. S. Pirie and M. V. Bell, *J. Exp. Mar. Biol. Ecol.*, *74*, 187 (1984).
37. P. Frank, R. M. K. Carlson, and K. O. Hodgson, *Inorg. Chem.*, *25*, 470 (1986).
38. P. Frank, R. M. K. Carlson, and K. O. Hodgson, *Inorg. Chem.*, *27*, 118 (1988).
39. H. Michibata, Y. Iwata, and J. Hirata, *J. Exp. Zool.*, *257*, 306 (1991).

40. A. L. Dingley, K. Kustin, I. G. Macara, and G. C. McLeod, *Biochem. Biophys. Acta*, *720*, 384 (1982).
41. M. I. Agudelo, K. Kustin, and G. C. McLeod, *Comp. Biochem. Physiol.*, *75*A, 211 (1983).
42. C. J. Hawkins, G. A. James, D. L. Parry, J. H. Swinehart, and A. L. Wood, *Comp. Biochem. Physiol.*, *76B*, 559 (1983).
43. S. G. Brand, C. J. Hawkins, and D. L. Parry, *Inorg. Chem.*, *26*, 627 (1987).
44. S. G. Brand, C. J. Hawkins, A. T. Marshall, G. W. Nette, and D. L. Parry, *Comp. Biochem. Physiol.*, *93B*, 425 (1989).
45. S. W. Taylor, C. J. Hawkins, D. L. Parry, J. H. Swinehart, and G. R. Hanson, *J. Inorg. Biochem.*, in press.
46. H. Winterstein, *Biochem. Z.*, *19*, 384 (1909).
47. I. G. Macara, G. C. McLeod, and K. Kustin, *Comp. Biochem. Physiol.*, *62A*, 821 (1979).
48. M. Kalk, *Quart. J. Micr. Sci.*, *104*, 483 (1963).
49. L. Botte, S. Scippa, and M. de Vincentiis, *Experientia*, *35*, 1228 (1979).
50. S. Scippa, K. Zierold, and M. de Vincentiis, *J. Submicrosc. Cytol. Pathol.*, *20*, 719 (1988).
51. H. Michibata, J. Hirata, M. Uesaka, T. Numakunai, and H. Sakurai, *J. Exp. Zool.*, *244*, 33 (1987).
52. D. A. Roman, J. Molina, and L. Rivera, *Biol. Bull.*, *175*, 154 (1988).
53. J. Gandeboeuf and P. Souchay, *J. Chim. Phys. Phys.-Chim. Biol.*, *56*, 358 (1959).
54. L. Pajdowski, *Rocz. Chem.*, *37*, 1351 (1963).
55. T. W. Newton and F. B. Baker, *Inorg. Chem.*, *3*, 569 (1964).
56. L. Pajdowski, *J. Inorg. Nucl. Chem.*, *28*, 433 (1966).
57. L. Pajdowski and B. Jezowska-Trzebiatowska, *J. Inorg. Nucl. Chem.*, *28*, 443 (1966).
58. F. A. Cotton, C. K. Fair, G. E. Lewis, G. N. Mott, F. K. Ross, A. J. Schultz, and J. M. Williams, *J. Am. Chem. Soc.*, *106*, 5319 (1984).
59. J. K. Beatti, S. P. Best, B. W. Skelton, and A. H. White, *J. Chem. Soc., Dalton Trans.*, 2105 (1981).

60. K. Wieghardt, M. Köppen, B. Nuber, and J. Weiss, *J. Chem. Soc., Chem. Commun.*, 1530 (1986).
61. M. Köppen, G. Fresen, K. Wieghardt, R. M. Llusar, B. Nuber, and J. Weiss, *Inorg. Chem.*, *27*, 721 (1988).
62. P. Knopp, K. Wieghardt, B. Nuber, J. Weiss, and W. S. Sheldrick, *Inorg. Chem.*, *29*, 363 (1990).
63. P. Knopp and K. Wieghardt, *Inorg. Chem.*, *30*, 4061 (1991).
64. K. Kanamori, Y. Ookubo, K. Ino, K. Kawai, and H. Michibata, *Inorg. Chem.*, *30*, 3832 (1991).
65. S. E. Lincoln and T. M. Loehr, *Inorg. Chem.*, *29*, 1907 (1990).
66. R. Meier, M. Boddin, S. Mitzenheim, and W. R. Fawcett, *J. Electroanal. Chem.*, submitted for publication.
67. M. Boddin, S. Mitzenheim, and R. Meier, *J. Chem. Soc., Dalton Trans.*, submitted for publication.
68. E. Deutsch and H. Taube, *Inorg. Chem.*. *7*, 1532 (1968).
69. E. Deutsch, PhD thesis, Stanford University, Stanford, 1967.
70. M. A. Hitchman, R. G. McDonald, P. W. Smith, and R. Stranger, *J. Chem. Soc., Dalton Trans.*, 1393 (1988).
71. S. C. Furman and C. S. Garner, *J. Am. Chem. Soc.*, *72*, 1785 (1950).
72. R. Meier, A. Jokuszies, M. Petrich, M. Hecht, and R. van Eldik, *Inorg. Chem.*, in press.
73. D. D. Perrin and I. G. Sayce, *Talanta*, *14*, 833 (1967).
74. D. W. Marquardt, *J. Soc. Ind. Appl. Math.*, *11*, 431 (1963).
75. H. Gampp, *J. Chem. Soc., Faraday Trans.*, *1*, *83*, 1719 (1987).
76. H. Gampp, *Anal. Chem.*, *59*, 2456 (1987).
77. H. Gampp, M. Maeder, C. J. Meyer, and A. D. Zuberbühler, *Talanta*, *32*, 257 (1985).
78. E. L. Yee, R. J. Cave, K. L. Guyer, P. D. Tyma, and M. J. Weaver, *J. Am. Chem. Soc.*, *101*, 1131 (1979).
79. S. W. Barr, K. L. Guyer, and M. J. Weaver, *J. Electroanal. Chem.*, *111*, 41 (1980).
80. M. J. Weaver, *J. Phys. Chem.*, *84*, 568 (1980).
81. M. J. Weaver and E. L. Yee, *Inorg. Chem.*, *19*, 1936 (1980).
82. J. T. Hupp and M. J. Weaver, *Inorg. Chem.*, *22*, 2557 (1983).

83. F. A. Cotton, M. W. Extine, L. R. Falvello, D. Beck Lewis, G. E. Lewis, C. A. Murillo, W. Schwotzer, M. Thomas, and J. M. Troup, *Inorg. Chem.*, *19*, 3505 (1986).

84. D. T. Richens, L. Helm, P.-A. Pittet, A. E. Merbach, F. Nicolo, and G. Chapius, *Inorg. Chem.*, *28*, 1394 (1989).

85. B.-L. Ooi and A. G. Sykes, *Inorg. Chem.*, *28*, 3799 (1989).

86. G. L. Lamprecht, M. Martinez, M. Nasreldin, C. A. Routledge, N. Al-Shatti, and A. G. Sykes, *J. Chem. Soc., Dalton Trans.*, 747 (1993).

87. S. J. Lippard, *Angew. Chem., Int. Ed. Engl.*, *27*, 344 (1988).

88. M. Petrich, Dissertation thesis, TH Leipzig, Leipzig, Germany, 1990.

89. S. Mitzenheim, M. Boddin, and R. Meier, *J. Chem. Soc., Dalton Trans*, submitted for publication.

90. J. B. Vincent, C. Christmas, J. C. Huffman, G. Christou, H.-R. Chang, and D. N. Hendrickson, *J. Chem. Soc., Chem. Commun.*, 236 (1987).

91. W. H. Armstrong, M. E. Roth, and S. J. Lippard, *J. Am. Chem. Soc.*, *109*, 6318 (1987).

92. S. M. Gorun and S. J. Lippard, *Inorg. Chem.*, *27*, 149 (1988).

93. P. Suesse, *Z. Kristallogr.*, *127*, 261 (1968).

94. R. A. Whiteker and N. Davidson, *J. Am. Chem. Soc.*, *75*, 3081 (1953).

95. NIST Standard Reference Database 46, NIST Critical Stability Constants and Related Thermodynamic Constants of Metal Complexes, National Institute of Standards and Technology, Gaithersburg, MD 20899, U.S.A.

96. L. Meites, *J. Am. Chem. Soc.*, *75*, 6095 (1953).

97. T. Kimura, K. Morinaga, and K. Nakano, *Nippon Kagaku Kaishi*, 664 (1972).

98. E. L. Martin and K. E. Bentley, *Anal. Chem.*, *34*, 354 (1962).

99. H.-J. Bielig, H.-D. Dell, H. Möllinger, and W. Rüdiger, *Justus Liebigs Ann. Chem.*, *662*, 206 (1963).

100. T. W. Newton and F. B. Baker, *J. Phys. Chem.*, *68*, 228 (1964).

101. R. Meier and M. Boddin, manuscript in preparation.

102. R. Endean, *Quart. J. Micr. Sci.*, *101*, 177 (1960).

103. S. Scippa, L. Botte, and M. de Vincentiis, *Acta Zool.*, *63*, 121 (1982).

104. R. C. Bruening, E. M. Oltz, J. Furukawa, K. Nakanishi, and K. Kustin, *J. Am. Chem. Soc.*, *107*, 5298 (1985).

105. E. Bayer, G. Schiefer, D. Waidelich, S. Scippa, and M. de Vincentiis, *Angew. Chem., Int. Ed. Engl.*, *31*, 52 (1992).

106. S. W. Taylor, PhD thesis, University of Queensland, St. Lucia, Australia, 1992.

107. S. W. Taylor, D. L. Parry, C. J. Hawkins, and J. H. Swinehart, *Comp. Biochem. Physiol.*, *106A*, 531 (1993).

108. J. H. Waite, *Comp. Biochem. Physiol.*, *97B*, 19 (1990).

109. S. W. Taylor, C. J. Hawkins, and D. J. Winzor, *Inorg. Chem.*, *32*, 422 (1993).

110. R. M. K. Carlson, PhD thesis, Stanford University, Stanford, 1977.

3

The Vanadyl Ion: Molecular Structure of Coordinating Ligands by Electron Paramagnetic Resonance and Electron Nuclear Double Resonance Spectroscopy

Marvin W. Makinen and Devkumar Mustafi

Department of Biochemistry and Molecular Biology,
The University of Chicago,
920 East 58th Street,
Chicago, Illinois 60637

1. INTRODUCTION

The oxovanadium(IV) or vanadyl (VO^{2+}) ion is one of the most stable diatomic ions known. In the ground state of V(IV), the unpaired electron occupies the $3d_{xy}$ orbital. VO^{2+} is, thus, orbitally nondegenerate with no nearby, low-lying excited states [1,2]. The paramagnetism of VO^{2+} is well behaved, having its origin almost entirely in spin angular momentum, and electron paramagnetic resonance (EPR) absorption is observed over a wide range of temperatures, including fluid solutions at room temperature. Reviews of the EPR spectroscopy of the VO^{2+} ion have been provided by McGarvey [3], Goodman and Raynor [4], and most recently by Mabbs and Collison [5].

There are special advantages in the use of VO^{2+} as a probe of biomolecular structure when compared with other paramagnetic metal ions. VO^{2+} can be substituted for a variety of divalent metal ions in metal-activated enzymes and metalloproteins. The free, solvated vanadyl ion, existing as the $[VO(H_2O)_5]^{2+}$ species at lower pH, ionizes in the neutral pH region to form an EPR-silent, polymeric $VO(OH)_2$ species, and in the pH 5–11 region the EPR signal intensity is proportional only to ligand- or macromolecule-bound VO^{2+}. Consequently, spectra are not complicated by equilibria between bound and free species with overlapping resonance absorption features. Reviews of the chemical properties of VO^{2+}, particularly as a paramagnetic probe of biological molecules, have been written by Chasteen [6,7; cf. also Chapters 7 and 19 in this volume]. Also, Butler and Carrano have reviewed the coordination chemistry of vanadium in biological systems [8].

A major limitation of EPR for probing structural details of a paramagnetic site is the failure to resolve the ligand hyperfine (hf) or superhyperfine (shf) couplings of nearby magnetic nuclei which interact with the unpaired electron spin, as illustrated in Fig. 1. In this diagram it is seen that the EPR absorption features in natural abundance solvent are broader than in perdeuterated solvent, reflecting the stronger but unresolved electron-nuclear hf couplings of protons. The change in linewidth can be analyzed to yield information about the symmetry and coordination environment of VO^{2+} [6,7]. However, analysis of the ligand hf interaction on the basis of comparison of EPR linewidths is restricted to equatorial ligands since the broadening influence of axial ligands is not sufficiently prominent to result in a comparable, measurable change. Nonetheless, quantitative evaluation of shf couplings of both axial and equatorial ligands is essential for structure determination. In order to assign nuclear coordinates and to estimate metal-nucleus separations, it is necessary to measure the strength of shf interactions and to determine their relationships with respect to magnetic axes in the molecule and with respect to the external (laboratory) magnetic field. This objective can be probably best achieved by electron nuclear double resonance (ENDOR) spectroscopy.

ENDOR spectroscopy was first developed by Feher [10,11]. Several monographs and review articles on the theory and application of ENDOR have been published [12–15], including a comparison of ENDOR studies of VO^{2+} in small-molecule systems [15]. Also, Eaton and Eaton reviewed EPR, ENDOR, and electron spin-echo envelope modulation (ESEEM) studies of VO^{2+} [16]. While ENDOR studies for purposes of structure determination have been carried out primarily with use of oriented single crystals, more recent studies, particularly from our laboratory, have demonstrated that ENDOR spectroscopy can be an incisive method to assign the relative coordinates of nuclei near paramagnetic sites in polycrystalline frozen solution or in glassy matrices. In this chapter we discuss how application of ENDOR spectroscopy, as a noncrystallographic method of structure determination, can lead to three-dimensional structure determination. We emphasize the application of ENDOR of VO^{2+} in polycrystalline or frozen solution systems, and we point out directions of future research in which ENDOR spectroscopy of VO^{2+} may be of advantage for biological structural studies.

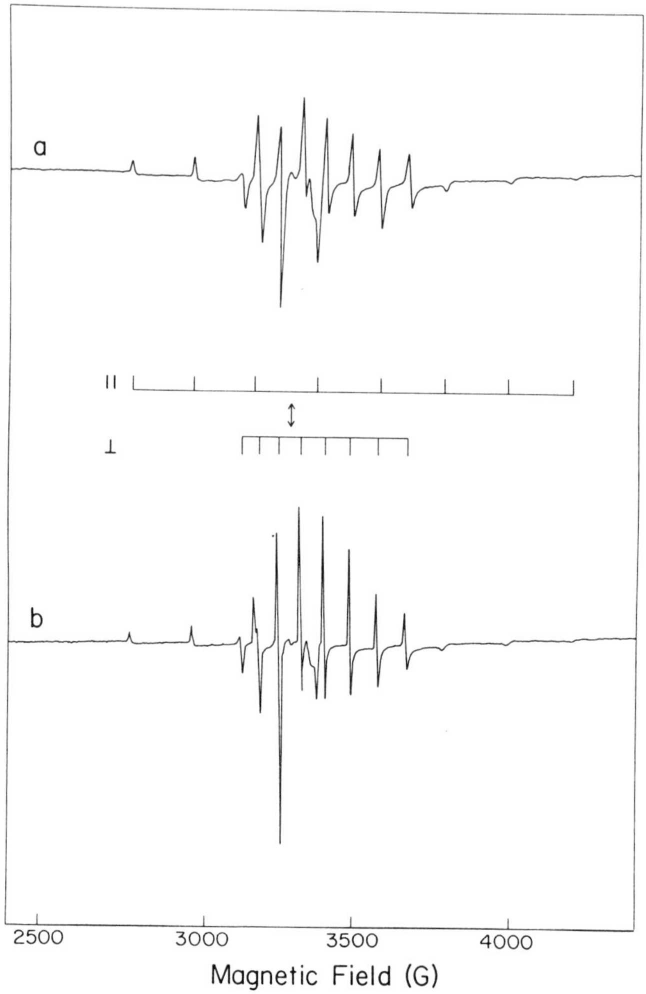

FIG. 1. First-derivative EPR absorption spectra of VO^{2+} in 50:50 (v/v) (a) H_2O:CH_3OH and (b) D_2O:CD_3OD solvent mixtures. The parallel (||) and perpendicular (⊥) EPR absorption components (−7/2, −5/2, . . . , 5/2, 7/2) are identified in sequence in the low- to high-field direction. The double-headed arrow indicates the magnetic field position corresponding to the forbidden $\Delta m_I = \pm 1$ transition. (Reproduced by permission from Ref. 9.)

2. EPR AND ENDOR OF VO^{2+}

2.1. EPR of VO^{2+} Complexes

The $3d^1$ electronic configuration of V(IV) belongs to a Kramers doublet ($S = 1/2$), corresponding in its lowest energy term to $^2D_{3/2}$. With an orbitally nondegenerate ground state, the orbital angular momentum is effectively quenched by the crystalline field, and the paramagnetism of VO^{2+} is due almost entirely to spin angular momentum [5,17]. For rapidly tumbling complexes of VO^{2+} in solution, eight sharp EPR transitions are observed because of the hf interaction of the unpaired electron with the 100% naturally abundant ($I = 7/2$) ^{51}V nucleus. Because the VO^{2+} ion in frozen solutions or polycrystalline systems exhibits an axially symmetric **g** tensor and an axially symmetric hf tensor due to the ^{51}V nucleus, there are eight parallel and eight perpendicular EPR absorption lines [18,19]. These features are illustrated in Fig. 1.

The EPR and ENDOR spectra of vanadyl complexes can be described by the spin Hamiltonian in Eq. (1) which includes the electronic

$$H_S = |\beta_e|\mathbf{H}_0\cdot\mathbf{g}_e\cdot\mathbf{S} - g_n\beta_n\mathbf{H}\cdot\mathbf{I} + \mathbf{S}\cdot\mathbf{A}\cdot\mathbf{I} \tag{1}$$
$$H_1 \qquad + \qquad H_2 \qquad + \quad H_3$$

Zeeman interaction (H_1) and the nuclear Zeeman interaction (H_2) while hf interactions of the unpaired electron with different classes of nuclei are contained within H_3. In Eq. (1), β_e and β_n represent the Bohr electron and nuclear magneton, $\mathbf{g}_e$ the electronic Zeeman interaction tensor, g_n the nuclear g factor, $\mathbf{H}_0$ the external (laboratory) magnetic field, **S** and **I** the electron and nuclear spin operators, respectively, and **A** the electron-nucleus hf tensor. The principal values of **A** for the vanadium nucleus and of $\mathbf{g}_e$ that characterize vanadyl complexes can be extracted from EPR spectra of frozen solutions or polycrystalline samples [6,7,18,19]. Representative examples are compared in Table 1.

2.2. ENDOR of VO^{2+} Complexes

The physical basis of ENDOR is illustrated in Fig. 2. ENDOR spectroscopy is performed by saturating the electronic transitions of the para-

TABLE 1

Comparison of Spectroscopic Parameters Characterizing EPR Spectra of Selected VO^{2+} Complexes[a]

Complex	$g_{\parallel}$	$g_{\perp}$	$A_{\parallel}$	$A_{\perp}$	Ref.
$[VO(H_2O)_5]^{2+}$	1.933	1.978	547.4	211.9	19
VO^{2+} in $Zn(NH_4)_2(SO_4)_2{\cdot}6H_2O$	1.933	1.981	548.1	215.3	20
VO^{2+} in $Mg(NH_4)_2(SO_4)_2{\cdot}6H_2O$	1.937	1.988	542.0	209.5	21
$[VO(acetylacetonate)_2]$ in tetrahydrofuran	1.945	1.981	506.7	183.3	22
[VO(TPP)] in tetrahydrofuran	1.964	1.989	477.0	162.4	22
VO^{2+} complexed to serum albumin	1.939	1.979	518.0	179.4	23
$[VO(ADP)_2(H_2O)]$[b]	1.928	1.970	540	195	—

[a]The principal components of the **g** and **A** tensors are defined as: $g_{\parallel} = g_{zz}$; $g_{\perp} = (g_{xx} + g_{yy})/2$; $A_{\parallel} = A_{zz}$; and $A_{\perp} = (A_{xx} + A_{yy})/2$. Hfc components are given in MHz.

[b]Calculated in association with studies in [24].

FIG. 2. Schematic illustration of the physical basis of ENDOR. The upper part of the figure illustrates energy level diagrams for a magnetic nucleus with $I = 1/2$ and an unpaired electron with $S = 1/2$ placed into a magnetic field. The nuclear and electronic transitions between the lower and upper levels are detected correspondingly by nuclear magnetic resonance (NMR) and EPR. Combining the two-energy-level schemes to correspond to a molecule with an unpaired electron interacting with a nearby magnetic nucleus results in the four-level scheme illustrated as **I** in which the levels are denoted by the magnetic spin quantum numbers $|m_S m_I\rangle$. In **I** the allowed transitions are indicated that are detected by EPR according to the selection rules $\Delta m_S = \pm 1$, $\Delta m_I = 0$ and by NMR for the selection rules $\Delta m_S = 0$, $\Delta m_I = \pm 1$. In **II** four particles in each of the two lower levels are shown to emphasize that the Boltzmann population of spins in the lower manifold is greater than that in the upper manifold. An electronic transition induced under low microwave power from a lower level to an upper level does not disturb the Boltzmann equilibrium significantly because of the much faster decay rate of an electron from an excited level to the ground state level, as indicated by W_e. This situation thus corresponds to detection of EPR absorption under nonsaturating conditions. In **III** the thick arrow represents EPR transitions induced under high microwave

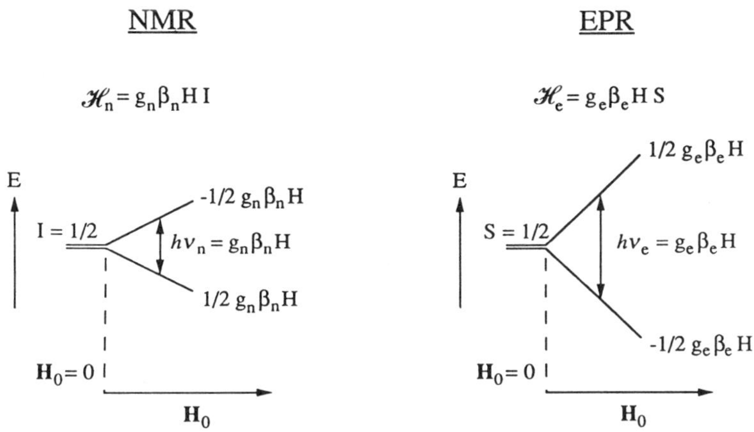

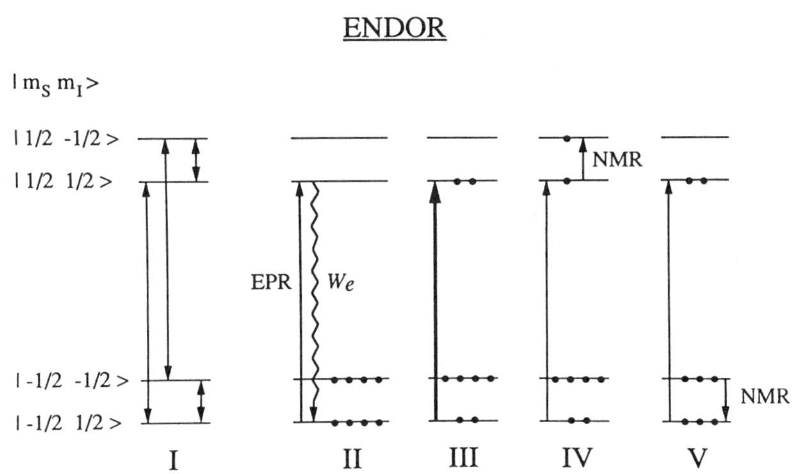

power, a condition corresponding to saturation which now perturbs the lower and upper levels away from their Boltzmann-defined populations. If a strong, saturating rf field is now applied, the populations of the upper or lower levels are altered through the rf-induced nuclear transitions, as illustrated in **IV** and **V**. This condition "desaturates" the system and restores the levels closer to the Boltzmann-defined equilibrium, and the intensity of the EPR transition will be increased. This situation corresponding to detection of EPR absorption by inducing an NMR transition thus depicts the physical processes which underlie ENDOR.

magnetic system under high microwave power and irradiating the system simultaneously with a strong radiofrequency (rf) field. The spectrometer is first adjusted to optimize the EPR signal intensity under low microwave power, and then the microwave power is increased, generally by several decibels, to achieve saturation. When the frequency of the rf field is scanned under these conditions and resonance of a nucleus interacting with the unpaired electron is reached, a forbidden transition involving simultaneous electron and nuclear "spin flips" is stimulated. The populations of the upper and lower levels are then "desaturated", giving rise to increased signal amplitude. Thus, the ENDOR method is equivalent to detection of nuclear resonance absorption by observing changes in the intensity of an EPR line.

ENDOR spectroscopy is optimally carried out with paramagnetic probes that are characterized by relatively long electronic spin-lattice relaxation times (T_{1e}), allowing microwave power saturation of the paramagnetic center to be easily achieved. Since the EPR absorption of VO^{2+} can be detected at room temperature, subsequent lowering of the temperature readily results in achieving an electronic spin-lattice relaxation time favorable for ENDOR spectroscopy. Also, in our experience, ENDOR absorption is optimally detected in systems characterized by narrow EPR absorption features with high peak-to-peak amplitudes. With respect also to this latter characteristic, we have found VO^{2+} to be an ideal paramagnetic probe for ENDOR spectroscopy [9,24].

ENDOR spectroscopy provides the most precise method of measuring the strength of electron-nucleus hf interactions. Consequently, its application results in very high resolving power for study of molecular structure. The high resolution of ENDOR spectroscopy is based on three factors:

1. The density of nuclear spectral lines for a multinuclear system is much smaller for ENDOR than for EPR.
2. The resonance absorption lines are much narrower for ENDOR than for EPR.
3. The direct interaction of the nucleus with the static magnetic field and the electron-nucleus hf interaction can be measured separately by ENDOR but not by EPR.

Since an ENDOR spectrum is a nuclear resonance spectrum, one must determine the total field at the nucleus to calculate the transition frequency. If we begin with the spin Hamiltonian in Eq. (1), it can be shown [25] that the first-order transition frequencies $\nu_{\pm}$ for a nucleus are given by Eq. (2) where the l_i are the direction cosines of $\mathbf{H}_0$ in the

$$\nu_{\pm} = [\sum_{i=1}^{3} [m_s/g_e(\sum_{j=1}^{3} g_j l_j A_{ji}) - l_i \nu_n]^2]^{1/2} \tag{2}$$

molecular axis system, A_{ji} is the orientation-dependent value of the hf coupling, ν_n is the nuclear Larmor frequency, m_s is the electron spin quantum number, and g_e represents the effective g value defined by the relation $g_e = [l_1^2 g_{xx}^2 + l_2^2 g_{yy}^2 + l_3^2 g_{zz}^2]^{1/2}$. In systems of low g-anisotropy, we assume the axis of quantization of the unpaired electron spin to be the direction of the applied field $\mathbf{H}_0$. If the applied field is oriented parallel to the principal axis j of the hf tensor $\mathbf{A}$, Eq. (2) then simplifies to that of Eq. (3):

$$\nu_{\pm} = \nu_n \pm |A_{jj}|/2 \tag{3}$$

where the separation of $\nu_{\pm}$ about ν_n is called the ENDOR shift. For symmetric separations, the hf coupling is thus twice the value of this frequency spacing. Equation (3) applies to the condition of $\nu_n > |A|/2$. For some nuclei the condition $\nu_n < |A|/2$ may apply, in which case Eq. (3) becomes $\nu_{\pm} = |A|/2 \pm \nu_n$.

Within the strong-field approximation the observed hf coupling A is given by Eq. (4) as a function of r and α where h is Planck's constant,

$$A = \frac{g_e|\beta_e|g_n|\beta_n|}{hr^3} \cdot (3\cos^2\alpha - 1) + A_{iso} \tag{4}$$

r is the modulus of the electron-nucleus position vector $\mathbf{r}$, and α is the angle between $\mathbf{H}_0$ and $\mathbf{r}$. For VO^{2+} systems, the observed principal hfc components $A_{\parallel}$ and $A_{\perp}$ correspond, respectively, to the maximum and minimum ENDOR shifts in the spectrum. The principal hfc components due to dipole-dipole interaction $A_{\parallel}^D$ and $A_{\perp}^D$ correspond to the first term of Eq. (4) for values of $\alpha = 0°$ and $90°$, respectively. Under the conditions that $|A_{iso}| \ll |A_{\parallel}|$ and $|A_{\perp}|$, and $A_{\parallel}^D > 0 > A_{\perp}^D$, the traceless dipolar hfc components $A_{\parallel}^D$ and $A_{\perp}^D$ can be calculated under the constraint $(A_{\parallel} + 2A_{\perp}) = 3A_{iso}$.

As seen in Table 1, VO^{2+} systems exhibit low g-anisotropy. Since the pseudo-contact contribution to the isotropic hfc, represented as the second right-hand term in Eq. (4), is negligible in cases of low g-anisotropy [26], a reasonable approximation is that A_{iso} arises entirely from the Fermi contact term and is thus determined by the extent of covalency in the metal-ligand bond. Because the unpaired electron is localized to the metal $3d_{xy}$ orbital in the VO^{2+} ion, the transfer of unpaired spin density to donor-ligand atoms either in the equatorial plane or in the axial position trans to the vanadyl oxygen atom is small. This aspect is of particular importance, for it ensures valid application of the point-dipole approximation, allowing structure determination with high precision. Quantitative aspects of the limits of application of Eq. (4) to VO^{2+} systems are discussed in Sec. 2.4.

2.3. Characteristics of ENDOR Spectra at Selected Molecular Orientations

To determine structural information by ENDOR one must (1) identify the surrounding nuclei contributing to the hf interactions and (2) determine the dependence of ENDOR spectra on the settings of the external magnetic field with respect to magnetic axes in the molecule. In VO^{2+} complexes the $g_{\parallel}$ or g_{zz} component of the $\mathbf{g}_e$ tensor is coincident with the V=O bond and the $g_{\perp}$ component is coincident with the molecular x,y plane. Therefore, separate application of $\mathbf{H}_0$ at parallel and perpendicular EPR absorption features should provide a means for selection of different molecular orientations for ENDOR with correspondingly distinguishable hfc components. This concept of angle selection for ENDOR yielding single-crystal-like spectra of molecules in polycrystalline matrices or frozen glassy solutions was first noted by Rist and Hyde for Cu^{2+} complexes [27]. Following this seminal observation, Kreilick and coworkers [25] and Hoffman et al. [28] provided insightful theoretical analysis of ENDOR spectra of polycrystalline or frozen glassy systems obtained at arbitrary field settings.

Figure 3 illustrates schematically the expected pattern of observed ENDOR splittings as a function of the positions of protons and the orientation of $\mathbf{H}_0$ with respect to the unique symmetry axis of g_{zz} of the VO^{2+} ion. When $\mathbf{H}_0$ is applied to g_{zz} at the low-field EPR absorption feature ($-7/2\,\|$), $A_{\parallel}$ of a proton located along the symmetry axis or $A_{\perp}$ of a proton located near or in the x,y plane should be observed. On the other hand, if $\mathbf{H}_0$ is set to an orientation perpendicular to g_{zz} and, therefore,

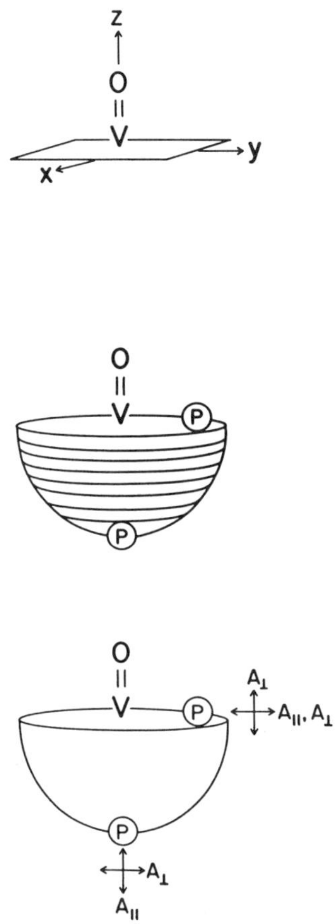

FIG. 3. Schematic illustration of the relationships of the symmetry axes of the VO^{2+} ion to the principal axes of the **g** tensor and of the hf tensor of nearby protons. The top diagram designates the direction of the molecular axes with respect to the V=O bond. The middle diagram illustrates the relative positions of protons near the molecular *x,y* plane and near or along the *z* axis. Each circle represents an orientation of $\mathbf{H}_0$ within the *g*-axis system. The bottom diagram illustrates the principal hfc components that are detected for equatorially or axially positioned protons according to whether $\mathbf{H}_0$ is aligned parallel or perpendicular to the *z* axis. (Reproduced by permission from Ref. 9.)

parallel to the x,y plane (i.e., $\mathbf{H}_0$ at $-3/2\ \perp$), $A_\perp$ of a proton located along the symmetry axis and both $A_\parallel$ and $A_\perp$ of a proton located near or in the x,y, plane should be observed. On this basis, the ENDOR splittings are classified into three categories:

1. Splittings that are observed only for the parallel orientation ($\mathbf{H}_0$ parallel to the g_{zz} or $g_\parallel$ axis).
2. Splittings that are observed only for the perpendicular orientation ($\mathbf{H}_0$ perpendicular to $g_\parallel$).
3. Splittings that are observed as common to both parallel and perpendicular orientations.

Figure 4 illustrates an EPR-selected orientation, the g axes, and the hf coordinate system for a proton located in the molecular (equatorial) plane of a paramagnetic metal ion with low g-anisotropy. The surface of the sphere represents the isotropic distribution of molecular

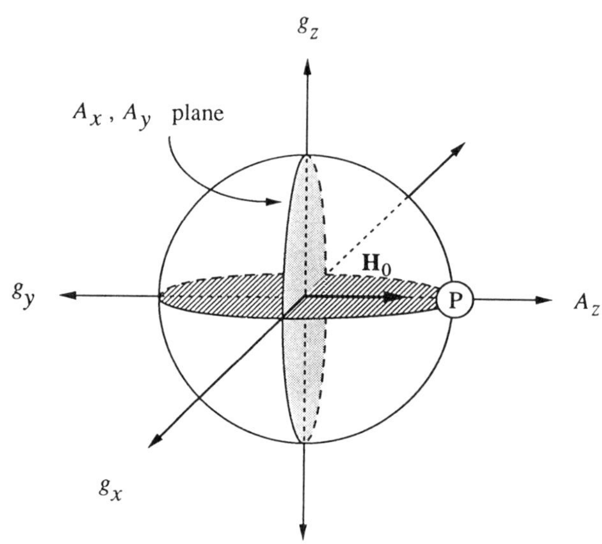

FIG. 4. Superposition of the principal axes of the **g** tensor and of the hf tensor. The proton is located in the molecular x,y plane. The electron-proton vector defines the principal hf axes: the A_z axis along the vector; and the A_x,A_y plane is perpendicular to the vector, indicated by dots. The circle represented by the hatched area indicates the orientation of the applied magnetic field $\mathbf{H}_0$ coincident with the molecular x,y plane.

orientations present in a randomly oriented sample. When g-anisotropy is small compared to the average g value, the maximum hf interaction energy occurs for a field oriented along the metal-nucleus vector (A_z axis), and the minimum occurs when the field is in the A_x,A_y plane. For a proton in the x,y plane, well-resolved features corresponding to $A_\perp$ are observed which persist with an essentially constant ENDOR shift at all $\mathbf{H}_0$ values because upon selection of any value of Θ, where Θ represents the angle between the g_{zz} axis and $\mathbf{H}_0$, an orientation set intersects the A_x,A_y plane. This relationship can be seen in Fig. 4 and provides the basis to analyze hf interactions in terms of nuclear coordinates with respect to the unique symmetry axis of the VO^{2+} ion.

2.4. Evaluation of the Quantitative Limits of Structure Determination by ENDOR

The structures obtained through ENDOR studies depend on estimates of the electron-nucleus dipolar separations which are ordinarily calculated on the basis of Eq. (4). Therefore, it is essential to develop an accurate working model of the spatial distribution of spin angular momentum in the metal ion complex. Although the spatial part of the wave function of the unpaired electron spin describes the region of space to which unpaired spin is localized, calculations to include the integrated wave function are generally not carried out, and the unpaired electron in the paramagnetic center is often considered as a point-dipole for purposes of simplification. To justify this simplification, two important questions must be carefully considered: How localized in space is the unpaired electron? What fraction of the unpaired spin density has been transferred to nearby atoms through chemical bonds?

For certain types of paramagnetic species, the unpaired spin density behaves very closely to that of a point-dipole. Hutchison and coworkers showed in ENDOR studies that the f electrons of the lanthanide ions Nd^{3+} [29] and Gd^{3+} [30] exhibit very close behavior to that of a point magnetic dipole. For the free radical of organic nitroxyl spin labels, we demonstrated that the unpaired electron localized to the $2p_z$ orbitals of the nitrogen and oxygen atoms behaves like an effective point-dipole centered near the midpoint of the N–O bond [31]. Atherton and Shackleton demonstrated through single crystal ENDOR studies that the hf interactions of VO^{2+} doped into crystals of $Mg(NH_4)_2(SO_4)_2{\cdot}6H_2O$ adhere closely to the point-dipole model [21,32]. In all of these cases, the traceless dipolar hfc components are shown to exhibit axial symmetry, as expected for point-dipole interactions.

For purposes of structural analysis, we assess here the quantitative limits of application of Eq. (4) to VO^{2+} complexes in polycrystalline or frozen solution systems. First, the accuracy of the strong-field approximation improves as the g-anisotropy decreases. Following Hutchison and McKay [29], we evaluate the expectation value of the electronic magnetic moment $\langle \mu_e \rangle_\pm$ associated with the VO^{2+} ion by Eq. (5) in which

$$\langle \mu_e \rangle_\pm = -|\beta_e| \mathbf{g}_e \cdot \langle S \rangle_\pm = -\frac{|\beta_e|}{2g_e} \begin{bmatrix} g_{xx}{}^2\, l_1 \\ g_{yy}{}^2\, l_2 \\ g_{zz}{}^2\, l_3 \end{bmatrix} \tag{5}$$

$\mathbf{g}_e$ is considered a diagonalized matrix, the quantity $\langle S \rangle_\pm$ is the expectation value of $\mathbf{S}$, and g_e is the effective value of $\mathbf{g}_e$, which in the case of axial symmetry ($g_{xx} = g_{yy} \neq g_{zz}$) reduces to $g_e = [g_\|{}^2 \sin^2\Theta + g_\perp{}^2 \cos^2\Theta]^{1/2}$. The principal values $g_\|$ and $g_\perp$ of VO^{2+} in Table 1 differ by $\leqslant$2.3%. While this corresponds to greater g-anisotropy than observed, for instance, in nitroxyl spin labels [33,34], it is sufficiently small to apply g as the average value of $g_\perp$ and $g_\|$. Then Eq. (5) reduces to $\langle \mu_e \rangle_\pm \simeq -|\beta_e| g/2$.

Second, the accuracy of the strong-field approximation increases with increasing r. For a spin system with $g \sim 2.0$, the maximum magnetic field $\mathbf{h}_e$ generated by an unpaired electron at an electron-nucleus separation of 5 Å is ~ 150 G. This value is small compared to that of the applied laboratory field ($\mathbf{H}_0$ of ~ 2800 G for the low-field resonance feature $-7/2\,\|$ or $\mathbf{H}_0$ of ~ 3300 G for the $-3/2\perp$ resonance feature; cf. Fig. 1) which generally corresponds to the field settings for microwave saturation of VO^{2+} in ENDOR experiments. Accordingly, the maximum angle between $\mathbf{H}_0$ and the effective field ($\mathbf{H}_0 \pm \mathbf{h}_e$) at a nucleus 5 Å from the metal center is ~ 2.4° for $\alpha = 45°$. This angle is 0° for $\alpha = 0°$ or 90°. The ratio of $\mathbf{h}_e$ to $\mathbf{H}_0$ will be even smaller for larger values of r.

Third, the accuracy of the point-dipole approximation improves with increasing r. Snetsinger et al. [35] showed that attributing the observed hfc of an unpaired electron in a metal $3d_{xy}$ orbital entirely to dipole-dipole interaction with a nuclear spin of $I = 1/2$ over a distance of 2.09 Å results in a calculated separation of 2.15 Å. This discrepancy will be correspondingly less for larger values of r, and for electron-nucleus distances $\geqslant$4.0 Å, it will be negligible. In our study of the solvation structure of VO^{2+}, we have found that electron-nucleus separations of equatorially positioned solvent protons or ^{13}C of inner sphere coordinated methanol ligands were underestimated by ~0.1 and 0.3 Å, respectively [9].

Furthermore, Snetsinger et al. [35] showed that for large values of Z (effective nuclear charge), as in the case of the vanadium nucleus, the hf couplings are relatively insensitive to small changes in Z of ± 0.3 as could occur through charge transfer from V(IV) to donor-ligand atoms. Thus, the point-dipole and strong-field approximations contribute less than 5–9% error to estimates of the principal dipolar hfc components for values of $r \sim 2.5$ Å. In view of the r^{-3} dependence of the electron-nucleus separation on the hfc, the error will be even less as r increases.

3. STRUCTURE DETERMINATION BY EPR AND ENDOR

3.1. Solvation Structure of the VO^{2+} Ion

Although the VO^{2+} ion in aqueous solution exists as the penta-aqua hydrated species $[VO(H_2O)_5]^{2+}$, it has not been observed hitherto by X-ray crystallography in any crystalline environment. The structure of the solvated complex was first formulated on the basis of its electronic absorption spectrum [1] and modeled according to the square-pyramidal geometry of the penta-coordinated complex of VO^{2+} with acetylacetonate [36] and of $VOSO_4 \cdot 5H_2O$ in crystals [37]. In the latter study, the structure of the coordination complex was found to correspond to $[VO^{2+}(H_2O)_4(SO_4^{2-})] \cdot H_2O$. Although the original X-ray study was carried out for $[VO^{2+}(H_2O)_4(SO_4^{2-})] \cdot H_2O$ only on the basis of Patterson projections [37], the complete X-ray study with refined crystallographic parameters determined subsequently has not been reported [38].

To elucidate the composition of the first hydration sphere of the VO^{2+} ion in aqueous solution, nuclear magnetic resonance (NMR) studies using oxygen-17-enriched water were carried out in two different laboratories [39–42]. These studies were based on the resonance linewidths of solvent molecules and their chemical shifts. Because of the long spin-lattice relaxation time of the metal ion, NMR parameters could not be obtained with high accuracy, particularly for solvent molecules in the first coordination sphere [39]. However, the results from these studies indicated that four water molecules were tightly bound in the equatorial plane and a fifth axially positioned water molecule was weakly bound to the VO^{2+} ion in the first coordination sphere. Although molecular geometry in terms of bond lengths and valence angles could not be derived, the results supported the intuitively adopted structure of the $[VO(H_2O)_5]^{2+}$ species as a square-pyramidal complex of tetragonal symmetry. From ligand hfc constants based on EPR linewidths of the VO^{2+} ion in frozen solutions of natural and perdeuterated water, Albanese and Chasteen showed that the anisotropic (dipolar) hf couplings are consistent with $[VO(H_2O)_5]^{2+}$ as a complex of square-pyramidal geometry [19].

The molecular structure of $[VO(H_2O)_5]^{2+}$ was first elucidated by Atherton and Shackleton from single-crystal ENDOR studies [21,32]. From the proton ENDOR spectra of vanadyl doped into single crystals of $Mg(NH_4)_2(SO_4)_2{\cdot}6H_2O$ [43] in which the VO^{2+} ion replaced the Mg^{2+} ion plus one water molecule, they determined the principal hfc components for all 10 protons in the $[VO(H_2O)_5]^{2+}$ species [21]. They showed that the traceless components of the principal hf tensors are nearly axially symmetric as expected for point-dipole interactions. Because the anisotropy of the various magnetic interactions could be explored by rotation of the applied magnetic field $\mathbf{H}_0$ with respect to crystal axes, the coordinates of each proton in the crystal could be assigned. The molecular structure of $[VO(H_2O)_5]^{2+}$ based on single-crystal ENDOR studies is shown in Fig. 5. In Table 2 the ENDOR based metal-proton distances are listed and compared with distances calculated on the basis of a complete hf tensor analysis for the interaction of the unpaired electron in the $3d_{xy}$ orbital of the V(IV) ion with each proton. The excellent agreement shown in Table 2 demonstrates that the VO^{2+}-proton hf couplings adhere closely to point-dipole interactions.

A more extensive study of the solvation structure of the VO^{2+} ion has been made by ENDOR spectroscopy of frozen solutions of VO^{2+} in methanol and water-methanol mixtures [9] on the basis of both ^1H and

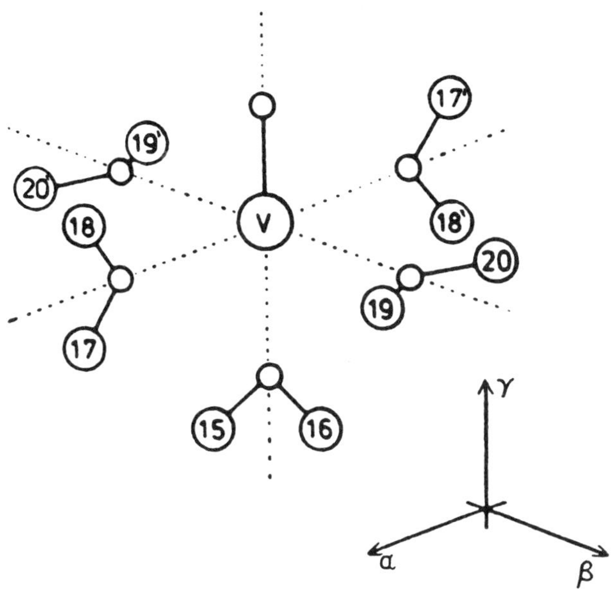

FIG. 5. Molecular structure and the numbering convention for protons of the $[VO(H_2O)_5]^{2+}$ ion in single crystals of $Mg(NH_4)_2(SO_4)_2 \cdot 6H_2O$. Protons i,i' are related by inversion through a center of symmetry at the divalent cation. The plane of the water molecule containing protons 19 and 20 is close to the molecular x,y plane while that for the molecule containing protons 17 and 18 is almost perpendicular to the plane. (Reproduced by permission from Ref. 32.)

^{13}C ENDOR, through which structures of complexes of the solvated VO^{2+} ion were assigned by molecular modeling constrained by ENDOR-determined electron-nucleus distances. Comparison of the spectra in Fig. 6 shows that resonance features arising separately from methyl and hydroxyl protons frequently overlap, and by selective deuteration there are a total of seven pairs of resonance features due to hydroxyl protons with five pairs due to methyl protons. The principal hfc components for each class of protons were identified from ENDOR spectra with $\mathbf{H}_0$ settings at the $-7/2\|$ and $-3/2\perp$ EPR absorption features corresponding to magnetic field orientations parallel or perpendicular to the V=O bond, respectively, as described in Figs. 3 and 4.

For each class of protons, the principal hfc components for the $[VO(CH_3OH)_5]^{2+}$ complex were axially symmetric. In Table 3 are

TABLE 2

Comparison of the ENDOR-Determined and Calculated Metal-Proton Distances of $[VO(H_2O)_5]^{2+}$ in $Mg(NH_4)_2(SO_4)_2 \cdot 6H_2O$ [21]

	Metal-proton distances (Å)	
Proton[a]	ENDOR	Calculated
15	2.772	2.892
16	2.777	2.867
17	2.681	2.671
17′	2.608	2.588
18	2.566	2.516
18′	2.619	2.599
19	2.600	2.530
19′	2.550	2.460
20	2.595	2.575
20′	2.539	2.519

[a]See Fig. 5.

listed the dipolar and isotropic hfc components, the metal-nucleus distances calculated according to Eq. (4), and brief comments for each class of nuclei to indicate the structural relationships of the ligand to the VO^{2+} ion. Because the unpaired electron is localized to the metal $3d_{xy}$ orbital, the isotropic contributions of ligand nuclei in the equatorial plane are seen as significantly higher than for the axially coordinated ligand. Nevertheless, the values of A_{iso} from this study [9] correspond very closely to values determined on the basis of single-crystal ENDOR studies [21].

Figure 7 illustrates the solvation structure of VO^{2+} with both inner and outer sphere coordinated methanol molecules, as determined by ENDOR-constrained molecular modeling. In this structure, the outer sphere coordinated molecules were assigned orientations that could account for plausible hydrogen-bonding interactions with inner

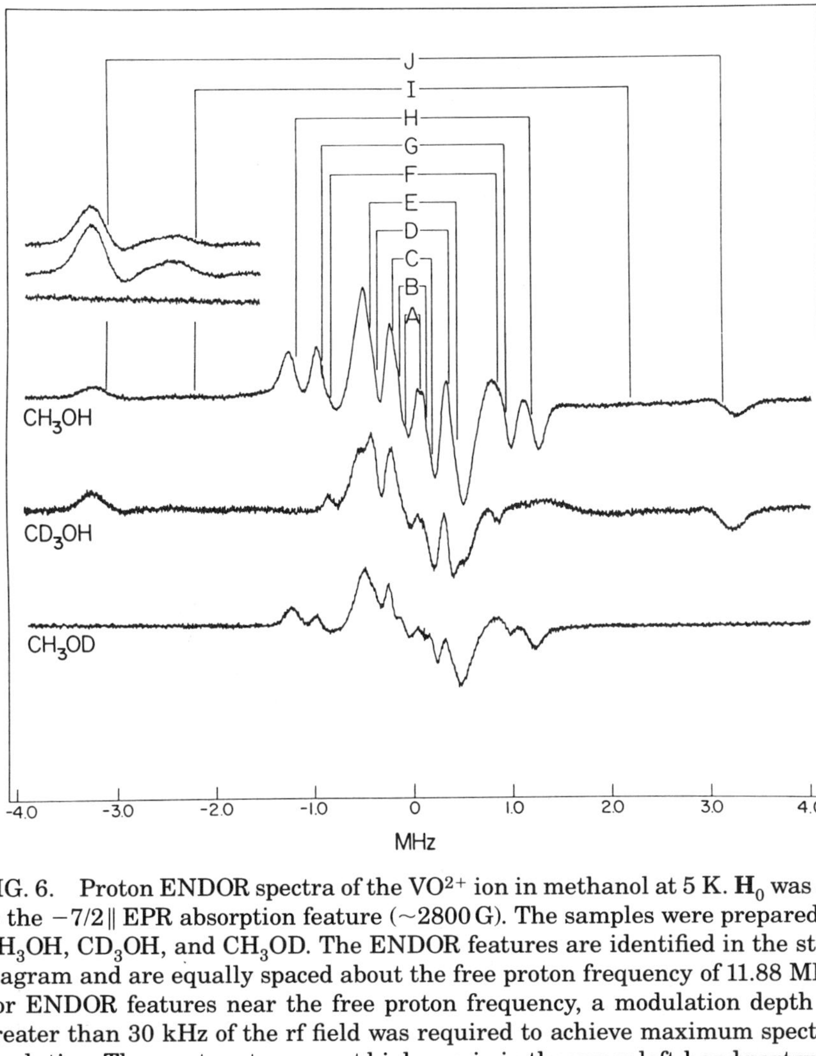

FIG. 6. Proton ENDOR spectra of the VO^{2+} ion in methanol at 5 K. $\mathbf{H}_0$ was set to the $-7/2\|$ EPR absorption feature (~2800 G). The samples were prepared in CH_3OH, CD_3OH, and CH_3OD. The ENDOR features are identified in the stick diagram and are equally spaced about the free proton frequency of 11.88 MHz. For ENDOR features near the free proton frequency, a modulation depth no greater than 30 kHz of the rf field was required to achieve maximum spectral resolution. The spectra at apparent higher gain in the upper left-hand part were recorded with a modulation depth of 90 kHz. (Reproduced by permission from Ref. 9.)

TABLE 3

Summary of the Principal hfc Components (in MHz) and Estimated Metal-Nucleus Distances of the VO^{2+} Ion in Frozen Methanol [9]

$A_{\parallel}$	$A_{\perp}$	A_{iso}	$A_{\parallel}^{D}$	$A_{\perp}^{D}$	r (Å)	Comment[a]
^{1}H ENDOR						
6.12	3.21	−0.10	6.22	−3.11	2.90	*ax.* OH
4.34	1.40	0.50	3.83	−1.91	3.42	V=O···H
15.67	0.70	4.76	10.91	−5.46	2.52	*eq.* OH (in plane)
13.38	1.03	3.77	9.61	−4.78	2.60	*eq.* OH (out-of-plane)
1.64	0.31	0.34	1.30	−0.65	4.80	outer sphere, *ax.* OH
7.61	0.12	2.45	5.16	−2.57	3.20	outer sphere, *eq.* OH
2.36	1.40	−0.15	2.51	−1.25	3.92	*ax.* CH_3
0.80	0.45	0.03	0.83	−0.42	5.60	V=O···$HOCH_3$
3.21	1.96	−0.23	3.44	−1.73	3.54	*eq.* CH_3
1.83	0.16	0.50	1.33	−0.66	4.85	outer sphere, *ax.* CH_3
^{13}C ENDOR						
0.99	0.44	0.04	0.95	−0.48	3.43	*ax.* CH_3
2.04	0.67	0.23	1.81	−0.90	2.78	*eq.* CH_3
0.23	0.13	−0.01	0.24	−0.12	5.45	outer sphere, *eq.* CH_3

[a]Axial and equatorial (abbreviated as *ax.* and *eq.*, respectively) protons of the hydroxyl and methyl groups, and ^{13}C nucleus of the methyl group were assigned on the basis of ENDOR spectra according to chemical origin and magnetic field ($\mathbf{H}_0$) settings.

sphere coordinated methanol molecules [9]. It was shown by analysis of ENDOR spectra that $[VO(CH_3OH)_5]^{2+}$ was a unique complex formed only in neat methanol. In aqueous methanol mixtures, two types of species were identified: one with an axially coordinated water molecule trans to the vanadyl oxygen and another with axially coordinated methanol. Both types of complexes were shown to have only equa-

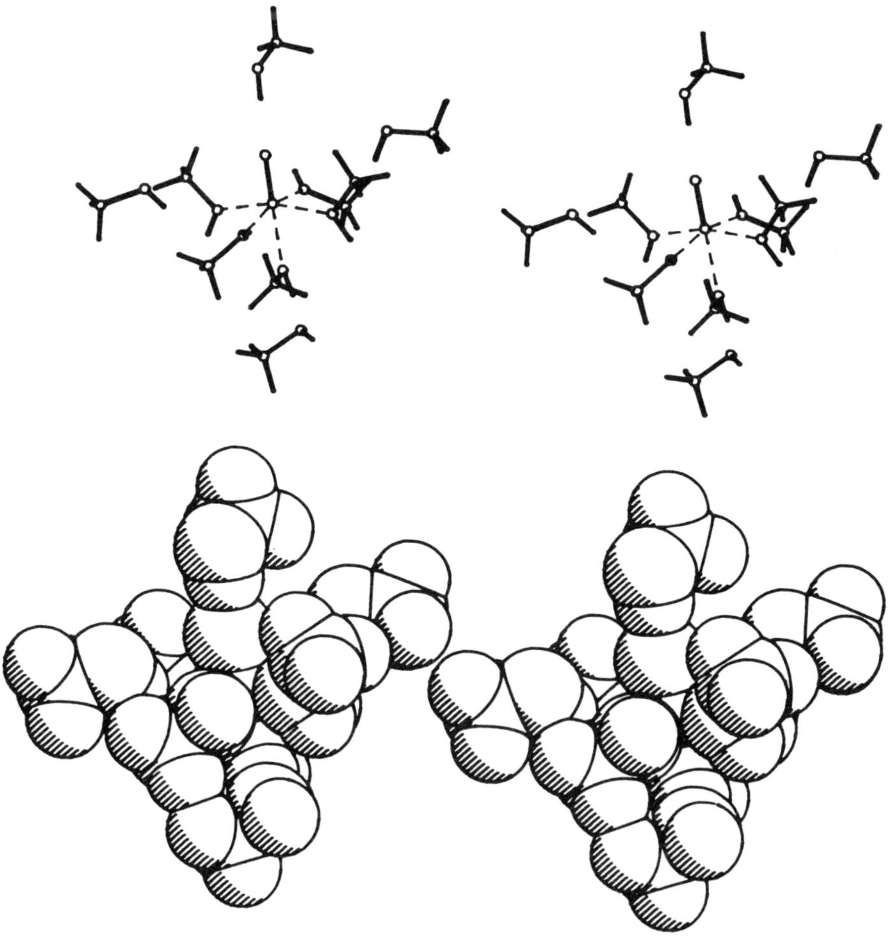

FIG. 7. Stereodiagram of the coordination structure of the VO^{2+} ion in methanol determined on the basis of ENDOR and molecular modeling. The upper diagram illustrates the complex in stick skeletal form. Broken lines connect the inner sphere methanol molecules coordinated to the vanadium. The lower diagram illustrates the complex in space-filling form (in the same projection as in the upper diagram) and was drawn to scale for van der Waals radii of 1.53 Å (C), 1.4 Å (O), 1.2 Å (H), and 1.35 Å (V). (Reproduced by permission from Ref. 9.)

torially coordinated water molecules. The coordination geometry of the $[VO(CH_3OH)_5]^{2+}$ species in neat methanol and of the $[VO(H_2O)_5]^{2+}$ and $[VO(H_2O)_4(CH_3OH)]^{2+}$ species in water-methanol cosolvent mixtures was best accounted for as square pyramidal with tetragonal symmetry. The structural detail obtained in this ENDOR study approaches the accuracy associated with small-molecule X-ray crystallographic studies.

3.2. Small Molecule Complexes

3.2.1. Nucleotides

The interactions of nucleotides in biological systems require a divalent metal ion, usually Mg^{2+}, as a cofactor. While metal-nucleotide interactions have been investigated by various spectroscopic and potentiometric techniques, there are only a small number of high-resolution X-ray crystallographic studies of ATP or its derivatives, primarily because of difficulties in obtaining suitable crystals, and relatively few provide detailed information on Mg^{2+}-ATP complexes [44,45]. In the case of ADP, the only crystal structure that has been hitherto determined to atomic resolution is that of the monorubidium salt [46]. While complexes of nucleotides with other divalent metal ions such as Co^{2+} and Ni^{2+} have been described [47], the ligand field stabilization effect of these transition metal ions promotes coordination with base nitrogens resulting in structures that may deviate significantly from those of nucleotide complexes of Mg^{2+} in solution. The affinity of base complexation relative to phosphate binding increases in the order $Mg^{2+} < Co^{2+} < Ni^{2+} < Mn^{2+} < Zn^{2+} < Cd^{2+} < Cu^{2+}$ [48].

Since Mg^{2+} complexation of nucleoside 5′-di- and triphosphates hinders their spontaneous hydrolysis and is generally required for phosphoryl transfer reactions in cells, it is important to determine structures of Mg^{2+}-nucleotide complexes or to employ spectroscopic probes that closely simulate the interactions of Mg^{2+} with phosphate groups. In our laboratory we have observed with EPR and ENDOR spectroscopy that VO^{2+} is coordinated only to the phosphate groups of nucleotides, similar to that expected in Mg^{2+}-nucleotide complexes, and no coordination is detected to atoms of the nucleic acid base [24,49]. Since it was also possible to demonstrate in these studies that VO^{2+}-phosphate binding was inhibited by Mg^{2+} with no evidence of nonspecific binding of the VO^{2+} ion to the ribose hydroxyl groups or to the base nitrogens, it is likely that VO^{2+} may serve as the most suitable spectroscopic probe to simulate Mg^{2+}-nucleotide interactions in solution.

Figure 8 illustrates the EPR spectrometric titration of the binding of VO^{2+} to adenine nucleotides [24]. For the $-3/2 \perp$ EPR absorption feature, there is a linear increase in signal peak-to-peak amplitude up to a VO^{2+}/nucleotide ratio of 1:2, after which the signal intensity remains constant. The results thus indicate that the stoichiometry of VO^{2+}/nucleotide binding is 1:2 for adenosine 5′-diphosphate (ADP), α,β-methylene adenosine 5′-diphosphate (AMP-CP), and adenosine 5′-triphosphate (ATP) in the presence of excess ligand. Figure 9 shows the ligand hf structure underlying the $-3/2 \perp$ resonance feature for a solution mixture containing excess ADP with respect to VO^{2+}. The resonance features in perdeuterated solvent exhibit a quintet pattern with 6.62 G interval spacings and approximate intensity ratio of 1:4:6:4:1. Since ligand hfc produces a total of $(2nI + 1)$ absorption lines under these circumstances, this hf pattern can be attributed only to four structurally equivalent ($I = 1/2$) ^{31}P atoms [24]. On this basis the conclusion must be made that VO^{2+} is coordinated to two pyrophosphate groups with the phosphate oxygens bound in the equatorial plane. Similar conclusions were drawn for VO^{2+} complexed to AMP-CP.

It was also demonstrated that for solutions in which the VO^{2+}/ADP ratio is 1:1 the shf coupling due to ^{31}P is observed as a three-line pattern with an approximate 6.74 G splitting and an intensity ratio of 1:2:1. This pattern is indicative of only two equivalent ^{31}P nuclei interacting with VO^{2+}. Thus, complexes of 1:1 and 1:2 metal/nucleotide stoichiometry can be formed with ADP. Since the proton ENDOR spectra were identical, reflecting metal-proton separations, the two nucleotides complexed to the central VO^{2+} ion must occupy equivalent positions and exhibit twofold rotational symmetry with respect to the unique symmetry axis defined by $g_{\parallel}$. Similar conclusions were reached for VO^{2+}-ATP complex formation except that binding occurs via the terminal β- and γ-phosphate groups [24].

Figure 10 shows the ENDOR spectrum of the $[VO(AMP\text{-}CP)_2(H_2O)]$ complex with $\mathbf{H}_0$ at the $-3/2 \perp$ setting to illustrate how resonance features were assigned and how structural analysis was performed. Since the spectrum of $[VO(ADP)_2(H_2O)]$ is identical to that in Fig. 10

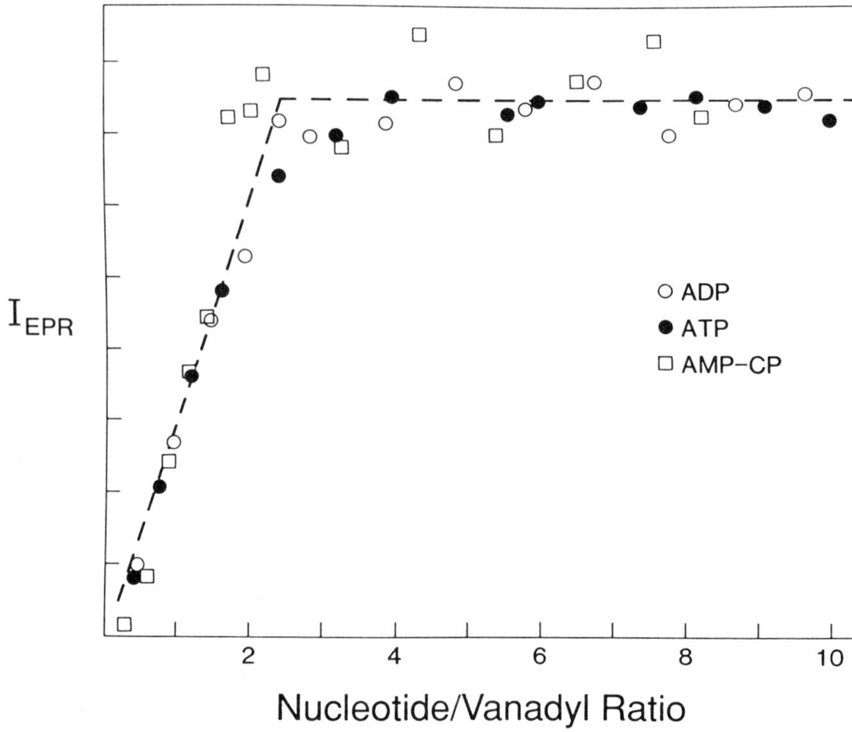

FIG. 8. EPR spectrometric titration of the VO^{2+} ion complexed to ADP (open circles), ATP (closed circles), and AMP-CP (open squares). The EPR signal intensity for the $-3/2\perp$ line of the VO^{2+} ion binding to nucleotides is plotted as a function of nucleotide/vanadyl molar ratio. The EPR spectra were recorded at 20 K (final VO^{2+} ion concentration of 0.01 M buffered to pH 6.5 with 1,4-piperazinediethanesulfonate). Dashed lines were drawn through the experimentally measured points. (Reproduced by permission from Ref. 24.)

except for the features labeled C1–C4, these line pairs were assigned to the methylene protons bridging the two phosphorus atoms. The line pairs A1–A4, also seen for the $[VO(ADP)_2(H_2O)]$ complex, are assigned to the H(8) proton of the adenine moiety and to the (averaged) H(5′) proton of the ribose moiety. All three VO^{2+}-nucleotide complexes exhibit the proton resonance features of an axially coordinated solvent molecule trans to the vanadyl oxygen [9]. With the ENDOR-determined electron-proton distances as constraints, summarized in Table 4, the

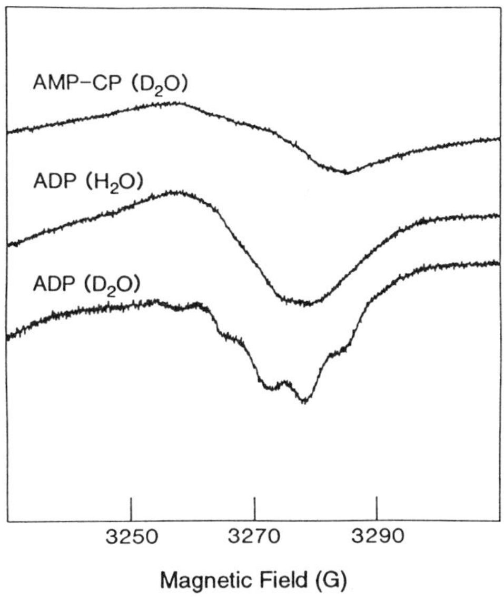

FIG. 9. Linewidths of the $-3/2 \perp$ EPR absorption feature and ^{31}P shf structure for the VO^{2+} ion complexed to ADP and AMP-CP in frozen solutions of natural abundance and perdeuterated aqueous methanol (50:50 v/v) cosolvent mixtures. The nucleotide/vanadyl ratio was in excess of 2:1 to ensure saturation of the VO^{2+} cation with nucleotide ligand. (Reproduced with permission from Ref. 24.)

conformation of the metal-nucleotide complex was then assigned by torsional angle search calculations. Figure 11 illustrates the structure of the complex of VO^{2+} with AMP-CP. The ENDOR-constrained structure shows that (1) the base moiety has an anti conformation with respect to the glycosidic C(1′)–N(9) bond, (2) the conformation about the C(4′)–C(5′) bond is gauche gauche, and (3) the conformation is trans about the C(5′)–O(5′) bond [24]. The ENDOR-determined dihedral angles around these three bonds are in excellent agreement with values obtained from X-ray studies of metal-nucleotide complexes [44–46,50].

Recently, Jiang and Makinen have similarly demonstrated by EPR and ENDOR that VO^{2+} is bound by the phosphate group of guanosine 5′-monophosphate (GMP) with 1:2 metal/nucleotide stoichiometry and have determined the structure of the $[VO(GMP)_2(H_2O)]$ complex [49]. Of the five types of nucleosides found in DNA and RNA,

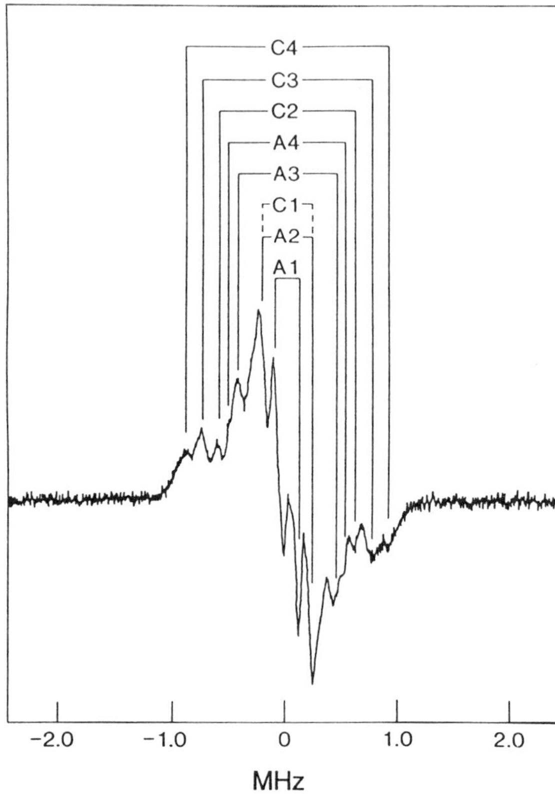

FIG. 10. Proton ENDOR spectrum of the $[VO(AMP\text{-}CP)_2(H_2O)]$ complex in perdeuterated cosolvent mixture at 20 K. The complex was prepared with a VO^{2+}/AMP-CP molar ratio of 1:4, buffered to pH 6.5. $\mathbf{H}_0$ was set to the $-3/2\perp$ EPR absorption line (~3270 G). The ENDOR line pairs denoted by symbols A and C belong to protons of the bound adenosine moiety and methylene group of AMP-CP, respectively. (Reproduced by permission from Ref. 24.)

guanosine is unique because it is capable of forming self-structured assemblies in solution through hydrogen bonding to give G–G base pairs and G-quartets. The G-quartet assemblies are formed through a square-planar array of hydrogen-bonded guanine bases, and equilibria controlling the formation of G-quartets and of stacked quartets to form octets and higher order assemblies are sensitive to pH and the presence of sodium and potassium ions [51–55].

TABLE 4

Summary of Principal Hyperfine Coupling Components (in MHz) and Estimated Metal-Proton Distances of the VO^{2+}-Adenine Nucleotide Complexes [24]

$A_{\parallel}$	$A_{\perp}$	A_{iso}	$A_{\parallel}^{D}$	$A_{\perp}^{D}$	r (Å)[a]	Proton assignment
1.81	1.22	−0.21	2.02	−1.01	4.23	$-CH_2-$ (AMP-CP)
1.52	0.44	0.21	1.31	−0.65	4.88	$-CH_2-$ (AMP-CP)
1.07	0.44	0.06	1.01	−0.50	5.32	Ade:H(8) (ADP/AMP-CP)[b]
0.81	0.24	0.11	0.70	−0.35	6.02	Rib:H(5′) (ADP/AMP-CP)[b]
6.07	3.13	−0.06	6.13	−3.07	2.92	axial OH (solvent)[c]
2.36	1.40	−0.15	2.51	−1.25	3.92	axial CH_3 (solvent)[c]

[a]Estimated error in r is 0.1–0.2 Å.
[b]The assignment of the H(8) and H(5′) protons is based on molecular modeling studies.
[c]These studies were carried out with aqueous methanol as the solvent.

Jiang and Makinen demonstrated on the basis of NMR studies that the $[VO(GMP)_2(H_2O)]$ complex forms self-structured quartet and octet assemblies through hydrogen bonding of the guanine bases like the free nucleotide [49]. They were able to show that the proton ENDOR resonances were identical for the monomeric form of $[VO(GMP)_2(H_2O)]$ as well as for the complex incorporated into quartet and octet assemblies. The ENDOR-determined electron-proton distances restricted the ribose conformation to a C3′-*endo* pucker. Modeling studies using the ENDOR distances as constraints were able to rule out the C2′-*endo* conformation, which is the other prevalent conformation found for monomeric ribonucleotides [48]. On the other hand, for G-quartets formed with $[VO(GMP)_2(H_2O)]$ in solution, the guanine base was restricted to the anti conformation.

In contrast, the X-ray structure of $d(G_4T_4G_4)$, which forms hydrogen-bonded arrays of G-quartets in crystals, shows the guanine bases to occupy both syn and anti conformations within each quartet [56]. The differences in base conformation with only the anti structure associated

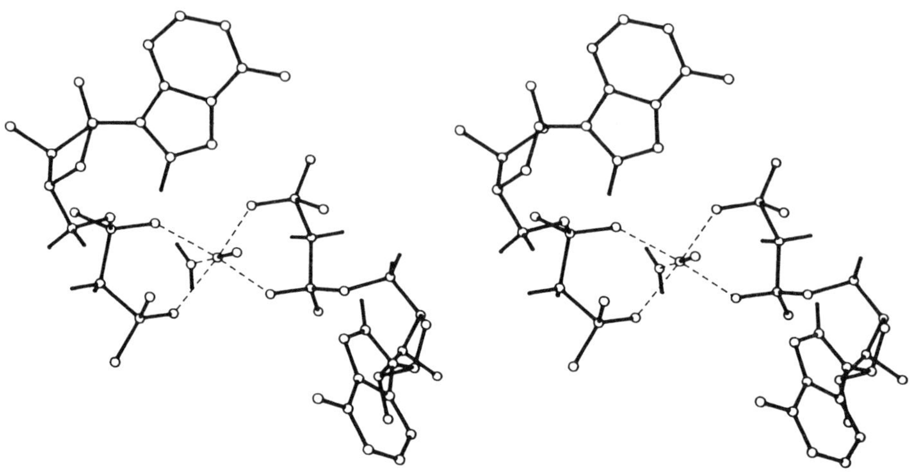

FIG. 11. Stereodiagram of [$VO(AMP\text{-}CP)_2(H_2O)$] determined on the basis of EPR and ENDOR spectroscopy and molecular modeling. Broken lines connect the vanadium atom both to the α- and β-phosphate oxygens of the two symmetry-related, equatorially positioned AMP-CP molecules and to the axial water molecule. The ENDOR-determined hydrogen atom positions of the two methylene protons, H(8) and H(5′), and the axially coordinated water protons are shown in addition to the nonhydrogen atoms. (Reproduced by permission from Ref. 24.)

with G-quartets formed with GMP may reflect the lack of packing constraints imposed by a polynucleotide strand, as is the case for $d(G_4T_4G_4)$ in crystals. However, the quality of the electron density maps did not allow an unambiguous assignment of ribose pucker, and this aspect was readily assessed on the basis of ENDOR measurements.

Modeling of VO^{2+}-nucleotide complexes in the two studies described above was based both on torsional angle search calculations constrained by ENDOR-determined electron-proton distances within van der Waals hard-sphere limits [57] and on prediction of the corresponding perpendicular or parallel hfc properties according to the assigned position of each proton. For instance, within the [$VO(GMP)_2(H_2O)$] complex, the position of the exchangeable proton of the 2′-hydroxyl group assigned through torsion angle search calculations predicted the correct hfc pattern only for a C3′-*endo* structure while a C2′-*endo* conformation did not. The same ENDOR-determined metal-proton distance was applied for each case. Thus, construction of a molecular model on the basis of ENDOR results is particularly exacting since the spectroscopic constraints specify not only metal-nucleus dipolar separations, which are scalar quantities, but also the dependence of hfc patterns on relative nuclear coordinates in the molecule, which reflect vector relationships.

Differences in the reactivity of VO^{2+} with nucleotides were observed in these studies. For instance, VO^{2+} binds readily to GMP in the pH 6–9 region [49] while binding of VO^{2+} to AMP was not detected under comparable conditions [24]. Also, both ADP and ATP form complexes of 1:1 and 1:2 metal-nucleotide stoichiometry, while only 1:2 stoichiometry is observed for GMP [49]. VO^{2+} does not react in an identical manner with methylene-substituted 5′-di- and triphosphate analogs of adenosine and guanosine. For instance, VO^{2+} forms identical complexes with ADP and with AMP-CP while the corresponding α,β- and β,γ-methylene analogs of ATP exhibit decreased affinity to VO^{2+} and do not form complexes that are isomorphous with that of ATP [24]. The different binding properties cannot be rationalized on the basis of the metal-binding affinity and altered pK_a values that are associated with methylene-substituted nucleotide analogs [58,59]. On the other hand, β,γ-methylene- and β,γ-imidoguanosine 5′-triphosphate derivatives form complexes of 1:2 stoichiometry with VO^{2+} identically to that of GTP [60]. These results, thus, indicate that one cannot assume that the methylene di- or triphosphate inhibitor analogs of different nucleotides behave identically with respect to metal-nucleotide interactions. Such differences could become important in the study of enzyme-nucleotide interactions if the nonhydrolyzable derivatives are employed as substrate inhibitor analogs.

3.2.2. Amino Acids

Mulks et al. reported ENDOR studies of vanadyl complexes of imidazole and histidine in frozen solution [61]. Using selectively deuterated and ^{15}N-labeled compounds, they determined the coordination environment of the metal ion. In the presence of a large excess of ligand, the ENDOR results showed that four imidazole molecules are coordinated to the VO^{2+} ion in equatorial positions while the fifth ligand is an axially coordinated water molecule. In the case of the VO^{2+}-histidine complex, no axial water is found [61]. The authors suggest that the amino nitrogens and possibly the carboxylate group might bind in such a manner that the VO^{2+} ion is completely shielded from solvent water. However, the coordination geometry of the VO^{2+}-histidine complex was not derived from ENDOR results.

EPR and ENDOR studies of the VO^{2+} ion in single crystals of triglycine sulfate have been reported by Böttcher and coworkers [62]. Strong ENDOR signals of the ^{51}V nucleus and of amino and methylene protons were observed over the 1–120 MHz frequency range. The authors also demonstrated that the point-dipole approximation was suitable for the VO^{2+}-triglycine sulfate complex. From EPR and ENDOR studies, it was concluded that VO^{2+} occupies an interstitial lattice site forming a $VO(glycine)_2$ complex of C_1 point-group symmetry.

3.2.3. *Porphyrins*

Mulks and van Willigen reported ENDOR studies of the vanadyl-tetraphenylporphyrin [VO(TPP)] complex in frozen solutions and diluted into polycrystalline [Zn(TPP)] [63,64]. With the magnetic field settings at the $-7/2\|$ and $-3/2\perp$ absorption features, their results yielded the hf and quadruple coupling components of the pyrrole nitrogens and the hfc components of pyrrole protons. In the [VO(TPP)] complex in frozen solutions, they also identified a solvent molecule coordinated to the VO^{2+} ion in the axial position trans to the V=O bond.

Comparison of the isotropic hfc of porphyrin pyrrole nitrogens with coordinated paramagnetic metal ions confirms that in [VO(TPP)] the unpaired electron is localized to the metal $3d_{xy}$ orbital. This orbital bisects the N-V-N valence angles, and the isotropic coupling of the pyrrole nitrogens directly coordinated to vanadium(IV) is small (6.8 MHz) [64]. On the other hand, the values of A_{iso} for the pyrrole nitrogens of Cu^{2+}- and Ag^{+}-TPP complexes are 47.0 and 67.7 MHz, respectively [65]. In the latter two complexes, the unpaired electron is localized to the metal $3d_{x^2-y^2}$ orbital which has strong overlap with the orbitals of the pyrrole nitrogens. In [VO(TPP)], Mulks and van Willigen also found axially symmetric principal hfc components for pyrrole protons with a very small isotropic contribution, confirming that the spin density on the coordinating nitrogen donor atoms was very small [64]. The ENDOR-determined dipolar hfc components yielded the metal-proton distance of 5.2 Å for the pyrrole protons, consistent with X-ray diffraction studies.

3.2.4. Other Organic Compounds

ENDOR studies of the vanadyl-acetylacetonate complex [VO(acac)$_2$] in frozen solutions of methanol and pyridine have been reported by Kirste and van Willigen [66]. Since the CH protons exchange with solvent deuterons, ENDOR assignments of CH and CH$_3$ protons of the bidentate acetylacetonate ligand were easily made. The ENDOR results indicated that for [VO(acac)$_2$] in methanol, two carbonyl oxygens of both acetylacetonate molecules were coordinated to VO^{2+} ion in equatorial positions in a twofold symmetric manner, while a methanol molecule was bound trans to the V=O bond. On the other hand, ENDOR results of the [VO(acac)$_2$] complex in pyridine showed that a solvent pyridine molecule could bind either cis or trans to the V=O bond.

ENDOR studies of vanadyl complexes of Schiff bases in frozen solutions have been reported by Attanasio [67,68]. The vanadyl complexes of the Schiff base ligands *N,N'*-ethylenebis(acetylacetonimine), *N,N'*-ethylenebis(salicylideneamine), and *N,N'*-*o*-phenylenebis(salicylideneamine) were prepared for structural characterization by ENDOR [67]. The principal hfc components were assigned to specific protons by deuterium substitution. ENDOR results showed that two nitrogen and two oxygen atoms of these tetradentate ligands were coordinated to VO^{2+} in equatorial positions. Axially coordinated water trans to the V=O bond and a second water molecule hydrogen-bonded to the vanadyl oxygen were also assigned by ENDOR. Using the point-dipole approximation, the ENDOR-determined electron-proton distances were used to define the conformation of the ethylenediamine bridge [67,68].

ENDOR studies of bis((1*R*)-3-(heptafluorobutyryl)camphorate)-oxovanadium(IV), a catalyst for asymmetric hetero Diels-Alder reactions, have been reported by Togni et al. [69]. EPR and ENDOR results indicate that four oxygen atoms of the camphorate derivative were coordinated to the VO^{2+} ion. By employing ^{13}C-benzaldehyde as a solvent, the ENDOR results showed that a benzaldehyde molecule is coordinated trans to the V=O bond and four oxygens of the two camphor ligands are equatorially coordinated to the vanadyl group with the fluorobutyryl groups located cis to the vanadyl oxygen. The ENDOR assigned distance for the O=V$\cdots$O(benzaldehyde) interaction in the complex was 2.2 Å, in excellent agreement with X-ray diffraction studies [69].

4. APPLICATION OF VANADYL AS A STRUCTURAL PROBE OF BIOLOGICAL SYSTEMS

ENDOR spectroscopy of VO^{2+} as a probe of macromolecular structure has received hitherto limited application. Since it is obvious from the above discussion that there are distinct advantages for use of VO^{2+} as a paramagnetic structural probe and that the accuracy of structural analysis, as developed in this laboratory, approaches that of X-ray diffraction of single crystals, we examine those factors that govern the resolving power of ENDOR for VO^{2+}-protein complexes in frozen solutions. In Fig. 12 are shown ENDOR spectra of the divanadyl complex of transferrin obtained in our laboratory [70], illustrating detailed spectral resolution of resonances of protein residues. The features of the ENDOR spectra in Fig. 12 clustered near the Larmor frequency highlight the importance of the spectral resolution that must be achieved to apply ENDOR to problems of biological importance. Two line pairs indicated by arrows in Fig. 12 have their origin in solvent-exchangeable protons. Since the calculated dipolar electron-proton distances are 3.7 and 4.0 Å, these resonance features cannot be ascribed to protons of inner sphere coordinated solvent molecules. If these protons are indeed from solvent molecules rather than from protein residues, they must be located in the outer coordination sphere. A similar conclusion was made by Eaton et al. in their ESEEM studies of VO^{2+}-transferrin complexes [71].

Because the hfc of distant nuclei is almost entirely dipolar in character, their resonance features are clustered about the Larmor frequency as a result of the r^{-3} dependence of the hfc on electron-nucleus separation, as shown through Eq. (4). The intrinsic linewidths of the ENDOR features will limit the accuracy of identifying different classes of nuclei on the basis of r. For r values of ~7 Å, it can be readily calculated that to resolve ENDOR lines of two closely spaced protons the absorption features must be separated by at least 20 kHz (often the ENDOR linewidth and the modulation depth of the rf field in our ENDOR experiments), for electron-nucleus separations differing by about 0.2 Å [34]. On the other hand, for r values of ~10 Å, the electron-nucleus separations must differ by ~1 Å for the same 20-kHz peak-to-peak separation.

In Fig. 12 it is seen that the features labeled $e1_{\perp}$ and $e2_{\perp}$ located

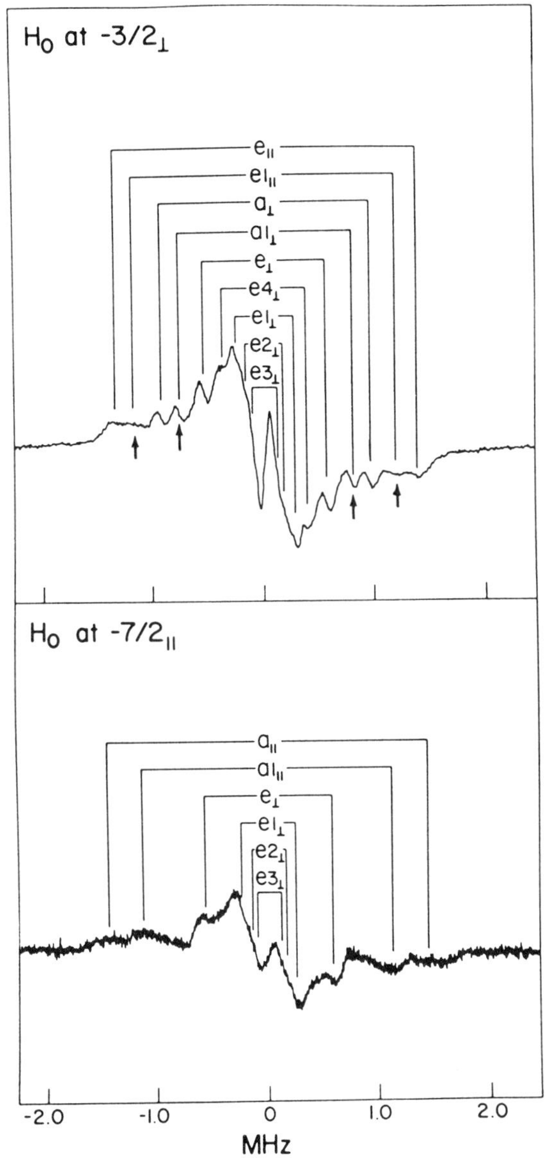

FIG. 12. Proton ENDOR spectra of the divanadyl complex of transferrin at 25 K with $\mathbf{H}_0$ at the $-3/2\perp$ (top) and $-7/2\|$ (bottom) settings. The sample was prepared with 1.74×10^{-3} M of VO^{2+} and 0.90×10^{-3} M of transferrin in 0.1 M NaCl and 0.01 M $NaHCO_3$ buffered to pH 8 with 0.03 M *N*-2-hydroxyethylpiperazine-*N'*-2-ethanesulfonate [70]. Tentative assignments of the ENDOR line pairs for axially and equatorially located protons are indicated by stick diagrams. The line pairs indicated by arrows are due to solvent exchangeable protons.

closest to the Larmor frequency exhibit hf couplings of 0.52 and 0.25 MHz, respectively, corresponding to dipolar electron-nucleus separations of 6.7 and 8.5 Å. These two features, separated from each other by ~0.27 MHz, appear only as weak absorption lines although the depth modulation of the rf field was ≤30 kHz. Within this spectral region, resolution is nonetheless partially obscured because of the overlapping contributions of the disordered protons in the macromolecule. Another factor that also limits spectral resolution through clustering of resonances derives from the significant isotropic hfc contribution that the protons of inner-coordinated or outer sphere ligands to the VO^{2+} ion may have, for instance, as seen for the equatorial hydroxyl protons in Table 3. While this is a relatively trivial factor in the case of H_2O molecules whose spectral contributions can be eradicated through ^2H_2O exchange, the problem is more complicated with amino acid side chains with nonexchangeable protons that occupy outer sphere sites.

Hitherto there is only one published ENDOR study of VO^{2+} complexed to a protein—in this case apoferritin [72]. In that study, spectral resolution near the proton Larmor frequency was very limited. There are, on the other hand, a number of ESEEM investigations to identify donor-ligand atoms of metal binding sites in proteins in which VO^{2+} was employed as a spectroscopic probe [71,72–75]. These results, in addition to the EPR-observed shf coupling of ^{31}P of VO^{2+} bound to *S*-adenosylmethionine synthetase [76,77], provide ample evidence that environmental effects can be characterized by measurement of the shf couplings of the VO^{2+} ion in macromolecules.

These observations highlight the need to develop methods for improved spectral resolution for structural analysis. In principle, this goal can be achieved through perdeuteration of the macromolecule with ^1H-enriched residues selectively incorporated at specific sites. By use of auxotrophic strains of bacteria engineered to overproduce a protein and grown on deuterated culture media, specific isotopically enriched amino acids can be supplemented for selective incorporation into the protein according to its amino acid sequence [78]. In addition to the application of chemical and biosynthetic methods, enhancement of spectral resolution may also be achievable through application of newer developments in instrumentation such as pulsed ENDOR [14,69] and stochastic ENDOR [79,80], which have been shown to provide substantive improvement in signal-to-noise and spectral resolution over that of conventional, continuous wave

ENDOR. These newer developments, we believe, will significantly extend detailed examination of structure-function relationships of biological macromolecules in solution by ENDOR spectroscopy at a level of accuracy that cannot be achieved at present by any other physical method.

ACKNOWLEDGMENT

This work was supported by grants from the National Institutes of Health (GM21900 and AA06374).

ABBREVIATIONS

acac	acetylacetonate
ADP	adenosine 5′-diphosphate
AMP-CP	α,β-methylene adenosine 5′-diphosphate
ATP	adenosine 5′-triphosphate
EPR	electron paramagnetic resonance
ENDOR	electron nuclear double resonance
ESEEM	electron spin-echo envelope modulation
GMP	guanosine 5′-monophosphate
hf	hyperfine
hfc	hyperfine coupling
NMR	nuclear magnetic resonance
rf	radiofrequency
shf	superhyperfine
TPP	α,β,γ,δ-tetraphenylporphyrin

REFERENCES

1. C. J. Ballhausen and H. B. Gray, *Inorg. Chem.*, *1*, 111 (1962).
2. C. D. Garner, I. H. Hillier, J. Kendrick, and F. E. Mabbs, *Nature*, *258*, 138 (1975).
3. B. R. McGarvey, in *Transition Metal Chemistry*, Vol. 3 (R. L. Carlin, ed.), Marcel Dekker, New York, 1966, pp. 89–201.

4. B. A. Goodman and J. B. Raynor, in *Advances in Inorganic Chemistry and Radiochemistry*, Vol. 13 (H. J. Emeléus and A. G. Sharpe, eds.), Academic Press, New York, 1970, pp. 135–362.
5. F. E. Mabbs and D. Collison, *Electron Paramagnetic Resonance of d Transition Metal Compounds*, Elsevier, Amsterdam, 1992, pp. 338–441.
6. N. D. Chasteen, *Coord. Chem. Rev.*, *22*, 1 (1977).
7. N. D. Chasteen, in *Structure and Bonding*, Vol. 53, Springer-Verlag, Berlin, 1983, pp. 105–138.
8. A. Butler and C. J. Carrano, *Coord. Chem. Rev.*, *109*, 61 (1991).
9. D. Mustafi and M. W. Makinen, *Inorg. Chem.*, *27*, 3360 (1988).
10. G. Feher, *Phys. Rev.*, *103*, 834 (1956).
11. G. Feher, *Phys. Rev.*, *114*, 1219 (1959).
12. A. L. Kwiram, *Adv. Phys. Chem.*, *22*, 133 (1971).
13. L. Kevan and L. D. Kispert, *Electron Spin Double Resonance Spectroscopy*, John Wiley and Sons, New York, 1976.
14. M. M. Dorio and J. H. Freed (eds.), *Multiple Electron Resonance Spectroscopy*, Plenum Press, New York, 1979.
15. A. Schweiger, in *Structure and Bonding*, Vol. 51, Springer-Verlag, Berlin, 1982, pp. 1–119.
16. S. S. Eaton and G. R. Eaton, in *Vanadium in Biological Systems* (N. D. Chasteen, ed.), Kluwer Academic, Dordrecht, 1990, pp. 199–222.
17. A. Abragam and B. Bleaney, *Electron Paramagnetic Resonance of Transition Ions*, Clarendon Press, Oxford, 1970.
18. H. R. Gersman and J. D. Swalen, *J. Chem. Phys.*, *36*, 3221 (1962).
19. N. F. Albanese and N. D. Chasteen, *J. Phys. Chem.*, *82*, 910 (1978).
20. R. H. Borcherts and C. Kikuchi, *J. Chem. Phys.*, *40*, 2270 (1964).
21. N. M. Atherton and J. F. Shackleton, *Mol. Phys.*, *39*, 1471 (1980).
22. D. Kivelson and S. K. Lee, *J. Chem. Phys.*, *41*, 1896 (1964).
23. N. D. Chasteen and J. Francavilla, *J. Phys. Chem.*, *80*, 867 (1976).
24. D. Mustafi, J. Telser, and M. W. Makinen, *J. Am. Chem. Soc.*, *114*, 6219 (1992).
25. G. C. Hurst, T. A. Henderson, and R. W. Kreilick, *J. Am. Chem. Soc.*, *107*, 7294 (1985).

26. H. M. McConnell and D. B. Chestnut, *J. Chem. Phys.*, *28*, 107 (1958).
27. G. Rist and J. Hyde, *J. Chem. Phys.*, *49*, 2449 (1968).
28. B. M. Hoffman, J. Martinsen, and R. A. Venters, *J. Magn. Reson.*, *59*, 110 (1984).
29. C. A. Hutchison, Jr., and D. B. McKay, *J. Chem. Phys.*, *66*, 3311 (1977).
30. R. A. Fields and C. A. Hutchison, Jr., *J. Chem. Phys.*, *82*, 1711 (1985).
31. D. Mustafi, H. Joela, and M. W. Makinen, *J. Magn. Reson.*, *91*, 497 (1991).
32. N. M. Atherton and J. F. Shackleton, *Chem. Phys. Lett.*, *103*, 302 (1984).
33. H. M. McConnell and B. Gaffney-McFarland, *Quart. Rev. Biophys.*, *3*, 91 (1970).
34. G. B. Wells and M. W. Makinen, *J. Am. Chem. Soc.*, *110*, 6343 (1988).
35. P. A. Snetsinger, N. D. Chasteen, and H. van Willigen, *J. Am. Chem. Soc.*, *112*, 8155 (1990).
36. R. P. Dodge, D. H. Templeton, and D. H. Zalkin, *J. Chem. Phys.*, *35*, 55 (1961).
37. M. B. Palma-Vittorelli, M. U. Palma, D. Palumbo, and F. Sgarlata, *Nuovo Cimento*, *3*, 718 (1956).
38. C. J. Ballhausen, B. F. Djurinskij, and K. J. Watson, *J. Am. Chem. Soc.*, *90*, 3305 (1968).
39. J. Reuben and D. Fiat, *Inorg. Chem.*, *6*, 579 (1967).
40. J. Reuben and D. Fiat, *Inorg. Chem.*, *8*, 1821 (1969).
41. K. Wüthrich and R. E. Connick, *Inorg. Chem.*, *6*, 583 (1967).
42. K. Wüthrich and R. E. Connick, *Inorg. Chem.*, *7*, 1377 (1968).
43. H. Montgomery and E. C. Lingafelter, *Acta Crystallogr.*, *17*, 1478 (1964).
44. M. Sabat, R. Cini, T. Haromy, and M. Sundaralingam, *Biochemistry*, *24*, 7827 (1985).
45. R. Cini, M. Sabat, M. Sundaralingam, M. C. Burla, A. Nanzi, G. Polidori, and P. F. Zanazzi, *J. Biomol. Struct. Dyn.*, *1*, 633 (1983).

46. M. A. Viswamitra, M. V. Hosur, Z. Shakked, and O. Kennard, *Nature, 262*, 234 (1976).
47. R. G. Gellert and R. Bau, in *Metal Ions in Biological Systems*, Vol. 8 (H. Sigel, ed.), Marcel Dekker, New York, 1979, pp. 1–55.
48. W. Saenger, *Principles of Nucleic Acid Structure*, Springer-Verlag, New York, 1984, pp. 556.
49. F. S. Jiang and M. W. Makinen, manuscript in preparation.
50. O. Kennard, N. W. Isaacs, W. D. S. Motherwell, J. C. Coppola, D. L. Wampler, A. C. Larson, and D. G. Watson, *Proc. Roy. Soc. London A, 325*, 401 (1971).
51. M. Gellert, M. N. Lipsett, and D. R. Davies, *Proc. Natl. Acad. Sci. USA, 48*, 2013 (1962).
52. V. Sasisekharan, S. Zimmerman, and D. R. Davies, *J. Mol. Biol., 92*, 171 (1975).
53. T. J. Pinnavaia, H. T. Miles, and E. D. Becker, *J. Am. Chem. Soc., 97*, 7198 (1975).
54. H. Klump, *Ber. Bunsenges. Phys. Chem., 80*, 121 (1976).
55. D. Sen and W. Gilbert, *Methods Enzymol., 211*, 191 (1992).
56. C. Kang, X. Zhang, R. Ratliff, R. Moyzis, and A. Rich, *Nature, 356*, 126 (1992).
57. H. Iijima, J. B. Dunbar, Jr., and G. R. Marshall, *Proteins: Structure, Function, and Genetics, 2*, 330 (1987).
58. R. G. Yount, D. Babcock, N. M. Ballantyne, and D. Ojala, *Biochemistry, 10*, 2484 (1971).
59. R. G. Yount, *Adv. Enzymol. Rel. Areas Mol. Biol., 43*, 1 (1975).
60. F. S. Jiang, H. Rütthard, and M. W. Makinen, manuscript in preparation.
61. C. F. Mulks, B. Kirste, and H. van Willigen, *J. Am. Chem. Soc., 104*, 5906 (1982).
62. R. Böttcher, D. Heinhold, and W. Windsch, *Chem. Phys. Lett., 65*, 452 (1979).
63. H. van Willigen, C. F. Mulks, A. Bouhaouss, M. Ferhat, and A. H. Roufosse, *J. Am. Chem. Soc., 102*, 4846 (1980).
64. C. F. Mulks and H. van Willigen, *J. Phys. Chem., 85*, 1220 (1981).
65. T. G. Brown and B. M. Hoffman, *Mol. Phys., 39*, 1073 (1980).
66. B. Kirste and H. van Willigen, *J. Phys. Chem., 86*, 2743 (1982).

67. D. Attanasio, *J. Phys. Chem.*, *90*, 4952 (1986).

68. D. Attanasio, *J. Chem. Soc., Faraday Trans. I*, *85*, 3927 (1989).

69. A.Togni, G. Rist, G. Rihs, and A. Schweiger, *J. Am. Chem. Soc.*, *115*, 1908 (1993).

70. D. Mustafi and M. W. Makinen, unpublished results.

71. S. S. Eaton, J. Dubach, K. M. More, G. R. Eaton, G. Thurman, and D. R. Ambruso, *J. Biol. Chem.*, *264*, 4776 (1988).

72. P. M. Hanna, N. D. Chasteen, G. A. Rottman, and P. Aisen, *Biochemistry*, *30*, 9210 (1991).

73. E. de Boer, C. P. Keijzers, A. A. K. Klaassen, E. J. Reijerse, D. Collison, C. D. Garner, and R. Wever, *FEBS Lett.*, *235*, 93 (1988).

74. P. A. Tipton, J. McCracken, J. B. Cornelius, and J. Peisach, *Biochemistry*, *28*, 5720 (1989).

75. G. J. Gerfen, P. M. Hanna, N. D. Chasteen, and D. J. Singel, *J. Am. Chem. Soc.*, *113*, 9513 (1991).

76. G. D. Markham, *Biochemistry*, *23*, 470 (1984).

77. G. D. Markham and T. S. Leyh, *J. Am. Chem. Soc.*, *109*, 599 (1987).

78. D. M. LeMaster, *Methods Enzymol.*, *177*, 23 (1989).

79. J. R. Niklas, *Habilitationschrift*, Universität Paderborn, Germany, 1983.

80. J. R. Niklas, R. U. Bauer, and J. M. Späth, *Phys. Stat. Sol.*, *B119*, 171 (1983).

4
Vanadyl(IV) Complexes of Nucleotides

Enrique J. Baran

Quimica Inorgánica (QUINOR), Facultad de Ciencias
Exactas, Universidad Nacional de La Plata, C. C. 962,
1900-La Plata, Argentina

1. INTRODUCTION

As is shown in various chapters of this volume, the similarities between the chemistry of vanadates(V) and phosphates play a central role in the biological activity of vanadium. On the other hand, it is clear that the first reduction product of vanadate(V), the vanadyl(IV) cation VO^{2+}, which is considered as the most stable oxo cation of the first-row transition metals [1] and, perhaps, the most stable diatomic ion known [2,3], can compete with other divalent cations for ligand binding sites in biological systems. These competitions are surely favored by the high flexibility of the VO^{2+} cation toward coordination geometries [1,4].

The coordination behavior of both vanadates(V) and VO^{2+} with respect to nucleotides and their constituents (nucleosides, nucleic bases, sugars, and phosphates) has a great significance in relation with competitive processes in the regulation of the activity of ATPases, ribonucleases, and similar systems, as well as with regard to the cancerostatic action of vanadates. In addition, it is well known that enzymatic reactions involving ATP and other nucleotides are often metal ion–dependent [5].

It was suggested that cytoplasmatic vanadium may be present as a VO^{2+}-ATP complex [6] and it is also known that ATP protects VO^{2+} from oxidation by H_2O_2 due to complex formation between the cation and the nucleotide [7].

The above indications suggest that a better knowledge of the interaction of VO^{2+} with nucleotides and their derivatives would be worthwhile at the present stage of development of vanadium biochemistry.

Interestingly, and despite these considerations, neither in a previous volume of this series devoted to nucleotide and related complexes

[8], nor in a review on metal complexes of nucleic acid constituents [9], was a single reference to a VO^{2+}-nucleotide complex found.

The present knowledge on VO^{2+}-nucleotide interactions and on closely related systems will be briefly reviewed and discussed in this chapter.

To facilitate this it is worthwhile remembering the structural characteristics of nucleotides and their constituents.

Nucleotides consist of three structural units: a purine or pyrimidine base, a ribose unit, and a phosphate group. The structures and numbering schemes of the common bases and the pentose rings are shown in Fig. 1. The bases are bound to the sugar moieties at C(1′) to N(9) (purines) or N(1) (pyrimidines), generating the nucleosides. Only thymine is bound to 2′-deoxyribose whereas all of the other bases are attached to ribose. The mono-, di- or triphosphate groups, if present, are usually attached to the 5′ position of the sugars and generate the

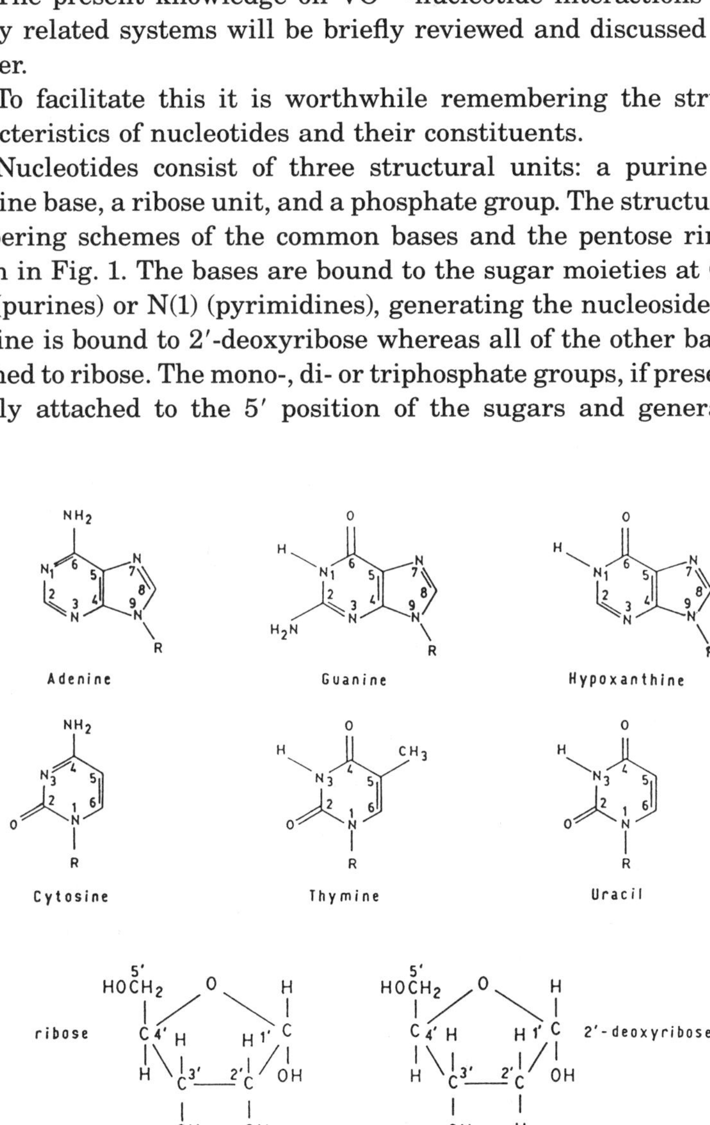

FIG. 1. Structures of the most common nucleic bases. For the base R = H, whereas for the nucleosides R is a sugar moiety.

monophosphate, diphosphate, and triphosphate nucleosides, respectively.

Abbreviations used along the text for the different discussed systems are given at the end of this chapter.

2. INTERACTION OF VANADYL(IV) WITH NUCLEOTIDES IN SOLUTION

2.1. Complexes with Nucleoside Di- and Triphosphates

A number of recent studies using different experimental approaches have been directed to attaining a better insight into the characteristics of the VO^{2+}-ATP interactions. Additionally, some information on the related VO^{2+}-ADP and VO^{2+}-AMP systems has also been published, whereas studies with other nucleotides are more limited. Interestingly, most of these studies have emphasized a somewhat different behavior of the monophosphate nucleosides in comparison with the di- and triphosphate analogous. Therefore, we shall discuss both groups of systems separately.

A qualitative study of the interaction of VO^{2+} with nucleotides using electron absorption spectroscopy in buffered (pH 4.9) solutions and with a great ligand excess (L:M ratios = 4 or higher) provides some general information which supports much of the following analysis and discussion [10].

The electron absorption spectrum of VO^{2+} in solution is very characteristic. As can be seen from Fig. 2A it presents a well-defined band located at 763 nm ($\epsilon = 16$ L mol^{-1} cm^{-1}) and a shoulder at $\sim$ 625 nm ($\epsilon = 8$ L mol^{-1} cm^{-1}). These two bands, originated in the complex $[VO(H_2O)_5]^{2+}$ cation, can be assigned, according to the well-known Ballhausen and Gray MO scheme to the $b_2 \rightarrow e$ and $b_2 \rightarrow b_1$ transitions, respectively [11]. Another band, found often in vanadyl(IV) complexes below 450 nm and usually overlapped by the tail of a strong charge transfer band centered in the near-UV region, has been assigned to the $b_2 \rightarrow a_1$ transition.

The interaction of VO^{2+} with the triphosphate nucleosides ATP, ITP, CTP, and UTP generates very similar spectral patterns. The main absorption band is displaced to lower energies (815–820 nm) and its

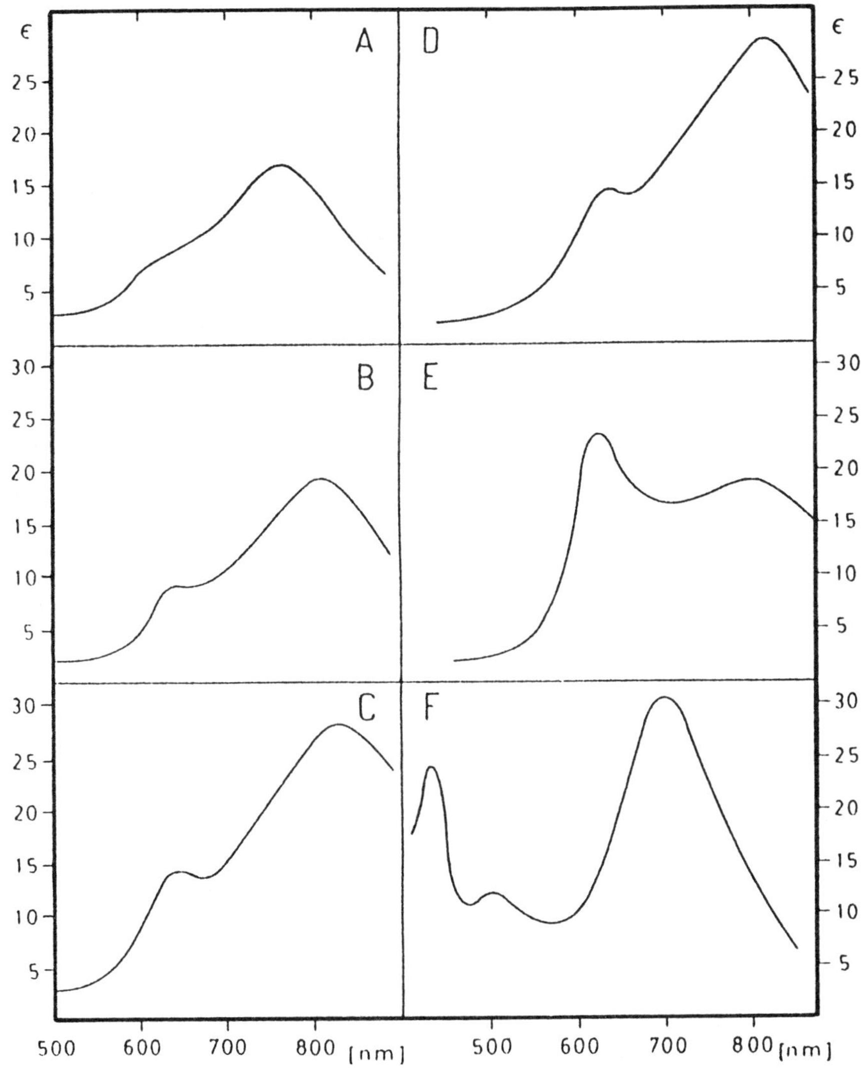

FIG. 2. Electronic absorption spectra of aqueous solutions at pH 4.9 of $[VO(H_2O)_5]^{2+}$ (A), VO^{2+}/ATP (B), $VO^{2+}/P_2O_7^{4-}$ (C), VO^{2+}/UDP (D), VO^{2+}/CMP (E), and VO^{2+}/CMP (at pH 11.5) (F). ε values in L mol^{-1} cm^{-1}. (Reproduced with permission from Ref. 10.)

intensity increases around 15%. The shoulder appears better defined and somewhat weaker in intensity. As an example, Fig. 2B shows the spectrum of the VO^{2+}-ATP solution. This spectrum is very similar to that obtained with a VO^{2+}-$P_2O_7^{4-}$ solution under similar experimental conditions (Fig. 2C).

In the case of the diphosphate nucleoside solutions (ADP, IDP, CDP, UDP, TDP), the spectra are very similar to those observed for the triphosphates as can be observed in Fig. 2D for the VO^{2+}-UDP system.

On the other hand, the monophosphate nucleosides (AMP, GMP, IMP, CMP, UMP, TMP) generate a different spectral pattern, as seen in Fig. 2E for the VO^{2+}-CMP system. Although their absorption remains in the same spectral range their relative intensities are clearly inverted, suggesting some differences in their bonding characteristics.

All the commented spectral results clearly show that at the investigated pH value interaction between VO^{2+} and the different nucleotides occurs essentially through the phosphate moieties.

Finally, if the pH value of the solutions is raised up to 11 or 12, all systems (except those derived from thymidine) show the same spectral pattern (Fig. 2F). In these cases, interaction takes place through the deprotonated 2′ and 3′ OH groups of the ribose moieties, as will be discussed later.

The interaction of phosphate groups with VO^{2+} in acidic media could also be confirmed by means of ^{31}P NMR measurements [12], although in the case of ATP it is not totally clear if all three [12] or only the two terminal phosphate groups [13–15] are involved in coordination. Such differences have also been found in the case of other transition metal-ATP complexes [16–18].

Also solution IR spectra clearly confirmed the VO^{2+}-phosphate interactions, as important changes were observed in the characteristic P-O stretching region [12].

More detailed solution studies for the VO^{2+}-ATP system [13,15] show the existence of different species in equilibria. $(VO)_2$ ATP seems to be one of the major components of all the investigated mixtures [13]. Complexes with this stoichiometry have also been found in other M(II)-ATP systems [19–21]. Other relevant species may be VO(ATP) and $VO(ATP)_2$ [13]. The latter seems to be the predominant one at high nucleotide/VO^{2+} ratios [14].

As the participation of the bases in bonding was controversial, a study of the VO^{2+}-ATP, VO^{2+}-ADP, and VO^{2+}-AMP systems was undertaken using UV-differential absorption spectroscopy at different pH values [22]. Measurements performed at pH 4.9 and 6.5 clearly suggest the binding of VO^{2+} to a ring position in ATP. Furthermore, at pH 6.5 the results indicate also the formation of two different complex species at different metal-to-ligand ratios, supporting the results of Cini et al. [13]. On the other hand, at pH 3 the spectra indicate that no interaction with the nucleobase occurs. This result refutes conclusively the formerly postulated bonding through adenine N(3) in acidic media [15].

For the VO^{2+}-ADP system, similar results are obtained: participation of adenine in bonding is evident at pH 4.9 and 6.5, but in this case no biphasic behavior is observed at the higher pH value. Also in this system no interaction occurs in acidic media.

On the other hand, in the case of AMP, the participation of adenine in bonding could be excluded in the pH range between 3.0 and 6.5.

Recent EPR and ENDOR measurements, performed at nearly neutral pH values and at high nucleotide/VO^{2+} ratios with ATP and ADP, establish the formation of $VO(nucl)_2$ species in which the equatorial positions are chelated by the terminal phosphate groups, whereas one solvent molecule occupies the axial position, trans to the V=O bond [14].

2.2. Complexes with Nucleoside Monophosphates

Most of the experimental evidence accumulated up to date suggests that monophosphate nucleosides behave in a different manner as the corresponding di- and triphosphate nucleosides in their interaction with the VO^{2+} cation.

The behavior of AMP, GMP, CMP, and UMP in relation to VO^{2+} was recently investigated using spectrophotometric techniques and working in the pH range between 2 and 11 [23]. In the pH range between 2.0 and 3.0 VO^{2+}-phosphate interactions could be unambiguously confirmed. At pH values between 3.0 and 5.0 precipitation of solid products occurs with all of the investigated ribonucleotides. These precipitates redissolve as the pH increases. In all cases at pH 5.0 the higher wavelength band suffers a new shift to lower energies, suggesting that the VO^{2+}-phosphate bond remains and an additional interaction through oxygen donors is started, probably involving OH groups of ribose, as supported by the similar behavior of *D*-ribose-5-phosphate [23]. At pH 7.0 the spectra change again, suggesting a marked symmetry diminution around the metal center. Finally, at higher pH values the interaction with the deprotonated OH groups of ribose takes place.

As expected, the behavior of the VO^{2+}-TMP system is markedly different, especially at higher pH values [23] at which, finally, hydrolysis of the cation is observed.

The results presented in these two sections allow the following conclusions:

1. For all three types of phosphate nucleosides:
   —In acidic media interactions with VO^{2+} occurs only through the phosphate groups.
   —At high pH values coordination takes place only through the deprotonated OH groups of ribose.
2. For tri- and diphosphate nucleosides:
   —In neutral or nearly neutral solutions, and at high ligand to metal ratios, $VO(nucl)_2$ complexes seem to be the predominant species.
   —Participation of base N atoms in bonding occurs probably only at lower ligand to metal ratios.
   —Coordination to one or two OH groups of ribose in acidic or neutral media also cannot be excluded (cf. [13,15]).
3. For monophosphate nucleosides:
   —Solid products precipitate at pH values between 3.0 and 5.0 and redissolve at higher pH.
   —Important ligand reaccommodation around the metal center takes place with raising pH values. The phosphate group together with the OH groups of ribose participate in coordination.

Finally, it should be noted that practically no stability constants for these VO^{2+}-nucleotide systems are known. A set of values for different VO^{2+}-ATP complexes has been reported [13]. Interestingly, in a recent review devoted to a critical evaluation of stability constants of nucleotide complexes [24] practically no references to VO^{2+} species are found and the few available data [25] are questioned. Therefore, this remains a very promising and fruitful field for future research and it is expected that in the coming years increasing information on this topic becomes available.

3. SOLID VANADYL(IV)-NUCLEOTIDE COMPLEXES

3.1. Simple VO^{2+}-Nucleotide Systems

Very few solid complexes of VO^{2+} with nucleotides have been reported and thoroughly characterized. Unfortunately, none of these complexes has been structurally investigated, either by X-ray or by EXAFS techniques.

Our own experience with these systems indicates that it is very difficult to obtain good crystalline samples, even in powdered form.

Two different solid VO^{2+}-ATP complexes have been described: $Na_2[VO(ATP)]\cdot 2H_2O)$ [26,27] and $[(VO)_2ATP]\cdot 6H_2O$ [13]. In the first case, chelation through the triphosphate group [27] and interaction with the adenine N(1) atom [26] were suggested from IR spectroscopy data and a magnetic moment of 1.70 BM close to the expected "spin-only" value was observed [26,27]. For the second complex, both IR data and reflectance measurements support metal coordination through the triphosphate chain [13].

A number of complexes with nucleoside monophosphates, $[VO(AMP)]\cdot 2H_2O$, $[VO(GMP)]\cdot 5H_2O$, $[VO(IMP]\cdot 5H_2O$, $[VO(CMP)]\cdot 5H_2O$, and $[VO(UMP)]\cdot 5H_2O$ were reported by Katsaros [26]. In these complexes, IR data clearly suggest participation of N donors of the bases in coordination. However, it was not clear if, in some cases, the phosphate group is also involved in coordination [26]. Notwithstanding our own work on some of these systems indicates the participation of both N donors and phosphate moieties in bonding [28]. On the other hand, these monophosphate complexes are probably polymeric [26] and their magnetic moments are also very close to the spin-only values [26].

An independent investigation of the VO^{2+}-AMP complex confirmed the participation of an adenine N atom and the phosphate group in bonding [29]. The metal-adenine interaction was also supported by fluorescence measurements [29].

Work in progress in our laboratory has also shown that it is possible to obtain solid complexes with diphosphate nucleosides. In the case of complexes with ADP, CDP, and UDP the infrared spectra also point to the participation of the bases and the phosphate groups in bonding [28,30].

3.2. Ternary Complexes Containing VO^{2+}, a Nucleotide, and a Second (Organic) Ligand

Ternary complexes involving a metal ion, a nucleotide (mainly ATP), and a second (usually organic) ligand have been the subject of extensive investigations in recent years [5,31–35]. Their importance arises from the fact that divalent metal cations usually catalyze the dephosphorylation of nucleoside di- and triphosphates, whereas ternary complexes show remarkable stability toward hydrolysis [5,33,36]. Therefore, these complexes can be considered as good models for the ATP transport mechanism [33]. On the other hand, they can also reveal the nature and extent of statistical, steric, and electronic restrictions imposed by other ligands to the metal-nucleotide bonding.

As the bonding of VO^{2+} to nucleotides in biological systems is surely influenced by the presence of other ligands, the characterization of ternary complexes of this cation appears extremely important. But despite this very few results are available.

Cini et al. reported the synthesis of a mixed VO^{2+}-ATP complex with bis(2-pyridyl)amine (bipyam) of the composition $[(VO)H_2ATP(bipyam)]\cdot 7H_2O$ [13], precipitated as a blue powder from hydroethanolic solutions. Its infrared spectrum as well as the EPR study of the analogous Zn(II) complex doped with VO^{2+} suggests coordination only through phosphate oxygens, but the reflectance spectrum also points to the presence of N donors in the coordination sphere of the metal. Bipyam is probably not involved in metal bonding, as suggested by structural studies of other similar systems [37].

More recently, two ternary VO^{2+} complexes containing ATP and 1,10-phenanthroline (phen) or 2,2′-bipyridyl (bipy) of the stoichiometry $[VO(H_2ATP)(phen)]\cdot 6H_2O$ and $[VO(H_2ATP)(bipy)]\cdot 6H_2O$ were prepared and characterized [38]. Both compounds appear to be isostructural but possess a structure different from that reported for similar complexes of other divalent first-row transition metal ions [33].

The electronic spectra of these vanadyl(IV) complexes clearly point to simultaneous interaction with O and N donors. The infrared spectra suggest that the α- and β-PO_2 moieties of ATP are involved in the formation of a chelate, whereas the terminal PO_3 group apparently acts as a bridging ligand, generating polymeric complexes [38].

Similar ternary complexes of VO^{2+} with phen and bipy can also be obtained with ADP and AMP. Preliminary spectroscopic results for ADP complexes suggest that in these cases no polymeric species were generated [39].

Recently, we could also obtain similar VO^{2+}-ATP, VO^{2+}-ADP, and VO^{2+}-AMP complexes, containing imidazole and benzimidazole as the second ligand [40].

4. VANADYL(IV) COMPLEXES OF NUCLEIC BASES, NUCLEOSIDES, SUGARS, PHOSPHATES, AND RELATED LIGANDS. MODELS FOR VO^{2+}/ NUCLEOTIDE INTERACTIONS

Studies on the interaction of the VO^{2+} cation with the individual constituents of nucleotides (nucleic bases, D-ribose, phosphates, and nucleosides) and with other closely related ligands are of great interest for a better understanding of the chemical behavior and structural characteristics of the vanadyl(IV)- nucleotide systems.

4.1. Complexes of Nucleic Bases

In buffered acid solutions no interaction between the VO^{2+} cation and the free nucleic bases could be established [10]. At pH values above 5.0, complex formation with some of these bases becomes apparent [41]. The electronic absorption spectra suggest coordination through N atoms for adenine and oxygen coordination for guanine and uracil [41].

In the cases of adenine [42] and guanine [43], adducts of the type $VO(HL)Cl_2$ (HL = neutral base) obtained by refluxing the ligand with $VOCl_2$ in an ethanol-triethylorthoformate mixture were reported. These complexes are probably linear chain polymers, with the bases acting as bridges through the N(7) and N(9) imidazole nitrogens. This suggests a structure similar to that reported for $[Cu(Hpu)(H_2O)_4]SO_4$ (Hpu = neutral purine) [44].

Both VO^{2+} complexes present a normal magnetic behavior at room temperature [42,43].

A similar complex has been obtained with purine [42]. Complexes of V(III) with purine, adenine, xanthine, and hypoxanthine have also been reported [45,46].

4.2. Complexes of Nucleosides

Nucleoside complexes of vanadyl(IV) have been reported several times in relation with inhibition studies in ribonuclease [4]. In particular, information is available on complexes of uridine [47], guanosine, and inosine [48].

Recent solution studies at different pH values and VO^{2+} to nucleoside ratios indicate that interactions begin in the pH range between 8.5 and 9.0, and occur through two adjacent deprotonated hydroxyl groups of the ribose moieties [49]. As expected, such an interaction is not possible in the case of thymidine [49].

Interestingly, no solid VO^{2+}-nucleoside complexes have so far been reported.

4.3. Complexes of Sugars and Related Ligands

It is well known that sugars can interact with metal ions by acting as either reductants or chelators [50]. Actually, most of them are able to reduce vanadates to VO^{2+} and to complex this cation.

Due to its strong hydrolytic tendency, the VO^{2+} ion usually needs the presence of additional donor groups (e.g., carboxylates) in the sugar molecule but, once bound to the ligand, it can easily deprotonate the hydroxyl groups and strongly coordinate to up to four of them [51–54].

The binding of VO^{2+} to simple sugars in neutral and basic aqueous solutions was recently investigated by EPR and electronic absorption spectroscopy. The results show that complexation is favored in basic media and occurs only with sugar molecules provided with cis couples of adjacent OH groups [54].

The particular behavior of D-ribose, relevant to the systems reviewed in this chapter, has been investigated independently and the generation of a 2:1 ribose-VO^{2+} complex could be confirmed [23].

In the cases of D-galacturonic and D-glucuronic acids the coordination is initiated through a metal-carboxylate bond at pH > 3 and one or two deprotonated sugar OH groups are involved in bonding [51,53]. Different complex species are generated in these systems at different pH values [51] and these studies were also extended to polygalacturonic acids [52]. In these cases, the VO^{2+} is bonded only to deprotonated monodentate carboxylate groups [52,55].

Vanadyl(IV) complexes of lactobionic acid show a behavior similar to that of the uronic acids but their stability constants are considerably higher [56].

Studies of the interaction of the VO^{2+} cation with the mucopolysaccharide chondroitin-sulfate A, which contains alternate units of D-glucuronic acid and *N*-acetyl-D-galactosamine, indicated the formation of a 2:1 ligand-to-metal complex, in which the coordination of the cation through the carboxylate group and the glycosidic oxygen of the D-glucuronate moieties is suggested [57].

The synthesis and characterization of a solid homoleptic glucose complex of VO^{2+} was also recently reported [58] and opens an interesting synthetic approach for other similar complexes.

Finally, it should be mentioned that interaction of metal ions, including VO^{2+}, with carbohydrates has been the subject of a recent review [59].

4.4. Complexes of Phosphates and Related Ligands

There are many old reports on solid vanadyl(IV) phosphates [2] and they shall not be discussed here.

With the orthophosphate ion, complex formation in aqueous acidic solution generates the following species [60,61]: $VO(HPO_4)$, $[VO(H_2PO_4)(HPO_4)]^-$, $[VO(HPO_4)_2]^{2-}$, $[VO(H_2PO_4)]^+$, and $[VO(H_2PO_4)_2]$, with the phosphate groups acting as bidentate ligands.

In the case of diphosphate, conductometric and potentiometric titration has shown the formation of three different species, namely, $[VO(P_2O_7)_2]^{6-}$, $[VO(P_2O_7)]^{2-}$, and $(VO)_2(P_2O_7)$ [62]. It was also found that at pH 5, the vanadyl(IV) cation is able to give a polynuclear species with diphosphate [63,64]. Its sodium salt, of stoichiometry $Na_6[(VO)_6(P_2O_7)_3]\cdot 18H_2O$, could be isolated and a cyclic structure with each $P_2O_7^{4-}$

ion acting as a bidentate ligand with each of two vanadyl ions has been suggested [63].

On the other hand, it has been found that the triphosphate chain, $P_3O_{10}^{5-}$, forms more stable complexes with the VO^{2+} cation than orthophosphate and diphosphate [13]. The stability constant for the complex $[VO(P_3O_{10})]^{3-}$ is higher than that for the $[VO(ATP)]^{2-}$ species [13].

The behavior of the system VO^{2+}/D-ribose-5-phosphate has also been investigated. It shows a behavior practically identical to that of the monophosphate nucleotides [23].

It is also interesting to comment that VO^{2+} is strongly adsorbed on the surface of calcium hydroxyapatite, $Ca_{10}(PO_4)_6(OH)_2$. The EPR spectra suggest an inhomogeneous distribution of the adsorbed species and the generation of $O{=}V(O)_4$ moieties with the four equatorial positions around the metal center probably occupied by phosphate oxygens [65]. The anchored cations show a high stability toward oxidation [65].

By direct reduction of V_2O_5 with hypophosphorous acid or by interaction of $VOSO_4$ and H_3PO_2 solutions, a vanadyl(IV) hypophosphite of the composition $VO(PO_2H_2)_2 \cdot H_2O$ could be prepared, in which the acid anion acts as a bidentate oxygen donor [66].

It has also been shown that VO^{2+} interacts very strongly with thiamine diphosphate (cocarboxylase). In the pH range 3–4 the formation of a 1:1 complex, involving only VO^{2+}-phosphate bonds, could be established. At higher pH values, the formation of another complex with the additional participation of an N atom of the pyrimidine ring was suggested [67].

Recently, we have also reported the formation of a soluble 1:1 complex of VO^{2+} with phytate (myoinositolhexaphosphate) at $pH < 1.0$. At slightly higher pH values, insoluble complexes of different stoichiometries are generated [68].

ACKNOWLEDGMENTS

It is a great pleasure to acknowledge the contributions of the collaborators whose names appear in the references. Work from this laboratory reported herein has been supported by the "Consejo Nacional de Investigaciones Científicas y Técnicas de la República Argentina," the "Comisión de Investigaciones Científicas de la Provincia de Buenos Aires," and the "Stiftung Volkswagenwerke" (FRG).

ABBREVIATIONS

ADP	adenosine 5′-diphosphate
AMP	adenosine 5′-monophosphate
ATP	adenosine 5′-triphosphate
CDP	cytidine 5′-diphosphate
CMP	cytidine 5′-monophosphate
CTP	cytidine 5′-triphosphate
GMP	guanosine 5′-monophosphate
IDP	inosine 5′-diphosphate
IMP	inosine 5′-monophosphate
ITP	inosine 5′-triphosphate
TDP	thymidine 5′-diphosphate
TMP	thymidine 5′-monophosphate
UDP	uridine 5′-diphosphate
UMP	uridine 5′-monophosphate
UTP	uridine 5′-triphosphate

REFERENCES

1. N. D. Chasteen in *Biological Magnetic Resonance*, Vol. 3 (L. Berliner and J. R. Reuben, eds.), Plenum Press, New York, 1981, p. 53 ff.
2. J. Selbin, *Chem. Rev., 65*, 153 (1965).
3. J. Selbin, *Coord. Chem. Rev., 1*, 293 (1966).
4. N. D. Chasteen, *Struct. Bond., 53*, 105 (1983).
5. H. Sigel, *Coord. Chem. Rev., 100*, 453 (1990).
6. I. G. Macara, K. Kustin, and L. C. Cantley, Jr., *Biochim. Biophys. Acta, 629*, 95 (1980).
7. G. M. Woltermann, R. A. Scott, and G. P. Haight, Jr., *J. Am. Chem. Soc., 96*, 7569 (1974).
8. H. Sigel (ed.), *Metal Ions in Biological Systems*, Vol. 8, Marcel Dekker, New York, 1979.

9. D. J. Hodgson, *Progress Inorg. Chem.*, *23*, 211 (1977).
10. G. Urretavizcaya and E. J. Baran, *Z. Naturforsch*, *42b*, 1537 (1987).
11. C. J. Ballhausen and H. B. Gray, *Inorg. Chem.*, *1*, 111 (1962).
12. S. B. Etcheverry, E. G. Ferrer and E. J. Baran, *Z. Naturforsch.*, *44b*, 1355 (1989).
13. R. Cini, G. Giorgi, F. Laschi, M. Sabat, A. Sabatini, and A. Vacca, *J. Chem. Soc. Dalton Trans.*, 575 (1989).
14. D. Mustafi, J. Telser, and M. W. Makinen, *J. Am. Chem. Soc.*, *114*, 6219 (1992).
15. H. Sakurai, T. Goda, and S. Shimomura, *Biochem. Biophys. Res. Commun.*, *108*, 474 (1982).
16. M. Cohn and T. R. Hughes, Jr., *J. Biol. Chem.*, *237*, 176 (1962).
17. M. Gabriel, D. Larcher, J. C. Boubel, A. A. Peguy, and J. Torreilles, *Inorg. Chim. Acta*, *26*, 77 (1978).
18. J. L. Bock, *J. Inorg. Biochem.*, *12*, 119 (1980).
19. H. Sigel and P. E. Amsler, *J. Am. Chem. Soc.*, *98*, 7390 (1976).
20. H. Sigel, F. Hofstetter, R. B. Martin, R. M. Milburn, V. Scheller-Krattiger, and K. H. Scheller, *J. Am. Chem. Soc.*, *106*, 7935 (1984).
21. T. A. Glassman, J. Suchy, and C. Cooper, *Biochemistry*, *12*, 243 (1973).
22. P. A. M. Williams and E. J. Baran, *J. Inorg. Biochem.*, *48*, 15 (1992).
23. P. A. M. Williams and E. J. Baran, *J. Inorg. Biochem.*, *50*, 101 (1993).
24. R. M. Smith, A. E. Martell, and Y. Chen, *Pure Appl. Chem.*, *63*, 1015 (1991).
25. E. E. Kriss and K. B. Yatsimirskii, *Russ. J. Inorg. Chem.*, *16*, 202 (1971).
26. N. Katsaros, *Transition Metal Chem.*, *7*, 72 (1982).
27. R. G. Bhattacharyya and I. Bhaduri, *J. Inorg. Nucl. Chem.*, *40*, 733 (1978).
28. E. G. Ferrer, PhD thesis, Universidad Nacional de La Plata, 1993.
29. A. Terrón and V. Moreno, *Inorg. Chim. Acta*, *56*, L57 (1981).
30. E. J. Baran et al., unpublished results.
31. H. Sigel, in *Metal Ions in Biological Systems* (H. Sigel, ed.), Marcel Dekker, New York, 1973, Vol. 2, p. 64 ff.

32. M. S. Mohan and M. M. Taqui Khan, *J. Coord. Chem.*, *8*, 207 (1979).
33. R. Cini and P. Orioli, *J. Inorg. Biochem.*, *14*, 95 (1981).
34. H. A. Azab, A. Hassan, A. M. El-Nady, and R. S. A. Azkal, *Monatsh. Chem.*, *124*, 267 (1993).
35. H. Sigel, K. Becker, and D. B. McCormick, *Biochim. Biophys. Acta*, *148*, 655 (1967).
36. H. Sigel, D. H. Buisson, and B. Prijs, *Bioinorg. Chem.*, *5*, 1 (1975).
37. R. Cini, M. C. Burla, A. Nunzi, G. P. Polidori, and P. F. Zanazzi, *J. Chem. Soc. Dalton Trans.*, 2467 (1984).
38. E. G. Ferrer and E. J. Baran, *Transit. Metal Chem.*, *16*, 599 (1991).
39. E. G. Ferrer and E. J. Baran, submitted for publication.
40. E. G. Ferrer and E. J. Baran, in preparation.
41. P. A. M. Williams, S. B. Etcheverry, and E. J. Baran, *Z. Naturforsch.*, *48b*, 1845 (1993).
42. C. M. Mikulski, S. Cocco, and N. De Franco, *Inorg. Chim. Acta*, *78*, L25 (1983).
43. C. M. Mikulski, L. Mattucci, L. Weiss, and N. M. Karayannis, *Inorg. Chim. Acta*, *92*, 275 (1984).
44. P. I. Vestues and E. Sletten, *Inorg. Chim. Acta*, *52*, 269 (1981).
45. C. M. Mikulski, S. Cocco, N. De Franco, and N. M. Karayannis, *Inorg. Chim. Acta*, *80*, L61 (1983).
46. C. M. Mikulski, K. Udell, D. L. Staley, and N. M. Karayannis, *Transit. Metal Chem.*, *17*, 159 (1992).
47. R. N. Lindquist, J. L. Lynn, Jr., and G. E. Lienhard, *J. Am. Chem. Soc.*, *95*, 8762 (1973).
48. Y. Georgalis, A. Zuoni, U. Hahn, and W. Saenger, *Biochim. Biophys. Acta*, *1118*, 1 (1991).
49. P. A. M. Williams, S. B. Etcheverry, and E. J. Baran, *An. Asoc. Quim. Argent.*, *82*, 13 (1994).
50. C. F. G. C. Geraldes and M. M. C. A. Castro, in *Metal Speciation in the Environment* (J. A. C. Broekaert, S. Gücer, and F. Adams, eds.), Springer-Verlag, Berlin, 1990, p. 105 ff.
51. G. Micera, A. Dessi, H. Kozlowski, B. Radomska, J. Urbanska, P. Decock, B. Dubois, and I. Oliver, *Carbohydr. Res.*, *188*, 25 (1989).

52. M. Branca, G. Micera, A. Dessi, and H. Kozlowski, *J. Chem. Soc. Dalton Trans.*, 1283 (1989).
53. M. Branca, G. Micera, D. Sanna, A. Dessi, and H. Kozlowski, *J. Chem. Soc. Dalton Trans.*, 1977 (1990).
54. M. Branca, G. Micera, A. Dessi, and D. Sanna, *J. Inorg. Biochem.*, *45*, 169 (1992).
55. S. Deiana, L. Erre, G. Micera, P. Piu, and C. Gessa, *Inorg. Chim. Acta*, *46*, 249 (1980).
56. H. Kozlowski, S. Bouhsina, P. Decock, G. Micera, and J. Swiatek, *J. Coord. Chem.*, *24*, 319 (1991).
57. S. B. Etcheverry, P. A. M. Williams, and E. J. Baran, *Biol. Trace Elem. Res.*, in press.
58. C. P. Rao and S. P.Kaiwar, *Inorg. Chim. Acta*, *186*, 11 (1991).
59. D. M. Whitfield, S. Stojkowski, and B. Sarkar, *Coord. Chem. Rev.*, *122*, 171 (1993).
60. J. E. Salmon and D. Whyman, *J. Chem. Soc.*, 980 (1966).
61. W. C. Copenhafer, M. W. Kendig, T. P. Russel, and P. H. Rieger, *Inorg. Chim. Acta*, *17*, 167 (1976).
62. L. I. Muschietti, S. B. Etcheverry, and E. J. Baran, *Monatsh. Chem.*, *113*, 1399 (1982).
63. C. C. Parker, R. R. Reeder, L. B. Richards, and P. H. Rieger, *J. Am. Chem. Soc.*, *92*, 5230 (1970).
64. A. Hasegawa, Y. Yamada, and M. Miura, *Bull. Chem. Soc. Jpn.*, *42*, 846 (1969).
65. G. E. Narda, E. D. Vega, J. C. Pedregosa, S. B. Etcheverry, and E. J. Baran, *Z. Naturforsch.*, *47b*, 395 (1992).
66. E. J. Baran, S. B. Etcheverry, and E. Diemann, *Polyhedron*, *4*, 1711 (1985).
67. P. A. M. Williams and E. J. Baran, *J. Inorg. Biochem.*, *38*, 101 (1990).
68. P. A. M. Williams and E. J. Baran, *Biol. Trace Elem. Res.*, *36*, 143 (1993).

5

Interaction of Vanadates with Biogenic Ligands

Debbie C. Crans

Department of Chemistry and Cell
and Molecular Biology Program
Colorado State University, Fort Collins,
Colorado 80523, USA

1. INTRODUCTION

An aqueous solution of vanadate contains a series of ionic oxovanadium(+5) species whose relative abundances depend on concentration, pH, ionic strength, temperature, and the presence of other substances. *Vanadate* is a term used indiscriminately to describe these vanadium(+5) species in aqueous solution, and throughout this chapter we will use the term vanadate generically and other terms such as *oxovanadate* when referring to specific species. Vanadate is known to induce a large number of biological responses including insulin-mimetic activity and cardiovascular effects (compiled in [1] and [2]), and accordingly, the interactions of vanadate with biogenic ligands are of interest. The aqueous chemistry of vanadium(+5) is particularly important because most of biology occurs in water-based environments such as blood plasma and intracellular fluids. This chapter will describe the chemistry of vanadate with essential organic and inorganic substances in aqueous environments of relevance to biological systems.

Proteins contain a wide range of reactive functionalities including hydroxyl, phenol, carboxylate, amine, amide, guanidine, and thiol groups. Other biogenic ligands can be either organic or inorganic. Proteins, organic acids, and carbohydrates total about 20 mM in blood plasma. The pH of plasma is 7.4 and a ~0.2 M ionic strength is maintained, mainly by sodium chloride. Intracellular fluids contain about 50 mM of organic substances at similar pH and ionic strengths; however, two important differences should be recognized. First, the intracellular potassium ion concentration is significantly higher than the sodium ion concentration; this is important for the transport of compounds in and out of cells. Second, the intracellular concentration of phosphate and phosphate-based species is significantly higher than the concentration of carboxylate-based species in blood plasma. The high intracellular levels of phosphate-based species reflect the central role of these compounds in metabolism. Inorganics, specifically phosphates, are an important group of biogenic ligands. Furthermore, it is important to recognize that tissue and cellular compartmentalization provide reducing environments which are maintained by millimolar concentrations of glutathione, varying levels of ascorbic acid and other reducing agents. Most biogenic ligands react with some form of vanadate to generate a variety of complexes, a fact which underscores the complexity of physiologically relevant vanadium(+5) chemistry.

Given the increasing interest in vanadium chemistry in the past decade, it is beyond the scope of this chapter to give a comprehensive review of all the recent vanadium(+5) chemistry. The readers are referred to previous compilations [1,2] and reviews for a detailed discussion of oxovanadate speciation [3], the vanadate-phosphate analogy [4,5], metal ion binding, transport, and redox chemistry [6,7], vanadium as a component in enzymes and vanadium-accumulating organisms [8,9], the bioinorganic chemistry of vanadium [10], vanadium(+5) chemistry as probed by ^{51}V nuclear magnetic resonance (NMR) spectroscopy [11], the coordination chemistry of vanadium [12], the classification and analysis of structural data [13], the chemistry of peroxovanadium compounds [14,15], complexes of polyoxovanadates with or without organic ligands [16], practical aspects of working with vanadates in biological systems [17], and interactions of vanadates with proteins [18]. In addition, the readers are referred to various chapters in this volume.

The chemistry of vanadates with biogenic ligands is often very complex, and in the absence of structural data, the chemistry is difficult to sort out even using the most sophisticated spectroscopic techniques. It is therefore important that model studies in organic solvents and with "man-made" biogenic-like ligands be used as an available tool to provide essential information concerning structure and reactivity. Information from model systems will create expectations based on chemical precedence and can assist in the interpretations of the observations made under biological conditions. This chapter describes studies on the real complex biological systems as well as model chemistry; together these approaches provide an understanding of how vanadates interact in biological systems.

2. VANADATES IN AQUEOUS SOLUTION

2.1. Oxovanadates-"Metavanadates"

In the neutral pH range aqueous colorless solutions of vanadate consist of a mixture of anionic species, oxovanadates, resulting from the various condensation and protonation equilibria [3]. The simplest oxovanadate, monomeric vanadate (V_1 in Fig. 1), is commonly recognized as a phosphate analog and can exist as a mono-, di-, or trianion ($H_2VO_4^-$, HVO_4^{2-}, VO_4^{3-}). The structural flexibility of the vanadium atom in monomeric vanadate is illustrated in crystallographic characterization of KVO_3, α-$NaVO_3$, NH_4VO_3 and $TlVO_3$, which contain tetrahedral vanadium (in infinite chains of corner-shared VO_4 tetrahedra) and $KVO_3{\cdot}H_2O$, $NaVO_3{\cdot}1.89H_2O$, $Ca(VO_3)_2{\cdot}2H_2O$, and $Sr(VO_3)_2{\cdot}4H_2O$ which contain trigonal bipyramidal vanadium (in chains of distorted VO_5 trigonal bipyramids) [3].

Other oxovanadates present in the physiological pH range include vanadate dimer (V_2 in Fig. 1), vanadate tetramer (V_4 in Fig. 1), and vanadate pentamer (V_5 in Fig. 1). These species are conveniently distinguished by ^{51}V NMR spectroscopy (Fig. 2) [19,20], and depending on pH can exist in various protonated forms ($H_3V_2O_7^-$, $H_2V_2O_7^{2-}$, $HV_2O_7^{3-}$, $V_2O_7^{4-}$, $V_4O_{12}^{4-}$, $HV_4O_{13}^{5-}$, $V_5O_{15}^{5-}$). Characterization of the speciation of

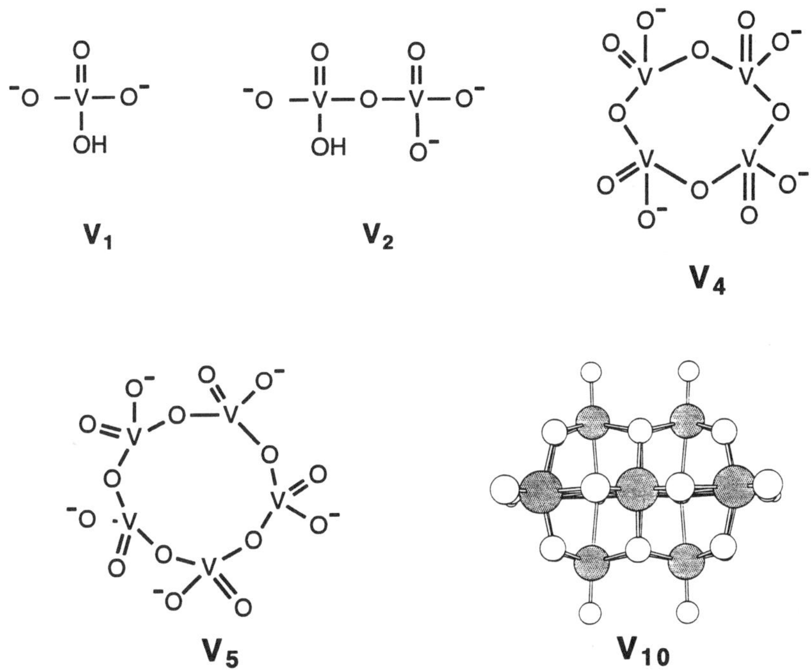

FIG. 1. Structural representations are shown for vanadate monomer (V_1), vanadate dimer (V_2), vanadate tetramer (V_4), vanadate pentamer (V_5), and vanadate decamer (V_{10}). The drawing for V_{10} was generated from the reported X-ray coordinates.

vanadate in solution was recently carried out using a combination of ^{51}V NMR spectroscopy and potentiometry, supporting some of the earlier studies identifying the major oxovanadate as V_4 [3,21]. The distribution of the oxovanadates is very sensitive to pH and total vanadium concentration, and the reader is referred to Chap. 1 of this volume and [3,19–21] for more information on distribution diagrams. The temperature as well as the presence of salts and other compounds can also affect the speciation [17]. Oxovanadate solutions reach equilibrium rapidly since the oxovanadates exchange vanadium atoms and oxygen atoms on a millisecond time scale in both aqueous and alcoholic solutions [22–24]. Despite characterization of crystalline salts isolated from aqueous and organic solutions [3,25–27], the rapid exchange makes structural assignments for these species more tentative. With these words of caution,

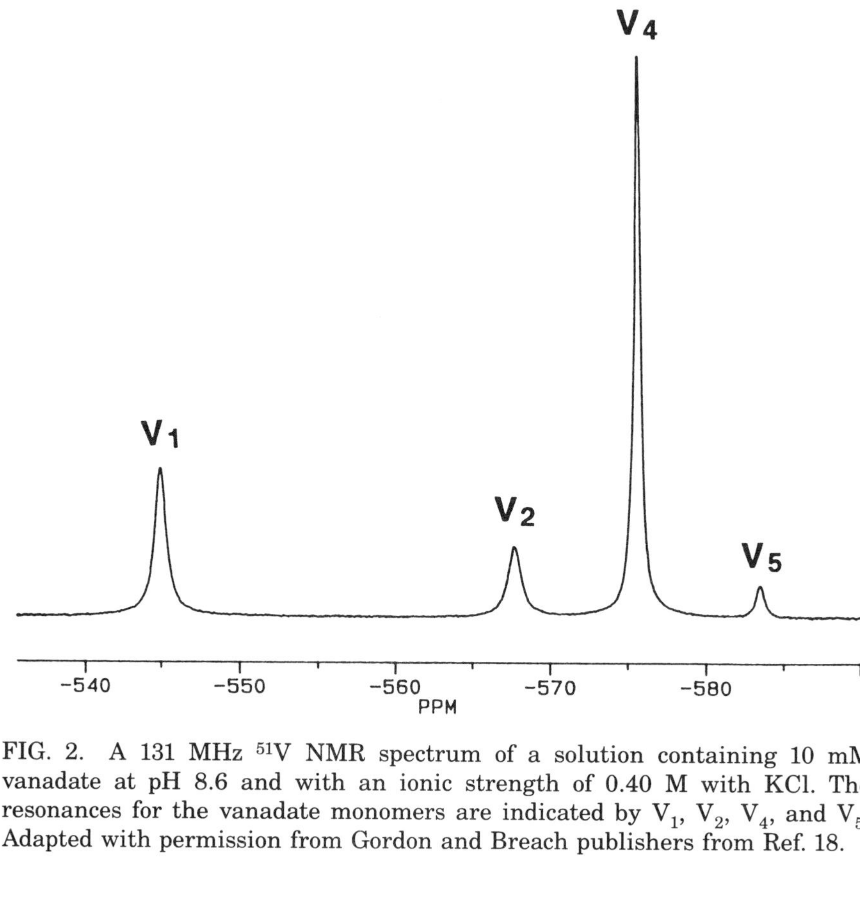

FIG. 2. A 131 MHz ^{51}V NMR spectrum of a solution containing 10 mM vanadate at pH 8.6 and with an ionic strength of 0.40 M with KCl. The resonances for the vanadate monomers are indicated by V_1, V_2, V_4, and V_5. Adapted with permission from Gordon and Breach publishers from Ref. 18.

V_2 is generally accepted to be a pyrophosphate analog [25], and V_4 and V_5 to be cyclical [26,27], although alternative structural suggestions have emerged in the literature [28,29].

The term *metavanadate* has been used to describe the colorless mixture of oxovanadates, with the empirical formula VO_3^-, in aqueous solution [3,20]. One of the vanadate salts is referred to as metavanadate (VO_3^-), a name that originates from the analogy with metaphosphate (PO_3^-). After adjustment of pH solutions containing the same mixture of oxovanadates (metavanadates) can be obtained by dissolution of metavanadate (in the form of $NaVO_3$ or NH_4VO_3) or by dissolution of orthovanadate (in the form of Na_3VO_4). The term metavanadate can cause confusion unless used carefully.

The chelate chemistry of vanadium(+5) is not limited to the chemistry of VO_2^+ but also includes the chemistry of $VOOH^{2+}$, VO^{3+}, and bare V^{5+} at physiological pH [12,30–32]. This chapter will briefly describe the chemistry of some of these systems in Secs. 5.4, 6.3, and 6.4.

Vanadate interacts with a modest to high affinity with a wide variety of proteins [18]. The structure and/or charge on the specific oxovanadate is important to the interactions with the protein and there is some evidence supporting sequence specificity. Under normal physiological conditions, the higher oxovanadates are not likely to be present in sufficient total vanadium concentration. However, recent studies suggest that these species have great potential as tools in the study of enzymes.

2.2. Other Oxovanadates

Decavanadate, V_{10} ($H_2V_{10}O_{28}^{4-}$, $HV_{10}O_{28}^{5-}$, $V_{10}O_{28}^{6-}$), is a complex oxovanadate [33] that forms upon addition of acid to vanadate solutions. Solutions of decavanadate are yellow, orange, or red, depending on concentration or countercation. At neutral and alkaline pH the decamer is thermodynamically unstable. In contrast to the labile oxovanadates in metavanadate solutions, decavanadate remains intact at neutral pH for limited time periods due to its slow hydrolysis rate [3]. Below pH 2, VO_2^+ is the favored oxovanadium species [3]. This cationic form of vanadium(+5) is found in many complexes.

Minor vanadate derivatives also form in aqueous solution and these species include a linear trimer, a linear tetramer, and hexamers [3,19,20]. New oxovanadates were recently characterized both in the solid state and in solution [34–37], and more oxovanadates are likely to be characterized in the future.

3. VANADATE REACTIONS WITH INORGANIC LIGANDS

3.1. Phosphate, Phosphate Esters, and Anhydrides

Aqueous solutions of vanadate will react with phosphate and phosphate derivatives to form a mixed 1:1 vanadate-phosphate anhydride [38]. In addition, a series of simple and condensed vanadate-phosphate compounds are known [3,39]. Most of these complexes were characterized in the solid state although some potentiometric studies were carried out [40]. The 1:1 V-P (vanadate-phosphate anhydride) complex that forms in aqueous solution is also expected to form in biological fluids containing millimolar concentrations of phosphate. At ambient temperature V_1 and the V-P complex (Structure **1**) are in rapid exchange, and only one signal is observed in the ^{51}V NMR spectrum for these two species. An exchange rate on the milliseconds time scale was deduced from the linewidths of the ^{51}V NMR spectra. The ^{51}V chemical shift of the V-P complex is pH-dependent but at pH 7.98 was deduced to be -582 ppm [38]. The apparent formation constant for the V-P complex ranges from 25 M^{-1} to 5 M^{-1} in the pH range from 6.69 to 7.98. The formation constant is 65 M^{-1} if it is calculated using the diprotonated vanadate and phosphate concentrations, respectively [38].

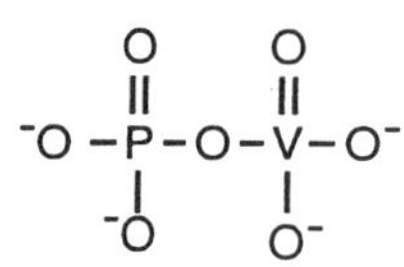

1

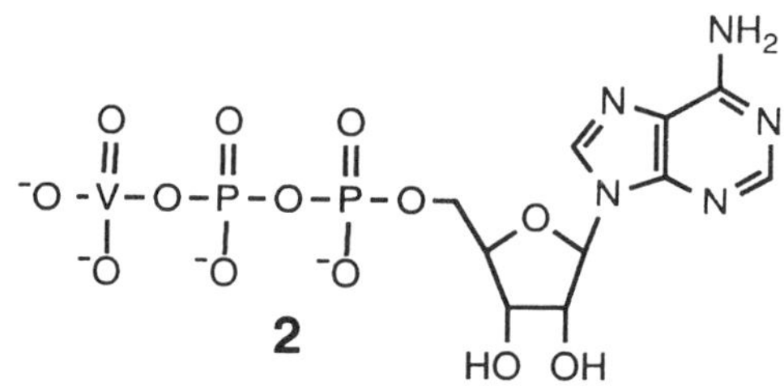

2

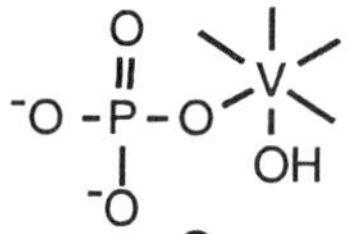

3

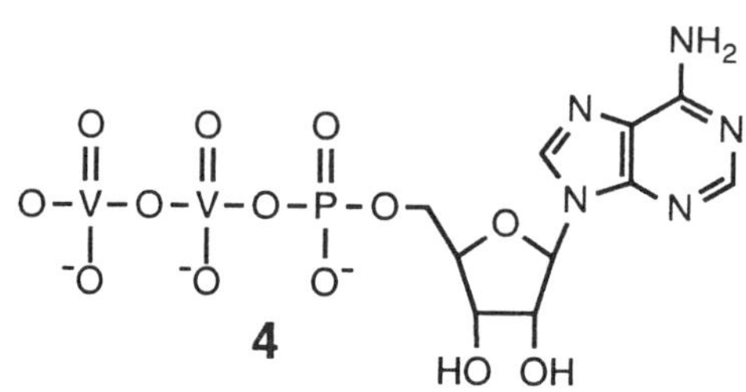

4

A similar complex forms between vanadate and pyrophosphate and has an apparent formation constant of 39 M^{-1} at pH 7.98. Formation constants for complexes from vanadate and other phosphate derivatives, including AMP, ADP, ATP, CTP, and PPP, are of the same order of magnitude, suggesting that these complexes all have similar structures [41,42]. Accordingly, the terminal phosphate group is presumed to coordinate to the vanadium atom. The coordination geometry around the vanadium atom in these species is not known although tetrahedral geometry is commonly preferred [42] (as shown for Structures **1** and **2**) over the alternative octahedral geometry (as shown for **3**) [41]. A second complex, probably cyclic, forms between pyrophosphate and vanadate [42].

When phosphate reacts with a vanadate monoester or vanadate diester an alkyl vanadate-phosphate or a dialkyl vanadate-phosphate derivative forms [28]. When phosphate reacts with V_2 a P-V-V unit forms [42]. In the reaction of AMP with vanadate the ATP analog AMP-V_2 (**4**) was observed by ^{51}V NMR spectroscopy. At pH 7.47 the apparent formation constant for the reaction between V_2 and AMP was 12 M^{-1} [42]. The potential for mimicking nucleotides with vanadate has been explored, and successful applications have been reported with myosin for the ADP-V analog (**2**) [43], with adenylate kinase for the AMP-V analog [38], and for adenylate cyclase with the GDP-V analog [44].

The VO_2^+-polyphosphate complexes were found to hydrolyze to inorganic phosphate (P_i) and a smaller polyphosphate at low pH [41]. Mechanistic studies suggested that the active form of the VO_2^+-polyphosphate complex had the monodentate polyphosphate ligated cis to H_2O or OH^-. The reaction rate was limited by the concentration of the VO_2^+-polyphosphate complex, and the complex had to form at acidic pH. At higher pH, when a mixture of oxovanadates was used as the source of vanadium(+5), no formation of P_i was observed ([41] and P. J. Stankiewicz, unpublished). Since the concentration of the VO_2^+-polyphosphate complex was rate limiting, the rate was increased significantly when a larger concentration of VO^{2+}-polyphosphate complex was generated first and then oxidized to the VO_2^+-polyphosphate complex using H_2O_2 or MnO_4^- [45].

Complex formation between vanadate and phosphate derivatives has not yet been appropriately recognized. Despite several reports describing the vanadate-phosphate complexes [28,41–45], phosphate buffers are still often used in biological studies with vanadate without acknowledgment of complex formation. Perhaps the most dramatic example is the vanadate-dependent NADH oxidation stimulated by plasma membranes, which is only observed in phosphate buffer [46]. In this system it seems likely that a vanadate-phosphate derivative plays a key role in inducing the NADH oxidation.

Interactions of vanadate with phosphates are likely to be extremely important for understanding how vanadium interacts in biological conditions [2]. The concentrations of phosphate-containing ligands in biological systems are significant and, combined with the affinity for vanadate (and vanadyl cation) and the properties of the resulting complexes as phosphate mimics, suggests that this group of compounds may be very important. All proteins, polynucleic acids, and other biological materials that interact with organic phosphates or phosphate anhydrides may respond to corresponding vanadate derivatives [43,44].

3.2. Carbonate, Arsenate, Chromate, and Molybdate

Vanadate may interact with carbonate and/or bicarbonate. Although quantitative information is not yet available, the interaction is likely to be weak (L. Pettersson, personal communication). Vanadate forms an anhydride with arsenate and the ^{51}V chemical shift is -575 ppm at pH 7.85. The situation is similar to the V-P case as only one resonance is observed since the V_1 and the V-As complex are in rapid exchange at ambient temperature. The formation constant for the 1:1 complex between vanadate and arsenate ranges from 13.0 to 4.5 M^{-1} over the pH range 6.47–7.85 [38].

Chromate forms anhydrides with vanadates to generate $[VO_2(OCrO_3)_2]^{3-}$ and $[V_2O_5(OCrO_3)_2]^{4-}$ [47]. In addition, chromate reacts with mono- and diprotonated decavanadate to form $[V_{10}O_{27}(OCrO_3)]^{6-}$ and $[HV_{10}O_{27}(OCrO_3)]^{5-}$. Both molybdate and tungstate react to replace the vanadium atom in the octahedral site in the decamer cluster rather than forming an anhydride linkage. The reactivity pattern of chromate thus resembles that of phosphate and arsenate rather than that of molybdate and tungstate.

Given the low formation constants of these types of complexes and the low natural occurrence of these biogenic ligands, these types of complexes are likely to play a limited role in biological systems.

3.3. Peroxide

V_1 reacts with H_2O_2 in aqueous solutions to generate complexes in which each of the O^- or OH groups have been substituted with an O_2^{2-} group [48–50]. The monoperoxo- ($[VO_2(OH)(OO)]^{2-}$) and diperoxovanadium derivatives ($[VO(OH)(OO)_2(OH_2)_2]^{2-}$ or expressed as $[VO(OH)(OO)_2]^{2-}$) have ^{51}V NMR chemical shifts at −627.6 and −770.6 ppm [48]. These and other species are conveniently monitored by ^{51}V NMR spectroscopy, although the chemical shifts of the ^{51}V NMR resonances change with pH and as the fraction of each species changes. The protonated forms of these species have chemical shifts at −602 and −686 ppm, respectively [50]. The speciation is sensitive to pH, concentrations of vanadium(+5) and H_2O_2, ionic strength, and other species in solution. At alkaline pH two species at −830 and −737.2 ppm in the ^{51}V NMR spectrum were attributed to $[VO(OO)_3]^{3-}$ and $[V(OH)(OO)_3]^{2-}$ [48]. A species attributed to $[V(O_2)_4]^{3-}$ has also been reported. VO_2^+ also reacts with the H_2O_2 and generates $[VO(OO)]^+$ with a chemical shift at −549 ppm below pH 2. H_2O_2 will also react with V_2 and thereby generates several binuclear derivatives which contain one or two types of vanadium atoms [48].

Peroxovanadium compounds will react with monodentate ligands to form a variety of complexes [14,15]. In general, the structurally characterized complexes are mononuclear with the vanadium atom in a pentagonal bipyramid with one or two peroxo groups in the equatorial plane (Fig. 3). In this context the peroxo group is formally taken to be bidentate. An X-ray structure of a triperoxovanadium compound has also been reported [51]. Exceptions to the above geometries are known and include dinuclear complexes represented by the citrate complex, $K_2[VO(O_2)(C_6H_6O_7)]_2 \cdot 2H_2O$ (see Fig. 18) [52], and complexes with other coordination environments at the vanadium atom represented by a picolinic acid complex, $VO_2(C_6H_4NO_2)(OPN_3C_6H_{18})$ [53]. A compilation of structural information has been published elsewhere [15].

The chemistry of the peroxovanadium complexes has been of prime interest given their relevance to the activity of haloperoxidases and the

FIG. 3. Schematic structures for monoperoxo- and diperoxovanadium complexes.

potent catalytic properties in the oxidation of organic and inorganic substrates [14,15]. Peroxovanadium species usually react with organic substrates by generating O_2 and some reduced form of the vanadium; and various synthetic approaches have been developed for the oxidation of alcohols, ketones, enolates, carbanions, thioethers, olefins, and the epoxidation of allylic alcohols. In the past most reactions have been described in organic solvents [14]; however, recent developments have been utilizing phase transfer systems [14] and aqueous solvents [54]. Accordingly, this type of chemistry in the biological environment of the bromo- and haloperoxidase enzymes is becoming more similar to the chemistry in aqueous solution (see Chap. 10). Studies designed to model the haloperoxidase activity have recently been described in detail elsewhere [55–57] and reviewed in [15].

The presence of the peroxo ligand generates complexes that are less labile than the oxovanadium complexes [24,58,59]. Accordingly, the peroxovanadium complexes would be able to survive transport in biological systems. Slow autolytic decomposition of H_2O_2 is probably facilitated by peroxovanadium complex decomposition. The catalytic applications of peroxovanadates take advantage of the increased oxidation rate of the peroxovanadium complex, which reacts after formation [14]. The activity of the peroxovanadium complexes as catalysts has been fine-tuned with ligands and various correlations have been made involving the electronics and other properties of the ligand [60]. In contrast, complexes formed between monoperoxovanadate and simple peptides reduced the rate of H_2O_2 decomposition [59].

Despite the sizable formation constants of peroxovanadate complexes [50], the low levels of H_2O_2 and related compounds maintained through the action of catalases, superoxide dismutases, and other detoxifying enzymes precludes the formation of peroxovanadium compounds in biological fluids at large. However, high local concentrations of peroxide derivatives are likely, and, given the lifetime of the compounds, some of the peroxovanadium compounds may be involved in biological reactions at sites remote from where they were formed. These properties are instrumental for the success of peroxovanadium compounds as insulin mimetics [61]. In addition, peroxo complexes and their chemistry are pertinent to the reactions catalyzed by bromo- and haloperoxidases, in which the vanadium cofactor is likely to be associated with a peroxo group during catalysis (see Chap. 10).

3.4. Other Inorganics

Vanadium(+5) derivatives with other inorganics containing bonds such as V–Cl, V–OSO_3^-, and V–ONO_2^- have been characterized in the solid state and/or in organic solvents [62–66]. In general, these derivatives are hydrolytically very unstable and will form oxovanadates in the presence of water. Studies of oxovanadates in aqueous solution do not form observable levels of complexes with these types of ligands; and, in fact, KCl, NaCl, and $NaClO_4$ are commonly used to maintain a constant ionic strength in aqueous studies since vanadate equilibria appear to be sensitive to the ionic strength.

4. VANADATE REACTIONS WITH MONODENTATE ORGANIC LIGANDS

Organic substances present in biological fluids commonly contain more than one functionality. The hydroxyl, carboxylate, amine, and thiol functionalities are abundant in all types of organic substances in biological fluids, and interaction of vanadate with this functionality is exceedingly important in many areas of vanadium(+5) biochemistry. In order to interpret and understand the interactions of such substances, the interaction of vanadate with each functionality and the properties of resulting simple complexes must be considered.

4.1. Alcohols and Phenols

The ethyl vanadate ester and the diethyl vanadate ester were first observed by Gresser and Tracey using ^{51}V NMR spectroscopy in 1985 (Fig. 4) [67,68]. The term alkyl vanadate ester was originally used for these oxovanadium alkoxides in order to stress the analogy with the alkyl phosphate esters, and this nomenclature will also be used in this chapter. The composition of the observed complexes was deduced by correlating the concentration of the respective vanadate esters with the ethanol and V_1 concentrations [67]. The formation constant (defined as $K = [RV][H_2O]/[ROH][V_1]$) was 10.4 for the monoester and 2.3 for the

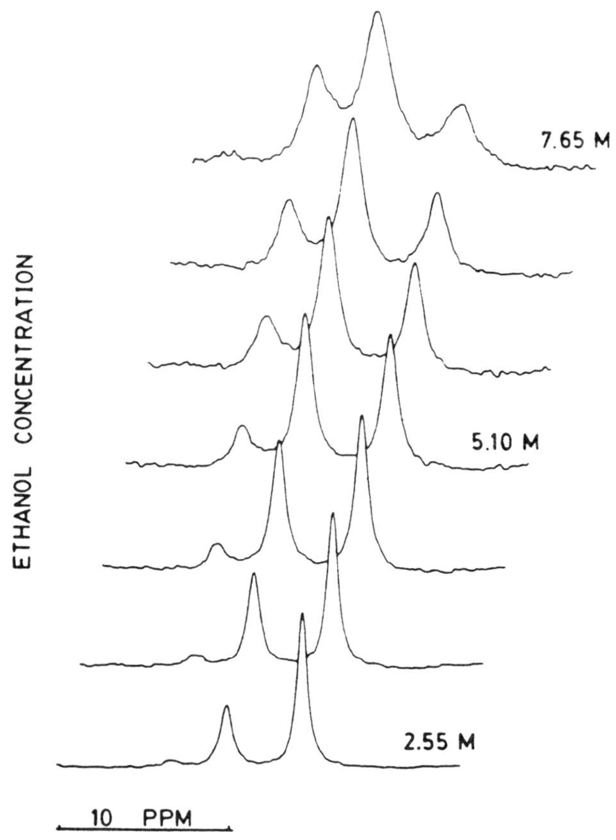

FIG. 4. 105 MHz ^{51}V NMR spectra of solutions containing vanadate and ethanol. The solutions contained 0.39 mM sodium vanadate, 50 mM Tris-acetate buffer at pH 7.5 and the indicated amounts of ethanol. Adapted with permission from the American Chemical Society from Ref. 67.

diester [68]. Despite the fact that these complexes are weak, the potential analogy of organic vanadate mono- and diesters (Fig. 5) with organic phosphate mono- and diesters provided the impetus to probe the chemistry and biochemistry of these compounds (see Chaps. 7 and 8 and [69,70]).

A series of ^{51}V NMR spectroscopic solution studies followed in which the formation constants for mono- and dialkyl vanadate esters were determined for methanol, ethanol, propanol, butanol, and other

FIG. 5. Schematic drawings for vanadate monoesters and diesters.

more complex alcohols (see [28,67,68,71,72] and the previous review in [11]). The formation constants for the esterification reaction change depending on the buffer used in the studies [17,22]. The formation constant for trialkyl esters in aqueous solutions is sufficiently low that these species have not been observed. No characterization by X-ray crystallography has yet been reported for any of the anionic vanadate esters. However, they are commonly referred to as tetrahedral vanadium complexes based on ^{51}V NMR chemical shift arguments (Fig. 5).

Given the importance of the vanadate ester in vanadium biochemistry and the lack of structural information, the limited structural characterization of trialkyl vanadate esters (commonly referred to as oxovanadium trialkoxides) is of interest. Oxovanadium trialkoxides in which the alcohol varied from methanol, ethanol, *i*-propanol, *t*-butanol, cyclohexanol, and various derivatives thereof were first synthesized in 1913 and at that time these materials were characterized using elemental analysis, UV and IR spectroscopy (reviewed in [10]). $VO(OCH_3)_3$ was the first oxovanadium trialkoxide to be characterized by X-ray crystallography [73]. Despite the twinning and disorder in this material, the structure was solved ($R = 0.168$) and showed the vanadium atom in an octahedral geometry in a dimeric unit arranged in a polymeric arrangement (Fig. 6, top left). Two additional compounds with the formula $VO(OR)_3$, namely, $(VO(OCH_2CH_2Cl)_3$ (Fig. 6, top right) [62] and $VO(OC_5H_9)_3$ [74], have now been characterized, and both place the vanadium atom in a distorted trigonal bipyramidal geometry. Despite the crystallographic characterization of the oxovanadium(+5) triphenylsilanol (Fig. 6, bottom) [75], the corresponding compound with triphenylmethanol as ligand has resisted structural characterization [76].

The solid state structural characterization described above may imply large differences between the trialkyl vanadate esters with coordination numbers larger than 4 and the aqueous anions with tetra-

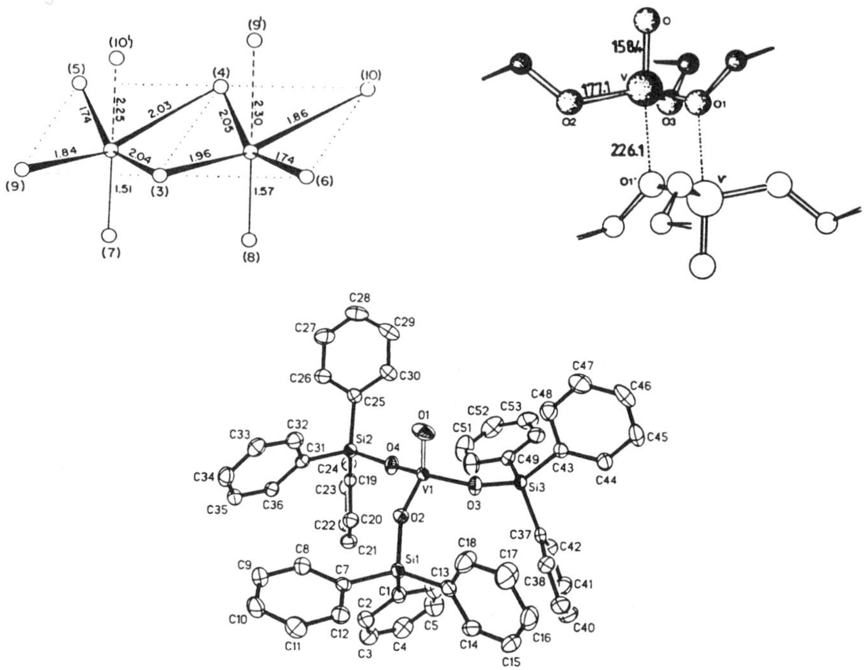

FIG. 6. Graphical representations of $VO(OCH_3)_3$ (top left), $VO(OCH_2CH_2Cl)_3$ (top right) and $VO(OSi(C_6H_5)_3)_3$ (bottom). Adapted with permission from the American Chemical Society from Refs. [62], [73], and [75].

hedral vanadium. Solution properties of the alkyl vanadate derivatives in aqueous [22,28,67,68,71,72] and organic solvents [62,65,66,76–82] also have several similarities. The lability of the mono- and dialkyl vanadate esters in aqueous solution is analogous to the lability of simple trialkyl vanadate esters in organic solution. The reactivity of vanadate esters in aqueous solutions is mirrored by the tendency of trialkyl vanadates to associate in organic solution. For example, a solution of $VO(OiPr)_3$ in toluene in the presence of excess KOiPr generated $[VO(OiPr)_4]^-$ and $[(iPrO)_3VO(OiPr)VO(OiPr)_3]^-$ [80]. In aqueous solutions the pK_a of the parent alcohol has been correlated with the formation constant of the alkyl monoester [71]. The sensitivity to the pK_a of the parent alcohol is much lower for the vanadate esters than for phosphate esters. Considering the sensitivity of the formation of vanadate esters to the presence of buffers and salts [22], the low sensitivity to the pK_a of the parent alcohol may be surprising. Similar correlations have not been described for alkyl vanadate esters in organic solvents although it has been possible to "tame" the reactivity of alkyl vanadate esters significantly by steric bulk [76]. Combined, these studies do suggest that appropriate electronic effects can stabilize the alkyl esters.

The reaction of vanadate with phenol in aqueous solution generates a monoester and a diester [83,84]. The chemical shift for the phenyl vanadate ester at −574 ppm is at a significantly higher resonance than the chemical shift for an alkyl vanadate ester (−554 ppm). The formation constant ($K = [PhOVO_3H^-][H_2O]/[H_2VO_4^-][PhOH]$) for the 1:1 monoester is 45, suggesting that esters formed from phenols are slightly more stable than esters formed from simple alcohols. The much larger difference in stability between the aryl phosphates and alkyl phosphates implies that much higher levels of seryl phosphates exist under biological conditions. Similar concentrations of aryl and alkyl vanadate esters could influence the tyrosine phosphatase activity but not the serine/threonine phosphatases in biological systems [5,83].

Serine, threonine, and tyrosine amino acid residues play a key role in biology. Given the interest in the effects of vanadate and other vanadium compounds on protein phosphatases [5], future characterization of vanadium complexes of these simple molecules with respect to structure, reactivity, and redox chemistry will continue to be very important. In addition, organic phosphate metabolites play an essential role in almost all aspects of biology, and the potential of organic vanadates as viable organic phosphate analogs has far-reaching implications for understanding some of the biological effects of vanadium (see Chaps. 7 and 8).

4.2. Carboxylic Acids

The reaction of vanadate with monodentate carboxylic acids has been examined both in aqueous and in organic solvents. Aqueous solutions containing vanadate and acetic acid form the bis(acetato)vanadate complex in the pH range 3.38–5.44 [85]. The chemical shift of this complex was deduced from the changes in the V_1 signal since the complex and V_1 are in rapid exchange at ambient temperature. *n*-Pentanoic acid is the only simple carboxylic acid that reacted with vanadate to form a complex (with a formation constant of 30 M^{-2}) which was observed separate from the V_1 signal at −536 ppm at pH 5.0 [85]. None of these complexes are stable in the neutral pH range. In contrast to the aqueous complexes, the material that forms from reaction of $VOCl_3$ in excess acetic acid has a molecular formula of VO_2OOCCH_3 [86]. This material contains the VO_2^+ cation and the acetate monoanion, and it presumably formed by losing acetic anhydride from $VO(OOCCH_3)_3$.

Structural characterization of materials with monodentate carboxylate ligands has provided a variety of different geometric arrangements [82,87–89]. The $[VO(OOCC(CH_3)_3)_3]$ complex has each of the three carboxylate groups chelated in a bidentate fashion and is currently the only mononuclear complex among this group of compounds (Fig. 7) [87]. Bidentate chelation is also observed for the tetranuclear complex with four ligating carboxylates ($[K\{V_4O_4(\mu\text{-}O)_4(\mu\text{-}O_2CCH_2t\text{Bu})_4\}][OOCCH_2t\text{-}Bu]\cdot t\text{BuCH}_2CO_2H$) [89] and $[V_4O_8(NO_3)(C_4H_3SCO_2)_4]^{2-}$ [88]. One example of a pentanuclear complex with four ligating carboxylates $[V_5O_9Cl(C_4H_3SCO_2)_4]^{2-}$ has also been reported [88]. Each carboxylate group is bidentate in all these complexes, but the two oxygen atoms in a carboxylate group span two different vanadium atoms. All four alkyl groups are pointing in the same direction in a cup-shaped fashion in all these three complexes [88,89].

None of the compounds that have been structurally characterized resemble those that presumably form in aqueous environments with

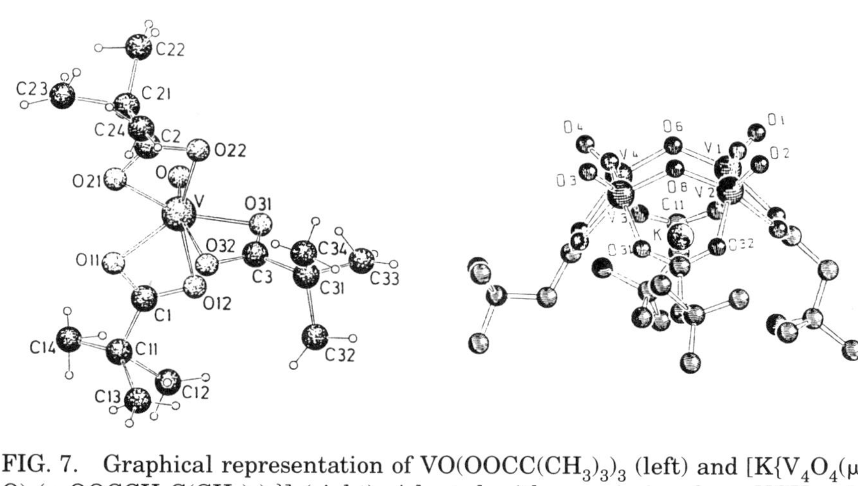

FIG. 7. Graphical representation of $VO(OOCC(CH_3)_3)_3$ (left) and $[K\{V_4O_4(\mu\text{-}O)_4(\mu\text{-}OOCCH_2C(CH_3)_3)_4\}]$ (right). Adapted with permission from VCH publishers, Inc. from Refs. [87] and [89].

simple carboxylic acids. In contrast, similarities have been obtained between compounds in organic solvents and in aqueous solution containing multidentate ligands. Carboxylate groups in general are components of substances prevalent in biological fluids. Interaction of vanadate with the carboxylate is likely to be essential to enzymes, such as the bromo- and haloperoxidases, that contain up to 25% acidic amino acid. Protein and organic substances in conjunction with carboxylate are, however, likely to interact with vanadate and derivatives; perhaps in this case, complexes with more than one functionality give better representation of the vanadate complexes that form in biological systems.

4.3. Amines, Imines, and Heterocyclic Amines

The reactions of monodentate nitrogen compounds with vanadate have been reported only in a few cases and result in weak complexes ([22,90] and L. Pettersson, results to be published). The 2:1 bis(pyridine)-vanadate complex gives a broad signal in the ^{51}V NMR spectrum (-467 ppm), whereas no resonance could be observed for the imidazole-vanadate complex. The formation of an imidazole-vanadate complex was inferred from the changes in the oxovanadate equilibria and the kinetics of oxovanadate exchange in the presence of imidazole [22]. The small formation constant for the bis(pyridine)-vanadate complex of 8.1×10^{-3} M^{-2} attests to the weakness of these types of complexes [90].

The interaction of vanadate with imidazole derivatives is important because the amino acid histidine contains an imidazole group. Histidine residues often play an essential role in enzymes of current interest, such as the haloperoxidases. At this time all of the systems that have been structurally characterized contain additional ligands and/or several functionalities to stabilize the complex. These complexes will be described below (Secs. 6.3 and 6.4).

4.4. Thiols

The reaction of vanadate with thiols results in the reduction of vanadium(+5) to vanadium(+4). The reactions of vanadate with cysteine, *N*-acetylcysteine, glutathione, and penicillamine have been characterized in some detail [91,92]. Vanadate reacts with cysteine with a stoichiometry of 1:3 to generate an oxovanadium(+4)-cystine complex [92]. It was proposed that the reaction involved a thiol radical [93] and electron paramagnetic resonance (EPR) studies of solutions containing vanadate and cysteine, *N*-acetylcysteine, glutathione, and penicillamine all show EPR signals attributable to thiol radicals [91]. It appears that the reaction is strongly dependent on pH.

The interaction of vanadate with glycerol-3-phosphate dehydrogenase and fructose 1,6-bisphosphate dehydrogenase provides some additional information on the reaction of vanadate with thiols since each of these enzymes contains at least one accessible thiol group. Two oxovanadates, specifically V_2 and V_4, inhibit glycerol-3-phosphate dehydrogenase [94]. An essential and accessible thiol group is present in the cofactor site of this enzyme, and oxidation of this thiol should have resulted in loss of enzyme activity. Since removal of the vanadium by EDTA or exclusion chromatography regenerated full enzyme activity, redox chemistry had not occurred with this thiol. It was suggested that thiol group oxidation may require the presence of two thiol groups in order to complete the oxidation reaction. Fructose 1,6-bisphosphate aldolase exhibits both reversible and irreversible (redox chemistry) interactions with vanadate [95]. The active enzyme consists of a tetrameric unit with a multitude of accessible thiol groups. A combination of enzyme kinetics and ^{51}V NMR spectroscopy was used to show that V_2 interacted reversibly with the enzyme whereas V_4 irreversibly deactivated the enzyme. Deactivation was preceded by formation of a protein-V_4 complex. Reactivation studies were performed using reactivation buffers containing dithiothreitol (DTT), and it was concluded that the vanadate slowly oxidized some of the thiol groups essential for enzyme activity. In any event, the results show that V_4 is competent to carry out redox chemistry and that the lack of oxidation of glycerol-3- phosphate dehydrogenase is not due to the form of the vanadium interacting with this enzyme. The formation of a protein-vanadate complex that is unable to complete the electron transfer is consistent with these observations. It remains to be determined as to whether complexes containing one or two thiol groups can undergo electron transfer.

The evidence for V-S coordination in the newly discovered vanadium nitrogenase has increased the interest in preparation of model compounds [96–98]. Most structurally characterized vanadium-sulfur complexes contain vanadium in an oxidation state less than +5. For example, a vanadium(+4) thiolate from cysteine methyl ester with the vanadium in a five-coordinate square-pyramidal geometry has been described [97]. Tetrathiovanadate is an analog of vanadate that contains vanadium(+5) [98]. The material is unstable in aqueous solution and it was prepared for development as a convenient precursor to thio-bridged vanadium-iron clusters.

When vanadate enters a cell it is reduced to vanadium(+4) [99–101]. The reduction has mainly been attributed to glutathione, although ascorbic acid [102] and other reducing agents can play a role. In the past, all the vanadate that entered a cell was believed to undergo reduction, but recent studies suggest that both vanadium(+5) and (+4) exist in biological fluids [103]. There is no doubt that this area of vanadium chemistry is essential to understanding the interactions of vanadium in biological systems and will grow significantly in the future.

5. VANADATE REACTIONS WITH BIDENTATE ORGANIC LIGANDS

Metabolites and biosynthetic precursors are often simple molecules containing two types of functionalities. Most of these types of substances will be prevalent in the biological environment. In addition, a few other bifunctional ligands have been used as effective ligands for vanadium(+5) and accordingly may be of relevance to the insulin-mimetic effects and other biological effects of vanadium compounds. In the reaction of vanadate with bifunctional ligands the possibilities for complex formation increase. First, the complex(es) with a monochelated ligand may form, and second, the complex(es) in which the ligand chelate in a bifunctional manner may form. In many cases the ligand strongly prefers to chelate in a bifunctional manner and, given the description of monodentate complexes above, less attention is given to these minor complexes in this section. In some cases, however, the complexes are of similar stability, and all complexes must be considered equally.

5.1. Diols

The reaction of vanadate with 1,2-ethanediol (ethylene glycol) generates a series of complexes with stoichiometries 1:1, 1:2, and 2:2 [104]. The minor complexes were assigned as vanadate monoester (1:1), vanadate diester (1:2), and esters of vanadate dimer (2:1 and 2:2). The major complex, a 2:2 complex with a characteristic chemical shift of −523 ppm, was originally proposed to have the structure shown in **5** based on

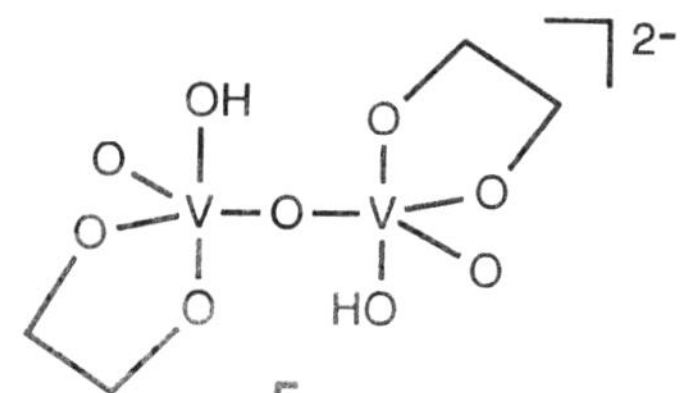

5

^{51}V NMR studies. The structure of complexes of this type are important because the analogous complex between vanadate and uridine has been recognized as a potent transition state analog for ribonuclease [105]. A series of studies have been published characterizing the reaction between four nucleosides and vanadate [42,106–110] as well as characterizing the complexes formed between vanadate and a series of related but less complex vicinal diols [72,111].

Solution and solid state structural characterization of various model complexes provides valuable information since the aqueous complexes have resisted solid state characterization. Replacement of one OR functionality in the complex with a chlorine atom allowed the first structural characterization of a cyclic oxovanadium 1,2-diolate (Fig. 8, top left) [112]. The five-coordinate distorted trigonal bipyramidal geometry clearly shows the expected structural features of the vanadium–transition state analog geometry. The X-ray structure of the uridine-vanadate-ribonuclease complex was reported in preliminary form from two groups, but due to some disorder in the ribose ring and low vanadium occupancy, the uridine-vanadate-ribonuclease structure did not refine to the point of being entered in the Cambridge Protein Data base [113–115]. The unpublished structure [113] of the vanadate-uridine complex inside ribonuclease is shown for comparison with the model compounds (Fig. 8, top right).

Substituting one $C(CH_3)_2$ group of the pinacol with a carbonyl group provided crystals with solvable diffraction patterns and a vanadium atom in trigonal bipyramidal geometry (Fig. 8, bottom left) [116]. The latter compound is stable in aqueous solution and thus amenable to spectroscopic characterization [29,111].

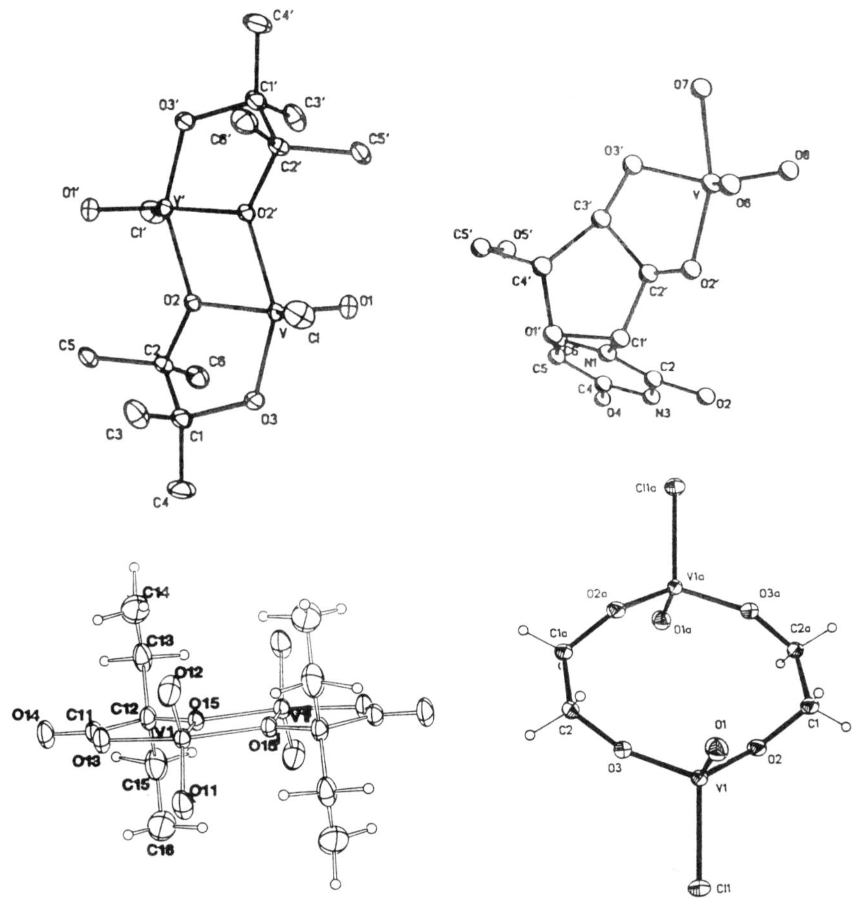

FIG. 8. Graphical representations of $[VOCl(OC(CH_3)_2C(CH_3)_2O)]_2$ (top left), the vanadate-uridine complex inside ribonuclease (top right), $[VO_2(OC(CH_2CH_3)_2\text{-}COO)]_2$ (bottom left) and $[VOCl(OCH_2CH_2O]_2$ (bottom right). Adapted with permission from American Chemical Society from Refs. [112], [116] and [117].

Replacing the pinacol (2,3-dimethylbutane-1,2-diol) ligand with ethylene glycol had a dramatic effect on the coordination geometry around the vanadium atom in the binuclear complex (Fig. 8, bottom right). The ethylene glycol complex contained two tetrahedral vanadium atoms; however, in solution this material forms a mixture of the four- and five-coordinate binuclear complexes [117].

The interactions of vanadate with *cis*- or *trans*-cyclohexane-1,2-diol in aqueous solution have been studied by ^{51}V NMR spectroscopy and the major complexes had, as anticipated, a 2:2 stoichiometry [72]. No differences were observed in ^{51}V chemical shifts of the *cis*- and *trans*-diol-derived complexes. The 2:2 complex formed from the *cis*-diol was found to be approximately 10 times more stable than the 2:2 complex formed from the *trans*-diol, suggesting that subtle geometric variations can have dramatic effects on the stability of the complexes. Potentiometric studies with linear polyols support the chelation of the vanadium atoms in a bidentate manner [118].

As more techniques were employed and other 1,2-diols were studied in aqueous solution, additional structural possibilities emerged: five distinct proposals (Structures **5** to **9**) have now surfaced in the literature. Structure **6** was proposed as a revised version of structure **5** to

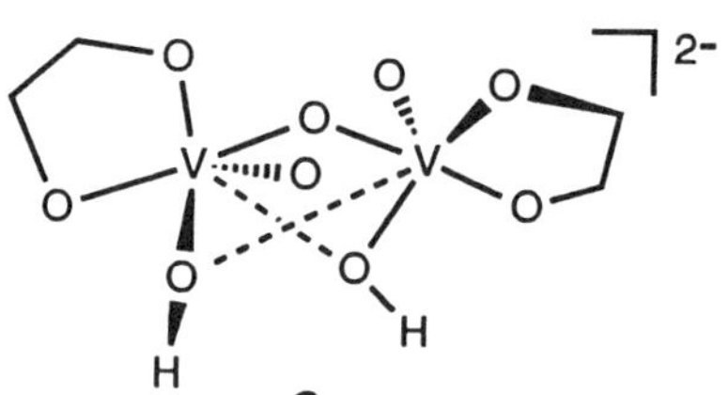

6

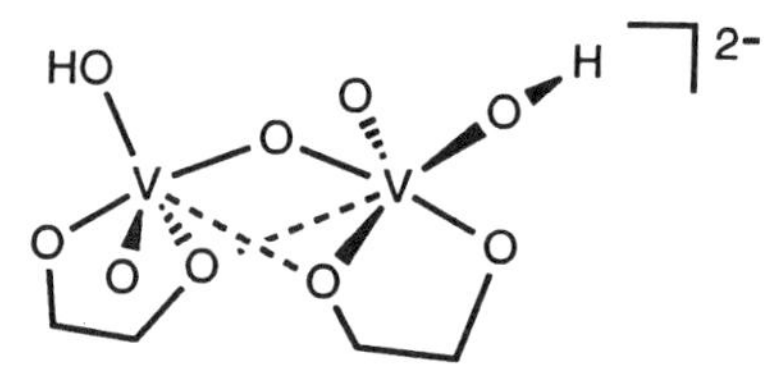

7

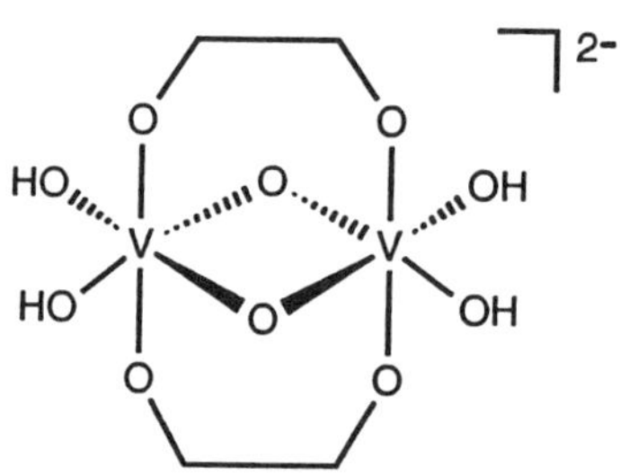

8

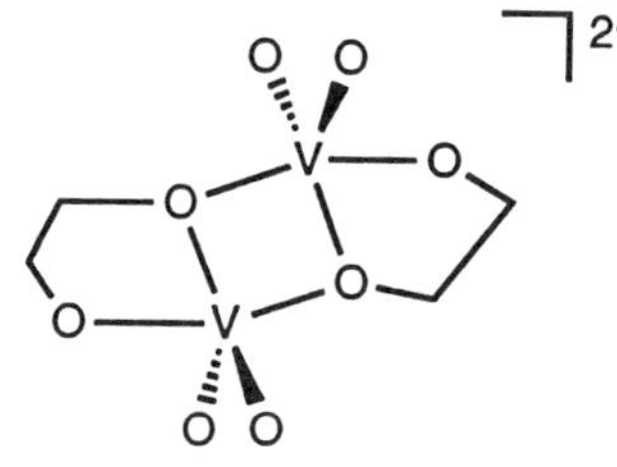

9

explain the large number of isomers in solution [108]. Structure **7** was based on the bridging of alkoxides in solution of various diols [119]. A very different geometry, structure **8**, was suggested based on the lack of circular dichroism (CD) spectra in conjunction with the observed magnetically induced circular dichroism (MCD) data of aqueous solutions of nucleoside-vanadate complexes [29,110]. Lastly, structure **9** was proposed based on a variety of data including Raman spectroscopy, IR spectroscopy, ^{51}V, ^{13}C, ^{1}H, and ^{17}O NMR spectroscopy, as well as model compounds [111].

The first aqueous complexes were characterized using only ^{51}V NMR spectroscopy [42,104]. Other techniques including ^{1}H, ^{13}C, and ^{17}O NMR spectroscopy, CD, MCD, UV-VIS and Raman spectroscopy have now been employed to examine the solution structure of this and related complexes [29,107–111]. It is beyond the scope of this chapter to describe all the available evidence and the reader is referred to the original studies for further information. Suffice it to note that the solution structure must satisfy all observations without undergoing physically and chemically unreasonable interconversions of isomers. Several of the structural proposals are no longer viable candidates based on the evidence that has accumulated in the past 5 years.

In contrast to the studies with 1,2-ethanediol, analogous studies with 1,3-propanediol showed only formation of the monoester and diester; no cyclic 2:2 complex formed with this ligand [104]. Multinuclear complexes of 1,3-diolate ligands have been structurally characterized but these complexes bear little resemblance to the complexes in aqueous solution [36,81,120]. As both solution and solid state information become available, differences and similarities are likely to emerge that should provide insight into the factors that govern formation of complexes with these types of ligands.

The current interest in the geometry of the solution structure of the vanadate-1,2-diol complexes is directed at understanding the high affinity of the vanadate-uridine complexes for ribonuclease. Transition state analogs have potential as inhibitors, and the vanadate complexes show very high affinities. In addition, diols are prevalent functionalities in biological fluids and thus will be important in the biochemistry of vanadium(+5).

5.2. Catechols

The study of reactions of vanadate with catechols has been stimulated by the interest in tunichrome complexes of vanadium [8], the abundance of this group in siderophores [121], and the catecholamine stimulated activation of Na,K-ATPases [6]. In addition, catechol derivatives have been used as metal ion indicators [122]. In most cases, structural characterization of vanadium(+5) complexes with catechol-derived ligands showed octahedral coordination about the vanadium atom [30, 123–125]. In some cases these compounds are dinuclear.

The reaction of catechol and other derivatives with VO_2^+ has been studied in great detail in acidic media [126–128] and in some detail at alkaline pH [129]. At acidic pH a vanadium(+5)-catechol complex is formed initially, which reacts with VO_2^+ with a half-life ranging from 5 to 15 msec to generate oxidized catechol and reduced vanadium(+4) [127]. The formation of the intermediate species is reversible and increases with increasing pH. It has been suggested that the loss of coordinated hydroxy and/or aqua ligands may be linked to the rate-determining step of intermediate formation [126]. The rate of the slow and rate-limiting oxidation-reduction reaction increases with decreasing pH [128].

At alkaline pH, the reaction could not be analyzed in such detail. However, three distinct stages could be identified corresponding to (1) complex formation,(2) reduction of V(+5) to V(+4), and (3) further reactions of the intermediate oxidation products [129]. The redox potential for the V(+5)/V(+4) couple at pH >6 for catechols implies that the redox reaction with catechols should not occur. This fact supports the experimental observations that a metal-ligand complex forms before the electron transfer occurs; it is the redox potential of the catechol complex that governs the reaction. In the reaction of vanadate with both catechols and thiols, redox chemistry is preceded by formation of a vanadium complex. Further information is needed to assess whether this is a general feature associated with redox chemistry of vanadate and vanadium(+5) complexes.

5.3. Dicarboxylates

The reaction between oxalic acid and vanadate has been examined using combined ^{51}V NMR spectroscopy and potentiometry, and speciation curves have been reported from pH 0 to 8 [130]. At low pH the major complex is mononuclear with one oxalate ligand and at micromolar concentrations of vanadate this complex continues to be the major complex at higher pH. The (2,1,1) species [see [131] for a brief definition of nomenclature and references to the original literature) has a formation constant of $\log \beta_{2,1,1} = 13.07$ [130] or expressed as an H^+-dependent apparent formation constant ($K = [V\text{-}Ox]/[V_1][Ox]$) a value of 13 M^{-1} at pH 7.04 [132]. At millimolar concentrations of vanadate and oxalate at pH 4, a mononuclear complex with two oxalate ligands becomes the major complex [130]. The (2,1,2) species [131] has a formation constant of $\log \beta_{2,1,2} = 17.11$ [130] or, expressed as the H^+-dependent apparent formation constant ($K = [V\text{-}Ox_2]/[V_1][Ox]^2$), a value of 2500 M^{-2} was measured at pH 7.04 [132]. Characterization using ^{13}C, ^{51}V, and ^{17}O NMR spectroscopy shows that the 1:2 complex in solution corresponds to a complex that has been characterized by X-ray crystallography [133]. This complex contains the *cis*-dioxo unit and a vanadium atom in octahedral geometry (Fig. 9) [134]. The 1:2 complex, the 1:1 complex, and the free ligand are in rapid exchange in aqueous solution [133]. A dissociative process has been suggested based on the positive entropies of activation measured for these processes. Although there has not been structural reports for the 1:1 complex in aqueous solutions, the dioxofluorooxalatovanadate complex has been structurally characterized and is shown in Fig. 9 [135].

The reactions of vanadate with other dicarboxylic acids have been examined at pH 5.0 [85]. Three different complexes have been observed

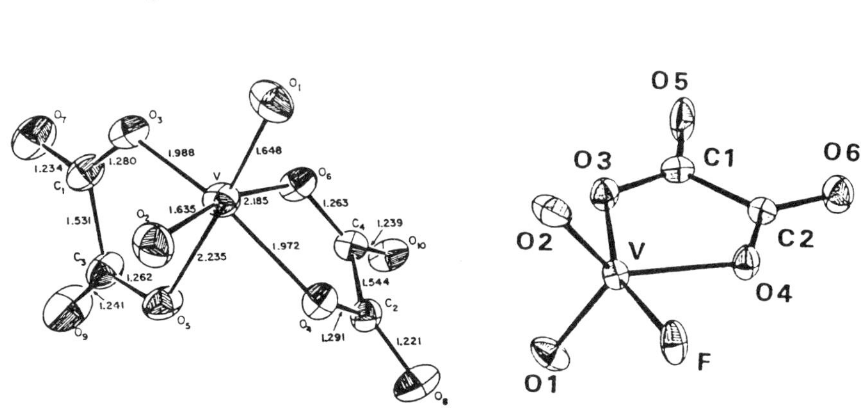

FIG. 9. Graphic representation of $VO_2(OOCCOO)_2$ (left) and $VO_2F(OOCCOO)$ (right). Adapted with permission from the American Chemical Society and Barth Verlagsgesellschaft mbH from Refs. [134] and [135].

in the vanadate–succinic acid system by ^{51}V NMR spectroscopy; two 1:1 complexes (at −526 and −548 ppm) and one 1:2 complex (at −541 ppm). Based on a series of solution ^{51}V NMR and selected ^{13}C NMR studies it was suggested that the two 1:1 complexes arose from differences in the coordination geometry around the vanadium atom. Succinic acid, methylsuccinic acid, glutaric acid (pentanedioic acid), and adipatic acid (hexanedioic acid) all form 1:1 complexes with H^+-dependent apparent formation constants in the range of 1–20 M^{-1} and 1:2 complexes with H^+-dependent apparent formation constants in the range of 0.2–12 M^{-2}. Malonic acid is the only simple dicarboxylic acid that forms a 1:2 complex with a large formation constant, 1130 M^{-2}. The results suggest that the stability of the vanadate complexes with dicarboxylic acids is greatest when fewer carbon atoms separate the carboxylate groups. With respect to interactions of vanadate with organic substances in biological fluids, only oxalic and malonic acids form complexes with sufficiently large formation constants to be biologically significant.

5.4. Simple Hydroxamates and Siderophores

Hydroxamates are important functionalities of two major groups of naturally occurring siderophores [121]. The hydroxamate functionality is also present in amavadin, a natural product containing vanadium(+4) [136]. In addition, hydroxamates have been used extensively as a reagent to determine vanadium concentrations [137–140]. The chemistry of hydroxamates and siderophores with vanadate is described here and in addition related chemistry with other oxidation states is briefly summarized.

In 1972 Bayer and Kneifel [141] isolated a vanadium containing compound from a toadstool variety of the *Amanita muscaria* that accumulates vanadate (reviewed in [136]). Upon isolation, the natural product named amavadin was found to contain dimethyl-*N*-hydroxyimino-α,α′-dipropionate and vanadium(+4) [141]. Model studies with the ligand *N*-hydroxyiminodiacetic acid lead to structural characterization of this vanadium(+4) complex [142]. Further characterization of spectroscopic studies of the model compound and amavadin led to a revisal of the previous proposal. It is now believed that both the oxygen and the nitrogen atoms in the hydroxylamine functionality chelate to the vanadium in amavadin [136,143,144].

Vanadate forms very stable complexes with hydroxamates and it appears that these complexes form rapidly in a wide pH range in aqueous solution [145,146]. The vanadium-hydroxamate complex is measured at low pH when used for analytic purposes, mostly because iron does not form complexes in this pH range [147]. At neutral pH, the vanadium(+5) complexes are also stable. For example, this is evidenced by the abstraction of the ATPase-bound vanadate and regeneration of all the vanadate-inhibited ATPase activity. A broad resonance at −515 ppm is observed in the ^{51}V NMR spectrum in solutions of vanadate and desferrioxamine B, the water-soluble polyhydroxamic acid siderophore originally isolated from *Streptomyces pilosus* [145].

Reaction of hydroxylamine with the dioxo(dipicolinato)vanadate in alkaline solutions generates a 1:2 complex which contained an axial nitrosyl ligand and a "side-on" hydroxylamine, coordinated through both the nitrogen and the oxygen atom [148]. In contrast, structural characterizations of several octahedral vanadium(+5) and +4 hydroxamate derivatives show that the hydroxamates coordinate through the carbonyl oxygen and the hydroxylamine oxygen atoms and not through the nitrogen atom [149–151]. In these hydroxamate complexes, the V=O bond is broken, as in the case of the catechol derivatives. However, structural characterization of a $[VO(salicylhydroxamate)(CH_3OH)]_3$ complex showed metal binding to the hydroxamate nitrogen [152]. This unusual binding array was in part generated by a trinuclear core consisting of $[-V-N-O]_3$.

Hydroxamic acids clearly provide an array of different types of interactions with vanadium and vanadates. Perhaps the range of naturally occurring siderophores may exhibit a variety of different interactions with vanadates.

5.5. Mixed Bidentate Organic Ligands

Ligands with two different functional groups further increase the number of complexes which vanadate can form. Two different complexes can form with a mixed bidentate ligand coordinated in a monodentate manner. In addition, the two functionalities increase the structural possibilities for complexes formed with the ligand coordinated in a bidentate manner. We will describe here the reaction of vanadate with bidentate ligands such as hydroxy aldehydes and ketones, hydroxy acids, amino alcohols, amino acids, pyridine carboxylates, 8-quinolinates, and pyridones.

The reaction of vanadate with aldehydes has not been reported although a significant affinity of vanadium(+5) for the carbonyl group has been observed in organic solvents [66] and in aqueous acetone solutions [17]. Several lines of evidence suggest that bifunctional ligands containing an aldehyde or a ketone interact with vanadate. An elegant set of experiments in an aqueous solution provided evidence that vanadate catalyzed the dehydration of hydrated glyoxalate [153]. Vanadate forms esters with the hydrated aldehyde functionality,which upon hydrolysis yield the free aldehyde. Utilizing the fact that lactate dehydrogenase only accepts the free aldehyde form of glyoxalate, the rate enhancements of NADH oxidation reflect the hydrolysis of hydrated glyoxalate vanadate ester. In general, compounds with the functionality HOH_2C-CO-R are less stable toward oxidation by vanadate than compounds with the functionality OHC-CHOH-R. For example, dihydroxyacetone, dihydroxyacetone phosphate, or fructose will fairly rapidly reduce vanadate at neutral pH, whereas the corresponding solutions containing glyceraldehyde, glyceraldehyde 3-phosphate, or glucose are significantly more stable [154].

The reactions of α-hydroxycarboxylic acids with vanadate have been examined by a number of workers. However, this literature is rife with disagreements concerning the stoichiometries and assignments of various complexes [29,116,118,132,155–158]. Perhaps part of the problem in examining these systems stems from the many products that form in aqueous solution compounded by the vanadate condensation reactions that complicate studies involving analyses of the ratios of ligand to vanadate. A recent careful ^{51}V NMR analysis showed, in addition to four vanadate oligomers, that four other signals with an additional fifth signal superimposed on the V_1 resonance are observed in the NMR spectrum of a solution of 1 mM vanadate, 70 mM lactate at pH 7.5 [132]. This suggests that up to five different complexes can form in these solutions at pH 7.5! It is not difficult to envision that studies relying exclusively on any one experimental approach may encounter significant problems in sorting out this complex system. We will not review here the various stoichiometries and geometries proposed for the complexes from various laboratories but will instead refer the reader to the original literature [29,116,118,132,155–158]. However, we will describe studies probing the stoichiometry and structure of one of the major complexes since several studies using various techniques at this time seem to be approaching an agreement.

The 2:2 complex of 2-ethyl-2-hydroxybutanoic acid and vanadate has been structurally characterized (Fig. 8, bottom left) [116]. Multinuclear ^{13}C, ^{1}H, and ^{51}V NMR spectroscopies suggest that the major complex remains a 2:2 species with the vanadium atom in a distorted trigonal bipyramidal geometry in aqueous solution [29]. The ^{51}V chemical shift of this complex is −550 ppm, whereas the chemical shifts of analogous complexes with 2-methyl-2-hydroxypropionic acid and 2-phenyl-2-hydroxypropionic acid are −547.9 and −545.7 ppm, respectively. It is well documented that the chemical shift of complexes with the same coordination geometry around the vanadium atom can be sensitive to substituents on the ligand, as reported previously for the simple vanadate esters [22,68,71], and for the structurally characterized oxovanadium chloroalkoxides [112,117]. Changes in the ^{51}V NMR shifts would then be anticipated for the series of complexes formed from glycolic acid (−540.8 ppm) [158], lactic acid (−544.1 ppm [158] or −536 ppm [132]), and 2-ethyl-2-hydroxybutanoic acid (−550 ppm) [29]. It seems reasonable to expect that these signals represent the 2:2 complexes with the vanadium atom in a trigonal bipyramidal geometry.

The above comments were made based on the assumption that the 2:2 complex of 2-ethyl-2-hydroxy butanoic acid and vanadate is truly representative of these types of complexes. In addition, one would also expect minor complexes to form upon mixing vanadate with 2-ethyl-2-hydroxybutanoic acid in aqueous solution and that such complexes would form upon dissolution of the 2:2 complex. Further information on the aqueous reaction and the kinetics of the complex formation may identify the source of the apparent discrepancies reported. There is no doubt that more structural information and careful correlations between solution and solid state structure are needed to sort out this very complex system.

The reaction of vanadate with a vicinal amino alcohol will result in vanadate ester formation. However, no new resonance in the ^{51}V NMR spectrum is observed for a cyclic product that would form should the ligand bind in a bidentate manner [159].

The reactions of vanadate with amino acids form complexes that can be observed by ^{51}V NMR spectroscopy and by other means [160–

164]. Complexes with a 1:1 stoichiometry and a ^{51}V NMR chemical shift about −550 to −560 ppm are expected to form a chelated amino acid. Obviously, either the amino group or the carboxylate group can form a complex as a monodentate ligand. Such 1:1 complexes may be in rapid exchange with V_1 or the resonances may be superimposed on the V_1 resonance. If so, no new resonance will be observed in the ^{51}V NMR spectrum, and a speciation analysis is necessary to determine the presence of these complexes. In the case of alanine, serine and arginine signals were observed at −556 ppm. However, in the case of glycine, asparagine and valine speciation analysis was required to infer complex formation. The apparent formation constants at pH 7.0 ranged from 0.13 to 0.94 M^{-1}. Since no complex was observed with glycinamide (which contains no carboxylate) it was concluded that the 1:1 complex involves an anhydride bond formation between the vanadate and the carboxylate group in the amino acid.

A second 1:1 complex was observed in the ^{51}V NMR spectra of glycine, alanine, serine, and asparagine at about −545 ppm. The signals were broad and presumably reflect involvement of both the carboxylate and the amine groups of the amino acid. Apparent formation constants ranged from 0.19 to 0.74 M^{-1} at pH 7.0. Lastly, a 1:2 complex formed with a chemical shift of −523 ppm in the reaction of vanadate with glycine. The apparent formation constant of this complex was 0.5 M^{-2} at pH 7.0. This complex presumably contains two chelated ligands and a vanadium atom in an octahedral geometry. The fact that only glycine formed this kind of complex suggests a very fine balance that is perturbed upon further substitution of the amino acid.

Neither the 3-pyridinecarboxylate (nicotinate) nor 4-pyridinecarboxylate (isonicotinate) undergoes significant reaction with vanadate until ligand concentrations above 2 M are reached [90]. At very high concentrations the pyridine begins to interact with vanadate. This is in contrast to 2-pyridine carboxylate, which reacts with vanadate at equimolar ligand concentrations to form four stable complexes at −550, −552, −529, and −523 ppm in the ^{51}V NMR spectrum. One of these complexes is a 1:1 complex (−550 ppm), and the other three are 1:2 complexes. The apparent formation constant at pH 7.0 is 200 M^{-1} for the 1:1 complex. The H^+-dependent apparent formation constants at the same pH are 1.5×10^4 M^{-2} for the −552 complex, 4.0×10^4 M^{-2} for the −529 ppm complex, and 8.4×10^4 M^{-2} for the −513 complex. Picolinic acid is frequently used for modification of the reactivity of vanadium for application in organic synthesis [60]. In addition, several species have been characterized crystallographically, and compounds with vanadium in trigonal bipyramidal [165], octahedral [166], and pentagonal bipyramidal [167] geometry have been reported. The two 1:2 vanadate-picolinato complexes at −513 and −529 ppm are of particular interest since they have chemical shifts in the same range of the unusual complex that forms between vanadate and glycine described above.

A few additional bifunctional ligands, including 8-quinolinates and the hydroxypyridones, deserve brief mentioning. These ligands are less abundant in biological fluids; however, the chemistry of the complexes is of relevance to the topic of this chapter. The 8-quinolinate (Q) formed a complex with VO_2^+ which could be protonated to $Q_2VO(OH)^{2+}$ (pK_a of 3.4) [168]. Floriani and coworkers [168] suggested that the $VO(OH)^{2+}$ moiety could be viewed as an inorganic analog of a carboxylic acid, although Bonadies and Carrano have suggested that the dipronated species is better formulated as L_2VO^{2+} + H_2O (L here carries no charge) [31]. 8-Quinolinates have been successfully used for preparation and characterization of model vanadate ester systems [169]. The hydroxypyridones are of interest because of the recent report of insulin-mimetic activity of the 1:2 vanadium(+4) complex of 2-hydroxypyridone [170]. This compound oxidized to vanadium(+5) in aqueous solution, and accordingly these solutions can be examined by NMR after oxidation and EPR spectroscopy prior to oxidation [170–173]. The hydroxypyridone complexes are related to the reagent $VO(ACAC)_2$ and have similar properties.

This group of ligands interacts either strongly or with modest affinity with vanadate. Many α-hydroxycarboxylic acids are prevalent metabolites and will be present at some level in the cellular fluids. Complexes of the type described in this section will form and can be important to the biochemistry of vanadium(+5).

6. VANADATE REACTIONS WITH POLYDENTATE LIGANDS

6.1. Carbohydrates

Carbohydrates make up a significant fraction of the organic substances present in biological fluids. The major functionalities in carbohydrates are hydroxyl and carbonyl groups. The structural considerations of complexes containing 1,2-diols suggest that sorting out the interactions of vanadate with various multifunctional carbohydrates is exceedingly complex. Several approaches to probe the interactions of vanadate with carbohydrates have been reported.

Cyclic α- or β-hemiacetals or cyclic α- or β-hemiketals are the major form of carbohydrates in aqueous solution but a small fraction exists in the extended form. Vanadate can interact with all of these molecules, and will form the most stable complexes. One approach to the study of how vanadate interacts with carbohydrates involves examination of ligands with limited geometric flexibilities, such as the α- or β-glycosides. The reaction of vanadate with α- or β-methyl glucoside forms vanadate monoesters as the major products [72]. The difference in the ^{51}V NMR spectra of the complexes formed by α- and β-methyl glucosides allows assignment of the 3- and 4-monoesters of the β-methyl glucoside (−558 ppm) and the 2-monoester (−562 ppm), respectively. The formation constants for these 1:1 vanadate complexes with the α- and β-glucosides are smaller than those observed for the cyclohexane-1,2-diol. The smaller constants were attributed to the ring flattening associated with the replacement of a carbon of the cyclohexane ring by an oxygen atom. Only the α-methyl glucoside formed a minor product at −520 ppm which is reminiscent of the diol complex at −523 ppm, and the authors assign this complex as having a structure akin to that shown in Structure **7**.

The reactions of vanadate with methyl galactosides and methyl mannosides generate cyclic vanadate diesters as the major products [72]. The reaction of vanadate with α- or β-methyl galactoside produces ^{51}V NMR spectra that are identical, with major complexes at −544, −517, and −502 ppm. The reaction of vanadate with α- or β-methyl mannoside shows only one major complex at −522 ppm with only small differences between the α and β isomers. In both of these systems other minor products form, and due to signal overlap the analysis is complex and cannot be completed solely by the use of ^{51}V NMR spectroscopy.

A study of the β-methyl riboside was completed using ^{1}H, ^{13}C, and ^{51}V NMR spectroscopy with the objective of exploring the aqueous structure of the 2:2 diol complex [111]. Accordingly, conditions were chosen to assure that the major complex in solution indeed was the 2:2 complex with the ^{51}V NMR chemical shift of −523 ppm. ^{13}C NMR spectroscopy shows two different types of α-hydroxy carbon atoms; however, the lines were very broad and variable-temperature studies showed that these signals were approaching coalescence at ambient temperature. Three different types of complexes exist in solution by ^{1}H NMR spectroscopy while only one complex was observed by ^{51}V NMR spectroscopy. Magnetization transfer experiments showed that two isomers exchange rapidly and that these two isomers and the third isomer exchange more slowly. The authors conclude that all of these observations are only consistent and chemically reasonable if this 2:2 complex has a geometry of the type shown in Structure **9**. Furthermore, the vanadate-adenosine (2:2) complex has recently been structurally characterized and an X-ray structure akin to Structure **9** will soon be forthcoming (A. S. Tracey, personal communication).

A study probing the interaction of vanadate with ribose, glucose, and mannose was particularly complicated by overlapping ^{51}V NMR signals of multiple complexes [106]. Although this study took advantage of the combination of ^{1}H, ^{13}C, and ^{51}V NMR spectroscopy to examine the nature of the complexes, the structural possibilities for complexes formed by these sugars in the α, β, and extended form presumably contributed to the overlap of resonances observed in these ^{51}V NMR spectra. D-Glucose in the presence of vanadate forms a major cyclic complex with a chemical shift at about −520 ppm, and a minor species with a chemical shift around −500 ppm. Given the differences in complex formation between the α- and β-methyl glucosides and that of free glucose one can conclude that the formation of the extended form is critical to generating this cyclic-vanadate-glucose complex. Glucose vanadate complexes were not observed by ^{1}H NMR spectroscopy, which led the authors to suggest that a tridentate complexation arrangement proposed for D-mannose [106] and tris(hydroxymethyl)aminoethane [174] is not occurring with glucose [106].

No structural information is available for any carbohydrate-vanadate derivative. Recent progress in characterization of vanadium complexes of tris(hydroxymethyl)ethane provides some information by showing these types of derivatives involve multiple ligands and oxovanadate clusters ([36] and D. C. Crans, unpublished). Structural characterization of vanadium(+5) carbohydrate derivatives is badly needed, particularly in view of the potential these types of compounds have for playing a role in the vanadium(+5) biochemistry.

There is no doubt that a variety of vanadate esters, including those of carbohydrates, are likely to form in the biological fluids. Carbohydrate mono- and divanadate esters are likely to be labile and not able to serve to store the vanadium compound for prolonged periods. This may not be the case for complexes with multidentate interactions. However, given the analogy of the simple vanadate esters with phosphate esters, it is likely that enzymes and/or receptors may respond to these derivatives as the corresponding phosphate derivative and, accordingly, trigger a particular response.

6.2. Nucleosides and Nucleotides

As described in Sec. 3.1 vanadate interacts with AMP and ADP to form vanadium analogs of ADP (AMP-V) and ATP (ADP-V, Structure **2**; AMP-V-V, Structure **3**) [38,42,175]. Reactions with the other nucleotides also form the corresponding vanadium nucleotides. With the exception of kinases, in general (see Chap. 9) enzymes appear to recognize these analogs as substrates and bind them [38,43,44]. The vanadium is presumed to be tetrahedral in these compounds.

Vanadate interacts with nucleosides to generate analogs of the corresponding monophosphates such as the AMP, AMV [42,106–108]. The major complex that forms in a solution of vanadate and nucleosides (see Sec. 5.1) is a 2:2 complex [106–108,110], and not the 1:1 complex that was the anticipated analog of cyclic 2′,3′-AMP. Given the fact that ribonuclease forms a vanadate-uridine-ribonuclease complex containing a 1:1 species [105,113,114] there is no doubt that such a species forms, but the aqueous chemistry clearly shows that this compound is not a major species.

No reports have yet emerged regarding a vanadium analog of cyclic 3′,5′-AMP. Aqueous studies suggest that simple bifunctional 1,3-diol cyclic derivatives with vanadate are not sufficiently stable for observation by NMR spectroscopy [104,175]. The ligands that have been proposed to form cyclic derivatives are always supported by other functionalities in a more complex structural arrangement [tris(hydroxymethyl)aminoethane, tris(hydroxymethyl)ethane, and mannose] [106,174]. Studies in nonaqueous solutions show that there are major differences between vanadium complexes containing 1,2-diol chelates and 1,3-diol chelates [81,112,117,119,169,176]. Formation of cyclic 3′,5′-

AMV in aqueous solution is likely to require additional stabilization perhaps provided by a polydentate ligand.

An analog of NADP, NADV, has been observed in aqueous solution [70,177]. The major complex in solutions containing NAD and vanadate is a 2:2 complex, as in the cases of other complexes formed from vicinal diols. Presumably, the minor 1:1 complex NADV is the species that shows the biological activity. NADV is a cofactor for enzymes such as glucose-6-phosphate dehydrogenase, 6-phosphogluconate dehydrogenase, and alcohol dehydrogenase [70,177].

In summary, vanadate is indeed able to mimic phosphate in various nucleotides and cofactors and thereby induce biological responses. The impact of these species in vivo remains to be determined.

6.3. EDTA and Other Ethylenediamine-Derived Ligands

The reactions of vanadate with EDTA are of interest because this ligand and related compounds are often additives in biological studies with vanadate. In addition, complexes with ethylenediamine-derived ligands show the types of chemistry that vanadate may participate in when coordinated to polydentate biogenic ligands such as proteins.

Crystallographic characterization of the monoanionic and trianionic complexes of ethylenediamine-*N,N,N',N'*-tetraacetatodioxovanadium(+5) (V-EDTA) have been described (Fig. 10) [178,179). ^{1}H and ^{13}C NMR studies confirm that the V-EDTA complex in solution has the same structure as that characterized in the solid state [180,181]. The kinetics for formation of the V-EDTA complex have been reported, and the second-order rate constant for reaction of the dianionic form of EDTA with $H_2VO_4^-$ is 10 times larger than when reacting with HVO_4^{2-} [126]. The first-order rate constant for V-EDTA hydrolysis was measured; the complex reacts 20 times faster with OH^- than with H_2O [182]. Given the kinetic lability of the V-EDTA complex, the solution stability directly reflects the great thermodynamic stability of the complex [183,184].

The oxidation of the vanadium(+4)-EDTA complex to the vanadium(+5)-EDTA complex has been examined from pH 3 to 4.3 and the reaction scheme for the oxidation with hexachloroiridate ($[IrCl_6]^{2-}$) is shown in Fig. 11 [185,186]. This particular reaction is of interest because some structural rearrangements take place when the vanadium(+4)

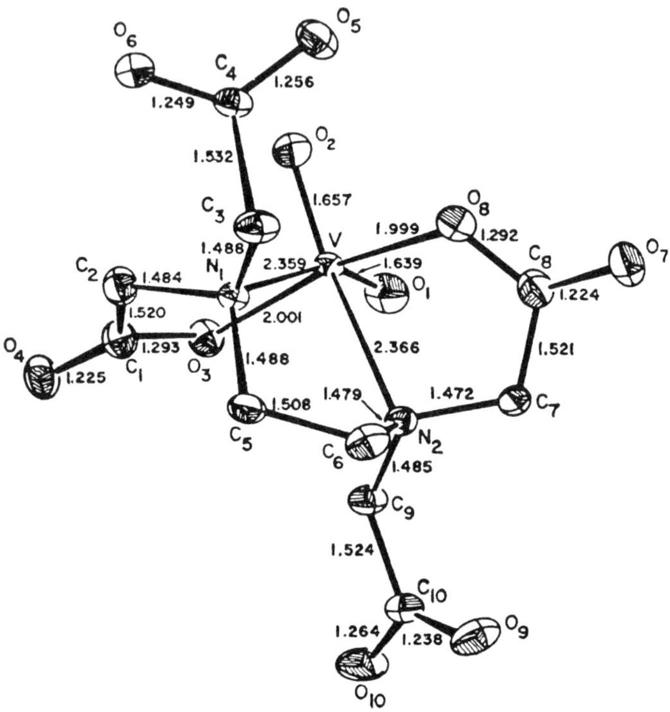

FIG. 10. Graphical representation of $VO_2(EDTA)^{3-}$. Adapted with permission from the American Chemical Society from Ref. 179.

complex with a pentadentate EDTA ligand [187] is converted to the vanadium(+5) complex with a tetradentate EDTA ligand. The overall rate of the redox reaction (k_a and k_b) is two orders of magnitude greater than the rate of the ligand substitution reaction in either of the vanadium-EDTA complexes. It was concluded that the outer sphere oxidation of the vanadium(+4)-EDTA complex by $[IrCl_6]^{2-}$ occurs in two steps transfer followed by the incorporation of the oxo ligand at the basal site in the complex [185,186].

When the vanadium atom is bound to a protein, several amino acid residues in the vicinity of the binding site are likely to participate, and these interactions will modify the reactivity of the metal. Polydentate ligands have been designed to probe the reactivity of complexes containing various functionalities including alkoxides, phenols, imidazoles, and

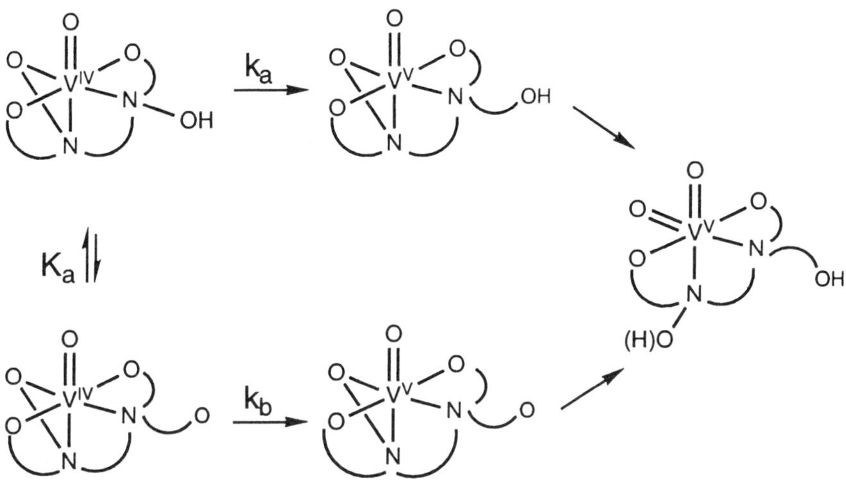

FIG. 11. Oxidation of $[VO(EDTA)]^{2-}$ with hexachloroiridate ($[IrCl_6]^{2-}$) to $[VO_2(EDTA)]^-$ (and other protonated forms) in aqueous solution at pH 3 to 4.5. Modified and redrawn from Ref. 185.

amino acids. Examples of such ligands include *N,N'*-ethylenebis[(*o*-hydroxyphenyl)glycine] (EHPG) [31,188], *N*[2-(*o*-salicylideneamino)ethyl](*o*-hydroxyphenyl)glycine (EHGS) [31], *N,N'*-disalicylideneethylenediamine (SALEN) [188], *N*-(3-methyl-6-*t*-butylsalicylaldehyde)-*N'*-methylethylenediamine [189], *N*-(salicylideneaminato)-*N'*-(2-hydroxyethyl)ethylenediamine (H_2SHED) [190,191], and *N*-(salicylideneaminato)ethylenediamine (HENSAL) [191]. These ethylenediamine-derived ligands have been found to generate vanadium(+5) complexes that are particularly amenable to characterization, although this list of ligands is not exhaustive [31,188,189,191].

The ligation, solvent, and pH play an important role in determining the stability of vanadium(+4) and (+5) complexes. In general, square-pyramidal, coordinatively unsaturated V(+4) complexes are more difficult to oxidize than their coordinatively saturated octahedral counterparts [188,192]. This observation reflects the behavior of the harder vanadium(+5) center, which strongly prefers octahedral coordination in many complexes with multidentate ligands.

Given the tendency the vanadate-1,2-diolates have for dimerization (Sec. 5.1), the nuclearity and coordination geometry around the vanadium(+5) nucleus was investigated using a polydentate ligand. The ligand *N*-(3-methyl-6-*t*-butylsalicylaldehyde)-*N'*-methylethylenediamine formed a complex which upon isolation and structural characterization proved to be mononuclear with the vanadium atom in a trigonal bipyramidal geometry (Fig. 12) [189]. A dimer with the vanadium atom in an octahedral geometry was obtained when the less sterically hindered ligand *N*-salicylaldehyde-*N'*-methylethylenediamine was used. The formation of the monomer and its tendency to dimerize clearly underscore the role of the protein in the interactions with the uridine-vanadate transition state analog.

The aqueous reaction of vanadate with catechol derivatives results in vanadium(+5) complexes in which the ligands undergo oxidation while reducing the vanadium. Polydentate ligands containing a catechol or a phenol moiety, EHPG, EHGS, and SALEN, were used to demonstrate that stable vanadium(+5) catecholates and phenolates can form [31,188]. Upon addition of chelating ligands, the rate of ligand oxidation has been reduced. The greater stability of the complexes formed from the catechol- and phenol-type ligands is in agreement with the observation that these types of ligands have stabilized the highest available oxidation state in complexes with other metals. Vanadium

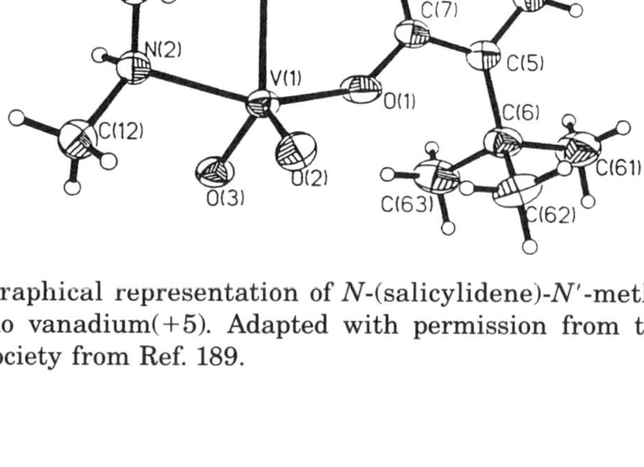

FIG. 12. Graphical representation of *N*-(salicylidene)-*N'*-methyl-ethylenediaminodioxo vanadium(+5). Adapted with permission from the American Chemical Society from Ref. 189.

complexes formed in biological fluids under aerobic conditions with these types of ligands should be most stable with the vanadium in oxidation state +5 [31].

Spectroscopic characterizations of complexes of a variety of these ligands illustrate that a very wide chemical shift range (up to 1200 ppm) in the ^{51}V NMR spectrum is to be expected for an enzyme-vanadium(+5) complex [191]. The observed chemical shifts for a series of model systems are linearly correlated with the inverse of the energy of the ligand-to-metal charge transfer (LMCT) bands in the visible and near-infrared spectra [191]. In addition, bare vanadium(+5) and VO^{3+} ions chelated to phenolate ligands all display LMCT bands in the visible region. Differences in the intensities of the electronic absorption bands provide a means to distinguish bare vanadium(+5) and +4 ions and VO^{3+} from the more common VO_2^+ and VO^{2+} [31,188].

Complexes containing this group of ligands often serve as excellent models for probing the interactions of vanadate with enzymes and other biological substances, even though many of the ligands do not exist in significant concentrations in biological systems. The ligands have been designed in such a manner that the chemistry of a particular functionality becomes tractable. In many cases, particularly when redox chemistry is involved, this approach provides an understanding of the chemistry when enzyme and receptor interact with vanadate.

6.4. Ethanolamine and Glycine-Derived Ligands

Interaction of vanadate with glycine is of interest since very little is known about the structural properties of complexes of amino acids and their derivatives. The interactions of vanadate with ethanolamine- and glycine-derived ligands are of further interest because these types of functionalities are often present in buffers [17]. In addition, vanadium complexes of these types of ligands have potential as chiral catalysts [193]. As described below, these types of complexes allow characterization of both the structural and redox properties of vanadium complexes.

Diethanolamine and related ligands form complexes with vanadate in aqueous solution with moderate formation constants [159,194–196]. Accordingly, in biological studies with buffers such as triethanolamine (TEA) and Tricine, most of the vanadate will be in a complexed form. Stable complexes form when the ligand contains the framework O-C-C-N-C-C-O in which various substitutions can be made on the carbon atoms [159]. Only one of the oxygen atoms can be replaced with other functionalities in the ligand if the stability of the vanadium(+5) complex is to remain high. Ligands in which one CH_2OH group has been replaced by a COOH group are amino acid derivatives and complexes formed from such ligands provide information on how simple amino acids coordinate to vanadate.

Few of the vanadium(+5) complexes have been structurally characterized [197,198], although several vanadium(+4) complexes are known (see, for example, Refs. [185,197]). A few examples of mixed-valent vanadium(+4) and (+5) systems have also been described [199]. The ligand *N*-[1,(2-pyridyl)ethyl]iminodiacetate is particularly interesting since the vanadium(+4), vanadium(+5), and mixed-valent complexes have all been structurally characterized (Fig. 13) [199].

Ethanolamine complexes and derivatives were first prepared from $VO(OCH_2CH_3)_3$ in organic solvents and characterized by elemental analysis [200]. Recently, the TEA-oxovanadium complex was characterized in the solid state and in organic solution (Fig. 14) [198,201]. Surprisingly, in the organic solution and in the solid state this complex was found to have a different structure than in aqueous solution (Fig. 14) [198]. The vanadium atom is in a distorted trigonal bipyramidal geometry in both structures; however, the TEA ligand is tetradentate in the solid state and organic solution, whereas in aqueous solution it is tridentate. Similar observations were made for the tri-2-propanolamine-oxovanadium complex in organic solution and the solid state compared to the complex in aqueous solution [198]. These findings show the importance of solution characterization of vanadium complexes in combination with solid state characterization; recent reports are addressing various aspects of solution chemistry and reactivity of these and related systems [24,181].

Kinetic studies have been carried out on vanadium(+5) complexes of a variety of ligands including EDTA, EDDA, iminodiacetic acid, *N*-methyliminodiacetic acid, and nitrilotriacetic acid using stopped-flow techniques [126,182,202–204] and diethanolamine, TEA, *N*-[tris-(hydroxymethyl)methyl]glycine (Tricine), *N*-hydroxyethyl iminodiacetic acid and EDTA using exchange spectroscopy (EXSY) [24,184,196]. Despite the thermodynamic stability of these complexes, they are found to

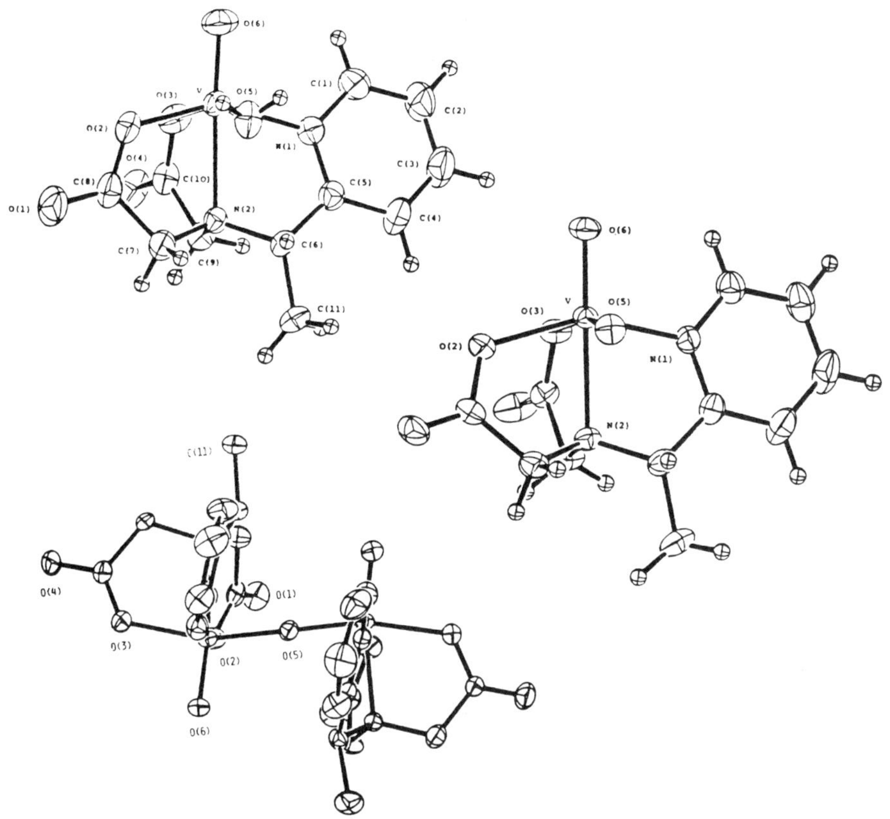

FIG. 13. Graphical representation of [$VO(peida)(H_2O)$] (top left), [$VO_2(peida)$] (right) and [$V_2O_3(peida)_2$] (bottom left). The abbreviation "peida" stands for *N*-[1-(2-pyridyl)ethyl]iminodiacetate. Adapted with permission from the American Chemical Society from Ref. 197.

be surprisingly labile in aqueous solution. This lability is illustrated by a ^{13}C EXSY spectrum showing the intra- and intermolecular exchange that is observed for the aqueous V-TEA complex (Fig. 15). The cross-peaks between the ^{13}C signals of the bound hydroxyethyl arms (C_{4b} and C_{3b}) in the complex and the ^{13}C signal for the free ligand (L_4 and L_3) describe the intermolecular exchange between the complex and the free ligand (connected with a solid line). In addition, cross-signals between the ^{13}C signals of the hydroxyethyl arms (C_{4b} and C_{4f}) illustrate a

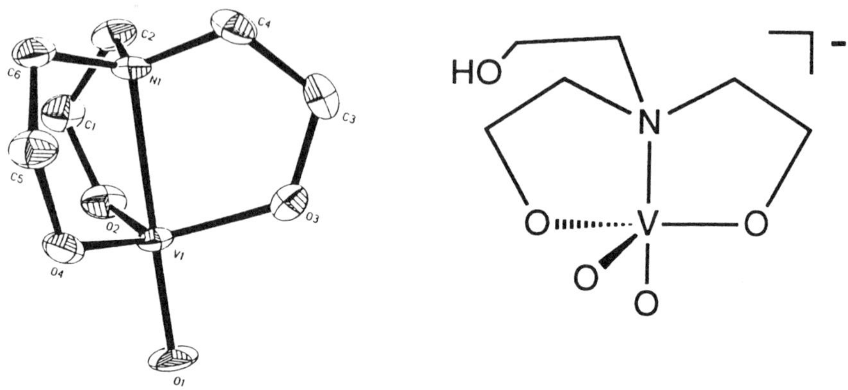

FIG. 14. Graphical representation of $VO(OCH_2CH_2)_3N$ (left) and the corresponding complex in aqueous solution $[VO_2(OCH_2CH_2)_2N(CH_2CH_2OH)]^-$ (right). Adapted with permission from the American Chemical Society from Ref. 198.

second mode of exchange (connected with a dotted line), namely, the intramolecular exchange within the complex [24].

Vanadium derivatives as well as other metal derivatives from this group of compounds have uses in organic synthesis, particularly when generated in a chiral manner [193]. The oxidation of *N*-(phosphonomethyl)iminodiacetic acid to generate *N*-(phosphonomethyl)glycine (the herbicide Roundup) was studied in detail and provides mechanistic information on the reactivity of these types of compounds [205]. The reaction at acidic pH was first order in *N*-(phosphonomethyl)iminodiacetic acid, total vanadium concentration, and H^+ concentration. Key mechanistic features of this reaction were determined beginning with the oxidation of the vanadium(+4) complex (by oxygen) to the vanadium(+5) complex. Oxidation of the carboxyl moiety by the vanadium(+5) followed by decarboxylation yields an intermediate carbon-centered *N*-methylene-*N*-(phosphonomethyl)glycine radical. Trapping of this radical with O_2 eventually results in the formation of *N*-formyl-*N*-(phosphonomethyl)glycine which hydrolyzes to *N*-(phosphonomethyl)-glycine. The reaction suffered from lower selectivities because the *N*-methylene-*N*-(phosphonomethyl)glycine radical abstracted H leading to *N*-methyl-*N*-(phosphonomethyl)glycine.

Variations in the ligand framework have resulted in complexes probing different aspects of vanadium(+5) chemistry. Ligands derived

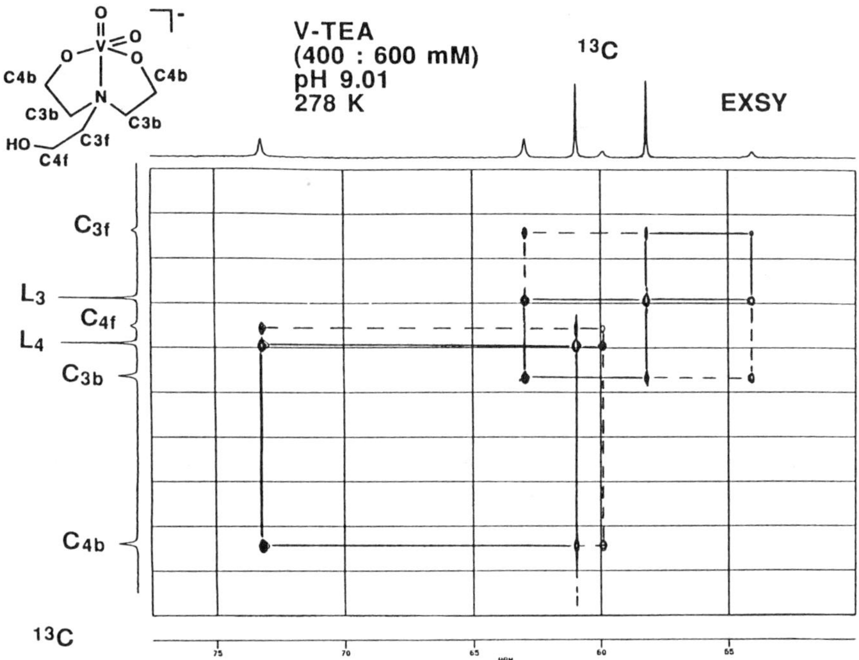

FIG. 15. 2D ^{13}C EXSY NMR spectrum recorded at 75.4 MHz of a solution containing 400 mM total vanadate and 600 mM TEA at pH 9.01. The labeling scheme for the carbon atoms is shown, and the corresponding carbon atoms in the free ligand are labeled L_3 and L_4. Exchange off-resonance signals reflecting intermolecular exchange are connected by solid lines, and exchange off-resonance signals reflecting intramolecular exchange are connected by dotted and solid lines. Adapted with permission from the American Chemical Society from Ref. 24.

from imine formation with ethanolamine or amino acids contain the framework O–C–C–C=N–C–C–O, and have been used to examine reactivity and structural aspects of vanadium complexes with the imidazole functionality [206,207]. A vanadium(+5) complex of [4-(2-salicylideneaminato)ethyl]imidazole (HSALIMH) shows the coordination of the imidazole to an octahedral vanadium atom [207] (Fig. 16). A Schiff base ligand generated between *o*-hydroxynaphthalenecarbaldehyde and histidine formed a vanadium complex in which the vanadium(+5) atom in a trigonal bipyramidal geometry with a pendant imidazole group

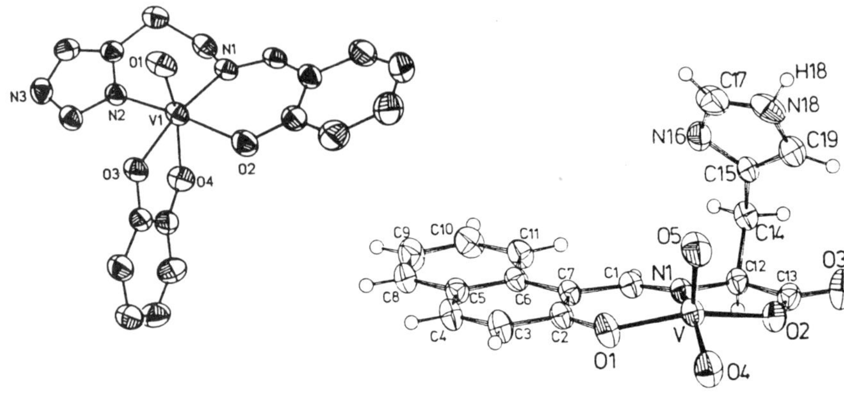

FIG. 16. Graphical representation of VO(cat)(HSALIMH) (left) and [VO_2{N-(2-oxidonaphthal)(H^+)HisO$^-$}] (right). See list of abbreviations for explanation of the ligands. Adapted with permission from the American Chemical Society from Refs. [206] and [207].

[206] (Fig. 16). The latter observation is particularly interesting when recognizing that the imidazole functionality was protonated and accordingly could not coordinate to the vanadium atom. These two model studies together point to the versatility of vanadium(+5) coordination chemistry and provide structural precedence for possible interactions with biogenic histidine-containing peptidic ligands.

In summary, not only have a series of compounds generated from this group of ligands been structurally characterized, but the stability, lability, reactivity, and mechanisms of reaction of selected complexes have been documented providing information important to interpretation of the solution studies between vanadate and peptidic ligands.

6.5. Peptides and Proteins

Proteins and peptides are polymers of amino acids, and all of the functionalities in amino acid side chains have the potential of forming complexes with vanadate. In addition, the protein backbone, perhaps in combination with side chain functionalities, has potential for forming complexes with vanadate.

The reaction between vanadate and simple dipeptides at neutral pH did indeed form a complex which was more stable than the complexes of both the individual amino acids [160–163]. Although a particular amino acid could affect the chemical shift somewhat, most vanadate-dipeptide complexes are in the range of −507 ppm and the formation constant at pH 7.0 ranges from 20 to 70 M^{-1}. The original report examined a series of dipeptides probing the possibility of various perturbations in the dipeptide structure [160]. The peptide amide was essential for complex formation since alkylation of the nitrogen eliminated complex formation at −507 ppm [160]. The terminal amine and the carboxylate groups were implicated in complex formation because protection of the free groups did not result in a signal at −507 ppm [160,162]. Evidence utilizing both ^{13}C and ^{15}N NMR spectroscopies support the above conclusions [164]. At this time only model compounds (see Secs. 6.3 and 6.4) have been structurally characterized. However, the literature favors complexes with octahedral vanadium atoms akin to that shown in Fig. 17 [160,162,181,208].

At alkaline pH a complex at −498 ppm is formed from vanadate and dipeptide. The latter complex generated an H^+ upon formation. Dipeptides containing serine and tyrosine form additional complexes that are observable in the ^{51}V NMR spectra. It appears that tripeptides can also from stable complexes, although details in this reaction have yet to be determined [161,162]. In addition, it appears that longer peptide sequences and even proteins show significant affinity for various forms of oxovanadates. This type of interaction is described in detail in Chap. 8 and [18].

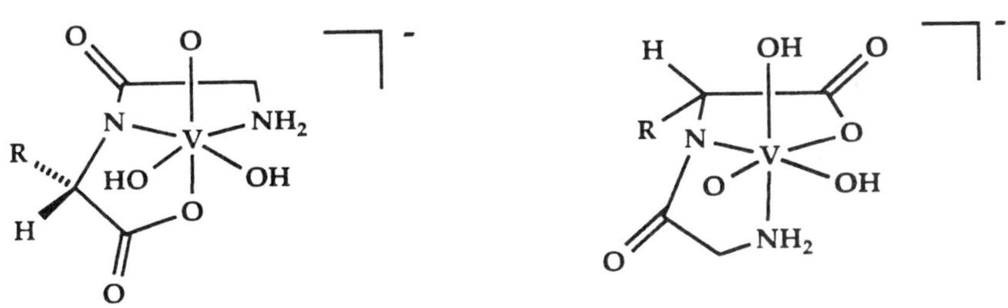

FIG. 17. Schematic representation of likely structural possibilities for the aqueous complex between vanadate and gly-X, where X could be any amino acid.

Studies characterizing the formation of the vanadate-gly-tyr complex show that these types of complexes distinguish themselves from the labile glycine complexes described in Secs. 6.3 and 6.4. The complex at −507 ppm formed from gly-tyr and vanadate forms over several hours and furthermore is fairly hydrolytically stable [164]. The reaction of vanadate and gly-ser to form the −507 ppm complex also needs a few hours to reach equilibrium. The complex at −498 ppm forms even more slowly, and accordingly caution must be taken to assure that such solutions have had sufficient time to reach equilibrium. It was suggested that the amide nitrogen is instrumental to the added stability of these vanadium complexes and shows potential for the future development of hydrolytically stable vanadium derivatives.

6.6. A Natural Chelator: Citrate

Throughout this chapter we have discussed a variety of naturally occurring organic substances grouped according to their specific functionalities. A few naturally occurring chelators are exceptional. Citrate is an important metabolite (citric acid cycle) and in addition is one of the most abundant natural chelators in cellular fluids. On the other hand, lower organisms produce specific chelators, siderophores, which aid the transport of metal ions. The interaction of vanadate with siderophores and their functionalities is described in Secs. 5.2 and 5.4. The interaction of vanadate with citrate is described here.

Vanadate interacts with citrate to form several products as would be expected based on the functionalities present in citric acid [209]. The three major complexes [131] are described as (1,2,1), (2,2,1), and (3,2,1) species and each contains two vanadium atoms and one citrate ligand. Their formation constants are $\log \beta_{1,2,1} = 12.84$, $\log \beta_{2,2,1} = 19.68$, and $\log \beta_{3,2,1} = 24.12$. At pH 7 the major species present are the (1,2,1) and the (2,2,1) species. At micromolar vanadate and citrate the species (3,1,1) becomes the major species in the system. Structural information is available for a peroxovanadium(+5) citrate complex [52], and a structure was proposed for the (1,2,1) species [209]. Both of these structures are shown in Fig. 18. The vanadate-citrate complex is very stable, and micromolar concentrations of citrate are sufficient to reactivate phosphatases inhibited by vanadate [161].

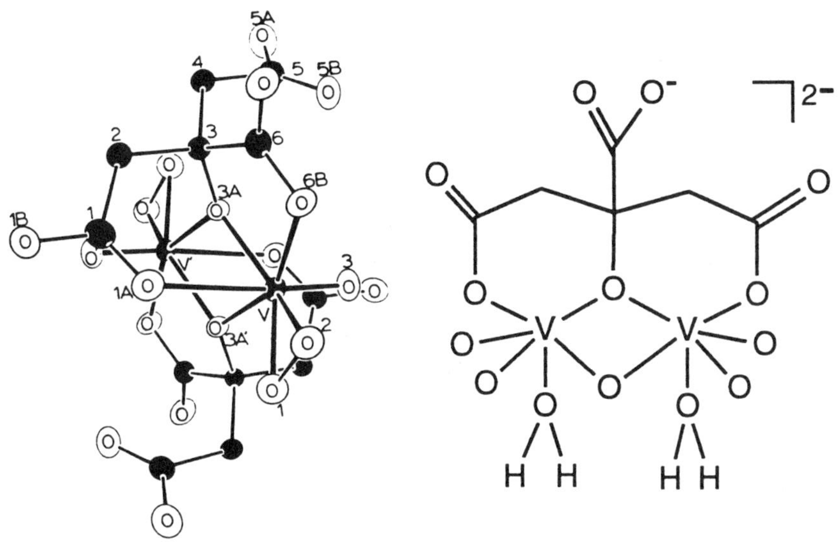

FIG. 18. Graphical representation of $[VO(O_2)(C_6H_6O_7)]_2^{2-}$ (left; Adapted with permission from the American Chemical Society from Ref. 52) and a schematic drawing of the structure proposed for the (1,2,1) vanadate-citrate complex [209].

7. FINAL COMMENTS

Interest in the interaction of vanadates with biogenic ligands in aqueous environments is fueled by the developments in vanadium biochemistry and biology, and selected topics are of particular importance at this time. The insulin-mimetic activities of vanadate and other vanadium compounds are currently under examination in human clinical trials as an oral substitute for insulin. Application of vanadate as a tool in studies of signal transduction continues to increase and provides mechanistic information that is otherwise unavailable. Application of vanadate derivatives as phosphate ground state and transition state analogs in enzymology and biochemistry is likely to continue spurred by the X-ray structures of enzymes with vanadate derivatives. Additional information concerning the interactions of vanadate with simple biogenic ligands will be needed in the future so that the most potent and biologically significant interactions can be identified. Not only are thermodynamic and structural properties of the compounds that form between vanadate and biogenic ligands important, but the lability and the redox properties of these complexes are essential to vanadium biochemistry.

Although vanadium may or may not be an essential element for higher mammals, enzymes have been identified from microorganisms and algae in which vanadium is a cofactor. Vanadium-dependent haloperoxidases and nitrogenases both utilize vanadium as a cofactor, albeit in a very different manner. Mimicking the vanadium chemistry of the haloperoxidase has led to functional mimics that probe the enzyme mechanism and may also have implications in organic synthesis. The interest in organisms that accumulate vanadium and various aspects of their chemistry will continue to intrigue chemists. Although the vanadium inside these species has oxidation states lower than +5, vanadate is the form that is absorbed by these organisms. The flexibility in the coordination chemistry of vanadium(+5) clearly contributes to the rich chemistry and biochemistry in all these cases.

Various aspects of vanadium chemistry described in this chapter will continue to develop at a very rapid pace. This chapter has brought together aqueous studies of systems that potentially can be observed in humans, animals, plants, and microorganisms with systems that are manmade and designed to discover and probe properties of the fundamental unit. In some cases the analogy may seem remote, although difficulties in modeling the biological systems may simply reflect the many subtle factors influencing and governing the dominant chemistry under a particular set of conditions. There is no doubt that interplay between the information obtained in aqueous and "natural" systems combined with model systems will be important to ultimately reach an understanding of the reactions of vanadates with biogenic ligands.

ACKNOWLEDGMENTS

DCC thanks the National Institutes of Health, the Sloan Foundation and the Colorado State Experiment Station for funds. DCC also thanks Drs. Christopher R. Roberts, Mohammad Mahroof-Tahir and Alan Tracey for reading this manuscript.

ABBREVIATIONS

ACAC	acetylacetonate
ADP	adenosine 5′-diphosphate
AMP	adenosine 5′-monophosphate
AMV	adenosine 5′-monovanadate
tBu	*tert*-butyl
cat	catechol
CD	circular dichroism
CTP	cytidine 5′-triphosphate
DTT	dithiothreitol
EDDA	ethylenediamine-*N*,*N*-diacetate
EDTA	ethylenediamine-*N*,*N*,*N*′,*N*′-tetracetate
EHGS	*N*-[2-(*o*-salicylideneamino)ethyl](*o*-hydroxy-phenyl)glycine
EHPG	*N*,*N*′-ethylenebis[(*o*-hydroxyphenyl)glycine]
EPR	electron paramagnetic resonance
EXSY	exchange spectroscopy
GDP	guanosine 5′-diphosphate
HSALIMH	[4-(2-salicylideneaminato)ethyl]imidazole
HENSAL	*N*-(salicylideneaminato)ethylenediamine
$HisO^-$	deprotonated histidine
H_2SHED	*N*-(salicylideneaminato)-*N*′-(2-hydroxyethyl)-ethylenediamine
IR	infrared
LMCT	ligand-to-metal charge transfer
MCD	magnetically induced circular dichroism
NAD	β-nicotinamide adenine dinucleotide
NADP	β-nicotinamide adenine dinucleotide phosphate
NADV	β-nicotinamide adenine dinucleotide vanadate
NMR	nuclear magnetic resonance
2-oxidonaphthal	*o*-hydroxynaphthalenecarbaldehyde or 2-hydroxy-1-naphthaldehyde
peida	*N*-[1-(2-pyridyl)ethyl]iminodiacetate
P_i	inorganic phosphate
PPP	linear triphosphate
iPr	isopropyl

Q	8-quinolinate
SALEN	*N,N'*-disalicylideneethylenediamine
TEA	triethanolamine
Tris	tris(hydroxymethyl)aminoethane
Tricine	*N*-[tris(hydroxymethyl)methyl]glycine
UV-VIS	ultraviolet-visible
V_1	vanadate monomer
V_2	vanadate dimer
V_4	vanadate tetramer
V_5	vanadate pentamer
V_{10}	vanadate decamer

REFERENCES

1. N. D. Chasteen, in *Vanadium in Biological Systems: Physiology and Biochemistry* (N. D. Chasteen, ed.), Kluwer Academic, Boston, 1990.
2. B. R. Nechay, L. B. Nanninga, P. S. E. Nechay, R. L. Post, J. J. Grantham, I. G. Macara, L. F. Kubena, T. D. Phillips, and F. H. Nielsen, *Fed. Proc.*, *45*, 123–132 (1986).
3. M. T. Pope, *Heteropoly and Isopoly Oxometalates*, Springer-Verlag, New York, 1983.
4. N. D. Chasteen, *Structure and Bonding*, *53*, 105–138 (1983).
5. M. J. Gresser, A. S. Tracey, and P. J. Stankiewicz, *Adv. Prot. Phosphatases*, *4*, 35–57 (1987).
6. D. W. Boyd and K. Kustin, *Adv. Inorg. Biochem.*, 311–365 (1984).
7. K. Kustin, G. C. McLeod, T. R. Gilbert, and L. B. R. Briggs IV, *Structure and Bonding*, *53*, 139–161 (1983).
8. R. Wever and K. Kustin, *Adv. Inorg. Chem.*, *35*, 81–115 (1990).
9. D. Rehder, *BioMetals*, *5*, 3–12 (1992).
10. D. Rehder, *Angew. Chem. Int. Ed. Engl.*, *30*, 148–167 (1991).
11. O. W. Howarth, *Prog. NMR Spec.*, *22*, 453–485 (1990).
12. A. Butler and C. J. Carrano, *Coord. Chem. Rev.*, *109*, 62–105 (1991).
13. C. E. Holloway and M. Melnik, *Rev. Inorg. Chem.*, *7*, 75–159 (1985).

14. V. Conte and F. Di Furia, in *Catalytic Oxidations with Hydrogen Peroxide as Oxidant* (G. Strukul, ed.), Kluwer Academic, Dordrecht, 1992, pp. 223–252.
15. A. Butler, M. J. Clague, and G. Meister, *Chem. Rev.*, *3*, 625–638 (1994).
16. Q. Chen and J. Zubieta, *Coord. Chem. Rev.*, *114*, 107–167 (1992).
17. D. C. Crans, *Comm. Inorg. Chem.*, *16*, 1–33 (1994).
18. D. C. Crans, *Comm. Inorg. Chem.*, *16*, 35–76 (1994).
19. E. Heath and O. W. Howarth, *J. Chem. Soc., Dalton Trans.*, 1105–1110 (1981).
20. M. A. Habayeb and O. E. Hileman, Jr., *Can. J. Chem.*, *58*, 2255–2261 (1980).
21. L. Pettersson, I. Andersson, and B. Hedman, *Chem. Scrip.*, *25*, 309–317 (1985).
22. D. C. Crans, S. M. Schelble, and L. A. Theisen, *J. Org. Chem.*, *56*, 1266–1274 (1991).
23. D. C. Crans, C. D. Rithner, and L. A. Theisen, *J. Am. Chem. Soc.*, *112*, 2901–2908 (1990).
24. D. C. Crans, P. K. Shin, and K. B. Armstrong, *Mechanistic Bioinorganic Chemistry* (H. Thorp and V. Pecoraro, eds.), American Chemical Society, Washington, DC, in press.
25. J. A. Konnert and H. T. Evans Jr., *Acta Crystallogr.*, *B31*, 2688–2690 (1975).
26. J. Fuchs, S. Mahjour, and J. Pickardt, *Angew. Chem. Int. Ed. Engl.*, *15*, 374–375 (1976).
27. V. W. Day, W. G. Klemperer, and A. Yagasaki, *Chem. Lett.*, 1267–1270 (1990).
28. A. S. Tracey, M. J. Gresser, and B. Galeffi, *Inorg. Chem.*, *27*, 157–161 (1988).
29. S.E. Harnung, E. Larsen, and E. J. Pedersen, *Acta Chem. Scand.*, *47*, 674–682 (1993).
30. S. R. Cooper, Y. B. Koh, and K. N. Raymond, *J. Am. Chem. Soc.*, *104*, 5092–5102 (1982).
31. J. A. Bonadies and C. J. Carrano, *J. Am. Chem. Soc.*, *108*, 4088–4095 (1986).
32. Y. Zhang and R. H. Holm, *Inorg. Chem.*, *29*, 911–917 (1990).

33. H. T. Evans Jr., *Inorg. Chem.*, *5*, 967–977 (1966).
34. L. Pettersson, I. Andersson, and O. W. Howarth, *Inorg. Chem.*, *31*, 4032–4033 (1992).
35. D. Hou, K. S. Hagen, and C. L. Hill, *J. Am. Chem. Soc.*, *114*, 5864–5866 (1992).
36. M. I. Khan, Q. Chen, H. Höpe, S. Parkin, C. J. O'Connor, and J. Zubieta, *Inorg. Chem.*, *32*, 2929–2937 (1993).
37. A. Müller, R. Rohlfing, E. Krickemeyer, and H. Bögge, *Angew. Chem. Int. Ed. Engl.*, *32*, 909–912 (1993).
38. M. J. Gresser, A. S. Tracey, and K. M. Parkinson, *J. Am. Chem. Soc.*, *108*, 6229–6234 (1986).
39. S. Ohashi, *Topics in Phosphorus Chemistry* (M. Grayson and E. J. Griffith, eds.), Interscience, New York, 1964, pp. 189–240.
40. R. U. Russell, J. E. Salmon, and H. R. Tietze, *J. Chem. Soc.*, 3211–3214 (1961).
41. T. Imamura, D. M. Hinton, R.L. Belford, R. I. Gumport, and G. P. Haight, Jr., *J. Inorg. Biochem.*, *11*, 241–259 (1979).
42. A. S. Tracey, M. J. Gresser, and S. Liu, *J. Am. Chem. Soc.*, *110*, 5869–5874 (1988).
43. C. C. Goodno, *Proc. Natl. Acad. Sci. USA*, *76*, 2620–2624 (1979).
44. W. L. Combest and R. A. Johnson, *Arch. Biochem. Biophys.*, *225*, 916–927 (1983).
45. G. M. Woltermann, R. A. Scott, and G. P. Haight Jr., *J. Am. Chem. Soc.*, *96*, 7569–7570 (1974).
46. S. I. Liochev and I. Fridovich, *Arch. Biochem. Biophys*, *279*, 1–7 (1990).
47. O. W. Howarth and D. Winter, *J. Chem. Soc. Dalton Trans.*, 399–402 (1991).
48. A. T. Harrison and O. W. Howarth, *J. Chem. Soc. Dalton Trans.*, 1173–1177 (1985).
49. N. J. Campbell, A. C. Dengel, and W. P. Griffith, *Polyhedron*, *8*, 1379–1386 (1989).
50. J. S. Jaswal and A. S. Tracey, *Inorg. Chem.*, *30*, 3718–3722 (1991).
51. R. E. Drew, F. W. B. Einstein, J. S. Field, and D. Begin, *Acta Crystallogr.*, *31A*, 5135 (1975).

52. C. Djordjevic, M. Lee, and E. Sinn, *Inorg. Chem.*, *28*, 719–723 (1989).

53. H. Mimoun, L. Saussine, E. Daire, M. Postel, J. Fischer, and R. Weiss, *J. Am. Chem. Soc.*, *105*, 3101–3110 (1983).

54. V. Conte, Vanadium Biochemistry Symposium, in Southeast Regional American Chemical Society Meeting, Johnson City, Tennessee, 1993.

55. R. I. de la Rosa, M. J. Clague, and A. Butler, *J. Am. Chem. Soc.*, *114*, 760–761 (1992).

56. M. J. Clague, N. L. Keder, and A. Butler, *Inorg. Chem.*, *32*, 4754–4761 (1993).

57. G. J. Colpas, B. J. Hamstra, J. W. Kampf, and V. L. Pecoraro, *Inorg. Chem.*, in press.

58. S. Funahashi, M. Inamo, and M. Tanaka, *Inorg. Chim. Acta*, *157*, 65–71 (1989).

59. A. S. Tracey and J. S. Jaswal, *J. Am. Chem. Soc.*, *114*, 3835–3840 (1992).

60. M. Bonchio, V. Conte, F. Di Furia, and G. Modena, *J. Org. Chem.*, *54*, 4368–4371 (1989).

61. A. Shaver, J. B. Ng, D. A. Hall, B. S. Lum, and B. I. Posner, *Inorg. Chem.*, *32*, 3109–3113 (1993).

62. W. Priebsch and D. Rehder, *Inorg. Chem.*, *29*, 3013–3019 (1990).

63. F. W. B. Einstein, E. Enwall, D. M. Morris, and D. Sutton, *Inorg. Chem.*, *10*, 678–686 (1971).

64. R. C. Hibbert, N. Logan, and O. W. Howarth, *J. Chem. Soc. Dalton Trans.*, 369–372 (1986).

65. W. Priebsch and D. Rehder, *Inorg. Chem.*, *24*, 3058–3062 (1985).

66. M. T. P. Viet, V. Sharma, and J. D. Wuest, *Inorg. Chem.*, *30*, 3026–3032 (1991).

67. M. J. Gresser and A. S. Tracey, *J. Am. Chem. Soc.*, *107*, 4215–4220 (1985).

68. A. S. Tracey and M. J. Gresser, *Can. J. Chem.*, *66*, 2570–2574 (1988).

69. M. J. Gresser and A. S. Tracey in *Vanadium in Biological Systems: Physiology and Biochemistry* (N. D. Chasteen, ed.), Kluwer Academic, Boston, 1990, pp. 63–79.

70. D. C. Crans, C. M. Simone, and J. S. Blanchard, *J. Am. Chem. Soc.*, *114*, 4926–4928 (1992).

71. A. S. Tracey, B. Galeffi, and S. Mahjour, *Can. J. Chem.*, *66*, 2294–2298 (1988).

72. A. S. Tracey and M. J. Gresser, *Inorg. Chem.*, *27*, 2695–2702 (1988).

73. C. N. Caughlan, H. M. Smith, and K. Watenpaugh, *Inorg. Chem.*, *5*, 2131–2134 (1966).

74. F. Hillerns, F. Olbrich, U. Behrens, and D. Rehder, *Angew. Chem. Int. Ed. Engl.*, *4*, 447–448 (1992).

75. F. J. Feher and J. F. Walzer, *Inorg. Chem.*, *30*, 1689–1694 (1991).

76. D. C. Crans, H. Chen, and R. A. Felty, *J. Am. Chem. Soc.*, *114*, 4543–4550 (1992).

77. A. Lachowicz and K.-H. Thiele, *Z. Anorg. Allgem. Chem.*, *434*, 271–276 (1977).

78. A. Lachowicz, W. Höbold, and K.-H. Thiele, *Z. Anorg. Allgem. Chem.*, *418*, 65–71 (1975).

79. C. Weidemann and D. Rehder, *Inorg. Chim. Acta*, *120*, 15–20 (1986).

80. O. W. Howarth and J. R. Trainor, *Inorg. Chim. Acta*, *127*, L27–L28 (1987).

81. F. Hillerns and D. Rehder, *Chem. Ber.*, *124*, 2249–2254 (1991).

82. D. Rehder in *Polyoxometalates: From Platonic Solids to Anti-Retroviral Activity* (A. Müller and M. T. Pope, eds.), Kluwer Academic, Dordrecht, 1993, pp. 157–166.

83. A. S. Tracey and M. J. Gresser, *Proc. Natl. Acad. Sci. USA*, *83*, 609–613 (1986).

84. B. Galeffi and A. S. Tracey, *Can. J. Chem.*, *66*, 2565–2569 (1988).

85. A. S. Tracey, H. Li, and M. J. Gresser, *Inorg. Chem.*, *29*, 2267–2271 (1990).

86. R. C. Paul and A. Kumar, *J. Inorg. Nucl. Chem.*, *27*, 2537–2547 (1965).

87. D. Rehder, W. Priebsch, and M. von Oeynhausen, *Angew. Chem. Int. Ed. Engl.*, *28*, 1221–1222 (1989).

88. D. D. Heinrich, K. Folting, W. E. Streib, J. C. Huffman, and G. Christou, *J. Chem. Soc., Chem. Commun.*, 1411–1413 (1989).

89. W. Priebsch, D. Rehder, and M. von Oeynhausen, *Chem. Ber.*, *124*, 761–764 (1991).

90. B. Galeffi and A. S. Tracey, *Inorg. Chem.*, *28*, 1726–1734 (1989).
91. Z. Shi, Z. Sun, and N. S. Dalal, *FEBS Lett.*, *271*, 185–188 (1990).
92. H. Sakurai, S. Shimomura, and K. Ishizu, *Inorg. Chim. Acta*, *55*, L67–L69 (1981).
93. R. J. Keller, R. A. Coulombe, R. P. Sharma, T. A. Grover, and L. H. Piette, *Arch. Biochem. Biophys.*, *271*, 40–48 (1989).
94. D. C. Crans and C. M. Simone, *Biochemistry*, *30*, 6734–6741 (1991).
95. D. C. Crans, K. Sudhakar and T. J. Zamborelli, *Biochemistry*, *31*, 6812–6821 (1992).
96. J. A. Kovacs and R. H. Holm, *Inorg. Chem.*, *26*, 702–711 (1987).
97. H. Sakurai, Z.-E. Taira, and N. Sakai, *Inorg. Chim. Acta*, *151*, 85–86 (1988).
98. Y.-P. Zhang and R. H. Holm, *Inorg. Chem.*, *27*, 3875–3876 (1988).
99. L. C. Cantley, Jr. and P. Aisen, *J. Biol. Chem.*, *254*, 1781–1784 (1979).
100. G. R. Willsky, D. A. White, and B. C. McCabe, *J. Biol. Chem.*, *259*, 13273–13281 (1984).
101. G. R. Willsky, in *Vanadium in Biological Systems: Physiology and Biochemistry* (N. D. Chasteen, ed.), Kluwer Academic, Boston, 1990, pp. 1–24.
102. K. Kustin and D. L. Toppen, *Inorg. Chem.*, *12*, 1404–1407 (1973).
103. N. D. Chasteen, J. K. Grady, and C. E. Holloway, *Inorg. Chem.*, *25*, 2754–2760 (1986).
104. M. J. Gresser and A. S. Tracey, *J. Am. Chem. Soc.*, *108*, 1935–1939 (1986).
105. R. N. Lindquist, J. L. Lynn, Jr., and G. E. Lienhard, *J. Am. Chem. Soc.*, *95*, 8762–8768 (1973).
106. C. F. G. C. Geraldes and M. M. C. A.Castro, *J. Inorg. Biochem.*, *35*, 79–93 (1989).
107. A. S. Tracey, J. S. Jaswal, M. J. Gresser, and D. Rehder, *Inorg. Chem.*, *29*, 4283–4288 (1990).
108. A. S. Tracey and C. H. Leon-Lai, *Inorg. Chem.*, *30*, 3200–3204 (1991).
109. X. Zhang and A. S. Tracey, *Acta Chem. Scand.*, *46*, 1170–1176 (1992).

110. D. C. Crans, S. E. Harnung, E. Larsen, P. K. Shin, L. A. Theisen, and I. Trabjerg, *Acta Chem. Scand.*, *45*, 456–462 (1991).
111. W. J. Ray, Jr., J. Zheng, H. Deng, J. W. Burgner III, M. Mahrood-Tahir, and D. C. Crans, submitted. (1994).
112. D. C. Crans, R. A. Felty, and M. M. Miller, *J. Am. Chem. Soc.*, *113*, 265–269 (1991).
113. T. Alber, W. A. Gilbert, D. R. Ponzi, and G. A. Petsko, *Ciba Found. Symp.*, *93*, 4–24 (1983).
114. A. Wlodawer, M. Miller, and L. Sjölin, *Proc. Natl. Acad. Sci.*, *80*, 3628–3631 (1983).
115. D. Kostrewa, H.-W. Choe, U. Heinemann, and W. Saenger, *Biochemistry*, *28*, 7592–7600 (1989).
116. T. W. Hambley, R. J. Judd, and P. A. Lay, *Inorg. Chem.*, *31*, 343–345 (1992).
117. D. C. Crans, R. A. Felty, O. P. Anderson, and M. M. Miller, *Inorg. Chem.*, *32*, 247–248 (1993).
118. M. Bartusek and U. Sustácek, *Collect. Czech. Chem. Commun.*, *48*, 2785–2797 (1983).
119. J. Richter and D. Rehder, *Z. Naturforschung*, 1613–1620 (1991).
120. D. C. Crans, R. A. Felty, H. Eckert, and N. Das, *Inorg. Chem.*, *33*, 2427–2438 (1994).
121. K. N. Raymond and C. J. Carrano, *Acc. Chem. Res.*, *12*, 183–190 (1979).
122. M. Siroki and C. Djordjevic, *Z. Anorg. Allgem. Chem.*, *430*, 277–282 (1977).
123. N. Galesic and M. Siroki, *Acta Crystallogr.*, *B35*, 2931–2937 (1979).
124. M. E. Cass, D. L. Greene, R. M. Buchanan, and C. G. Pierpont, *J. Am. Chem. Soc.*, *105*, 2680–2686 (1983).
125. A. R. Bulls, C. G. Pippin, F. E. Hahn, and K. N. Raymond, *J. Am. Chem. Soc.*, *112*, 2627–2632 (1990).
126. K. Kustin and D. L. Toppen, *J. Am. Chem. Soc.*, *95*, 3564–3568 (1973).
127. K. Kustin, S.-T. Liu, C. Nicolini, and D. L. Toppen, *J. Am. Chem. Soc.*, *96*, 7410–7415 (1974).
128. K. Kustin, C. Nicolini, and D. L. Toppen, *J. Am. Chem. Soc.*, *96*, 7416–7420 (1974).

129. J. H. Ferguson and K. Kustin, *Inorg. Chem.*, *18*, 3349–3357 (1979).

130. P. M. Ehde, I. Andersson, and L. Pettersson, *Acta Chem. Scand.*, *A40*, 489–499 (1986).

131. The (*p*, *q*, *r*) nomenclature describes complexes with the stoichiometry *p*, *q*, and *r* as defined by $(H^+)_p(H_2VO_4^-)_q(L)_r$. See Refs. 130 and 209 for further description of the uses of this nomenclature.

132. A. S. Tracey, M. J. Gresser, and K. M. Parkinson, *Inorg. Chem.*, *26*, 629–638 (1987).

133. P. M. Ehde, L. Pettersson, and J. Glaser, *Acta Chem. Scand.*, *45*, 998–1005 (1991).

134. W. R. Scheidt, C.-C. Tsai, and J. L. Hoard, *J. Am. Chem. Soc.*, *93*, 3867–3872 (1971).

135. V. H. Rieskamp and R. Mattes, *Z. Anorg. Allgem. Chem.*, *419*, 193–199 (1976).

136. J. J. R. Fraústo da Silva, *Chem. Spec. Bioavail.*, *1*, 139–150 (1989).

137. G. D. Lutwick and D. E. Ryan, *Can. J. Chem.*, *32*, 949–955 (1954).

138. R. R. Nanewar and U. Tandon, *Talanta*, *25*, 352–354 (1978).

139. Y. K. Agrawal, *Bull. Soc. Chim. Belg.*, *89*, 261–265 (1980).

140. S. Luterotti and V. Grdinic, *Analyst*, *111*, 1163–1165 (1986).

141. H. Kneifel and E. Bayer, *Ang. Chem.*, *85*, 542–543 (1973).

142. M. A. A. de C. T. Carrondo, M. T. L. S. Duarte, J. C. Pessoa, J. A. L. Silva, J. J. R. Fraústo da Silva, M. C. T. A. Vaz, and L. F. Vilas-Boas, *J. Chem. Soc. Commun.*, 1158–1159 (1988).

143. H. Kneifel and E. Bayer, *J. Am. Chem. Soc.*, *108*, 3075–3077 (1986).

144. E. Bayer, E. Koch, and G. Anderegg, *Angew. Chem. Int. Ed. Engl.*, *26*, 545–546 (1987).

145. A. Butler, S. M. Parsons, S. K. Yamagata, and R. I. De La Rosa, *Inorganica Chimica Acta*, *163*, 1–3 (1989).

146. I. Batinic, M. Birus, and M. Pribanic, *Croatica Chem. Acta*, *60*, 279–284 (1987).

147. D. C. Bhura and S. G. Tandon, *Anal. Chim. Acta*, *53*, 379–386 (1971).

148. K. Wieghardt, U. Quilitzsch, B. Nuber, and J. Weiss, *Angew. Chem. Int. Ed. Engl.*, *17*, 351–352 (1978).

149. D. C. Fisher, S. J. Barclay-Peet, C. A. Balfe, and K. N. Raymond, *Inorg. Chem.*, *28*, 4399–4406 (1989).

150. A. Dessì, G. Micera, D. Sanna, and L. S. Erre, *J. Inorg. Biochem.*, *48*, 279–287 (1992).

151. A. Abu-Dari and K.N. Raymond, *Inorg. Chem.*, *19*, 2034 (1980).

152. V. L. Pecoraro, *Inorg. Chim. Acta*, *155*, 171–173 (1989).

153. M. M. Craig, M. Baff, and M. J. Gresser, *J. Am. Chem. Soc.*, *110*, 3302–3303 (1988).

154. D. G. Drueckhammer, J. R. Durrwachter, R. L. Pederson, D. C. Crans, L. Daniels, and C.-H. Wong, *J. Org. Chem.*, *54*, 70–77 (1989).

155. D. K. Mullick, S. P. Moulik, and B. N. Ghosh, *J. Ind. Chem. Soc.*, *40*, 137–143 (1963).

156. B. Viossat, *Bull. Soc. Chim. France*, *1–2*, 113–116 (1975).

157. A. D. Kelkar, G. V. Gholap, P. V. Khadikar, and S. G. Harmalkar, *Science*, *4*, 68–72 (1977).

158. M. M. Caldeira, M. L. Ramos, N. C. Oliveira, and V. M. S. Gil, *Can. J. Chem.*, *65*, 2434–2440 (1987).

159. D. C. Crans and P. K. Shin, *Inorg. Chem.*, *27*, 1797–1806 (1988).

160. D. Rehder, *Inorg. Chem.*, *27*, 4312–4316 (1988).

161. D. C. Crans, R. L. Bunch, and L. A. Theisen, *J. Am. Chem. Soc.*, *111*, 7597–7607 (1989).

162. J. S. Jaswal and A. S. Tracey, *Can. J. Chem.*, *69*, 1600–1607 (1991).

163. K. Knüttel, A. Müller, D. Rehder, H. Vilter, and V. Wittneben, *FEBS. Lett.*, *302*, 11–14 (1992).

164. D. C. Crans, H. Holst, and D. Rehder, submitted.

165. B. Nuber, J. Weiss, and K. Wieghardt, *Z. Naturforschung*, *33b*, 265–267 (1978).

166. B. H. Bersted, R. L. Belford, and I. C. Paul, *Inorg. Chem.*, *7*, 1557–1562 (1968).

167. R. E. Drew and F. W. B. Einstein, *Inorg. Chem.*, *12*, 829–835 (1973).

168. A. Giacomelli, C. Floriani, A. Ofir de Souza Duarte, A. Chiesi-Villa, and C. Guastini, *Inorg. Chem.*, *21*, 3310–3316 (1982).

169. C. Weidemann, W. Priebsch, and D. Rehder, *Chem. Ber.*, *122*, 235–243 (1989).

170. J. H. McNeill, V. G. Yuen, H. R. Hoveyda, and C. Orvig, *J. Med. Chem.*, *35*, 1489–1491 (1992).

171. J. J. Habeeb, D. G. Tuck, and F. H. Wasters, *J. Coord. Chem.*, *8*, 27–33 (1978).

172. H. E. Jungnickel and W. Klinger, *Z. Anal. Chem.*, *203*, 257–260 (1964).

173. W. Bechmann, E. Uhlemann, R. Kirmse, and K. Köhler, *Z. Anorg. Allg. Chem.*, *544*, 215–224 (1987).

174. A. S. Tracey and M. J. Gresser, *Inorg. Chem.*, *27*, 1269–1275 (1988).

175. C. F. G. C. Geraldes and M. M. C. A. Castro, *J. Inorg. Biochem.*, *37*, 213–232 (1989).

176. D. C. Crans, R. W. Marshman, M. S. Gottlieb, O.P. Anderson, and M. M. Miller, *Inorg. Chem.*, *31*, 4939–4949 (1992).

177. D. C. Crans, R. W. Marshman, R. Nielsen, and I. Felty, *J. Org. Chem.*, *58*, 2244–2252 (1993).

178. W. R. Scheidt, D. M. Collins, and J. L. Hoard, *J. Am. Chem. Soc.*, *93*, 3873–3877 (1971).

179. W. R. Scheidt, R. Countryman, and J. L. Hoard, *J. Am. Chem. Soc.*, *93*, 3878–3882 (1971).

180. L. W. Amos and D. T. Sawyer, *Inorg. Chem.*, *11*, 2692–2697 (1972).

181. D. C. Crans and P. K. Shin, *J. Am. Chem. Soc.*, *116*, 1305–1315 (1994).

182. K. Zare, J. Lagrange and P. Lagrange, *Inorg. Chem.*, *148*, 568–571 (1979).

183. L. Przyborowski, G. Schwarzenbach, and T. Zimmerman, *Helv. Chim. Acta*, *48*, 1556–1565 (1965).

184. P. K. Shin, Multinuclear NMR elucidation of the structure of vanadium(V) complexes with amino acid derivatives, PhD thesis, Colorado State University, 1993.

185. K. Saito and Y. Sasaki, *Pure Appl. Chem.*, *60*, 1123–1132 (1988).

186. Y. Sasaki, M. Kanesato, K. Okazaki, A. Nagasawa, and K. Saito, *Inorg. Chem.*, *24*, 772–775 (1985).

187. Z. Kangjing and L. Xiaoping, *J. Struct. Chem.*, *6*, 14–16 (1987).

188. J. A. Bonadies and C. J. Carrano, *Inorg. Chem.*, *25*, 4358–4361 (1986).

189. L. Mokry and C. J. Carrano, *Inorg. Chem.*, *32*, 6119–6121 (1993).

190. X. Li, M. S. Lah, and V. L. Pecoraro, *Inorg. Chem.*, *27*, 4657–4664 (1988).

191. C. R. Cornman, G. J. Colpas, J. D. Hoeschele, J. Kampf, and V. L. Pecoraro, *J. Am. Chem. Soc.*, *114*, 9925–9933 (1992).

192. A. Butler, in *Vanadium in Biological Systems: Physiology and Biochemistry* (N. D. Chasteen, ed.), Kluwer Academic, Boston, 1990, pp. 25–50.

193. W. A. Nugent, T. V. RajanBabu, and M. J. Burk, *Science*, *259*, 479–483 (1993).

194. S. Yamada, J. Hagase, S. Funahashi, and M. Tanaka, *J. Inorg. Nucl. Chem.*, *38*, 617–621 (1976).

195. K. Zare, P. Lagrange, and J. Lagrange, *J. Chem. Soc., Dalton Trans.*, 1372–1376 (1979).

196. D. C. Crans, P. M. Ehde, P. K. Shin, and L. Pettersson, *J. Am. Chem. Soc.*, *113*, 3728–3736 (1991).

197. A. Kojima, K. Okazaki, S. Ooi, and K. Saito, *Inorg. Chem.*, *22*, 1168–1174 (1983).

198. D. C. Crans, H. Chen, O. P. Anderson, and M. M. Miller, *J. Am. Chem. Soc.*, *115*, 6769–6776 (1993).

199. M. Nishizawa, K. Hirotsu, S. Ooi, and K. Saito, *Chem. Soc. Chem. Commun.*, 707–708 (1979).

200. R. K. Mittal, *Z. Anorg. Allgem. Chem.*, *351*, 309–312 (1967).

201. Z. Yin-Zhuang, J. Xiang-Lin, and L. Shun-Cheng, *J. Struct. Chem.*, *12*, 48–51 (1993).

202. J. Lagrange and P. Lagrange, *Bull. Soc. Chim. France*, 19–23 (1972).

203. J. Lagrange and P. Lagrange, *Bull. Soc. Chim. France*, 1460–1462 (1975).

204. S. Yamada, Y. Ukei, and M. Tanaka, *Inorg. Chem.*, *15*, 964–967 (1976).

205. D. P. Riley, D. L. Fields, and W. Rivers, *Inorg. Chem.*, *30*, 4191–4197 (1991).
206. V. Vergopoulos, W. Priebsch, M. Fritzsche, and D. Rehder, *Inorg. Chem.*, *32*, 1844–1849 (1993).
207. C. R. Cornman, J. Kampf, and V. L. Pecoraro, *Inorg. Chem.*, *31*, 1981–1983 (1992).
208. D. Rehder, H. Holst, W. Priebsch, and H. Vilter, *J. Inorg. Biochem.*, *41*, 171–185 (1991).
209. P. M. Ehde, I. Andersson, and L. Pettersson, *Acta Chem. Scand.*, *43*, 136–143 (1989).

6

Use of Vanadate-Induced Photocleavage for Detecting Phosphate Binding Sites in Proteins

Andras Muhlrad[1] *and Israel Ringel*[2]

[1]Department of Oral Biology, The Hebrew University–Hadassah School of Dental Medicine and
[2]Department of Pharmacology, The Hebrew University–Hadassah Medical School, P.O.B. 1172, 91010 Jerusalem, Israel

1. INTRODUCTION

1.1. Vanadate-Induced Photocleavage of the Polypeptide Chain

Vanadate (Vi) is considered to act as a phosphate analog on proteins [1,2]. Irradiation in the presence of vanadate covalently modifies proteins and may lead to a specific cleavage of the polypeptide chain at one site or more. It has been suggested that some of those sites represent the phosphate-binding locus of the ATP binding site [3–5]; thus photocleavage by vanadate may serve as a tool to identify the location of ATP binding sites and conformational change in the vicinity of those sites. Site-directed photolysis mediated by vanadate was first described by Lee-Eiford et al. [6]. Irradiation at 254 nm of dynein ATPase from sperm flagella cleaved the α and β heavy-chain polypeptides at a specific site. The cleavage was accompanied by an additional nonspecific damage that was due to absorption by the aromatic amino acids. The obstacle of nonspecific damage was solved by changing the irradiation wavelength to 365 nm where, aside from the specific cleavage, no nonspecific damage was detected [3]. The photocleavage of dynein by vanadate is probably a one-step event whereby the photoactivation of vanadate generates a localized free radical. The vanadate is presumably bound at the γ-phosphate locus of the hydrolytic ATP binding site in the inhibited dynein-ADP-vanadate complex [7]. On the other hand, the cleavage of myosin (M) or myosin subfragment-1 (S1) in the presence of Vi was reported to proceed in a well-controlled two-step manner where the first irradiation resulted in a photomodified enzyme which upon second irradiation cleaves into fragments [8]. The first step may involve photoexcitation of the electronic structure of Vi while it is complexed with the divalent ion (i.e., Mg^{2+}) and one or more amino acid side chains on the protein as well as to a nucleotide when present. The oxidized radical group attached to the protein participates, with or without further irradiation, in additional radical reactions that lead to the cleavage of the polypeptide backbone [9]. The cleavage reaction depends on the divalent cation present, since quenching of the photocleavage was observed by transition metals—Mn(II), Fe(II), and Co(II)—whose divalent cation state is in the lower of the two accessible oxidation states. The quenching by these divalent cations presumably occurs through a redox reaction which oxidizes the cation and finally dissipates the energy of the excited vanadate to the solvent. The sites of cleavage and their intensity also depend on the solution conditions (ionic strength, concentration, pH, temperature). For instance, performing the cleavage of S1 by vanadate in Tris buffer will significantly inhibit the V2 cleavage [10]. The cause for this outcome is probably the fact that vanadate(+5) (V^{5+}) is reduced to V^{4+}, which is then protected from rapid reoxidation by O_2 by complexation with the Tris buffer [10].

Irradiation of the myosin-MgADP-Vi complex covalently modifies the protein by oxidizing the hydroxymethyl side chain of Ser-180 to an aldehyde [4,11], the aldehyde can then tautomerize to an enol which is ionized to form a highly chromophoric enolate anion which is the source of the large increased UV absorbance of the photomodified S1 [4,12]. When adenylate kinase (AK) was irradiated in the presence of vanadate, the polypeptide chain was cleaved near the N terminus. The amino terminus of this fragment was found to contain an acetylated serine, while its carboxy terminus was modified by a cyclic γ-aminobutyric acid which originated from photooxidation and decarboxylation of proline-17 by vanadate [13].

1.2. ^{51}V NMR as a Probe for Vanadate-Protein Interactions

^{51}V NMR studies have proved to be highly informative in systems of biological relevance. The ^{51}V isotope has exceptionally high nuclear magnetic resonance (NMR) sensitivity due to its high natural abundance (99.76%), large magnetic moment, and rapid quadrupolar relaxation in solutions. Although its spin is 7/2, the linewidths are relatively narrow and resolvable spectra can be obtained in several minutes.

Concentrations of various vanadium species can be determined accurately by integration of the signals.

^{51}V NMR studies on the binding of mono- and oligovanadate oligomers to sarcoplasmic reticulum [14], peroxidase [15], apotransferrin [16], 6-phosphogluconate dehydrogenases [17], and other biologically related molecules reveal a delicate equilibrium which depends on kinetic parameters and solution conditions (ionic strength, concentration, pH, temperature). ^{51}V NMR is a noninvasive and reliable technique that allows accurate analysis of the various oligomeric forms and their concentrations. Moreover, the shape and linewidth of each form in the presence of a macromolecule or organelle is indicative of the magnitude of binding of the vanadate species. Several studies, based on NMR data, concluded that tetravanadate is the major oligomer associated with UV photocleavage of polypeptides [10,11,18–20]. NMR experiments analyzing the effect of ATP on the photocleavage of myosin and S1 in the presence of various ionic forms of vanadate reveal that more monomeric and less oligomeric vanadate binds to the protein in the presence than in the absence of ATP [18].

2. DETECTION OF PHOSPHATE BINDING SITES IN SPECIFIC PROTEINS BY VANADATE-INDUCED PHOTOOXIDATION OR PHOTOCLEAVAGE

2.1. Dynein

2.1.1. Cleavage at a Single Site (V1 Cleavage)

Dynein is a cytoplasmatic motor protein ATPase that interacts with tubulin, the major component of the microtubules. This interaction is the molecular basis of the flagellar and ciliary movement in eukariotes and partially responsible for cytoplasmic vesicle transport and chromosome movement [21]. Dynein is a giant heteromultimer molecule with a molecular mass of more than 1000 kDa. It contains, among others, two or three heavy chains (α, β, and γ) with molecular masses between 400 and 500 kDa. By using photoaffinity analogs of ATP, it was demonstrated that each heavy chain contains an ATP binding site [22] and a distinct microtubule binding site [23]. Vanadate together with MgADP forms a complex at the ATP binding site and inhibits the dynein ATPase. Vanadate-induced photocleavage has been first detected on the outer arm dynein from sea urchin sperm flagella in the presence of 5–10 μM vanadate and MgATP by irradiation with far-UV light at 254 nm [6]. The UV irradiation in the presence of a low micromolar concentration of vanadate cleaves both the α and the β-heavy chain of dynein at a single site resulting in the formation of two peptides with 228- and 200-kDa molecular masses. The cleavage is accompanied by the irreversible loss of the ATPase activity of dynein [6]. Irradiation by UV light at 365 nm considerably improves the specificity of the photocleavage because it completely eliminates the nonspecific effects, which were observed upon irradiation by far-UV light [3]. No detectable cleavage occurred in the absence of vanadate or ATP or when ATP was substituted with AMP. Substitution of Mg^{2+} with any of the transition metal cations Mn^{2+}, Co^{2+}, or Fe^{2+} inhibited both the photocleavage and the loss of the ATPase activity. Since the cleavage at a single site, which is called V1, was always accompanied by the loss of ATPase activity, it is assumed that the cleavage occurred at the hydrolytic ATP binding site.

In solution, vanadate (Vi) exists in either monomer or various polymer forms, depending on the Vi concentration and the pH. At low micromolar Vi concentrations necessary for the photocleavage at the V1 site and at the near-neutral pH, the monomeric Vi is the predominant form. Therefore, it is assumed that the monomeric Vi forms an inhibitory dynein-MgADP-Vi complex at the ATP hydrolytic site and the photosensitization of Vi, which has an absorbance peak at 365 nm, leads to the cleavage of the polypeptide chain at this site [3]. It is assumed that the cleavage mechanism also involves a divalent cation (which is Mg^{2+}, under physiological conditions) to this cation the excitation energy is directly transferred from Vi, possibly because the bonding of Pi or Vi at the ATP γ-phosphate locus includes coordination to the divalent cation. On the other hand, the cleavage does not involve the action of oxygen-free radicals which may form as a consequence of the photoreduction of Vi, since free radical scavengers, like cysteamine, do not affect the extent of cleavage or the irreversible loss of ATPase activity [3].

The V1 cut was described also in dyneins from other sources including inner arm dynein from sea urchin sperm flagella [24], outer arm dynein from *Chlamydomonas* flagella [25,26], *Tetrahymena* [27], and *Paramecium tetraurelia* cilia [28]. Cytoplasmic dyneins from *Paramecium* [29], chicken brain [30], sea urchin egg [31], and nematodes [32] were also cleaved at the same site. The amino acid sequence of the β heavy chain of urchin sperm flagella outer arm dynein was recently deduced from its cDNA, and the site of the V1 cut was proposed at the "consensus" ATP binding site, in the sequence GPAGTGKT at residues 1852–1859 [33]. By analogy to the known location of the vanadate-mediated photocleavage in myosin [4] and in adenylate kinase [13], the most probable candidate of the V1 cut is proline 1853. In all described cases the V1 cleavage of dynein occurred in the presence of Mg^{2+}, ATP, and low micromolar concentration of Vi. However, in the case of *Tetrahymena* cilia outer arm dynein β heavy chain [27] and of *Chlamydomonas* outer arm dynein α and γ heavy chains [25,26], the V1 cleavage occurred also in the presence of Mg^{2+} and Vi only, without the nucleotide.

As a consequence of the loss of ATPase activity because of the V1 cut, all cellular movements in which dynein has a force-generating role are severely inhibited. The effects on cellular movements include, among others, (1) a progressive decrease in sea urchin sperm flagella beat frequency up to 50% cleavage (more cleavage causes irreversible loss of motility) [27]; (2) an over 90% inhibition of organelle motility directed to the minus end of microtubules following the cleavage of chick brain dynein [30]; (3) loss in "motility-inducing activity" of *Nematoda C. elegans* [32]. Since according to our present knowledge dynein is the only motor protein, which can be cleaved in the presence of 5 μM Vi by irradiation with near-UV light, the loss of the biological movement upon irradiation under these conditions indicates the *involvement of dynein* in the affected movement, which points to the biological significance of the Vi-mediated photocleavage. The cut at the V1 site was proved to be useful in assigning the hydrolytic ATP binding site of the five "consensus" ATP binding site motifs located on the β heavy chain of the outer arm dynein from sea urchin sperm flagella [33]. It was shown earlier that a single V1 cut in the β heavy chain leads to complete loss of ATPase activity [3]. Since only one of the ATP binding sites is affected by this cut, which involves the loss of ATPase activity, it is obvious that only the cleaved ATP binding site is catalytically active in the β heavy chain of dynein, while the rest may have a regulatory function.

2.1.2. Cleavage at a Second Site (V2 Cleavage)

Both the α and β chains of the sea urchin sperm flagella outer arm dynein are cleaved by near-UV irradiation at 365 nm at an additional and distinct site from Vi, called V2, in the presence of 50–200 μM Vi and in the absence of ATP [34]. The V2 cut is not affected by substituting Mg^{2+} with other metal divalent cations or by scavengers of free radicals but strongly inhibited by nucleotide triphosphates, especially ATP. Because the V1 cut, unlike V2, is strongly inhibited by the Mg^{2+} to Mn^{2+} substitution, the V2 cut is generally carried out in the presence of 1 mM Mn^{2+} to avoid V1 cleavage. The cleavage at the V2 site requires much higher Vi concentrations, 50–200 μM, than the cut at V1. Since at these Vi concentrations at neutral pH various forms of vanadate exist side by side, the Vi concentration dependence of the V2 cut was studied. The concentration dependence was found to be sigmoidal, with 2.7 sigmoidicity, which according to Tang and Gibbons [34] indicates that the trimeric form of Vi is active in the V2 cleavage, while the monomeric Vi is inactive. This assumption is supported by the finding that the half-maximal V2 cleavage rate is at 58 μM Vi, which is in good agreement with the calculated 65 μM Vi concentration necessary for the half-maximal formation of trimeric Vi as obtained from the equilibrium constants [35].

The variability in the V2 cleavages of the various dyneins is much greater than that in the Vi cuts. Generally, the V2 cut is inhibited by nucleotides, but no inhibition was observed during the V2 cleavage of the α, β, and γ chains of the *Tetrahymena* cilia outer arm dynein [27]. The V2 cut has been found distinct from the V1 cut in all dynein heavy chains studied. In the majority of cases only a single V2 cut was observed per heavy chain, e.g., in the outer arm dynein α and β heavy chains from the sea urchin sperm flagella [33], in the α, β, and γ heavy chains of *Tetrahymena* cilia outer arm dyneins [27], and in the β heavy chain of the *Chlamydomonas* flagella outer arm dynein [25]. On the other hand, there are three or two distinct V2 cleavage sites in the α and γ heavy chains of the *Chlamydomonas* dynein, respectively [25,26].

In spite of the findings that the V2 cleavage does not cause loss in the ATPase activity or in the biological activity of various dyneins [9], it is believed that this cut occurs at the noncatalytic ATP binding sites on the basis of the following arguments: (1) the V2 cleavage is inhibited by nucleotides in most of the cases; (2) dynein sequences, which could be labeled with photoaffinity analogs of nucleotides before the V2 cleavage, cannot be labeled after the cleavage, possibly because the cut damaged the nucleotide binding sites [36]. Since the noncatalytic ATP binding sites might have regulatory functions, it will be necessary to study how the regulation of the dynein-dependent cellular motility is affected by the V2 cleavage. Finally, it is worth mentioning that near-UV irradiation of microtubule-bound *Tetrahymena* cilia outer arm dynein under V1 conditions in the absence of ATP resulted in V2 cleavage on all three heavy chains [27]. This observation indicates that microtubule binding induces conformational changes in the dynein structure, which are manifested in an altered Vi-dependent photo-cleavage pattern. The finding also shows the usefulness of the Vi-dependent photocleavage in detecting conformational changes in the dynein molecule.

2.2. Myosin

2.2.1. Vanadate-Mediated Photomodification

Myosin (M) is the major motor protein in all eukaryotes. It possesses ATPase activity and interacts with actin. This interaction, which is coupled with the myosin-catalyzed ATP hydrolysis, powers muscle contraction and a large number of various cellular movements. The molecular mass of myosin is about 500 kDa, and it is a heterohexamer, composed of two heavy chains and four light chains. The N-terminal segment of both heavy chains, called S1, is a globular head, and contains the distinct ATP and actin binding sites. The predominant intermediate during the myosin-catalyzed ATP hydrolysis is the M**-MgADP-Pi complex and the dissociation of the phosphate (Pi) from the complex is the relatively slow, rate-limiting step of the process. The substitution of Pi with Vi, which is a good Pi analog [1], dramatically stabilizes the complex [37]. The M-MgADP-Vi complex, which has no ATPase activity, remains stable for several days ($t_{1/2}$ is 3–4 days) even after the removal of free Vi and MgADP. The stability of the complex is not affected by substituting the whole myosin molecule with its active and soluble "head segment," S1, and most of the experiments involving Vi were carried out by S1. Attempts were made to exploit the high stability of the M-MgADP-Vi complex for localization of the ATP binding site of myosin by substituting ADP with photoaffinity label nucleotide analogs [38]. However, the irradiation of the complex with near-UV light led to its destabilization and the dissociation of ADP and Vi from S1. The decomposition of the complex is caused by Vi-mediated photomodification, which is accompanied by the appearance of an absorbance difference spectrum, peak 270 nm, and an increase in the Ca^{2+}-activated ATPase activity of S1 from rabbit skeletal muscle myosin [8].

The photomodification can be at least partially reversed by reduction with $NaBH_4$, which restores the spectral properties and ATPase activity of unmodified S1 [11]. Amino acid analysis of photomodified S1, which was reduced by tritiated $NaBH_4$, showed incorporation of the tritium label only to serine, indicating that this was the photomodified residue. These findings together with the pH dependence of the absorbance spectrum of the photomodified S1 led to the assumption that the result of the Vi-mediated photoreaction is the oxidation of the β-hydroxymethyl group of a serine to an aldehyde, which can tautomerize to an enol. This enol is ionized at neutral pH to form the chromophoric enolate ion, which is the source of the observed UV difference spectrum of the photomodified S1 (see Figure 5 in Ref. 11). Finally, it has been shown that only a single serine, Ser-180, was modified during the reaction [4]. This serine is part of the "consensus ATP binding site" of myosin (residues 178–185 in rabbit skeletal muscle myosin [39]) which is conserved in all myosin heavy chains. Since Vi replaces Pi in the stable S1-nucleotide complex, the photomodification of Ser-180 is direct evidence that this residue is at the γ-phosphate binding site of ATP or at the Pi binding site of the M**-MgADP-Pi complex. It has been shown that essentially the same serine is modified both in rabbit muscle skeletal myosin [4] and chicken gizzard smooth muscle myosin [5], which further supports the putative presence of serine at the nucleotide binding site.

2.2.2. Vanadate-Mediated Photocleavage at Ser-180

Photomodified S1 is still capable of forming an M-MgADP-Vi complex, which remains stable after the removal of the excess of MgADP and Vi [8]. The irradiation of this complex with near-UV light at 365 nm results in the cleavage of the heavy chain at 23 kDa from the N terminus and the loss of both the Ca^{2+} and the K^+ (EDTA)-activated ATPase activities. The N-terminal Edman sequencing of the photocleavage site was not successful because the N terminus of the newly formed peptide has been blocked during the photocleavage. However, following intensive proteolysis, the peptides containing the new C and N termini formed in the photocleavage were isolated and analyzed by several methods including mass spectrometry and capillary electrophoresis. The analysis showed that Ser-180 was cleaved during the photoreaction and its amino group was transferred to Glu-179 blocking the newly formed C terminus [40]. This result supported the finding of the Vi-mediated photomodification, i.e., that Ser-180 is located at the Pi-binding loop of the nucleotide binding site of myosin. The NaB^3H_4 treatment of the photocleaved S1 revealed an additional photomodification, which occurred at Ser-243 during the second irradiation by near-UV light [41]. This indicates that Ser-243 is proximal to the nucleotide binding site through a loop in the S1 structure. Both photomodification and photocleavage could be prevented by substituting Mg^{2+} with transition metal cations Co^{2+}, Mn^{2+}, or Ni^{2+} in the stable M-ADP-Vi complex. The existence of the light-insensitive myosin-metal-ADP-Vi complexes made it possible to label the nucleotide binding site by photoaffinity label ATP analogs in Vi containing myosin-nucleotide complexes [42].

2.2.3. *Tetravanadate-Mediated Photocleavage of the Myosin Heavy Chain*

Myosin heavy chain from rabbit skeletal muscle was found to be cleaved in the presence of 0.5 mM Vi and absence of nucleotide by near-UV irradiation at three specific sites, located at 23 kDa and 74 kDa from the N terminus in the globular S1 segment and at about 10 kDa from the C terminus in the helical rod segment [43]. From the sigmoid Vi concentration dependence, sigmoidicity 2.7, it was assumed that oligomeric, trimeric, or tetrameric vanadate was responsible for the photocleavage. The cleavage affected the ATPase characteristics of myosin: it increased its Ca^{2+}-activated while decreasing the K^+ (EDTA)- and actin-activated ATPase activities. The presence of MgATP strongly inhibited the cleavage at 23 kDa and to a lesser degree at 74 kDa from the N terminus. On the basis of these results and on the basis of Pi-Vi analogy, it was assumed that the cleavage occurred in the proximity of specific phosphate binding sites of myosin [43].

^{51}V NMR studies revealed, by monitoring the dependence of the fraction of monomer and various oligomer vanadate species and the dependence of the extent of photocleavage on Vi concentration, that tetrameric vanadate is responsible for the photocleavage of myosin and S1 [10,18]. The ^{51}V NMR studies also showed that all vanadate species bind to S1 and the binding of the tetrameric vanadate is specifically inhibited by MgATP. A good correlation was found between the inhibition of tetrameric vanadate binding and the inhibition of the cut at 23 and 31 kDa from the N terminus, which further supports the assumption that the tetrameric vanadate binds to the nucleotide binding site and its binding is a prerequisite of the photocleavage [18]. 0.2 mM was the lowest Vi concentration where photocleavage and formation of tetravanadate were observed. At this concentration two cleavages appeared in the S1 heavy chain, located at 23 and 31 kDa from the N terminus, while the cleavage at 74 kDa appeared only at higher Vi concentrations. The 23- and 31-kDa cuts are inhibited simultaneously by Pi and several polyphosphates, including ATP, ADP, and pyrophosphate (PPi) [44] and by phosphate analogs BeF_3^- and AlF_4^- [45] and sulfate [46], which is a competitive inhibitor of the binding of ATP to myosin [47]. From the simultaneous appearance and disappearance (inhibition) of the two cuts, it is assumed that both participate in the formation of the γ-phosphate binding site or the Pi binding site of the M**-MgADP-Pi complex. The 23-kDa site is most probably located at Ser-180, while the 31-kDa is at Ser-243. Both serines were shown to be located at or in the proximity of the ATP binding site of myosin [4,41]. The photocleavage at the 31-kDa site is pH-dependent with pK_a 6.8 [44]. The probable reason for this pH dependence is that only the enol and not the ionized enolate form of the oxidized Ser-243 is susceptible for the photocleavage. By studying the divalent cation requirement of the tetravanadate-dependent photocleavage, it was found that this reaction, unlike the formation of the M-MgADP-Vi complex, does not require the presence of divalent cations [46].

The cleavage site at 74 kDa from the N terminus of the myosin heavy chain is near (linear distance of about 1 kDa) to a highly positively charged, lysine-rich stretch of the sequence, which is part of the actin binding site of myosin [48]. This cleavage is completely prevented by the presence of actin. The actin-activated ATPase activity of myosin is strongly inhibited by the cleavage and there is a good correlation between the extent of the cleavage and the inhibition of the actin-activated ATPase when the photocleavage was performed in the presence of ATP with the 23- and 31-kDa sites protected [44]. These findings show a close correlation between actin binding and tetravanadate-dependent photocleavage at the 74-kDa site. There is probably a competition between actin and tetravanadate—both of them are negatively charged—for the positively charged actin binding site, which is located in the proximity of the 74-kDa cleavage site. In the absence of actin tetravanadate binds to the actin binding site, which is accessible through a loop to the 74-kDa site 1 kDa upstream in the sequence. The near-UV irradiation causes a cleavage at this site, which is mediated by excitation of Vi. Since phosphate and polyphosphate also inhibit the photocleavage at the 74-kDa site, one may assume that this is a second phosphate binding site, which may have an important regulatory role in the myosin-actin interaction.

2.3. Adenylate Kinase

Adenylate kinase (AK) is a relatively small enzyme with a molecular mass of 21 kDa, which catalyzes the transfer of γ-phosphate of ATP to AMP. It contains a glycine-rich phosphate-binding loop near its N terminus. Vanadate interacts with AK and inhibits its enzymatic activity [49]. Near-UV light irradiation in the presence of 0.25 mM vanadate was found first to inactivate the enzyme and then to cleave the polypeptide chain near the N terminus [13]. The inactivation and the cleavage were prevented by the multisubstrate analog, P^1, P^5-bis(5′-adenosyl)-pentaphosphate, which binds to both the ATP and AMP binding sites. ATP, but not AMP, also inhibited both inactivation and the cleavage in a saturable manner, indicating that these reactions take place as a consequence of the photomodification of the ATP binding site. The small N-terminal peptide produced in the photocleavage was isolated and analyzed. This peptide contained the first 16 residues of the enzyme and a γ-aminobutyric acid which derived from the photooxidized Pro-17 [13]. Pro-17 is located in the phosphate-binding loop of AK in a position equivalent to that of Ser-180 in the ATP binding site of myosin.

2.4. Isocitrate Lyase

Isocitrate lyase catalyzes the reversible cleavage of isocitrate to glyoxalate and succinate. Near-UV irradiation of the enzyme in the presence of 0.4 mM Vi leads to the loss of enzymatic activity. The presence of substrates, e.g., glyoxalate or isocitrate and, a lesser extent, succinate, prevents the Vi-mediated photoinactivation [50]. This indicates that the photomodification occurs at the active site of the enzyme. The activity can be partially restored by reduction with $NaBH_4$. Upon reduction of the photomodified enzyme with tritiated $NaBH_4$, Ser-319 and Ser-321 were found to be differentially labeled in the absence and in the presence of substrates. The irradiation in the presence of Vi also causes a small degree (about 4%) of photocleavage. The photocleavage produces a 36- to 37-kDa peptide and the size of the peptide is in good agreement with the assumed cleavage site at Ser-319 and Ser-321. These residues are highly conserved and suggested to be located at the active site of the enzyme [50].

2.5. Phosphofructokinase

Phosphofructokinase catalyzes the transfer of phosphate from ATP to fructose-6-phosphate with the formation of fructose-1,6-bisphosphate. ATP is both a substrate and an allosteric inhibitor of the enzyme. Decavanadate is also an allosteric inhibitor of the enzyme through an interaction with the ATP inhibitor site [51], while monomeric vanadate does not have any inhibitor effect [52]. UV irradiation in the presence of Vi led first to the inhibition of the enzyme, then to the cleavage of its polypeptide chain [53]. In the presence of decavanadate a progressive photodegradation was observed to a 78-kDa fragment, while in the presence of monovanadate a number of smaller peptides were produced by the irradiation.

2.6. Ca^{2+}-ATPase of Sarcoplasmic Reticulum

Sarcoplasmic reticulum Ca^{2+}-ATPase, molecular mass 110 kDa, is a membrane ATPase which couples the hydrolysis of ATP to the transfer of Ca^{2+} from the cytosol to the sarcoplasmic reticulum of muscle. The activity of the enzyme is strongly inhibited by Vi [54]. A broad-spectrum (maxima at 253, 309, and 363 nm) UV light irradiation of the enzyme leads to the irreversible loss of ATPase activity and to the cleavage of the polypeptide chain, which is accompanied by formation of sodium dodecyl sulfate (SDS)-insoluble aggregates [55]. Illumination by near-UV only, 366 nm, proved to be more effective and eliminated aggregate formation [56]. The photocleavage occurred at 22 and 71 kDa from the N terminus in the presence and absence of Ca^{2+}, respectively, indicating that Ca^{2+} induced conformational changes in the enzyme structure. The 22-kDa and 71-kDa cleavage sites are near the phosphate acceptor residue Asp-351 and the fluorescein isothiocyanate (FITC) binding site (Lys-515), respectively. Both regions are components of the active site of the enzyme. In the presence of Ca^{2+} the irreversible loss of ATPase activity is faster than the photocleavage, which points to possible photomodification prior to the cleavage. Half of the enzyme molecules are resistant to photocleavage even after a prolonged (2-hr) exposure to irradiation. Na^+,K^+-ATPase, and gastric H^+,K^+-ATPase, which have a similar structure of sarcoplasmic reticulum Ca^{2+}-ATPase [57], are essentially insensitive to the photocleavage, although partially inactivated during illumination in the presence of Vi [56].

2.7. Aldolase

Aldolase is a glycolytic enzyme, consisting of two monomers of a molecular mass of 41-kDa, which catalyzes the reversible cleavage of fructose-1,6-bisphosphate to triose phosphates. The activity of the enzyme is inhibited by various oxoanions including vanadate. The various vanadate species, which exist in equilibrium at neutral pH, affect the activity differently. Monovanadate has no effect, while dimeric, tetrameric and decameric vanadate are reversible competitive, irreversible competitive and reversible non-competitive inhibitors, respectively [17]. The authors believe, but do not prove, that the irreversible inhibition of the enzyme by tetrameric vanadate is caused by a redox reaction between aldolase thiols and tetravanadate. The oxoanions, including the inhibitory vanadate species, probably bind to the subunit interface of aldolase, which regulates the enzymatic activity. Irradiation of aldolase at 365 nm in the presence of tetravanadate causes photocleavage with formation of a 36-kDa major peptide. The photocleavage is accompanied by irreversible loss of enzymatic activity. The cleavage is assumed to take place at the oxoanion binding site [17].

2.8. Glycogen Phosphorylase

Phosphorylase is the key enzyme of glycogen degradation, and comprises one (phosphorylase b) or two (phosphorylase a) subunits of 95-kDa molecular mass. The activity of the enzyme is strongly inhibited by oxoanions of transition metals, including vanadate. Near-UV light irradiation in the presence of 1 mM Vi progressively degrades phosphorylase to smaller peptides. The first product of the photocleavage is an 82-kDa peptide, which is further cleaved to a 76-kDa fragment. When phosphorylase a labeled with ^{32}Pi on Ser-14 is subjected to photocleavage, the resulting 82- and 76-kDa fragments contain the radiolabel which shows that the cleavages took place in the C-terminal region [58]. The photocleave has a sigmoidal dependence of Vi concentration, which indicates, contrary to the authors' claim, that tetrameric Vi is responsible for the photocleavage.

2.9. Ribulose 1,5-Bisphosphate Carboxylase/Oxygenase

This enzyme is a hexadecamer consisting of eight large (55-kDa) and eight small (15-kDa) subunits. It catalyzes the carboxylation of ribulose-1,5-bisphosphate during photosynthesis. Irradiation of the enzyme in the presence of 0.3 mM Vi by near-UV light results in rapid loss of carboxylase activity. The presence of substrate, substrate analog, or phosphate during the irradiation substantially prevents the activity loss, which indicates that a Pi-binding loop has been affected in the active site of the enzyme by the Vi-modulated photomodification. Activity can be partially restored by reduction with $NaBH_4$. Following treatment of the photomodified enzyme with NaB^3H_4, tritium has been found to be incorporated only to serine, which places one yet unidentified serine residue to the active site of the enzyme. Subsequent to photomodification a small-scale Vi-dependent photocleavage was also observed in the large subunit of the enzyme [59].

2.10. Tubulin and Microtubule Proteins

Microtubules are a major component of the spindle fibers and the cytoskeleton of eukaryotic cells. They participate in a variety of cellular functions, such as mitosis, morphogenesis, motility, and intracellular organelle transport. Microtubules are assembled mainly from tubulin, a dimeric protein of ~110 000 MW consisting of two polypeptide chains (α and β) of approximately identical molecular weight [60]. Microtubules purified by repeated cycles of polymerization and depolymerization contain tubulin and several copurified proteins collectively referred to as microtubule-associated proteins (MAPs). A number of the MAP proteins are important for microtubule stability and participate significantly in both the nucleation and elongation phases of the assembly reaction. Tubulin has two binding sites for guanine nucleotides, one exchangeable and one nonexchangeable [61–63]. Generally, GTP bound at the exchangeable site of the protein is required to initiate the assembly of microtubules, and in the reaction process this GTP is hydrolyzed to GDP. Studies reveal that at high molar ratios of vanadate/tubulin (×13–×90), the assembly of tubulin is inhibited up to 65% and the effect of vanadate is independent of GTP concentration at which the polymerization assays were performed [64]. In a more elaborate study performed recently, Lobert et al. [65] used ^{51}V NMR analysis and purified tubulin and MAPs. The study revealed the presence of various vanadate oligomers in the assembly buffer. The various forms of vanadate have diverse effects on tubulin assembly. Monomeric vanadate does not interact with tubulin or MAPs and does not bind to the exchangeable nucleotide binding site on tubulin. The assembly of microtubule protein is inhibited by decavanadate, which strongly interacts with tubulin and with MAP2 and τ as well, while tetravanadate inhibits the assembly through weak binding to MAPs. Photoirradiation of tubulin dimers in the presence of vanadate cleaved both α and β subunits, under conditions where tetravanadate, pentavanadate, and decavanadate are present in solution [19]. While most cleavage sites on dynein and myosin are associated with the MgADP-Pi binding site, the cleavage reaction of tubulin is independent of MgGTP or MgGDP and cleavage occurs at two or more sites per subunit [19]. Depending on the buffer used and pH, the major species associated with the cleavage are tetravanadate at pH 6.9 in Pipes buffer and V′ (oligomeric form appearing at −582 ppm) at pH 6.0 in phosphate buffer [19].

ABBREVIATIONS

ADP	adenosine 5′-diphosphate
AK	adenylate kinase
AMP	adenosine 5′-monophosphate
ATP	adenosine 5′-triphosphate
EDTA	ethylenediamine-*N,N,N′,N′*-tetraacetate
FITC	fluorescein isothiocyanate
GDP	guanosine 5′-diphosphate
GTP	guanosine 5′-triphosphate
M	myosin
M**	myosin with increased tryptophan fluorescence
MAP	microtubule-associated protein
NMR	nuclear magnetic resonance
Pi	phosphate
PPi	pyrophosphate
S1	myosin subfragment-1
SDS	sodium dodecyl sulfate
Vi	vanadate

REFERENCES

1. R. N. Lindquist, J. L. Lynn, and G. E. Lienhard, *J. Am. Chem. Soc.*, *95*, 8762 (1973).
2. D. Rehder, *Biometals*, *5*, 3 (1992).
3. I. R. Gibbons, A. Lee-Eiford, G. Mocz, C. A. Phillipson, W-J. Y. Tang, and B. H. Gibbons, *J. Biol. Chem.*, *262*, 2780 (1987).
4. C. R. Cremo, J. C. Grammer, and R. G. Yount, *J. Biol. Chem.*, *264*, 6608 (1989).

5. D. G. Cole and R. G. Yount, *Biochemistry*, *31*, 6186 (1992).
6. A. Lee-Eiford, R. A. Ow, and I. R. Gibbons, *J. Biol. Chem.*, *261*, 2337 (1986).
7. B. H. Gibbons, W-J. Y. Tang, and I. R. Gibbons, *Cell Motil. Cytoskel.*, *11*, 188 (1988).
8. J. C. Grammer, C. R. Cremo, and R. G. Yount, *Biochemistry*, *27*, 8408 (1988).
9. I. R. Gibbons and G. Mocz, *Meth. Enzymol.*, *196*, 428 (1991) and references therein.
10. R. C. Cremo, G. T. Long, and J. C. Grammer, *Biochemistry*, *29*, 7982 (1990).
11. C. R. Cremo, J. C. Grammer, and R. G. Yount, *Biochemistry*, *27*, 8415 (1988).
12. C. R. Cremo, J. C. Grammer, and R. G. Yount, *Meth. Enzymol.*, *196*, 442 (1991).
13. C. R. Cremo, J. A. Loo, C. G. Edmons, and K. M. Hatleid, *Biochemistry*, *31*, 491 (1992).
14. P. Csermely, A. Martonosi, G. C. Levy, and A. J. Ejchart, *Biochem. J.*, *230*, 807 (1985).
15. H. Vilter, and D. Rehder, *Inorg. Chim. Acta*, *136*, L7 (1987).
16. A. Butler, and M. J. Danzitz, *J. Am. Chem. Soc.*, *109*, 1864 (1987).
17. D. C. Crans, K. Sudhakar, and T. J. Zamborelli, *Biochemistry*, *31*, 6812 (1992).
18. I. Ringel, Y. M. Peyser, and A. Muhlrad, *Biochemistry*, *29*, 9091 (1990).
19. J. J. Correia, L. D. Lipscomb, J. C. Dabrowiak, N. Isern, and J. Zubieta, *Arch. Biochem. Biophys.*, *309*, 94 (1994).
20. D. C. Crans, E. M. Willging, and S. R. Butler, *J. Am. Chem. Soc.*, *112*, 427 (1990).
21. R. B. Vallee and H. S. Shpetner, *Annu. Rev. Biochem.*, *59*, 909 (1990).
22. K. K. Pfister, B. E. Haley and G. B. Witman, *J. Biol. Chem.*, *259*, 8499 (1984).
23. R. D. Vale, D. R. Soll, and I. R. Gibbons, *Cell*, *59*, 915 (1989).
24. B. H. Gibbons and I. R. Gibbons, *J. Biol. Chem.*, *262*, 8354 (1987).
25. S. M. King and G. B. Witman, *J. Biol. Chem.*, *262*, 17596 (1987).

26. S. M. King and G. B. Witman, *J. Cell. Biol., 107*, 1799 (1988).
27. S. P. Marchese-Ragona, K. C. Facemyer, and K. A. Johnson, *J. Biol. Chem., 264*, 21361 (1989).
28. J. Larsen, K. Barkalow, T. Hamasaki, and P. Satir, *J. Protozool., 38*, 55 (1991).
29. C. C. Schroeder A. K. Fok, and R. D. Allen, *J. Cell. Biol., 111*, 2553 (1990).
30. T. A. Schroer, E. R. Steuer, and M. P. Sheetz, *Cell, 56*, 937 (1989).
31. M. E. Porter, P. M. Grissom, J. M. Scholey, E. D. Salmon, and J. R. McIntosh, *J. Biol. Chem., 263*, 6759 (1988).
32. R. J. Lye, M. E. Porter, J. M. Sholey, and J. R. McIntosh, *Cell, 51*, 309 (1987).
33. I. R. Gibbons, B. H. Gibbons, G. Mocz, and D. I. Asai, *Nature, 352*, 640 (1991).
34. W. J. Y. Tang and I. R. Gibbons, *J. Biol. Chem., 263*, 17728 (1987).
35. N. D. Chasteen, *Struct. Bond., 53*, 105 (1983).
36. S. M. King, B. E. Haley, and G. B. Witman, *J. Biol. Chem., 264*, 10210 (1988).
37. C. C. Goodno, *Proc. Natl. Acad. Sci. USA, 76*, 2620 (1979).
38. Y. Okamoto, T. Sekine, J. C. Grammer, and R. G. Yount, *Nature, 324*, 78 (1986).
39. S. W. Tong and M. Elzinga, *J. Biol. Chem., 21*, 13100 (1983).
40. J. C. Grammer and R. G. Yount, *Biophys. J., 64*, A142 (1993).
41. J. C. Grammer and R. G. Young, *Biophys. J., 59*, 226a (1991).
42. T. E. Garabedian and R. G. Yount, *Biochemistry, 30*, 10126 (1991).
43. G. Mocz, *Eur. J. Biochem., 179*, 373 (1989).
44. A. Muhlrad, Y. M. Peyser, and I. Ringel, *Biochemistry, 30*, 958 (1991).
45. M. M. Werber, Y. M. Peyser, and A. Muhlrad, *Biochemistry, 31*, 7190 (1992)
46. A. Muhlrad, Y. M. Peyser, and I. Ringel, *Eur. J. Biochem., 201*, 409 (1991).
47. C. Tesi, T. Barman, and F. Travers, *FEBS Lett., 236*, 256 (1988).
48. D. Mornet, P. Pantel, E. Audemard, and R. Kassab, *Biochem. Biophys. Res. Commun., 89*, 925 (1979).

49. D. W. Boyd, K. Kustin, and M. Niwa, *Biochim. Biophys. Acta*, *827*, 472 (1985).
50. Y. H. Ko, C. R. Cremo, and B. A. McFadden, *J. Biol. Chem.*, *267*, 91 (1992).
51. G. Choate and T. E. Mansour, *J. Biol. Chem.*, *254*, 11457 (1979).
52. F. Climent, R. Bartrons, G. Pons, and J. Carreras, *Biochem. Biophys. Res. Commun.*, *101*, 570 (1981).
53. M. Signorini and C. M. Bergamini, *Biochem. Biophys. Res. Commun.*, *172*, 919 (1990).
54. U. Pick and S. J. D. Karlish, *Biochim. Biophys. Acta*, *626*, 255 (1980).
55. M. Vegh, E. Molnar, and A. Martonosi, *Biochim. Biophys. Acta*, *1023*, 168 (1990).
56. E. Molnar, S. Varga, and A. Martonosi, *Biochim. Biophys. Acta*, *1068*, 17 (1991).
57. D. H. McLennan, C. J. Brandl, B. Korczak, and N. M. Green, *Nature*, *316*, 696 (1985).
58. C. M. Bergamini and M. Signorini, *Biochem. Int.*, *25*, 143 (1991).
59. S. N. Mogel and B. A. McFadden, *Biochemistry*, *28*, 5428 (1989).
60. P. Dustin, *Microtubules*, Springer-Verlag, New York, 1984.
61. J. E. Allende, *FASEB J.*, *2*, 2356 (1988).
62. M.-F. Carlier, D. Didry, and D. Pantaloni, *Biochemistry*, *26*, 4428 (1987).
63. E. T. O'Brien, W. A. Voter, and H. P. Erickson, *Biochemistry*, *26*, 4148 (1987).
64. E. P. Kirazov and D. G. Weiss, *Cell Motil. Cytoskel.*, *6*, 314 (1986).
65. S. Lobert, N. Isern, B. S. Hennington, and J. J. Correia, *Biochemistry*, *33*, 6244 (1994).

7
Vanadium-Protein Interactions

N. Dennis Chasteen

Department of Chemistry, University of New Hampshire, Parsons Hall, Durham, New Hampshire 03824-3598, USA

1. INTRODUCTION

1.1. Coordination Chemistry of Vanadium(IV) and Vanadium(V)

The coordination chemistry of vanadium in biological systems is largely that of vanadium(IV) and vanadium(V), although there are organisms such as tunicates where the principal oxidation state of vanadium is +3 in the blood cells (Chap. 13). The coordination and redox chemistry of vanadium(III) is discussed in Chap. 2.

The chemistries of vanadium(IV) and vanadium(V) are rich and varied. Complexes of vanadium(IV), such as the VO^{2+} cation, with ligands containing oxygen, nitrogen, and thiol donor atoms are well known [1–6]. More recently, an octahedral *tris*-catecholate complex of "bare" V(IV) lacking the oxo group has been prepared and characterized [7]. Vanadium(IV) coordination geometries include square-pyramidal, square-bipyramidal, trigonal-bipyramidal, and octahedral [1,3,4]. This flexibility in coordination undoubtedly accounts for the large number of VO^{2+}-protein complexes which have been observed over the years. The VO^{2+} cation appears to readily accommodate the geometric requirements imposed by the structure of the protein. Unfortunately, with the exception of ribonuclease-vanadium(V)-nucleoside complex, which has a trigonal bipyramidal coordination about V [8], crystal structures of vanadium-protein complexes are lacking at this point in time. In most instances our knowledge is limited to the identity of the protein ligands and not their locations within the first coordination sphere of the metal.

Vanadium(V) chemistry is dominated by mononuclear and oligonuclear anionic vanadate species with tetrahedral coordination [1,4–6]. Such complexes are well-known analogs of phosphate and have been studied extensively (Chaps. 1 and 5). Oftentimes vanadium(V), when acting as a cation, exists as a *cis*-dioxo group, VO_2^+. Thus, both anionic and cationic binding sites for vanadium(V) must be considered. Some of the more recent literature on the complexes of vanadium(IV) and vanadium(V) is reviewed in Chaps. 1, 3, 4, and 5. The reader is referred to previous reviews for citations to the earlier literature [1–6].

1.2. Physical Methods

Electron paramagnetic resonance (EPR) and nuclear magnetic resonance (NMR) are the most widely employed techniques for studying vanadium(IV) and vanadium(V), respectively. VO^{2+} is a $3d^1$, $S = ½$ system with a nondegenerate ground state and no excited states nearby in energy. These are the requirements to observe EPR signals in both room temperature and frozen solutions. The hyperfine coupling to the ^{51}V nucleus and the *g* factor have long been used to infer information about the coordination environment of the metal [3]. More recently, electron nuclear double resonance (ENDOR) and electron spin-echo envelope modulation (ESEEM) have been used to probe the ligand nuclei directly [9,10] (see also Chaps. 3 and 19).

There is an extensive literature on ^{51}V NMR spectroscopy of vanadates and vanadate esters and complexes. Because of its modest quadrupole moment and large nuclear moment, the ^{51}V nucleus is relatively easily observed by NMR spectroscopy. Much of the chemistry of small vanadate chelates as well as some protein complexes has been elucidated by NMR [11,12]. Magnetic resonance techniques as well as several other physical methods for studying vanadium are reviewed in Chap. 19.

2. VANADIUM(IV)-PROTEIN COMPLEXES

There is a considerable literature on VO^{2+} complexes with proteins. Several key proteins are discussed below with leading references to the literature. The proteins selected for coverage in this chapter are not all inclusive, however. Space does not permit citation of the many contributors to our understanding of vanadium(IV) and vanadium(V) interactions with proteins. Here we simply review some of the salient features of several of the most extensively studied protein complexes.

2.1. Insulin

VO^{2+} bonded to insulin was the first protein complex studied with the vanadyl(IV) electron spin probe method [13]. This early work laid the groundwork for later studies with other proteins [3]. The insulin work demonstrated that VO^{2+} binds at the two symmetry related Zn^{2+} sites in the hexameric solid state structure. The native metal is displaced by VO^{2+}. In the hexamer, the metal is bound to the three imidazole groups of the B10 His residues of adjacent subunits, the residues being related by trigonal symmetry. EPR spin Hamiltonian parameters in conjunction with additivity relationships were used to assess the equatorial ligand field of the vanadium. A square-bipyramidal metal site containing two histidyl nitrogens and two water molecules in the basal plane and the third histidyl and the vanadyl oxygen in the apical positions was inferred from the data. Additional weaker complexation at protein carboxylate groups was observed, a phenomenon subsequently found with other proteins as well (e.g., [14,15]).

2.2. Carboxypeptidase A and Carbonic Anhydrase B

In further development of the use of VO^{2+} as a spin probe, VO^{2+} was substituted for the native Zn^{2+} in the enzymes carboxypeptidase A [14] and carbonic anhydrase B [16]. Zn^{2+} is essential for activity in both of these enzymes. VO^{2+} was found to restore both peptidase and esterase activities in carboxypeptidase A with maximal activity achieved when vanadium was bound 1:1 with the protein as for the native Zn^{2+} protein. The VO^{2+} derivative of carbonic anhydrase B, however, was inactive toward the ester substrate *p*-nitrophenyl phosphate. The EPR parameters suggested that, in the case of carboxypeptidase A, the equatorial environment about the VO^{2+} ion consisted of the imidazole nitrogens of His-69 and His-196 and two water molecules with the axial position possibly occupied by Glu-72. A similar coordination was inferred for carbonic anhydrase. Similar metal site structures in both solution and solid state was suggested by the similarity in the EPR spin Hamiltonian parameters for the two physical states of both proteins [14,16]. The ability to carry out metal ion exchange and displacement reactions involving vanadium was demonstrated for carbonic anhydrase B [16].

2.3. Serum Albumin

The binding of vanadium by serum albumin is of special interest since this protein may play a role in transport of vanadium in the blood [17–19]. Moreover, significant vanadium contamination in the form of vanadium(V) has been observed in commercial preparations of serum albumins from various species [20,21]. Vanadate(V) binds only weakly to albumin [18] and can be removed by dialysis against chelators such as EDTA or *o*-phenanthroline [20,21]. In serum samples studied in vitro, VO^{2+} binds appreciably to albumin as well as to transferrin whereas vanadate(V) is primarily bound to transferrin [18]. The redox cycling of vanadium between the +4 and +5 states occurs on a short time scale relative to the residence time of vanadium in the circulation, implying that both oxidation states are probably present to some degree in blood.

One VO^{2+} binds tightly to bovine serum albumin at the specific sites for Cu^{2+} located in the N terminus of the polypeptide chain (Asp-Thr-His-Lys) of the protein [15]. The nature of this Cu^{2+} binding site has been reviewed [22]. There are four to five additional weaker binding sites for VO^{2+} at carboxylate groups on the protein. At pH 4.8 and above, binding at the strong site is complete. An association constant of 2.6×10^6 M^{-1} was measured at pH 5 [15] but is probably four or more orders of magnitude higher at pH 7. The EPR parameters suggest that most of the ligands are oxygen donors with perhaps one nitrogen donor; however, the nature of the metal coordination is not well established.

2.4. Collagen

Binding of VO^{2+} to tropocollagen in room temperature solution results in an isotropic EPR spectrum characterized by a rotational correlation time of about 6×10^{-9} sec, which is short relative to the correlation time of $\sim 10^{-7}$ sec for the protein as a whole [23]. A similar situation has also been found for weak binding to the carboxylate groups of serum albumin [15]. In both proteins there is local motion of the VO^{2+} ion at its binding site. This observation plus the EPR parameters suggest that the interaction of VO^{2+} with collagen occurs in the nonhelical end regions of the protein through monodentate coordination with either an amino or imidazole nitrogen in liquid solution. In frozen samples, binding in the equatorial plane of the VO^{2+} appears to involve two imidazole groups of histidyl residues and two water molecules as for insulin [13], carbonic anhydrase [16], and carboxypeptidase A [14]. Thus, this type of coordination motif is common to several vanadyl(IV) protein complexes.

2.5. Transferrin, Lactoferrin, and Ovotransferrin

The transferrins are perhaps the most extensively studied of the VO^{2+}-protein complexes. Interest in these iron transport and sequestering proteins stems from the early use of VO^{2+} as a spin probe of their metal sites and also from the potential role of serum transferrin in the transport of vanadium [17–19,24]. Binding studies and EPR spectra have been reported for human serum transferrin [25–28], ovotransferrin [29,30], and lactoferrin [31,32]. Although the Fe^{3+} ligands of all of these proteins are known to be the same [33], they exhibit somewhat different VO^{2+} EPR spectra, suggesting some variations in metal site geometries due to conformational and structural differences among the different proteins in the vicinity of the two metal binding sites.

The obligate anion carbonate, which is required for Fe^{3+} binding at the specific sites of these proteins, is also required for VO^{2+} binding [27,31]. Ovotransferrin is an exception in that VO^{2+} will bind in the absence of carbonate but the coordination has changed as reflected in the EPR spectrum [30]. The binding stoichiometry is 2 VO^{2+} per protein as it is for iron. The ligands to Fe^{3+} are 2 Tyr, 1 Asp, 1 His, and bidentate CO_3^{2-} to give a six-coordinate complex [33]. In the case of VO^{2+}, ESEEM [34] and EPR [27] measurements have demonstrated that carbonate is a ligand. ESEEM of $(VO^{2+})_2$-transferrin has also shown nitrogen couplings from equatorially coordinated imidazole of histidine [34], a result confirmed by ENDOR spectroscopy [35]. ESEEM suggests the lack of water or hydroxide bound to the vanadium [34], whereas the coordination of two phenolate groups of tyrosine residues has been demonstrated by UV difference spectroscopy [36]. No information regarding whether Asp, a ligand in the Fe^{3+} complex, is present in the first coordination sphere has been obtained. No combination of the known or potential ligands in the equatorial plane of the vanadium predicts the observed spin Hamiltonian parameters based on additivity relationships and model compounds [3]. This finding suggests a distorted geometry from an idealized square bipyramid about the vanadium.

2.6. Ferritin

The complexes of VO^{2+} with the iron storage protein ferritin have been used to investigate the binding of Fe^{2+} and Fe^{3+} to this iron storage protein [37–40]. Vanadyl ferritin complexes are also of interest since such complexes are observed in the tissues of rats on vanadium supplemented diets [17]. High and low pH forms of VO^{2+} binding are observed in vitro which appear to be related to proton dissociation from metal bound H_2O to form a hydroxo complex at higher pH [37,40]. ENDOR [39] and ESEEM [40] reveal the presence of nitrogen coordination, presumably from histidine. This putative histidyl site is also involved in binding of iron to the protein since both Fe^{2+} and Fe^{3+} compete with VO^{2+} for the site. The stoichiometry of VO^{2+} binding is 2 VO^{2+} per 3 subunits in the 24-subunit protein, implying that the binding occurs in the vicinity of each of the eight threefold rotation axes [38]. Hydrophilic channels are located along the axis and are thought to be avenues by which iron gains entry to the protein [38,41].

2.7. D-Xylose Isomerase

VO^{2+} has been used as a probe of the two types of binding sites (A and B) on each subunit in tetrameric D-xylose isomerase [44]. Both types of sites must be occupied by divalent cations for the enzyme to be active. A key question is which site contains the histidine residue. EPR and ENDOR data of the VO^{2+} proteins show that the low-affinity A site consists of oxygen donor ligands without nitrogen ligation. Exchangeable H_2O ligands in the vicinity of the VO^{2+} are observed with this site. In contrast, the high-affinity B site was demonstrated to contain a histidine ligand as evidenced by ^{14}N resonances in the ENDOR spectrum and an exchangeable proton assigned to an N-H group.

2.8. *S*-Adenosylmethionine Synthetase

S-Adenosylmethionine synthetase is an enzyme which has been fruitfully studied using VO^{2+} as a spin probe [45–47]. It catalyzes the transfer of the adenosyl group of ATP to the sulfur atom of L-methionine to form *S*-adenosylmethionine (AdoMet). The enzyme requires two divalent metal ions such as Mg^{2+}, Mn^{2+}, or Ca^{2+} and a monovalent cation K^+, Na^+, or Tl^+ for activity. The two divalent and the monovalent cations are associated in a cluster which has not been fully defined. VO^{2+} can be substituted for one of the divalent cations with Ca^{2+} or Mg^{2+} as the second metal, resulting in an active enzyme which can be probed by EPR and ESEEM spectroscopies. A superhyperfine coupling of 23 G from ^{203}Tl + ^{205}Tl, both $I = ½$ nuclei, are cleanly resolved in the VO^{2+} EPR spectrum of the enzyme-VO^{2+}-methionine-AMPPNP-Tl^+ complex (AMPPNP ≡ 5′-adenylylimidodiphosphate ≡ β,γ-imidoadenosine 5′-triphosphate, an ATP analog), demonstrating that the VO^{2+} and Tl^+ are in close proximity [46]. Superhyperfine coupling of 8.5 G from two equivalent ^{31}P nuclei are observed in complexes with polyphosphate (PP_i) substrates or products: enzyme-VO^{2+}-ATP-methionine and enzyme-VO^{2+}-PP_i-Mg^{2+}. That the diphosphate group is a bidentate ligand to the VO^{2+} was confirmed by [^{17}O]pyrophosphate experiments [45]. Narrowing of the EPR linewidths in D_2O/H_2O exchange experiments suggest the presence of coordinated water [45].

ESEEM measurements of selectively labeled ^{15}N of the ε-amino nitrogen of lysine residues in the enzyme demonstrates that lysine is a ligand to the vanadium [48]. Lysine coordination remains in the intermediate complex enzyme-VO^{2+}-AdoMet·PPP_i·K^+ and in the initial product complex enzyme-VO^{2+}-AdoMet·PP_i·K^+ but is lost in the subsequent product enzyme-VO^{2+}-AdoMet·K^+, where PP_i dissociates and the methionyl amino nitrogen of AdoMet becomes a ligand [47]. Thus through EPR/ESEEM measurements on a series of complexes, information regarding the rearrangement of the coordination sphere of the vanadium is obtained. Based on the EPR and ESEEM data and additivity relationships, the most reasonable assignment for the metal coordination prior to catalysis is two phosphate oxygens, a water molecule and a ϵ-amino nitrogen of lysine in the equatorial plane of the VO^{2+} group. The identity of the axial ligand opposite the vanadyl oxygen is unknown.

2.9. Pyruvate Kinase

Pyruvate kinase is another example of an enzyme which has been very successfully probed using VO^{2+} [48]. Pyruvate kinase, although not a metalloprotein, requires metal ions for the catalytic formation of ATP from ADP and phosphoenolpyruvate. However, VO^{2+} does not impart appreciable activity to the enzyme in the physiological reaction but is an activator of the enzyme-catalyzed phosphorylation of glycolate by ATP [49]. Like *S*-adenosylmethionine synthetase, pyruvate kinase requires two divalent metal ions (Mg^{2+}) and a monovalent cation (K^+) for activity. Several different complexes of VO^{2+}, the enzyme, pyruvate, and monovalent cation have been studied by ESEEM [48]. Reporter molecules [1-^{13}C]pyruvate, [2-^{13}C]pyruvate, and pyruvate-d_3, and reporter ions $^{23}Na^+$, $^{39}K^+$, and $^{133}Cs^+$ were employed. Magnetic couplings between the unpaired electron of VO^{2+} with all the isotopically labeled pyruvate molecules were observed, demonstrating that pyruvate is bidentate to the VO^{2+}, a result confirmed by EPR measurements on the [^{17}O]pyruvate complex [49]. Also, lysine coordination from the protein is observed [48]. Couplings with $^{23}Na^+$, $^{39}K^+$, and $^{133}Cs^+$ in the ESEEM [48] and $^{203,205}Tl$ in the EPR [50] indicated that these ions are in close proximity to the vanadium.

2.10. Calmodulin

The regulatory protein calmodulin binds four Ca^{2+} ions, two at high-affinity and two at low-affinity sites. The protein has been the subject of intensive study for many years. VO^{2+}, like Ca^{2+}, binds to calmodulin with an approximate stoichiometry of 4:1 in two classes of sites (A and

B) whose EPR spectra are resolved in frozen solution [51,52]. The A sites are of higher affinity for VO^{2+} than the B sites. Ca^{2+} competes with VO^{2+} for binding at the weaker B sites but under the conditions of the experiment does not compete with VO^{2+} for binding at the stronger A sites. These latter sites may bind VO^{2+} much more strongly than Ca^{2+} or may simply be Ca^{2+}-independent sites. The EPR spectra of mixed Ca^{2+}, VO^{2+}-calmodulin complexes compared with VO^{2+}-calmodulin suggest that the A and B sites are relatively independent, i.e., the spin Hamiltonian parameters of the A class of sites change little when Ca^{2+} is bound to the B sites.

3. VANADIUM(V)-PROTEIN COMPLEXES

3.1. Bromoperoxidase

The haloperoxidases were the first bona fide enzymes in which vanadium was shown an essential component for activity [42]. The vanadium is in the +5 state in the catalytically active native form of the enzyme but can be reduced to EPR-observable VO^{2+} (see [43] and Chap. 10 for reviews). A number of physical techniques suggest a coordination sphere comprising oxygen and nitrogen donor atoms. The role of the vanadium in catalysis is unclear but it may be simply acting as a Lewis acid. The reader is referred to Chap. 10 for further information concerning this very interesting enzyme.

3.2. Ribonuclease

Ribonuclease was the first enzyme in which vanadate(V) [as well as vanadyl(IV)] was shown to be a potent inhibitor, in this case of the hydrolysis reaction of uridine 2′,3′-cyclic phosphate [53]. Here the inhibitor was in the form of a vanadium(V)-uridine complex [53]. Lindquist and coworkers [53] postulated that the vanadate uridine chelate formed a trigonal bipyramidal pentacoordinate complex with the protein, mimicking the transition state of the enzyme-substrate complex. Subsequent neutron/X-ray diffraction and ^{51}V and ^{1}H NMR studies a decade later confirmed this hypothesis [54,55], arriving at a structure remarkably close to that postulated by Lindquist et al. [53]. In the vanadate(V)-uridine complex with ribonuclease A, one of the two apical positions of the trigonal bipyramid is occupied by the 2′ oxygen of ribose which is hydrogen-bonded to Lys-41. The other apical position is occupied by a vanadate oxygen which is hydrogen-bonded to His-119. One of the three basal plane oxygens is the 3′ oxygen of ribose and the other two are vanadate oxygens, one of which is hydrogen-bonded to His-12. Studies of these complexes have provided new insights into the mechanism of action of RNase A [55]. The structures of model compounds with trigonal bipyramidal structures have also been reported [56,57].

The X-ray structure of $H_2VO_4^-$ complexed to ribonuclease T_1 has been determined to 1.8 Å [58]. In this guanosine-free complex, the vanadate anion is seen to have a distorted tetrahedral geometry with two vanadate oxygens slightly shorter than the other two. The vanadate is fixed in place by a complex network of 11 hydrogen bonds involving 8 amino acid side chains and 3 water molecules. Comparison between the structures of the guanosine 2′-phosphate RNase T_1 and vanadate RNase T_1 complexes has revealed changes in protein structure thought to be due to conformational adjustments associated with substrate binding.

3.3. Phosphoglucomutase

Phosphoglucomutase (PGM) catalyzes the phosphoryl transfer reaction between α-D-glucose-1-phosphate (Glc-1-P) and α-D-glucose-6-phosphate. The 6-vanadate ester of Glc-1-P, namely, V-6-Glc-1-P, is a competitive inhibitor vs. 1,6-bisphosphate (Glc-P_2), an effect which has been ascribed to binding at the V-6-Glc-1-P to the active site of the dephospho form of the enzyme (E) to form an E-V-6-Glc-1-P which rapidly interconverts to another species E*-V-6-Glc-1-P [59]. It is postulated that this latter species represents a transition state for the interconversion between E-P·Glc-1-P and E·P-6-Glc-1-P in which the vanadium is most likely in a trigonal bipyramidal pentacoordinate complex [59]. Extensive kinetics and UV and NMR spectroscopic measurements by Ray and coworkers [60–62] have shown that the vanadate group binds where P_i transfer normally takes place. UV spectroscopy indicates that the vanadium(V) is in an unusual environment. However, the spectrum is most consistent with a decrease in "net bonding" to the vanadium in the transition state analog even though the coordination number may have increased from 4 to 5 [61,62] as proposed by Gresser and coworkers [59].

3.4. Transferrin

Transferrin represents a special case in that both vanadium(V) and vanadium(IV) (see above) bind to the two specific Fe^{3+} sites of the protein. In contrast to vanadium(IV), vanadium(V) binding does not require the presence of bicarbonate [63]. Binding of vanadium(V) has been probed by UV difference spectroscopy which measures tyrosine ligation [18,63], $^{35}Cl^-$ NMR spectroscopy which measures protein folding [18], and ^{51}V NMR spectroscopy which measures bound vanadium directly [64,65]. It is interesting that the vanadium(V) ions bound in the two sites are seen by ^{51}V NMR spectroscopy and are spectroscopically distinguishable [64,65]. In this case only the central $m_I = -1/2 \leftrightarrow +1/2$ transition of the $I = 7/2$ nucleus is observed with the $(V^{5+})_2$-protein complex being in the motional narrowing regime [65]. The vanadium(V) is most likely five or six-coordinate and in the cationic *cis*-dioxovanadium(V) form, VO_2^+, with two tyrosine ligands from the protein.

3.5. Glucose-6-Phosphate Dehydrogenase

Glucose-6-phosphate dehydrogenase catalyzes the oxidation of glucose-6-phosphate to D-glucono-δ-lactone-6-phosphate which then rapidly hydrolyzes to 6-phosphogluconate. $NADP^+$ or NAD^+ is a cofactor for the reaction. Gresser and coworkers [66] showed that vanadate activates the catalysis of the oxidation of glucose itself by the enzyme. Here a glucose-6-vanadate ester is formed in solution which then serves as the substrate for the enzyme. However, vanadate itself is an inhibitor of the native reaction, namely, the oxidation of glucose-6-phosphate [67]. Inhibition does not appear to be a simple substitution for phosphate at a binding site on the protein. From species correlations using ^{51}V NMR spectroscopy, Crans and Schelble [67] showed that it is the vanadate dimer and/or tetramer which is responsible for the inhibition and not the monomer. The data suggest that oligomeric forms of vanadate may be important inhibitors of other enzymes in which organophosphates are substrates, especially at high concentrations of vanadate. The mode of binding to the enzyme is unknown.

3.6. Other Proteins

There are numerous other enzymes which are either activated or inhibited by vanadate(V), including acid and alkaline phosphatases, various ATPases, kinases, and phosphoglycerate mutase among others. The literature on these enzymes is extensive. The reader is referred to various reviews [1–6,68] and other chapters in this volume for further information.

4. CONCLUDING REMARKS

Over the past 15 years the interest in the biochemistry of vanadium has grown enormously. Vanadyl(IV) is now being used routinely as an electron resonance spectroscopic probe of metal binding sites. Recent examples employing ENDOR and ESEEM as well as EPR spectroscopy, including pyruvate kinase, D-xylose isomerase, *S*-adenosylmethionine synthetase, ferritin, and transferrin, have been briefly discussed above. Vanadate(V) has been and continues to be extensively used as a probe of phosphate metabolizing enzymes. While there is no shortage of papers on vanadium- protein interactions, the number of crystal structures is limited and none exists for vanadium(IV). This situation will undoubtedly change as the level of interest in these systems continues to increase. For the moment, we must be satisfied with the insight into their properties which is provided by spectroscopic, kinetic, and model compound studies.

ACKNOWLEDGMENT

This work was supported by NIH grant R37 GM20194.

ABBREVIATIONS AND DEFINITIONS

AdoMet	*S*-adenosylmethionine
ADP	adenosine 5′-diphosphate
AMPPNP	5′-adenylylimidodiphosphate

ATP	adenosine triphosphate
E	enzyme
EDTA	ethylenediamine-*N,N,N′,N′*-tetraacetate
ENDOR	electron nuclear double resonance
EPR	electron paramagnetic resonance
ESEEM	electron spin-echo envelope modulation
Glc-1-P	α-D-glucose-1-phosphate
Glc-P_2	1,6-bisphosphate
NAD^+	nicotinamide adenine dinucleotide
$NADP^+$	nicotinamide adenine dinucleotide phosphate
NMR	nuclear magnetic resonance
PGM	phosphoglucomutase
PP_i	pyrophosphate
PPP_i	triphosphate
RNase	ribonuclease
UV	ultraviolet-visible

REFERENCES

1. A. Butler and C. J. Carrano, *Coord. Chem. Rev., 109*, 61 (1991).
2. D. Rehder, *Angew. Chem. Int. Ed. Engl., 30*, 148 (1991).
3. N. D. Chasteen, in *Biological Magnetic Resonance*, Vol. 3 (J. Reuben and L. Berliner, eds.), Plenum Press, New York, 1981, pp. 53–117.
4. N. D. Chasteen, *Struct. Bond., 53*, 105 (1983).
5. A. Butler, in *Vanadium in Biological Systems* (N. D. Chasteen, ed.), Kluwer Academic, Dordrecht, 1990, pp. 25–49.
6. D. W. Boyd and K. Kustin, *Adv. Inorg. Biochem., 6*, 311 (1984).
7. M. Branca, G. Micera, A. Dessi, D. Sanna, and K. N. Raymond, *Inorg. Chem., 29*, 1586 (1990).
8. B. Borah, C.-W. Chen, W. Egan, M. Miller, A. Wlodawer, and J. S. Cohen, *Biochemistry, 24*, 2058 (1985).
9. S. S. Eaton and G. R. Eaton, in *Vanadium in Biological Systems* (N. D. Chasteen, ed.), Kluwer Academic, Dordrecht, 1990, pp. 199–222.

10. N. D. Chasteen, in *Methods in Enzymology: Metallobiochemistry*, Part C (J. F. Riordan and B. L. Vallee, eds.), Academic Press, Orlando, 1993, pp. 232–244.
11. D. Rehder, in *Vanadium in Biological Systems* (N. D. Chasteen, ed.), Kluwer Academic, Dordrecht, 1990, pp. 173–198.
12. O. W. Howarth, *Progr. NMR Spectrosc.*, *22*, 453–485 (1990).
13. N. D. Chasteen, R. J. DeKoch, B. L. Rogers, and M. W. Hanna, *J. Am. Chem. Soc.*, *95*, 1301 (1973).
14. R. J. DeKoch, D. J. West, J. C. Cannon, and N. D. Chasteen, *Biochemistry*, *13*, 4347 (1974).
15. N. D. Chasteen and J. Francavilla, *J. Phys. Chem.*, *80*, 876 (1976).
16. J. J. Fitzgerald and N. D. Chasteen, *Biochemistry*, *13*, 4338 (1974).
17. N. D. Chasteen, E. M. Lord, H. J. Thompson, and J. K. Grady, *Biochim. Biophys. Acta*, *884*, 84 (1986).
18. N. D. Chasteen, J. K. Grady, and C. E. Holloway, *Inorg. Chem.*, *25*, 2754 (1986).
19. W. R. Harris, S. B. Friedman, and D. Siberman, *J. Inorg. Biochem.*, *20*, 157 (1984).
20. H. Sakurai, M. Nishida, K. Kida, M. Koyama, and J. Takada, *Inorg. Chim. Acta*, *138*, 149 (1987).
21. H. Sakurai, M. Nishida, M. Koyama, and J. Takada, *Biochim. Biophys. Acta*, *924*, 562 (1987).
22. B. Sarkar, *Progr. Food Nutr. Sci.*, *11*, 363 (1987).
23. R. P. Ferrari, *Inorg. Chim. Acta*, *176*, 83 (1990).
24. E. Sabbioni and E. Marafante, *Bioinorg. Chem.*, *9*, 389–407 (1978).
25. J. C. Cannon and N. D. Chasteen, *Biochemistry*, *14*, 4573 (1975).
26. N. D. Chasteen, L. K. White, and R. F. Campbell, *Biochemistry*, *16*, 363 (1977).
27. R. F. Campbell and N. D. Chasteen, *J. Biol. Chem.*, *252*, 5996 (1977).
28. L. K. White and N. D. Chasteen, *J. Phys. Chem.*, *82*, 279 (1979).
29. J. D. Casey and N. D. Chasteen, *J. Inorg. Biochem.*, *13*, 111 (1980).
30. J. D. Casey and N. D. Chasteen, *J. Inorg. Biochem.*, *13*, 127 (1980).
31. J. Mazuerier, J.-M. Lhoste, J. Montreuil, and G. Spik, *Biochim. Biophys. Acta*, *745*, 44–49 (1983).
32. A. Carmichael and J. S. Vincent, *FEBS Lett.*, *105*, 349 (1979).

33. E. N. Baker and P. F. Lindley, *J. Inorg. Biochem.*, *47*, 147 (1992).
34. S. S. Eaton, J. Dubach, K. M. More, G. R. Eaton, G. Thurman, and D. R. Ambruso, *J. Biol. Chem.*, *264*, 4776 (1989).
35. P. M. Hanna, N. D. Chasteen, G. Rottman, and P. Aisen, *Biochemistry*, *30*, 9210 (1991).
36. J. C. Cannon and N. D. Chasteen, in *Proteins of Iron Storage and Transport in Biochemistry and Medicine* (R. R. Crichton, ed.), North Holland, Amsterdam, 1975, pp. 67–72.
37. N. D. Chasteen and E. C. Theil, *J. Biol. Chem.*, *257*, 7672 (1982).
38. J. G. Wardeska, B. Viglione, and N. D. Chasteen, *J. Biol. Chem.*, *261*, 6677 (1986).
39. P. M. Hanna, N. D. Chasteen, G. A. Torrman, and P. Aisen, *Biochemistry*, *30*, 9211 (1991).
40. G. J. Gerfen, P. M. Hanna, N. D. Chasteen, and D. J. Singel, *J. Am. Chem. Soc.*, *113*, 9513 (1991).
41. A. Treffry, E. R. Bauminger, D. Hechel, N. W. Hodson, I. Nowik, S. J. Yewdall, and P. M. Harrison, *Biochem. J.*, *296*, 721 (1993).
42. H. Vilter, *Phytochemistry*, *23*, 1387 (1984).
43. R. Wever and B. E. Krenn, in *Vanadium in Biological Systems* (N. D. Chasteen, ed.), Kluwer Academic, Dordrecht, 1990, pp. 63–97.
44. R. Bogumil, J. Hüttermann, R. Kappl, R. Stabler, C. Sudfeldt, and H. Witzel, *Eur. J. Biochem.*, *196*, 305 (1991).
45. G. D. Markham, *Biochemistry*, *23*, 470 (1984).
46. G. D. Markham and T. S. Leyh, *J. Am. Chem. Soc.*, *109*, 599 (1987).
47. C. Zhang, G. D. Markham, and R. LoBrutto, *Biochemistry*, *32*, 9866 (1993).
48. P. A. Tipton, J. McCracken, J. B. Cornelius, and J. Peisach, *Biochemistry*, *28*, 5720 (1989).
49. K. A. Lord and G. H. Reed, *Arch. Biochem. Biophys.*, *281*, 124 (1990).
50. K. A. Lord and G. H. Reed, *Inorg. Chem.*, *26*, 1464 (1987).
51. J. Nieves, L. Kim, D. Puett, and L. Echegoyen, *Biochemistry*, *26*, 4523 (1987).
52. R. H. Ahmed, J. Nieves, L. Kim, L. Echegoyen, and D. Puett, *J. Protein Chem.*, *6*, 431 (1987).

53. R. N. Lindquist, J. L. Lynn, Jr., and G. E. Lienhard, *J. Am. Chem. Soc.*, *95*, 8762 (1973).
54. A. Wlodawer, M. Miller, and L. Sjölin, *Proc. Natl. Acad. Sci. USA*, *80*, 3628 (1993).
55. B. Borah, C.-W. Chen, W. Egan, M. Miller, A. Wlodawer, and J. S. Cohen, *Biochemistry*, *24*, 2058 (1985).
56. D. C. Crans, R. A. Felty, and M. M. Miller, *J. Am. Chem. Soc.*, *113*, 265 (1991).
57. T. W. Habley, R. J. Judd, and P. A. Lay, *Inorg. Chem.*, *31*, 343 (1992).
58. D. Kostrewa, H.-W. Choe, U. Heinemann, and W. Saenger, *Biochemistry*, *28*, 7593 (1989).
59. M. D. Percival, K. Doherty, and M. J. Gresser, *Biochemistry*, *29*, 2764 (1990).
60. W. J. Ray, Jr., J. W. Burgner, II, and C. B. Post, *Biochemistry*, *29*, 2770 (1990).
61. W. J. Ray, Jr., and C. W. Post, *Biochemistry*, *29*, 2779 (1990).
62. W. J. Ray, Jr., and J. M. Puvathingal, *Biochemistry*, *29*, 2790 (1990).
63. W. R. Harris and C. J. Carrano, *J. Inorg. Biochem.*, *22*, 201 (1984).
64. A. Butler, M. J. Danzitz, and H. Eckert, *J. Am. Chem. Soc.*, *109*, 1864 (1987).
65. A. Butler and H. Eckert, *J. Am. Chem. Soc.*, *111*, 2802 (1989).
66. A. F. Nour-Eldeen, M. M. Craig, and M. J. Gresser, *J. Biol. Chem.*, *260*, 6836 (1985).
67. D. C. Crans and S. M. Schelble, *Biochemistry*, *29*, 6698 (1990).
68. N. D. Chasteen (ed.), *Vanadium in Biological Systems*, Kluwer Academic, Dordrecht, 1990.

8

Stimulation of Enzyme Activity by Oxovanadium Complexes

Paul J. Stankiewicz and Alan S. Tracey

Department of Chemistry, Simon Fraser University,
Burnaby, British Columbia, V5A 1S6, Canada

1. INTRODUCTION

Many of the important discoveries about vanadate chemistry that provide the basis for our understanding of the activation of enzyme function by vanadium anions were made only during the past decade. As a result, much of our knowledge concerning the stimulation of enzyme activity by vanadium is highly limited. At the present time, the process of direct activation is thought to be reasonably well understood. Indirect activation can be exceedingly complicated and its basis may involve a cascade of effects. For the situation in which vanadium directly activates enzyme activity, the activation, in most cases, is directly caused by the formation of vanadate complexes that mimic the structure of the normal substrate and are then utilized as such by the enzyme. This substrate, in all cases that have been reported, is a phosphate derivative.

The situation is much more complicated when there is indirect activation of the enzymic function. A relatively simple process that could lead to such an activation might involve an enzyme function that is normally regulated in a feedback loop. If the behavior of the loop is influenced by the presence of vanadium, then the regulation may be impaired and enzyme catalysis could proceed in an unrestrained fashion. Clearly, much more complicated interaction networks are to be expected.

In this chapter, both of the above types of enzyme activation will be examined but this examination will be preceded by a description of some of the chemistry that is relevant to the biological function that leads to enzyme activation. The various chapters of this volume discuss a number of other biochemical effects of vanadium and there are also a number of recent review articles and compendiums that discuss much of the chemistry and biochemistry of vanadium [1–3].

2. RELEVANT CHEMISTRY

One of the rather unique properties of vanadium is the facile manner in which its oxo compounds undergo oxidation and reduction. Throughout the range of pH that is of principal interest for biochemical work, about pH 4–10, the major oxidation states are vanadate [vanadium(V)] and vanadyl [vanadium(IV)]. The interplay between these states can be mediated by reactive extraneous groups leading to oxidation or reduction of the group itself. Although the vanadium(III) oxidation state may be involved in some redox reactions, a discussion of its properties is not included here (see Chap. 2).

In aqueous solution, both vanadate and vanadyl undergo a number of structural modifications and self-condensations that are dependent both on the pH of the solution of interest and on the concentration of the reactants. Furthermore, both ions can undergo additional reactions with a wide range of ligands to generate many products of varying complexity. The formation of a number of such complexes is relevant to understanding vanadium biochemistry. A more detailed description of vanadium chemistry is provided in earlier chapters of this book.

2.1. Vanadate Aqueous Chemistry

2.1.1. Oligomerization Reactions

At concentrations of about 1 mM and at near-neutral pH, the predominate vanadate species in solution are the monomer, dimer, and tetramer with the pentamer being a minor component. The trimer is also a minor component that, although present, is observed most clearly at pH values above ~9. Other oligomers are also formed but are present in low concentrations relative to the major components [4,5]. Upon mild acidification of the solution, decavanadate becomes the major component [4,6].

The relative concentrations of the various vanadate derivatives are strongly dependent on the ionic strength and the pH of the medium. This is shown very clearly in Fig. 1 where the formation constant of the dimer from the monomer ($[V_2]/[V_1]^2$, Fig. 1A) and the tetramer from the monomer ($[V_4]/[V_1]^4$, Fig. 1B) are shown as a function of both ionic strength and pH.

Except for those situations that involve decavanadate, equilibrium conditions are rapidly established [4,7]. Decavanadate forms and decomposes relatively slowly and the kinetics of decavanadate decomposition are slow enough that problems are routinely encountered if vanadate solutions are acidified, even transiently. The formation of decavanadate is strongly favored by mildly acidic conditions, pH 3–5. Below this pH the cationic species, VO_2^+, is formed and with decreasing pH becomes the predominant product [5].

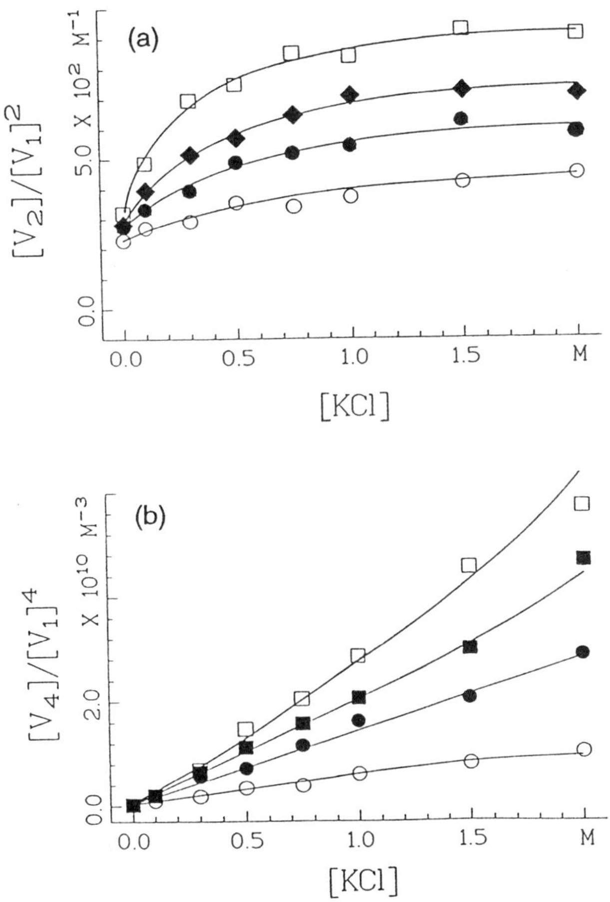

FIG. 1. The variation in the formation constants for the formation of divanadate (a) and tetravanadate (b) are shown as a function of added KCl for pH values of: □ pH 6.5; ♦ pH 7.0; • pH 7.5; ○ pH 8.0. The solid lines have no theoretical significance.

2.1.2. Reactions with Alcohols

Vanadate undergoes a rapid and reversible condensation reaction with both primary and secondary aliphatic alcohols to form vanadate esters. Although equilibrium conditions are established within a few milliseconds [8,9], the formation constants are quite small, being in the order of about 0.2 M^{-1} for formation of the monoester and about one-half to one-fourth of this for incorporation of a second alkoxo group to form the diester. These two reactions are, of course, dependent on the pK_a values of both vanadate and its monoester [9,10]. The chemistry is not significantly dependent on whether the alcohols are cyclic or acyclic, and esters are readily formed with the hydroxyl groups of cyclohexane diols [11], carbohydrates [11,12], or nucleosides [13–16]. The formation of neither aliphatic nor aromatic esters is particularly sensitive to the pK_a of the starting alcohol [10,17].

Aromatic alcohols behave similarly to their aliphatic counterparts but the formation constants of the vanadate esters tend to be slightly larger, in the order of 1 M^{-1} [17,18]. Although this value is not large, it is about five orders of magnitude larger than the formation constants for structurally similar phosphate esters, which are comparatively high-energy species. This suggests that, even at low vanadium concentrations, there may be a significant biological activity associated with aromatic vanadate esters if the corresponding phosphate ester is biochemically active.

2.1.3. Anhydride Formation with Carboxylates and Phosphates

It is well known that vanadate undergoes strong complexation reactions with carboxylate-containing bidentate ligands. Complexation is accompanied by a change in coordination geometry, generally to five coordination or six coordination or a mixture of both, dependent on the ligand [19]. In addition to such reactions, in aqueous solution vanadate spontaneously forms anhydrides with very simple carboxylic acids, such as acetic acid. These products are in fairly rapid dynamic equilibrium and both mono and bis ligand products are formed and are thought to have octahedral coordination [20].

Phosphate spontaneously forms an anhydride that, like the acetate product, is in rapid dynamic equilibrium. Unlike with acetate, only a monoanhydride was observed to form and this product has been assigned a tetrahedral coordination [21]. The phosphovanadate formation constant is an order of magnitude smaller than that for divanadate formation (3.5×10^2 M^{-1}) but is six orders of magnitude larger than the formation constant (4×10^{-5} M^{-1}) of pyrophosphate from two phosphate ions [21]. A phosphate derivative is also formed with divanadate to give a product that is assumed to be linear, like triphosphate. Products similar to the above have been observed with pyrophosphate and adenosine monophosphate, the latter reactant forming an ADP analog with monovanadate and an ATP analog with divanadate [22]. In addition to the above, pyrophosphate forms a bis ligand bidentate complex.

^{51}V nuclear magnetic resonance (NMR) spectroscopy indicates that there is an important difference between divanadate and pyrophosphate in that divanadate does not bind Mg^{2+} particularly well [23]. The cause of this poor binding may well derive from the structure of divanadate itself. This structure is not known with certainty although it is generally assumed to be, and most probably is, close to linear.

2.1.4. Reactions with Amino Acids and Peptides

Although amino acids form a number of complexes with vanadate, none are highly favored [24–26]. This contrasts with the case of dipeptides where tridentate products are readily formed [24,26]. Formation of these products requires the presence of the terminal amino, the terminal carboxyl, and the peptido nitrogen. Appropriately substituted side chains that can replace one of the above groups yield similar derivatives. Reaction of vanadate with tyrosyl and imidazole side chains is comparable to that with phenol or imidazole. Imidazole, in addition, can replace the peptido nitrogen in the above complexes.

Tripeptides yield complexes of a probable tetradentate coordination while simple tetrapeptides do not provide products [26]. Although it has not yet been clearly shown, it seems likely that with the appropriate side chains, multidentate complexes of vanadate with polypeptides that do not involve the main chain groups will form readily. Many common buffers that have a di- or tripeptide-like structure, such as diethanolamine, also form complexes with vanadate [27,28] and must be used with circumspection [29].

2.1.5. Oligomer Chemistry

On reading this chapter, and the companion Chap. 9 that follows, a clear demarcation of biological function between vanadate and its oligomers will be evident. Although to a large extent much of this selectivity is a function of size, there are also significant chemical differences between the various species and this is also important in delineating the function of the various oligomers.

Oligomers of vanadate that do not have proton-bearing oxygens have not been observed to form esters or anhydride-type derivatives, presumably because of a requirement for the production of water as the product forms. Thus, divanadate reacts spontaneously with alcohols to afford the corresponding divanadate esters [8], whereas a similar reaction with tetravanadate has not been observed. In fact, it appears that, except for possible electrostatic-type binding, tetravanadate does not form complexes easily.

Decavanadate, unlike tetravanadate, carries protons [30] and reacts readily with alcohols to yield ester-type products. The kinetics for the formation of these materials is quite slow but they are easily observable in vanadium NMR spectra. Figure 2 shows a spectrum of an equilibrated system. By comparing the number and intensities of the signals observed in the spectrum to those expected from different

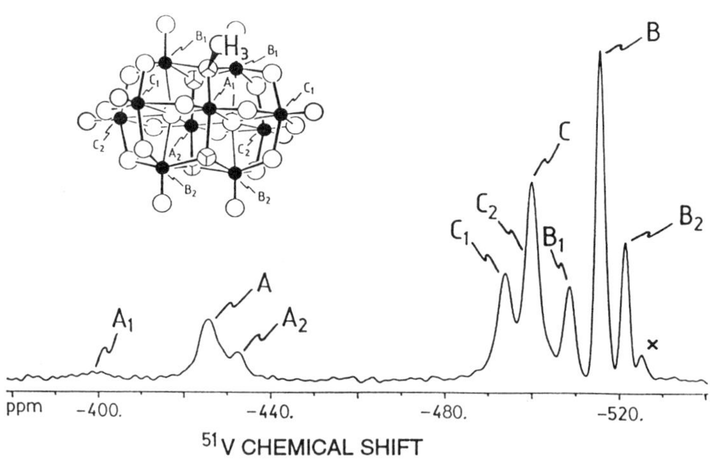

FIG. 2. A ^{51}V NMR spectrum showing the incorporation of methanol into the decavanadate structure. The signals labeled A, B, and C derive from unsubstituted decavanadate. The remaining signals arise from the positions indicated in the inset. Conditions of the experiment: 3.0 mM total vanadate; 3.7 M methanol, 20 mM MES buffer, 1.0 M KCl, pH 5.4. The signal marked with the X derives from a minor product.

positions of substitution, it seems evident that substitution occurs as indicated in the figure.

2.1.6. Reactions with Hydrogen Peroxide

That hydrogen peroxide is an exceedingly good ligand for vanadates has been known for many years. Only with the advent of the multinuclear capabilities and spectral dispersion of modern NMR spectrometers has it proven possible to study in detail the aqueous equilibria established in this system. There are a number of peroxovanadate products formed, including monoperoxovanadate, diperoxovanadate, triperoxovanadate, tetraperoxodivanadate, and others [31,32]. Of these, the two of predominant importance are the mono- and diperoxovanadates, and these are readily obtained in solution under the conditions of interest here [32].

The structures in solution of the latter two derivatives are not known with certainty and six, seven [32], and eight [31] coordination has been suggested. However, many X-ray structure determinations of crystalline peroxovanadate complexes reveal seven coordination and this seems to be the most likely solution structure. If this is true and the peroxide ligand is formally treated as being unidentate, then the structures are trigonal bipyramidal, the coordination often associated with enzyme inhibition by vanadate. Both mono- and diperoxovanadate undergo complexation reactions with a large number of ligands and may enter into biochemical processes by means of such products. Furthermore, peroxovanadates tend to be good oxidants and may influence some biochemical reactions via an oxidative pathway.

2.2. Peroxovanadate Aqueous Chemistry

Both mono- and diperoxovanadate undergo reactions with a number of metabolites and related materials to form heteroligand complexes of varying complexity. The chemistry of the two peroxovanadates is quite different and it seems unwise to extrapolate the chemistry of one complex to that of the other.

Diperoxovanadate undergoes a favorable condensation reaction with the histidyl imidazole ring ($K_f = 4.1 \times 10^2\ M^{-1}$) to form a monodentate product. This is quite different from the situation with monoperoxovanadate where a similar product has not been observed [33]. There is evidence, however, that side chain imidazoles can enter into the formation of multidentate complexes involving monoperoxovanadate [33].

Both mono- and diperoxovanadate form complexes with the aromatic hydroxyl of the tyrosyl side chain. Tyrosines aside, there is an interesting dichotomy in the chemistry of these two peroxovanadates that may be particularly relevant to the biological function of these complexes. Diperoxovanadate interacts in a moderate to strong monodentate fashion with amino, carboxyl, and imidazole ligands of amino acids [34] and simple peptides [33] but forms comparatively weak multidentate products. Monoperoxovanadate, on the other hand, tends toward strong multidentate complexation and forms comparatively weakly complexed monodentate products. It is not yet known as to whether this chemistry is reflected in the enzymology of these two peroxides.

2.3. Vanadium(IV) Aqueous Chemistry

Physiological reducing agents such as glutathione [35], cysteine [36], ascorbic acid [37], and catechols [38] can convert vanadium(V) to vanadium(IV). The commonly utilized form of vanadium(IV) is VO^{2+} and it is most stable below pH 5 [1]. At higher pH vanadyl incorporates two hydroxide ions with pK_b values of 5.3 and 6.7, respectively. At pH 7, the dihydroxide is essentially fully formed and is exceedingly insoluble. It seems likely that any significant amount of soluble vanadyl observed at pH 7 is chelated to a solubilizing ligand. Biological ligands which are known to bind vanadium(IV) include cysteine [36], glutathione, citrate, pyrophosphate, ATP, catechols, amino acids [1], and heme groups [39] or hemoglobin [40]. Because the complexation interaction is reversible, vanadyl, at a pH value above about 5.0, will eventually undergo air oxidation due to reaction with dissolved molecular oxygen. The superoxide formed in this reaction can remain associated to vanadate. The vanadate-superoxide complex is a strong oxidizing agent capable of oxidizing cysteine and NADH [41]. Oxidation-reduction cycles may occur in aqueous aerobic vanadyl or vanadate solutions in the presence of a reducing agent.

3. DIRECT ACTIVATION OF ENZYME ACTIVITY BY VANADIUM(V)

3.1. Subclass I: Vanadate Generates Functional Substrate Analogs

3.1.1. Indirect Participation in Catalysis

Certain classes of enzymes, including dehydrogenases, epimerases, and isomerases, catalyze redox reactions or structural changes that do not involve phosphate transfer. In some cases, the organic substrates for these reactions are phosphorylated compounds. The role of the phosphate group in these reactions is to increase substrate specificity by increasing the k_{cat}/K_m ratio by providing favorable binding interactions. It is reasonable to assume, and has been observed [42], that the dephosphorylated analog will act as an alternative substrate but with a higher K_m. It is possible to convert the dephosphorylated substrate into a high-affinity substrate by including vanadate in the reaction medium. Under such conditions, the spontaneously formed vanadate ester will provide a satisfactory analog to the phosphorylated compound [42,43]. For example, the glucose dehydrogenase activity of glucose-6-phosphate (glucose-6-P) dehydrogenase is activated by vanadate. A mixture of vanadate esters is formed with the different hydroxyl groups of glucose, but presumably only the species with the correct substrate specificity will bind in the active site of the enzyme. At a given concentration of glucose, vanadate, and NAD^+, the oxidation rate increased linearly as a function of enzyme concentration until the observed rate of reaction approaches the observed on-rate for glucose esterification by vanadate. At a sufficiently high enzyme concentration the rate will reach a constant value that is related to the second-order rate constant for vanadate ester formation. For the example here, the rate constant of glucose-6-V formation was 2.4 $M^{-1} s^{-1}$. The 6-vanadate ester of gluconic acid that is the product of this enzymic reaction rapidly and spontaneously hydrolyzes to give gluconic acid.

Other enzymes can function in a manner similar to glucose-6-P dehydrogenase. For instance, glycerol-3-phosphate dehydrogenase utilizes the dihydroxyacetone vanadate ester together with NADH to synthesize glycerol plus vanadate plus NAD^+ [43].

Epimerases and isomerases can function in a similar manner. The enzyme ribulose-5-phosphate epimerase converts a solution of ribulose plus vanadate to xylulose plus vanadate [43] while phosphoglucose isomerase converts fructose plus vanadate to glucose plus vanadate [43]. Ribose-5-phosphate isomerase converts ribose plus vanadate to ribulose plus vanadate [43]. Utilization of these and similar enzymes in organic chemistry has potential for the synthesis of required compounds that otherwise would require the chemical synthesis of the appropriate phosphate derivative, frequently a difficult task.

3.1.2. Vanadate Directly Participates in Catalysis

There are a number of examples in which vanadate can act as a surrogate substrate for enzymes that catalyze phosphotransferase reactions. The enzyme glyceraldehyde-3-phosphate dehydrogenase normally uses glyceraldehyde-3-phosphate and inorganic phosphate as its substrates. However, the enzyme will convert glyceraldehyde-3-phosphate and vanadate in the presence of NAD^+ to 3-phosphoglycerate plus vanadate plus NADPH [44]. The initial product 1-vanadio-3-phosphoglycerate rapidly hydrolyzes to provide 3-phosphoglycerate. A similar behavior has been observed with arsenate [45].

A somewhat similar reaction has been observed with nucleoside phosphorylase. This enzyme normally catalyzes the phosphorolysis of inosine to ribose-1-phosphate plus hypoxanthine. When inosine in the presence of vanadate is utilized as the substrate, the observed reaction products are ribose and hypoxanthine [43].

Adenylate kinase normally catalyzes the synthesis of two adenosine-diphosphate molecules from AMP plus ATP. Adenosine, in the presence of vanadate, will form adenosine 5′-vanadate as one of the reaction products. This material acts as a surrogate for AMP. In the presence of Mg^{2+}-ATP, the enzymic reaction provides adenosine plus vanadate plus inorganic phosphate [23]. The reaction products imply that the ADP analog, adenosine-V-P, was initially formed then hydrolyzed to the observed products. In an effort to catalyze the reverse reaction it was found that the ADP analog AMP-V was not an alternate substrate for adenylate kinase. For this reaction to occur it was presumed that there is a requirement that the ADP analog binds Mg^{2+}. Apparently the AMP-V is unable to complex Mg^{2+} at the active site of the enzyme thus effectively preventing catalysis [23]. The proposed inability of nucleotide monophosphate vanadate anhydride to form catalytic complexes with Mg^{2+} was also invoked in order to explain the inability of AMP-V to substitute for ADP in reactions catalyzed by hexokinase or pyruvate kinase [23].

3.2. Subclass II: Vanadate Generates Functional Enzymes

It is well known that cyclic AMP–dependent protein kinases phosphorylate a large number of proteins on serine or threonine hydroxyl groups. These phosphorylations can result either in kinetic activation or inhibition. Classic cAMP-activated systems include activation of hepatic fructose-6-P-2 kinase [46], which stimulates glycolysis, activation of glycogen degradation [47], and inhibition of glycogen synthesis [48]. Exposure of these enzymes to vanadate should generate an equilibrium mixture of specific vanadoserine esters which would mimic the kinetics normally associated with the phosphorylated enzyme. No activations or inhibitions of this type have been reported. The reason for this is probably that, because of the low vanadate concentrations normally employed, the equilibrium solution contains only a very small fraction of the specific vanadio enzyme species. This conclusion is based on the application of an equilibrium constant of 0.2 M^{-1} for the esterification of a hydroxyl group, a value typically observed [49]. However, there does exist a specific class of phosphoenzymes which may reasonably be expected to be activated by this mechanism. This class consists of the autophosphorylating protein kinases.

Glycogen phosphorylase kinase is an example of the above type of enzyme and its activity is sustained and amplified by autophosphorylating serine protein kinase activity [47]. Protein kinases that have this autophosphorylation activity are reasonable candidates for vanadate-dependent activation because vanadate may generate small quantities of an active vanadio enzyme species which could then catalyze a stable ATP-dependent activation cascade. As yet there are no known reports in the literature which have postulated or even tested the ability of vanadate to stimulate autophosphorylating serine kinases. In vivo these kinases are typically activated by cAMP-dependent kinases.

The dominant physiological effect of vanadate is to activate proteins of the insulin-dependent metabolic pathways. Insulin antagonizes many of the effects of cAMP by stimulating serine protein phosphatases [50]. This may dampen the ability of vanadate to activate cAMP-dependent pathways in vivo. There are nevertheless a few examples in which vanadate causes cAMP-like effects in vivo. These include activation of glycogen degradation in the liver [51] and positive inotropic effects in cardiac muscle [52,53]. In order to avoid complications caused by the presence of other hormonal systems, it is prudent to test the ability of vanadate to activate autophosphorylating serine kinases using purified enzyme systems.

A second type of autophosphorylating protein kinase exists for which there is significant support for a vanadate-dependent mechanism. This is the autophosphorylating tyrosine kinase. Autophosphorylating tyrosine kinases are found among the insulin receptor tyrosine kinases, insulin-like growth factor (IGF) receptors, epidermal growth factor (EGF) receptor, platelet-derived growth factor receptors, and proto-oncogene tyrosine kinases [54,55]. Using the insulin receptor as a model, the first step in signal transduction involves the binding of insulin to the receptor α subunit. This induces a conformational change which stimulates the β subunit to catalyze ATP-dependent phosphorylation at one or more sites near the carboxy terminus of the β subunit. The receptor may also use neighboring insulin receptor β subunits as substrates for tyrosine phosphorylation. Once activated by autophosphorylation, the receptor may use heterogeneous proteins as substrates for tyrosine phosphorylation. Examples in this class include activation of a cAMP phosphodiesterase by insulin receptor [56], activation of phospholipase C_γ by both platelet-derived growth factor receptor [57] and EGF receptor [58], and platelet-derived growth factor–dependent phosphorylation of proto-oncogene raf-1 protein serine kinase [59].

The equilibrium constant for nonenzymatic esterification of tyrosine by vanadate is close to 1.0 M^{-1} and is about five orders of magnitude greater than the corresponding equilibrium constant for esterification by phosphate [60]. This comparatively large formation constant may be the key to some of the physiological significance of vanadate esterification of tyrosine. For instance spontaneous vanadiotyrosine formation can be postulated to explain the vanadate-dependent activation of the cell-free purified insulin receptor protein [61]. The insulin-like stimulation of glycogen synthase in vanadate-treated cell-cultured adipocytes has been attributed to vanadate stimulated insulin receptor tyrosine kinase activity [61].

4. FUNCTION OF VANADATE OLIGOMERS

4.1. Divanadate

In aqueous solution, at pH 7, divanadate is an important species whose concentration varies from 5.0 μM to 43 μM as total vanadate varies from 0.1 to 1.0 mM. The only enzyme reported to be activated by divanadate is the 2,3-bisphosphoglycerate phosphatase activity of phosphoglycerate mutase. This enzyme converts 3-phosphoglycerate (3PGA) to 2-phosphoglycerate (2PGA) in glycolysis. In addition, it also regulates erythrocyte 2,3-BPG levels. For the catalytic reaction to occur, 2,3-BPG is required to phosphorylate the enzyme on an active site histidine residue [62]. 3PGA reacts with the phosphoenzyme to form an enzyme bound 2,3-BPG intermediate. A rearrangement occurs in which the substrate rephosphorylates the enzyme from the 3 position and 2-PGA is released. Divanadate acts as a mimic of monophosphoglycerate and binds to the phosphoenzyme. The two anionic moieties of divanadate are postulated to occupy the phosphate and carboxylate binding sites in the enzyme [63]. This type of binding normally induces the enzyme to transfer phosphate to the hydroxyl group of the substrate. In this case, there is no hydroxy group and only phosphate transfer to water is possible. Thus, since 2,3-BPG is required to phosphorylate the enzyme, divanadate stimulates net hydrolysis of 2,3-BPG.

The dissociation constant for divanadate from phosphoglycerate mutase, determined by ^{51}V NMR, is 4.0 μM [63]. The K_m of divanadate in the phosphatase reaction is 25 μM and this K_m is three orders of magnitude lower than the apparent K_m of pyrophosphate, 15 mM [64]. Under conditions of low vanadate concentration it was also found that phosphate augmented the phosphatase activity. This led to the conclusion that the phosphovanadate is also an activator of this function. The greater affinity of divanadate over pyrophosphate for the phosphoenzyme suggests that at least one vanadium of the divanadate is bound in a pentacoordinate configuration, a hypothesis supported by the function of phosphovanadate.

4.2. Decavanadate

Decavanadate is a closed spheroidal structure containing two internal octahedral vanadate moieties. Because of its large size, it is more likely that decavanadate will bind at an allosteric polyanionic binding site rather than at the active site. A 5′-nucleotidase from rat kidney is activated 16-fold by 100 nM decavanadate. Half-maximal activation is observed at 1.4 nM decavanadate [65]. This enzyme is allosterically activated by either ATP or 2,3-BPG. The V_m of the reaction is essentially the same regardless of whether decavanadate, ATP, or 2,3-BPG is used as activator. It was postulated, therefore, that decavanadate bound at the ATP/2,3-BPG binding site [65].

Decavanadate is also postulated to activate several G-protein-dependent systems (see Sec. 7 of this chapter).

5. VANADATE ACTIVATION OF PHYSIOLOGICAL PATHWAYS REGULATED BY TYROSINE PHOSPHORYLATION

In the following discussion, the physiological function of vanadium is attributed to vanadate since the V(V) form of the metal was used in these studies. It must be recognized that in a biological system complex redox equilibria will be established. Consequently, the influence of vanadyl [V(IV)] cannot and should not be ruled out. The discussion is in accordance with the known chemistry of vanadate. Work based on the use of vanadyl instead of vanadate is discussed in Sec. 10 of this chapter.

There are a wide range of enzyme activities and coordinated physiological responses stimulated in cells or in vivo by vanadate. Most of these enzymes could not be activated by direct binding by vanadate at the active site. Two viable hypotheses are presently available that might explain the cellular action of vanadate and that are also consistent with the reasonable proposition that one or a few key enzymes exist which are sensitive to low vanadate concentrations and which are also control sites for coordinated control of a family of enzyme activities. The first hypothesis proposes that there is direct activation of tyrosine kinase activity because of spontaneous vanadate esterification of tyrosine. The second proposal suggests that there is strong inhibition of phosphotyrosyl phosphatase activity by vanadate. At micromolar concentrations, vanadate is a good inhibitor of phosphotyrosyl phosphatases [66].

5.1. Mitogenic Action of Vanadium

Vanadate stimulates mitogenic processes in cell culture. Generally the concentration of vanadate required for mitogenesis is 10–50 μm and this is about 20-fold less than the amount required for the insulin-mimetic effects. Vanadate also causes reversible transformations in fibroblast culture [67]. The effects include (1) morphological changes, (2) increased glucose transport, (3) inhibition of density-dependent growth, and (4) growth in the absence of a solid support. Transformation of chondrocytes is promoted by 50 μM vanadate [68]. The fibroblast growth factor receptor is a tyrosine kinase which stimulates translation of mRNA. Vanadate mimics fibroblast growth factor action in fibroblasts by stimulating the translation of actin and creatine phosphokinase transcription of mRNA for c-fos proto-oncogene and promoting a growth-stimulating transition from the G_0 to G_1 phase of the cell cycle [69]. Vanadate was found to activate the pp60$^{\nu\text{-src}}$ tyrosine kinase of Rous sarcoma virus (RSV) transformed cells two- to threefold but could not activate the purified kinase [70].

DNA synthesis was stimulated in mouse fibroblasts at concentrations of 5–50 μM vanadate [71] and vanadate increased DNA synthesis in the presence of suboptimal amounts of EGF [72]. Vanadate also stimulated collagen and DNA synthesis in prebone tissue [73]. Vanadate mimicked EGF by stimulating the plasma membrane Na^+/H^+ exchange protein which promotes cell alkalinization [74–76]. Vanadate (0.2 mm) or EGF stimulated Ca^{2+} influx in epidermal carcinoma cells [77].

Increased levels of inositol monophosphate, inositol biphosphate, and inositol triphosphate are found in fibroblasts treated with 100 μm vanadate. Vanadate was found to stimulate inositol phosphate accumulation by both a genistein-sensitive tyrosine kinase-dependent mechanism and a tyrosine kinase-independent mechanism [78]. Vanadate or EGF also stimulates phosphatidylinositol 4-kinase in rat liver plasma membranes [79,80]. Vanadate also caused altered microtubule arrangements in kidney cells [81]. The altered microtubules showed increased separation to microfilament structures. Vanadate was reported to activate phospholipase by a pertussis toxin–sensitive G-protein mechanism rather than tyrosine phosphorylation [82]. This effect was seen using a vanadate concentration of 10 mM which is 100-fold higher than the concentration needed to activate the mitogenic tyrosine kinase pathway.

5.2. Insulin-like Action of Vanadium

The largest and best characterized group of physiologically related activations by vanadate can be accounted for via increased tyrosine phosphorylation of the insulin receptor and associated proteins. The insulin-mimetic effect of vanadate can be observed both in vitro and in vivo.

Work with live animals showed 0.8 mg/mL (approximately 4.0 mM) of vanadate administered in the drinking water caused a normalization of hyperglycemia and improved cardiac output in streptozotocin treated diabetic rats [83]. Other laboratories [84,85] have confirmed and extended these results to show that vanadate produces other downstream metabolic insulin-like effects in diabetic rats. Orally administered vanadium activated glycogen synthase and restored glycogen levels in streptozotocin diabetic rats [86]. Oral vanadate decreased the K_m of glycogen synthase by stimulating an insulin-like dephosphorylation of the enzyme [85]. Vanadate restored depressed levels of fructose-2,6-biphosphate by stimulating an insulin-like activation of fructose-6-P-2 kinase in streptozotocin-diabetic rats [86,87] and also restored hepatic glucokinase activity [87]. The reactions described are metabolically associated with increased glycolytic rates.

Livers from sacrificed oral vanadate–treated diabetic rats showed increased activity of glucose-6-P dehydrogenase and malic enzyme [88]. These activities are metabolically related to increased rates in the pentose phosphate pathway and fatty acid synthesis, respectively. Syringe injection (12 mg/kg of vanadate) enhanced glycogen content in rat diaphragm muscle and was found to be 50% as potent as insulin [89].

A 12-week course of 4 mM oral vanadate treatment in genetically diabetic rats stimulated a return to normal weight from obesity, decreased plasma insulin levels, normoglycemia, and an insulin-like decrease in glycogen phosphorylase and increase in glucose-6-P dehydrogenase activity [90]. Oral treatment with 0.25 mg/mL vanadate for 3 weeks in non-insulin-dependent diabetic rats also produced a classical insulin-like recovery [91]. The animals in this latter study had very few insulin receptor molecules and thus it could be argued that the effects of vanadium were not mediated by a tyrosine phosphorylation mechanism in this case. This suggests that vanadate may act by postreceptor mechanisms and perhaps not by, or in addition to, receptor-dependent mechanisms. Normal Wistar rats drinking ~3 mM vanadate had approximately normal rates of glycolysis and glycogen levels [92].

The insulin-like effects of vanadate in tissue culture are similar to effects seen in animal studies [93]. The vanadate concentration typically employed in these studies is about 1.0 mM. Vanadate caused an increase in tyrosine kinase in rat adipocytes [94]. Proteins that are phosphorylated include a 180-kDa protein that is known to be phosphorylated on tyrosine by the insulin receptor [95]. However, the complete pattern of protein tyrosine phosphorylation by vanadate is somewhat different than that observed with insulin [96]. Vanadate at 0.2 mM stimulated down-regulation of the insulin receptor but inhibited the degradation of the internalized receptor [97], while an increase in vanadate concentrations to 1.0 mM was required to stimulate insulin receptor tyrosine kinase activity. Vanadate caused increased glucose transport [98,99], glycolysis, and glucose oxidation through the pentose phosphate pathway and the citric acid cycle [98–100]. Vanadate inhibited lipolysis [100] and stimulated glycogen synthase and casein kinase II activities while antagonizing the effects of cAMP [101]. Fatty acid synthesis and acetyl-CoA carboxylase activity were increased 1.5-fold by 3 mM vanadate in hepatocytes [102]. Vanadate elevated the level of fructose-2,6-bisphosphate in hepatocytes [103]. This molecule functions as an allosteric activator of glycolytic phosphofructokinase [46]. Vanadate causes an insulin-like activation of a cGMP inhibited cAMP phosphodiesterase [104]. Vanadate stimulates amino acid transport into cells [105] and stimulates ribosomal protein S6 kinase in epidermal cells which actuates protein synthesis [106]. Vanadate also inhibited a cellular ATP-dependent mechanism for protein catabolism in liver cells [107].

6. VANADATE MIMICS ATRIAL NATRIURETIC HORMONE

Injection of 1.5 μmol of vanadate in saline solution into rats produced a 15-fold increase in urinary volume output and a 25-fold increase in Na^+ output [108]. This diuretic/natriuretic effect is characteristic of atrial natriuretic factor (ANF). The plasma membrane receptor which binds ANF has intrinsic guanyl cyclase activity [109]. The cyclic guanosine monophosphate that is generated causes relaxation of vascular smooth muscle which is related to the diuretic effect. A cGMP-dependent serine protein kinase is known to catalyze the opening of the Na^+ channel in the renal inner medullary duct [110] and this is related to the natriuretic effect. Some evidence indicates that a cGMP kinase is able to phosphorylate cardiac phospholamban [111] and this would augment the positive inotropic effect. Such a process could mediate restoration of the cardiac contractile force seen in vanadate-treated diabetic rats [83].

7. G-PROTEIN ACTIVATION BY VANADATE

7.1. The Visual Receptor System

Vanadate activates a phosphodiesterase which hydrolyzes cGMP in the rod outer segment [112]. In the normal system, a light-excited rhodopsin molecule binds a G protein. The G-protein binds GTP and remains capable of transducing the signal until the GTPase activity produces hydrolysis of the GTP. The excited rhodopsin/G-protein-GTP complex allosterically binds and activates the phosphodiesterase. Hydrolysis of cGMP causes closing of the Na^+ channel, which triggers the visual action potential. The mechanism of action of vanadate seems to be at the level of thc G protein. Decavanadate was reported to allosterically bind the G-protein-GDP complex [113]. This complex apparently mimics the GTP-bound form of the G protein and transduces the signal. Vanadate, at a minimum concentration of 0.1–0.2 mM, or the nonhydrolyzable GTP analog, GDPγS, produced action potentials in crab photoreceptor cells [114]. The mechanism, as for the visual receptor system, was proposed to involve activation at the G-protein level.

7.2. Adenylate Cyclase

Adenylate cyclase is activated by a G-protein mechanism in all known cases. In the general model, a hormone binds and causes a conformational change in its receptor. The hormone-receptor complex binds a G-protein and induces the binding of GTP. The G-protein-GTP complex binds and activates adenylate cyclase. The G protein is a GTPase. The signal level is attenuated by hydrolysis of GTP.

Vanadate is reported to activate adenylate cyclase and elevate cAMP levels in several cell systems. In fat cells, the activation of adenylate cyclase by vanadate was postulated to involve inhibition of GTPase activity [115]. Vanadate increased cAMP levels and activated cAMP-dependent protein kinases in liver slices [116]. Vanadate caused a cAMP-like effect and stimulated phosphorylation and inhibition of glycogen synthase in rat livers [51]. Pancreatic adenylate cyclase was activated by 10 mM vanadate [117] and this action was pharmacologically similar to the effect of G-protein-activating GTP analogs. Turkey erythrocyte adenyl cyclase was activated by 3 mM vanadate [118]. Vanadate did not interfere with isopreterinol binding to the β-adrenergic adenyl cyclase–linked receptor and vanadate acted in a different manner than fluoride. Fluoride seems to bind at the active site and stabilizes the G-protein-GTP complex [119]. In colon epithelial cells, vanadate acted as a vasoconstrictor and stimulated extracellular chloride secretion in a manner similar to that of cAMP [120].

Vanadate has been reported to cause both positive and negative inotropic effects in cardiac tissue [121]. The positive inotropic effects are observed more frequently than the negative inotropic effects. Vanadate stimulated positive inotropic effects in ventricular cardiac muscle in rat, rabbit, guinea pig, and cat at concentrations of 0.01–1.0 mM [122]. Adenyl cyclase was shown to be stimulated simultaneously. In cat ventricular muscle vanadate activated adenyl cyclase and caused positive inotropic effects with half-maximal activation at 0.2 mM vanadate [52].

The pharmacological actions of vanadate which mimic cAMP can be explained by intervention at the G-protein level by some polymeric vanadate species. For instance, preformed decavanadate activated G-proteins in the visual system [113]. The concentration ranges employed in these studies are usually in the millimolar range and at such concentrations the di- and tetravanadate oligomers are the predominant forms of vanadate.

An alternate model of G-protein activation by vanadate has been proposed and is based on the formation of a GDP-V ester which is an analog of GTP [123]. This model has some merit by analogy to the GDP-V_1 complex which inhibits myosin ATPase [124]. But the experimental system on which the model is based required 3 mM vanadate. Furthermore G proteins bind the Mg^{2+} complex of GTP in the active state [119]. Analogously, ADP-V complexes are usually not inhibitors of enzymes which normally bind the Mg^{2+}-ATP complex such as the ATP kinases [125]. Because adenyl cyclase activation requires high vanadate concentrations and because it is antagonized by the insulin-like effect, activation is usually observed only in vitro and is unlikely to be observed at low pharmacological concentrations in vivo.

8. VANADATE AS A COFACTOR FOR HALOPEROXIDASES

Nonheme vanadate is a required cofactor for several iodoperoxidases [126] and bromoperoxidases [127,128] from marine algae and terrestrial lichens. Substrates for this reaction are the halogen anion, hydrogen peroxide, and an organic molecule. Products are hydroxide, water, and the halogenated organic compound [129]. These enzymes are discussed in detail in Chap. 10.

9. PHYSIOLOGICAL EFFECTS OF VANADIUM(V) PEROXIDES

Peroxovanadate is typically more potent or effective at lower concentrations than vanadate with regard to the insulin-like effect. Peroxovanadate is usually prepared as an equimolar mixture of vanadate and hydrogen peroxide in the presence of catalase and used within 30 min. Peroxovanadate (1 mM) was found to be 10-fold more potent than vanadate alone in stimulation of both glycolysis and glycogen synthesis in rat skeletal muscle [130]. Peroxovanadate at 0.5 mM concentrations increased the rates of lactate formation and glycogen synthesis approximately twofold in skeletal muscle preparations [131]. Solutions containing 2 mM H_2O_2 and 10–100 μM vanadate were ~100-fold more potent than vanadate alone and ~5-fold more potent than hydrogen peroxide in stimulating protein tyrosine phosphorylation and protein tyrosyl phosphatase activity in rat hepatoma cells [132]. In rat adipocytes, 1.0 mM peroxovanadate was found to be ~100-fold more effective than either 1.0 mM vanadate or 1.0 mM hydrogen peroxide in stimulating tyrosine kinase activity [133]. The extent of lipid synthesis was about equal to that of insulin and about 100-fold more potent than vanadate when activated by 10 μm peroxovanadate [134]. Peroxovanadate also antagonized epinephrin-stimulated lipolysis. These effects were attributed to increased tyrosine kinase activity. Translocation of IGF-II receptor tyrosine kinase to the plasma membrane was stimulated by 100 μM peroxovanadate in rat adipocytes [135]. This resulted in enhanced binding of IGF-II. The effect of peroxovanadate was fourfold more potent than 100 μM concentrations of either vanadate or hydrogen peroxide alone [133,136]. Protein synthesis was stimulated by 1 μM peroxovanadate in rat adipocytes [134]. The effect was ~100-fold more potent than an equal concentration of vanadate but about the same as from stimulation by optimal insulin concentrations.

Coordinated increases in tyrosine phosphorylation and inositol triphosphate (IP_3) synthesis were stimulated by the presence of 3 mM hydrogen peroxide plus 1 mM vanadate [137]. The effect was approximately threefold greater than either 3 mM hydrogen peroxide or 1 mM vanadate. Vanadate, at a 50 μM concentration, stimulated a respiratory burst of phagocytic neutrophils [138]. The effects were augmented by a combination of vanadate with the respiratory burst products superoxide and hydrogen peroxide. In another study, 0.1 mM peroxovanadate activated phospholipase D and the respiratory burst NADPH oxidase, which are thought to be controlled by tyrosine kinase in phagocytic neutrophils [139]. Combination of 0.1 mM vanadate and 1.0 mM hydrogen peroxide stimulated coordinated tyrosine phosphorylation and aggregation in platelets [140]. Tyrosine phosphorylation reached maximal levels after 2 min. In a further study it was found that platelets treated with 0.1 mM peroxovanadate underwent an increase in tyrosine phosphorylation. Increases were also seen in intracellular Ca^{2+}, protein kinase C activity, secretion, actin polymerization, and aggregation. These processes putatively are all under the control of tyrosine kinase. Inhibition of the appropriate phosphotyrosyl phosphatase should lead to an increased lifetime and also to autostimulation of the tyrosine phosphorylated state of the respective kinases. There are at least two possibilities for the specific inhibition of phosphotyrosyl phosphatase. Peroxovanadate may act as a phosphate analog or as an oxidizing agent. As a phosphate analog peroxovanadate may bind competitively at the enzyme active site, while as an oxidizing agent it may oxidize the cysteine residue required for activity [66,141].

10. ENZYME ACTIVATION BY VANADIUM(IV)

Because of the redox chemistry of vanadium and the difficulties in maintaining an oxygen-free environment, it is often difficult to be confident that the chemical observations recorded are actually attributable to vanadium(IV). Complimentary studies utilizing vanadate are often necessary in order to properly credit results. The problem is much more severe in biological systems. In general, the reported biological influences of vanadium(IV) parallel that of vanadium(V). Examples of this include formation of surrogate substrates as well as insulin-like and mitogen-like effects at the physiological level.

Vanadium(IV) cannot substitute for vanadium(V) in all cases and vanadium(IV) performs one unique function that is not possible with V(V) and that is to satisfy the divalent cation requirement of some enzymes.

10.1. Enzyme Activation by Surrogate Substrates

In most cases, vanadyl is not known to form esters analogous to those of vanadate. Apparently, however, a complex formed between vanadium(IV) and dihydroxyacetone can act as a substitute for dihydroxyacetone-3-P. This material serves as a substrate for glycerol-3-P dehydrogenase [43]. Vanadyl is only 60% as effective as vanadate in activating this enzyme.

10.2. Enzyme Activation by Oxidation

A vanadyl complex with reduced glutathione was reported to activate a liver microsomal cAMP phosphodiesterase [142]. This enzyme was also reported to be activated allosterically by an insulin-mediated oxidation [143]. Therefore it was proposed that a vanadyl-glutathione complex oxidized the phosphodiesterase, perhaps generating a disulfide that in turn activated the enzyme function.

10.3. Vanadium(IV) as a Divalent Cation

Vanadium(IV) is capable of substituting at the Mg^{2+} binding site of pyruvate kinase [144] where it is coordinated in an octahedral environment. Interestingly, the vanadyl-substituted enzyme could not catalyze the normal transfer of phosphate from ATP to pyruvate but did allow the ATP-dependent phosphorylation of glycolate.

Vanadyl occupies one of the two Mg^{2+} binding sites of *S*-adenosylmethionine synthase [145]. The enzyme is active only if Mg^{2+} or Ca^{2+} occupies the second site. Electron spin resonance (ESR) studies indicated that there are two water molecules and two or three protein ligands in the coordination group. In the absence of enzyme, it was shown by ESR that vanadyl formed a bidentate complex with pyrophosphate. Pyrophosphate is a product of the *S*-adenosylmethionine synthase reaction and may function as one of the vanadyl ligands in the enzyme complex.

The Zn^{2+} binding site of carboxypeptidase A can be occupied by vanadium(IV) [146]. In this complex, vanadyl is surrounded by four equatorial ligands. Two ligands are water molecules and two are enzyme carboxylate groups. Maximal velocity of the vanadyl-substituted enzyme was 22% of the native reaction and the K_m was threefold lower.

Both of the Mg^{2+} binding sites in xylose isomerase could be substituted with vanadyl [147]. The tight binding site has four carboxylate groups and two water molecules in a distorted octahedron. The low-affinity site contains three carboxylate ligands, one histidine, and one water molecule in a pentacoordinate structure. The activity of the vanadyl-substituted enzyme was apparently significantly lower than the Mg^{2+}-activated enzyme.

10.4. Insulin-like Effect of Vanadium(IV)

Vanadyl sulfate was shown to be an effective pharmacological agent for the treatment of streptozotocin-diabetic rats. Vanadyl sulfate normalized plasma glucose levels when given as either a daily injection of 50 mg/kg of vanadyl sulfate hexahydrate [148] or at 1 mg/mL in drinking water [149]. Vanadyl is less toxic than vanadate. Also, submaximal doses of vanadyl acted synergistically with insulin by decreasing the amount of externally added insulin required to normalize blood glucose levels [150]. The organ levels of vanadium were obtained from diabetic rats drinking 1 mg/mL vanadyl sulfate for 2 months. The concentration found by ^{51}V NMR, given in nmol vanadium per g wet tissue, were: kidney, 185; spleen, 55; liver, 40; plasma, 15; lung, 15; heart, 5; erythrocyte, 5 [151]. Vanadium levels were slightly higher in rats drinking vanadate solution.

It is reasonable that the same pharmacological effects should be exhibited by vanadium(IV) as by vanadium(V) because a cellular redox equilibrium is established. In vitro, glutathione and cysteine can reduce vanadium(V) [152]. Vanadate is reduced in adipocytes at the expense of glutathione [153]. Vanadate is reduced in erythrocytes by glutathione and vanadyl is found predominantly in the complex with hemoglobin [40]. Microsomes also reduce vanadate [154]. Consequently, it has been argued that vanadium(IV) rather than vanadium(V) is the active insulin-mimetic species [155]. The bismalto-oxovanadium(IV) complex is 50% more potent than free vanadium(IV) [156]. This complex could act as a delivery agent by preventing oxidation of vanadium(IV) in the blood.

It may well be that V(IV) and V(V) acting together are responsible for the observed pharmacological effects. However, as indicated by the very effective insulin-mimetic properties of peroxovanadate, the function is certainly very complicated and probably involves a variety of vanadium complexes. Such complexes, if they had no other function, could serve as transfer agents (see also Chap. 17 of this volume).

10.5. Mitogenic Effects of Vanadium(IV)

Concentrations of 5–50 μM vanadyl stimulated DNA synthesis fourfold in mouse fibroblast 3T3 cells [71]. In this case vanadyl and vanadate were equally effective. The increase in [^{3}H]thymidine incorporation into DNA stimulated by insulin, EGF, chloratoxin, or phorbol myristate was synergistically augmented by vanadium.

11. UNUSUAL MECHANISMS OF ENZYME ACTIVATION BY VANADATE

11.1. Lactate Dehydrogenase

The NADH-dependent reduction of pyruvate is catalyzed by lactate dehydrogenase. Glycolate is accepted as an alternative substrate and vanadate specifically stimulates the enzymatic reduction of glycolate [23]. Kinetic double-reciprocal plots gave a family of parallel lines with higher V_m values at higher vanadate concentrations. Glycolate is an aldehyde that exists in solution predominantly in a hydrated form which is not an enzymatic substrate. It was proposed that vanadate stimulates the reaction by generating a higher concentration of glycolate aldehyde. This occurs because vanadate forms esters with the hydrated glycolate and the product, in turn, rapidly hydrolyzes. The hydrolysis rate is approximately 1.2×10^3 sec^{-1} using the rate of formation of ethyl vanadate esters as a model. Product breakdown generates higher concentrations of the enzymatically competent aldehyde.

11.2. Vanadate Promotes Protein Association

Vanadate and molybdate catalyze the formation of a specific protein-protein complex between pp60$^{\nu\text{-src}}$ and its chaperone protein hsp90

[157]. It is speculated that the oxoanion binds one of the two proteins at what may be a phosphate binding site. This could stabilize a conformation which promotes protein-protein association.

11.3. Phosphoglucomutase

In addition to catalyzing the interconversion of glucose-1-phosphate and glucose-6-phosphate, phosphoglucomutase catalyzes the synthesis of glucose-1,6-bisphosphate. In the reaction sequence, either fructose-1,6-bisphosphate or glycerate-1,3-bisphosphate may phosphorylate the enzyme. The phosphoenzyme reacts with glucose-1-phosphate and generates glucose-1,6-bisphosphate, which dissociates to provide the product. This synthase activity was stimulated two- to threefold by 100 μM vanadate [158]. Mechanistically, it was postulated that vanadate bound at a unique position in the active site simultaneously with glucose-1,6-bisphosphate and this promoted product dissociation.

ABBREVIATIONS AND DEFINITIONS

acetyl-CoA	acetyl coenzyme A
ADP	adenosine 5′-diphosphate
AMP	adenosine 5′-monophosphate
cAMP	adenosine cyclic 2′,3′-monophosphate
ANF	atrial natriuretic factor
ATP	adenosine 5′-triphosphate
2,3-BPG	2′,3′-bisphosphoglycerate
DNA	deoxyribonucleic acid
EGF	epidermal growth factor
ESR	electron spin resonance
GDP	guanosine 5′-diphosphate
cGMP	guanosine cyclic 2′,3′-monophosphate
GTP	guanosine 5′-triphosphate
IGF-II	insulin-like growth factor II
IP_3	inositol triphosphate
MES	2-[*N*-morpholino]ethanesulphonic acid

mRNA	messenger ribonucleic acid
NAD^+	nicotinamide adenine dinucleotide
NADH	nicotinamide adenine dinucleotide, reduced form
NADPH	nicotinamide adenine dinucleotide phosphate, reduced form
NMR	nuclear magnetic resonance
2PGA	2-phosphoglyceric acid
3PGA	3-phosphoglyceric acid
RNA	ribonucleic acid
RSV	Rous sarcoma virus

REFERENCES

1. N. D. Chasteen, *Struct. Bond.*, *53*, 105–138 (1983).
2. N. D. Chasteen (ed.), *Vanadium in Biological Systems*, Kluwer Academic, Dordrecht, 1990.
3. R. Wever and K. Kustin, *Adv. Inorg. Chem.*, *35*, 81–115 (1990).
4. E. Heath and O. W. Howarth, *J. Chem. Soc., Dalton Trans.*, 1105–1110 (1981).
5. L. Pettersson, I. Andersson, and B. Hedman, *Chem. Script.*, *25*, 309–317 (1985).
6. M. T. Pope and B. W. Dale, *Chem. Soc. (Lond) Quart. Rev.*, *22*, 527–549 (1968).
7. D. C. Crans, C. D. Rithner, and L. A. Theisen, *J. Am. Chem. Soc.*, *112*, 2901–2908 (1990).
8. A. S. Tracey, M. J. Gresser, and B. Galeffi, *Inorg. Chem.*, *27*, 157–161 (1988).
9. M. J. Gresser and A. S. Tracey, *J. Am. Chem. Soc.*, *107*, 4215–4220 (1985).
10. A. S. Tracey, B. Galeffi, and S. Mahjour, *Can. J. Chem.*, *66*, 2294–2298 (1988).
11. A. S. Tracey and M. J. Gresser, *Inorg. Chem.*, *27*, 2695–2702 (1988).
12. C. F. G. C. Geraldes and M. M. C. A. Castro, *J. Inorg. Biochem.*, *35*, 79–93 (1989).

13. A. S. Tracey and C. H. Leon-Lai, *Inorg. Chem.*, *30*, 3200–3204 (1991).
14. A. S. Tracey, J. S. Jaswal, M. J. Gresser, and D. Rehder, *Inorg. Chem.*, *29*, 4283–4288 (1990).
15. D. C. Crans, S. M. Schelble, and L. A. Theisen, *J. Org. Chem.*, *56*, 1266–1274 (1991).
16. C. F. G. C. Geraldes and M. M. C. A. Castro, *J. Inorg. Biochem.*, *37*, 213–232 (1989).
17. B. Galeffi and A. S. Tracey, *Can. J. Chem.*, *66*, 2565–2569 (1988).
18. A. S. Tracey and M. J. Gresser, *Proc. Natl. Acad. Sci. (USA)*, *83*, 609–613 (1986).
19. C. E. Holloway and M. Melnik, *Rev. Inorg. Chem.*, *7*, 75–159 (1985).
20. A. S. Tracey, H. Li, and M. J. Gresser, *Inorg. Chem.*, *29*, 2267–2271 (1990).
21. M. J. Gresser, A. S. Tracey, and K. M. Parkinson, *J. Am. Chem. Soc.*, *108*, 6229–6234 (1986).
22. A. S. Tracey, M. J. Gresser, and S. Liu, *J. Am. Chem. Soc.*, *110*, 5869–5874 (1988).
23. M. M. Craig, *Selective inhibition by vanadate of phosphoryl transfer enzymes*, PhD thesis, Simon Fraser University, Burnaby, 1991.
24. D. Rehder, *Inorg. Chem.*, *27*, 4312–4316 (1988).
25. D. C. Crans, R. L. Bunch, and L. A. Theisen, *J. Am. Chem. Soc.*, *111*, 7597–7607 (1989).
26. J. S. Jaswal and A. S. Tracey, *Can. J. Chem.*, *69*, 1600–1607 (1991).
27. D. C. Crans and P. K. Shin, *Inorg. Chem.*, *27*, 1797–1806 (1988).
28. D. C. Crans, P. M. Ehde, P. K. Shin, and L. Pettersson, *J. Am. Chem. Soc.*, *113*, 3728–3736 (1991).
29. A. S. Tracey and M. J. Gresser, *Inorg. Chem.*, *27*, 1269–1275 (1988).
30. V. W. Day, W. G. Klemperer, and D. J. Maltbie, *J. Am. Chem. Soc.*, *109*, 2991–3002 (1987).
31. A. T. Harrison and O. W. Howarth, *J. Chem. Soc., Dalton Trans.*, 1173–1177 (1985).

32. J. S. Jaswal and A. S. Tracey, *Inorg. Chem.*, *30*, 3718–3722 (1991).
33. J. S. Jaswal and A. S. Tracey, *J. Am. Chem. Soc.*, *115*, 5600–5607 (1993).
34. A. S. Tracey and J. S. Jaswal, *Inorg. Chem.*, *32*, 4235–4243 (1993).
35. I. G. Macara, K. Kustin, and L. C. Cantley, Jr., *Biochim. Biophys. Acta*, *629*, 95–106 (1980).
36. H. Sakurai, S. Shimomura, and K. Ishizu, *Inorg. Chim. Acta*, *55*, L67–L69 (1981).
37. K. Kustin and D. L. Toppen, *Inorg. Chem.*, *12*, 1404–1407 (1973).
38. L. C. Cantley, J. H. Fergusen, and K. Kustin, *J. Am. Chem. Soc.*, *100*, 5210–5212 (1978).
39. M. Tamura, N. Shimidzu, and K. Hayashi, *Biochem. Biophys. Res. Commun.*, *75*, 1029–1036 (1977).
40. L. C. Cantley and P. Aisen, *J. Biol. Chem.*, *254*, 1781–1784 (1979).
41. P. J. Stankiewicz, P. J. Stern, and A. Davison, *Arch. Biochem. Biophys.*, *287*, 8–17 (1991).
42. A. F. Nour-Eldeen, M. M. Craig, and M. J. Gresser, *J. Biol. Chem.*, *260*, 6836–6842 (1985).
43. D. G. Drueckhammer, J. R. Durrwachter, R. L. Pederson, D. C. Crans, L. Daniels, and C.-H. Wong, *J. Org. Chem.*, *54*, 70–77 (1989).
44. E. G. De Master and A. Mitchell, *Biochemistry*, *12*, 3616–3621 (1973).
45. E. A. Bramhall, *The effects of vanadate and arsenate on mitochondrial ATP synthase*, MSc thesis, Simon Fraser University, Burnaby, 1987.
46. L. Hue and M. H. Rider, *Biochem. J.*, *245*, 313–324 (1987).
47. C. A. Pickett-Gies and D. A. Walsh, in *The Enzymes*, Vol. 17A (P. D. Boyer, ed.), Academic Press, New York, 1986, pp. 395–459.
48. P. Cohen, in *The Enzymes*, Vol. 17A (P. D. Boyer, ed.), Academic Press, New York, 1986, pp. 461–497.
49. M. J. Gresser and A. S. Tracey, *J. Am. Chem. Soc.*, *107*, 4215–4220 (1985).
50. G. J. Sale, *Int. J. Biochem.*, *20*, 897–908 (1988).
51. F. Bosch, J. Arino, A. M. Gomez-Foix, and J. J. Guinovart, *J. Biol. Chem.*, *262*, 218–222 (1987).

52. W. Krawietz, K. Werdan, and E. Erdmann, *Biochem. Pharmacol.*, *28*, 2517–2520 (1979).
53. G. Grupp, I. Grupp, C. L. Johnson, E. T. Wallick, and A. Schwartz, *Biochem. Biophys. Res. Commun.*, *88*, 440–447 (1979).
54. J. Hunter and J. A. Cooper, *Annu. Rev. Biochem.*, *54*, 897–930 (1985).
55. Y. Yarden and A. Ullrich, *Annu. Rev. Biochem.*, *57*, 443–478 (1988).
56. N. J. Pyne, W. Cushley, H. G. Nimmo, and M. D. Houslay, *Biochem. J.*, *261*, 897–904 (1989).
57. C. Kanik-Ennulat and N. Neff, *Mol. Cell. Biol.*, *10*, 898–909 (1990).
58. S. Nishibe, M. I. Wahl, P. B. Wedegaertner, J. J. Kim, S. G. Rhee, and G. Carpenter, *Proc. Natl. Acad. Sci. USA*, *87*, 424–428 (1990).
59. K. S. Kovacina, K. Yonezawa, D. L. Brautigan, N. K. Tonks, V. R. Rapp, and R. A. Roth, *J. Biol. Chem.*, *265*, 12115–12118 (1990).
60. A. S. Tracey and M. J. Gresser, *Proc. Natl. Acad. Sci. USA*, *83*, 609–613 (1986).
61. M. Tamura, T. A. Brown, J. H. Whipple, Y. Fujito-Yamaguchi, R. E. Dubler, K. Cheng, and J. Larner, *J. Biol. Chem.*, *254*, 6650–6658 (1984).
62. Z. B. Rose, *Adv. Enzymol.*, *51*, 211–253 (1980).
63. P. J. Stankiewicz, M. J. Gresser, A. S. Tracey, and L. F. Hass, *Biochemistry*, *26*, 1264–1269 (1987).
64. D. R. Harkness, W. Thompson, S. Roth, and V. Grayson, *Arch. Biochem. Biophys.*, *138*, 208–219 (1970).
65. M. Le Hir, *Biochem. J.*, *273*, 795–798 (1991).
66. K.-H. W. Lau, J. R. Farley, and D. J. Baylink, *Biochem. J.*, *257*, 23–36 (1989).
67. J. K. Klarlund, *Cell*, *41*, 707–717 (1985).
68. M. K. Owada, M. Iwamoto, T. Koike, and Y. Kato, *J. Cell Physiol.*, *138*, 484–492 (1989).
69. B. Wice, J. Milbrandt, and L. Glaser, *J. Biol. Chem.*, *262*, 1810–1817 (1987).
70. D. J. Brown and J. A. Gordon, *J. Biol. Chem.*, *259*, 9580–9586 (1984).

71. J. B. Smith, *Proc. Natl. Acad. Sci. USA*, *80*, 6162–6167 (1983).
72. G. Carpenter, *Biochem. Biophys. Res. Commun.*, *102*, 1115–1121 (1981).
73. E. Canalis, *Endocrinology*, *116*, 855–862 (1985).
74. S. Paris, J.-C. Chambard, and J. Pouysségur, *J. Biol. Chem.*, *262*, 1977–1983 (1987).
75. P. Rothenberg, L. Glaser, P. Schlesinger, and D. Cassel, *J. Biol. Chem.*, *258*, 12644–12653 (1983).
76. D. Cassel, Y. X. Zhuang, and L. Glaser, *Biochem. Biophys. Res. Commun.*, *118*, 675–681 (1984).
77. I. G. Macara, *J. Biol. Chem.*, *261*, 9321–9327 (1986).
78. P. A. Randazzo, J. S. Olshan, A. A. Bijivi, and L. Jarett, *Arch. Biochem. Biophys.*, *292*, 258–265 (1992).
79. Y. Chen, D. C. Yang, A. B. Brown, Y. Jeng, A. Tatoyan, and T. M. Chan, *Arch. Biochem. Biophys.*, *283*, 184–192 (1990).
80. D. C. Yang, A. B. Brown, and T. M. Chan, *Arch. Biochem. Biophys.*, *274*, 659–662 (1989).
81. E. Wang and P. W. Choppin, *Proc. Natl. Acad. Sci. USA*, *78*, 2363–2367 (1981).
82. S. Paris and J. Pouysségur, *J. Biol. Chem.*, *262*, 1970–1976 (1987).
83. C. F. Heyliger, A. G. Tahiliani, and J. H. McNeill, *Science*, *227*, 1474–1477 (1985).
84. J. Meyerovitch, Z. Farfel, J. Sack, and Y. Schechter, *J. Biol. Chem.*, *262*, 6658–6662 (1987).
85. L. Rossetti and M. R. Laughlin, *J. Clin. Invest.*, *84*, 892–899 (1989).
86. M. Bollen, M. Miralpeix, F. Ventura, B. Toth, and R. Bartrons, *Biochem. J.*, *267*, 269–271 (1990).
87. J. Gil, M. Miralpeix, J. Carreras, and R. Bartrons, *J. Biol. Chem.*, *263*, 1868–1871 (1988).
88. S. Pugazhenti, R. L. Khandelwal, and J. F. Angel, *Biochim. Biophys. Acta*, *1083*, 310–312 (1991).
89. H. V. Strout, P. P. Vicario, R. Saperstein, and E. E. Slater, *Endocrinology*, *124*, 1918–1924 (1989).
90. S. Pugazhenti, J. F. Angel, and R. L. Khandelwal, *Metab. Clin. Exp.*, *40*, 941–946 (1991).

91. J. Meyerovitch, P. Rothenberg, Y. Schechter, S. Bonner-Weir, and C. R. Kahn, *J. Clin. Invest.*, *87*, 1286–1294 (1991).
92. R. A. J. Challiss, B. Leighton, F. Lozeman, L. Budohoski, and E. A. Neusholme, *Biochem. Pharmacol.*, *36*, 357–361 (1987).
93. Y. Schechter, J. Meyerovitch, and S. Amir, *Biochem. Pharmacol.*, *37*, 1891–1896 (1988).
94. M. Tamura, T. A. Brown, R. E. Dubler, and J. Larner, *Biochem. Biophys. Res. Commun.*, *113*, 80–86 (1983).
95. D. Heffetz and Y. Zick, *J. Biol. Chem.*, *264*, 10126–10132 (1989).
96. R. A. Moonly, K. L. Bordwell, S. Luhowskyi, and J. E. Casnellie, *Endocrinology*, *124*, 422–429 (1989).
97. K. Torrossian, D. Freedman, and I. G. Fantus, *J. Biol. Chem.*, *263*, 9353–9359 (1988).
98. G. R. Dubyak and A. Kleingeller, *J. Biol. Chem.*, *255*, 5306–5312 (1980).
99. A. S. Clark, J. M. Fagan, and W. E. Mitch, *Biochem. J.*, *232*, 273–276 (1985).
100. W. L. Duckworth, S. S. Solomon, J. Liepnieks, F. G. Hamel, S. Hand, and D. E. Peavy, *Endocrinology*, *122*, 2285–2289 (1988).
101. C. Villar-Palasi, J. J. Guinovart, A. M. Gómez-Foix, J. E. Rodriquez-Gil, and F. Bosch, *Biochem. J.*, *262*, 563–567 (1989).
102. M. Guzmán and J. Castro, *Arch. Biochem. Biophys.*, *283*, 90–95 (1990).
103. A. M. Gómez-Foix, J. E. Rodriquez-Gil, C. Fillat, J. J. Guinovart, and F. Bosch, *Biochem. J.*, *255*, 507–512 (1988).
104. H. Ueki, R. Okuhama, M. Sera, T. Inoue, N. Tominaga, and T. Morita, *Endocrinology*, *131*, 441–446 (1992).
105. P. Muñoz, A. Gumà, M. Furriols, X. Testar, M. Palacín, and A. Zoranzo, *J. Biol. Chem.*, *267*, 10381–10388 (1992).
106. I. Novak-Hofer and G. Thomas, *J. Biol. Chem.*, *260*, 10314–10319 (1985).
107. M. Desautels and A. L. Goldberg, *Proc. Natl. Acad. Sci. USA*, *79*, 1869–1873 (1982).
108. W. E. Balfour and J. J. Grantham, *Nature*, *275*, 768 (1978).
109. D. L. Vesely, *Atrial Natriuretic Hormones*, Prentice-Hall, Englewood Cliffs, NJ, 1992.

110. D. B. Light, J. D. Corbin, and B. A. Stanton, *Nature*, *344*, 336–339 (1990).
111. L. Ralymakers, F. Hofmann, and R. Casteels, *Biochem. J.*, *252*, 269–273 (1988).
112. M. Grzeskowiak, V. D. Bianca, P. De Togni, E. Papini, and F. Rossi, *Biochim. Biophys. Acta.*, *845*, 81–90 (1985).
113. Y. Kanaho, P. P. Chang, J. Moss, and M. Vaughan, *J. Biol. Chem.*, *263*, 17584–17589 (1988).
114. A. Fein and D. W. Corson, *Science*, *212*, 555–557 (1981).
115. V. Schwabe, C. Puckstein, H. Hanneman, and E. Söchtig, *Nature*, *277*, 143–145 (1979).
116. R. E. Catalán, A. M. Martínez, and M. D. Aragonés, *Biochem. Biophys. Res. Commun.*, *96*, 672–677 (1980).
117. A. Da Luz Duque, P. Dieryck, M. Lambert, M. Svoboda, and J. Christophe, *Arch. Int. Physiol. Biochim.*, *88*, B127–B128 (1980).
118. W. Krawietz, R. W. Downs, Jr., A. M. Spiegel, and G. D. Aurbach, *Biochem. Pharmacol.*, *31*, 843–848 (1982).
119. A. G. Gilman, *Annu. Rev. Biochem.*, *56*, 615–649 (1987).
120. M. Hatch, R. Freel, A. M. Goldner, and D. L. Earnest, *Biochim. Biophys. Acta*, *732*, 699–704 (1983).
121. K. Takeda, K. Temma, and T. Akera, *J. Pharmacol. Exp. Therapeut.*, *222*, 132 (1982).
122. G. Grupp, I. Grupp, C. L. Johnson, E. T. Wallick, and A. Schwartz, *Biochem. Biophys. Res. Commun.*, *88*, 440–447 (1979).
123. W. L. Combest and R. A. Johnson, *Arch. Biochem. Biophys.*, *225*, 916–927 (1983).
124. C. C. Goodno, *Proc. Natl. Acad. Sci. USA*, *76*, 2620–2624 (1979).
125. D. W. Boyd, K. Kustin, and M. Niwa, *Biochim. Biophys. Acta*, *827*, 472–475 (1985).
126. J. A. Manthey and L. D. Hager, *J. Biol. Chem.*, *256*, 11232–11238 (1981).
127. B. E. Krenn, H. Plat, and R. Wever, *Biochim. Biophys. Acta*, *912*, 287–291 (1987).
128. B. E. Krenn, M. G. M. Tromp, and R. Wever, *J. Biol. Chem.*, *264*, 19287–19289 (1989).
129. R. Wever and K. Kustin, *Adv. Inorg. Chem.*, *35*, 81–115 (1990).

130. B. Leighton, G. J. S. Cooper, C. Da Costa, and E. A. Foot, *Biochem. J.*, *276*, 289–292 (1991).

131. D. J. MacEwan, J. Simpson, R. Mitchell, M. S. Johnson, F. J. Thompson, and G. Finl, *Biochem. Soc. Trans.*, *20*, 133s (1992).

132. D. Heffetz, I. Bushkin, R. Dror, and Y. Zick, *J. Biol. Chem.*, *265*, 2896–2902 (1990).

133. S. Kadota, I. G. Fantus, G. Deragon, H. J. Guyda, B. Hersh, and B. I. Posner, *Biochem. Biophys. Res. Commun.*, *147*, 259–267 (1987)

134. I. G. Fantus, S. Kadota, G. Deragon, B. Foster, and B. I. Posner, *Biochemistry*, *28*, 8864–8871 (1989).

135. S. Kadota, I. G. Fantus, G. Deragon, H. J. Guyda, and B. I. Posner, *J. Biol. Chem.*, *262*, 8252–8256 (1987).

136. S. Kadota, I. G. Fantus, B. Hersh, and B. I. Posner, *Biochem. Biophys. Res. Commun.*, *138*, 174–178 (1986).

137. Y. Zick and R. Sagi-Eisenberg, *Biochemistry*, *29*, 10240–10245 (1990).

138. S. Trudel, M. R. Pâquet and S. Grinstein, *Biochem. J.*, *276*, 611–619 (1991).

139. S. Borgoin and S. Grinstein, *J. Biol. Chem.*, *267*, 11908–11916 (1992).

140. T. Inazu, T. Taniguchi, S. Yanagi, and H. Yanamura, *Biochem. Biophys. Res. Commun.*, *170*, 259–263 (1990).

141. K. Guan and J. E. Dixon, *J. Biol. Chem.*, *266*, 17026–17030 (1991).

142. W. J. Thompson, B. H. Tan, and S. J. Strada, *J. Biol. Chem.*, *266*, 17011–17019 (1991).

143. E. G. Loten and J. C. Redshaw-Loten, *Int. J. Biochem.*, *18*, 847–851 (1986).

144. K. A. Lord and G. H. Reed, *Arch. Biochem. Biophys.*, *281*, 124–131 (1990).

145. G. D. Markham, *Biochemistry*, *23*, 470–478 (1984).

146. R. J. De Koch, D. J. West, J. C. Cannon, and N. D. Chasteen, *Biochemistry*, *13*, 4347–4354 (1974).

147. R. Bogumil, J. Hütterman, R. Kappl, R. Stabler, C. Sudfeldt, and H. Witzel, *Eur. J. Biochem.*, *196*, 305–312 (1991).

148. H. Sakurai, K. Tsuchiya, M. Nukatsuka, M. So Fue, and J. Kawada, *J. Endocrinol.*, *126*, 451–459 (1990).

149. S. Ramanadham, J. J. Mongold, R. W. Brownsey, G. H. Cros, and J. H. McNeill, *Am. J. Physiol.*, *257*, H904–H911 (1989).

150. S. Ramanadham, G. H. Cros, J. J. Mongold, J. J. Serrano, and J. H. McNeill, *Can. J. Physiol. Pharmacol.*, *68*, 486–491 (1990).

151. S. Ramanadham, C. F. Heyliger, M. J. Gresser, A. S. Tracey, and J. H. McNeill, *Biol. Trace Element Res.*, *30*, 119–124 (1991).

152. X. Shi, X. Sun and N. S. Dalal, *FEBS Lett.*, *271*, 185–188 (1990).

153. H. Degani, M. Gochin, S. J. D. Karlish, and Y. Shechter, *Biochemistry*, *20*, 5795–5799 (1981).

154. X. Shi and N. S. Dalal, *Arch. Biochem. Biophys.*, *295*, 70–75 (1992).

155. J. Gil, M. Miralpeix, J. Carreras, and R. Bartrons, *J. Biol. Chem.*, *263*, 1868–1871 (1988).

156. J. H. McNeill, V. G. Yuen, H. R. Hoveyda, and C. Orvig, *J. Med. Chem.*, *35*, 1489–1491 (1992).

157. K. A. Hutchinson, L. F. Stancato, R. Joves, and W. B. Pratt, *J. Biol. Chem.*, *267*, 13952–13957 (1992).

158. M. Carreras, A. M. Bassols, J. Carreras, and F. Climont, *Arch. Biochem. Biophys.*, *264*, 155–159 (1988).

9

Inhibition of Phosphate-Metabolizing Enzymes by Oxovanadium(V) Complexes

Paul J. Stankiewicz,[1] *Alan S. Tracey,*[1] *and Debbie C. Crans*[2]

[1]Department of Chemistry, Simon Fraser University, Burnaby, British Columbia V5A 1S6, Canada

[2]Department of Chemistry and Cell and Molecular Biology Program, Colorado State University, Fort Collins, Colorado 80523, USA

1. INTRODUCTION

The observation and subsequent identification of the vanadium(V) oxoanion, vanadate, as a potent inhibitor of Na^+,K^+-ATPase activity provided much of the original impetus for the study of the biochemical properties of vanadium [1,2]. However, this early work did not represent the first report of the inhibition of enzyme function by vanadate as it had previously been shown that vanadate in the presence of uridine inhibited ribonuclease A. The possibility that the ribonuclease inhibition might arise because vanadate could readily adopt a pentacoordinate structure that resembled the transition state structure of phosphate was raised in this work [3].

In this chapter, the role of pentacoordinate vanadate in enzyme inhibition will be examined. This type of inhibition can reasonably be expected to be limited mostly to enzymic systems that catalyze reaction at the phosphorus of a phosphate group. Other processes may provide alternative possibilities for inhibition of enzyme activity and, as a consequence, allow for inhibition by vanadate oligomers, complexes of vanadate, or vanadium in other oxidation states.

This discussion of enzyme inhibition is preceded by a brief description of some of the chemistry relevant to the possible roles of vanadate as an inhibitor of the function of a variety of enzymes. Earlier chapters in this book discuss much of the aqueous chemistry of vanadium; other chapters discuss various other biochemical effects of vanadium as do a number of recent review articles and compendiums [4–9].

2. RELEVANT AQUEOUS CHEMISTRY OF VANADATE

Perhaps the greatest source of the biochemical function of vanadate derives from the ease in which the coordination geometry about the vanadium nucleus can be altered. Such changes are easily effected by a number of different ligands. However, often one ligand will give rise to products of differing coordination. Tetrahedral, trigonal bipyramidal, and octahedral coordination are the geometries most frequently encountered in aqueous solution. The molecular requirements necessary for the formation of such products is relevant to understanding vanadium biochemistry.

Vanadium oxides undergo facile oxidation-reduction reactions and these reactions also can lead to biochemical activity. The interplay between oxidation states is mediated by a number of materials produced by metabolic processes. Many aldehydes and sulfides readily reduce vanadium while peroxides oxidize it. In addition, the oxidative potential of vanadium(V) is quite sensitive to pH with vanadium(V) becoming a better oxidizing agent as the pH is lowered [10]. Oxidation by vanadium peroxides may account for some of the biochemical activity of these vanadium complexes.

2.1. Oligomerization Reactions

Throughout the range of concentrations most frequently employed for biochemical studies, a few micromolar to a few millimolar under near-neutral conditions, the predominant forms of vanadate are the monomer, dimer, and tetramer [10–12] and these compounds undergo exchange reactions in the millisecond time scale [13]. Other oligomers are also formed, probably the most biochemically important of which is decavanadate. This latter compound is the predominant species present in moderately acidic solution when vanadate is present in concentrations higher than about 100 μM. If the pH of the solution is lowered below about pH 3, the cationic ion VO_2^+ is formed and, with continued decrease in pH, becomes the predominant material [14]. When the pH is raised substantially, the monomeric vanadate trianion becomes the dominant species. It finally becomes the only observable product as the pH approaches pH 14, even at high concentrations.

2.2. Reactions with Alcohols and Diols

Under the range of conditions suitable for most biochemical reactions, monomeric vanadate reacts in a rapid and reversible manner with hydroxyl groups to form vanadate complexes [15–17]. These products are generally thought to be structurally analogous to phosphate esters and are frequently referred to as vanadate esters. These materials can act as substrates or as inhibitors depending on the type of reaction being catalyzed. If the enzyme uses the phosphate of the normal substrate simply as a prosthetic group to enable a reaction at a remote site, then the corresponding vanadate derivative can also act as the enzyme substrate and the reaction will proceed. In contrast, if the reactive site is normally at the phosphorus atom, the vanadate ester can be a potent inhibitor of the enzyme function as found, for instance, in the case of a number of phosphatases and mutases.

1,2-Diols also form acyclic esters but, in addition, give rise to cyclic products [18]. These products are readily formed and those that have been well characterized, without exception, are dimeric materials containing two metal centers and two ligands [19–22]. Some evidence for a monomer has been obtained that suggests it is undergoing fast equilibration with its precursors [21]. The exact stereochemistry about the vanadium in the dimeric product is not known but it almost certainly is a trigonal bipyramidal type of structure [21–24]. The formation constant for these products tends to be quite large, particularly if the geometry of the ligand itself is favorable for reaction, as in nucleosides [21,25,26]. The formation constant for reaction with adenosine (4×10^7

M^{-1}), for instance, is more than 10 000 times larger than for reaction with ethylene glycol [18]. It seems reasonable to expect that the dimerization constants are not particularly sensitive to the ligands, since the ligands are chemically quite similar. On this basis, large overall changes in the formation constants are a reflection of the ability of the monomeric unit to form. Somewhat surprisingly, cyclic products formed from 1,3-diols have not yet been observed [27,28]. It seems unlikely that such products are not formed, albeit with small formation constants.

2.3. Oxidation-Reduction

The studies of the effects of vanadium on ribonuclease and other enzymes have often been directed to both the V(IV) and V(V) oxoanions and complexes that they form [3,29]. Because of the problems associated with maintaining the V(IV) oxidation state, it is often exceedingly difficult to be confident that an observed effect arises from vanadyl, particularly when it is found that vanadate gives a similar response. Unless it is clear that this problem has been adequately addressed during the experimental work or that the vanadyl ion has different activity than vanadate, it is not possible to be confident that all or any of the observed effects should be accredited to vanadyl or its complexes.

3. ACTIVE SITE INHIBITION BY MONOMERIC VANADATE

The ability of vanadate to inhibit enzymes is linked to its physical and chemical properties. Structurally, vanadate is similar in both shape and size to phosphate, but in addition it has a flexible coordination chemistry. In order for an enzyme to generate the pentacoordinate structure associated with the transition state structure of phosphate it is necessary to balance the energy requirements of the phosphate group with an energy source inherent in the catalyst itself. This stabilization energy may derive from conformational changes, the generation of salt bridges, the formation of hydrogen bonds, or other interactions of the enzyme. Unlike phosphate, vanadate readily forms stable pentacoordinate structures. The energy normally needed to stabilize the pentacoordinate phosphate structure is not required because the material formed is not a transition state complex but a product. This product, to the extent that pentacoordinate vanadate resembles pentacoordinate phosphate, will be structurally analogous to the transition state structure associated with the normal catalytic reaction. Transition state complexes involving coordination somewhat as represented in Fig. 1 will be generated. Of course, in some cases, other coordinating groups will be involved.

Molecular orbital calculations have also been carried out to explore the similarities between pentacoordinate vanadate and pentacoordinate phosphate [30]. The authors concluded that the electron density contours of pentacoordinate vanadate were satisfactorily similar to those of phosphate with the minor qualification that vanadium-oxygen bonds are slightly longer and less polar and would form less favorable hydrogen bonds and charge interactions.

3.1. Vanadate Esters Mimic Intermediates in Phosphoryl Transfer Reactions

3.1.1. Ribonuclease

2′,3′-Cyclic uridine monophosphate is formed as an intermediate during the ribonuclease-catalyzed hydrolysis of RNA. Ribonuclease catalyzes both the formation and hydrolysis of the cyclic phosphate intermediate and both reactions involve pentacoordinate phosphate intermediates. Early investigations of the inhibition of ribonuclease A by vanadate showed that it was only effective as an inhibitor when in the presence of uridine [3]. Recent work showed that the major product of the reaction

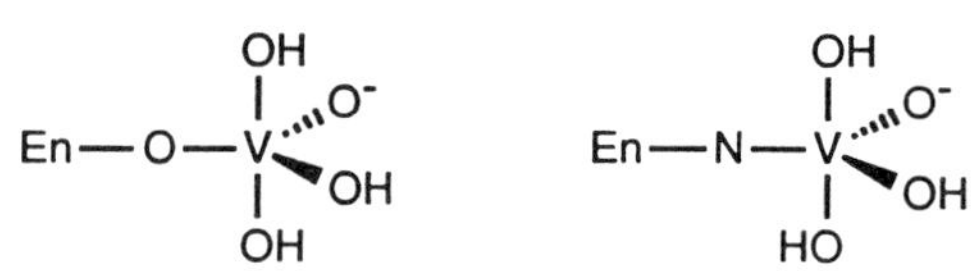

FIG. 1. Putative covalent enzyme intermediates depicting vanadate analogs to possible transition states corresponding to phosphate ester hydrolysis. Linkages are depicted as occurring through oxygen as for serine, threonine, or tyrosine residues and through nitrogen as in histidine, lysine, or arginine amino acid residues.

of vanadate with uridine is a dimeric compound with the monomer being a minor product under the conditions of the studies. Detailed measurements, taking this into account, provided a value for the inhibition constant of 0.3 μM at pH 7.0 [29]. This inhibition constant is a factor of 10^4 smaller than the K_i for the corresponding 2′,3′-cyclic phosphate and its value demonstrates the effectiveness of the vanadate complex as a transition state analog of the normal substrate.

A detailed picture of the structure of the vanadate-uridine complex attached to the active site of ribonuclease was obtained from a combination of X-ray and neutron diffraction studies and a distorted trigonal bipyramidal structure about vanadium was revealed [31–33]. These data provided very firm evidence for the role of vanadate as a mimic of the phosphate transition state structure. In view of this, it is not surprising that vanadate-nucleoside complexes inhibit a number of other ribonucleases, including human plasma ribonuclease and ribonuclease T_1 [34,35]. An interesting facet of ribonuclease T_1 is that it forms a crystalline adduct with vanadate ion where the vanadate occupies the active site as a tetrahedrally coordinated species [36]. Although intriguing, such binding is unlikely to be particularly relevant to the role of vanadate as a phosphate analog.

3.1.2. *Other Nucleic Acid–Related Enzymes*

Application of vanadate complexes as transition state analogs has been extended to a variety of other enzymes that catalyze reactions of ribonucleic acid [34]. Enzymes were assayed qualitatively by the occurrence of ^{32}P-labeled products using gel electrophoresis. As expected, the family of ribonucleases including types A, T_1, N_1, and V_2 were strongly inhibited. Other enzymes that were inhibited included polynucleotide phosphorylase (3′,5′-exonuclease), poly A polymerase, polynucleotide kinase (phosphorylates 5′ terminus of RNA or DNA), RNA ligase (joins 5′-phosphate terminus to 3′-hydroxyl ends), and micrococcal nuclease which degrades DNA or RNA to generate phosphate ends. Reverse transcriptase was partially inhibited. These studies were carried out using high concentrations (2 mM); thus little or no information concerning mechanism-based specificity was obtained.

Polynucleotide kinase and reverse transcriptase were inhibited competitively against nucleotide 5′-triphosphate substrates while enzymes which operate on DNA such as δ_1 nuclease and deoxyribonuclease I were not inhibited. However, there was some apparent nonspecific inhibition of the DNA restriction endonucleases.

3.1.3. Phosphoglycerate Mutase

Glycolytic phosphoglycerate mutase (PGM) operates by a ping-pong mechanism with a phosphohistidine enzyme intermediate [37–39]. Reaction with the substrate, 2,3-bisphosphoglycerate (2,3-BPG), generates the phosphorylated enzyme (EP), which catalyzes the interconversion of 2- and 3-phosphoglycerate (PGA) via an intermediate PGM/2,3-BPG complex; EP(3PGA) $\rightleftharpoons$ E(2,3-BPG) $\rightleftharpoons$ EP(2PGA). Vanadate specifically inhibits this and related enzymes comprising the class of mutases that require 2,3-BPG and have stable phosphoenzyme intermediates [40]. Nearly complete inhibition of rabbit muscle phosphoglycerate mutase can be achieved at 50–100 μM vanadate concentrations [41] and this led to the suggestion [41] that vanadate could be used to distinguish between the vanadate-sensitive ping-pong phosphoenzyme phosphotransferases and vanadate-insensitive sequential type phosphotransferase such as the kinases. Vanadate was also shown to inhibit erythrocyte phosphoglycerate mutase in a competitive manner with 2,3-BPG with an apparent K_i value of 2.5 μM [42]. Vanadate is known to stimulate hydrolysis of the phosphoenzyme intermediate [43] at concentrations between 0.1 and 1.0 mM but this is irrelevant to the inhibition of the mutase activity which is effective at micromolar concentrations. The vanadate-promoted hydrolysis is, however, thought to involve an active site divanadate species that activates the catalytic cycle and causes a transfer of phosphate to water [44].

Kinetics studies showed that the onset of inhibition of phosphoglycerate mutase was time dependent in the second to minute time range [40,45]. The rate of onset and the extent of inhibition increased both with vanadate and with 3-phosphoglycerate concentration. This suggested that the ester, 2-vanadio-3-phosphoglycerate (2-V-3-PGA) was the true transition state analogue inhibitor species. Figure 2 gives a diagrammatic representation of the inhibitor in the active site of phosphoglycerate mutase. The binding of vanadate to the enzyme could be quantitated by observing the loss of the ^{51}V NMR signal as vanadate was bound by the enzyme. Using this approach, the dissociation con-

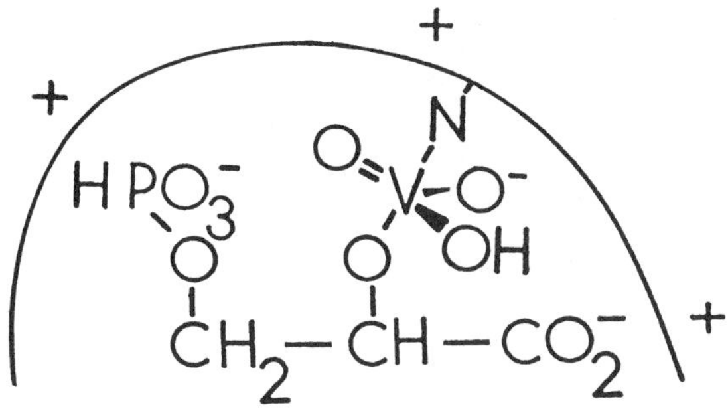

FIG. 2. Diagrammatic representation of the binding of 2-vanadio-3-phosphoglycerate in the active site of phosphoglycerate mutase. Binding is thought to involve a covalent interaction with an active site histidine [44] and noncovalent interactions with the carboxylate group [169].

stant of 2-V-3-PGA was found to be 1.0×10^{-11} M [44]. This constant is a factor of 10^{-5} of the K_m for 2,3,-BPG and its value illustrates the exceedingly tight binding of this vanadate complex.

The 2,3-BPG synthase activity of erythrocyte phosphoglycerate mutase assists in the synthesis of 2,3-BPG from 1,3-BPG. This activity is also inhibited by vanadate [43]. The structure of the inhibitor species is presumably the same as that described for the mutase activity of this enzyme.

3.1.4. Phosphoglucomutase

Following the reports of vanadate inhibition of ribonuclease and the ATPases, a survey of the susceptibility of glycolytic enzymes to vanadate inhibition was carried out. Only phosphoglucomutase and phosphoglycerate mutase were sensitive to vanadate and both were completely inhibited at concentrations between 50 and 100 μM vanadate [41].

Phosphoglucomutase is active in its phosphorylated form and operates by a ping-pong mechanism [46]. The enzyme is phosphorylated by glucose-1,6-bisphosphate on an active site serine and transfers phosphate between the 1 and 6 positions of glucose monophosphate. An early kinetic study of this enzyme indicated that the vanadate inhibitor was competitive toward Glu-1,6-BP and noncompetitive toward Glu-1-P with apparent K_i values of 0.74 and 1.1 μM, respectively [47]. A similar study of phosphoglucomutase from erythrocytes also found competitive inhibition toward the bisphosphate with an apparent K_i value of 0.15 μM [42]. Vanadate did not stimulate hydrolysis of the phosphoenzyme intermediate [43].

The earlier studies were extended by work that showed that the extent of inhibition of the enzyme in the presence of Glu-1-P and vanadate increased as a function of time. Furthermore, the rate at which an equilibrium level of inhibition could be achieved was increased by increasing either Glu-1-P or vanadate concentration [48]. This implicated both species in the inhibition and the true inhibitor was assigned as the Glu-1-P-6-V ester and indicated a complex interaction scheme. Detailed studies led to a model that invoked the formation of an enzyme-inhibitor complex that subsequently underwent an isomerization according to the model $E + I \rightleftharpoons EI \rightleftharpoons E^*I$, where EI and E*I represent the two enzyme complexes. The K_i value related to formation of the EI complex was found to be 1.0×10^{-9} M. The rate constants for the formation and breakdown of E*I from EI were found to be 4×10^{-2} sec^{-1} and 2.5×10^{-4} sec^{-1}, respectively, and the overall dissociation constant of the E*I complex was 2.0×10^{-12} M. The EI complex represents an equilibrium mixture of E·V-6-Gl-1-P and E·V-1-Glu-6-P and contains an Mg^{2+} ion. The rate of dissociation of Mg^{2+} from the enzyme-vanadate·Glu-1-P complex was found to be much slower than the dissociation from the enzyme-substrate complex and the affinity of the inhibitor for the enzyme was decreased by three orders of magnitude when Li^+ was substituted for Mg^{2+}.

Recent Raman spectroscopy studies shed considerable light on the mechanism of action of this enzyme [23,49] and revealed structural details of how vanadate is bound. The results are in full accord with a pentacoordinate vanadium in the active site. More recently, an unpublished X-ray structure has confirmed the pentacoordinate structure of vanadate when complexed to this enzyme [50].

3.2. Influence of Vanadate on ATPases

3.2.1. *E_1E_2 ATPases*

An inhibitor of Na^+,K^+-ATPase in commercial ATP preparations was reported in 1975 [51]. The material responsible for the inhibition was isolated from commercial equine ATP and identified as vanadate [52,53]. A detailed kinetic study showed, for the porcine kidney membrane Na^+,K^+-ATPase, that vanadate was capable of binding at two sites with equilibrium binding constants of 4.0 nM and 0.5 μM [54]. The high-affinity site for vanadate corresponded to the low-affinity site for ATP in the flip-flop kinetic model. In this work, vanadate was assumed to bind at the aspartyl residue that is phosphorylated [55] during the normal catalytic cycle. The results of the inhibition studies were confirmed and extended for additional Na^+,K^+-ATPases by other workers [1,2]. Vanadate was also shown to inhibit enzyme phosphorylation by ATP in electric eel ATPase for which 10 μM vanadate gave 98% inhibition [56].

Vanadate binds in concert with Mn^{2+} and it was therefore deduced that vanadate was bound in a site adjacent to the divalent cation site of the enzyme [57]. This conclusion was supported by other work that showed that the highest affinity of vanadate for human erythrocyte ATPase (K_1 = 0.1 μM) occurred when both the Mg^{2+} and K^+ binding sites were occupied [58]. This supported the previous conclusion that vanadate was bound at the site from which phosphate is normally released. Vanadate also inhibited the *Streptococcus faecalis* Na^+,K^+-ATPase and it was proposed that vanadate inhibits all E_1E_2-ATPases which are characterized by a two-state alternate sites kinetic model and an acyl phosphate intermediate [59]. Other members of the E_1E_2-ATPases are the Ca^{2+}-ATPases of sarcoplasmic reticulum and plasma membrane, K^+,H^+-ATPase of gastric mucosa, and H^+-ATPases of fungal membranes.

In the sarcoplasmic reticulum Ca^{2+}-ATPase, vanadate inhibited the enzyme function with an apparent K_i of 10 μM. The activity of the vanadate inhibited enzyme could, however, be restored by the addition of Ca^{2+} [60]. The H^+-ATPase of *Neurospora crassa* membrane was inhibited by vanadate with a K_i = 0.75 μM and the inhibition was accompanied by a change in kinetics [61].

3.2.2. Other ATPases

The mitochondrial ATPase which synthesizes ATP is inhibited weakly by vanadate. In this system, vanadate acts as a mixed type of inhibitor with K_i values of 0.18 and 0.8 mM with respect to phosphate. In this case, the vanadate could be acting as an analog to tetrahedral phosphate [62]. In contrast to this behavior, vanadate is a potent inhibitor of myosin ATPase. Unlike for the Na^+,K^+-ATPases, the vanadate-induced inhibition of myosin requires ADP. In the presence of ADP and 0.5 mM vanadate, the half-time for onset of inhibition is 1.5 hr [63]. The E·ADP·V_i complex is stable enough to withstand purification procedures [64]. The remarkable stability of the ternary complex was attributed to its isomerization to a stable intermediate which is analogous to the isomerization of the E·ADP·P_i complex which occurs during the catalytic mechanism. Vanadate was released from the E·ADP·V_i complex of myosin subfragment 1 (SF1) with a rate constant of 5×10^{-6} sec^{-1} [65] as indicated by loss of tryptophan fluorescence concomitant with an enzyme conformational change. The binding of actin to SF1 accelerates the release of vanadate 10^4–10^5 fold [65].

A bis-ADP derivative composed of a 22-atom spacer which connected the ribose hydroxyl groups of two ADP molecules acted as a bivalent ligand for SF1. The bis-ADP•V_i complex which bound two molecules of SF1 was stable enough to withstand chromatographic analysis and gave information about the topography of the active sites of the enzyme [66].

Vanadate is a potent inhibitor of the dynein ATPase system; 1 μM vanadate caused complete paralysis [67]. The ATPase of sea urchin mitotic apparatus was inhibited 40% by 10 μM vanadate [67] and similar concentrations inhibited anaphase movement of chromosomes and spindle elongation in lysed cells [68]. Vanadate at 1.0 mM could not inhibit chromosome movement in unlysed cells. Vanadate at 0.5 μM inhibited dynein ATPase of sperm cilia [69] but vanadium(IV) at concentrations up to 25 μM was ineffective in inhibiting dynein ATPase [67].

3.3. Inhibition of Phosphatases

3.3.1. Phosphotyrosyl Protein Phosphatases

The structures, mechanisms of action, and regulation of cellular phosphotyrosyl protein phosphatases are not known to a high level of refinement. This topic has been the subject of a number of recent reviews [70,71]. It is a widely accepted premise that the most salient biological action of vanadate is the inhibition of phosphotyrosyl protein phosphatases. It is by this mechanism that vanadate is thought to increase cellular tyrosine phosphorylation levels and augment the signal transduction pathways of insulin and growth factors. Despite this acceptance, vanadate has been shown to be only a moderately specific inhibitor of phosphotyrosyl protein phosphatase activities associated with purified enzymes or cell extracts. For instance, in rat spleen, 50 μM vanadate inhibited only one of three purified proteins with phosphatase activity [72]. An acid phosphatase from bone required 2.0 mM vanadate to achieve 70% inhibition [73]. Other examples of weak inhibition of phosphatase activity have been reported [72,74,75]. These results clearly indicate that vanadate is not a specific inhibitor of all protein phosphatase activities. However, vanadate can be a potent inhibitor of such activity [1,8].

The mechanism of action is not well understood and the requisite detailed studies have not been done. In many cases it seems possible that vanadate could inhibit the tyrosine phosphatase activity by forming a substrate analog by reacting with the dephosphorylated substrate to form a tyrosyl vanadate derivative [76]. This analog would then inhibit the phosphatase activity by forming a stabilized transition state structure during the catalytic cycle [8]. It is also possible that vanadate could inhibit these enzymes by forming a covalently bound enzyme complex. Evidently, much more work that utilizes purified enzymes and purified phosphoprotein substrates is required to establish the function of vanadate.

There are other processes that could lead to inhibition of the phosphatases. Vanadate might cause active site oxidation of cysteine. Many phosphotyrosine protein phosphatases require thiol-reducing agents for optimal activity [71,77,78]. It is not known how effectively vanadate will oxidize a thiol group if there is no possibility of forming a sulfur-sulfur bond. For instance, the active site thiol group of glycerol-3-phosphate dehydrogenase is not reduced by vanadate even though vanadate inhibits this enzyme [79]. Another factor that may be important here is the steric requirements of the active site itself, perhaps preventing access to the thiol group. In addition, and more speculatively, partial cellular reduction of vanadate to vanadyl (VO^{2+}) may generate a cation which can mimic the Zn^{2+} inhibitor effect. Vanadyl cation is known to substitute at the Zn^{2+} binding site of carboxypeptidase A [80] and most, but not all, phosphotyrosyl protein phosphatases are inhibited by 10–100 μM Zn^{2+} [71,72,81,82].

3.3.2. Phosphoseryl Protein Phosphatases

Cell physiologists normally add vanadate to cell extracts in order to stabilize protein phosphotyrosine content [81]. However, the assumption that vanadate does not inhibit phosphoryl protein phosphatases has not been well documented. This assumption is based mostly on the observed selectivity of vanadate inhibition of phosphotyrosyl proteins in cell extracts from fibroblast [83,84] and astrocytoma cell lines [85,86]. A careful kinetic study of the ability of vanadate to inhibit each of the four classes of protein seryl phosphatases is needed. There is no obvious mechanistic reason why vanadate should not inhibit these phosphatases if they operate by a trigonal bipyramidal transition state for phosphoryl transfer. One speculation which may be put forward to explain the apparent inability of vanadate to inhibit phosphoseryl protein phosphatases is that the protein or peptide substrate concentration is above its K_m. The action of vanadate acting as a competitive inhibitor could be suppressed by high substrate concentration [8]. In most cases the K_m of the substrate is not known.

In one case [87], vanadate was tested as an inhibitor of the seryl protein phosphatase, calcineurin. This phosphatase can be activated by Ni^{2+} in place of the normal activator, calmodulin. In the form of the Ni^{2+} phosphatase, vanadate at 0.1 mM produced no inhibition, although the same concentration of vanadate caused 60% inhibition of an Mn^{2+}-activated form.

3.3.3. Alkaline Phosphatase

E. coli alkaline phosphatase is inhibited by either vanadium(IV) or vanadium(V). The K_i values at pH 8.0 are 4.0×10^{-7} M and 2.2×10^{-6} M, respectively [88]. Vanadyl, in most cases, is a better inhibitor of alkaline phosphatases than vanadate [89]. Other potent inhibitors of alkaline phosphatase include molybdate ($K_i = 7 \times 10^{-6}$ M) [90] and periodate ($K_i = 1 \times 10^{-7}$ M) [91]. Because these last two ions can form expanded coordination fields it was proposed that they inhibit as octahedral complexes. However, it is highly likely that vanadate inhibits as a trigonal bipyramidal species and vanadyl may function similarly. The catalytic mechanism of alkaline phosphatase operates by a phosphorylserine intermediate [92]. Thus the trigonal bipyramidal and octahedral structures probably involve an active site serine as a ligand as suggested by Fig. 1. The crystal structure of alkaline phosphatase shows that the active site is near the surface [93], but kinetic evidence suggests that there are enough binding interactions to provide some selectivity toward the organic moiety of the substrate. Electron-withdrawing functions decreased the K_m of synthetic substrates [94]. Phenol, which can form the phenylvanadate ester in solution, decreased the apparent K_i of vanadate by an order of magnitude. This observation lead to the proposal that the protein tyrosyl vanadate ester would selectively enhance the inhibition of vanadate toward phosphotyrosyl protein phosphatases [90].

3.3.4. *Acid Phosphatase*

Vanadate inhibits both human liver and potato acid phosphatase with K_i values of 2×10^{-7} M [90,95]. This value is three orders of magnitude lower than the inhibition constant for phosphate [95]. The enhanced affinity for vanadate is attributed to its ability to adapt a pentacoordinate complex at the enzyme active site which resembles the transition state for phosphoryl transfer [90,95]. This intermediate may be somewhat similar to that depicted in Fig. 1. The wheat germ acid phosphatase has been reported to have a phosphorylhistidine as a catalytic intermediate [96]. The rate-limiting step of acid phosphatase has been proposed to be the hydrolysis of the phosphoenzyme intermediate [97], and since the rate-limiting step is also the highest energy point on the reaction coordinate, this observation argues in favor of a pentacoordinate phosphoryl enzyme transition state.

Human prostatic acid phosphatase is also inhibited by vanadium, but in a pH-dependent manner. Monomeric vanadate was found to inhibit this enzyme at pH 7.0 while at lower pH (5.5) divanadate was found to be the inhibiting species [98]. The monomeric vanadate was, however, found to be the more efficient inhibitor, $K_i = 0.013$ mM, compared to divanadate, $K_i = 0.7$ mM. The change in mode of binding was attributed to a conformational change in the enzyme brought about by protonation of a carboxylate group [98].

3.3.5. Fructose-2,6-Bisphosphatase

Fructose-2,6-bisphosphate is a signal molecule which allosterically activates glycolytic phosphofructokinase and inhibits gluconeogenic fructose-1,6-bisphosphatase. Vanadate was a noncompetitive inhibitor of the purified hepatic phosphatase with a K_i = 0.15 mM [99]. The physiological effect of inhibition would be an insulin-like stimulation of glycolysis. This effect has been seen in hepatocytes [100] and diabetic rats drinking water containing the vanadyl ion [101].

The catalytic mechanism of the enzyme involves an intermediate phosphorylation of histidine$_{258}$ [102] and this provides a rational for an inhibition mechanism involving a vanadioenzyme intermediate. However, in contrast to observation, this mechanism should produce competitive inhibition. A mechanism in which vanadate catalyzes hydrolysis of the phosphoenzyme has been excluded [99].

This unusual enzyme also has a fructose-6-P 2-kinase activity at a separate site on the same polypeptide chain that is inhibited by decavanadate [103]. At low vanadate concentrations the observed noncompetitive inhibition might not be caused by decavanadate but by smaller vanadate oligomers.

3.3.6. Glucose-6-phosphatase

The generation of free glucose from hepatic glucose-6-P is the last step of gluconeogenesis. Glucose is able to diffuse across the plasma cell membrane and this is the major metabolic mechanism for the increase in plasma glucose. The inhibition of this activity by vanadate is coherent with its insulin-like effect. In addition to the phosphohydrolase activity, this enzyme can catalyze transfer of phosphate from carbonyl phosphate to glucose. The second activity is strongly suggestive of a ping-pong mechanism involving a phosphohistidine intermediate and there is some evidence in support of this [104]. Vanadate competitively inhibits the hydrolase activity with a K_i value of 2.2 μM and the transferase activity with a K_i of 0.8 μM [105].

3.4. Kinases, Lyases, and Synthases

3.4.1. ATP-Dependent Kinases

There are no reports in the literature that show that vanadate significantly inhibits ATP-dependent kinases by an active site mechanism. Attempts to inhibit hexokinase, adenylate kinase, or glycerol kinase by monomeric vanadate produced little or no inhibition [106–108] although it would be expected that vanadate could form an ADP-vanadate transition state complex. The apparent inability to inhibit by this mechanism may be based on the inability of vanadate to complex Mg^{2+} [107]. A second possibility is that ADP•V would not be a transition state analog in the cases where the β-phosphate rather than the γ-phosphate becomes pentacoordinate during the reaction. However, attempts to study this last possibility by preparing AMPVP in the reaction solution did not reveal significant activity [107].

3.4.2. ATP-Citrate Lyase, Succinyl-CoA Synthase, and Phosphatidic Acid Hydrolase

ATP-citrate lyase [109] and succinyl-CoA synthase [110] are both influenced by vanadate. The two enzymes function via a phosphoenzyme intermediate. Vanadate was found to inhibit the incorporation of $[^{32}P]$phosphate into the enzyme to generate its active form. The ATP-citrate lyase is known to form a phosphorylhistidine during the catalytic cycle [111] and the succinyl-CoA synthase is thought to function similarly [112]. It seems quite probable that the vanadate inhibits these two enzymes by forming an enzyme vanadiohistidine complex, perhaps somewhat similar to that proposed for the alkaline phosphatases. Another reasonable candidate for a phosphoenzyme mechanism is phosphatidic acid hydrolase and this enzyme is strongly inhibited by vanadate [113].

3.4.3. HMG-CoA Reductase

3-Hydroxy-3-methylglutaryl-CoA reductase is a microsomal enzyme which controls the rate-limiting step in cholesterol synthesis. The enzyme is inhibited in microsomes by 10 mM vanadate [114]. The mechanism of inhibition is unknown.

3.4.4. *Horseradish Peroxidase*

The known effects of vanadate on cellular redox systems provoked a study of the direct inhibitor action on enzymes which quench cellular oxidizing agents. Vanadate did not inhibit catalase or glutathione peroxidase but was a competitive inhibitor of horseradish peroxidase with a K_i value of 41 μM [115]. Vanadate bears no obvious homology to the substrate or transition states of these reactions. The most reasonable explanation for the inhibition of horseradish peroxidase would be the complexation of hydrogen peroxide by vanadate [116,117].

3.5. Protein Catabolism

Proteolytic degradation of exogenous proteins or endocytosed proteins degraded by lysosomal proteases in hepatocytes was inhibited by 10 mM vanadate [118]. The inhibition of proteolysis in rat liver mitochondria required only 100–200 μM vanadate [119,120]. Protein degradation is controlled by a multistep pathway. It is not likely that vanadate is a mechanism-based inhibitor of the serine protease but probably acts at an earlier step in the transduction pathway, possibly by inhibiting ATP hydrolysis which is somehow linked to protein degradation [118]. The action of vanadate is consistent with the insulin-like effect of vanadate since cAMP antagonists stimulate and insulin inhibits protein degradation [118].

An *E. coli* proteolysis is carried out by protease La. A single protein appears to catalyze peptide bond hydrolysis by an ATP-dependent mechanism. This protein was inhibited 50% by 50 μM vanadate [121]. This suggests that protease La forms a phosphoprotein intermediate during the proteolytic cycle.

4. INHIBITION BY OLIGOMERIC VANADATE

At total vanadate concentrations of about 1 mM or less and at neutral pH, monomeric vanadate is the principal oxovanadate in solution. At concentrations of about one-tenth this, the oligomers are essentially unobservable by nuclear magnetic resonance (NMR) spectroscopy.

However, even though they are not observable, they are still present in solution and can influence enzyme activity. For instance, at a residual vanadate concentration of 60 μM at pH 7.0 and 1.0 M KCl, it can be calculated from the known formation constants [25,122] that there is about 1.2 μM divanadate and 0.3 μM tetravanadate free in solution. This is not sufficient divanadate to strongly influence phosphoglycerate mutase but enough tetravanadate to almost half-saturate phosphoglucomutase. Evidently, even at low vanadate concentrations, observation of inhibition cannot be taken as evidence that monomeric vanadate is the inhibiting species. Rather, titration studies are necessary in order to identify the inhibitor. Figure 3 illustrates the use of this procedure

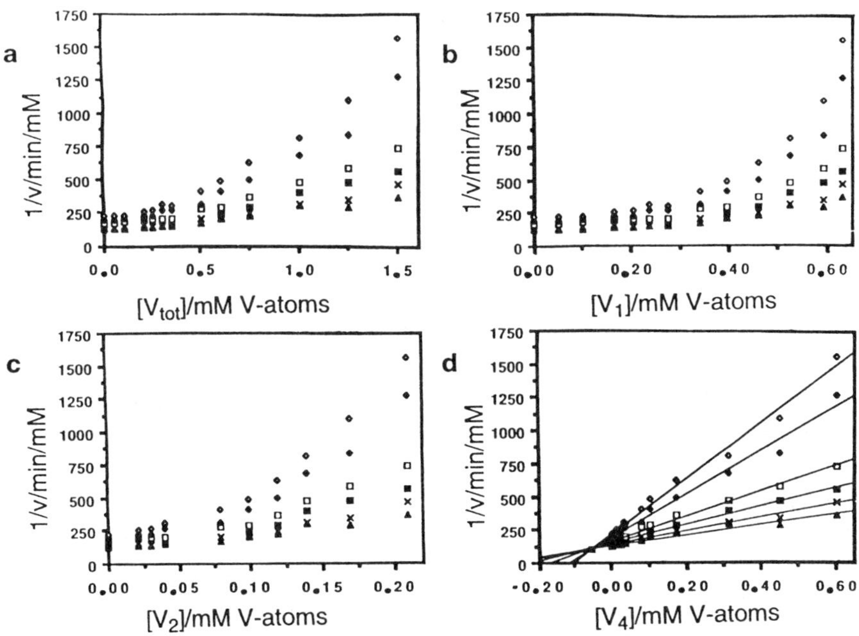

FIG. 3. The reciprocal rates of 6-phosphogluconate dehydrogenase catalyzed oxidation of 6-phosphogluconate are plotted as a function of the following concentrations: (a) total vanadate, V_{tot}; (b) vanadate monomer, V_1; (c) vanadate dimer, V_2; and (d) vanadate tetramer, V_4. Concentrations are in millimolar units and are given as vanadium atom concentrations. The results were obtained at pH 7.1 as described in [127].

for the identification of tetravanadate as an inhibitor of the function of 6-phosphogluconate dehydrogenase.

4.1. Divanadate

4.1.1. Dehydrogenases and Aldolases

Divanadate has been found to be a reasonably good competitive inhibitor against NADP, $K_i \approx 0.12$ mM, but only a weak inhibitor against NAD, $K_i \geqslant 15$ mM, for the enzyme glucose-6-phosphate dehydrogenase [123,127]. Divanadate also is a competitive inhibitor of 1,6-bisphosphate fructose aldolase from rabbit muscle, $K_i = 0.23$ mM [124], and a noncompetitive inhibitor of rabbit muscle glycerol-3-phosphate dehydrogenase [79]. In this last example, dianionic divanadate, $K_i = 0.3$ mM, was a much better inhibitor than the trianionic species, $K_i = 1.5$ mM [79].

In these studies, no obvious correlation was observed between the effects of divanadate and those of pyrophosphate or pyrophosphate-containing cofactors. This suggests that the inhibition derives, at least in part, from a change in coordination about one or both of the vanadiums of divanadate. Further evidence for this possibility has been provided from studies of glycogen phosphorylase.

4.1.2. Glycogen Phosphorylase

Oligomeric forms of molybdate, tungstate, and vanadate are inhibitors of rabbit muscle glycogen phosphorylase [125]. Measurement of the rate of glucose-1-P linkage to the nonreducing end of glycogen with the release of inorganic phosphate showed that the inhibitors were competitive against glucose-1-P. Significant inhibition required millimolar quantities of decavanadate. However, with a modified active form of the enzyme in which the phosphate moiety of the pyridoxal phosphate cofactor was replaced by phosphite ion, decavanadate was a more effective inhibitor, having K_i values of 6 μM and 4 μM toward phosphite and glucose-1-P, respectively. Thus, the inhibitors preferentially encompassed the phosphate site of pyridoxal phosphate as well as the substrate. Smaller vanadate polymers were observed to inhibit the enzyme but their kinetics were not characterized.

In our laboratory (unpublished data) we have observed that both divanadate and phosphovanadate are good inhibitors of glycogen phosphorylase. Both inhibitors bind to the free form of the enzyme with K_i values of 2.5 and 7 μM, respectively. Divanadate was found to bind to the enzyme-glycogen complex with fivefold weaker affinity. The similar but low K_i values of both divanadate and phosphovanadate suggest that one vanadium of divanadate is bound in a pentacoordinate fashion. Previously, a mechanism has been proposed in which the phosphorus of pyridoxal phosphate becomes pentacoordinate during the catalytic cycle [126]. The strong inhibition by both divanadate and phosphovanadate support this proposal.

4.2. Tetravanadate

4.2.1. Dehydrogenases, Mutases, and Others

The vanadate tetramer has been found to be a reasonably good competitive inhibitor of 6-phosphogluconate dehydrogenase obtained from a number of sources with inhibition constants ranging in value from 0.013 mM for the torula yeast enzyme through 0.63 mM (sheep liver) and 0.08 mM (human erythrocyte) to 0.24 mM for a bacterial enzyme [127].

Glucose-6-phosphate dehydrogenase is not as strongly inhibited by tetravanadate and is inhibited in a noncompetitive manner. Inhibition constants for the enzymes from the various sources are all in the order of 0.5 mM [127–129]. Inhibition of this enzyme is, however, competitive against the cofactor [123]. Glycerol-3-phosphate dehydrogenase is also inhibited by tetravanadate with an inhibition constant of 0.7 mM [79]. The mechanism by which tetravanadate inhibits this type of enzyme is not known but it may be that the shape, charge distribution, and size of tetravanadate are sufficiently close to that of the phosphate-containing substrates to allow it to be bound. On this basis, the sugar phosphates might be expected to be more tightly bound than glycerol phosphate and this is observed. Other examples that can be explained by this postulate are the inhibition of phosphoglucose isomerase [130], fructose-1,6-bisphosphate aldolase [124], phosphatidylinositol phospholipase C [131], and dephosphophosphoglucomutase [108]. Only the last enzyme tightly binds tetravanadate and does so with a dissociation constant of 0.5 μM. However, this is not too surprising, considering that the transition state analog, 1-vanadio-6-phosphoglucose, is bound with a dissociation constant of 2×10^{-12} M [48]. This exceedingly tight binding of the transition state analog suggests that in these systems the latter type of interaction has greater physiological relevance than the reaction with tetravanadate.

4.2.2. *Superoxide Dismutase*

^{51}V NMR studies of vanadium binding to superoxide dismutase (SOD) revealed independent and tight binding (K_{diss} = 0.1 μM) of two tetravanadate ions. Chemical modification of this enzyme suggested that lysine residues were required for binding [132]. Tetravanadate carries four negative charges and in this case the evidence suggests that binding is predicted mostly by electrostatic interactions. Equilibrium dialysis studies of the bovine erythrocyte superoxide dismutase did not reveal a similarly tight binding of tetravanadate to this enzyme [115]. If there is no difference in the response of the bovine and erythrocyte superoxide dismutases to vanadate, then tetravanadate in the SOD·V_4 complex must be labile. In this event, tetravanadate would not remain bound to the SOD during the dialysis treatment but would undergo hydrolysis. Hydrolysis of the product would also occur during the NMR experiments but the product would be regenerated because the vanadate is not removed from the solution. It is perhaps useful to reiterate one problem arising from studies similar to this and that is the role of the buffer used for the studies. Many buffers react with vanadate [28,133,134] and therefore can have a significant effect on the results. Phosphate is one such buffer [122] and it also catalyzes vanadate ester hydrolysis [15]. Whether it can similarly catalyze the hydrolysis of some enzyme-vanadate complexes is not known but it is possible.

4.3. Decavanadate

In all of the examples known, decavanadate binds at polyphosphate binding sites of enzymes or receptors either in the substrate domain or in an allosteric effector site. Decavanadate is a covalently closed polyanionic species of ellipsoidal shape [135]. It is readily formed in mildly acidic medium and is kinetically stable at neutral pH for the time periods required for most kinetic assays [136,137].

4.3.1. Kinases

Decavanadate, unlike monomeric vanadate, is a good noncompetitive inhibitor of several kinases. These include hexokinase (K_i = 62 μM), adenylate kinase (K_i = 11 μm), and bacterial phosphofructokinase (K_i = 20 μM) [138].

Adenylate kinase may be susceptible to decavanadate due to the tetraphosphate domain over the AMP/ATP binding site on the enzyme. Electron density maps of decavanadate diffused into adenylate kinase crystals were thought to show an oligovanadate bound to a cationic domain in the active site cleft of the protein [139]. Further structural refinement was unable to confirm this conclusion [140].

Mammalian phosphofructokinase was inhibited by decavanadate with an IC_{50} of approximately 300 nM [141,142]. It was postulated that decavanadate bound to the inhibitor sites of the polyanions: ATP, citrate, glycerate-3-P, or phosphoenolpyruvate. This proposition was supported by the relief of decavanadate inhibition by physiological antagonists of ATP inhibition: cAMP and glucose-1,6-DP.

Mammalian fructose-6-P 2-kinase accommodates four phosphate moieties within the active site and is allosterically activated by inorganic phosphate [143]. Nearly millimolar concentrations of pyrophosphate, molybdate, sulfate, tungstate, or selenate are required to inhibit this enzyme, whereas decavanadate is a good noncompetitive inhibitor toward both substrates (K_i = 27 μM). This suggests that decavanadate may bind at both substrate and allosteric sites. However, decavanadate does not inhibit all kinases since galactokinase, creatine kinase, glycerol kinase, pyruvate kinase, and inorganic pyrophosphatase are not inhibited by this material [138].

Decavanadate was found to inhibit four reactions related to inositol phosphate metabolism [144]. Decavanadate inhibited inositol-1,4,5-triphosphate-dependent Ca^{2+} release from microsomes (K_i = 0.25 μM for the triphosphate) [145]. The K_i of decavanadate was 1.2 μM. Heparin, the polysulfated glycosaminoglycan, was also capable of inhibiting Ca^{2+} release [146]. Other enzymes of inositol phosphate metabolism inhibited by decavanadate include inositol-1,4,5-triphosphate 3-kinase (K_i = 5 μM), inositol-1,4,5-triphosphate 5-phosphatase (K_i = 1.5 μM), and inositol-1,3,4,5-tetraphosphate 5-phosphatase (K_i = 0.6 μM).

Further evidence for the interaction of decavanadate at polyphosphate binding sites is the allosteric activation of 5′-nucleotidase at the allosteric activator site which recognizes ATP and 2,3-diphosphoglycerate [147] and the inhibition of the sarcoplasmic reticulum ATPase [136,148].

4.3.2. cAMP-Dependent Protein Kinase

Cyclic AMP-dependent protein kinase is noncompetitively inhibited vs. ATP (K_i = 1.7 mM) and competitively inhibited vs. the substrate, kemptide (K_i = 1.9 mM), by decavanadate. The mechanism of inhibition is somewhat novel since the evidence suggests that decavanadate itself is bound by kemptide, a heptapeptide containing two arginines and a serine (LeuArgArgAlaSerLeuGly). This binding is thought to prevent the substrate from binding to the protein. This result also shows that specific amino acid sequences can lead to binding of oligonuclear polyanionic materials [149].

5. INFLUENCE OF PEROXOVANADATE COMPLEXES

5.1. Reversible Inhibition by Peroxovanadate Complexes

Both ^{51}V NMR [7,116,117] and X-ray diffraction [150,151] show that the addition of hydrogen peroxide to vanadate increases the coordination number. Mono- and diperoxovanadate both tend to adopt seven-coordinate structures in a distorted pentagonal bipyramid. This suggests that they may not be good inhibitors for trigonal bipyramidal transition states. This is supported by unpublished observations in our laboratory that show that peroxovanadates do not inhibit calf thymus and *E. coli* alkaline phosphatase or potato acid phosphatase, phosphoglycerate mutase, or phosphoglucomutase. However, other work suggests that peroxovanadate may inhibit some osteoblast and bovine intestinal alkaline phosphatases [152].

Peroxovanadates were found to inhibit three enzymes. Two of these enzymes, glycogen phosphorylase and glyceraldehyde-3-P dehydrogenase, employ an inorganic phosphate substrate which does not become pentacoordinate during the reaction mechanism. This suggests that the peroxovanadate may bind at enzymatic binding sites for tetrahedral phosphate. In glycogen phosphorylase, inhibition was competitive with phosphate and it was also found that monoperoxovanadate (K_i = 0.34 μM) had a higher affinity than diperoxovanadate (K_i = 1.9 μM) for the binding site. With glyceraldehyde-3-P dehydrogenase, it was not possible to determine the dissociation constant because of concomitant oxidation of the active site cysteine by peroxovanadate. The third enzyme inhibited, aryl sulfatase, also does not have a pentacoordinate intermediate. Here monoperoxovanadate also had higher affinity (K_i = 0.11 μM) than diperoxovanadate (K_i = 0.52 μM) [153].

5.2. Oxidation by Peroxovanadate Complexes

The most important physiological effect of vanadate may be its role as an oxidant of cellular thiols [154,155]. Peroxovanadates are more potent oxidizing agents than vanadate and enzymes with active site cysteine residues may be the most susceptible. Both vanadium(IV) and vanadium(V) catalyze the oxidation of cysteine by hydrogen peroxide. The half-time for oxidation of 5 mM cysteine by vanadium peroxides under an argon atmosphere was a few seconds while the half-time for hydrogen peroxide alone was 1–2 min [153]. Vanadium(IV)-hydrogen peroxide catalyzes the oxidation of two molecules of cysteine per molecule of hydrogen peroxide. This reaction involves a two-electron reduction of hydrogen peroxide to water concomitant with the one-electron oxidation of one cysteine molecule to generate vanadate and one-half equivalent of cystine. Oxidation by the vanadium(V)-peroxide complex is accompanied by the loss of only one molecule of cysteine per molecule of hydrogen peroxide. This is explained by a mechanism involving a two-electron oxidation of cysteine by one molecule of hydrogen peroxide. In this case the reaction products are water and sulfenic acid (R-S-OH). The latter reaction has similarities to the vanadate-dependent bromoperoxidase reaction [156–159]. In this case the vanadate enzyme catalyzes the conversion of hydrogen peroxide and X^- to water and the hypohalous acid (X-OH).

In our laboratory we have observed that a combination of 0.1 mM vanadate and 0.1 mM hydrogen peroxide caused irreversible inactivation of the enzyme glyceraldehyde-3-phosphate dehydrogenase, which has an active site cysteine. Vanadate and hydrogen peroxide cause a significant kinetic enhancement over the rate of inactivation by hydrogen peroxide alone [153]. Other evidence exists for oxidation of erythrocyte glyceraldehyde-3-P dehydrogenase [160].

A protease (bromalein) with cysteine in the active site was equally well inactivated by either hydrogen peroxide or peroxovanadate. The selective action of peroxovanadate in the case of glyceraldehyde-3-P dehydrogenase compared to that of bromalein may be a result of the presence of a substrate binding site for phosphate in glyceraldehyde-3-P dehydrogenase. As mentioned previously, sites for binding tetrahedral phosphate have affinity for peroxovanadate. Tyrosyl phosphatase is known to have a phosphate binding site and an active site Cys-215 [161]. Therefore, this enzyme may also be susceptible to inhibition by peroxovanadate.

The oxidation of hydrocarbons by peroxovanadate in nonaqueous solvents has been shown to occur [150]. Peroxovanadate ligand complexes were shown to oxidize alkanes to alcohols and aromatics to phenols. The aromatic reactants were more easily oxidized than alkanes. These reactions have potential implications for the oxidation of enzymatic tyrosyl residues since such oxidations may lead to irreversible enzyme inactivation.

6. PHOSPHATE ESTER HYDROLYSIS BY VANADIUM(IV)

It is not known as to whether the vanadium(IV)-catalyzed hydrolysis of phosphate esters is biochemically significant. However, this function does represent a potentially important source of activity which, at least for now, cannot be ignored.

6.1. Vanadium(IV) Oxide

Certain types of phosphate derivatives are susceptible to hydrolysis by vanadyl (VO^{2+}) but not vanadate. These include compounds with high-energy phosphate bonds and phosphate esters adjacent to a carboxylic acid. The pH optimum for hydrolysis is in the acidic range, at pH 4–5.

At pH 4, in the presence of 4 mM vanadium(IV) and 4 mM phosphate ester, rates of hydrolysis are: acetyl phosphate, 540 μM/min; ADP, 23 μM/min; phosphocreatine, 17 μM/min; phosphoenolpyruvate, 5 μM/min; and 2,3-bisphosphoglycerate, 220 μM/min [153]. *O*-Phosphobenzoic acid has also been shown to be susceptible to V(IV)-stimulated hydrolysis [162]. This type of reaction may make a significant contribution to the observed lowering of erythrocyte 2,3-bisphosphoglycerate levels after incubation with vanadate [160]. Vanadate is partially reduced to vanadium(IV) in erythrocytes [163–166].

6.2. Vanadium(IV) plus Hydrogen Peroxide

Ordinary phosphate esters rather than high-energy acyl phosphates or pyrophosphate linkages are apparently the preferential targets for vanadium(IV)-hydrogen peroxide. Hydrolysis is complete within a few minutes while no reaction was observed with vanadate alone or with vanadate in the presence of hydrogen peroxide. Phosphoamino acids and most of the glycolytic sugars are hydrolyzed by this reagent. At pH 4 and 4 mM each of vanadium(IV), hydrogen peroxide, and ATP, no inorganic phosphate was detected by enzymatic assay within 5–10 min [153]. Other workers reported significant hydrolysis of ATP at pH 2.5 using higher concentrations and longer reaction times [167,168].

7. SUMMARY

The chemical similarities between vanadate and phosphate combined with the ability of vanadate to readily undergo changes in coordination geometry allows this ion to strongly influence the function of a large variety of phosphate-metabolizing enzymes. As transition state analogs, spontaneously formed vanadate complexes are potent inhibitors of a number of enzymes, including some ribonucleases, mutases, and phosphatases. In addition, vanadate is an effective inhibitor of many ATPases, kinases, lyases, and synthases. Vanadate oligomers tend to be weaker inhibitors than vanadate but do influence the function of dehydrogenases, mutases, aldolases, kinases, and others. Of the oligomers, decavanadate is unique in that it seems to bind only in polyphosphate binding domains. Peroxovanadate has not yet been well studied but it seems to inhibit enzymes that do not utilize a pentacoordinate vanadate in the catalysis cycle. Additional detailed studies of vanadate-initiated inhibition of enzymes will expand our understanding of the various mechanisms of action of vanadate and its derivatives that have been briefly described here and will doubtless provide insight into other functions of this unique material.

ABBREVIATIONS AND DEFINITIONS

ADP	adenosine 5′-diphosphate
ADP·V	adenosine 5′-phosphophosphatovanadate
AMP	adenosine 5′-monophosphate
AMPVP	adenosine 5′-phosphato[vanadio]phosphate
ATP	adenosine 5′-triphosphate
cAMP	adenosine 2′,3′-cyclic monophosphate
1,3-BPG	1,3-bisphosphoglycerate
2,3,-BPG	2,3-bisphosphoglycerate
CoA	coenzyme A
DNA	deoxyribonucleic acid
E	enzyme
EP	phosphorylated enzyme
Glu-1-P-6-V	1-phospho-6-vanadioglucose
I	inhibitor
NAD	nicotinamide adenine dinucleotide
NADP	nicotinamide adenine dinucleotide phosphate
NMR	nuclear magnetic resonance
2-PGA	2-phosphoglyceric acid
3-PGA	3-phosphoglyceric acid
PGM	glycolytic phosphoglycerate mutase
RNA	ribonucleic acid
SF-1	myosin subfragment 1
SOD	superoxide dismutase
V-1-Glu-6-P	1-vanadio-6-phosphoglucose
V-6-Glu-1-P	1-phospho-6-vanadioglucose
2-V-3-PGA	2-vanadio-3-phosphoglyceric acid

REFERENCES

1. N. D. Chasteen, *Struct. Bond.*, *53*, 105–138 (1983).
2. T. Ramasarma and F. L. Crane, *Curr. Top. Cell. Regul.*, *20*, 247–301 (1981).
3. R. N. Lindquist, J. L. Lynn, Jr., and G. E. Lienhard, *J. Am. Chem. Soc.*, *95*, 8762–8768 (1973).
4. N. D. Chasteen (ed.), *Vanadium in Biological Systems*, Kluwer Academic, Dordrecht, 1990.
5. D. Rehder, *Angew. Chem. Int. Ed. Engl.*, *30*, 148–167 (1991).
6. R. Wever and K. Kustin, *Adv. Inorg. Chem.*, *35*, 81–115 (1990).
7. O. W. Howarth, *Prog. Nucl. Magn. Reson. Spectrosc.*, *22*, 453–485 (1990).
8. M. J. Gresser, A. S. Tracey, and P. J. Stankiewicz, *Adv. Protein Phosphatases*, *4*, 35–57 (1987).
9. N. K. Tonks, C. D. Diltz, and E. H. Fischer, *Adv. Protein Phosphatases*, *4*, 431–451 (1987).
10. C. F. Baes and R. E. Mesmer, *The Hydrolysis of Cations*, Wiley-Interscience, New York, 1976.
11. M. T. Pope and B. W. Dale, *Chem. Soc. (Lond) Quart. Rev.*, *22*, 527–548 (1968).
12. E. Heath and O. W. Howarth, *J. Chem. Soc., Dalton Trans.*, 1105–1110 (1981).
13. D. C. Crans, C. D. Rithner, and L. A. Theisen, *J. Am. Chem. Soc.*, *112*, 2901–2908 (1990).
14. L. Pettersson, I. Andersson, and B. Hedman, *Chem. Script.*, *25*, 309–317 (1985).
15. A. S. Tracey, M. J. Gresser, and B. Galeffi, *Inorg. Chem.*, *27*, 157–161 (1988).
16. M. J. Gresser and A. S. Tracey, *J. Am. Chem. Soc.*, *107*, 4215–4220 (1985).
17. D. C. Crans, S. M. Schelble, and L. A. Theisen, *J. Org. Chem.*, *56*, 1266–1274 (1991).
18. M. J. Gresser and A. S. Tracey, *J. Am. Chem. Soc.*, *108*, 1935–1939 (1986).

19. A. S. Tracey and M. J. Gresser, *Inorg. Chem.*, *27*, 2695–2702 (1988).

20. A. S. Tracey, J. S. Jaswal, M. J. Gresser, and D. Rehder, *Inorg. Chem.*, *29*, 4283–4288 (1990).

21. A. S. Tracey and C. H. Leon-Lai, *Inorg. Chem.*, *30*, 3200–3204 (1991).

22. J. Richter and D. Rehder, *Z. Naturforsch.*, *46b*, 1613–1620 (1991).

23. H. Deng, W. J. Ray, Jr., J. W. Burgner, II, and R. Callender, *Biochemistry*, *32*, 12984–12992 (1993).

24. W. J. Ray, Jr., J. W. Burgner, II, H. Deng, and R. Callender, *Biochemistry*, *32*, 12977–12983 (1993).

25. A. S. Tracey, M. J. Gresser, and S. Liu, *J. Am. Chem. Soc.*, *110*, 5869–5874 (1988).

26. A. S. Tracey, J. S. Jaswal, M. J. Gresser, and D. Rehder, *Inorg. Chem.*, *29*, 4283–4288 (1990).

27. A. S. Tracey and M. J. Gresser, *Can. J. Chem.*, *66*, 2570–2574 (1988).

28. A. S. Tracey and M. J. Gresser, *Inorg. Chem.*, *27*, 1269–1275 (1988).

29. C. H. Leon-Lai, *Chemical Equilibrium Studies of Vanadium(V) Complexes and Their Effects on the Catalytic Activity of Ribonuclease A*, M.Sc. thesis, Simon Fraser University, Burnaby, British Columbia, 1990.

30. M. Krauss and H. Basch, *J. Am. Chem. Soc.*, *114*, 3630–3634 (1992).

31. A. Wlodawer, M. Miller, and L. Sjolin, *Proc. Natl. Acad. Sci. USA*, *80*, 3628–3631 (1983).

32. B. Borah, C.-W. Chen, W. Egan, M. Miller, A. Wlodawer, and J. S. Cohen, *Biochemistry*, *24*, 2058–2067 (1985).

33. T. Alber, W. A. Gilbert, D. Ringe Ponzi, and G. A. Petsko, *Mobility and Function in Proteins and Nucleic Acids*, CIBA Foundation Symposium, vol. 93, Pitman, London, 1983, pp. 4–24.

34. R. S. Puskas, N. R. Manley, D. M. Wallace, and S. L. Berger, *Biochemistry*, *21*, 4602–4608 (1982).

35. D. Rehder, H. Holst, R. Quaas, W. Hinrichs, U. Hahn, and W. Saenger, *J. Inorg. Biochem.*, *37*, 141–150 (1989).

36. D. Kostrewa, H.-W. Choe, U. Heinemann, and W. Saenger, *Biochemistry*, *28*, 7592–7600 (1989).

37. L. A. Fothergill-Gilmore and H. C. Watson, *Adv. Enzymol.*, *62*, 227–313 (1989).

38. P. J. Stankiewicz and L. F. Hass, *J. Biol. Chem.*, *261*, 12715–12721 (1986).

39. Z. B. Rose, *Adv. Enzymol.*, *51*, 211–253 (1980).

40. J. Carreras, R. Bartrons, and S. Grisolia, *Biochem. Biophys. Res. Commun.*, *96*, 1267–1273 (1980).

41. F. Climent, R. Bartrons, G. Pons, and J. Carreras, *Biochem. Biophys. Res. Commun.*, *101*, 570–576 (1981).

42. P. Ninfali, A. Accorsi, A. Fazi, F. Palma, and G. Fornani, *Arch. Biochem. Biophys.*, *226*, 441–447 (1983).

43. J. Carreras, F. Climent, R. Bartrons, and G. Pons, *Biochim. Biophys. Acta*, *705*, 238–242 (1982).

44. S. Liu, M. J. Gresser, and A. S. Tracey, *Biochemistry*, *31*, 2677–2685 (1992).

45. P. J. Stankiewicz and L. F. Hass, *Fed. Proc.*, *45*, 1649 (1986).

46. C. B. Post, W. J. Ray, Jr., and D. G. Gorenstein, *Biochemistry*, *28*, 548–558 (1989).

47. M. Carreras, A. M. Bassols, J. Carreras, and F. Climent, *Arch. Biochem. Biophys.*, *264*, 155–159 (1988).

48. M. D. Percival, K. Doherty, and M. J. Gresser, *Biochemistry*, *29*, 2764–2769 (1990).

49. W. J. Ray, Jr., J. W. Burgner, II, and C. B. Post, *Biochemistry*, *29*, 2770–2778 (1990).

50. Y. Liu and W. J. Ray, Jr., unpublished result, 1994.

51. A. N. Charney, P. Silva, and F. H. Epstein, *J. Appl. Physiol.*, *39*, 156–158 (1975).

52. L. Josephson and L. C. Cantley, Jr., *Biochemistry*, *16*, 4572–4578 (1977).

53. L. C. Cantley, Jr., L. Josephson, R. Warner, M. Yanagisawa, C. Lechene, and G. Guidotti, *J. Biol. Chem.*, *252*, 7421–7423 (1977).

54. L. C. Cantley, Jr., L. G. Cantley, and L. Josephson, *J. Biol. Chem.*, *253*, 7361–7368 (1978).

55. R. L. Post, A. K. Sen, and A. S. Rosenthal, *J. Biol. Chem.*, *240*, 1437–1445 (1965).

56. A. S. Hobbs, J. P. Frochlich, and R. U. Albers, *J. Biol. Chem.*, *255*, 5724–5727 (1980).

57. R. L. Smith, K. Zinn, and L. C. Cantley, Jr., *J. Biol. Chem.*, *255*, 9852–9859 (1980).

58. G. H. Bond and P. M. Hudgins, *Biochemistry*, *18*, 325–331 (1979).

59. P. Fürst and M. Solioz, *J. Biol. Chem.*, *260*, 50–52 (1985).

60. S. G. O'Neal, D. B. Rhodes, and E. Racker, *Biochem. Biophys. Res. Commun.*, *89*, 845–850 (1979).

61. B. J. Bowman and C. W. Slayman, *J. Biol. Chem.*, *254*, 2928–2934 (1979).

62. E. A. Bramhall, *The effects of vanadate and arsenate on mitochondrial ATP synthase*, M.Sc. thesis, Simon Fraser University, Burnaby, British Columbia, 1987.

63. C. C. Goodno, *Proc. Natl. Acad. Sci. USA*, *76*, 2620–2624 (1979).

64. T. Kawamura and K. Tawada, *J. Biochem.*, *91*, 1293–1298 (1982).

65. C. C. Goodno and E. W. Taylor, *Proc. Natl. Acad. Sci. USA*, *79*, 21–25 (1982).

66. K. D. Munson, M. J. Smerdon, and R. G. Yount, *Biochemistry*, *25*, 7640–7650 (1986).

67. T. Kobayaski, T. Martensen, J. Nath, and M. Flavin, *Biochem. Biophys. Res. Commun.*, *81*, 1313–1318 (1978).

68. W. Z. Cande and S. M. Wolniak, *J. Cell Biol.*, *79*, 573–580 (1978).

69. C. R. Gibbons, M. P. Casson, J. A. Evans, B. H. Gibbons, B. Houk, K. H. Martinson, W. S. Sale, and W-J. Y. Tang, *Proc. Natl. Acad. Sci. USA*, *75*, 2220–2224 (1978).

70. K. M. Walton and J. E. Dixon, *Annu. Rev. Biochem.*, *62*, 101–120 (1993).

71. K.-H. W. Lau, J. R. Farley, and D. J. Baylink, *Biochem. J.*, *257*, 23–36 (1989).

72. A. M. Brunati and L. A. Pinna, *Biochem. Biophys. Res. Commun.*, *133*, 929–936 (1985).

73. K.-H. W. Lau, J. R. Farley, and D. J. Baylink, *J. Biol. Chem.*, *260*, 4653–4660 (1985).

74. P. Boivin and C. Galand, *Biochem. Biophys. Res. Commun.*, *134*, 557–564 (1986).

75. S. S. Imes, N. O. Kaplan, and A. F. Knowles, *Anal. Biochem.*, *161*, 316–322 (1987).

76. A. S. Tracey and M. J. Gresser, *Proc. Natl. Acad. Sci. USA*, *83*, 609–613 (1986).

77. N. K. Tonks, C. D. Diltz, and E. H. Fischer, *Adv. Protein Phosphatases*, *4*, 431–451 (1987).

78. C. L. Shriver and D. L. Brautigan, *J. Biol. Chem.*, *259*, 11383–11390 (1984).

79. D. C. Crans and C. M. Simone, *Biochemistry*, *30*, 6734–6741 (1991).

80. R. J. De Koch, D. J. West, J. C. Cannon, and N. D. Chasteen, *Biochemistry*, *13*, 4347–4354 (1974).

81. J. A. Gordon, *Meth. Enzymol.*, *201*, 477–482 (1991).

82. D. L. Brautigan, P. Bernstein, and B. Gallis, *J. Biol. Chem.*, *256*, 6519–6522 (1981).

83. G. Swarup, K. V. Speeg, Jr., S. Cohen, and D. L. Garbers, *J. Biol. Chem.*, *257*, 7298–7301 (1982).

84. G. Swarup, S. Cohen, and D. L. Garbers, *Biochem. Biophys. Res. Commun.*, *107*, 1104–1109 (1982).

85. J. F. Leis, A. F. Knowles, and N. O. Kaplan, *Arch. Biochem. Biophys.*, *239*, 320–326 (1985).

86. J. F. Leis and N. O. Kaplan, *Proc. Natl. Acad. Sci. USA*, *79*, 6507–6511 (1982).

87. C. J. Pallen, K. A. Valentine, J. H. Wang, and M. D. Hollenberg, *Biochemistry*, *24*, 4727–4730 (1985).

88. V. Lopez, T. Stevens, and R. N. Lindquist, *Arch. Biochem. Biophys.*, *175*, 31–38 (1976).

89. C. Simone, *Vanadium and Protein Interactions: Inhibition, Redox Chemistry and Conversion of Cofactors*, PhD thesis, Colorado State University, Fort Collins, 1992.

90. P. J. Stankiewicz and M. J. Gresser, *Biochemistry*, *27*, 206–212 (1988).

91. J. T. Ohlsson and I. B. Wilson, *Biochim. Biophys. Acta*, *350*, 48–53 (1974).

92. M. Cocivera, J. McManaman, and I. B. Wilson, *Biochemistry*, *19*, 2901–2907 (1980).

93. J. M. Sowadski, M. D. Handschumacker, H. M. K. Murthy, B. A. Foster, and H. W. Wyckoff, *J. Mol. Biol.*, *186*, 417–433 (1985).

94. A. Williams and R. A. Naylor, *J. Chem. Soc. B*, 1973–1979 (1971).

95. R. L. Van Etten, P. P. Waymack, and D. M. Rehkop, *J. Am. Chem. Soc.*, *96*, 6782–6785 (1974).

96. M. E. Hickey and R. L. Van Etten, *Arch. Biochem. Biophys.*, *152*, 423–425 (1972).

97. M. E. Hickey, P. P. Waymack, and R. L. Van Etten, *Arch. Biochem. Biophys.*, *172*, 439–448 (1976).

98. D. C. Crans, C. M. Simone, A. K. Saha, and R. H. Glew, *Biochem. Biophys. Res. Commun.*, *165*, 246–250 (1989).

99. M. H. Rider, R. Bartrons, and L. Hue, *Eur. J. Biochem.*, *190*, 53–56 (1990).

100. M. Miralpeix, J. Gil, J. L. Rosa, J. Carreras, and R. Bartrons, *Life Sci.*, *44*, 1491–1497 (1989).

101. J. Gil, M. Miralpeix, J. Carreras, and R. Bartrons, *J. Biol. Chem.*, *263*, 1868–1871 (1988).

102. H. B. Stewart, M. R. El-Maghrabi, and S. J. Pilkis, *J. Biol. Chem.*, *260*, 12935–12941 (1985).

103. P. D. Kounty, R. W. McCain, M. R. El-Maghrabi, and S. J. Pilkis, *Arch. Biochem. Biophys.*, *251*, 104–113 (1986).

104. R. C. Nordlie, *Curr. Top. Cell. Regul.*, *8*, 33–112 (1974).

105. J. Singh, R. C. Nordlie, and R. A. Jorgenson, *Biochim. Biophys. Acta*, *678*, 477–482 (1981).

106. D. W. Boyd, K. Kustin, and M. Niwa, *Biochim. Biophys. Acta*, *827*, 472–475 (1985).

107. M. M. Craig, *Selective Inhibition by Vanadate of Phosphoryl Transfer Enzymes*, PhD thesis, Simon Fraser University, Burnaby, British Columbia, 1991.

108. M. J. Gresser and A. S. Tracey, in *Vanadium in Biological Systems* (N. D. Chasteen, ed.), Kluwer Academic, Dordrecht, 1990, pp. 63–79.

109. J. Krivanek and L. Novakova, *FEBS Lett.*, *282*, 32–34 (1991).

110. J. Krivanek and L. Novakova, *FEBS Lett.*, *254*, 121–123 (1989).

111. S. Ramakrishna, S. K. Murthy, and W. B. Benjamin, *Biochemistry, 28*, 856–860 (1989).

112. W. A. Bridges, in *The Enzymes* (P. D. Boyer, ed.), Academic Press, New York, 1974, pp. 581–607.

113. H. P. G. Swarts, M. Moes, F. M. A. H. Schuurmans Stekhoven, and J. J. H. H. M. De Pont, *Biochim. Biophys. Acta, 1107*, 143–149 (1992).

114. A. S. Menon, M. Rau, T. Ramasarma, and F. L. Crane, *FEBS Lett., 114*, 139–141 (1980).

115. M. A. Serra, A. Pintar, L. Casella, and E. Sabbioni, *J. Inorg. Biochem., 46*, 161–174 (1992).

116. A. T. Harrison and O. W. Howarth, *J. Chem. Soc., Dalton Trans.*, 1173–1177 (1985).

117. J. S. Jaswal and A. S. Tracey, *Inorg. Chem., 30*, 3718–3722 (1991).

118. M. F. Hopgood, M. G. Clark, and F. J. Ballard, *Biochem. J., 186*, 71–79 (1980).

119. M. Desautels and A. L. Goldberg, *Proc. Natl. Acad. Sci. USA, 79*, 1869–1873 (1982).

120. M. Desautels and A. L. Goldberg, *J. Biol. Chem., 257*, 11673–11679 (1982).

121. F. S. Larimore, L. Waxman, and A. L. Goldberg, *J. Biol. Chem., 257*, 4187–4195 (1982).

122. M. J. Gresser, A. S. Tracey, and K. M. Parkinson, *J. Am. Chem. Soc., 108*, 6229–6234 (1986).

123. D. C. Crans and S. M. Schelble, *Biochemistry, 29*, 6698–6706 (1990).

124. D. C. Crans, K. Sudhakar, and T. J. Zamborelli, *Biochemistry, 31*, 6812–6821 (1992).

125. G. Soman, Y. C. Chang, and D. J. Graves, *Biochemistry, 22*, 4994–5000 (1983).

126. S. G. Withers, N. B. Madsen, S. R. Sprang, and R. J. Fletterick, *Biochemistry, 21*, 5372–5382 (1982).

127. D. C. Crans, E. M. Willging, and S. R. Butler, *J. Am. Chem. Soc., 112*, 427–432 (1990).

128. D. C. Crans, *Comments Inorg. Chem., 16*, 35–76 (1994).

129. M. D. Cohen, A. C. Sen, and C.-I. Wei, *Inorg. Chim. Acta*, *138*, 179–186 (1987).

130. D. C. Crans, *Mol. Eng.*, *3*, 277–284 (1993).

131. A. S. Campbell and G. R. J. Thatcher, *Bioorg. Med. Chem. Lett.*, *2*, 655–658 (1992).

132. L. Wittenkeller, A. Abraha, R. Ramasamy, D. Mota de Freitas, L. A. Theisen, and D. C. Crans, *J. Am. Chem. Soc.*, *113*, 7872–7881 (1991).

133. D. C. Crans, *Comments Inorg. Chem.*, *16*, 1–33 (1994).

134. D. C. Crans and P. K. Shin, *Inorg. Chem.*, *27*, 1797–1806 (1988).

135. V. W. Day, W. G. Klemperer, and D. J. Maltbie, *J. Am. Chem. Soc.*, *109*, 2991–3002 (1987).

136. P. Csermely, A. Martinosi, G. C. Levy, and A. J. Ejchart, *Biochem. J.*, *230*, 807–815 (1985).

137. R. K. Murmann and K. C. Giese, *Inorg. Chem.*, *17*, 1160–1166 (1978).

138. D. W. Boyd, K. Kustin, and M. Niwa, *Biochim. Biophys. Acta*, *827*, 472–475 (1985).

139. E. F. Pai, W. Sachsenheimer, R. H. Schirmer, and G. E. Schultz, *J. Mol. Biol.*, *114*, 37–45 (1977).

140. U. Egner, A. G. Tomasselli, and G. E. Schulz, *J. Mol. Biol.*, *195*, 649–658 (1987).

141. G. L. Choate and T. E. Mansour, *J. Biol. Chem.*, *254*, 11457–11462 (1979).

142. G. L. Choate and T. E. Mansour, *Fed. Proc.*, *37*, 1433 (1978).

143. P. D. Kounty, R. W. McCain, M. R. El-Maghrabi, and S. J. Pilkis, *Arch. Biochem. Biophys.*, *251*, 104–113 (1986).

144. J. Struplish, R. J. H. Wojcikiewicz, R. A. J. Challiss, S. T. Safrany, A. L. Willcocks, B. V. L. Potter, and S. R. Nahorski, *Biochem. J.*, *277*, 294 (1991).

145. K. J. Föhr, J. Scott, G. Aknert-Hilger, M. Gratzl, and J. M. Senard, *Biochem. J.*, *262*, 83–89 (1989).

146. H. Paris, J. Galitzky, and J. M. Senard, *Mol. Pharmacol.*, *35*, 339–344 (1989).

147. M. Le Hir, *Biochem. J.*, *273*, 795–798 (1991).

148. C. Coan, D. C. Scales, and A. J. Murphy, *J. Biol. Chem.*, *261*, 10394–10403 (1986).

149. S. Pluskey, M. Mohammad-Tahir, D. C. Crans, and D. S. Lawrence, unpublished work.

150. H. Mimoun, L. Saussine, E. Daire, M. Postel, J. Fischer, and R. Weiss, *J. Am. Chem. Soc.*, *105*, 3101–3110 (1983).

151. D. Begin, F. W. B. Einstein, and J. Field, *Inorg. Chem.*, *14*, 1785–1790 (1975).

152. A. M. Cortizo, V. C. Salice, and S. B. Etcheverry, *Biol. Trace Element Res.*, in press.

153. P. J. Stankiewicz, unpublished result, 1994.

154. I. G. Macara, K. Kustin, and L. C. Cantley, *Biochim. Biophys. Acta*, *629*, 95–106 (1980).

155. H. Sakurai, S. Shimomura, and K. Ishizu, *Inorg. Chim. Acta*, *55*, L67–L69 (1981).

156. E. de Boer, M. G. M. Tromp, H. Plat, B. E. Krenn, and R. Wever, *Biochim. Biophys. Acta*, *872*, 104–115 (1986).

157. B. E. Krenn, H. Plat, and R. Wever, *Biochim. Biophys. Acta*, *912*, 287–291 (1987).

158. H. Plat, B. E. Krenn, and R. Wever, *Biochem. J.*, *248*, 277–279 (1987).

159. R. Wever and K. Kustin, *Adv. Inorg. Chem.*, *35*, 81–115 (1990).

160. J. E. Berabe, L. A. Echegoyen, B. Pastrana, and M. Martinez-Malonado, *J. Biol. Chem.*, *262*, 9555–9560 (1987).

161. K. Guan and J. E. Dixon, *J. Biol. Chem.*, *266*, 17026–17030 (1991).

162. R. Hofstetter, Y. Murakami, G. Mont, and A. E. Martell, *J. Am. Chem. Soc.*, *84*, 3041–3046 (1962).

163. G. L. Mendz, S. J. Hyslop, and P. W. Kuchel, *Arch. Biochem. Biophys.*, *276*, 160–171 (1990).

164. P. J. Stankiewicz, *Arch. Biochem. Biophys.*, *270*, 489–494 (1989).

165. X. Shi, X. Sun, and N. S. Dalal, *FEBS Lett.*, *271*, 185–188 (1990).

166. L. C. Cantley, Jr. and P. Aisen, *J. Biol. Chem.*, *254*, 1781–1784 (1979).

167. T. Imamura, D. M. Hinton, R. L. Belford, R. I. Gumport, and G. P. Haight, Jr., *J. Inorg. Biochem.*, *11*, 241–259 (1979).

168. G. M. Woltermann, R. Scott, and G. P. Haight, Jr., *J. Am. Chem. Soc.*, *96*, 7569–7570 (1974).

169. P. J. Stankiewicz, M. J. Gresser, A. S. Tracey, and L. F. Hass, *Biochemistry*, *26*, 1264–1269 (1987).

10
Vanadium-Dependent Haloperoxidases

Hans Vilter*

Gesellschaft für Biotechnologische Forschung mbH
Mascheroder Weg 1, D-38124 Braunschweig, Germany

---

*Present address: Zurmaienerstrasse 16, D-54292 Trier, Germany.

1. INTRODUCTION

1.1. Discovery of Vanadium-Dependent Haloperoxidases

Peroxidases are oxidoreductases acting on peroxide as electron acceptor (EC 1.11.1.X) [1]. The detection of peroxidases has been dated back to the beginning of the nineteenth century [2]. Most of the peroxidases have been reported to be heme-containing enzymes and these have been divided into three superfamilies: plant peroxidases, animal peroxidases, and catalases [3]. Those from higher plants are some of the most studied enzymes, especially the peroxidases from horseradish, and are used in large amounts for diagnostic kits [4]. Plant peroxidases are involved in the biosynthesis of plant hormones [5,6] and in lignin polymerization [7]. There is evidence that heme-containing peroxidases from fungi are involved in lignin degradation, although overproduction does not enhance lignin degradation [8]. Lignin mineralization is reduced in peroxidase-deficient [9] strains.

In comparison to the peroxidases from higher plants and mammalians, little was known about peroxidases from seaweeds until the 1970s. There were only a few reports on the detection of peroxidases in tissues and crude extracts from brown algae [10–17]. No spectroscopic studies had been performed, as nobody was able to purify peroxidases from these algae due to the presence of alginates [16]. However, from the overwhelming presence of heme-containing peroxidases in higher plants and mammalians, it was concluded that the peroxidases in seaweeds would also probably be heme-containing peroxidases. Hence it was reasoned that time should not be wasted in attempting the purification of these enzymes. An anomalous substrate specificity of peroxidases from algae was reported [15] and was assumed to be the reason for seaweeds not being lignificated. On the other hand, some of the brown algae contain tannins in large amounts [18], as phloroglucinol polymers, and it was assumed that peroxidases are involved in the polymerization process.

Interest in the biogenesis of these compounds resulted eventually in the successful isolation and purification of peroxidases from brown algae [19,20]. The electron absorption spectrum (Fig. 1) lacked the Soret band in the range of ~410 nm. During the purification process there was a partial loss of enzyme activity, although after handling the enzyme preparations with a Hamilton syringe an increase of activity was observed. Dialyzing the enzyme preparations, using ethylenediamine-*N,N,N′,N′*-tetraacetate (EDTA)-containing citrate/phosphate buffers, yielded inactive preparations. This prompted the study of the effect of metal ions on enzyme activity, which eventually showed that inactive

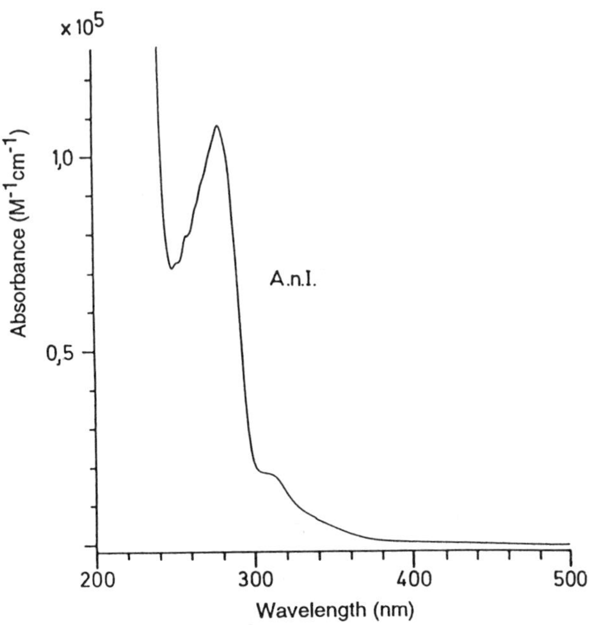

FIG. 1. Electron absorption spectrum of peroxidase A.n.I. from *Ascophyllum nodosum*.

apoenzyme could be reconstituted to active holoenzyme using vanadate. There was no reconstitution of the enzyme activity with iron or other metal ions [21]. This observation was confirmed [22] and is compatible with the observation that vanadium is an essential trace element for algae and other organisms [23–28].

The reactivation of the apoperoxidase by vanadate was inhibited by phosphate and phosphate analogs [29,30], and is due to the structural similarity of these compounds. This was the first study to show that vanadium is essential for the activity of an enzyme [31]. The isoenzyme first isolated is called A.n.I., using the initials from the source, the brown alga *Ascophyllum nodosum*. The reconstitution from the apoenzyme to the active holoenzyme using the isoenzyme A.n.I. is a slow process (Fig. 2) compared to other isoenzymes that are reconstituted in a few minutes. The reconstitution of the activity is correlated with the increase of the UV absorption at 310 nm (Fig. 3).

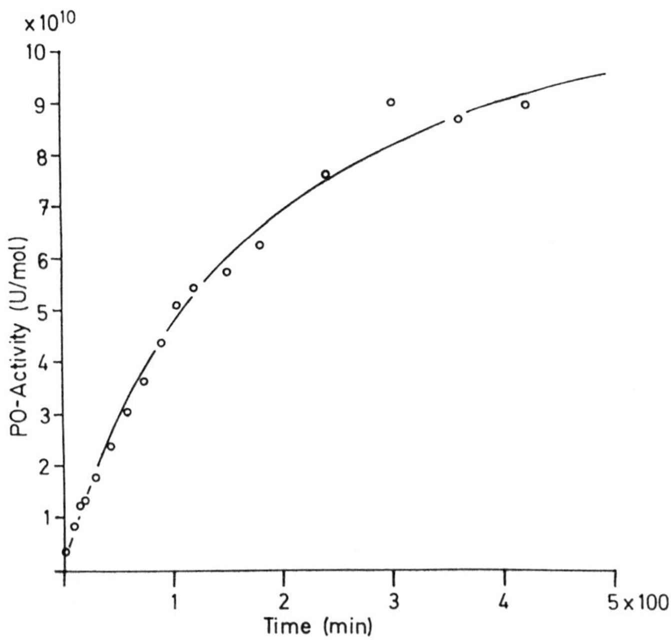

FIG. 2. Reconstitution of peroxidase A.n.I. from *Ascophyllum nodosum*. Increase of activity after the addition of 0.2 mM Na-vanadate to 0.7×10^{-6} M apoenzyme A.n.I. in 0.05 M Tris/HCl, pH 9.0 at 25°C.

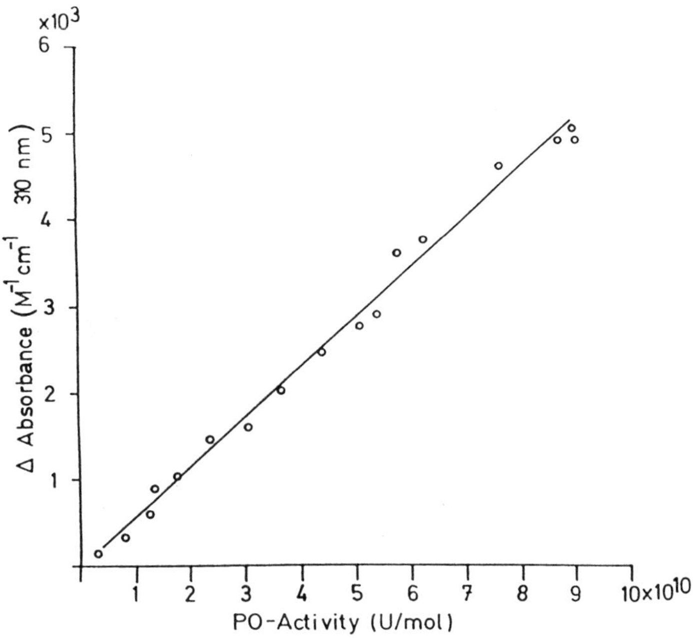

FIG. 3. Reconstitution of peroxidase A.n.I. from *Ascophyllum nodosum*. Increase of activity correlated to the absorbance at 310 nm after the addition of 0.2 mM Na-vanadate to 0.7×10^{-6} M apoenzyme A.n.I. in 0.05 M Tris/HCl, pH 9.0.

Vanadium-containing nonheme peroxidases have been isolated from several other sources (Table 1), although as yet the isoenzyme A.n.I. is the most studied vanadium-containing peroxidase. Nonheme, nonvanadium haloperoxidases have been isolated from bacteria [32–37] and a flavin-containing haloperoxidase has been isolated from the marine worm *Notomastus lobatus* [38]. The NADH-peroxidase (EC 1.11.1.1.) is also a flavoprotein [1,39], while the selenoproteid glutathione-peroxidase (EC 1.11.1.9) is a well-known nonheme peroxidase [1,40,41].

1.2. Sources of Vanadium-Dependent Haloperoxidases

The detection of the first vanadium-dependent enzyme stimulated the search for other vanadium-dependent peroxidases. It turned out that

TABLE 1

Sources of Vanadium-Dependent Haloperoxidases

Source	Ref.
Brown Algae	
Phaeophyta	
Fucales	
Fucaceae	
Ascophyllum nodosum	21,22,46
Fucus distichus	47–49
Fucus serratus	49,50
Fucus spiralis	49,50
Fucus vesiculosis	49–51
Pelvetia canaliculata	49,50
Pelvetioptis litoralis	49
Sargassaceae	
Sargassum undulatum	49
Sargassum spinuligerum	49
Carpophyllum marschalocarpum	49
Carpophyllum plumosum	49
Carpophyllum flexuosum	49
Carpophyllum angustifolium	49
Cystophora retroflexa	49
Cystophora torulosa	49
Durvilleaceae	
Xiphophora chondrophylla	49
Hormoseiraceae	
Hormoseira banksii	49
Laminariales	
Alariaceae	
Alaria esculenta	48,49
Alaria marginata	49
Alaria nana	49
Agarum cribosum	49
Eckonia radiata	49
Eckonia cava	52
Eckonia maxima	49

TABLE 1

Continued

Source	Ref.
Laminariales	
Alariaceae (continued)	
Eisenia arborea	49
Eisenia bicyclis	52
Pterygophora californica	49
Undaria pinnatifida	52
Chordaceae	
Chorda filum	49
Laminariaceae	
Cymathera triplicata	49
Hedophyllum sessile	49
Laminaria digitata	53,54
Laminaria groenlandica	49
Laminaria hyperborea	49,55
Laminaria ochroleuca	49
Laminaria saccharina	51,54,55
Lessoniaceae	
Lessionia variegata	49
Lessioniopsis litoralis	49
Macrocystis integrifolia	49
Macrocystis pyrifera	45,49
Nerocystis lutkeana	49
Postelsia palmaeformis	49
Phyllariacae	
Sacchoriza polyschides	56
Chordariales	
Leathesia difformis	49
Splachnidium rugoum	49
Dictyotales	
Glossophora harvei	49

(Continued)

TABLE 1

Continued

Source	Ref.
Red Algae	
Rhodophyta	
Ceramium rubrum	57
Corallina pilulifera	58
Corallina officinalis	59,60
Corallina vancouverensis	61
Plocamium hamatum	51
Green Algae	
Chlorophyta	
Halimeda sp.	62
Fungi	
Curvularia inaequalis	63,64
Lichen	
Xantoria parietina	65

peroxidases isolated from other brown algae [19] belonging to Fucales and Laminariales are also vanadium-dependent peroxidases. Similar peroxidases have been isolated from several brown algae, some red algae, one green alga, one lichen, and one fungus (see Table 1). The peroxidases from the red algae *Corallina pilulifera* [42–44] and the fungus *Curvularia inaequalis* [64,65] have been reported before to be nonheme iron-containing enzymes. Initially there was some doubt as to the source from which the peroxidase had been isolated. Although determination of the raw material collected at shore is not difficult, it has to be kept in mind that in the raw material the algae are colonized by bacteria, diatoms, and other epiphytes, which are detectable using a microscope. Further there are endophytes. In *Ascophyllum nodosum*, for example, there is the obligate endosymbiont *Mycospherella ascophylli*, a fungus belonging to the ascomycetes [66,67], which uses carbohydrates produced by the alga. It is not known if the alga benefits from the presence of the fungus; it does gain from the presence of bacteria on the surface. In the absence of these bacteria the alga is killed by the fungus [68]. The diatom *Navicula endophytia* lives in the intercellulares [69], while the red alga *Nemalion vermiculare* contains a symbiotic blue-green alga [70]. Such contaminations of the raw material have caused problems for phytochemists. For example, chlorinated compounds isolated from sorghum were in fact produced by algae [71]. Therefore the raw material has to be controlled by using the histochemical localization of the peroxidases in the tissues or by control experiments using axenic culture material. These controls have been performed on *Ascophyllum nodosum*. Histochemical peroxidases were located in the cell walls. Due to differences in stability two isoenzymes could be distinguished, one located in cell walls between cortex and medulla and the other located in the cell walls at the surface [20,72,73]. Snails, *Litorina* sp., graze on the surface of the algae and the surface isoenzyme was detectable in their excrements. The endosymbiont *Mycospherella ascophylli* was known to produce brominated compounds [74].

Peroxidases from this fungus have been isolated from axenic cultures [75] and have been proved to be non-vanadium-dependent peroxidases. Isoenzyme patterns have been studied from several brown algae [76]. Using protoplasts from *Laminaria saccharina*, *Laminaria digitata* [77], and *Macrocystis pyrifera* [78], it was demonstrated that different isoenzymes are expressed in different cells. However, most of the peroxidases were found in the cell wall lysates. The patterns of the intercellular peroxidases are different from the extracellular peroxidase patterns and may be due to translatoric modification during the excretion of the peroxidases from the protoplast into the cell wall. The biosynthesis of peroxidases by the protoplasts was demonstrated during the synthesis of the new cell wall [77].

Axenic culture material form *Ascophyllum nodosum* yielded the same isoenzymes as those isolated from the raw material [79]. Using axenic red algae only peroxidases were isolated from *Nemalion helminthoides*; however, they did not show the characteristics of vanadium-dependent peroxidases [79]. Heme-containing peroxidases are known from the red algae *Cystocloneum purpureum* [80], *Rhodomela larix* [81], and the green algae *Penicillus capitatus* [82,83], *Penicillus lamourouxii* [83], and *Rhipocephalus phoenix* [83]. The lichen *Xanthoria paretina* represents a symbiosis between the green alga *Trebouxia* sp. and an ascomycete. To date localization experiments have not been successful

[84] and thus it is not known as to whether the vanadium-dependent peroxidases are produced by the green alga or the fungus.

Further candidates, containing vanadium-dependent peroxidases, may be diatoms and other unicellular algae, and last but not least blue-green algae. The blue-green alga *Fischerella ambigua* has been tested without success [85], but further studies are needed. Polychlorinated compounds have been isolated from this blue-green alga [86].

2. EXTRACTION AND PURIFICATION

2.1. Conventional Extraction Process

The isolation of enzymes from brown algae is attended by several problems. The disruption of plant tissue is difficult because of the rigidity of the cell walls. The rigidity is due to cellulose fibers embedded in a matrix of anionic phycocolloids such as alginates and fucansulfates. The alginates are used as gel-forming or thickening agents and emulsifiers, e.g., in food and pharmaceutical products. The viscosity of the extracts from algae hinders the chromatographic processes; in addition, there is the difficulty of using liquid-solid separations. The presence of anionic polysaccharides may convert an anion exchanger to a mixed-bed ion exchanger, thus causing still more problems during the separation process. Cell wall vacuoles contribute to the bulk of most algal tissues and contain secondary algal products which can affect the stability of enzymes during extraction.

Most of the latter problems are caused by the presence of phenolic compounds [18] which can modify and even denature proteins. Additional pigments are frequently located in the plastids, e.g., lipophilic compounds such as chlorophylls and fucoxanthin, and these become attached to the protein by hydrophobic interactions.

No general method for the isolation of enzymes from algal material can be given. However, there is a general approach. Fresh algae can be disrupted with a meat cutter. Even when the disruption is not very efficient [19,46], this causes no problems as the algae are available in large amounts. For most purposes an efficient disruption and extraction process is not necessary. The tissues can be disrupted in large amounts more easily, after freeze-drying, using a centrifuge mill [19], which is convenient for large-scale processes. On an analytic scale small quantities of tissue can be cut with a razor blade into small pieces and then be disrupted very efficiently with a French press. Several phenol scavengers are known. Primarily insoluble polyvinylpyrrolidone (trade name Polyclar AT, Serva, Heidelberg, Germany) is used for this purpose as well as caffeine.

Several liquid-solid separation steps are necessary. The presence of insoluble polyvinylpyrrolidone makes the homogenization medium containing phycocolloids reminiscent of soft cheese. The extract is obtained by pressing the homogenate through muslin. Scale-up of the process is limited. Tris buffers [19,46,87] are preferred as extraction mediums. As most peroxidases are located with the anionic mucilages in the cell walls, all attempts to increase the solubility of the enzymes also lead to increased alginate content. Therefore all conventional extraction procedures are not optimal. A crude fractionation of the extracts can be achieved by using ammonium sulfate, ethanol, or acetone to precipitate the enzymes. The alginates can be removed by dialysis after the digestion using alginate lyasis [88]. The enzymes can be purified to homogeneity using anion exchange, size exclusion, hydrophobic, and affinity chromatography. Scale-up of conventional extraction procedures is limited and therefore enzyme production for commercial applications is not convenient by these procedures.

2.2. Liquid-Liquid Extraction Process

Recently, the advantage of applying aqueous two-phase systems to enzyme purification from plants was described [53,54,89] and liquid-liquid extractions have been developed for industrial applications [90].

Vanadium-dependent peroxidases were isolated from brown algae using two-phase systems generated by polyethylene glycol (PEG) (Hüls, Marl, Germany) and salt solutions. Pilot experiments were performed to analyze the partitioning of peroxidases from different sources: alginates, fucan sulfates, tannins, and plant pigments. Based on the results from these pilot experiments, the most promising PEG/salt systems were selected. The peroxidases were extracted either as active holoenzyme or as inactive apoenzyme. The inactivation of peroxidases resulting from treatment with potassium phosphate or potassium carbonate, up to pH 12, was reversible. Transfer of the apoenzymes into Tris buffer containing vanadate reconstituted the holoenzymes. No artifacts from this treatment were observed on PAGE.

The extraction process was developed on the mL scale and was easily scaled up to 300 L. In general, the process is started with 50 kg of fresh algae. Using a centrifugal mill (Retsch, Haan, Germany), freeze-dried algae were easily and efficiently converted in large quantities to powder. Further homogenization was achieved by passing the extraction mixture through a colloid mill and then a Mauton-Gaulin high-pressure homogenizer. Due to the efficient disruption of the cell wall there is a drastic increase of the viscosity of the bottom phase. Extraction of the peroxidases occurs in the top phase and the viscosity of the bottom phase causes no problems. From a technological viewpoint the viscosity is an advantage, since the sedimentation of particles is reduced and prevents blocking in the flow-through separator. Two-phase systems offer a potential for further scale-up.

Two methods have been developed. The first method is suitable for extracting peroxidases from the brown alga *Ascophyllum nodosum* and other brown algae belonging to the Fucaceae. It is a three-step liquid-liquid extraction process. In the first step the peroxidases are extracted into the top-phase using a PEG/potassium phosphate system. Adding magnesium sulfate solution causes the peroxidases to be transferred to the second bottom phase. Adding PEG/ammonium sulfate the enzymes were then extracted into the top phase of the third system. From the third top phase the peroxidases are recovered by precipitation using acetone and the precipitate dissolved using Tris buffer. After dialysis the peroxidases are purified using hydrophobic interaction and anion exchange chromatography. A pure isoenzyme A.n.I. was obtained in quantity and quality sufficient for crystallization [91].

The second method was developed to extract peroxidases from the brown algae *Laminaria digitata* and *Laminaria saccharina*, as results with the first method were unsatisfactory in these cases. The method is also suitable for the isolation of peroxidases from other brown algae belonging to the Laminariales. It is not suitable for isolating peroxidases from Fucaceae. The peroxidases are extracted into the top phase using a PEG/potassium carbonate system and they could then be recovered by addition of ammonium sulfate and acetone. It has been confirmed that the PEG/potassium carbonate system is not efficient for isolating peroxidases from *Ascophyllum nodosum* [87].

3. FUNCTION

3.1. Substrate Specificity and Kinetic Properties

The preferential electron acceptor in haloperoxidases is hydrogen peroxide, although there was initially some confusion about the electron donors. In the past the substrate specificity was tested using crude extracts; as a consequence brown algae, which accumulate iodide, have extracts that contain bromide and chloride from the seawater in addition to iodide. A broad range of compounds has been reported as substrates of the peroxidases from brown algae [10–17], but in the end it turned out that these enzymes were using none of the common peroxidase substrates. The enzymes are haloperoxidases. Iodide is oxidized preferentially and can be detected directly due to the formation of triiodide. The reaction product is visible, with an absorption maximum at 353 nm [19,92]. From the practical point of view the differences are not important in the range of ±5 nm. Important are several reverse reactions. Most important are the equilibria:

$$I_2 + I^- \rightleftharpoons I_3^- \qquad K_c = 1.3 \times 10^{-3}\ \text{M}$$

and

$$I_2 + 2OH^- \rightleftharpoons I^- + IO^- + H_2O \qquad K_c = 30\ \text{M}$$

Polyhalide anions are formed in the presence of low iodide concentration, while disproportionation of the hypohalide takes place in alkaline solutions. The assay can be used below pH 7. The iodide concentration should be in the range 2–10 mM iodide and standard conditions are 6 mM iodide. In this iodoperoxidase assay the substrate concentration is below saturation, as at this concentration the nonenzymatic oxidation of iodide by hydrogen peroxide is sufficiently low. Therefore the enzymatic reaction can be measured easily using a double-beam photometer. If this is not available, the difference between the nonenzymatic and the nonenzymatic-plus-enzymatic reaction has to be calculated. The vanadium-dependent peroxidase is remarkably stable (Fig. 4), in contrast to lactoperoxidase or other heme-containing peroxidases. The heme-containing peroxidases are inactivated by the reaction products.

Several substances are oxidized by iodine, triiodide, and hypoiodide, and thereby iodide is regenerated. These coupled reactions can be used to monitor the enzymatic reaction. The formation of oxidation

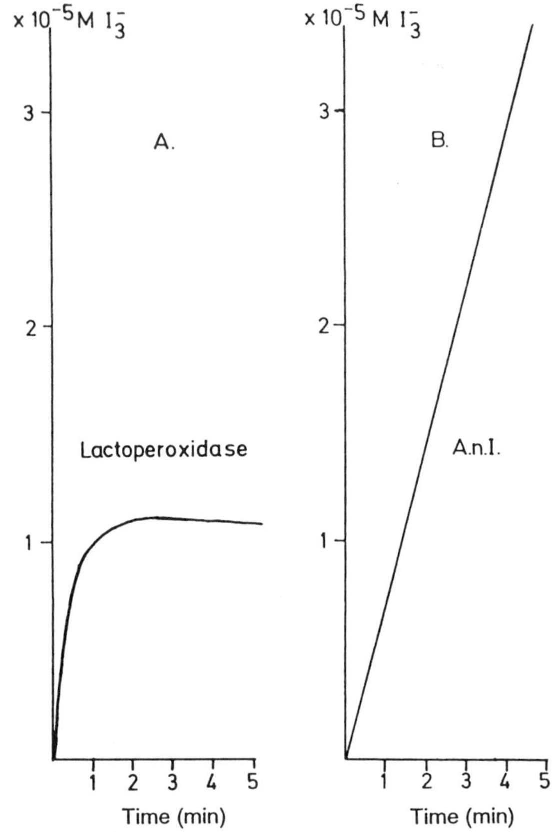

FIG. 4. Formation of triiodide catalysed by peroxidases. (A) Inactivation of lactoperoxidase due to the product. (B) Stability of peroxidases A.n.I. from *Ascophyllum nodosum*.

products from *p*-phenylenediamine or *o*-dianisidine can be used to measure the activity using a photometer or to detect the enzyme activity in polyacrylamide (PA) gels after electrophoresis or isoelectric focusing [76]. In order to localize peroxidase activity these oxidation products are more convenient than reaction products which are more soluble, such as bromphenol blue. In the micromolar range the iodination of organic substances can be monitored using an iodide electrode [93], although the response characteristics of the electrode are not suitable for steady-state kinetic experiments. Iodinate compounds are frequently labile compounds and thus spectroscopic studies of iodinations are less convenient than the monitoring of brominations.

In the absence of organic compounds the formation of tribromide was detected by using a rapid-scan stopped-flow photometer [94] as it is not possible to monitor the formation of tribromide with a conventional photometer. In the absence of organic compounds the light is scattered by the formation of bubbles in the cuvettes and could be detected using an oxygen electrode. This indicates the formation of hypobromide, which is well known to be reduced by hydrogen peroxide with the formation of oxygen [95]. The formation of dioxygen in the singlet exited state has been studied using the near-infrared emission at 1268 nm [96,97]. There is evidence that organic compounds are bound to the active holoenzyme as well as at the inactive apoenzyme [98]. In the presence of suitable organic compounds the bromination is favored over the hydrogen peroxide oxidation. Preferential bromination of monochlordimedone (MCD) has been monitored [87,93,94]. The formation of the dihalogenated compound causes a decrease of the UV absorption. In the standard test 50 μM MCD is used and the decrease at 290 nm is measured, which corresponds to the absorption maximum of the MCD enolate anion [99]. The pH dependence of the absorption spectrum of MCD caused by the enol/enolate anion equilibrium has to be kept in mind. Corrections are necessary below pH 5 [94] but it is more convenient to use the isobestic wavelength between the enol and the enolate anion at 275 nm [99]. The assay is less sensitive than that involving the formation of triiodide. The advantage of the assay is that there is no significant nonenzymatic oxidation of bromide by hydrogen peroxide. Impurity with high UV absorption in crude extracts and the presence of iodide in brown algae interferes with the assay [93]. Steady-state kinetic analysis has been performed using MCD bromination [94].

Different K_M values have been reported for different isoenzymes [72,87,94,100] although some of the results may be within the experimental and methodological errors. It has been reported that the rate of the MCD bromination in the range above pH 6 is lower than the rate of the bromide-dependent dioxygen formation in the absence of MCD [96]. Therefore results deduced from MCD bromination studies may be complicated by side reactions. Nonphysiological high bromide concentrations cause substrate inhibition. There is a disagreement as to whether this is a competitive [94] or a noncompetitive inhibition [96] with respect to hydrogen peroxide binding. A noncompetitive substrate inhibition caused by hydrogen peroxide has been reported [87]. The storing and freezing of isoenzyme A.n.I. is possible without loss of activity even in the presence of hydrogen peroxide [101]. Different reaction mechanisms have been proposed. Preferential formation of a vanadium(V) peroxo complex is assumed [87,100,102] that then reacts with halides to give an intermediate which may release hypohalides. These are in equilibrium in solution with the halogen and trihalides, which may react in solution with several compounds. However, even the enzyme-intermediate complex may react directly with some compounds. Stereo- and/or regioselective halogenations should be possible with organic compounds when these are bound to the active center of the enzyme. Chiral indoles are known [103]. The vanadium-dependent peroxidases catalyze the bromination of indoles [98] although there is no report on the stereochemical structure of the products. Nonstereospecificity has been reported for the bromination of barbituric acid and its derivatives [104]. Although stereoselective halogenations with haloperoxidase are still unknown, the enzymes are useful catalysts because they act in a much smoother way than the conventional halogenating agents [105]. Regiospecificity has been reported for the bromination of cinnamyl substrates [106].

Furthermore, thiocyanide can be oxidized [107] to CN^+, which is detected by the König reaction [108–111]. CN^+ cleaves pyridine derivatives which form colored polymethine compounds with, for example, *p*-phenylenediamine or barbituric acid derivatives. The vanadium-dependent peroxidases are not inhibited by cyanide [46,93] and CN^+ was not detected in the presence of hydrogen peroxide and cyanide, while the enzyme is irreversibly inactivated [112] in the presence of azide and hydrogen peroxide.

There are reports on chloroperoxidase activity using a peroxidase from the fungus *Curvularia inaequalis* [63] and the peroxidase from *Ascophyllum nodosum* [113]. In the case of the peroxidase from *Ascophyllum nodosum* the chloroperoxidase activity is very low. In comparison to the iodide concentration used in the standard iodoperoxidase assay the chloride concentration was 250-fold. However, the reported chloroperoxidase activity is less than 0.1% compared to the iodoperoxidase activity. Chlorinated compounds were isolated [113], in contrast to experiments on the chlorination of anisole [114]. Even using analytic grade sodium chloride, which contains less than 0.005% bromide, the products were predominantly brominated anisole derivatives. Enzyme preparations from brown algae are contaminated by iodide and bromide, not only in the crude extract, but also in the purified isoenzyme where there are still traces of iodide and bromide [115].

Even after preparation of the vanadium-free apoenzyme, some iodide is detectable [115], e.g., iodinated compounds are labile and reactive. Maybe in the presence of high concentrations of chloride, iodinated compounds are converted to chlorinated compounds. Whether it is an iodide- and/or bromide-mediated reaction or not, the chloroperoxidase activity from *A. nodosum* is not sufficient to give chlorinated products for commercial applications. In comparison to the chloroperoxidase activity of the peroxidase from *Ascophyllum nodosum*, the specific activity of the peroxidase from the *Curvularia inaequalis* is more pronounced. However, no explanation exists as to why the specific activity is only 5% that of the first report of this enzyme [45]. A complete reconstitution with vanadate of the holoenzyme has been performed [63].

3.2. Stability

As demonstrated (Fig. 4) the vanadium-dependent peroxidases are, in contrast to heme-containing peroxidases, not inactivated by the reaction products. The enzyme is stable against freeze drying and X-ray radiation. At room temperature the isoenzyme A.n.I. is stable even in the presence of 6 M urea [20]. The denaturation by guanidine hydrochloride is more efficient than in other cases [29]. The activity is still detectable after sodium dodecyl sulfate (SDS) electrophoresis [46] if the enzyme is not denatured by boiling. Vanadium-dependent peroxidases are reasonably stable in the presence of several organic solvents. Therefore acetone [19] or ethanol [46] can be used in the purification process and there is no problem in storing the enzymes in the presence of organic solvents. At room temperature bromination in a membrane reactor has been performed [116]. The enzyme was stable for 20 days in a continuous process, but then the efficiency decreased drastically. This may be due to irreversible inactivation or to a problem with the stability of the membrane.

3.3. Biological and Ecological Relevance

The brown and red algae are known to accumulate iodide in large amounts. In 1811 Courtois discovered iodine in seaweed ash and until 1873 seaweeds were the only source for iodine [117]. Brown algae, belonging to the Laminariales, accumulate up to 30 000-fold iodide from the seawater. Some studies on iodide accumulations have been done [118–129] and most of the iodine has been found as inorganic iodide [124]. Only traces of monoiodotyrosine and diiodotyrosine were detected. The transport and distribution in the algae have been studied [129]. Iodine and bromine (maybe the halide or halogenated substances) have been detected in the outer cell wall, in the middle lamella, and in physodes in the cells [130].

Furthermore most of the phlorotannins, which are dehydropolymerizates based on the phloroglucinol units, are located in physodes [131] inside the cells of the brown algae. Halogenated phenols have been isolated [132–136] in small amounts, although it is possible that these are artefacts produced during storage and extraction of the algae [137]. Some chlorinated compounds isolated from plants are known to be artefacts [138]. As halide-mediated formation of singlet oxygen is possible, it has been assumed that the peroxidases may be involved in the oxidation of phlorotannins [61]. Iodination, but not oxidation, has been detected in the presence of phloroglucinol, iodide, and hydrogen peroxide [93]. Hence it is unlikely that haloperoxidases located in the cell walls are involved in the oxidation of phlorotannins located in the physodes. It is more likely that cells are protected against singlet oxygen and other reactive substances caused the oxidation of phlorotannins. Until now there has been no proof that haloperoxidases are located in the physodes. It is possible that phlorotannins excreted from the cells are modified through the activity of haloperoxidases.

Little is known about the physiological effect of iodide on algae. From culture experiments it is obvious that iodide is an essential growth factor for the brown algae [139–142] and the liberation of free elementary iodine from seaweeds has been reported [143,144]. More recent studies have investigated the production of halomethanes from seaweeds and from unicellular algae [65,145,146]. Haloperoxidases are assumed to be involved in the production of halomethanes, which are detectable in water near the algae as well as in the air. The halomethanes probably protect the algae through their antibiotic activity.

Calculations of the production of halomethanes by macroalgae range from 10 000 to 26 000 tons/year [147–153]. The calculations are based on the assumptions about biomass and production rates of halomethanes from certain algae, while contributions from microalgae and other biological sources are not included. Determinations of the methyl iodide in seawater and the atmosphere using a two-phase model suggest a global flux from sea to air of about 800 000 tons/year methyl iodide [151]. This is in the range of other calculations [154]. Halomethanes contribute to the ozone degradation in the lower atmosphere [155–158]. There is evidence that iodinated and brominated halomethanes are nonenzymatically converted in seawater into chlorinated compounds [159,160]. Besides haloperoxidases, methyltransferases are involved in the biosynthesis of halomethane [154,161]. A third route for production of methyl iodide has been suggested from in vitro experiments, i.e., the reaction of dimethylsulfoniumpropinate (DMSP) produced by algae and phytoplankton with iodide [162]. In nature, however, no correlation between DMSP and methyl iodide was found and this is probably due to the production of methyl iodide via several pathways. Furthermore, it has been proposed that chlorine-containing prostaglandin-like compounds from the brown alga *Egregia menziesii* may be generated due to the activity of a lipoxygenase. A cation intermediate may react with chloride addition [163].

Inorganic iodide is not taken up in the reduced form by brown algae. Enzymatic oxidation is a prerequisite for the uptake [128] and an iodide oxidase located in the cell walls has been postulated. However, this enzyme has not been isolated as yet. It is possible that the haloperoxidases and hydrogen peroxide–producing oxidases are involved in the iodine uptake. Iodide is taken up by tissue but not by protoplasts from *Laminaria saccharina* [164]. Iodine uptake has been achieved by the oxidation by hydrogen peroxide. No loss of iodine from the incubation medium caused by liberation of either free elementary iodine or iodomethane was observed [164]. The experiments have been performed using natural algae from the ocean and hence bacteria were present in the incubation media, which may cause deiodination of iodomethane. The observation is in agreement with studies on the production of halomethanes by *Laminaria saccharina* and show that in the presence of bacteria the algae only release small amounts of iodomethane. In the absence of bacteria the release of iodomethane is enhanced [145]. To date the effect is unexplained, but it may be caused by bacterial deiodination as several dehalogenating bacteria are known [165,166].

4. STRUCTURE

4.1. The Active Site and Substrate Binding

The vanadium content has been found to range from 0.4 to 2 V per enzyme molecule for different peroxidase preparations [96,115,167,168]. The variation can probably be explained by a partial loss of vanadium during the purification, while further unspecific binding of vanadium is possible [115]. Several other trace elements are detectable in the purified enzyme [115] whose content decreases during the preparation of the inactive apoenzyme, except for iron, which shows an increase compared to the active holoenzyme.

Vanadium(V) is proposed to be the active site in the holoenzyme as vanadate in the presence of hydrogen peroxide forms vanadium(V) peroxide [169–174]. The oxidation of iodide [175] and bromide [176] by peroxovanadate has been reported. As shown in Fig. 3, the binding of vanadium(V) during the reconstitution of the active holoenzyme has an effect on the electron absorption spectrum, which is also affected by the addition of hydrogen peroxide (Fig. 5). The changes in the extinction coefficients are very low. Vanadium(V) can act as an oxidase. During the reaction vanadium(V) is reduced and then reoxidized by oxygen [177]. As the vanadium center was not accessible for electrochemical studies [178], no redox potentials are known for these enzymes. The reconstitution conditions indicate that vanadium(V) is present in the holoenzyme, which is in agreement with ^{51}V NMR studies [168,179,180] and indicates that vanadium(V) is surrounded by oxygen ligands, probably bidentate carboxylato groups. ^{51}V NMR spectra from the vanadate/dipeptide complexes are quite different from the ^{51}V NMR spectrum of the native holoenzyme [115,180].

X-ray absorption near-edge structure spectroscopy (XANES) was used to study the coordination charge and the surrounding of the vanadium center in the native and reduced enzyme. The coordination

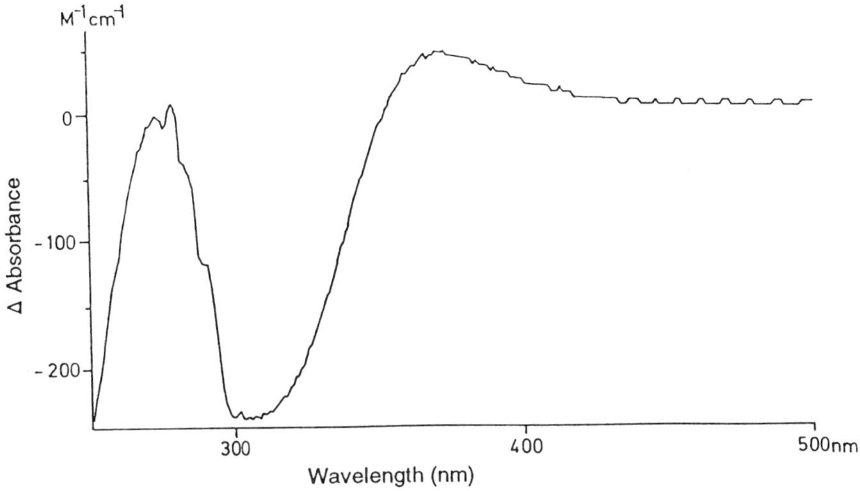

FIG. 5. Electron absorption difference spectrum of peroxidase from *Ascophyllum nodosum*. 1×10^{-5}M A.n.I. with 20 mM H_2O_2 against A.n.I. without H_2O_2 in 0.1 M phosphate, pH 6.2.

charge can be deduced from the energy position of the pre-edge peak. In the native enzyme the vanadium center is in the oxidation state (V) [101,102]; this is in agreement with ^{51}V NMR studies. The oxidation state was affected neither by the substrate bromide nor by hydrogen peroxide. This is in agreement with ^{51}V NMR and electron paramagnetic resonance (EPR) studies. Vanadium(V) is EPR-silent; hence no signal was detectable in the native enzyme or in the presence of the substrates. In order to detect an EPR signal the enzyme has to be reduced by dithionite to vanadium(IV) [22,115] and this has been confirmed by the change of the energy position of the pre-edge peak in the presence of dithionite [101,102]. Furthermore, there was a drastic decrease of the pre-edge peak due to the reduction indicating a change in the coordination sphere.

It has been proposed that there is an octahedral coordination of the ligands around the vanadium(V) center in the native enzyme and this is compatible with the extended X-ray absorption fine structure (EXAFS) spectra [181]. According to these spectra it was concluded that one ligand at a distance of 1.61 Å should be an oxygen ligand, three ligands at a distance of 1.72 Å may be oxygen or nitrogen ligands, and two at a distance of 2.11 Å are postulated to be nitrogen ligands. Changes in the distances were observed upon reduction of vanadium(V) to vanadium(IV). Information concerning the distance may be an indication but is not a proof for certain ligands. According to XANES spectra from native enzyme it has been suggested, that the coordination sphere consists of oxygen ligands. Electron spin-echo envelope modulation (ESEEM) spectroscopic studies have shown that a nitrogen ligand is coordinated to the vanadium(IV) center in the equatorial plane [182]. Whether the nitrogen ligand was coordinated to the vanadium center before reduction could not be proven by this method.

Reduced peroxidases show axially symmetric vanadyl-type EPR spectra and these are very similar from different sources [64,183]. The spectra are also similar to vanadyl EPR spectra obtained after the reduction of vanadate/dipeptide by dithionite [115], although there are some differences. After the removal of dithionite the reduced holoenzyme reoxidized by air is active [115]. Thus it is probable that in the reduced enzyme, vanadium(IV) is still in or near the active site. In the presence of dithionite and vanadium(IV) only weak and nonspecific binding and/or rapid exchange between vanadyl bound at the apoenzyme and free vanadyl takes place [115]. From the pH dependence of EPR spectra it has been concluded that protonation of a group of pK_a 5.4 takes place and corresponds to either histidine or aspartate/glutamate residues near vanadium(IV) [167]. In contrast, using the native holoenzyme no pH dependence of the vanadium(V) XANES spectra was observed [102]. This is in accordance with the reconstitution experiments which failed to demonstrate an ionizable group in this range [29].

It is an open question as to whether the protonation of protein residues or the protonation of vanadate prevents the reconstitution of the holoenzyme from the apoenzyme by vanadate at low pH. The pH effect is quite different in the case of the vanadium-containing chloroperoxidase from *Curvularia inaequalis* [65], where the protonation of a group with pK_a 6.0 was detected. Histidine derivatization in the apoenzyme hinders the reconstitution by vanadate. The modification was reversible. Even the EPR spectra of vanadium(IV) in the fungal and the algal enzymes are similar; there are differences in the environment of the vanadium center and the catalytic properties.

The binding sites of substrates have not been elucidated from kinetic or spectroscopic studies. Changes observed in the XANES spectrum upon the addition of hydrogen peroxide is indicative of changes in the coordination sphere of the vanadium center. However, whether hydrogen peroxide binds to the vanadium(V) center or elsewhere near the vanadium(V) center is an open question. From ^{51}V NMR studies it is concluded that the iodide is not bound to the vanadium center, but binding affects the coordination sphere of the vanadium center. Binding of the substrate iodide was detectable by ^{127}I NMR [184]. In respect to pH there was no correlation between activity and substrate binding. However, NMR studies need substance concentrations in the millimolar range. From the catalytic point of view substrate concentrations in the micromolar range are physiologically more relevant. Assuming that there may be a competition between the binding of azide and the substrate iodide ^{15}N NMR studies were performed. By adding trace amounts of iodide the binding of azide is more effective than in the absence of iodide [112]. This cooperative effect indicates that iodide and azide are binding at different places.

To date much effort has been spent on kinetic and spectroscopic studies but still little is known about the active site. Vanadium(V) is essential for activity, but it is still possible that it is not the place where the reaction takes place. Vanadium(V) may be essential for the structure of the active site. The structure of a nonheme nonmetal haloperoxidase is known and in this case the active site has a lyase-like structure [37]. This indicates that at least there is no need to have a metal in haloperoxidases for substrate activation.

4.2. Molecular Weight

Precise knowledge of the molecular weight of the protein is important. The results using denaturation conditions, e.g., SDS–polyacrylamide gel electrophoresis (PAGE) are assumed to give the molecular weight of subunits, whereas nondenaturation conditions should give the total weight of the protein aggregates. Each method has its particular advantage as well as disadvantage. SDS electrophoresis is commonly used and simple. Variable binding of SDS, e.g., to acidic proteins, results in inaccurate molecular weight determinations [185]. The results for peroxidase A.n.I. range from 40 000 to 97 000 Da [46,74]. Molecular weight determination using gel chromatography or ultracentrifugation is more laborious and in both methods errors may rise because the density of the protein may differ from that of typical proteins. The results are in the range of 95 000 ± 5000 Da for the native enzyme [20,22,29].

The results corresponded poorly with the assumption of one or two subunits. More precise determination was possible using matrix-assisted UV laser desorption/ionization mass spectroscopy (MALDI-MS) [186] (Fig. 6). The molecular weight of the native enzyme A.n.I. was 120 400 ± 200 Da and there are two subunits which are bridged by a disulfide [187]. The difference between the molecular weight of the native enzyme and the dimeric acylated derivative is in the range expected from the cysteine content [29].

4.3. Primary Structure

The knowledge of the primary structure is important for the further elucidation of the enzyme structure as well as for the enzyme application. Even when the enzymes can be produced in large amounts from the raw material, overexpression and genetic engineering of these enzymes are of interest.

As long as the protein is not a small peptide, protein sequence analysis is combined with cDNA sequence analysis encoding the protein of interest. Sufficient sequence information from the protein sequence can be used to design gene probes in order to screen cDNA libraries. After sequence analysis of several cDNA clones, the comparison of protein sequence information and protein sequence deduced from the cDNA allows selection of the proper sequence. The peroxidase A.n.I. was cleaved chemically and also by several proteases. The resulting peptides were separated by reverse high-performance liquid chromatography (HPLC) and sequenced [188].

The next important aspect was the availability of a cDNA library for which the isolation of highly purified mRNA is necessary. The presence of anionic polysaccharides hampers the isolation of RNA from brown algae, as the RNA is also an anionic polymer. Furthermore there is a need to inactivate RNases. Purified mRNA was prepared and used to produce a cDNA library using suitable phages. The library was screened and the inserts from positive phages were subcloned into a

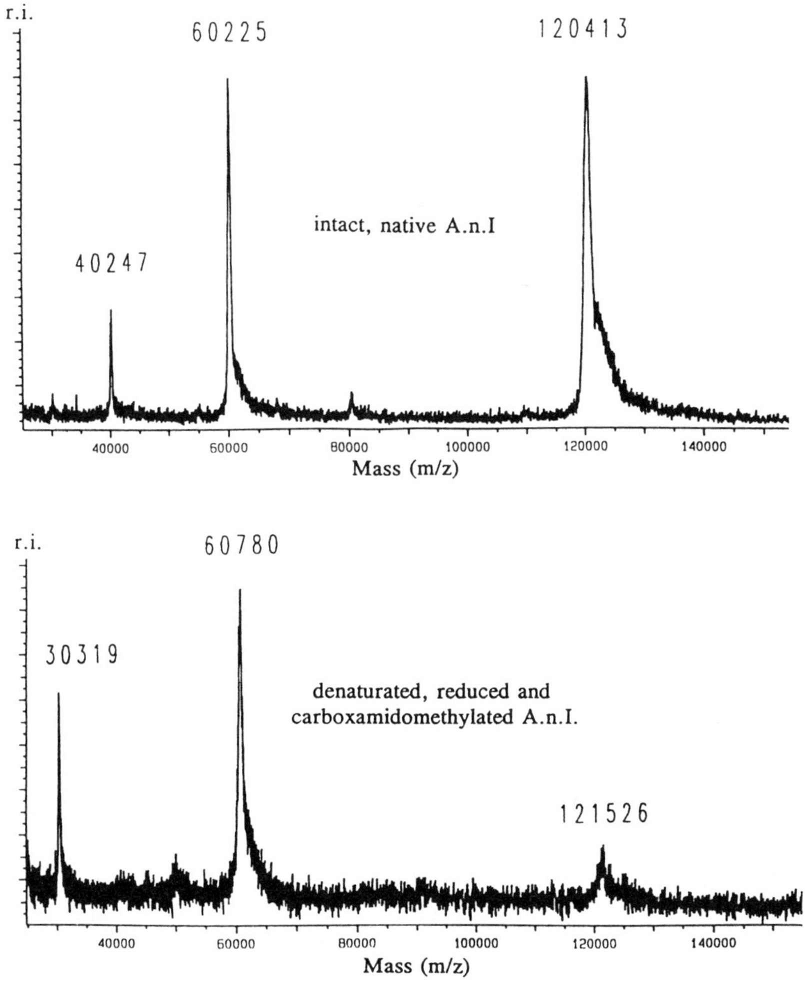

FIG. 6. Mass spectra of haloperoxidase A.n.I.

TABLE 2

Part of the Sequence of the Vanadium-Dependent Peroxidase A.n.I. from *Ascophyllum nodosum*

```
Peptid :                          L  I  G  A  A  E  L  A  Q  R  S  S
frame 0:  I  D  H  Y  F  P  F  R  L  I  G  X  A  E  L  A  Q  R  S  S
       5' ATCGATCACTACTTTCCCTTTAGATTGATAGGTGTNGCCGAGTTGGCGCAGCGTTCCTCG 3'
                1508      1518      1528      1538      1548      1558

Peptid :  W  Y  Q  K
frame 0:  W  Y  Q  N  W  Q  V  H  R  F  A  R  P  E  A  L  G  G  T  L
       5' TGGTACCAAAACTGGCAGGTGCATCGATTCGCACGCCCCGAGGCCCTCGGAGGTACCCTC 3'
                1568      1578      1588      1598      1608      1618

Peptid :
frame 0:  H  L  T  I  K  G  E  L  N  A  D  F  D  L  S  L  L  E  N  A
       5' CACCTCACCATCAAGGGGGAACTAAATGCAGACTTCGACCTCTCCCTCCTTGAAAATGCT 3'
                1628      1638      1648      1658      1668      1678

Peptid :                 V  A  A  I  N  A  A  Q  N  P  N  N  E  V  T
frame 0:  E  L  L  K  R  V  A  A  I  N  A  X  X  N  P  N  N  E  V  T
       5' GAGCTCTTGAAACGTGTGGCGGCCATAAATGCGGCNNAGAACCCCAACAACGAGGTCACC 3'
                1688      1698      1708      1718      1728      1738

Peptid :  Y  L  L  P  Q  A  I  Q  E  G  T  S  T  H  P  S  Y  P  S  G
frame 0:  Y  L  L  P  Q  V  Y  Q  E  G  T  P  T  H  P  S  Y  P  S  G
       5' TACCTTCTTCCACAGGTCTACCAAGAGGGTACACCAACGCACCCTTCGTACCCGTCTGGC 3'
                1748      1758      1768      1778      1788      1798

Peptid :  G  A  T  Q  N  G  A  F  A
frame 0:  H  A  T  X  N  G  A  F  A  T  V  L  K  A  L  I  G  L  D  R
       5' CACGCTACCCNNAACGGAGCATTTGCCACAGTTCTCAAGGCCCTCATTGGCCTAGATCGG 3'
                1808      1818      1828      1838      1848      1858

Peptid :
frame 0:  G  G  D  A  I  R  P  V  Y  P  D  D  D  G  L  K  L  I  D  F
       5' GGGGGTGATGCTATCCGCCCCGTGTACCCAGACGATGACGGCCTGAAACTAATCGACTTC 3'
                1868      1878      1888      1898      1908      1918

Peptid :                                      K  L  A  V  N  V  A  F
frame 0:  R  G  S  C  L  T  F  E  G  E  I  K  K  L  A  V  N  V  A  F
       5' AGAGGGAGTTGCCTCACGTTTGAGGGAGAGATCAAGAAGCTCGCGGTCAACGTCGCATTC 3'
                1928      1938      1948      1958      1968      1978

Peptid :  G     Q  M  L  G  I  H  Y  R  F  D  G  I  Q  G  L  L  L  G
frame 0:  G  R  Q  M  L  G  I  H  Y  R  F  D  G  I  Q  G  L  L  L  G
       5' GGGAGGCAGATGCTAGGCATCCACTATCGGTTTGACGGTATTCAAGGCCTGCTCCTCGGA 3'
                1988      1998      2008      2018      2028      2038

Peptid :  E  T  I  T  V  R  T  L  H  Q  E  L  M  T  F  A  E  E  S  T
frame 0:  E  T  I  T  V  R  T  L  H  Q  E  L  M  T  F  A  E  E  S  T
       5' GAGACAATTACCGTACGAACACTTCACCAGGAGCTGATGACTTTCGCCGAGGAATCCACC 3'
                2048      2058      2068      2078      2088      2098

Peptid :  F  E  F  R  L  F  T  G  E  V  I  K  L  F  Q  D  G  T  F  T
frame 0:  F  E  F  R  L  F  T  G  E  V  I  K  L  F  Q  D  G  T  F  T
       5' TTTGAGTTCCGCCTGTTCACCGGCGAGGTCATCAAACTTTTCCAGGACGGCACATTCACC 3'
                2108      2118      2128      2138      2148      2158

Peptid :  I  D  G  F  K  .  P  G  L  V  Y  T  G  V  E  N  .  V
frame 0:  I  D  G  F  K  C  P  G  L  V  Y  T  G  V  E  N  C  V
       5' ATCGATGGATTTAAATGTCCCGGTTTGGTCTACACTGGCGTGGAGAATTGCGTG 3'
                2168      2178      2188      2198      2208

Linear DNA sequence: clone 1s, length 2289, 5' strand gene from:  1499:2212
Peptid : protein sequence analysis
Frame 0: protein sequence deduced from cDNA
```

plasmid. Several clones have been sequenced [189] and proven to be encoding for vanadium-dependent peroxidases. Part of one sequence is given in Table 2.

4.4. Secondary Structure

Secondary structure prediction is possible from the sequence and global information can be obtained from CD spectra (Fig. 7). The results indicate more than 70% helical structure in agreement with Ref. 29.

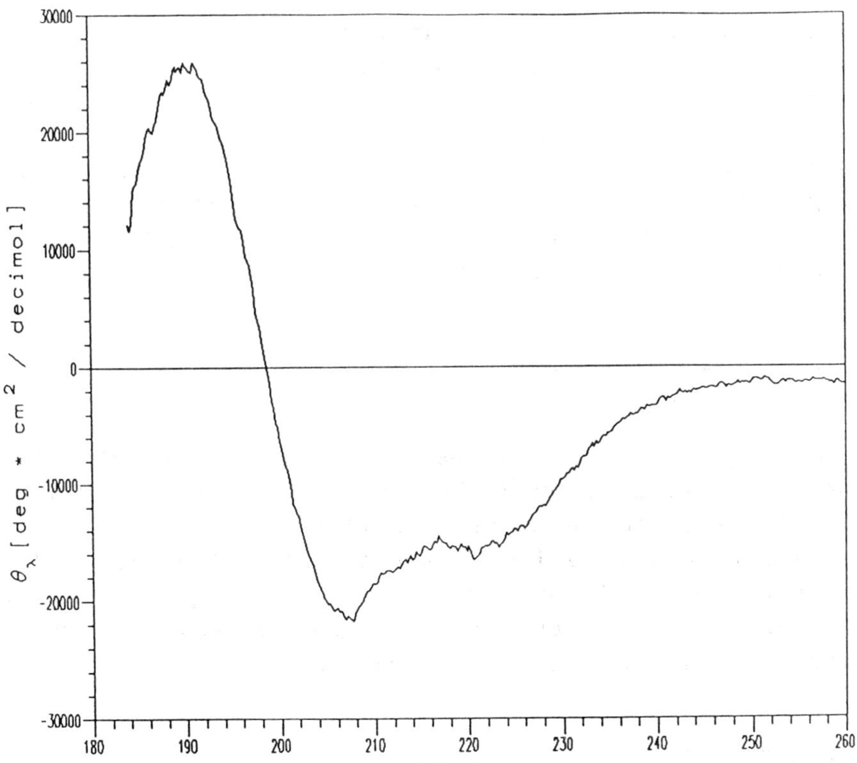

FIG. 7. Circular dichroism spectrum of peroxidase A.n.I. from *Ascophyllum nodosum*. It is concluded that the protein contains 71 ± 2.1% helical structure, 12 ± 2.4% β sheet, and 17 ± 4.0% other structures.

4.5. Tertiary Structure

The tertiary structure analysis by NMR techniques is not possible because of the high molecular weight while X-ray crystallography requires suitable crystals. Fortunately, two different crystal forms are now available. One crystal form is monoclinic ($0.5 \times 0.15 \times 0.15$ mm, space group $P2_1$) [91,190] and the other one is tetragonal ($0.4 \times 0.4 \times 0.3$ mm, space group $P4_{1/3}2_12$). Currently the structure elucidation is in progress [191].

4.6. Posttranslatoric Protein Modification and Protein Surface

Little is known about the posttranslatoric protein modification and protein surface. Some of the peroxidases have been shown to interact with concanavalin A [74,192], which has an affinity for terminal α-D-mannosyl and α-D-glucosyl residues. The vanadium-dependent peroxidases are acidic proteins with isoelectric points in the range 3.0–5.0 [77,100]. There are significant differences in the protein surfaces of different isoenzymes, which is evident from partitioning in the two-phase systems as well as from cross-immuno experiments [54,55,59,89,192].

ACKNOWLEDGMENTS

The project was started at the Institute of Pharmaceutical Biology at the University of Bonn and has been continued at the Gesellschaft für Biotechnologie mbH and the Institute of Genetics at the University of Braunschweig. I thank Prof. Glombitza (Bonn), Prof. Flohé, Prof. Schomburg, and Prof. Cerff (Braunschweig) for the opportunity to do this work. I thank the Deutsche Forschungsgemeinschaft for financial support and all those who have collaborated.

ABBREVIATIONS

A.n.I.	*Ascophyllum nodosum* isoenzyme
CD	circular dichroism

cDNA	copy deoxyribonucleic acid
DMSP	dimethylsulfoniumpropinate
EDTA	ethylenediamine-*N*,*N*,*N′*,*N′*-tetraacetate
EPR	electron paramagnetic resonance
ESEEM	electron spin-echo envelope modulation spectroscopy
EXAFS	extended X-ray absorption fine structure
HPLC	high-performance liquid chromatography
MALDI-MS	matrix-assisted ultraviolet laser desorption/ionization mass spectroscopy
MCD	monochlordimedone = 1,1-dimethyl-4-chloro-3,5-cyclohexadione
mRNA	messenger ribonucleic acid
NADH	nicotinamide adenine dinucleotide (reduced)
NMR	nuclear magnetic resonance
PA	polyacrylamide
PAGE	polyacrylamide gel electrophoresis
PEG	polyethylene glycol
RNA	ribonucleic acid
SDS	sodium dodecyl sulfate
Tris	tris(hydroxymethyl)aminomethane = 2-amino-2-(hydroxymethyl)-1,3-propanediol
XANES	X-ray absorption near-edge structure

REFERENCES

1. E. C. Webb, *Enzyme Nomenclature*, Academic Press, San Diego, 1992.
2. K. G. Paul, in *Molecular and Physiological Aspects of Plant Peroxidases*, University of Geneva, 1986, pp. 1–14.
3. K. G. Welinder, in *Molecular and Physiological Aspects of Plant Peroxidases*, University of Geneva, 1991, pp. 3–13.
4. H. W. Krell, in *Molecular and Physiological Aspects of Plant Peroxidases*, University of Geneva, 1991, pp. 469–478.
5. A. Campa, in *Peroxidases 1 in Chemistry and Biology*, Vol. 2 (J. Everse, K. Everse, and M. B. Grisham, eds.), CRC Press, Boca Raton, 1991, p. 25.
6. S. F. Yang, *Arch. Biochem. Biophys.*, *122*, 481 (1967).
7. G. G. Gross, *Rec. Adv. Phytochem.*, *11*, 141–148 (1977).

8. M. Tien and S. B. Myer, *Appl. Environ. Microbiol.*, *56*, 2540–2544 (1990).
9. S. Boominathan, D. S. Balachandra, T. A. Randall, R. J. Kelly, and C. A. Reddy, *Bacteriology*, *172*, 260–265 (1990).
10. W. R. G. Atkins, *Sc. Proc. R. Dublin Soc.*, *14*, 199–206 (1914).
11. G. B. Reed, *Bot. Gazette*, *59*, 407–409 (1915).
12. S. Petersson, *Kgl Fys. Sällsk. Förh.*, *10*, 171–182 (1940).
13. S. Rönnerstrand, *Akademische Abhandlung*, Lindström, Lund, 1943.
14. H. Krienke, *Wiss. Z. Univ. Rostock*, *16*, 455–456 (1967).
15. B. Z. Siegel and S. M. Siegel, *Am. J. Bot.*, *53*, 285–287 (1970).
16. M. J. Murpy and C. Ó hEocha, *Phytochemistry*, *12*, 2645–2648 (1973).
17. L. Oliveria and T. Bisalputra, *Can. J. Bot.*, *54*, 913–922 (1976).
18. M. A. Ragan and K.-W. Glombitza, in *Progress in Phycological Research*, Vol. 4 (F. E. Round and D. J. Chapman, eds.), Biopress Ltd., Bristol, 1986, p. 129.
19. H. Vilter, K.-W. Glombitza, and A. Grawe, *Bot. Mar.*, *26*, 331–340 (1983).
20. H. Vilter, *Bot. Mar.*, *26*, 451–455 (1983).
21. H. Vilter, *Phytochemistry*, *23(7)*, 1387–1390 (1984).
22. E. De Boer, Y. van Kooyk, M. G. M. Tromp, H. Plat, and R. Wever, *Biochim. Biophys. Acta*, *869*, 48–53 (1986).
23. L. Fries, *Planta*, *154*, 393–396 (1982).
24. H.-U. Meisch and H. J. Bielig, *Arch. Microbiol.*, *105*, 77–82 (1975).
25. H.-U. Meisch, H. Benzschawel, and H. J. Bielig, *Arch. Microbiol.*, *114*, 67–70 (1977).
26. H.-U. Meisch and H. Brenzschawel, *Arch. Microbiol.*, *116*, 91–95 (1978).
27. H.-U. Meisch and J. Bauer, *Arch. Microbiol.*, *117*, 49–52 (1978).
28. F. H. Nielsen and E. O. Uthus, in *Vanadium in Biological Systems: Physiology and Biology* (N.D. Chasteen, ed.), Kluwer Academic, Doordrecht, 1990, pp. 51–62.
29. M. G. M. Tromp, G. Olafsson, B. E. Krenn, and R. Wever, *Biochim. Biophys. Acta*, *1040*, 192–198 (1989).

30. M. G. M. Tromp and R. Wever, *Biochim. Biophys. Acta, 1079*, 53–56 (1991).

31. N. D. Chasteen, in *Vanadium in Biological Systems: Physiology and Biology* (N. D. Chasteen, ed.), Kluwer Academic, Doordrecht, 1990, pp. VII–VIII.

32. W. Wiesner, K.-H. van Pee, and F. Lingens, *FEBS Lett.*, *209*, 321–324 (1986).

33. K.-H. van Pee, G. Sury, and F. Lingens, *Biol. Chem. Hoppe-Seyler*, *368*, 1225–1232 (1987).

34. B. E. Krenn, H. Plat, and R. Wever, *Biochim. Biophys. Acta*, *952*, 255–260 (1988).

35. M. Weng, O. Pfeifer, S. Krauss, F. Lingens, and K.-H. van Pee, *J. Gen. Microbiol.*, *137*, 2539–2546 (1991).

36. H. Sobek, T. Haag, D. Schomburg, F. Lingens, and K.-H. van Pee, *J. Mol. Biol.*, *221*, 35–37 (1991).

37. H. J. Hecht, Abstracts of the 3rd European Workshop on Crystallography of Biological Macromolecules, Centro di Cultura Scientifica Alessandro Volta–Como (Italy), O41 (1993).

38. Y. P. Chen, D. E. Lincoln, S. A. Woodin, and C. R. Lovell, *J. Biol. Chem.*, *266*, 23909–23915 (1991).

39. M. L. Dohlin, in *The Bacteria, Vol. 2, Metabolism* (I. C. Cunsalus and R. Y. Stanier, eds.), Academic Press, New York, 1961, pp. 437–444.

40. L. Flohé, W. A. Günzler, and R. Ladenstein, in *Glutathione: Metabolism and Function* (I. M. Arias and W. B. Jakoby, eds.), Raven Press, New York, 1976, pp. 115–138.

41. L. Flohé, in *Glutathione: Chemical, Biochemical and Medical Aspects*, Part A (D. Dolphin, R. Poulson, and O. Avramovic, eds.), John Wiley and Sons, New York, 1989, pp. 643–731.

42. N. Itoh, Y. Izumi, and H. Yamda, *Biochem. Biophys. Res. Commun.*, *131*, 428–435 (1985).

43. N. Itoh, Y. Izumi, and H. Yamda, *J. Biol. Chem.*, *261*, 5194–5200 (1986).

44. N. Itoh, Y. Izumi, and H. Yamda, *J. Biol. Chem.*, *262*, 11982–11987 (1987).

45. T. N. E. Lui, T. M'Timkulu, J. Geigert, R. Wolf, S. L. Neidelman,

D. Silva, and J. C. Hunter-Cevera, *Biochem. Biophys. Res. Commun., 142*, 329–333 (1987).

46. R. Wever, H. Plat, and E. De Boer, *Biochim. Biophys. Acta, 830*, 181–186 (1985).
47. H. S. Soedjak and A. Butler, *Biochemistry, 29*, 7974–7981 (1990).
48. R. Wever, *Nature, 335*, 501 (1988).
49. H. Vilter, *Plant Peroxidase Newsletter*, submitted for publication.
50. E. De Boer, M. G. M. Tromp, H. Plat, B. E. Krenn, and R. Wever, *Biochim. Biophys. Acta, 872*, 104–115 (1986).
51. R. Wever, M. G. M. Tromp, B. E. Krenn, A. Marjani, and M. van Tol, *Environ. Sci. Technol., 25*, 446–449 (1991).
52. M. Miyata, H. Vilter, and T. Hosoya, *Annu. Rep. Mar. Ecosystem Res. Ctr., 11*, 27–30 (1991).
53. H. Vilter, *Bioseparation, 1*, 283–292 (1990).
54. P. Jordan and H. Vilter, *Biochim. Biophys. Acta, 1073*, 98–106 (1991).
55. R. Wever, J. W. P. M. Van Schijndel, E. Vollenbroek, R. L. Olsen, and E. Folgelqvist, in *The Biogeochemistry of Global Change: Radiative Trace Gases* (R. S. Oremland, ed.), Chapman and Hall, New York, 1992, pp. 811–824.
56. A. Almeida, M. Humanes, and H. Vilter, Communicated at the European Research Conference: Chemistry of Metals in Biological Systems, Praia da Falésia, Portugal; details to be published.
57. B. E. Krenn, H. Plat, and R. Wever, *Biochim. Biophys. Acta, 912*, 287–291 (1987).
58. B. E. Krenn, Y. Izumi, H. Yamada, and R. Wever, *Biochim. Biophys. Acta, 998*, 63–68 (1989).
59. H. Yu and J. W. Whittaker, *Biochem. Biophys. Res. Commun., 160*, 87–92 (1989).
60. D. J. Shieffield, T. Harry, A. J. Smith, and L. J. Rogers, *Phytochemistry, 32*, 21–26 (1993).
61. R. R. Everett, J. R. Kanofsky, and A. Butler, *J. Biol. Chem., 265*, 4908–4914 (1990).
62. H. S. Soedjak, PhD dissertation, University of California–Santa Barbara, 1991.
63. J. W. P. M. Van Schijndel, E. G. M. Vollenbroek, and R. Wever, *Biochim. Biophys Acta, 1161*, 249–256 (1993).

64. J. W. P. M. Van Schijndel, L. H. Simons, E. G. M. Vollenbroek, and R. Wever, *FEBS Lett.*, *336*, 239–242 (1993).
65. H. Plat, B. E. Krenn, and R. Wever, *Biochem. J.*, *248*, 277–279 (1987).
66. J. Kohlmeyer and E. Kohlmeyer, *Bot. Mar.*, *15*, 109–112 (1972).
67. N. Fries and K. Thorén-Tolling, *Bot. Mar.*, *21*, 409–411 (1978).
68. L. Fries, *J. Phycol.*, *24*, 333–337 (1988).
69. E. Baardseth and M. A. Cuesta, *Sarsia*, *68*, 165–169 (1983).
70. S. Mitsuyoshi and M. Tomitaro, *Jpn. J. Phycol.*, *33*, 239–244 (1985).
71. T. H. Haines, *Annu. Rev. Microbiol.*, *27*, 403–411 (1973).
72. B. E. Krenn, H. Plat, M. G. M. Tromp, H. Vilter, and R. Wever, Abstracts of the 14th International Congress of Biochemistry, Th160, Prague, 1988.
73. B. E. Krenn, M. G. M. Tromp, and R. Wever, *J. Biol. Chem.*, *264*, 19287–19292 (1989).
74. M. Pedersén and N. Fries, *Z. Pflanzenphysiol.*, *82*, 363–366 (1977).
75. H. Vilter and N. Fries, unpublished results.
76. H. Vilter and K.-W. Glombitza, *Bot. Mar.*, *26*, 341–344 (1983).
77. P. Jordan, B. Kloareg, and H. Vilter, *J. Plant Physiol.*, *137*, 520–524 (1991).
78. A. Butler, H. S. Soedjak, M. Polne-Fuller, A. Gibor, C. Boyen, and B. Kloareg, *J. Phycol.*, *26*, 589–592 (1990).
79. H. Vilter and L. Fries, unpublished results.
80. M. Pedersén, *Physiol. Plant.*, *37*, 6–11 (1976).
81. T. J. Ahern, G. G. Allan, and D. G. Medcalf, *Biochim. Biophys. Acta*, *616*, 329–339 (1980).
82. J. A. Manthey and L. P. Hager, *J. Biol. Chem.*, *256*, 11232–11238 (1981).
83. D. G. Baden and M. D. Corbett, *Biochem. J.*, *187*, 205–211 (1980).
84. M. Kirchgessner, personal communication.
85. H. Vilter, B. S. Falch, and G. M. König, unpublished results.
86. B. S. Falch, G. M. König, A. D. Wright, O. Sticher, H. Ruegger, and G. Bernardinelli, *J. Org. Chem.*, *58*, 6570–6575 (1993).

87. A. Butler and J. V. Walker, *Chem. Rev.*, *93*, 1937–1944 (1993).

88. H. Vilter, *Planta Medica*, *10*, 438 (1986).

89. H. Vilter, in *Methods in Enzymology*, Vol. 228 (G. Johansson and H. Walter, eds.), Academic Press, San Diego, 1994, pp. 665–672.

90. H. Hustedt, K. H. Kroner, H. Schütte, and M.-R. Kula, *Ausarbeitung biologischer Medien. Schriftenreihe der GBF*, Vol. 7 (K. Schugerl, M.-R. Kula, and U. Omken, eds.), Verlag Chemie, Weinheim, 1984, pp. 83–93.

91. A. Müller-Fahrnow, W. Hinrichs, W. Saenger, and H. Vilter, *FEBS Lett.*, *239*, 292–294 (1988).

92. F. Björkstén, *Eur. J. Biochem.*, *5*, 133–142 (1968).

93. H. Vilter, *Bot. Mar.*, *26*, 429–435 (1983).

94. E. De Boer and R. Wever, *J. Biol Chem.*, *263*, 1226–1233 (1988).

95. C. F. Schönbein, *Lieb. Ann. Suppl.*, *2*, 217 (1863).

96. E. E. Everett, H. S. Soedjak, and A. Butler, *J. Biol. Chem.*, *265*, 15671–15679 (1990).

97. H. S. Soedjak and A. Butler, *Biochim. Biophys. Acta*, *1079*, 1–7 (1991).

98. R. A. Tschirret-Guth and A. Butler, *J. Am. Chem. Soc.*, *116*, 411–412 (1994).

99. W. D. Hewson and L. P. Hager, *J. Phycol.*, *16*, 304–345 (1980).

100. R. Wever and B. E. Krenn, in *Vanadium in Biological Systems: Physiology and Biology* (N. D. Chasteen, ed.), Kluwer Academic, Doordrecht, 1990, pp. 81–97.

101. J. Hormes, U. Kuetgens, R. Chauvistre, W. Schreiber, N. Anders, H. Vilter, D. Rehder, and C. Weidemann, *Biochem. Biophys. Acta*, *956*, 293–299 (1988).

102. U. Küsthardt, B. Hedman, K. O. Hodgson, R. Hahn, and H. Vilter, *FEBS Lett.*, *329*, 5–8 (1993).

103. R. S. Norton and R. J. Wells, *J. Am. Chem. Soc.*, *104*, 3628–3635 (1982).

104. M. C. R. Franssen, J. D. Jansma, H. C. Van der Pas, E. De Boer, and R. Wever, *Bioorg. Chem.*, *16*, 352 (1988).

105. M. C. R. Franssen and H. C. van der Plas, *Adv. Appl. Microbiol.*, *37*, 41–99 (1992).

106. P. Coughlin, S. Roberts, C. Rush, and A. Willetts, *Biotechnol. Lett.*, *15*, 907–912 (1993).
107. H. Vilter, *Nachr. Chem. Lab.*, *39*, 686–689 (1991).
108. J. L. Lambert, J. Ramasamy, and J. V. Paukstells, *Anal. Chem.*, *47*, 916–918 (1975).
109. W. König, *J. Prak. Chem.*, *69*, 105 (1904).
110. W. König, *Z. Angew. Chem.*, *69*, 115 (1905).
111. G. Schwarzenbach and R. Weber, *Helv. Chim. Acta*, *25*, 1628–1639 (1942).
112. H. Vilter, H. Rüterjans, and H. Hanssum, Abstracts of the 27th International Conference on Coordination Chemistry, M62, Brisbane, 1989.
113. H. S. Soedjak and A. Butler, *Inorg. Chem.*, *29*, 5017–5025 (1990).
114. B. Walter, Diplomarbeit Universität Ulm, 1989.
115. K. Knüttel, A. Müller, D. Rehder, H. Vilter, and V. Wittneben, *FEBS Lett.*, *302*, 11–14 (1992).
116. E. De Boer, H. Plat, M. G. M. Tromp, R. Wever, M. C. R. Franssen, H. C. Van der Pas, E. M. Meijer, and H. E. Schoemaker, *Biotech. Bioeng.*, *30*, 607–610 (1987).
117. B. Courtois, *Ann. Chim. Phys.* (Paris), *88*, 314 (1813).
118. W. A. P. Black, *J. Soc. Chem. Ind.*, *67*, 165–176 (1948).
119. N. A. Baily and S. Kelly, *Biol. Bull.*, *109*, 13–21 (1955).
120. W. Tong and I. L. Chaikoff, *J. Biol. Chem.*, *215*, 473–484 (1955).
121. H. G. Klemperer, *Biochem. J.*, *67*, 381–390 (1957).
122. J. Roche and M. Lafon, *C. Roy. Acad. Sci.*, *229*, 481–482 (1949).
123. S. André, *C. R. Soc. Biol. Paris*, *159*, 2327–2332 (1965).
124. S. André, *C. R. Soc. Biol. Paris*, *165*, 2293–2298 (1971).
125. J. Roche and Y. Yagi, *C. Roy. Acad. Sci.*, *146*, 642–645 (1952).
126. T. I. Shaw, *Proc. Roy. Soc., Ser. B*, *150*, 356–371 (1959).
127. T. I. Shaw, *Proc. Roy. Soc., Ser. B*, *152*, 109–117 (1960).
128. T. I. Shaw, in *Physiology and Biochemistry of Algae* (R.A. Lewin, ed.), Academic Press, New York, 1962, pp. 247–253.
129. M. A. Amat and L. M. Srivastava, *J. Phycol.*, *21*, 330–333 (1985).
130. M. Pedersén and G. M. Poomans, *Bot. Mar.*, *26*, 113–118 (1983).
131. M. A. Ragan, *Bot. Mar.*, *19*, 145–154 (1976).

132. K.-W. Glombitza, M. Koch, and G. Eckhardt, *Phytochemistry*, *16*, 796–798 (1977).

133. M. Koch, K.-W. Glombitza, and G. Eckhardt, *Phytochemistry*, *19*, 1821–1823 (1980).

134. M. Koch and P. R. Gregson, *Phytochemistry*, *23*, 2633–2637 (1984).

135. K.-W. Glombitza and G. Gerstberger, *Phytochemistry*, *24*, 543–551 (1985).

136. K.-W. Glombitza and G. Zieprath, *Planta Med.*, *55*, 171–175 (1989).

137. K.-W. Glombitza, personal communication.

138. K. C. Engvild, *Phytochemistry*, *25*, 781–791 (1986).

139. K. Fries, *Physiol. Plant*, *19*, 800–808 (1966).

140. H. A. Von Stosch, *Naturwissenschaften*, *49*, 42–43 (1962).

141. M. Pedersén, *Physiol. Plant*, *22*, 680–685 (1969).

142. S. J. C. Hsiao, *Can. J. Bot.*, *47*, 1611–1616 (1969).

143. P. Dangeard, *C. R. Acad. Sci.*, *192*, 500–501 (1931).

144. H. Kylin, *Hoppe-Seyler's Z. Physiol. Chem.*, *203*, 58–65 (1931).

145. F. Laturnus, *Ber. Polarforsch.*, *132*, 1–188 (1993).

146. J. E. Lovelock, *Nature*, *156*, 193–194 (1975).

147. P. M. Gschwend, J. K. MacFarlane, and K. A. Newman, *Science*, *227*, 1033–1035 (1985).

148. P. M. Gschwend and J. K. MacFarlane, *Org. Mar. Geochem.*, ACS Symp. Ser. No. 305, 314–323 (1986).

149. G.-R. Helz and R. Y. Hsu, *Limnol. Oceanogr.*, *23*(5), 858–869 (1978).

150. T. Class and K. Ballschmiter, *J. Atmos. Chem.*, *6*, 35–46 (1988).

151. W. Reifenhäuser and K. G. Heumann, *Atmos. Environ.*, *24*(16), 2905–2912 (1992).

152. R. Moore, R. Tokarcyk, and C. Geen, in *The Tropospheric Chemistry of Ozone in Polar Regions*, NATO ASI Series, Subserie I, Global Environmental Changes, Vol. 7, 1993, pp. 235–250.

153. N. Itoh and M. Shinya, *Mar. Chem.*, in press.

154. D. B. Harper, in *Metal Ions in Biological Systems*, Vol. 29 (H. Sigel and A. Sigel, eds.), Marcel Dekker, New York, 1993, pp. 345–388.

155. L. A. Barrie, J. W. Bottenheim, R. C. Schnell, P. J. Crutzen, and R. A. Rasmussen, *Nature*, *334*, 138–141 (1988).

156. J. W. Bottenheim, L. A. Barrie, E. Altas, L. E. Heidt, H. Niki, R. A. Rasmussen, and P. B. Shepson, *J. Geophys. Res.*, *95*, 18555–18568 (1990).

157. B. J. Finlayso-Pitt, F. E. Livingston, and H. N. Berko, *Nature*, *343*, 622–625 (1990).

158. J. C. McConnell, G. S. Henderson, L. A. Barrie, J. W. Bottenheim, H. Niki, C. H. Langford, and E. M. J. Templeton, *Nature*, *355*, 150–152 (1992).

159. O. C. Zafiriou, *J. Mar. Res.*, *33*, 75–81 (1975).

160. Th. Class and K. Ballschmiter, *Fresenius Z. Anal. Chem.*, *327*, 40–41 (1987).

161. A. M. Wuosmaa and L. P. Hager, *Science*, *249*, 160–162 (1990).

162. R. H. White, *J. Mar. Res.*, *40*, 529–536 (1982).

163. J. S. Todd, P. J. Proteau, and W. H. Gerwick, *Tetrahedron Lett.*, *34*, 7689–7692 (1993).

164. H. Vilter and B. Kloareg, unpublished results.

165. S. L. Neidleman and J. Geigert, *Biohalogenation: Principles, Basic Roles and Application*, Ellis Horwood, London, 1986, pp. 156–169.

166. W. W. Mohn and J. M. Tiedje, *Microbiol. Rev.*, *56*, 482–507 (1992).

167. E. de Boer, K. Boon, and R. Wever, *Biochemistry*, *27*, 1629–1635 (1988).

168. H. Vilter and D. Rehder, *Inorg. Chim. Acta*, *136*, L7–L10 (1987).

169. N. J. Campbell, A. C. Dengel, and W. P. Griffith, *Polyhedron*, *8*, 1379 (1989).

170. A. T. Harrison and O. W. Howarth, *J. Chem. Soc., Dalton Trans.*, 1173 (1985).

171. A. P. Makarov, A. E. Gekhman, Y. M. Nekidelov, E. P. Talzi, O. Y. Polotnyuk, K. I. Zamarev, and I. I. Moiseev, *Bull. Akad. Sci. USSR, Div. Chem. Sci.*, *34*, 1764 (1985).

172. H. Mimoun, L. Saussine, E. Daire, M. Postel, J. Fischer, and R. Weiss, *J. Am. Chem. Soc.*, *105*, 3101–3110 (1983).

173. O. Borttolini, F. Di Furia, and G. Modena, *J. Am. Chem. Soc.*, *103*, 3924–3926 (1981).

174. O. W. Howarth and J. R. Hunt, *J. Chem. Soc., Dalton Trans.*, 1388–1391 (1979).

175. S. Celsi, F. Secco, and M. Venturini, *J. Chem. Soc., Dalton Trans.*, 793–795 (1974).

176. R. de la Rosa, M. J. Clague, and A. Butler, *J. Am. Chem. Soc., 114*, 760–761 (1992).

177. B. Galeffi, M. Postel, A. Grand, and P. Rey, *Inorg. Chim. Acta, 129*, 1–5 (1987).

178. P. Peters, H. Vilter, and A. O. Hill, unpublished results.

179. D. Rehder, H. Vilter, A. Duch, W. Priebsch, and C. Weidemann, *Rec. Trav. Chimiques, 106*, 408 (1987).

180. D. Rehder, H. Holst, W. Priebsch, and H. Vilter, *J. Inorg. Biochem., 41*, 171–185 (1991).

181. J. M. Arber, E. de Boer, C. D. Garner, S. S. Hasnain, and R. Wever, *Biochemistry, 28*, 7968–7973 (1989).

182. E. de Boer, C. D. Keijzers, A. A. K. Klassen, E. J. Reijerse, D. Collison, C. D. Garner, and R. Wever, *FEBS Lett., 235*, 93–97 (1988).

183. R. Wever and K. Kustin, *Adv. Inorg. Chem., 35*, 81–115 (1990).

184. H. Vilter, S. Takahashi, K. Sato, M. R. Wälchi, and T. Hosoya, Abstracts of SAMBAS III, Bosen (Saarland, Germany), 1990, p. 46.

185. P. J. Blackshear, in *Methods in Enzymology*, Vol. 104 (W. B. Jakoby, ed.), Academic Press, San Diego, 1984, p. 244.

186. F. Hillenkamp and M. Karas, in *Methods in Enzymology*, Vol. 193 (J. A. McClosky, ed.), Academic Press, San Diego, 1990, pp. 280–285.

187. A. Severin and H. Vilter, unpublished results.

188. M. Kiess and H. Vilter, unpublished results.

189. H. Vilter and M. F. Liaud, unpublished results.

190. V. A. Erdmann, C. Lippmann, C. Betzel, Z. Dauter, K. Wilson, R. Hilgenfeld, J. Hoven, A. Liesum, W. Saenger, A. Müller-Fahrnow, H. Hinrichs, M. Düvel, E. Georg, G. E. Schulz, C. W. Müller, H. G. Wittmann, A. Yonath, G. Weber, K. Stegen, and A. Plaas-Lin, *FEBS Lett., 259*, 194–198 (1989).

191. M. Weiyand, H. Vilter, and D. Schomburg, unpublished results.

192. P. Jordan and H. Vilter, *Electrophoresis, 11*, 653–655 (1990).

11
Vanadium Nitrogenases of *Azotobacter*

Robert R. Eady

AFRC-IPSR Nitrogen Fixation Laboratory,
The University of Sussex, Falmer, Brighton BN1 9RQ,
United Kingdom

1. INTRODUCTION

It is only since 1986 that vanadium has been known to have a role in biological nitrogen fixation, when a vanadium-containing nitrogenase was isolated from mutants of two species of *Azotobacter* unable to synthesize Mo-nitrogenase. Subsequent genetic and physiological studies showed that although all nitrogen-fixing organisms possess Mo-nitrogenase, V-nitrogenases are widely distributed.

Nitrogenases catalyze the conversion of N_2 to NH_3, and are responsible for the cycling of some 10^8 tonnes of N per year from the atmosphere to the soil, thereby making it potentially available to higher organisms. This process of biological nitrogen fixation compensates for the loss of fixed nitrogen in the soil as N_2 and N_2O, the gaseous products produced as a consequence of the reduction of NO_2^- by denitrifying bacteria.

Since the pioneering study of the metal requirement for dinitrogen-dependent growth of the soil bacterium *Azotobacter vinelandii* by the microbial physiologist Bortels in the period 1930–1936, Mo had been accepted to have an essential role in nitrogen fixation [1,2]. Although these and other early studies [3] also showed that V was partially effective in supporting growth, these findings were generally ignored. Subsequent biochemical studies up to 1986 showed that all N_2-fixing organisms had a nitrogenase system based on Mo. Even when isolated from quite different taxonomic groups, this enzyme consisted of two oxygen-sensitive metalloproteins: an Fe protein which acts as a specific ATP-dependent electron donor to a MoFe protein, containing a Fe- and Mo-containing cofactor center (FeMoco), the probable site of binding and reduction of N_2 [4–6]. Genetic studies since 1970 showed that the structural genes of Mo-nitrogenases were highly conserved in different N_2-fixing organisms, and were often clustered with genes involved in FeMoco biosynthesis. Parallel developments in the characterization of defined chemical systems—many based on Mo or W and capable of reducing dinitrogen—fitted well with the dogma that Mo was intimately involved in the enzymic reduction of N_2 [7].

It was a 1980 study of the genetics and physiology of diazotrophic growth of *A. vinelandii*, the organism studied some 50 years earlier by Bortels, which changed this view of the unique role for Mo in nitrogen fixation. The experimental observations were simple: the inability of mutants with defective structural genes for nitrogenase, and therefore unable to use N_2 as a source of nitrogen for growth, surprisingly regained this ability when cultured in media lacking added Mo. The proposal by Bishop and coworkers [8] that an Mo-independent route of N_2 fixation existed was not generally accepted until 1985 when the N_2-dependent growth in chemostat culture of a mutant strain lacking the structural genes encoding Mo-nitrogenase (*nifHDK*) was reported [9]. The growth of this strain was inhibited by the addition of 1 nM Mo to the medium, explaining the long delay in establishing the reality of an Mo-independent route for nitrogen fixation, since following the 1930 observations of Bortel's, Mo had been routinely added to the selective growth media for diazotrophs, thereby preventing synthesis of the Mo-independent nitrogenases. A year later, the V stimulation of growth of a strain of the closely related organism *Azotobacter chroococcum* also lacking the structural genes of Mo-nitrogenase was reported, and V-nitrogenases were purified from this [10] and the equivalent strain *A. vinelandii* [11,12]. A surprising development in 1988 was the purification of a third nitrogenase system which apparently lacks Mo and V, from *A. vinelandii* [13], grown in the absence of added Mo or V. It should be emphasized that it is now known that the three systems, although related, are genetically distinct and are not simply the substitution of different metals into the same enzyme. This early work on the involvement of metals in nitrogen fixation is extensively reviewed in [14].

In this chapter, my intention is to indicate that the V-nitrogenase and Mo-nitrogenases of *Azotobacters* share a great deal in common. The available information for the third nitrogenase system, which is found in *A. vinelandii* and several other species of bacteria, but not *A. chroococcum*, is more restricted, but since the expression of this system is regulated by the availability of V it is discussed in context here. In providing the background of the better characterized Mo-nitrogenase, review references are cited, except for the most recent and directly relevant work where primary citations are made.

1.1. Mo-Dependent Nitrogen Fixation

Diazotrophy, the ability of an organism to use N_2 as a source of nitrogen for growth, is a property shared by many different prokaryotes [15]. Despite the sensitivity of the enzyme to irreversible inactivation by dioxygen and the requirement of Mo-nitrogenase for a low potential electron donor ($E_m \sim -400$ mV) for activity, a wide range of different physiologies are compatible with diazotrophy. Aerobes, anaerobes, photosynthetic bacteria, and cyanobacteria (which conduct an oxygenic plant-type photosynthesis) all have N_2-fixing representative species, and their genetics and biochemistry has been studied extensively. All species which have been investigated have been shown to possess a Mo-nitrogenase system, the presence of which is readily measured in vivo using the alternative substrate C_2H_2 which is reduced specifically to C_2H_4.

Most diazotrophs are free living, but some are capable of forming specific symbiotic associations with plants and make a positive contribution to the N nutrition to the plant symbiont. A product of nitrogenase function is H_2, which is often not detectable in vivo owing to an uptake hydrogenase which recycles the H_2 and conserves energy which would otherwise be lost (see [16]).

1.2. Mo-Independent Nitrogen Fixation

Although the distribution of Mo-independent diazotrophy is apparently quite wide, (see Sec. 1.3) the most detailed information available arises from the study of two species of *Azotobacter* in which the process was discovered. The overall physiology of this system appears to be similar to Mo-nitrogenase activity, in terms of its response to the effect of the ambient dioxygen concentration on activity. There is no information as to the contribution N_2 fixation by this route makes to global N cycling, although some conditions where its contribution by soil microorganisms may be significant, i.e., low pH or low temperature are discussed in Sec. 4.1. A study of the symbiotic association of several legume species with *Rhizobium* and *Bradyrhizobium*, which resulted in the formation of Mo-deficient legume root nodules, found no evidence for their presence under these conditions [17].

1.3. Distribution of Mo-Independent Nitrogenases in Diazotrophs

Soon after the discovery of V-nitrogenase it was found that, in addition to C_2H_4, C_2H_6 was a minor product of the reduction of C_2H_2 [18] and that this product was readily detected in vivo, but only during growth under Mo-deficient conditions. It was subsequently shown that C_2H_6 was also formed by the third nitrogenase [19]; thus C_2H_6 formation from C_2H_2 provides an assay for the presumptive presence of Mo-independent nitrogenases [18]. This assay, coupled with continued growth on subculture in Mo-deficient media and the possible stimulatory effect of added

V on the rate of growth, was initially used to probe the distribution of Mo-independent nitrogenases in diazotrophs. By these criteria *Clostridium pasteurianum* [18] and the filamentous cyanobacterium *Anabaena variabilis* were positive [20,21]. More recently, these approaches have been supplemented by DNA hybridization tests.

TABLE 1

Distribution of V-Nitrogenase and the Third Nitrogenase

Genus/species	*vnfDGK*	*anfDGK*	Ref.
Aerobes			
Azotobacter vinelandii	Sequenced	Sequenced	31
Azotobacter chroococcum	Sequenced	n.d.	32
Azotobacter beijerinckii	Hybridization	n.d.	22
Azotobacter nigricans	Hybridization	n.d.	22
Azotobacter paspali	Hybridization	Hybridization	22,103
Azomonas agilis	Hybridization	Hybridization	22
Azorhizobium caulinodans	Hybridization	n.d.	22
Anaerobes			
Clostridium pasteurianum	n.d.	Sequenced	26
Archaebacteria			
Methanosarcina barkeri[a]	—	—	
Photosynthetic bacteria			
Rhodobacter capsulatus	n.d.	Sequenced	25
Rhodospirillum rubrum[a]	—	—	106
Heliobacterium gestii[a]	—	—	
Cyanobacteria			
Anabaena variabilis	Sequenced	n.d.	23

[a]V stimulates growth of *M. barkeri* under Mo-deficient conditions [105], and growth of *H. gestii* is stimulated by Fe under Mo-deficient conditions and C_2H_6 is a product of C_2H_2 reduction [104]. In addition to the DNA hybridization data, V stimulates growth and C_2H_6 is a product of C_2H_2 reduction by *A. paspali* [103], the third nitrogenase has been purified from *R. rubrum* [106]; n.d. = not detected.

The genes for the Mo- and V-nitrogenases in *A. chroococcum* and those of the Mo-, V-, and third nitrogenase systems of *A. vinelandii* had been cloned and sequenced (see Sec. 2.1). The availability of the DNA of Mo-independent nitrogenases of *Azotobacter* enabled various organisms to be screened by hybridization of their genomic DNA to *vnf* and *anf* DNA, probes of V- and third nitrogenase, respectively [22]. V-nitrogenase appears to be widespread among the family Azotobacteriaceae, but not in a range of other aerobic and microaerobic bacteria including several strains of *Pseudomonas*, *Bradyrhizobium*, and *Rhizobium*. Of the species with a presumptive V-nitrogenase, half have DNA which hybridizes to *anf* DNA.

Genomic DNA of the filamentous cyanobacterium *A. variabilis* hybridized to *vnf DGK* probe of *A. vinelandii* [23], a finding which enabled the cloning and sequencing of the *vnfDGK* genes. Despite the growth experiments in Mo- and V-deficient media which suggested the presence of a third nitrogenase [24], no hybridization to an *anf* DNA probe was detected. Table 1 summarizes the distribution and criteria used to assign the presence of Mo-independent nitrogenase in a range of organisms which have been screened.

The *anf* genes of the photosynthetic bacterium *Rhodobacter capsulatus* [25] and the anaerobe *C. pasteurianum* [26] have also been cloned and sequenced. The availability of three gene sequences for V-nitrogenase and also of the third nitrogenase from different organisms make comparisons with the large number of Mo-nitrogenase sequences more meaningful as discussed in Sec. 2.1. The unambiguous identification of these Mo-independent nitrogenases from DNA sequences now appears possible and establishes the widespread distribution of these nitrogenase systems.

2. GENETICS OF NITROGEN FIXATION

In this section, the genetics of nitrogen fixation is considered from (1) the viewpoint of the organization and sequencing of the structural genes of the three nitrogenases, (2) other genes required for all three systems, and (3) the regulation of the expression of nitrogenases by the availability of Mo and V.

The organization of the nitrogen fixation genes of five free-living diazotrophs—*Klebsiella pneumoniae*, *Enterobacter agglomerans*, *A. vinelandii*, *A. chroococcum*, and *Rhodobacter capsulatus*—are now defined in some detail. Comparative studies indicate that some 14 common genes form the basic set required for Mo-dependent nitrogen fixation (see [27]). Many of these genes have been assigned general roles, although the details of their biochemical function remain obscure, being defined at the level of the biochemical complementation pattern of extracts of mutants defective in nitrogenase synthesis [28]. Such studies have shown that *nifH*, *E*, *N*, and *B* are essential for FeMoco biosynthesis, since mutation of any one of them results in the synthesis of an inactive apoMoFe protein lacking FeMoco, and mutation of *nifV* results in the formation of a species of MoFe protein with an altered substrate specificity, due to the incorporation of citrate rather than homocitrate into FeMoco [29].

2.1. Organization of the Structural Genes of Nitrogenases

The α and β subunits of the MoFe proteins are encoded by *nifD* and *nifK* genes, and the γ subunit of the Fe protein by *nifH*. The *nifH* genes have been cloned and sequenced from some 20 diazotrophs and show a high degree of conservation of sequence, even when the eubacteria and the methanogenic archaebacteria are compared [30].

The genes of V-nitrogenase have been cloned and sequenced from *A. vinelandii* [31], *A. chroococcum* [32], and *An. variabilis* [23]. The genes of the α and β subunits of VFe protein *vnfDK* are separated by a gene (*vnfG*) in *A. vinelandii* and *A. chroococcum* which encodes a third small (δ) subunit of the VFe protein in these organisms [32]. In *An. variabilis* this coding region is fused with the α subunit encoded by *vnfD* [23]. Although the function of the small subunit is not known, it has no counterpart in the MoFe proteins. Mutations of *vnfG* result in the inability of V to support N_2 fixation [33].

The genes required for a functional Mo-nitrogenase in *A. vinelandii* are located in two unlinked regions of the genome, the major region is 29 kb of DNA containing *nifHDKTY*, *nifENX*, *nifUSVP*, *nifWZ*, *nifM*, and *nifF* and a smaller 6 kb section containing *nifBQ* and *nifAL*

[34]. A detailed discussion of the regulation of these genes was recently reviewed [27] and will not be reiterated here. In the context of this review, the role of NifA as a positive activator of *nif* gene expression is summarized in Sec. 2.4 and compared with VnfA and AnfA, the activators required for the expression of V-nitrogenase and the third nitrogenase, respectively.

Genes specific for V-nitrogenase are located on a 14-kb region containing *vnfA*, *vnfENX* separated by *vnfH*, and *vnfDGK*, and those for the third nitrogenase on a different fragment of DNA containing *anfA*, and *anfHDGK* [33]. As discussed below in Sec. 2.3, some *nif* genes are required to form functional V- and third nitrogenases.

2.2. Amino Acid Sequence Comparisons of MoFe and VFe Proteins

When DNA sequence data became available for a number of MoFe proteins, one of the most striking features in the derived amino acid sequences was the conservation of a number of invariant cysteine residues, five in the α subunit and three in the β subunit. These were clearly candidates for being potential ligands to the FeS centers that these proteins contain and were the subject of early directed mutagenesis studies [35,36]. Such studies implicated seven of these residues, an assignment confirmed subsequently by the X-ray crystal structure. Six of the cysteine residues, three from each subunit, are involved in binding the P clusters and a single cysteine residue together with a single histidine residue bind FeMoco to the MoFe proteins Av1 [37] and Cp1 [38].

These advances are directly relevant to the evaluation of data on the amino acid sequences of VFe proteins of *A. vinelandii* and *A. chroococcum*, and the redox centers that they contain. Comparison of the amino acid sequences of the VFe proteins with those of the MoFe proteins shows the α subunits to be 32% and the β subunits 35% identical in sequence. In particular, the cysteine residues which ligate the P clusters and FeMoco and the cysteine and histidine residues which bind FeMoco which are invariant in all MoFe protein sequences are also conserved in the VFe proteins [39]. As discussed in Sec. 3.2, spectroscopic and extrusion data indicate the presence of homologous

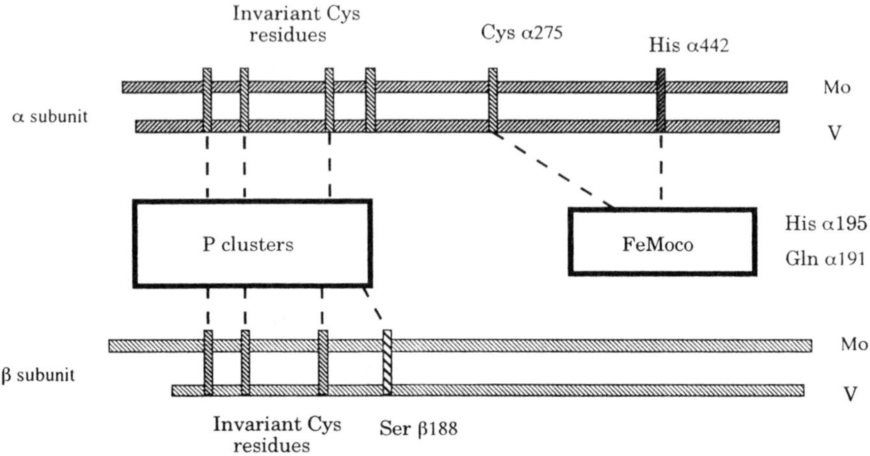

FIG. 1. Amino acids conserved in MoFe proteins, VFe proteins and FeFe proteins known to bind FeMoco and P clusters in the MoFe proteins. The horizontal bars represent the subunit polypeptides of MoFe and VFe proteins and the vertical bars those amino acid residues which the X-ray structure shows to bind FeMoco and P clusters in Av1 [56] and Cp1 [68]. The His195 and Gln191 residues of the α chain are conserved in both proteins and mutation of these residues changes the substrate specificity of the MoFe protein [28].

centers in the VFe proteins and Fig. 1 shows the suggested mode of binding of these clusters.

An additional feature which comparison of the α-subunit sequences reveals is that the amino acids around the conserved histidine residue which ligates to FeMoco are different in the VFe proteins and also in the corresponding component of the third nitrogenase. In each of the three organisms for which sequences have been determined (see Table 1) these residues are differently conserved (Table 2). As a consequence, this region appears to provide a fingerprint for differentiation between *nif*, *vnf*, or *anf* genes, which otherwise can be difficult. An exception to this high degree of conservation is observed in the case of Cp1, which shows substitutions in four out of eight of the flanking residues. However, the clostridial *anf* sequences do conform to the consensus sequence.

TABLE 2

Conserved Amino Acid Sequences in MoFe Proteins, VFe Proteins, and the Third Nitrogenase Flanking His442 of the α Subunit

Genus/species	Sequence	Ref.
Nif		
A. vinelandii	FRQM**H**SWDY	
R. capsulatus	FRQM**H**SWDY	
K. pneumoniae	FRQM**H**SWDY	
P. boryanum	FRQM**H**SWDY	
C. pasteurianum	SKQL**H**SYDY	
Anf		
A. vinelandii	YLNA**H**AYH	46
R. capsulatus	YLNA**H**AYH	25
C. pasteurianum	YLNA**H**AYH	26
Vnf		
A. vinelandii	YVNG**H**GYH	31
A. chroococcum	YVNG**H**GYH	32
An. variabilis	YVNG**H**GYH	23

The sequences shown for the MoFe protein α subunits are representative of a larger number of sequences which have been determined; see [34].

2.3. Genes Involved in Both Systems

The potential role for some *nif* genes in the formation of active Mo-independent nitrogenases has been tested by mutational analysis. This approach established that *nifM* [40], *U*, *S*, *V* [41], and *B* [42,43] were required for activity. Although the biochemical roles of the products of these genes is largely unknown, as discussed above in Sec. 2.1, the effect of their mutation on Mo-nitrogenase has been established.

The *nifM* gene product is involved in the maturation of the Fe protein, and given the similarity of structure and function of the Fe proteins associated with the Mo-independent systems (see Sec. 3.2.1), this involvement is not unexpected [40].

It has been proposed that *nifS* and *nifU* are involved in FeS cluster biosynthesis [44]. The *nifS* gene product has been purified and shown to be a pyridoxal phosphate-containing L-cysteine desulfurase which activates sulfur by the formation of an enzyme-bound persulfide [45]. Since mutation of this gene resulted in a lowered activity of both MoFe protein and Fe proteins, it is probable that it serves as the sulfide donor in the formation of FeS clusters of Mo-nitrogenase components. An analogous role in the formation of active Mo-independent nitrogenases is understandable given that these systems contain FeS centers which appear homologous to those of Mo-nitrogenase.

The involvement of two genes, *nifB* [42] and *nifV* [43,44], in the formation of active Mo-independent nitrogenases is more instructive. Both are involved in the formation of active FeMoco, *nifB* in some unknown role, and *nifV* (which amino acid sequence homology indicates almost certainly encodes a homocitrate synthase) to catalyze the condensation of α-oxoglutarate with acetyl CoA to form homocitrate, a ligand to Mo in FeMoco (see Sec. 3.1.3). The involvement of these genes in the formation of an active V- and third nitrogenases provides very strong genetic evidence for the presence of cofactors analogous to FeMoco, and which has homocitrate as a component, in these enzymes.

Other genetic data suggesting the presence of cofactors analogous to FeMoco in the Mo-independent nitrogenases is provided by the presence of a reiteration of *nifEN*-like genes in *A. vinelandii* [46] and *A. chroococcum* [47]. *NifEN* are essential for FeMoco biosynthesis, and the products of these genes, which show homology with *nifDK*, have a suggested role as a template on which FeMoco is synthesized [44]. The reiterated genes are part of the *vnfAEN* operon, and are essential for the function of V- and the third nitrogenase [48]. In addition, mutation of *nifEN* of *R. capsulatus* indicates that they are not required for the functioning of the third nitrogenase [49], suggesting the presence of as yet undetected reiterations of *EN* genes in this organism.

2.4. Regulation of Nitrogen Fixation Genes by Mo and V in *Azotobacter*

The regulation of the expression of Mo-nitrogenase by the availability of fixed N and dioxygen is understood in considerable detail and surprisingly, given the metabolic diversity of many of the diazotrophs investigated, has common features in a range of organisms (see [27]). In contrast, studies on the regulatory role of Mo and V in the expression of nitrogen fixation genes are in their infancy.

It has been established that a Mo concentration of 25 nM is required for the expression of the structural genes of Mo-nitrogenase [50]. At this level, the onset of repression of V-nitrogenase is observed, although in the range of Mo concentrations of 25–100 nM mRNA transcripts of both Mo- and V-nitrogenase are detectable. Only under conditions of limiting Mo does the presence of V allow synthesis of V-nitrogenase and prevent the expression of the third nitrogenase [51,52].

Transcriptional activation of Mo-nitrogenase regulons is regulated by the availability of NH_3, and *nif* promoters have a consensus sequence recognized by RNA polymerase with a σ subunit (σ^{54}) associated with the core polymerase. The model for the regulation of *nif* expression is briefly summarized below since some features appear to be common to the regulation of V- and the third nitrogenase.

Transcription of *nif* genes is modulated by an upstream activator binding site some 100 bp upstream from the point at which transcription is initiated. The *nifA* gene encodes a specific activator which binds to the upstream activator site. DNA bending in this region is enhanced by an integration host factor protein which enables nifA bound to the activator site to interact with the σ^{54} subunit of RNA polymerase bound to a *nif* promoter and thereby initiating transcription. Since expression of the σ subunit of RNA polymerase is constitutive, regulation of *nif* expression by NH_3 and dioxygen is controlled by modulation of *nifA* gene expression, or by modulating the activity of the NifA protein (see [27,53]).

NifA is homologous with a number of σ^{54}-dependent activator proteins comprising a C-terminal sensing domain, a central ATP binding domain, and an N-terminal DNA binding domain with a characteristic helix-turn-helix motif, the sequence of which determines the specificity of the interaction with DNA [53]. NifA binding sites are absent in the promoter sequences of the structural genes of alternative nitrogenases and the expression of V-nitrogenase is activated by the product of *vnfA* and the third nitrogenase by *anfA* [39,52]. These proteins have central domains which are highly homologous with NifA but have different DNA recognition sequences. The binding site of VnfA to DNA upstream of the *vnf* promoters has been identified and is different from that of the NifA binding site.

In wild-type *A. vinelandii* expression of the structural genes of Mo-nitrogenase does not occur in the absence of Mo. However, mutation of *vnfA* results in the synthesis of Mo-nitrogenase in the absence of Mo, whether V is present or not, indicating a role for VnfA in the repression of *nif* genes when Mo is absent [54]. In the presence of Mo, *nifHDK* is transcribed due to activation by NifA, but *vnfH* and *vnfDGK* are not, since *vnfA* is not transcribed. In some way Mo modulates the expression of these genes mediated through the transcriptional activators NifA and VnfA [39,55].

In the absence of Mo, *vnfH* is expressed irrespective of the presence or absence of V. However, V is required for maximum levels of transcription of *vnfDGK* and for the accumulation of VFe protein polypeptides.

V prevents the transcription of *anfHDGK*, but full repression requires the presence of an active VFe protein. The full expression of this operon requires *vnfH* to form an active Fe protein. It has been proposed that AnfA is reduced by the Fe protein when VFe protein is absent and that reduced AnfA activates transcription of the third nitrogenase.

3. BIOCHEMISTRY OF NITROGEN FIXATION

3.1. Mo-Nitrogenase

The properties of Mo-nitrogenase described in this section are not a comprehensive review, having been selected for purposes of comparison with the more restricted range of data available for V-nitrogenase. The recent X-ray crystal structures of the MoFe proteins Av1 and Cp1 at 2.2 Å resolution [56,57] revealed the tertiary structure of the $\alpha_2\beta_2$ tetramer and provided structures for the redox centers that they contain. This information enables a more meaningful comparative interpretation to be made with data obtained for V-nitrogenase than was possible before these structures were known. In particular, it enables the structure of the V-containing cofactor of the VFe proteins to be predicted with some certainty.

3.1.1. Fe Protein: Structure and Function

The Fe proteins of Mo-nitrogenase are encoded by *nifH*, a gene which has been cloned and sequenced from some 20 diazotrophs. The sequences are remarkably homologous and indicate that the Fe proteins are a highly conserved family. Several Fe proteins have been purified and characterized. They are γ_2 homodimers of $M_r \sim 64000$ and have a single 4Fe4S cluster which typically has a midpoint potential of −300 mV, normally cycling between the +1 and +2 oxidation levels. The cluster is unusual, in existing as a physical mixture of two spin states of $S = 1/2$ and $S = 3/2$, detectable by electron paramagnetic resonance (EPR) spectroscopy. The proteins are extremely sensitive to irreversible inactivation by dioxygen with a $t_{1/2}$ of 45 sec. The binding of MgADP (K_d = 10 μM) or MgATP (K_d = 100 μM) results in a conformational change of the protein and a decrease in the midpoint potential of the cluster by ~100 mV. In studies of the ATP-dependent electron transfer from the Fe protein to the MoFe protein, the changes in the absorbance of the 4Fe4S cluster at 420 nm and in the $g = 1.94$ EPR spectrum of the reduced Fe protein have been used extensively.

The structure of the Fe protein from *A. vinelandii* has been determined at 2.9 Å resolution by X-ray crystallography [58]. This showed each of the subunits to be folded in a single αβ domain and confirmed the location of the cluster at the subunit interface, as had been inferred from chemical modification studies [59]. The cluster is symmetrically coordinated by two cysteines from each subunit and located at one end of the molecule. It is exposed to the solvent to a much greater extent than is found in the ferredoxins, and may account for the extreme sensitivity of the Fe proteins to inactivation by dioxygen.

Comparison of the amino acid sequence and the secondary structure of the folding of the polypeptide chain of the Fe protein subunits with $p21^{ras}$, a GTPase which couples nucleotide hydrolysis to stabilize different conformational states, has led to the proposal that these nucleotide-binding proteins utilize a common signal transducing system triggered by nucleotide binding and hydrolysis [58]. This proposal is consistent with the location of the two ATP/ADP binding sites in the cleft formed between the two subunits, some 20 Å from the 4Fe4S center. A separation of this magnitude implies that the decrease in redox potential of the cluster observed when ATP or ADP bind is mediated by a conformational change in the protein. However, it should be noted that in the case of the Fe proteins ATP hydrolysis does not occur in the absence of the MoFe protein.

A number of amino acid residues on the face of the Fe protein where the 4Fe4S cluster is located have been implicated, from chemical crosslinking studies and the effects of covalent modification on activity [60,61], as forming part of the contact surface involved in complex formation with the MoFe protein. The crystal structures of Av2 and Av1 have been used to model complex formation between the two proteins, and the docking model developed from these studies envisages the surface of the Fe protein surrounding the 4Fe4S cluster being a contact surface to the exposed surface of the pseudosymmetric $\alpha\beta$ interface of the MoFe protein [62].

An unexpected role for the Fe protein as an essential component in the assembly of FeMoco has recently been established [28]. Studies on the effect of mutations on this requirement have revealed that the Fe protein does not have to be active as an electron donor or in binding MgATP. To effectively participate in this biosynthetic role the 4Fe4S cluster appears to be required, since mutation of one of the cysteinyl ligands which normally binds the cluster results in a mutant unable to form FeMoco.

3.1.2. MoFe Protein: Structure and Function

MoFe proteins have been purified from a number of different organisms and are $\alpha_2\beta_2$ tetramers of M_r ~240 000. They are complex proteins which contain, in preparations with highest activity, 2 Mo, and 30–34 Fe, and 30 acid-labile sulfur atoms (Table 3). The arrangement of these atoms and the types of redox centers present in MoFe proteins were subject to extensive spectroscopic and chemical studies during the 1970s and 1980s [4–6]. As isolated, in the dithionite-reduced state, MoFe proteins are characterized by an EPR signal arising from an $S = 3/2$ spin system. This EPR signal, with g values close to 4.3, 3.7, and 2.01, is a sensitive spectroscopic fingerprint of FeMoco. Mössbauer spectroscopy indicated that approximately half the 30–34 Fe atoms present in MoFe proteins were associated with this center and the remainder associated with the P clusters. Such studies showed that MoFe proteins isolated from different organisms all contained two types of metal sulfur cluster which were unique to nitrogenase: the P clusters which are Fe/S clusters with unusual spectroscopic properties, and FeMoco, an extractable iron- and molybdenum-containing cofactor.

The spectroscopic properties of P clusters in the MoFe protein have no counterpart in other Fe/S proteins studied. Recent low-temperature

TABLE 3

Comparison of the Physicochemical Properties of MoFe Proteins and VFe Proteins

	Kp1	Ac1^{v}	Av1^{v}
Property			
Native M_r	225000	240000	240000
Subunit structure	$\alpha_2\beta_2$	$\alpha_2\beta_2\delta_2$	$\alpha_2\beta_2\delta_2$
Subunit M_r	2 × 54100	2 × 53800	2 × 53800
	2 × 58300	2 × 52700	2 × 52700
		2 × 13300	2 × 13300
Metal and S^{2-} content (g atom/mol)			
V	0.06	2 ± 0.3	0.7 ± 0.3
Mo	2	0.06	0.05
Fe	32 ± 3	21 ± 1	9 ± 2
S^{2-}	n.d.	19 ± 0.2	21 ± 1
EPR *g* values for dithionite-reduced protein	4.32, 3.73, 2.018	5.6, 4.35, 3.77, 1.93	5.31, 4.34, 2.01, 1.93
Specific activity (nmol product formed/min/mg of protein)			
NH_3 formation	990	350	660
H_2 under N_2	648	928	n.d.
C_2H_4 from C_2H_2	1693	608	220
H_2 evolution under C_2H_2	304	998	n.d.
H_2 evolution under Ar	2100	1348	1400
Ref.	[106]	[74]	[11]

n.d. = not determined.

magnetic circular dichroism (MCD) and Mössbauer spectroscopy coupled with EPR studies have defined characteristic properties of these unique centers which can achieve more than the two redox levels accessible to simpler Fe/S centers [63–66]. Useful comparative features are the unusually steep magnetization curves of the MCD transitions of P centers in dye-oxidized proteins, and a range of EPR signals which arise from oxidized species in samples of MoFe- and VFe proteins poised at redox potentials in the range −500 mV to +100 mV. The consensus view is that the oxidation of the P clusters occurs in a series of one-electron steps leading to species with characteristic paramagnetic properties. A scheme for the redox states of P centers based on these studies is shown below:

$$P^{N}\ (S = 0) \leftrightarrow P^{\text{semi-ox}}\ (S = 1/2;5/2) \leftrightarrow P^{\text{ox1}}\ (S = 3 \text{ or } 4) \leftrightarrow P^{\text{ox2}}\ (S = 1/2;7/2)$$

P^N is the diamagnetic species present in dithionite reduced protein, $P^{\text{semi-ox}}$ is a physical spin mixture of $S = 1/2$ and $S = 5/2$ [64], P^{ox1} is a non-Kramer's paramagnetic state with a $g = 12$ excited state [63,65,66], most easily detected in parallel mode EPR, and P^{ox2} is a physical spin mixture of $S = 1/2$ and $S = 7/2$ [66,67].

As discussed below (Sec. 5), EPR signals arising from some of these oxidation levels are detectable during enzyme turnover and presumably arise from catalytically competent species. For many years the consensus was that the P centers of MoFe proteins were four unusual 4Fe4S clusters [4–6]. However, in 1987, when the $S = 7/2$ EPR signals arising from P centers of dye-oxidized MoFe protein were first detected, they were proposed to arise from larger centers containing 6–8 Fe atoms [67]. Subsequently, the first low-resolution X-ray crystallographic data for any MoFe protein showed that in Cp1 the P clusters were too large to be isolated 4Fe4S clusters [38]. The P clusters and the FeMoco centers were also shown to be some 19 Å apart and the two cofactor centers 70 Å apart. A subsequent 2.7-Å resolution X-ray structure of Av1 confirmed this separation of the redox centers and allowed structural models of P clusters and FeMoco to be developed [37]. The structure of P clusters and the FeMoco centers in Av1 and Cp1 have now been refined to 2.2 Å [56,68], a resolution which allows a definitive assignment to atoms in these centers.

The X-ray structure shows that the P clusters are located at the α/β subunit interface, buried some 12 Å from the surface, and ligated by cysteines from both subunits. They are made up of two 4Fe4S clusters linked by bridging cysteine thiolate groups (see Fig. 2A). In addition, a bridging sulfur-sulfur bond in the P clusters is observed in a 2.2-Å-resolution structure of Av1. In the case of Cp1 a similar structure has been proposed [57], but also a structure whereby a single sulfur atom is shared by the cubanes [68]. The bridged double-cubane structure of the

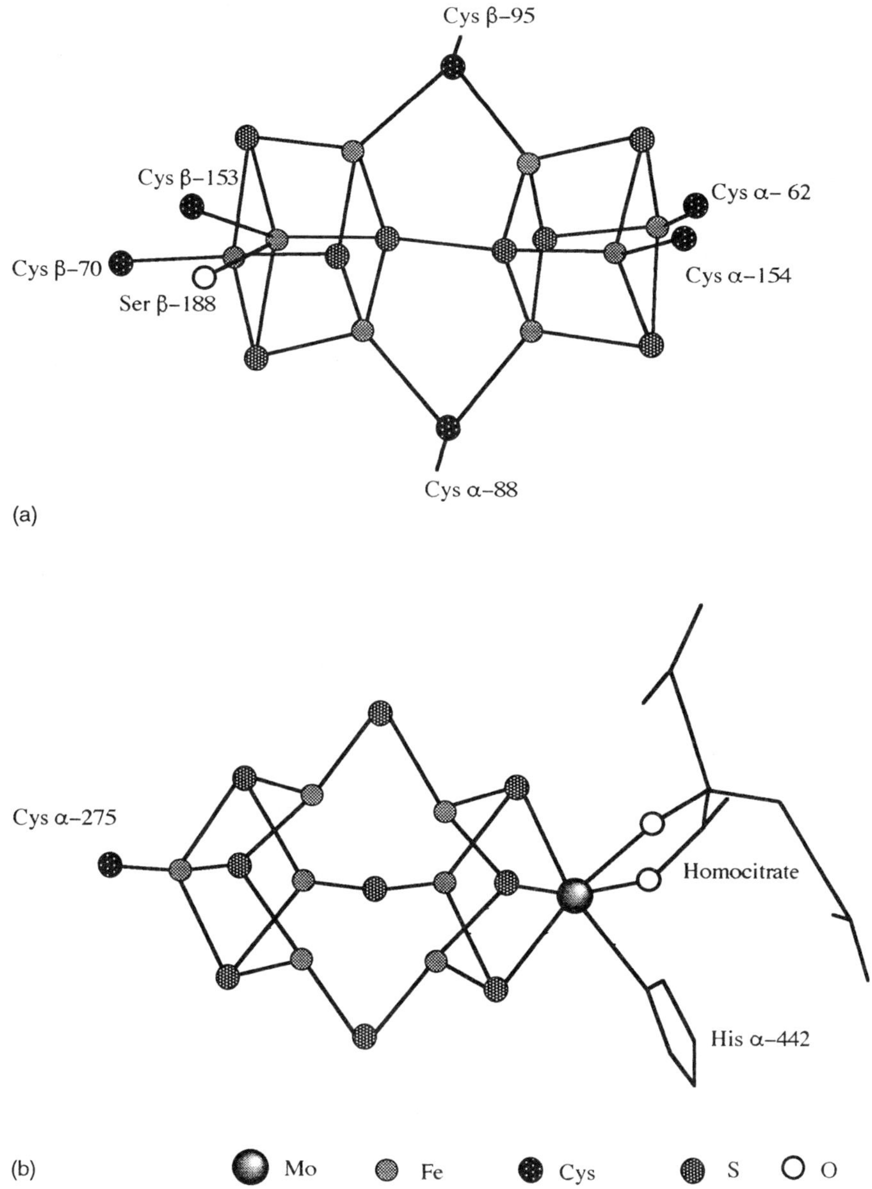

FIG. 2. (a) Schematic representation of the P cluster of Av1 [56]. (b) Schematic representation of the FeMoco center of Av1 [56]. (Generated from coordinates kindly provided by Prof. D. Rees.)

P clusters, and the possible structural variant of a shared sulfur atom readily accounts for their unusual spectral properties and their ability to undergo multielectron redox processes. These centers in Kp1 were recently shown to become transiently oxidized during the reduction of N_2 [69], providing an experimental basis for their long-suggested role as electron capacitors.

3.1.3. FeMoco: Structure and Role in N_2 Reduction

Mo is present in the MoFe proteins as a component of an iron- and molybdenum-containing cofactor (FeMoco) which gives rise to the characteristic $S = 3/2$ EPR spectrum of the dithionite-reduced protein. In 1977 FeMoco was shown to be extractable into *N*-methylformamide (NMF) in a form capable of activating the inactive apoMoFe protein isolated from mutant strains unable to synthesize FeMoco [70]. Subsequent studies showed that in this type of reconstitution experiment FeMoco transferred the *nifV* phenotype of forming an acetylene reductase but a poor nitrogenase to the MoFe protein, providing the strongest evidence that FeMoco formed the active site of nitrogenase [71].

Genetic studies established that MoFe protein was not necessary for the synthesis of FeMoco, a fact that enabled an in vitro assay for the gene products and factors necessary for its synthesis to be developed [44,72]. One important discovery arising from such studies was that (*R*)-homocitrate (2-hydroxy-1,2,4-butanetricarboxylic acid) was a component of FeMoco, and that analogs or fluorinated derivatives led to the assembly of an MoFe protein with altered substrate specificity.

Before the X-ray structure of MoFe protein was known, the chemical environment of the Fe and Mo atoms had been investigated by X-ray absorption spectroscopy. These data are presented in Table 3, where they are compared with the data for V in VFe proteins. They provide the most accurate estimate of the nearest-neighbor distances of atoms in the first coordination sphere of Mo and V. It is clear from these data that Mo in FeMoco of the MoFe proteins occupies a very similar environment to V in the FeVaco center of the VFe proteins.

The X-ray crystal structures of Av1 and Cp1 show that FeMoco is anchored to the protein by α-Cys-275, which is a thiolate ligand to an Fe atom at one end of the cofactor, and by α-His-442 where the imidazole N atom is coordinated to the Mo atom. The initial models for FeMoco at 2.7

Å resolution [37] were constrained by atomic for metal-sulfur distances derived from extended X-ray absorption fine structure (EXAFS) studies of the MoFe protein and isolated FeMoco, and also by model cluster data. The structures of Av1 [56] and Cp1 [68] were subsequently refined to a 2.2-Å-resolution difference map which confirms that FeMoco is made up of two M Fe3 S3 (M = Mo or Fe) clusters, linked by three bridging sulfides with (*R*)-homocitrate coordinated to the Mo atom by its 2-hydroxy and 2-carboxy groups (Fig. 2B). This higher resolution refinement gives average Fe--Fe distances between the bridged Fe sites of 2.5 Å suggestive of some Fe--Fe bonding interactions that could provide a fourth coordination interaction for the bridging iron atoms [56]. The essentially identical structure proposed for the FeMoco center of Cp1 [38] is based on the analysis of anomalous dispersion data where the interpretation of the electron density maps of the clusters was not constrained by other spectral information.

It has been suggested that N_2 binds to the center of FeMoco replacing the weak FeFe bonds with multiple Fe--N bonds [38]. To account for the cavity being some 0.5 Å too small to allow this mode of binding it is proposed that the more reduced states of the cofactor, generated during enzyme turnover, and to which N_2 binds, has an increased separation distance. However, other alternative substrates and reduction intermediates of N_2 are too large to be bound in this manner. The X-ray crystal structure determination of two MoFe proteins at 2.2 Å resolution is a major achievement which will change the direction of future research, but it has not yet enabled the nature of subtle interactions of FeMoco and the polypeptide, essential for N_2 reduction, to be identified.

3.2. V-Nitrogenase

The biochemical data for V-nitrogenases arises from the study of the enzymes from *A. vinelandii* and *A. chroococcum* and, as expected from the DNA sequence homology of their structural genes, these two V-nitrogenases have very similar properties (see [73]). Since their discovery, much of the work on these enzymes has focused on the VFe protein to define how the presence of V changes the structure and function compared with the more extensively studied Mo-nitrogenase.

3.2.1. Fe Protein: Structure and Function

The *vnfH* genes of *A. vinelandii* and *A. chroococcum* have been cloned and sequenced [31,32]. They are 90% identical and show a high degree of conservation with the *nifH* sequences of other organisms, and they retain the cysteine residues which ligate the 4Fe4S cluster of the Fe protein of Mo-nitrogenases as well as the N-terminal ATP/ADP binding motif. The Fe proteins of V-nitrogenase have been purified and shown to be typical members of this class of highly conserved proteins as described in Sec. 2.3.3. They contain a 4Fe4S center which EPR spectroscopy shows to exhibit mixed spin states. The midpoint potential of the cluster in the $Ac2^{V}$·MgADP species at −463 mV compares with −450 mV for $Ac2^{Mo}$·MgADP. When oxidized, these nucleotide-bound species are reduced in a one-electron reaction by dithionite at very similar rates. The binding of MgATP results in a conformational change of the protein which makes the Fe atoms of the cluster more accessible to the chelator 2,2′-bipyridyl [73].

3.2.2. VFe Protein: Structure and Function

VFe proteins have been purified from only two species of *Azotobacter*; they are $\alpha_2\beta_2\delta_2$ hexamers of $M_r \sim 250\,000$. They are complex proteins which contain, in preparations with highest activity, 0.7–2 V atoms, 9–19 Fe atoms, and 20 acid-labile sulfur atoms [11,74] (Table 3). Although these metal contents are generally lower than is currently found in a typical preparation of MoFe protein, this difference is probably not significant since in the case of MoFe proteins comparable values were common 15 years ago, but have increased with time.

The conservation of the amino acid residues involved in binding P clusters and the FeMoco center of the MoFe proteins (see Sec. 2.2), and the requirement for *nifV* and *nifB* for activity (see Sec. 2.3) together provide a very strong presumptive case for the presence of homologous redox centers in the VFe proteins. The types of redox centers present in VFe proteins have been investigated by EPR [74–76], MCD [77], Mössbauer [78], and X-ray absorption spectroscopies [79–83], in order to compare them with those of MoFe proteins.

3.2.2.1. EPR Spectroscopy. As isolated in the dithionite-reduced state, MoFe proteins are characterized by a rhombic EPR signal arising from an $S = 3/2$ spin system with g values close to 4.3, 3.7, and 2.01, which has proved to be a sensitive spectroscopic fingerprint for FeMoco. In addition to this signal, a spin $S = 1/2$ with $g_{av} = 1.92$ is often observed and has been assigned to arising from a de-molybdo species of the protein which is catalytically inactive. The EPR spectra of the VFe proteins are more complex since several paramagnetic species are present but exhibit signals assignable to a putative cofactor center. When isolated from *A. vinelandii* the VFe protein shows a broad, poorly resolved EPR signal with g values at 5.8 and 5.4 which has been assigned to transitions from the ground state of a spin $S = 3/2$ Kramer's doublet which integrate to 0.89 spins per V atom. The protein from *A. chroococcum* is more complex and difficult to analyze, since the g values at 5.6, 4.3, and 3.77 appear to arise from a mixture of $S = 3/2$ species, presumably reflecting different environments of the cofactor center. Oxidative titration showed this center to have an E_m of -125 mV.

Both proteins also exhibit an axial EPR signal with g values of 2.04 and 1.93 assigned to a $S = 1/2$ ground state spin system, which integrates to ~0.2 spin/mol. In dye oxidation redox titrations this signal is bleached before the $S = 3/2$ signals indicating that this center has a lower redox potential. In the case of the protein from *A. chroococcum* the intensity of this signal does not correlate with specific activity, being low in samples of high activity [84]. It has been assigned to an FeS center present in an inactive de-vanadio species. However, this signal in the protein from *A. vinelandii* does remain at a constant ratio to the $S = 3/2$ signal on further purification which resolved two species with polypeptide patterns, and it is thought to be an integral part of this VFe protein [85]. Given the similarity of these proteins at the amino acid sequence level, this difference is surprising and further work will be required to resolve this question.

A third signal, detected in rapid adiabatic passage dispersion EPR, is present in the spectra of the protein of *A. vinelandii* [75]. At 2 K and high power to saturate the signals a broad structureless absorption envelope dominates the spectrum, and it has been suggested to arise from a paramagnetic center coupled to a metal center. This signal is not seen in samples of MoFe protein under the same conditions.

Oxidative redox titrations of MoFe proteins studied by EPR spectroscopy have defined a number of possible oxidation states of the P clusters in these proteins (see Sec. 3.1.3). Similar titrations of VFe proteins have shown that EPR signals characteristic of the P^{ox1} [64] (a physical spin state mixture of $S = 1/2$ and 3/2) and P^{ox2} states [86] (with spin $S = 3$ or 4) of the P clusters of the MoFe protein. The differences in the E_m at which these different signals arise is presumably due to small variations in environment of the cluster, since they show species variation even among the MoFe proteins. The $g = 12$ EPR signal arising from the P^{ox2} state, which was most easily detected in the parallel mode, appears to be a unique and characteristic spectroscopic fingerprint for P clusters.

In summary, EPR studies of the VFe proteins in the dithionite-reduced state provide evidence for the presence of a cofactor center containing vanadium (FeVaco), and studies on the proteins in different oxidation states, for their containing P clusters. The significance of other signals, at least one of which has no counterpart in the MoFe proteins, remains to be established.

3.2.2.2. Mössbauer Spectroscopy. Low-temperature Mössbauer spectra of the $S_2O_4^{2-}$-reduced ^{57}Fe-substituted VFe protein from *A. vinelandii* [78] show features due to a paramagnetic species assigned to the $S = 3/2$ spin species of reduced FeVaco and three other species of iron assigned to P clusters in a diamagnetic state. The data are very similar to those for MoFe proteins, and the magnetic spectrum of the VFe cofactor can be decomposed into seven components, consistent with there being a close similarity between the structures of the FeMoco and FeVaco centers.

3.2.2.3. Magnetic Circular Dichroism Spectroscopy. Low-temperature MCD monitors the magnetically induced dichroism of the Fe--S charge-transfer band of paramagnetic Fe/S centers which dominate spectra measured at low temperature. Studies of the redox centers of MoFe proteins were facilitated by the fact that the paramagnetism of FeMoco resulted in its dominating the spectrum of $S_2O_4^{2-}$-reduced protein, and that of P clusters spectra of dye-oxidized protein. The spectra of $S_2O_4^{2-}$-reduced VFe protein of *A. vinelandii* [77] show transitions arising from the ground state of an $S = 3/2$ species. Although there are differences in the intensity and frequency of the temperature-dependent MCD transitions when compared with the MoFe proteins, such differences have been attributed to changes in the zero field splitting parameters of the system. These findings supported the EPR data which suggested the presence of a similar type of cofactor center in the MoFe and VFe proteins.

Early MCD studies on the P clusters of oxidized MoFe protein and VFe protein of *A. vinelandii* used the dye thionine as an oxidant. It is now known from EPR and Mössbauer studies that although oxidation can proceed in one-electron steps under the conditions of the early experiments, the proteins would have been in the P^{ox1} state, as was subsequently confirmed by combined MCD/EPR study [63]. The MCD spectra were similar in form, with very steep magnetization curves, and were interpreted as arising from a paramagnetic site of $S = 5/2$ or 7/2 spin state. These unusually steep magnetization curves, when compared with data for other FeS containing proteins, may be a property of larger FeS clusters which the X-ray data show to be present in MoFe proteins, and which the spectroscopic and sequence data reviewed above strongly suggest are present in the VFe proteins.

3.2.3. FeVaco: Structure and Function

The chemical environment of Mo in the MoFe proteins and V in the VFe proteins has been extensively studied by X-ray absorption spectroscopy. The shape and position of the absorption edge, and the X-ray absorption near-edge structure (XANES), are influenced by the coordination geometry and oxidation level of the absorbing atom. At higher energies, the extended X-ray absorption edge fine structure (EXAFS region) provides information as to the distance, type, and number of atoms in the first and second coordination sphere of the primary absorbing atom. This technique provides very precise atomic distances (1% out to 4 Å from the absorbing atom), only approximate coordination numbers, and cannot distinguish the light elements C, N, or O. Despite these limitations, the assignment for the chemical environment of Mo in MoFe proteins, based on this technique, have been confirmed by the X-ray crystal structures.

The VFe proteins of *A. chroococcum* [70] and *A. vinelandii* [80] have been studied by this technique at the V K-edge, and the data and assignments are in excellent agreement (Table 4). The features of the absorption edge are consistent with V^{II} or V^{IV} in a distorted octahedral coordination. Analysis of the EXAFS region indicate Fe, S and C, O or N as nearest-neighbor atoms to the V atom. These assignments of the types of atom and their distances from the V atom are essentially the same as those of Mo in FeMoco in the MoFe protein. In the case of the MoFe proteins, the X-ray crystal structure has removed the ambiguity

TABLE 4

Comparison of the Environment of Fe and V in the Cofactor Center of VFe Proteins and Fe and Mo in the Cofactor Center of MoFe Proteins Determined from EXAFS Measurements

	Type of interaction	*N*	*R* (Å)
MoFe protein of *A. vinelandii*	Mo---O(N)	2	2.12
	Mo---S	4.5	2.37
	Mo---Fe	3.5	2.67
	Fe---Fe	1.4	3.78
FeMoco	Mo---O(N)	2	2.09
	Mo---S	3	2.36
	Mo---Fe	3.5	2.68
	Fe---Fe	1.3	3.68
VFe protein of *A. vinelandii*	V---O(N)	2 − 3	2.15
	V---S	3 − 4	2.33
	V---Fe	3 ± 1	2.76
	Fe---Fe	1	3.76
VFe protein of *A. chroococcum*	V---O(N)	3 ± 1	2.15
	V---S	3 ± 1	2.31
	V---Fe	3 ± 1	2.75
FeVaco	Fe---S	3	2.24
	Fe---V	1	3.69
	Fe---Fe	2	2.65
	Fe---Fe	1	3.69

Data for the VFe proteins of *A. vinelandii* are from [80, 83] and those for the VFe protein [79] and FeVaco of *A. chroococcum* from [82]; those for MoFe protein and FeMoco are from [83].

as to the nature of the light atoms. The Mo is octahedrally coordinated to two O atoms of homocitrate and a N atom of the imidazole ring of His-442. As discussed above, since homocitrate is a likely component of FeVaco, and His-442 is conserved in the sequences of the VFe proteins (Table 2), the light atoms detected in V EXAFS are very probably O and N atoms of homocitrate and the histidine residue.

As has been shown for MoFe proteins, where the Mo edge and

XANES are unchanged on dye oxidation, or superreduction of the MoFe protein during enzyme catalysis, only small of angular rather than radial changes in the environment of the V atom are observed in the VFe protein [81].

Fe K-edge EXAFS measurements have been made on the VFe protein of *A. vinelandii* [83] and the extracted cofactor of the protein of *A. chroococcum* [82]. The EXAFS is dominated by Fe--S and Fe--Fe interactions, as was found for the MoFe protein and extracted FeMoco (Table 4). The two Fe--Fe separations observed are consistent with the structure proposed for FeMoco from the X-ray crystallographic study, and again emphasize the structural similarity between the cofactor centers of MoFe and VFe proteins.

The comparative biochemical, genetic, and spectroscopic data are consistent with the presence of an iron- and V-containing cofactor (FeVaco) in the VFe proteins. This cofactor center has been extracted from the VFe protein of *A. chroococcum* [87], using methods developed for the extraction of FeMoco from the MoFe proteins. Preparations of extracted FeVaco in NMF contained V-, Fe- and acid-labile sulfide in the ratio ~1:6:5, comparable to those of FeMoco with V replacing Mo. The EPR spectra exhibited weak signals characteristic of a spin $S = 3/2$ system.

FeVaco is capable of binding to the FeMoco binding site in MoFe protein since when recombined with the inactive MoFe protein isolated from a nifB mutant of *K. pneumoniae* an active protein was formed. The hybrid protein had the characteristic property of V-nitrogenase [88] in producing some C_2H_6 as well as C_2H_4 as a product of C_2H_2 reduction. Although C_2H_2 and H^+ were substrates, N_2 was not reduced either to NH_3 or to N_2H_4 suggesting that specific interactions of the cofactor center with the polypeptide required for the reduction of N_2 were missing.

4. SUBSTRATE REDUCTION

Both Mo- and V-nitrogenases require MgATP, the absence of dioxygen, and a low potential reductant (~−400 mV) for activity [73]. In most biochemical studies dithionite ($S_2O_4^{2-}$) is used as an electron donor, and extensive mechanistic studies of Mo-nitrogenase have been made using this system. Such studies have shown that the Fe protein acts as a specific ATP-dependent electron donor to the MoFe protein, a reaction which is subject to strong product inhibition by MgADP.

The strong structural similarities between the protein components of the Mo- and V-nitrogenases is shown by the ability of $Ac2^v$ to form a fully functional enzyme with the same specific activity when titrated with either Ac1 or $Ac1^v$. The corresponding hybrid $Ac1^v \cdot Ac2$ has 70% of the activity of the homologous V-nitrogenase [74]. Such ability to form functional nitrogenases is not always shown when components of Mo-nitrogenase isolated from different organisms are recombined. This indicates that the components of Mo- and V-nitrogenases of *A. chroococcum* have sufficiently conserved contact surfaces so as to allow effective heterologous electron transfer and catalytically competent complexes to be formed.

4.1. Reduction of N_2 and H^+

The apparent K_m for N_2 as a reducible substrate for Mo- and V-nitrogenases is very similar (14 kPa). However, under conditions of optimum electron flux [89], the distribution of electrons under N_2 to the alternative substrates N_2 and protons are very different. Mo-nitrogenase transfers 75% of the electron flux to N_2 [see reaction (1)], but in contrast 50% of the electron flux through V-nitrogenase is utilized for the reduction of protons [see reaction (2)].

$$N_2 + 8e^- + 8H^+ = 2NH_3 + H_2 \quad (1)$$

$$N_2 + 12e^- + 12H^+ = 2NH_3 + 3H_2 \quad (2)$$

The higher proportion of electrons allocated to proton reduction is not due to restricted electron transfer between the two proteins, since the presteady state rate of electron transfer between the two proteins of V-nitrogenase is very similar to that of Mo-nitrogenase, and the ratio of H_2/NH_3 does not change as the electron flux through the system is increased at high Fe protein–to–VFe protein ratios [74]. As discussed below, the lower efficiency appears to be an intrinsic property of V-dependent nitrogen fixation since it is also observed in vivo.

The relative efficiency of the Mo- and V-nitrogenases of *A.*

chroococcum was recently assessed using chemostat cultures [90] (which readily allow the rate of N_2 fixation in vivo to be quantified). This study used mutant strains of *A. chroococcum* which carried only the genes for a single type of nitrogenase and in which the uptake hydrogenase was inactive. Consequently, any H_2 generated by nitrogenase activity was not recycled and was detectable in the gas headspace over the culture. The data showed that the ratio of H_2 evolved to N_2 reduced at 30°C (the H_2/N_2 ratio) was 1 for Mo-nitrogenase and 3 for V-nitrogenase [see reactions (1) and (2)]. Similar findings had been reported earlier for mutant strains of *A. vinelandii* which lacked the structural genes for Mo-nitrogenase, when grown under Mo-deficient conditions in the absence of added V in the presence of an inhibitor of uptake hydrogenase [9]. This stoichiometry is observed with purified V- and Mo-nitrogenases during N_2 reduction under optimum conditions, and indicates that the inefficiency of the V system when compared with Mo-nitrogenase is not due to the use of suboptimum assay conditions.

The lower efficiency of V-dependent nitrogen fixation was also indicated by a higher requirement for carbon energy source to achieve the same yield of biomass during V-dependent growth of *A. chroococcum* [90]. These studies also provided support for the suggestion that V-nitrogenase, despite its lower efficiency, may have been retained during evolution as a nitrogenase capable of functioning at lower temperatures than Mo-nitrogenase. Comparative studies of the isolated enzymes showed that as the assay temperature is lowered V-nitrogenase continues to reduce N_2, whereas Mo-nitrogenase diverts an increasing proportion of electron flux into the reduction of protons rather than N_2. For example, over the temperature range 30°C to 5°C the specific activity of Mo-nitrogenase for N_2 decreases 10-fold more than that of V-nitrogenase [91]. Studies of the temperature dependence of nitrogenase activity of cultures of *A. chroococcum* showed that at pH 6.4 V-nitrogenase, but not Mo-nitrogenase, functioned at temperatures below 10°C. In addition, V-nitrogenase supported growth at 30°C at pH values as low as 6.2, but Mo-nitrogenase did not [90]. The indication from these studies is that the possession of a V-nitrogenase would confer a selective advantage to nitrogen-fixing organisms under conditions of low temperature or in an acidic environment. In this context, it has been shown that the repression of the expression of V-nitrogenase by Mo at 30°C (see Sec. 2.3) does not occur at lower temperatures [92].

The reduction of N_2 by V-nitrogenase also differs from that by Mo-nitrogenase in that a small amount of N_2H_4 is formed as a product by V-nitrogenase [93]. N_2H_4 formation is a property of the VFe protein and the amount formed increases electron flux and temperature [89,93]. The possible significance of these differences are discussed below in Sec. 5.

To date, there have been no reports on the contribution that V-dependent nitrogen fixation makes to global nitrogen cycling, but it is clear from the wide distribution of this system in the *Azotobacters* and other organisms that it has the potential for being significant.

4.2. Reduction of C_2H_2

C_2H_2 is a widely used substrate for the assay of Mo-dependent nitrogenase activity both in vivo and in vitro, since the product of the reduction C_2H_4 is readily detected and quantified by standard gas chromatographic techniques. The apparent K_m for C_2H_2 reduction by purified Mo-nitrogenases range from 0.3 to 2.0 kPa, and at saturating concentrations of C_2H_2 the evolution of H_2 is effectively suppressed [5].

The reduction of C_2H_2 by V-nitrogenase results in the formation of C_2H_6 as well as C_2H_4. Although the amount of C_2H_6 is small (typically 0.02–7.4% of the electrons utilized for C_2H_2 reduction), it is readily detected and is applicable to whole organisms [94]. The apparent K_m for C_2H_2 for the formation of both products is 6 kPa, some threefold higher than the upper range for Mo-nitrogenases. A striking difference for Mo-nitrogenase is the inability of C_2H_2 to effectively suppress the reduction of protons to H_2, even when present at saturating concentrations (see Table 3). As is observed with Mo-nitrogenase, the reduction of C_2H_2 is stereospecific; when the reaction is carried out in ^2H_2O [*cis*-$^2H_2$]C_2H_4 is formed consistent with side-on bonding of C_2H_2 to a metal site in both enzymes [88].

The route by which C_2H_6 is formed is not clear since there is a lag of several minutes before its rate of formation becomes linear, whereas the rates of concomitant C_2H_4 and H_2 formation show no detectable lag. Free C_2H_4 has been shown not to be involved as an intermediate in C_2H_6 formation. The proportion of C_2H_6 formed as a product increases with both the assay temperature and the rate of electron flux at high ratios of Fe to MoFe protein. The reduction of C_2H_2 by cultures of microorganisms grown under conditions of Mo deficiency, to form C_2H_6 in addition to C_2H_4, is presumptive evidence for the functioning of an Mo-independent nitrogenase [94]. This test was the first indication that *A. variabilis* had such a system, and subsequently the genes for V-nitrogenase were cloned and sequenced from this organism (see Table 1).

4.3. Inhibitors of Substrate Reduction

No extensive systematic study of mutual competition by alternative substrates of V-nitrogenase has been reported, as has been done for Mo-nitrogenase, except for the inhibition of H_2 evolution by C_2H_2 discussed above. In the case of inhibition by nonsubstrates, in the Mo-nitrogenase system, H_2 is a specific competitive inhibitor of N_2 reduction but does not affect the reduction of any other substrates, whereas CO inhibits the reduction of all substrates except protons [4]. The specific competitive inhibition of N_2 reduction by H_2 has been interpreted as arising from them binding to a common site on the enzyme (see Sec. 5).

Data for V-nitrogenase are broadly similar, but no analysis of kinetic data has been presented [88]. CO inhibits C_2H_2 reduction by V-nitrogenase less strongly than it does Mo-nitrogenase, as might be expected (assuming binding occurs at V) from the correlation of the bond energies of metal carbonyls which increase with the atomic number of the metal. H_2 at 85 kPa completely inhibits the reduction of N_2 but not the reduction of C_2H_2; the nature of the inhibition by H_2 remains to be established.

5. CATALYTIC CYCLE OF N_2 REDUCTION

The most comprehensive scheme for the function of Mo-nitrogenase in reducing N_2 is that of Thorneley and Lowe [95], based on computer modeling of pre-steady-state stopped-flow and rapid-quench data, for the rates of electron transfer and for the time courses of the generation of enzyme-bound intermediates and product formation. In the context of this review, only an outline is presented in order to indicate where the differences in the mechanism and efficiency of reduction of N_2 by V-nitrogenase might arise. In the case of V-nitrogenase, only a limited range of pre-steady-state data has been collected, but these are very similar to those for Mo-nitrogenase. However, the ability of the components of V-nitrogenase to form catalytically competent enzymes with the complementary protein of Mo-nitrogenase, even for N_2 as a substrate, indicates that inferences as to the broad similarity with regard to the mechanism between the two systems (as described in this scheme) are probably valid.

In the reactions of this scheme, the FeMoco center is reduced by successive transfer of eight electrons from the Fe protein to the MoFe protein in an MgATP-dependent reaction. Following the transfer of each electron, the protein complex dissociates in the rate-limiting reaction of nitrogenase turnover. MgATP then displaces MgADP from oxidized Fe protein-MgADP species which is then reduced before another round of complex formation and electron transfer. The cycle of reactions involving the Fe protein has been measured for the V-nitrogenase of *A. chroococcum*. The rate of electron transfer from the Fe protein to the VFe protein was measured by stopped-flow spectrophotometry utilizing the increase in absorbance of the Fe protein as it is oxidized [96]. In the reactions forming the first two steps of the catalytic cycle of nitrogenase turnover, V-nitrogenase components were shown to form a weaker electron transfer complex, but the rate of electron transfer between the two proteins and the dependence on MgATP concentration were very similar to those of Mo-nitrogenase. The inhibition of the rate of this electron transfer reaction by MgADP was also very similar.

The sequence of events which occur on the MoFe protein following each electron transfer from the Fe protein is shown in Fig. 3. Three electrons are required to be transferred before N_2 can bind by the displacement of H_2, before this stage, hydric species are proposed to be formed at the substrate binding site on FeMoco. Under conditions of low electron flux through the enzyme, or when the reaction is carried out under argon, these species can relax to the resting state of the enzyme by evolving H_2. This binding of N_2 in this way explains the specific competitive nature of the inhibition of N_2 reduction by H_2 and the stoichiometry of reaction 1. During the reduction of N_2, a dinitrogen hydride intermediate is believed to be generated, which produces N_2H_4 upon quenching of the reaction mixture in acid or alkali. In this scheme the dinitrogen is assumed to bind end-on to the metal at the active site at one end only. Protonation of the terminal N atom of the bound N_2 then proceeds, with electron flow from the metal to the dinitrogen.

The nature of the binding site of N_2 on FeMoco has not been clarified by the X-ray structure of the MoFe protein, and many binding

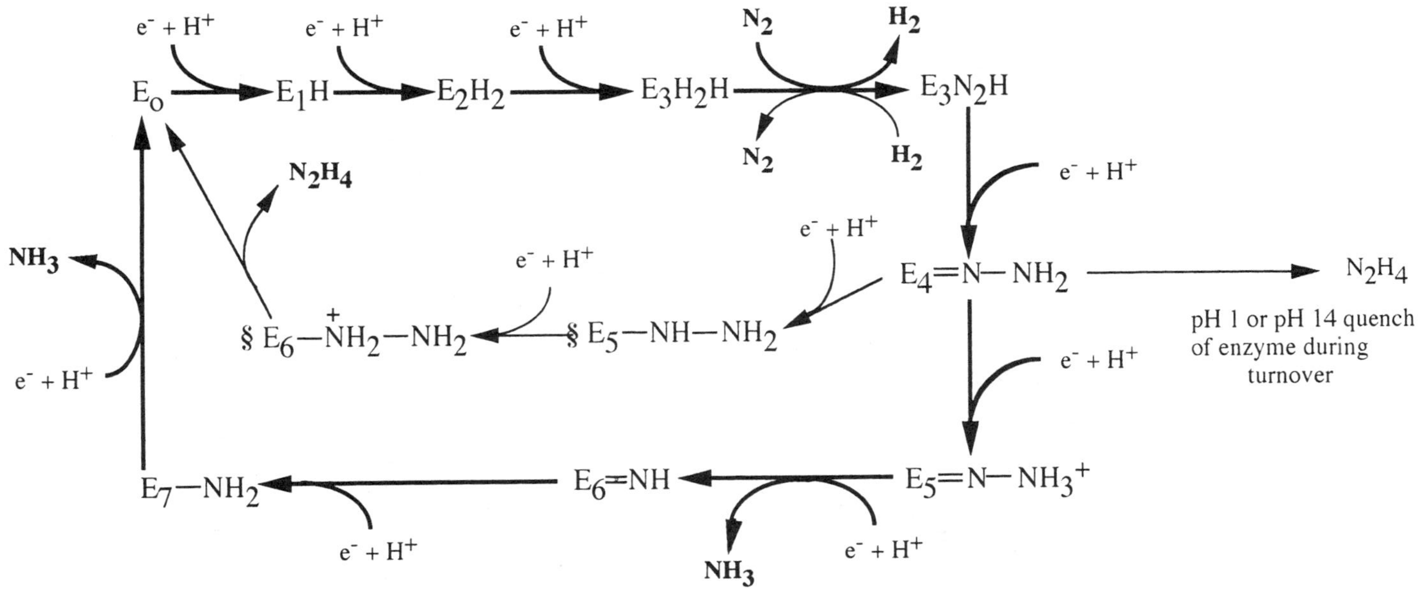

FIG. 3. The catalytic cycle for the reduction of N_2. This scheme as indicated by the bold arrows was developed from studies on Mo-nitrogenase of *K. pneumoniae* and is discussed in detail in [95]. E_0 represents the resting state of the enzyme, and the later subscripts the number of times the MoFe protein has been reduced by the Fe protein in an ATP-dependent reaction. The arrows between each state represent complex formation, electron transfer, and protonation, followed by the rate-limiting step in the overall reaction, that of dissociation of the protein complex. The side reaction at E_4 shows where an enzyme-bound intermediate gives rise to N_2H_4 on acid or alkaline quenching of the enzyme. The light arrows leading to states §E_5 and §E_6 are the proposed route leading to the formation of N_2H_4 as a product of N_2 reduction by V-nitrogenase [93] and do not form part of the cycle of the reactions of Mo-nitrogenase.

modes have been suggested [62,97], including binding to Fe atoms. In order to bind to the Mo atom loss of a ligand to the metal appears to be necessary, and very recent electrochemical model studies have indicated how this might be achieved [98]. A carboxylate functions as a leaving group on reduction of the electrochemically generated Mo^{IV} molybdenum dihydride center in $[MoH_2(\eta^2\text{-}OOCCH_3)(dppe)_2]^+$ exposing a site at which N_2 (and H_2, CO, or C_2H_2) can bind. In this system the stoichiometric loss of H_2 occurs after N_2 binds, as a consequence of the instability of the hydride in the species $[MoH_2(N_2)(dppe)_2]$. The possibility that in FeMoco the homocitrate (which effectively is a tethered carboxylate) functions as a leaving group from the Mo atom to enable N_2 to bind was suggested by these workers [98].

The reduction of N_2 by V-nitrogenase differs in three aspects from N_2 reduction by Mo-nitrogenase: (1) the stoichiometry of H_2/N_2 reduced [reaction (2)]; (2) the continued reduction of N_2 to lower temperatures by V-nitrogenase; (3) the formation of N_2H_4 as a minor product of the reduction of N_2 by V-nitrogenase. These differences are considered below.

1. Under conditions of saturating N_2, limiting values for the H_2/N_2 ratio of 1.13 ± 0.13 for Mo-nitrogenase and 3.5 ± 0.03 for V-nitrogenase were obtained. These data suggest that the early hydridic intermediates shown in Fig. 3 relax to form H_2 even under conditions of maximum electron flux through V-nitrogenase, or a fundamentally different mode of binding of N_2 to the two enzymes exists.
2. Studies with heterologous nitrogenases formed by components of Mo- and V-nitrogenases have shown that the effect of temperature on N_2 reduction is conferred by the Fe protein of V-nitrogenase and therefore is not a property of the VFe protein [91].
3. The formation of N_2H_4 as a product of N_2 reduction is consistent with a four-electron-reduced dinitrogen hydride species being formed on the VFe protein which dissociates to form free hydrazine (Fig. 3). Although quench experiments with Mo-nitrogenase result in the detection of hydrazine released from an enzyme intermediate, it has never been detected as a product on N_2 reduction. If a common intermediate is involved, then its stability in the case of V-nitrogenase is such

that its rate of dissociation compared with further reduction enables it to be detected without denaturation of the reacting enzyme. Assuming that both enzymes do follow a common route to NH_3 formation, then the reactivity of an Mo-NH_2-NH_2 type of intermediate on Mo-nitrogenase prevents the release of hydrazine as a product.

As the temperature is raised to 50°C from 42°C, V-nitrogenase reversibly loses the ability to reduce N_2 to NH_3, but continues to form N_2H_4. A reversible change in conformation of the protein which prevents interaction of FeVaco with residues of the polypeptide, required for the reduction of N_2 to NH_3, but not to N_2H_4, has been proposed [89]. This would suggest that such residues are involved in the final protonation step resulting in cleavage of the N-N bond to yield NH_3.

6. CHEMICAL STUDIES ON N_2 REDUCTION BY VANADIUM COMPLEXES

Aqueous chemical systems capable of reducing N_2 have been known for some time. Heterogeneous V^{II}·Mg^{II} alkaline gels [99], and the soluble V^{II}-catechol complex [100], exhibit some similarities with N_2 reduction by V-nitrogenase. However, such systems have proved difficult to characterize, and mechanistic interpretations have often been controversial, but it is clear that the reduction of N_2 by V^{II} has highly specific requirements.

Among the transition metals, V was one of the last for which stable characterized N_2 complexes were reported. It was only in 1989 that the isolation and structure of the bridging N_2 complex $[\{V(C_6H_4CH_2NMe_2)_2$-$(C_5H_5N)\}_2N_2]$ was reported [101] (Me = CH_3), which reacts with acid to yield NH_3. Subsequently, vanadium N_2 complexes, with singly bound end-on N_2, such as $[V(N_2)_2(Me_2PCH_2CH_2PMe_2)_2]^-$ were isolated [102]. These were isoelectronic and isostructural with the well-established complexes of Mo and W, and gave NH_3 and some N_2H_4 on reaction with acid. The isolation of the latter complex indicates that mononuclear V--N_2 complexes are stable and provide a basis for studying the chemistry of N_2 reduction at sites that are potentially relevant to the mechanism of V-dependent biological nitrogen fixation.

7. OUTLOOK

The area reviewed here, i.e., the involvement of vanadium in biological nitrogen fixation and by the isolation of V-containing bromoperoxidases from marine algae described elsewhere in this volume, shows that this element, which for many years had no known biological role, plays an important role in two discreet enzymic processes. Although the quantitative importance of V-nitrogenase in global nitrogen cycling will take many years to establish, it is already clear that this system is widely distributed in nature, and some clues as to why this apparently inefficient system may have persisted along with the more efficient Mo-dependent nitrogenase have emerged. Its discovery has stimulated chemical research on modeling biological nitrogen fixation which had for so long been dominated by a perceived essential role for Mo.

ACKNOWLEDGMENTS

I thank the many coworkers who made significant contributions to the work on the V-nitrogenase of *A. chroococcum* reviewed here, and to my colleagues at the nitrogen fixation laboratory who were involved in different aspects of this work, in particular Drs. Richard Pau, Antonio Perick, David Lowe, Geoff Yates and Profs. G. Jeffery Leigh and Barry Smith.

ABBREVIATIONS AND DEFINITIONS

Cp1, Av1, and Kp1 denote the MoFe proteins of the Mo-nitrogenase of *Clostridium pasteurianum*, *Azotobacter vinelandii*, and *Klebsiella pneumoniae*, respectively, and Av2 the Fe protein of *A. vinelandii* Mo-nitrogenase. $Ac1^{v}$ and $Ac2^{v}$ denote the VFe protein and Fe protein of the V-nitrogenase of *Azotobacter chroococcum*, $Av1^{v}$ and $Av2^{v}$, the corresponding proteins of the V-nitrogenase of *A. vinelandii*. The structural genes of the nitrogenase systems are denoted by *nif* for Mo-nitrogenase, *vnf* for V-nitrogenase, and *anf* for the third Fe-nitrogenase.

acetyl CoA	acetyl coenzyme A
ADP	adenosine 5′-diphosphate
ATP	adenosine 5′-triphosphate
dppe	$Ph_2PCH_2CH_2PPh_2$
EPR	electron paramagnetic resonance
EXAFS	extended X-ray absorption fine structure
FeMoco	Fe- and Mo-containing cofactor
FeVaco	Fe- and V-containing cofactor
GTP	guanosine 5′-triphosphate
MCD	magnetic circular dichroism
NMF	*N*-methylformamide
XANES	X-ray absorption near-edge structure

REFERENCES

1. H. Bortels, *Arch. Mikrobiologie, 1*, 333 (1930).
2. H. Bortels, *Zentralbl. Bakteriol. Parasitenkd. Infektioskr. Abt., 2, 95*, 193 (1936).
3. D. Burk, *Ergebnisse Enzymforschung, 3*, 23 (1934).
4. W. H. Orme-Johnson, *Annu. Rev. Biophys. Chem., 14*, 419 (1985).
5. R. R. Eady, in *Nitrogen Fixation, Vol. 4, Molecular Biology* (W. J. Broughton and A. Pühler, eds.), Clarendon Press, Oxford, 1986, p. 1 ff.
6. B. E. Smith and R. R. Eady, *Eur. J. Biochem., 205*, 1 (1992).
7. V. K. Shah, R. A. Ugalde, J. Imperial, and W. J. Brill, *Annu. Rev. Biochem., 53*, 231 (1984).
8. P. E. Bishop, D. M. L. Jarlenski, and D. R. Heatherington, *Proc. Natl. Acad. Sci. USA, 74*, 7342 (1980).
9. P. E. Bishop, M. E. Hawkins, and R. R. Eady, *Biochem. J., 238*, 437 (1986).
10. R. L. Robson, R. R. Eady, T. H. Richardson, R. W. Miller, M. Hawkins, and J. R. Postgate, *Nature, 322*, 388 (1986).
11. B. J. Hales, E. E. Case, J. E. Morningstar, M. F. Dzeda, and L. A. Mauterer, *Biochemistry, 25*, 7251 (1986).
12. B. J. Hales, D. J. Langosch, and E. E. Case, *J. Biol. Chem., 261*, 15301 (1986).

13. J. R. Chisnell, R. Premakumar, and P. E. Bishop, *J. Bacteriol.*, *170*, 27 (1988).
14. R. D. Joerger and P. E. Bishop, *CRC Crit. Rev. Microbiol.*, *16*, 1 (1988).
15. R. R. Eady, in *The Prokaryotes*, 2nd ed. (A. Balows, H. G. Trüper, M. Dworkin, W. Harder, and K.-H. Schleifer, eds.), Springer-Verlag, New York, 1992, p. 534 ff.
16. D. J. Arp, in *Nitrogen Fixation: Objectives and Achievements* (P. M. Gresshoff, L. E. Roth, G. Stacey, and W. E. Newton, eds.), Chapman and Hall, London, 1990, p. 67 ff.
17. M. J. Dilworth and J. F. Lonergan, *New. Phytol.*, *118*, 303 (1991).
18. M. J. Dilworth, R. R. Eady, R. L. Robson, and R. W. Miller, *Nature*, *327*, 167 (1987).
19. R. N. Pau, L. A. Mitchenall, and R. L. Robson, *J. Bacteriol.*, *171*, 124 (1989).
20. T. Kentemitch, G. Danneberg, B. Hundshagen, and H. Bothe, *FEMS Microbiol. Lett.*, *51*, 19 (1988).
21. A. F. Yakunin, C. V. Ni and I. N. Gogatov, *Microbiology*, *60*, 52 (1991).
22. E. Fallik, Y.-K. Chan, and R. L. Robson, *J. Bacteriol.*, *173*, 356 (1991).
23. T. Theil, *J. Bacteriol.*, *175*, 6276 (1993).
24. H. Bothe, G. Haverkamp, and T. Kentemitch, in *New Horizons in Nitrogen Fixation* (R. Palacios, J. Mora, and W. E. Newton, eds.), Kluwer Academic, Dordrecht, 1993, p. 505 ff.
25. K. Schüdderkopf, S. Hennecke, U. Liese, M. Kutsch, and E. W. Klipp, *Mol. Microbiol.*, *8*, 673 (1993).
26. F. R. Zinoni, R. M. Robson, and R. L. Robson, *Biochim. Biophys. Acta*, *174*, 83 (1993).
27. M. J. Merrick, in *New Horizons in Nitrogen Fixation* (R. Palacios, J. Mora, and W. E. Newton, eds.), Kluwer Academic, Dordrecht, 1993, p. 43 ff.
28. D. R. Dean, J. T. Bolin, and L. Zheng, *J. Bacteriol.*, *175*, 6737 (1993).
29. J. Liang, M. Madden, V. K. Shah, and R. H. Burris, *Biochemistry*, *29*, 857 (1990).
30. P. B. Normand and J. Bousquet, *J. Mol. Evol.*, *29*, 436 (1989).

31. R. D. Joerger, T. M. Loveless, R. N. Pau, L. A. Mitchenal, B. H. Simon, and P. E. Bishop, *J. Bacteriol.*, *172*, 3400 (1990).
32. R. L. Robson, P. R. Woodley, R. N. Pau, and R. R. Eady, *EMBO J.*, *8*, 1217 (1989).
33. P. E. Bishop, S. I. MacDougal, E. D. Wolfinger, and C. L. Shermer, in *Nitrogen Fixation: Objectives and Achievements* (P. M. Gresshoff, L. E. Roth, G. Stacey, and W. E. Newton, eds.), Chapman and Hall, London, 1990, p. 117 ff.
34. D. R. Dean and M. R. Jacobson, in *Biological Nitrogen Fixation* (G. Stacey, R. H. Burris, and H. J. Evans, eds.), Chapman and Hall, New York, 1992, p. 763 ff.
35. H. M. Kent, M. Baines, C. Gormal, B. E. Smith, and M. Buck, *Mol. Microbiol.*, *4*, 1497 (1990).
36. D. R. Dean, D. J. Scott, and W. E. Newton, in *Nitrogen Fixation: Objectives and Achievements* (P. M. Gresshoff, L. E. Roth, G. Stacey, and W. E. Newton, eds.), Chapman and Hall, London, 1990, p. 95 ff.
37. J. Kim and D. C. Rees, *Science*, *257*, 1677 (1992).
38. J. T. Bolin, A. E. Ronco, L. E. Mortenson, T. V. Morgan, and N. H. Xuong, *Proc. Natl. Acad. Sci. USA*, *90*, 1078 (1993).
39. R. N. Pau, in *Biology and Biochemistry of Nitrogen Fixation* (M. J. Dilworth and A. R. Glenn, eds.), Elsevier, Amsterdam, 1991, p. 37 ff.
40. C. Kennedy, R. Gamal, R. Humphry, J. Ramos, K. Brigle, and D. R. Dean, *Mol. Gen. Genet.*, *205*, 318 (1986).
41. C. Kennedy, and D. R. Dean, *Mol. Gen. Genet.*, *231*, 494 (1992).
42. R. D. Joerger, R. Premakumar, and P. E. Bishop, *J. Bacteriol.*, *168*, 673 (1986).
43. R. D. Joerger and P. E. Bishop, *J. Bacteriol.*, *170*, 1475 (1988).
44. D. R. Dean, J. T. Bolin, and L. Zheng, *J. Bacteriol.*, *173*, 6737 (1993).
45. L. Zheng, R. H. White, V. L. Cash, R. F. Jack, and D. R. Dean, *Proc. Natl. Acad. Sci. USA*, *90*, 2754 (1993).
46. R. D. Joerger, M. R. Jacobson, R. Premakumar, E. D. Wolfinger, and P. E. Bishop, *J. Bacteriol.*, *171*, 1075 (1989).
47. D. J. Evans, R. Jones, P. R. Woodley, and R. L. Robson, *J. Gen. Microbiol.*, *134*, 931 (1988).

48. E. D. Wolfinger and P. E. Bishop, *J. Bacteriol.*, *173*, 7565 (1991).
49. K. Schüddekopf, S. Hennecke, U. Liese, M. Kutsche, and W. Klipp, *Mol. Microbiol.*, *8*, 673 (1993).
50. M. R. Jacobson, R. Premakumar, and P. E. Bishop, *J. Bacteriol.*, *167*, 490 (1986).
51. F. Luque and R. N. Pau, *Mol. Gen. Genet.*, *227*, 481 (1991).
52. P. E. Bishop and R. Premakumar, in *Biological Nitrogen Fixation* (G. Stacey, R. H. Burris, and H. J. Evans, eds.), Chapman and Hall, New York, 1992, p. 736 ff.
53. M. J. Merrick, in *Biological Nitrogen Fixation* (G. Stacey, R. H. Burris, and H. J. Evans, eds.), Chapman and Hall, New York, 1992, p. 835 ff.
54. R. D. Joerger, M. R. Jacobson, and P. E. Bishop, *J. Bacteriol.*, *171*, 3258 (1989).
55. R. N. Pau, in *New Horizons in Nitrogen Fixation* (R. Palacios, J. Mora, and W. E. Newton, eds.), Kluwer Academic, Dordrecht, 1993, p. 117 ff.
56. M. K. Chan, J. Kim, and D. C. Rees, *Science*, *260*, 792 (1993).
57. J. Kim, D. Woo, and D. C. Rees, *Biochemistry*, *32*, 7104 (1993).
58. M. M. Georgiadis, H. Komiya, P. Chakrabarti, D. Woo, J. J. Kornuc, and D. C. Rees, *Science*, *257*, 1653 (1992).
59. J. B. Howard, G. L. Anderson, and T. L. Deits, in *Nitrogen Fixation Research Progress* (H. Evans, P. J. Bottomley, and W. E. Newton, eds.), Martinus Nijhoff, Dordrecht, 1985, 559 ff.
60. J. Kim and D. C. Rees, *Nature*, *360*, 553 (1992).
60. A. Willing and J. B. Howard, *J. Biol. Chem.*, *265*, 6596 (1990).
61. S. A. Murrel, R. G. Lowery, and P. W. Ludden, *Biochem. J.*, 609 (1988).
62. J. Kim and D. C. Rees, *Nature*, *360*, 553 (1992).
63. S. Marritt, A. J. Pierik, A. J. Thomson, W. R. Hagen, D. J. Lowe, S. P. J. Albrecht, and B. E. Smith, in *New Horizons in Nitrogen Fixation* (R. Palacios, J. Mora, and W. E. Newton, eds.), Kluwer Academic, Dordrecht, 1993, p. 153 ff.
64. R. C. Tittsworth and B. J. Hales, *J. Am. Chem. Soc.*, *115*, 9763 (1993).
65. K. K. Surerus, M. P. Hendrich, P. D. Christie, D. Rottgardt, W. H.

Orme-Johnson, and E. J. Münck, *J. Am. Chem. Soc.*, *114*, 8579 (1992).

66. A. J. Pierik, H. Wassink, H. Haaker, and W. R. Hagen, *Eur. J. Biochem.*, *212*, 51 (1993).
67. W. R. Hagen, H. Wassink, R. R. Eady, B. E. Smith, and H. Haaker, *Eur. J. Biochem.*, *169*, 457 (1987).
68. J. T. Bolin, N. Campobasso, S. W. Muchmore, W. Minor, T. V. Morgan, and L. E. Mortenson, in *New Horizons in Nitrogen Fixation* (R. Palacios, J. Mora, and W. E. Newton, eds.), Kluwer Academic, Dordrecht, 1993, p. 89 ff.
69. D. J. Lowe, K. Fisher, and R. N. F. Thorneley, *Biochem. J.*, *292*, 93 (1993).
70. V. K. Shah and W. J. Brill, *Proc. Natl. Acad. Sci. USA*, *74*, 3249 (1977).
71. T. R. Hawkes, P. A. McLean, and B. E. Smith, *Biochem. J.*, *217*, 317 (1983).
72. J. R. Hoover, J. Imperial, P. W. Ludden, and V. K. Shah, *Biofactors*, *1*, 199 (1988).
73. R. R. Eady, in *Vanadium in Biological Systems* (N. D. Chasteen, ed.), Kluwer Academic, Dordrecht, 1990, p. 99 ff.
74. R. R. Eady, R. L. Robson, T. H. Richardson, R. W. Miller, and M. Hawkins, *Biochem. J.*, *244*, 167 (1986).
75. B. J. Hales, A. E. True, and B. M. Hoffman, *J. Am. Chem. Soc.*, *111*, 8519 (1989).
76. J. E. Morningstar and B. J. Hales, *J. Am. Chem. Soc.*, *109*, 6854 (1989).
77. J. E. Morningstar, M. K. Johnson, E. E. Case, and B. J. Hales, *Biochemistry*, *26*, 1795 (1987).
78. N. Ravi, V. Moore, B. J. Hales, and B. H. Huynh in *New Horizons in Nitrogen Fixation* (R. Palacios, J. Mora, and W. E. Newton, eds.), Kluwer Academic, Dordrecht, 1993, p. 148.
79. J. M. Arber, B. R. Dobson, R. R. Eady, P. Stevens, S. S. Hasnain, C. D. Garner, and B. E. Smith, *Nature*, *325*, 327 (1987).
80. G. N. George, C. L. Coyle, B. J. Hales, and S. P. Cramer, *J. Am. Chem. Soc.*, *110*, 4057 (1988).
81. J. M. Arber, B. R. Dobson, R. R. Eady, S. S. Hasnain, C. D. Garner,

T. Matsushita, M. Nomura, and B. E. Smith, *Biochem. J.*, *258*, 733 (1989).

82. I. Harvey, J. M. Arber, R. R. Eady, B. E. Smith, C. D. Garner, and S. S. Hasnain, *Biochem. J.*, *266*, 929 (1990).

83. J. Chen, J. Christiansen, R. C. Tittsworth, B. J. Hales, S. J. George, D. Coucouvanis, and S. P. Cramer, *J. Am. Chem. Soc.*, *115*, 5509 (1993).

84. R. R. Eady, R. N. Pau, D. J. Lowe, and F. J. Luque, in *Nitrogen Fixation: Objectives and Achievements* (P. M. Gresshoff, L. E. Roth, G. Stacey, and W. E. Newton, eds.), Chapman and Hall, London, 1990, p. 125 ff.

85. R. C. Tittsworth, C. Z. Blanchard, and B. J. Hales, *J. Inorgan. Biochem.*, *51*, 366 (1993).

86. A. J. Pierik, D. J. Lowe, M. E. Eldridge, S. Marritt, J. A. Farrar, A. J. Thomson, and R. R. Eady, in *New Horizons in Nitrogen Fixation* (R. Palacios, J. Mora, and W. E. Newton, eds.), Kluwer Academic, Dordrecht, 1993, p. 147.

87. B. E. Smith, R. R. Eady, D. J. Lowe, and C. Gormal, *Biochem. J.*, *250*, 299 (1988).

88. M. J. Dilworth, R. R. Eady, and M. E. Eldridge, *Biochem. J.*, *249*, 745 (1988).

89. M. J. Dilworth, M. E. Eldridge, and R. R. Eady, *Biochem. J.*, *289*, 359 (1993).

90. J. Kahindi, D. Phil. thesis, University of Sussex (1994).

91. R. W. Miller and R. R. Eady, *Biochem. J.*, *256*, 429 (1988).

92. J. Walmsley and C. Kennedy, *Appl. Environ. Microbiol.*, *57*, 522 (1991).

93. M. J. Dilworth and R. R. Eady, *Biochem. J.*, *277*, 465 (1991).

94. M. J. Dilworth, R. R. Eady, R. L. Robson, and R. W. Miller, *Nature*, *327*, 167 (1987).

95. R. N. F. Thorneley and D. J. Lowe, in *Molybdenum Enzymes* (T. G. Spiro, ed.), John Wiley and Sons, New York, 1985, p. 221 ff.

96. J. Bergström, R. R. Eady, and R. N. F. Thorneley, *Biochem. J.*, *251*, 165 (1988).

97. H. Deng and R. Hoffmann, *Angew. Chem. Int. Ed. Engl.*, *32*, 1062 (1993).

98. D. L. Hughes, S. K. Ibraham, G. Querne, A. Laouenan, J. Talmarin, A. Fonseca, and C. J. Pickett, *J. Chem. Soc. Chem. Commun.*, submitted.
99. N. T. Denesov, N. I. Shuvalova, and A. E. Shilov, *Kinet. Katal.*, *18*, 1606 (1977).
100. L. A. Nikonova, A. G. Ovcharenko, O. N. Efimov, V. A. Avilov, and A. E. Shilov, *Kinet. Katal.*, *13*, 1602 (1972).
101. J. H. H. Edema, A. Meelsma, and S. Gambarotta, *J. Am. Chem. Soc.*, *111*, 6867 (1989).
102. C. Woitha and D. Rehder, *Angew. Chem. Int. Ed. Eng.*, *29*, 1438 (1990).
103. E. Fallik, P. G. Hartel, and R. L. Robson, *Appl. Environ. Microbiol.*, *59*, 1883 (1993).
104. L. K. Kimble and M. T. Madigan, *FEMS Microbiol. Lett.*, *100*, 255 (1992).
105. P. Scherer, *Arch. Microbiol.*, *151*, 44 (1989).
106. P. W. Ludden, R. Davis, R. Petrovitch, and G. P. Roberts in *New Horizons in Nitrogen Fixation* (R. Palacios, J. Mora, and W. E. Newton, eds.), Kluwer Academic, Dordrecht, 1993, p. 143.

12
Amavadin, the Vanadium Compound of Amanitae

Ernst Bayer

Institut für Organische Chemie der Universität,
Auf der Morgenstelle 1, D-72076 Tübingen, Germany

1. INTRODUCTION

Amavadin is a naturally occurring vanadium compound found in *Amanita* species. So far amavadin was not yet detected in other organisms. Amavadin is the first natural vanadium compound for which the structure has been completely elucidated. The biological function of amavadin is still not known.

2. OCCURRENCE OF AMAVADIN

The mushroom *Amanita muscaria* shows a surprisingly high capability to accumulate vanadium. Ter Meulen [1] described in 1931 the vanadium enrichment in *A. muscaria*. Later investigations [2,3] reported less vanadium in other *Amanita* species. However, Meisch et al. [4] found high amounts of vanadium in *A. regalis*, a closely related mushroom to *A. muscaria*. A very comprehensive study of the vanadium levels in *A. muscaria* from different locations and related species was reported by Koch et al. [5]. The vanadium concentrations vary between 36 and 250 mg V/kg dry weight. Stem and root of the mushroom show a higher vanadium accumulation in comparison to the hat. Table 1 shows the vanadium accumulation in a variety of mushrooms related to *A. muscaria*.

Whereas in most species vanadium accumulation is below 2 mg V/kg dry weight, only in *A. velatipes*, *A. wellsii*, and *A. regalis* is the vanadium content above 100 mg/kg dry weight, which is comparable to *A. muscaria*. Aqueous extracts of fly agaric show a characteristic electron spin resonance spectrum of tetravalent vanadium [6,7], which can be used to guide the isolation of the vanadium compound and to detect the presence of amavadin in extracts of biological samples. The electron spin resonance (ESR) spectrum of the purified compound, which was named *amavadin*, is the same as the one of the extracts before isolation (Fig. 1).

3. ISOLATION OF AMAVADIN AND STRUCTURE DETERMINATION OF THE ORGANIC LIGAND

The isolation of amavadin [5,8] can be achieved by high-performance liquid chromatography (HPLC): 100–200 mg of the dried and pulverized

TABLE 1

Vanadium Content of Different Mushrooms (mg/kg dry weight)

Species	Location	Content	
A. citrina (Schff.)	Schönbuch,	1.9 ± 0.6	(H)
S. F. Gray	Germany	1.2 ± 0.7	(S + K)
A. citrina (Schff.)	Netherlands	<1	(H)
S. F. Gray var. *alba* (Gill.)		<1	(S + K)
A. citrina (Schff.)	Netherlands	<1	(H)
S. F. Gray var. *citrina*		2.4 ± 0.3	(S + K)
A. crocea (Quél.) Singer	Wales	<1	(H)
	U.K.	<1	(S + K)
A. eliae (Quél.)	—	<1	(H)
		1.7 ± 0.1	(S + K)
A. frostiana (Peck)	New Hampshire	1.4 ± 0.8	(H)
	USA	1.2 ± 0.2	(S + K)
A. frostiana (Peck)	USA	<1	(H)
A. gemmata (Fr.) Gill	Netherlands	<1	(H)
		<1	(S + K)
A. gemmatoides		<1	(H)
		1.4 ± 0.4	(S + K)
A. hyperborea P. Karst.	Finland	<1	(H)
		3.9 ± 1.7	(S + K)
A. parcivolvata (Peck)	Alabama/USA	1.1 ± 0.2	(H)
E. J. Gilb.		6.1 ± 0.9	(S + K)
A. regalis (Fr.)	Finland	98 ± 0.9	(H)
		122 ± 7	(S + K)
A. rubescens (Pers. ex Fr.)		8.5 ± 0.5	(H)
S.F. Gray	Netherlands	1.1 ± 0.1	(S + K)
A. rubescens (Pers. ex Fr.)	France	<0.5	(H)
S. F. Gray	France	1.7 ± 0.9	(H)
		4.5 ± 0.8	(S)
A. velatipes Atk.	Michigan/USA	257 ± 4	(H)
		377 ± 20	(S + K)
A. wellsii (Murr.) Sacc.	Massachusetts	2.5 ± 2.5	(H)
	USA	5.0 ± 1.0	(S + K)
	Massachusetts	1.6 ± 0.2	(H)
	USA	6.1 ± 1.7	(S + K)

H = hat; S = stem; K = root.

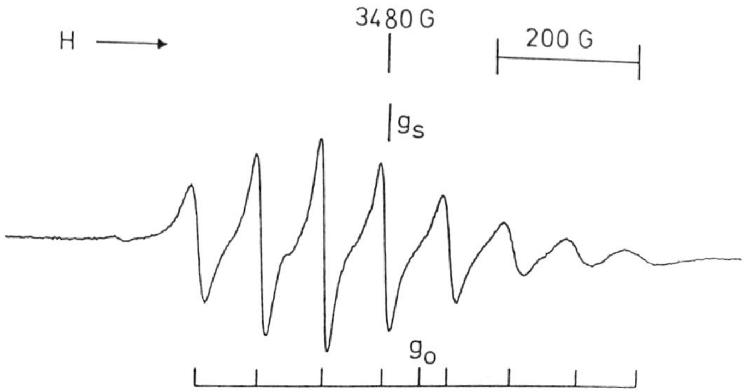

FIG. 1. ESR spectrum of amavadin.

mushroom is suspended in 5 mL distilled water and extracted for 60 min at 60°C. Thereafter the solution is filtered off over glass wool and over a C_{18} cartouche. This solution is submitted to an HPLC column (ODS-1, LATEK). Isocratic elution with a solution of 10 mM tetrabutylammonium hydrogen sulfate in acetonitrile-water (20:80) results in separation, which is shown in Fig. 2.

The isolated compound is pale blue. Molecular weight determination reveals that amavadin is a low molecular weight compound of 420 ± 20 Da [6]. Since the ESR spectra before and after isolation are identical it is clear that no binding to a protein is present in the native amavadin in the fungus cells. ESR, IR, and UV spectra clearly indicate that amavadin contains tetravalent vanadium.

The structure elucidation was difficult because standard methods could not be utilized. The paramagnetism of amavadin caused extreme peak broadening of nuclear magnetic resonance (NMR) spectra. Due to the low volatility of the vanadium complex no useful mass spectra in the electron impact (EI), chemical ionization (CI), and fast atom bombardment (FAB) mode could be obtained [6].

All attempts to crystallize amavadin isolated from natural samples for X-ray structure determination have failed so far. It is also not possible to remove the vanadium without degradation or change of the ligand due to the extreme stability of the vanadium complex.

Tetravalent vanadium salts and coordination compounds normally occur in the vanadyl form VO^{2+} and a IR band at 985 cm^{-1} in amavadin suggested first that it is also a VO^{2+} species [6].

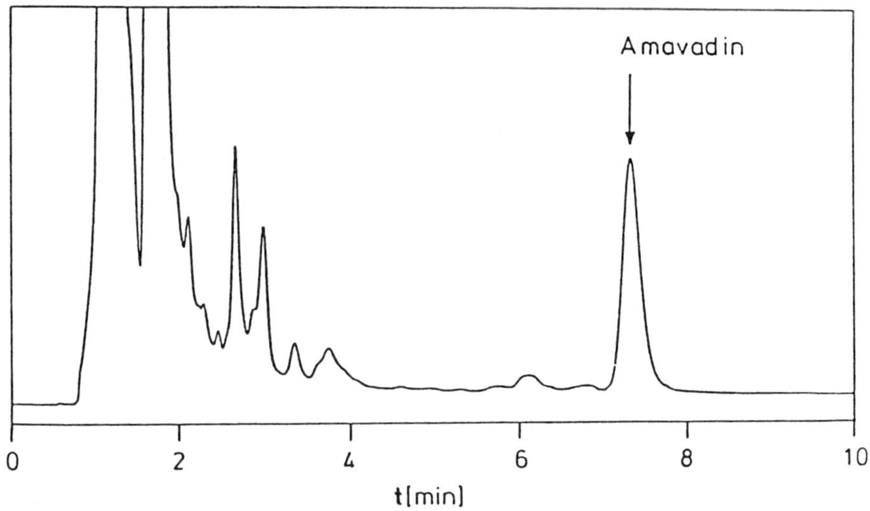

FIG. 2. Separation of amavadin by HPLC of mushroom extracts. Column: Latek ODS-1,5 μm; 150 × 4 mm. Solvent 80% acetonitrile, 20% tetrabutylammonium hydrogen sulfate. Flow 1 mL/min; pressure 1700 psi; temperature 23°C; detection: UV 235 nm.

The structure of the amavadin ligand had to be determined by chemical degradation (Fig. 3) and revealed a peptidic nature [7,8].

Strongly acidic and alkaline hydrolysis yielded L-alanine, pyruvic acid, and acetaldehyde as the only products, indicating a deeper degradation.

Under reducing conditions with zinc/HCl α,α′-iminodipropionic acid (**2**) (IDP) was isolated. IDP was already described in the literature by Karrer and Appenzeller [9], and occurs as RR and SS enantiomer and meso form. All forms were synthesized and compared with the IDP obtained by degradation of amavadin. Since the amount of material was restricted a micromethod had to be devised for this purpose. Therefore capillary gas chromatography on Chirasil-Val [10,11], a chiral polysiloxane peptide stationary phase, was chosen. Very well resolved are the chiral forms of IDP as derivatization products with isopropyl isocyanate on Chirasil-L-Val, as is shown in Fig. 4. The IDP degradation product of amavadin is identical with *SS*-IDP.

IDP, which is closely related to the classical sequestering agent ethylenediaminetetraacetate (EDTA), belongs to a group of natural

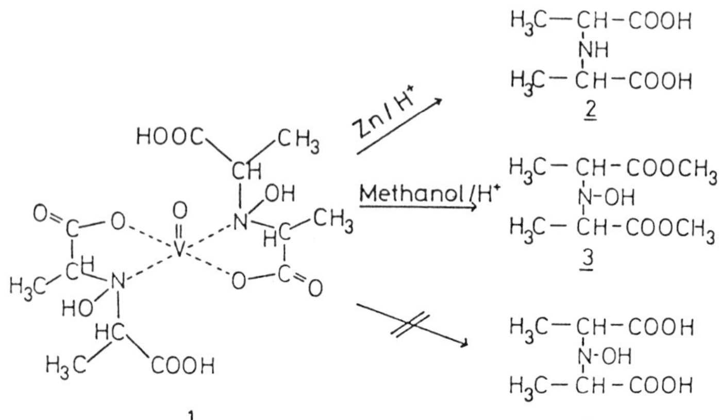

FIG. 3. Degradation reactions of amavadin.

products ("opines") which can be found in crown gall tumors [12]. IDP itself was found in mussels and squids where it occurred in the optically inactive meso form [13]. However L-IDP is not the organic ligand of amavadin because it does not produce amavadin when reacted with vanadyl sulfate.

Upon treatment with diluted sodium hydroxide in air a change of the characteristic ESR spectrum of amavadin can be observed. First the signals disappear and subsequently a new spectrum occurs which can be assigned to a relatively stable nitroxyl radical, suggesting that the ligand of amavadin contains an NOH group.

Finally by reacting amavadin with concentrated sulfuric acid in methanol at elevated temperature a compound was isolated which contained one oxygen more than IDP. The NMR and mass spectra suggested that the structure may be the dimethyl ester of *N*-hydroxy-α,α'-iminodipropionic acid (**4**) (Fig. 3). Under these reaction conditions the formation of the dimethyl ester of the ligand (**4**) (HIDP) is to be expected. Since this compound was not yet described in the literature it had to be synthesized, including all stereoiosmers (LL, DD, meso) in order to assign the structure and stereochemistry.

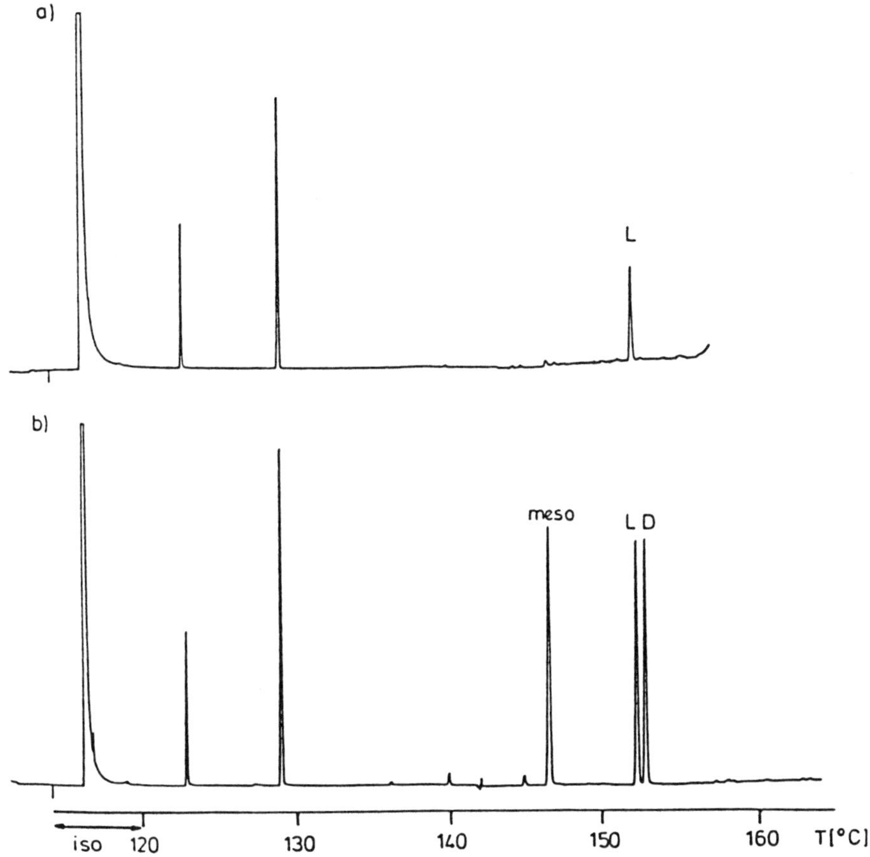

FIG. 4. Gas chromatogram on Chirasil-Val of IDP from amavadin (a) and of synthetic meso-, L- and D-IDP (b) after derivatization with isopropyl isocyanate.

4. STEREOCHEMISTRY OF THE ORGANIC LIGAND

For the assignment of the stereochemistry of the ligand (**4**) of amavadin two compounds are available (Fig. 3): the reduction of amavadin by zinc dust resulted in the formation of α,α′-iminodipropionic acid (**2**). After methanolysis of amavadin, the methyl ester (**3**) of the ligand can be obtained. Both compounds isolated from natural amavadin contain the two asymmetric carbon atoms and can exist in the form of the L, D, and meso isomers. The imino compound (**2**) has already been described in the literature [9]. As mentioned previously, all three isomers were synthesized and compared with the product obtained from amavadin and their structure assigned by gas chromatography using isopropylisocyanate as the derivatizing agent. By comparison of the retention times, the degradation product of amavadin was shown to be the L enantiomer. This indicated that also in amavadin the asymmetric carbon atoms have an L configuration. Also the isolation of L-alanine as a degradation product by strongly acid or alkaline treatment further support this.

The stereoisomers of the ligand (**4**) had not been described in the literature prior to this work. Earlier reports [6,14] on the synthesis of (**4**) did neither discuss the occurrence of isomers nor were they isolated.

5. TOTAL SYNTHESIS OF AMAVADIN

For the synthesis of (**4**) either of two ways seemed feasible: the direct oxidation of the imino compound (**2**) or a synthesis by starting from hydroxylamine and D-2-bromopropanoic acid. In the latter case a stereospecific synthesis of the L enantiomer could be expected, avoiding racemate separation.

First attempts to obtain (**4**) by direct oxidation of (**2**) were not successful and, therefore, the condensation reaction of 2-bromopropanoic acid with hydroxylamine in the presence of sodium carbonate was chosen [15]. The separation of (**4**) from the reaction mixture was easily achieved by precipitation after addition of zinc acetate and the zinc ion removed by cation exchange, yielding pure (**4**). By starting from D-2-bromopropanoic acid the product contained mostly the L enantiomer of (**4**) but also considerable amounts of the meso form. Purification could be achieved by preparative ion pair high-performance liquid chromatography (HPLC). The structure was confirmed by NMR spectra. The purity of the synthesized L enantiomer was analyzed by capillary gas chromatography of its dimethyl ester on Chirasil-Val [11,12]. The recrystallized product contained less than 10% of the D enantiomer and about 2% of the meso isomer. Retention times in gas chromatography (GC) of the dimethyl ester of the synthesized L enantiomer of (**4**) and of the corresponding compound obtained from amavadin were identical.

To prove that the synthesized *N*-hydroxyiminodipropionic acid (HIDP) had indeed the L configuration in both asymmetric centers, it was reduced with retention of its configuration. The iminodipropionic acid (**2**), after derivatization with isopropylisocyanate, gave GC retention times on Chirasil-Val identical with that of authentic iminodipropionic acid.

From (**4**) the synthesis of amavadin could be performed easily by adding the stoichiometric amount of vanadyl sulfate. The synthesized product and amavadin isolated from *A. muscaria* showed excellent agreement [15] in all analytic and spectroscopic data like IR, ESR, UV, and the circular dichroism (CD) spectra (Fig. 5).

On the basis of these results, and the assumption of a vanadyl group structure (**1**) was first proposed for amavadin. However, despite the fact that a total synthesis was achieved the arrangement of the ligand and the exact structure of the complex had to be determined.

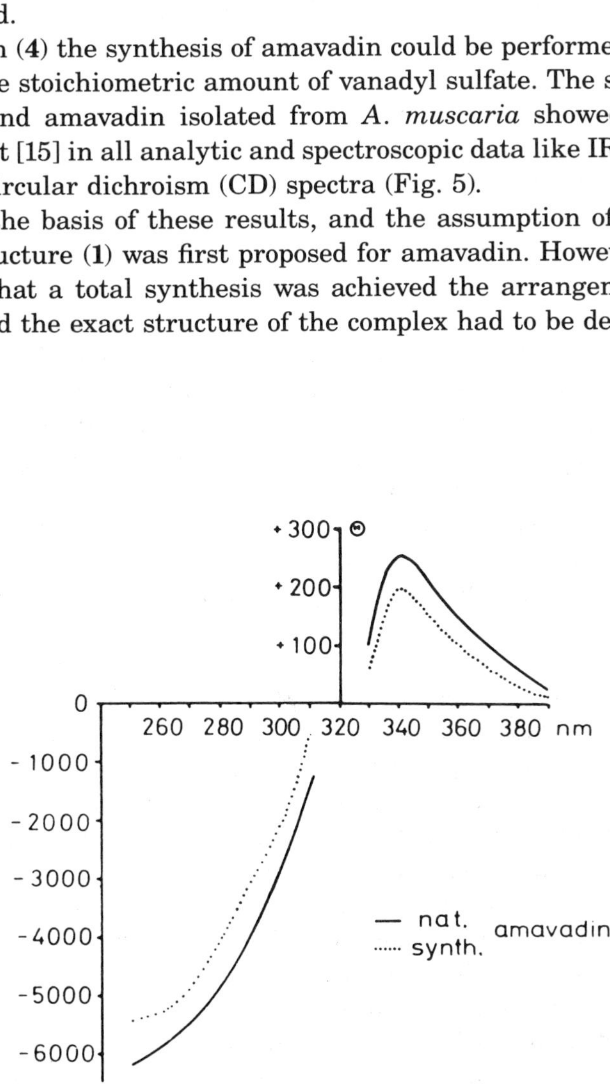

FIG. 5. CD spectra of isolated (−) and synthesized (....) amavadin.

6. COMPLEX STABILITIES OF AMAVADIN AND RELATED CHELATES

Since the nonesterified ligand of amavadin could not be isolated from amavadin directly by hydrolysis due to the extreme stability of the natural complex, the synthetic ligand was used to investigate in general the chelating properties of HIDP and to compare it with the very well-known sequestering capability of IDP.

Table 2 shows the $\log K$ values of IDP and HIDP for different metal ions [18,19], which have been investigated together with Anderegg [16]. The stability of the 2:1 complex of HIDP is remarkable, and so far amavadin reveals itself as the most stable V(IV) complex ever reported in the literature. It is more than a billion times more stable than the copper complex. Also remarkable is the favored formation of the 2:1 complex. Independent from the stoichiometric ratio of HIDP/vanadyl salt we never could observe a 1:1 complex, contrary to the observation of a stepwise complex formation in the case of other metals. This is in sharp contrast to the complex formation of IDP with vanadyl ions. IDP forms only a 1:1 complex. The stability of this complex is much smaller than that of amavadin. In addition, the vanadium selectivity disappears in the case of IDP. Copper forms the more stable chelate. So nature has created a unique vanadium selectivity only by inserting an oxygen into an NH bond.

However, if one looks at Table 2 in more detail, it is difficult to understand why the vanadium complex of the *N*-hydroxy compound should be more stable than the vanadium compound of IDP on the basis of the structure (**1**) proposed with a bond to nitrogen. Normally one would assume that the nitrogen in a hydroxylamine derivative is less basic than in an amine, and therefore also the complex stability should decrease. Table 2 shows this expected behavior for calcium, nickel, copper, and zinc complexes of both ligands. With the exception of vanadium all other metal complexes of IDP are more stable than those of HIDP by approximately one order of magnitude.

Therefore the complex stability constants of a larger variety of ligands related to HIDP were determined, which are shown for the V(IV) and Cu(II) chelates in Table 3 [20–23]. Only *N*-hydroxyimino-diacetic acid forms a vanadium compound with similar stability to that

TABLE 2

Logarithms of the Stability Constants K_n ($[ML_n]/([ML_{n-1}][L])$) for Various Metal Ions[a]

Metal ion		Ligand			
		N-Hydroxy-α,α′-iminodiacetic acid (HIDA)	*N*-Hydroxy-α,α′-iminodipropionic acid (HIDP)	α,α′-Iminodiacetic acid (IDA)	α,α′-Iminodipropionic acid (IDP)
Ca^{2+}	$\lg K_1$	3.0 (1)	2.7 (1)	2.6[b]	2.0 (1)
Ni^{2+}	$\lg K_1$	6.4 (1)	5.7 (1)	8.19[b]	7.6 (1)
	$\lg K_2$	4.65 (10)	5.0 (1)	6.11[b]	5.9 (1)
Cu^{2+}	$\lg K_1$	8.45 (10)	9.2 (1)	10.63[b]	10.6 (2)
	$\lg K_2$	4.10 (5)	3.45 (10)	6.05[b]	4.4 (1)
Zn^{2+}	$\lg K_1$	5.5 (5)	5.05 (10)	7.27[b]	6.6 (1)
	$\lg K_2$	3.96 (1)	4.05 (10)	5.33[b]	4.5 (1)
VO^{2+}	$\lg K_1$	—	—	9.00[c]	9.45[c]
	$\lg K\beta_2$[d]	21.9 ± 1.0	23.0 ± 1.0		

[a]The threefold standard deviation in the last digit(s) is shown in parentheses. Measurements at 25°C (HIDA, HIDP, IDP) and 20°C (IDA) and ionic strength $I = 0.1$ (KNO_3).
[b]From [18].
[c]From [19], $T = 25°C$.
[d]$\beta_2 = [VL_2^{2-}]/[VO^{2+}][L^{2-}]^2$.

TABLE 3

Complex Formation Constants K_n ($[ML_n]/([ML_{n-1}][L])$) for Various Amino and N-Hydroxyamino Acid Metal Complexes [$T = 25°C$, $I = 0.1$ (KNO_3)]

	VO^{2+}			Cu^{2+}	
Ligand	lg K_1	lg K_2	lg K_3	lg K_1	lg K_2
H_2N-O-CH_2-COOH	$\bar{n}_{max} < 0.15$			5.02[a]	
H_2N-CH_2-CH_2-COOH	<7.5			7.10[b]	5.40[b]
HO-NH-CH_2-COOH	6.4	6.2	5.1	Cu^{2+} reduction	
H_2N-CH_2-COOH	<6.6			8.46[b]	6.83
HO-N(CH_3)-CH_2-COOH	6	5	4	Cu^{2+} reduction	
H_3C-NH-CH_2-COOH	<7.5			7.94[c]	6.65[c]
HOOC-CH_2-CH(NHOH)-COOH	7.24	5.4		6.54	4.3
HOOC-CH_2-CH(NH_2)-COOH	9.2			8.57[d]	6.78[d]
HON(CH_2CH_2COOH)$_2$	5.8	4.2		7.3	

[a]From [20].
[b]From [21].
[c]From [22].
[d]From [23].

of amavadin. All other ligands like the isomeric compound N-hydroxy-β,β′-iminodipropionic acid or the bidentate ligand N-hydroxy-α-alanine form much weaker complexes. This means that all ligands, especially the NO group as such, and not only the nitrogen, as well as both carboxylic groups of HIDP, are involved in the complex formation, and that the inner sphere of the coordination depicted in (**1**) does not agree with the real structure.

Since no X-ray investigations could be obtained, large-angle X-ray scattering (LAXS) experiments were performed. They clearly indicated that no metal-ligand distance corresponding to a V=O distance (0.157–0.165 nm) was found, the shortest metal-ligand distance being larger than 0.19 nm. In addition, the IR signal at 985 cm^{-1} does not appear in the analogous stable vanadium complex of N-hydroxyiminodiacetic acid [17]. This excludes the presence of a vanadyl group and indicated a

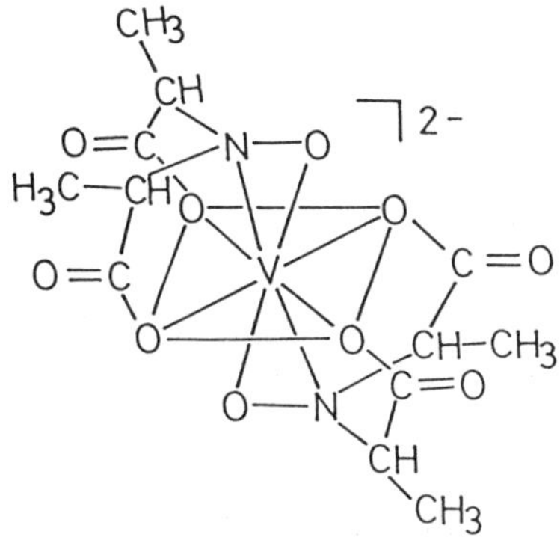

FIG. 6. Structure of amavadin.

V(IV) cation without oxygen. The high acidity of amavadin (pK = 0) excludes also the free carboxylic groups. On the basis of LAXS and EXAFS investigations we have to assume that in amavadin the donating groups of the ligands are arranged as depicted in Fig. 6 [7].

This arrangement is also supported by the X-ray structure determination of the vanadium complex of *N*-hydroxy-2,2′-iminodiacetic acid [24]. This compound also forms an exceptionally stable vanadium(IV) compound like the natural ligand.

7. CONCLUSION

With amavadin nature has designed the most stable among all known V(IV) complexes. However it remains still a question if amavadin has a biological function and by which mechanism just the few *Amanita* species accumulate vanadium. Up to now it remains a mystery as to why just *Amanita* species have designed such an efficient chemistry for vanadium enrichment.

ABBREVIATIONS

CD	circular dichroism
CI	chemical ionization

EDTA	ethylenediamine-*N,N,N′,N′*-tetraacetate
EI	electron impact
ESR	electron spin resonance
EXAFS	extended absorption fine structure
FAB	fast atomic bombardment
GC	gas chromatography
HIDP	*N*-hydroxy-α,α′-iminodipropionic acid
HPLC	high-performance liquid chromatography
IDP	α,α′-iminodipropionic acid
IR	infrared
LAXS	large-angle X-ray scattering
NMR	nuclear magnetic resonance
UV	ultraviolet

REFERENCES

1. H. ter Meulen, *Revl. Trav. Chim. Pays Bas*, *50*, 491 (1931).
2. D. Bertrand, *Bull. Am. Museum Nat. History*, *94*, 409 (1950).
3. H. U. Meisch, J. A. Schmitt, and W. Reinle, *Z. Naturforsch.*, *33C*, 1 (1978).
4. H. U. Meisch, W. Reinle, and J. A. Schmitt, *Naturwissenschaften*, *66*, 620 (1979).
5. E. Koch, H. Kneifel, and E. Bayer, *Z. Naturforsch.*, *42C*, 873 (1987).
6. E. Bayer and H. Kneifel, *Angew. Chem.*, *85*, 542 (1973).
7. P. Krauss, E. Bayer, and H. Kneifel, *Z. Naturforsch.*, *39B*, 829 (1984).
8. E. Bayer and H. Kneifel, *Z. Naturforsch.*, *27B*, 207 (1972).
9. P. Karrer and R. Appenzeller, *Helv. Chim. Acta*, *25*, 595 (1942).
10. E. Bayer, *Z. Naturforsch.*, *38B*, 1281 (1983).
11. H. Frank, G. Nicholson, and E. Bayer, *Angew. Chem. Int. Ed. Engl.*, *17*, 363 (1978).
12. J. Tempe, A. Petit, M. Holsters, M. v. Montaga, and J. Schell, *Proc. Natl. Acad. Sci. USA*, *74*, 2848 (1977).
13. M. Sato, Y. Sato, and Y. Tsuchiya, *Bull. Jpn. Soc. Sci. Fish.*, *43*, 1441 (1977).

14. J. Felcman, M. Candida, T. A. Vaz, and J. J. R. Frausto da Silva, *Inorg. Chim. Acta*, *93*, 101 (1984).
15. H. Kneifel and E. Bayer, *J. Am. Chem. Soc.*, *108*, 3075 (1986).
16. G. Anderegg, E. Koch, and E. Bayer, *Inorg. Chim. Acta*, *127*, 183 (1987).
17. E. Bayer, E. Koch, and G. Anderegg, *Angew. Chem. Int. Ed. Engl.*, *26*, 545 (1987).
18. L. G. Sillen and A. E. Martell, in *Stability Constants of Metal Complexes*, Chemical Society, London, Special Publications, Vol. 17 (1964); Vol. 25 (1971).
19. T. Vogt, E. Koch, G. Folkers, and E. Bayer, unpublished results.
20. Z. Warnke, C. Trojanoska, and A. Liwo, *J. Coord. Chem.*, *14*, 31 (1985).
21. V. S. Sharma, H. B. Mathur, and P. S. Kilharni, *Ind. J. Chem.*, *3*, 1461 (1965).
22. F. Basolo and Y. T. Chen, *J. Am. Chem. Soc.*, *76*, 953 (1954).
23. S. Chabarek and A. E. Martell, *J. Am. Chem. Soc.*, *74*, 6021 (1952).
24. M. A. A. F. de C. T. Carrando, M. T. L. S. Duarte, J. C. Pessoa, J. A. L. Silva, J. J. R. Frausto da Silva, M. C. T. A. Vaz, and L. F. Vilas-Boas, *J. Chem. Soc. Chem. Commun.*, *1158* (1988).

13
Vanadium in Ascidians and the Chemistry of Tunichromes

Mitchell J. Smith,[1] Daniel E. Ryan,[2] Koji Nakanishi,[2] Patrick Frank,[3] and Keith O. Hodgson[3]

[1]Natural Products Division, CFSAN, Food and Drug Administration (HFS-345), 200 C Street, SW, Washington, DC 20204, USA

[2]Department of Chemistry, Columbia University, New York, NY 10027, USA

[3]Department of Chemistry, Stanford University, Stanford, CA 94305-5080, USA

1. INTRODUCTION

Despite nearly a century of scientific scrutiny [1], the manifest presence of vanadium in many animals has eluded virtually all attempts to disclose a natural purpose. To this end, several convergent, bioinorganic trends are elaborated on below. In particular, ongoing studies worldwide are investigating just how and why ascidians assimilate and reduce unprecedented quantities of this metal. In Sec. 2, Patrick Frank and Keith Hodgson begin at the atomic and molecular levels. Here, the results from vanadium K-edge X-ray absorption spectroscopic studies on blood cells of *Ascidia ceratodes* suggest that endogenous vanadium is maintained in a single major state, as ${V^{III}}_{aq}$, which may perhaps represent a metabolic terminus. Thereafter, in Sec. 3, Daniel Ryan and Koji Nakanishi cover the formidable complexation and redox chemistries of the tunichromes (Fig. 1) [2–12]. An extensive array of *indirect* evidence indicates that dissimilar ascidians may actually utilize this type of ligand to drive earlier stages of transition metal assimilation, chelation, and reduction. In fact, the original isolation and structure determination of these elusive yellow (blood) pigments met with success through the realization that they could not withstand comingling with the constituents of green, vanadium-laden blood cells [13]. It is equally remarkable that the vanadium cycle of such creatures, including redox, is surprisingly rapid. Many if not most of their blood cells appear to be renewed within several weeks [14–22]. In Sec. 4, this author (M.J.S.)

approaches the proposed purposes of vanadium for ascidians via evolutionary and ecophysiological findings which, serendipitously enough, afford several new corollaries involving organismal homeostasis. In light of the sheer abundance and distribution of vanadium in ascidians, one can anticipate a variety of important biological roles for this metal.

In view of the daunting number of relevant reports on vanadium, the tunichromes, and ascidians proper, we have sought to minimize any reiteration of readily available information. For such background material, we direct your attention to one of several topical overviews on the biochemistry of vanadium [22–30], the tunichromes and analogous ligands [2,31–36], and ascidian biology [37–39]. Although we have endeavored to cover each issue objectively, several curious discrepancies [3,22,40,41] continue to confound this exceptionally broad, interdisciplinary field. Consequently, an ever-present challenge is to distinguish the transient and mutually exclusive vanadium chemistries that comprise this convoluted bioinorganic cycle from outright incongruities. Accordingly, an apt *modus vivendi* for ascidians (a.k.a. tunicates) was offered by N. J. Berrill several decades ago [42]: "A tunicate could not be itself if it were not ambiguous." Our point of departure has thus been to transcend this twilight truism through the assistance of several novel findings and strong inferences.

2. X-RAY SPECTROSCOPIC STUDIES ON *ASCIDIA CERATODES* BLOOD CELLS

2.1. Background

X-ray absorption spectroscopy (XAS) has become an important tool for the structural elucidation of metal sites not only within solutions or amorphous solids, but also within noncrystalline biological materials [43,44]. Recent advances in the intensity of available synchrotron X-ray sources and improved detector sensitivity and data analysis methods now permit the resolution of metrical details to intermediate ranges (3–5 Å) in extended metal-containing complexes of biological importance [45,46].

Figure 2 illustrates the origin and appearance of a K-edge transmission X-ray absorption spectrum. At X-ray energies below the ionization energy, promotion of a 1s electron into unoccupied valence levels often produces spectral features which reflect the symmetry of the

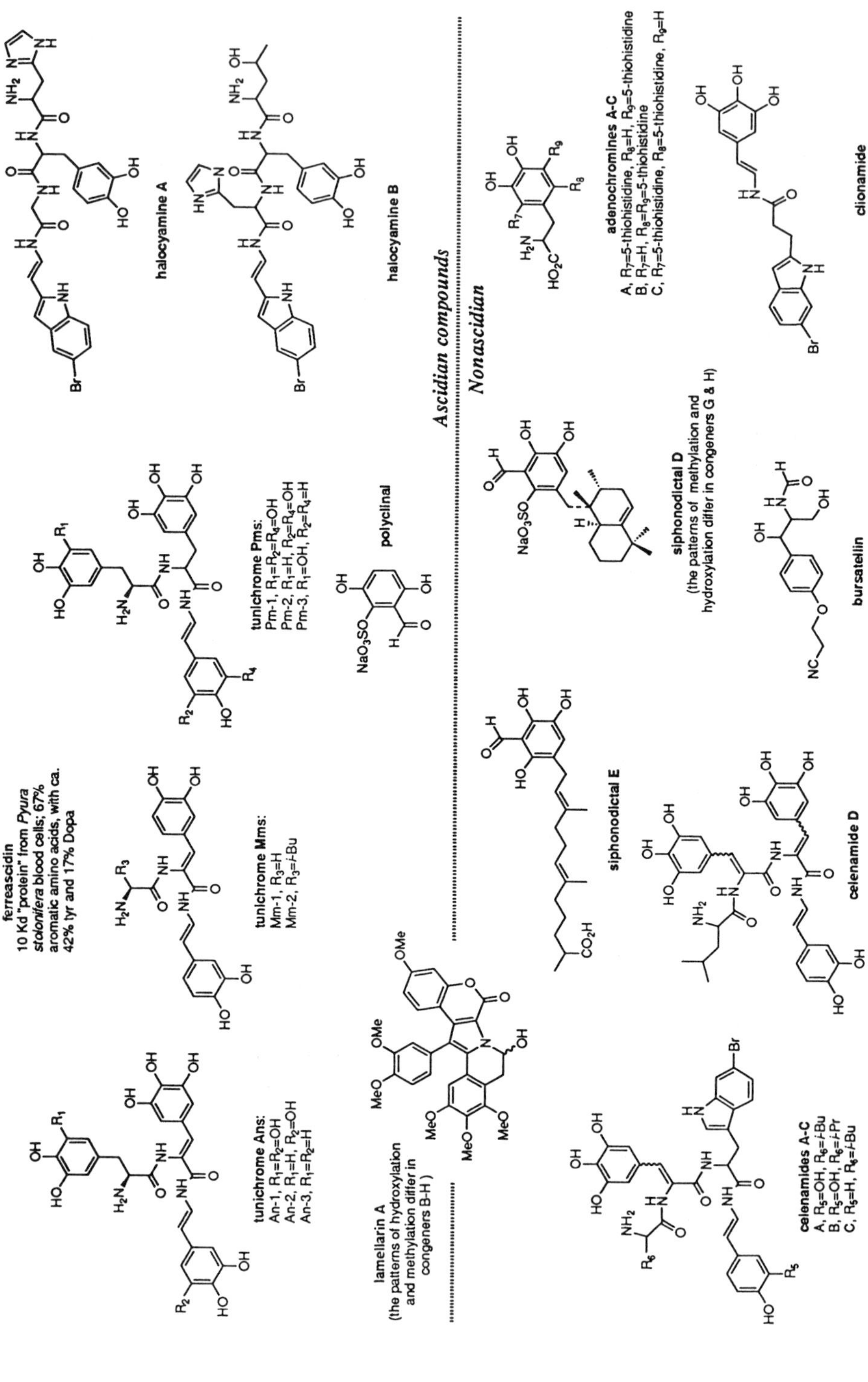
halocyamine A
halocyamine B
Ascidian compounds
Nonascidian
adenochromines A-C
A, R_7=5-thiohistidine, R_8=H, R_9=5-thiohistidine
B, R_7=H, R_8=R_9=5-thiohistidine
C, R_7=5-thiohistidine, R_8=5-thiohistidine, R_9=H
clionamide
ferreascidin
10 Kd "protein" from Pyura stolonifera blood cells; 67% aromatic amino acids, with ca. 42% tyr and 17% Dopa
tunichrome Pms:
Pm-1, R_1=R_2=R_4=OH
Pm-2, R_1=H, R_2=R_4=OH
Pm-3, R_1=OH, R_2=R_4=H
polyclinal
siphonodictal D
(the patterns of methylation and hydroxylation differ in congeners G & H)
bursatellin
tunichrome Mms:
Mm-1, R_3=H
Mm-2, R_3=i-Bu
siphonodictal E
celenamide D
tunichrome Ans:
An-1, R_1=R_2=OH
An-2, R_1=H, R_2=OH
An-3, R_1=R_2=H
lamellarin A
(the patterns of hydroxylation and methylation differ in congeners B-H)
celenamides A-C
A, R_5=OH, R_6=i-Bu
B, R_5=OH, R_6=i-Pr
C, R_5=H, R_6=i-Bu

FIG. 1.

absorber. For example, 1s→*n*d transitions are forbidden under normal selection rules governing octahedral complexes. However, in a tetrahedral first-row transition metal complex, 3d mixing with p-symmetry orbitals can produce significant intensity in this transition. Transitions to bound atomic p states, at energies above the lowest unoccupied molecular orbital (LUMO), become Rydberg-like at energies just below the ionization threshold. Virtual states can give rise to so-called continuum resonances at somewhat higher energies [43,44,47–50].

In the extended X-ray absorption fine structure (EXAFS) region the quasi-periodic oscillations in the intensity profile arise because the wavelength of the outgoing photoelectron is alternatively in phase or out of phase with the absorber-ligand distance. This results in respective resonant constructive or nonresonant destructive interference of the photoelectron wave with itself. Analysis of the interference pattern yields the absorber-scatterer distance, usually to within ± 0.02 Å. The phase of the EXAFS and the shape of the amplitude envelope carry information about the atomic number (Z) of the ligand, and the amplitude per se of the EXAFS signal carries information about the number of ligands [51].

2.2. Vanadium K-Edge X-ray Absorption Spectroscopy

In this section we discuss the application of synchrotron XAS to studies on whole blood cells from the solitary tunicate *A. ceratodes*. This includes both vanadium EXAFS and K-edge XAS, addressing questions regarding the oxidation state, symmetry, and ligand arrangement of

---

FIG. 1. The tunichromes [2,3] and related natural products: the halocyamines of ascidian morula cells [4]; iron-binding ferreascidin of ascidian blood cells [5]; the celenamides and clionamides from burrowing sponges [6]; the lamellarins from ascidians, as well as from molluscs, some of which feed on ascidians [7]; bursatellin from opisthobranch molluscs [sea hares], which possess circulatory systems seemingly devoid of respiratory pigments [8,9]; and the Fe^{III}-adenochromines from the branchial heart of octopus [10]. These and numerous other, more distantly related redox agents, i.e., the sulfated siphonodictals from burrowing sponges [11] and sulfated polyclinal from an ascidian [12], are *air-sensitive* and potent *metal chelators*. The compounds above the dashed line are from ascidians.

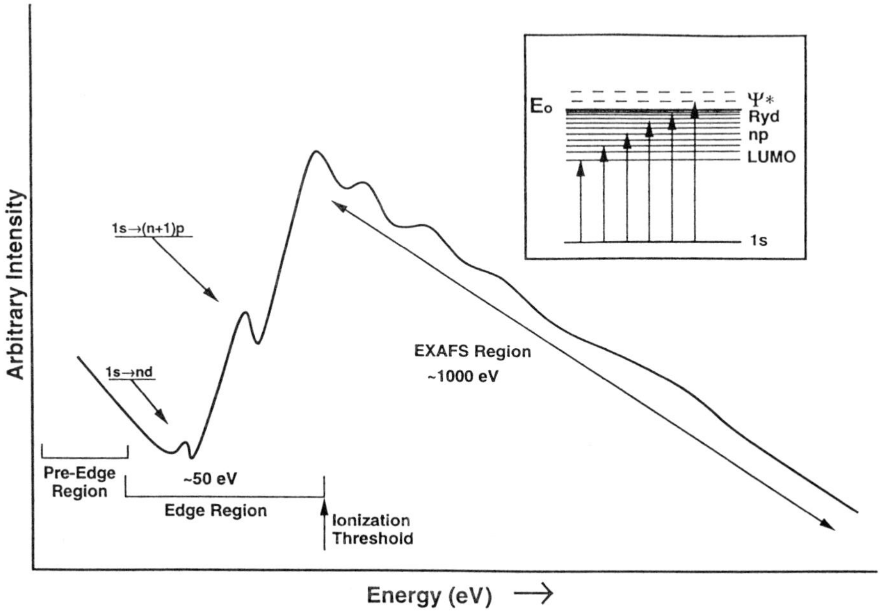

FIG. 2. A schematic representation of a K-edge transmission X-ray absorption spectrum for a transition metal with an unfilled d-orbital manifold. The features have been exaggerated for clarity. Below the ionization threshold, transitions occur from the 1s core state to the lowest unoccupied orbital, in this case an unfilled "d" orbital (*inset*). Considerable intensity can be acquired through d-p mixing (see text), yielding information about site symmetry. Successive transitions can occur to higher unfilled orbitals, typically of "p" symmetry for K-edge transitions. These become more atomic (Rydberg-like) at energies just below the ionization energy (Eo). At still higher energies, excitations to unbound continuum resonance states (Ψ^*) are possible. In the EXAFS region, scattering of the photoelectron from the atomic potentials around the absorber yield absorber-scatterer distance information, as well as the identity and number of scattering atoms.

this metal within living whole blood cells. In addition, we have been investigating the products of a surprisingly rich sulfur biochemistry within *A. ceratodes* blood cells, through use of XAS at the sulfur K edge.

The average concentration of blood cell vanadium within *A. ceratodes* is 0.1–0.2 M [21,52,53], which represents a concentration gradient against seawater of about 2–5 million fold. Work in the field has established that this metal is not in a protein matrix, plays no part in oxygen transport [9,54–56], and among phlebobranch ascidians is maintained as V^{III}, an air-sensitive oxidation state [56–64]. The collection and storage of V^{III} is considered to be a primitive characteristic among ascidians [21,54,55,64–67]. The finding that lysis of ascidian blood cells releases remarkable quantities of acid along with vanadium [53,55–57,60,67–75] implies either that acid is present somewhere within the cells [53,63,68–71] or that the acid found is an artifact of lysis, arising from hydrolysis and/or oxidation of blood cell V^{III} [72–75]. The corollary to the latter hypothesis is that the intracellular V^{III} is managed as a tunichrome chelate within a pH neutral regime. The tunichromes (Fig. 1) are polyphenolic secondary metabolites that contain gallate- or catecholate-like aromatic molecular units, which have so far been found to be ubiquitous to ascidian blood cells [2,34,76], although this has been disputed [41]. Tunichrome would likely chelate V^{III} in a planar or octahedral array [34,41,77] of oxygen ligands. The XAS experiments we describe here will have bearing on all these aspects of *A. ceratodes* blood cell inorganic biochemistry.

In Fig. 3 we compare the vanadium K-edge X-ray absorption spectrum arising from frozen (10 K) whole blood cells from *A. ceratodes* [78] with the spectra of authentic potassium-*tris*-(catecholato)-V^{III} and 0.1 M V^{III} [as $V_2(SO_4)_3$] in a frozen 9 M sulfuric acid glass. The catecholate complex is likely to be a good model for any biological *tris*-chelate involving V^{III} and the tunichromes [2,34,76,77]. Of the two inorganic model X-ray absorption spectra, the blood cell spectrum is much closer to that of acidic V^{III}_{aq}. This includes the shapes, intensities, and energy positions of the respective 1s→3d transition features (Fig. 3, inset), the maximum at 5484.7 eV, and the broad smooth multiple scattering feature around 5494 eV [79,80]. By the same token, the dissimilarity between the blood cell X-ray absorption spectrum and that of V^{III}-*tris*-catecholate argues against a significant presence ($\leqslant$10%) of a *tris*-tunichrome-like V^{III} chelate within the blood cells. The low intensities of the 1s→3d XAS transitions for both the blood cell sample and the inorganic V^{III} solution sample are comparable with that of the strictly octahedral [81] *tris*-catecholate V^{III} model complex. This implies a ligand field of analogous symmetry about V^{III} in the two former samples. The blood cell X-ray absorption spectrum also exhibits the same line shape as observed in the X-ray absorption spectra of known aquo-transition metal ions (Fig. 3) [61,80,82,83].

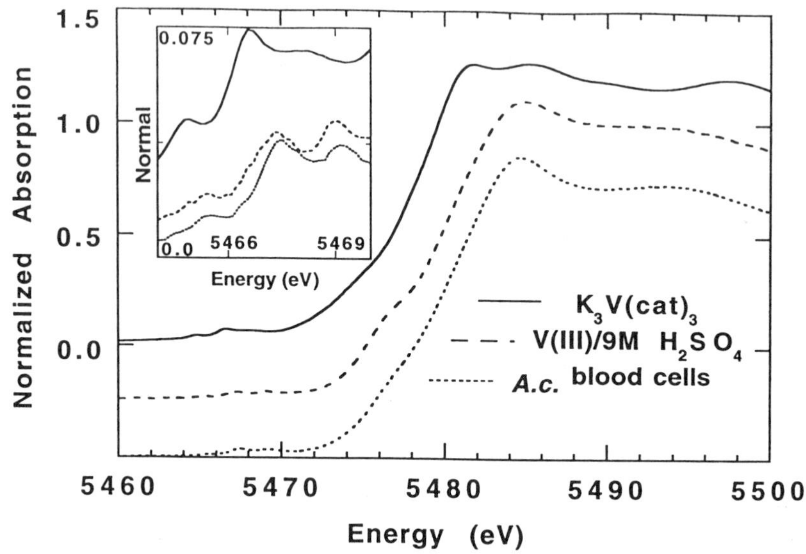

FIG. 3. Vanadium K-edge X-ray absorption spectra of $K_3V(cat)_3$, 0.050 M $V_2(SO_4)_3$ dissolved in 9 M H_2SO_4 (offset by -0.25), and whole blood cells from *A. ceratodes* (offset by -0.50). The two inorganic models test the two hypothetical states of trivalent vanadium within the blood cells of *A. ceratodes* (see text). *Inset*: Detailed view of the K-edge 1s→3d transition for vanadium in each of the above environments. The energy position, shape, and multiplicity (which reflects the splitting in the d manifold) of the X-ray absorption spectrum of the ascidian blood cell vanadium is virtually coincident with that of V^{III} in acidic aqueous solution.

Following the arguments above, the ligand type and ligand field symmetry for a preponderance of blood cell V^{III} are each likely to be very similar to those characterizing V^{III} in acidic aqueous solution, i.e., a set of water ligands arranged in a high-symmetry array. These findings corroborate previous vanadium K-edge XAS studies [61] performed at room temperature on blood cells, from *A. ceratodes*, suspended in blood plasma.

None of the Fig. 3 spectra contain evidence for the presence of V^{IV} as the vanadyl ion, VO^{2+}. Due to d-p mixing in the unique V=O bond, the XAS K-edge spectrum of vanadyl ion exhibits a strong transition at 5469 eV [61], equal in intensity to about 50% of the total vanadium edge jump. Thus, $\geqslant$5% of the endogenous vanadium as vanadyl ion would be readily detectable. However, electron paramagnetic resonance (EPR) examination of *A. ceratodes* blood cell samples has, in our hands, invariably revealed a small amount of pentaaquovanadyl ion [53,68,69]. Quantitation of the EPR spectra of the Fig. 3 samples indicated a [VO^{2+}] of about 0.3 mM within the inorganic V^{III} solution and about 1 mM VO^{2+} within the whole blood cell sample. These findings [61] again provide explicit evidence that routine in-air handling of blood cells results in little or no detectable artifactual oxidation of the endogenous V^{III}.

Any *tris*-tunichrome chelate of V^{IV} (i.e., lacking a V=O moiety) would be minimally evident as a shift to higher X-ray energies, relative to V^{III}_{aq}, of the rising edge of the blood cell vanadium K-edge spectrum. Since this shift was not observed, such complexes, if present, can account for no more than about 10% of the total blood cell vanadium. Endogenous *tris*-tunichrome V^{IV} complexes are unlikely to give an EPR signal at 77 K [81].

Lysis of blood cell samples, even in air, does not greatly change the shape or energy position of the vanadium K-edge X-ray absorption spectrum [61] (unpublished results). This finding clearly argues against extensive oxidation of V^{III} when this procedure is followed. Therefore, the acid produced on lysis cannot arise from oxidation of V^{III} to VO^{2+} or VO_2^+, with concomitant liberation of protons [68,74]. The prior presence of acid in the vanadium-bearing cells is not thereby demonstrated, however [84,85]. Nevertheless, the clear XAS evidence noted above for intracellular V^{III}_{aq} is not consistent with a neutral pH. Rather, an endogenous pH $\leqslant$ 2 is indicated, since above this pH, deprotonation of V^{III}_{aq} becomes extensive [24,86]. Such chemistry would have X-ray absorption spectroscopic consequences, which are absent from all current data. This important point is expanded on in Sec. 2.3.

2.3. Vanadium K-Edge EXAFS Spectroscopy

In Fig. 4 we compare the EXAFS spectra of the Fig. 3 samples, also acquired at 10 K [78]. The spectrum of V^{III}-*tris*-catecholate is comparatively much more complicated because it includes components not only from the first shell of six catecholate oxygens, but also from each of the more distant three shells of aromatic carbon atoms. The chelate structure of the complex means that scattering from these carbon shells is

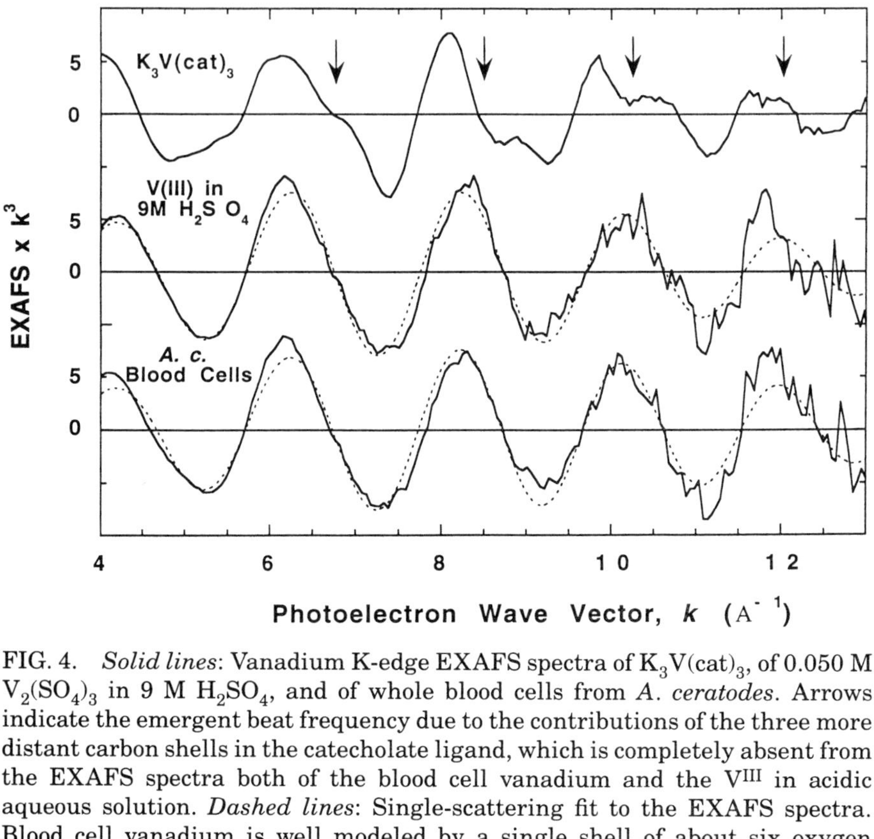

FIG. 4. *Solid lines*: Vanadium K-edge EXAFS spectra of $K_3V(cat)_3$, of 0.050 M $V_2(SO_4)_3$ in 9 M H_2SO_4, and of whole blood cells from *A. ceratodes*. Arrows indicate the emergent beat frequency due to the contributions of the three more distant carbon shells in the catecholate ligand, which is completely absent from the EXAFS spectra both of the blood cell vanadium and the V^{III} in acidic aqueous solution. *Dashed lines*: Single-scattering fit to the EXAFS spectra. Blood cell vanadium is well modeled by a single shell of about six oxygen scatterers at 1.99 Å. In acidic $V^{III}{}_{aq}$, a single shell of about seven oxygens at 1.99 Å is found.

particularly strong because the position of the chelate carbon is rigidly correlated with respect to the central vanadium absorber. Any blood cell V^{III}-tunichrome complex (or other carbon-containing chelate structure) would yield a similarly complicated EXAFS spectrum. In single-scattering curve-fitting experiments (Fig. 4, dashed lines) both the blood cell X-ray absorption spectrum and the $V^{III}{}_{aq}$ X-ray absorption spectrum can be well represented by a single shell of oxygen scatterers at 1.99 Å [78]. No contributions from any more distant carbon shells, as would be expected from a chelate complex, were detected. For the *A.*

ceratodes blood cell V^{III}, about *six oxygen atom scatterers at an average distance of 1.99 Å* were found [78]. The presence of first-shell nitrogen scatterers is not strictly ruled out, but is considered much less likely [61], particularly in the absence of any second-shell carbon. *The conclusion best supported from the combined K-edge XAS and EXAFS analysis is that within A. ceratodes blood cells, the* V^{III} *is surrounded by six oxygen ligands (most likely from water) in an octahedral array with* $(V\text{–}O)_{av} = 1.99$ *Å*. Any *tris*-chelated V^{III} which might be present could account for no more than about 10% of the total blood cell vanadium. The results from the earlier room temperature V-K EXAFS and XAS studies [61] and nuclear magnetic resonance (NMR) experiments [60] are thus completely corroborated. Six water ligands at 1.99 Å is a result consistent with the crystallographic coordination number and distance found in hexaaquo V^{III}. This complex exhibits ligation by six waters in an octahedral array with $V\text{–}O_{av} = 1.992 \pm 0.02$ Å [87,88].

Given the ecological restraints imposed on V^{III} by an aerobic environment, the conclusions from XAS analysis make sense. Any V^{III}-*tris*-tunichrome complex is likely to be highly reducing at neutral pH, with the (III,IV) redox couple of a *tris*-catecholate-like complex estimated to be −370 mV vs. NHE [77]. The difficulty of maintaining such an air-sensitive complex intracellularly, in the presence of any significant concentration of dissolved blood plasma dioxygen, is formidable. Alternatively, the endogenous pH 1.8 milieu, found by EPR to characterize the intracellular region containing vanadyl ion, can be reasonably assigned to the blood cell region containing V^{III} aquo ion [53,68]. The redox potential of $V(H_2O)_6^{3+}$ in pH 1.8 sulfuric acid is calculated to be about +260 mV vs. NHE [53], thus conferring resistance against autoxidation. These suggestions have been strengthened by results from XAS experiments at the sulfur K edge [69,89] (vide infra).

The room temperature K-edge EXAFS spectrum of vanadium from aerobically lysed whole blood cells [61] showed only small changes in the coordination sphere of V^{III}, as compared to intact blood cells. This experiment was recently repeated at 10 K, and the results will be reported elsewhere. In an interesting contrast, single-scattering fits to the EXAFS spectrum of V^{III} in 9 M sulfuric acid solution invariably revealed seven to eight oxygen ligands at an average distance of 1.99 Å [78]. Complexes of V^{III} within a seven-coordinate oxygen ligation sphere are known [90,91]. It thus appears possible that inorganic V^{III} in strongly acid solution supports a larger sphere of water ligands than is found in *A. ceratodes* blood cells or within the crystalline hexaaquo V^{III} complex. This may explain the subtle difference in the vanadium K edges of blood cells V^{III} and inorganic acidic V^{III} (Fig. 3).

2.4. Sulfur K-Edge X-ray Absorption Spectroscopy

Sulfur K-edge XAS is sensitive to variations in both the oxidation state and the bonding array of sulfur within molecules, even within a single molecule such as thiosulfate [92]. A monotonic shift in sulfur K-edge position of 1.6 ± 0.2 eV per unit oxidation state change has been found for a series of related sulfur-containing compounds [89]. Sulfur and sulfate have been identified within ascidian blood cells by X-ray microprobe and other analytic methods [53,60,67,84,93,94]. Figure 5 shows the sulfur K-edge X-ray absorption spectrum of a sample of whole blood cells from *A. ceratodes*. In prior work, the XAS features were demonstrated to be consistent with the intracellular presence of a disulfide unlike cystine (at 2473.5 eV), an aliphatic sulfonic acid (at 2481.1 eV), and sulfate (at 2482.5 eV) [69,89,92]. The sulfur K-edge X-ray absorption spectra of all four *A. ceratodes* whole blood cell collections thus far examined have been similar in general but quite dissimilar in detail.

Sulfate features in all the *A. ceratodes* sulfur K-edge X-ray absorption spectra observed to date have been typically broadened relative to the purely inorganic ion dissolved in neutral aqueous solution [69,89]. Figure 6 shows the effects of pH and [V^{III}] on the linewidth of the sulfate sulfur K-edge X-ray absorption spectrum [69]. Isosbestic points in each case argue for the interconversion of only uncomplexed and complexed sulfate [Eq. (1): $M^{n+} = H^+$ or V^{3+}].

$$M^{n+} + SO_4^{2-} \rightarrow (MOSO_3)^{(n-2)+} \tag{1}$$

The observed increase in sulfate XAS linewidth can be explained by the conversion of T_d sulfate to C_{3v} symmetry following protonation or complexation. The spectroscopic consequence of new incompletely resolved $1s \rightarrow a_1$ and $1s \rightarrow e$ transitions would appear as line broadening [69]. A similar effect has been observed in the resonance Raman spectrum of acidified sulfate [95]. Curve-fitting experiments [69] allowed the

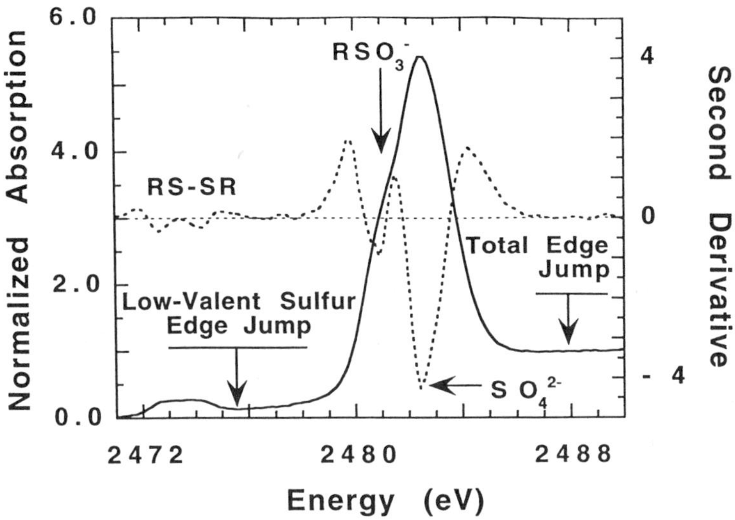

FIG. 5. *Solid line*: Sulfur K-edge X-ray absorption spectrum of a whole blood cell sample from *A. ceratodes*. The relative height of each edge jump is a direct measure of the relative quantities of sulfur types present. *Dashed line*: Second derivative of the same spectrum. The doublet near 2473 eV is characteristic of disulfide sulfur. The shoulder at 2481 eV, due to an aliphatic sulfonic acid, is now resolved from the sulfate transition at 2482.5 eV.

ratios of intracellular sulfate:sulfonate:disulfide to be obtained from the X-ray absorption spectra of three independent *A. ceratodes* blood cell samples collected over a period of 18 months. The results are summarized in Table 1. In all cases the ratios reflect an examination of whole blood. Thus the sulfur ratios are the average for the population of cells within each whole blood sample.

The finding of strongly broadened sulfate or sulfonate from the blood cell sulfur K-edge X-ray absorption spectra must mean that a dominant fraction, if not the entire quantity, of intracellular sulfate ions was maintained in acid, and where appropriate, with V^{III}. Indeed, in the case of the 0- and 18-month collections, the blood cell sulfate XAS features were so strongly broadened as to render unlikely not only the possibility that much of the endogenous sulfate could have been in neutral, V^{III}-free solution, but also unlikely that all the broadening could have been due only to endogenous pH 1.8 acid. That is, the excess

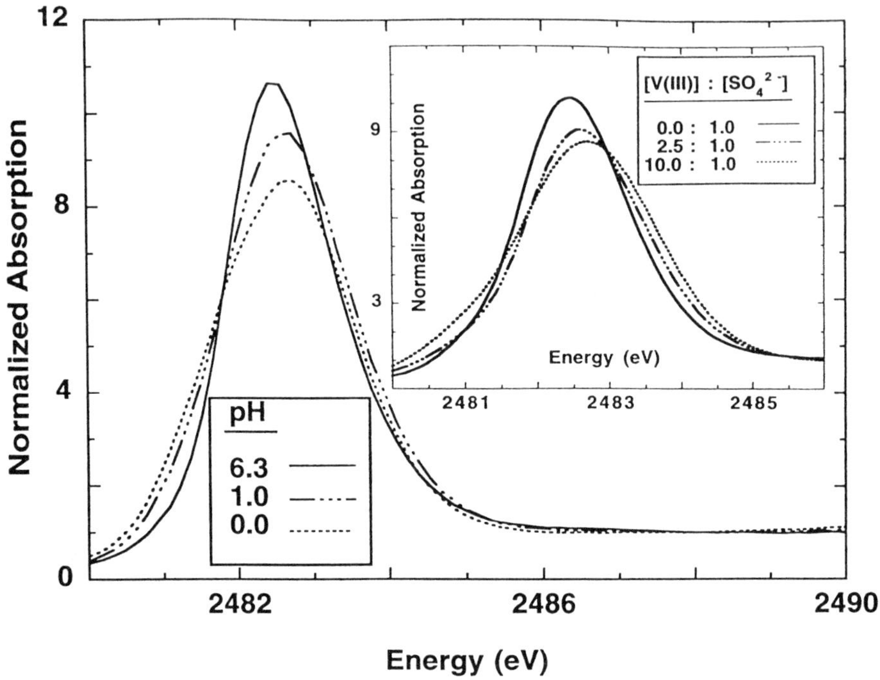

FIG. 6. Sulfur K-edge X-ray absorption spectrum of 0.1 M sulfate titrated with acid. The pH of each solution is given on the face of the figure. The isosbestic point of 2481.8 eV evidences the interconversion of sulfate and bisulfate. *Inset*: Sulfur K-edge X-ray absorption spectrum of 0.1 M sulfate in pH 1.5 sulfuric acid, on titration with V^{III}. The mole ratios are given on the face of the inset. Here the isosbestic point at 2483.0 eV indicates the interconversion of sulfate and the VSO_4^+ complex ion. The observed line broadening in each case reflects the symmetry change from T_d to C_{3v} about sulfur (see text).

broadening must have been caused by intracellular V^{III}. Other potential sulfate-complexing transition metals are in low concentration within *A. ceratodes* blood cells [53]. The fact that sulfate and V^{III} can interact intracellularly necessitates the assignment of these species, at least on occasion, to the *same intracellular locale* [53,60,68,69].

Interaction between sulfate and V^{III} implies that this vanadium is probably within a low molecular weight complex, with at least one exchangeable coordination position. Proton NMR experiments on whole blood cell packs have yielded evidence for four to five water-exchangeable

TABLE 1

Sulfur in Three Whole-Blood Cell Samples from *Ascidia ceratodes*

Ratio of:	Sulfate ($\pm V^{3+}, \pm H^{+}$)[a]	: Sulfonate ($\pm V^{3+}$)[b]	: Disulfide
Sample at:			
0 months	1.0 (+,+)	: 0.90 (+) :	0.37
12 months	1.0 (−,+)	: 0.50 (−) :	1.5
18 months	1.0 (+,+)	: 0.67 (−) :	0.24

[a]Plus or minus a deduced interaction between sulfate and V^{III} and/or acid protons.
[b]Aliphatic sulfonic acid S-K edge spectral linewidths are not sensitive to acid above pH 0.

sites on endogenous V^{III} [60,68,96]. Therefore, these results taken together suggest that blood cell sulfate is virtually indistinguishable from aqueous, acidic V^{III} sulfate. Though in general agreement with all the vanadium K-edge XAS results discussed in Secs. 2.2 and 2.3, these considerations do not include any vanadium which may be stored apart from sulfate.

The question [67,69,84] as to the choice of sulfate rather than chloride as counterion for intracellular V^{III} can be potentially answered by a consideration of ionic strength [69]. Complexation of V^{III} by sulfate [97] yields, for 0.5 M $V_2(SO_4)_3$, an ionic strength [69] of 2.0 M, compared with 6 M for a fully dissociated [60] 1 M VCl_3 solution. Thus the osmotic pressure gradient which must be maintained by the blood cell against the external blood plasma is greatly lowered with sulfate as endogenous counterion.

2.5. Conclusions and Summary

We have taken an XAS approach to elicit a view of the status of vanadium and sulfur within the blood cells of the tunicate *A. ceratodes*. These experiments have the advantage of an in situ measurement, yielding information about the native state of these elements. Whole blood cell packs were examined. Therefore the results must be taken to be averaged over all the blood cell environments. However, since all environments are represented, the data are not compromised by the high, possibly systematic, differential mortalities which can occur from losses during cell sorting, perhaps even within a single blood cell type [63,70,98–100].

The results from vanadium K-edge XAS experiments are unambiguous. The blood cell vanadium is ≥95% V^{III}. Vanadyl and vanadate ions cannot in combination account for more than 5% of the total blood cell vanadium. Comparative vanadium K-edge X-ray absorption spectra clearly imply that at least 90% of the blood cell V^{III}, despite the heterogeneous blood cell population, is in a highly uniform ligand environment that differs only slightly from that of V^{III} in 9 M sulfuric acid.

The results from vanadium K-edge EXAFS analysis are equally uniform and equally convincing. The strong beat frequencies from the carbon shells of V^{III}-*tris*-catecholate are completely undetectable within the EXAFS spectrum of blood cell V^{III}. Single-scatterer curve fitting of the blood cell vanadium X-ray absorption spectrum revealed only approximately six oxygen atoms, consistent with water oxygens, at 1.99 Å. This coordination environment is that expected from hexaaquo V^{III}. These experiments were carried out at 10 K. Thus, the possibility has been eliminated that any chelate-carbon contribution to the EXAFS could have been lost because of thermal noise. Any endogenous *tris*-chelated V^{III} complex which might be present can constitute only ≤10% of blood cell vanadium.

Sulfur K-edge X-ray absorption spectra show that there is a rich and unexpected variation of sulfur compounds within *A. ceratodes* blood cells and that endogenous sulfate can interact with dissolved V^{III} within an acidic intracellular region. Aside from possible differences among phlebobranch species, the sulfur K-edge XAS results may be qualified by *intra*specific differences arising from locale [69], seasonality, or other factors, especially variable blood cell population [15,39,101,102]. The results discussed here are certainly indicative of a highly variable blood sulfur chemistry. Overall these results are strictly relevant only to the undifferentiated whole blood cell population of the Monterey Bay *A. ceratodes*.

The problem of possible alternative vanadium storage regimens within the several blood cell types and the variations possibly pertaining thereto are obviously cautionary. However, the uniform vanadium XAS results obtained for two whole blood cell collections separated so widely in time (13 years) suggest that *A. ceratodes* maintains blood V^{III} in only a single major state, as ${V^{III}}_{aq}$.

3. TUNICHROME COMPLEXATION AND REDOX CHEMISTRY

3.1. Background

One of the most unusual attributes of tunicates of the suborder Phlebobranchia is their ability to reduce V^V, collected from seawater, to the V^{III} oxidation state in vivo [52]. In seawater (pH ~ 8.3), vanadium exists as monomeric vanadate ($H_2VO_4^-/HVO_4^{2-}$, pK = 8.3). In *Ascidia nigra* and *A. ceratodes*, for example, the vanadium stored in the blood cells is at least 90% V^{III} with small amounts of V^{IV} usually observed [57,61,62]. The vanadium oxidation states in ascidian blood cells have been determined by a superconducting quantum interference device (SQUID) [62] and other magnetic susceptibility measurements [57], EXAFS [61], EPR [52,53], and cytological staining [58,64]. Oxygen-sensitive V^{III} must be kept in a reducing environment in the blood cells; however, the blood plasma is aerobic to some extent and contains V^V [21].

Such intriguing vanadium redox chemistry must be mediated by a potent reducing agent. The standard reduction potentials (volts vs. NHE) for the V^V/V^{IV} couple and the V^{IV}/V^{III} couple are 1.00 and 0.337 V, respectively [103]. The prevalence of tunichromes in ascidian blood cells (up to 50% of the dry weight of morula blood cells [98]) and their electron-rich catechol and pyrogallol rings suggest that tunichromes are viable candidates for the native reducing agent.

Of relevance to this discussion is the well-studied redox chemistry between vanadium and catechols. This topic was reviewed and significantly clarified in a paper from Ken Raymond's group [81]. The standard reduction potential for the catechol/quinone couple is 0.795 V; that for the pyrogallol/hydroxyquinone couple is 0.680 V [104]. In aqueous media, reactions between V^V species and catechol generally involve complexation followed by reduction of the metal to the V^{IV} state [105–110]. However, stable *tris*-catecholate complexes of V^{V} have been prepared in methanol by using triethylamine [81].

The reactions observed between vanadium species and catechols are strongly influenced by the type of reaction medium (aqueous vs. nonaqueous) and by pH. For example, catechol and V^{V} or V^{IV} have generated semiquinones and/or V^{III} species, both potent reducing agents, in nonaqueous media (vide infra); however, such products have not been found in aqueous media. For comparison to catechol, the standard reduction potentials for *p*-hydroquinone/semiquinone (1.08 V) and semiquinone/*p*-benzoquinone (0.32 V) in aqueous media were estimated from electrochemical and spectrophotometric data [111]. In acetonitrile, the reduction potentials (volts vs. SCE) were measured to be 1.14 and 0.51 V, respectively [112]. The reports of catechols reducing vanadium to the +3 state are especially relevant to studies of vanadium chemistry in ascidians. In anhydrous tetrahydrofuran (THF) under anaerobic conditions, pyrogallol reduced $V^{IV}O(acac)_2$ to produce a pyrogallol-bridged V^{III} dimer with acetylacetonate ligands (Fig. 7) [113]. X-ray diffraction analysis of the dimer shows that the pyrogallol ligands are not oxidized. The portion of pyrogallol that reduced the V^{IV} starting material was apparently removed during filtration of the crystals. The reaction above occurs under neutral conditions. In strongly acidic, nonaqueous media (2 eq $HClO_4$ in acetonitrile), the $V^{IV}O$(salen) complex disproportionates to V^{V} and V^{III} complexes under anaerobic conditions [salen = *N,N'*-ethylenebis(salicylideniminato)] [114]. Although $V^{IV}O$(salen) is not a catecholate complex, the salen ligand has two *ortho*-substituted phenolic groups that are similar to catechols. The formation of V^{III} in these examples is strongly supported. The other examples of catechols reducing vanadium to V^{III} are based on uncertain charge distributions within the complexes produced. In toluene or anhydrous methanol, reactions between $V(CO)_6$ and 3,5-di-*tert*-butyl-1,2-benzoquinone and between $V^{IV}O(acac)_2$ and 3,5-di-*tert*-butylcatechol give the same product formulated as $V^{III}(DBSQ)_3$ (acac = acetylacetonate; DBSQ = 3,5-di-*tert*-butylsemiquinone) [115,116]. Shifting electronic charge from V^{III} to the semiquinone ligands would give catecholate ligands and a metal center in a higher oxidation state. Although a semiquinone ligand is clearly evident, proof of structure is not available. A similar

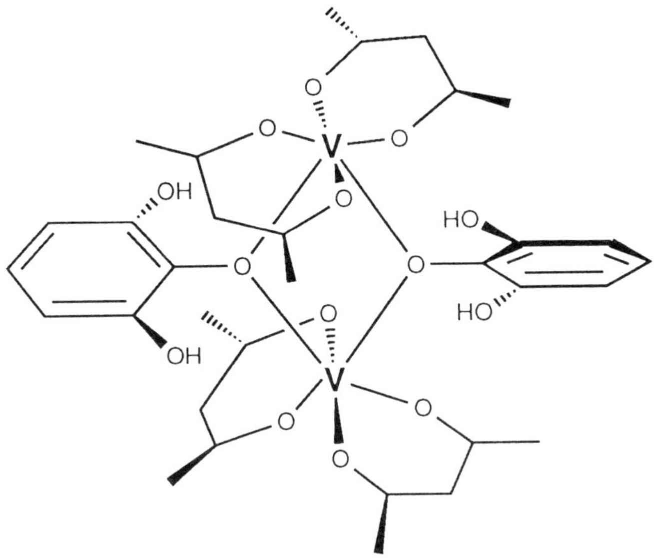

FIG. 7. Structure of the pyrogallol-bridged V^{III} dimer, $(acac)_4V_2[\mu\text{-}OC_6H_3(OH)_2]_2$.

semiquinone complex of V^{III} was postulated from the reaction of a heteropolyvanadate, $[PV^{V}_{14}O_{42}]^{9-}$, with 3,5-di-*tert*-butylcatechol in anaerobic acetonitrile [117]. The product was formulated as the dimer $[V^{III}(DBSQ)(DBcatH)_2]_2$; however, proof of structure is not available for this complex either.

In aqueous media, the reduction of V^V to V^{IV} by catechol and substituted catechols has been studied using stopped-flow spectroscopy. For a series of catechols with V^V perchlorate in acidic aqueous media (0.1–1.0 M $HClO_4$), the kinetics of the reactions indicate an initial complexation of V^V followed by reduction of V^V to V^{IV} followed by further reaction of the initial oxidation product [107,108]. The generality of this sequence was noted for a second series of substituted catechols in basic media (pH 8–9) [110]. It was noted that the rate of initial complex formation increases with increasing pH, whereas the rate of redox reactions increases with decreasing pH. In polarographic studies of catechol reactions over a pH range of 0–9, reduction of V^V to V^{IV} was observed from pH 3 to 9 [106]. At lower pH, however, reduction of V^V by catechol was clearly observed in the stopped-flow studies by the characteristic appearance of *ortho*-benzoquinone (λ_{max} 380–390 nm [118]) and by isolation of the reduced metal, V^{IV} [108].

Closer to the tunichrome structures are models comprised of three pyrogallol groups tethered by flexible linkages. Kime-Hunt et al. [28] studied the reactions of such models with V^V. The models reduce approximately 2 mol eq of V^V to V^{IV} according to UV-visible and EPR monitoring. The first study of reactions between tunichrome and vanadium was performed by Swinehart et al. [55] (although the term tunichrome was subsequently introduced by Macara et al. [119]). The yellow chromogen isolated from *A. ceratodes* blood cells showed the same absorption curve as more recently characterized tunichrome from this species (λ_{max} 340 nm) [2,13,120]. The chromogen is reportedly oxidized by V^V; however, no chemical evidence is given.

More extensive investigations conducted by Macara et al. [119,121] found that tunichrome readily reduces V^V and Fe^{III}. Reactions between tunichrome and vanadium (V^V, V^{IV}, or V^{III}) were monitored spectrophotometrically in the regions of tunichrome absorption (325 nm, ϵ = 30 000 M^{-1} cm^{-1}) and V^{IV} absorption (760 nm, ϵ = 17 M^{-1} cm^{-1}). No reaction was observed for additions of V^{IV} or V^{III}. However, additions of V^V produced a transient red-brown color (due to V^V complexation) and an attenuation of the tunichrome absorption at 325 nm. Simultaneously a peak appeared for V^{IV} at 760 nm. The redox behavior of tunichrome treated with Fe^{III} is interesting because tunicates of the suborder Stolidobranchia accumulate Fe^{III} from seawater and store the reduced metal, Fe^{II}, in their blood cells [121]. The reduction of Fe^{III} to Fe^{II} by tunichrome was monitored colorimetrically. Pink complexes of Fe^{II} with bipyridine or phenanthroline absorb at 520 or 510 nm, respectively. The initial redox reaction was found to be first order with respect to tunichrome and Fe^{III} [119]. Slower reactions were observed in which additional V^V and Fe^{III} were reduced. Tunichrome's ability to reduce Fe^{III} places its standard reduction potential below that of the Fe^{III}/Fe^{II} couple (0.770 V) [103].

More accurate studies of the redox reactivity of tunichromes were possible after tunichromes were isolated and their structures determined [2,13,120]. Pure tunichromes can be obtained by synthesis [122,123] or more readily from tunicate blood via a protection-isolation-deprotection method [124].

3.2. Spectroscopic Investigations of Tunichrome-Vanadium Chemistry

The initial studies of tunichrome treated with vanadium or iron were conducted in aqueous media at pH$\leq$3 [2,119,121], which may not approximate cellular conditions. The intracellular pH of vanadium-rich blood cells has been the subject of some controversy [53,68,70,73,75]. As the redox and complexation equilibria for reactions of vanadium and catechols are pH-dependent, the reaction medium strongly influences which reactions are favored. Therefore, prior to studying redox reactivity, we screened reactions between tunichrome and vanadium or iron in various media: deionized water, methanol (HPLC grade), and aqueous buffers at pH 2 (sulfate) and pH 7 (phosphate). These reactions were conducted under argon using cuvettes sealed with septum screw caps, thoroughly deoxygenated solutions, and airless techniques.

In the various reaction media, synthetically pure tunichrome Mm-1 (Fig. 1) was treated with up to 8 mol eq of V^V, V^{IV}, Fe^{III}, or Fe^{II} by stepwise additions. Reactions were monitored spectrophotometrically over the 900- to 200-nm range. In the various media, the absorption curve for Mm-1 is composed of a peak (λ_{max} 336–42 nm) and a slight shoulder around 300 nm. For all metals tested except V^V, no reaction with Mm-1 was observed in water, methanol, or pH 2 buffer. For additions of V^V to Mm-1 in water or methanol, the absorption curves are slightly attenuated in the presence of V^V relative to controls lacking V^V. For additions of V^V to Mm-1 in pH 2 buffer, the tunichrome peak became strongly attenuated, and the product showed two peaks at 285 and 318 nm and a weak shoulder at ~380 nm. In contrast to water, methanol, or pH 2 buffer, reactions were apparent for all metals tested with Mm-1 in pH 7 buffer. For V^{IV}, the addition of 2 mol eq (or more) of V^{IV} generates a product (λ_{max} 382–3 nm) that is quite air-sensitive (air exposure produces a spectrum similar to V^V treatment). For V^V, the addition of 2 mol eq (or more) of V^V produces a peak at 372–3 nm that is attenuated by air exposure. The broad absorption features produced by V^V with air exposure are strikingly similar to previously reported spectra of the following: a chromatographed An-2/V complex, signet ring blood cells lysed under argon, and lysed morula blood cells exposed to air [2]. It appeared from the recent experiments that reactions between tunichrome and vanadium or iron proceed to the greatest extent in pH 7 buffer relative to the other conditions screened.

The reactions of V^V and V^{IV} with tunichrome Mm-1 in pH 7 buffer were studied using EPR spectroscopy to determine the oxidation state(s) of product vanadium [125]. Considering the V^V, V^{IV}, and V^{III} oxidation states of interest, only the V^{IV} state is observable by EPR at room temperature [126]. V^V is diamagnetic, and paramagnetic V^{III} is generally EPR-silent above 4 K because of its non-Kramers' d^2, $S = 1$ electronic configuration [127]. The EPR method for determining the relative amounts of V^V, V^{IV} and V^{III} is illustrated in Fig. 8. The V^{IV} signal is first measured for a product mixture that has been acidified to pH 2. The acidification dissociates the EPR-silent, spin-coupled V^{IV} dimers and hydroxides present in situ at pH 3–11 (predominant at neutral pH) [53,128,129]. At pH 2, vanadium catecholate complexes are

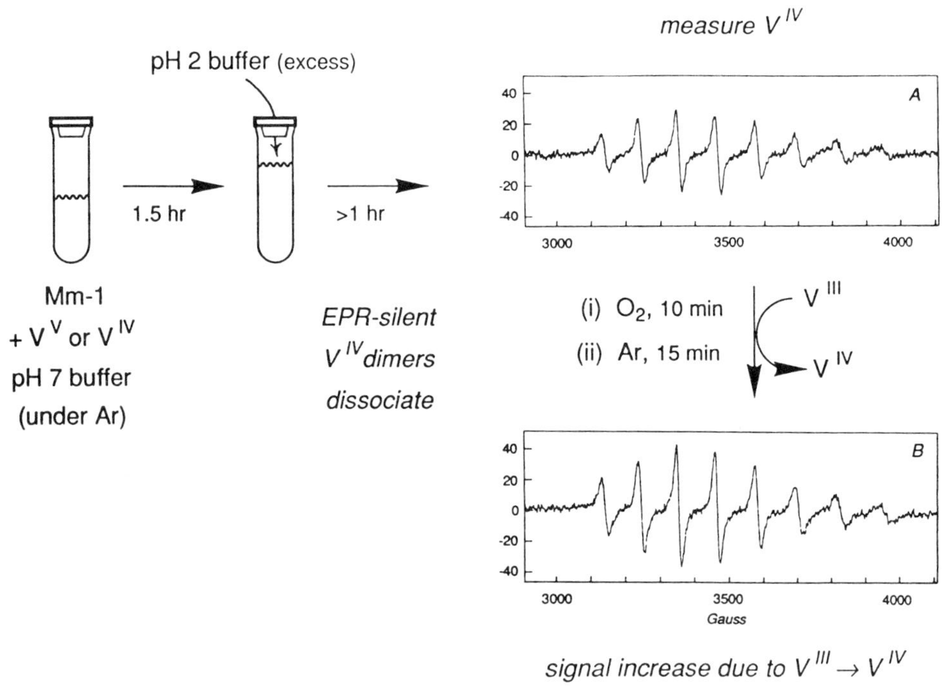

FIG. 8. EPR method for the analysis for V^{IV} and V^{III}.

also dissociated [105,106]; therefore only free pentaaquavanadyl ions are observed by EPR for the tunichrome reactions. After the V^{IV} signal is measured, the sample is bubbled with oxygen to convert V^{III} to V^{IV}. This oxidation of V^{III} to V^{IV} does not progress to V^{V} because of the pH of the acidified sample [24]. The sample is purged with argon to expel paramagnetic oxygen, and another EPR measurement is taken. The increase in the V^{IV} signal intensity reflects the quantity of V^{III} that was converted to V^{IV}. To calibrate the EPR signal intensity for the total vanadium in the sample, a control sample is prepared at the same vanadium concentration by using V^{IV} and no tunichrome. Using the measurements of V^{IV} and V^{III} (in a product) and total vanadium (in the control), one can calculate $\%V^{IV}$ and $\%V^{III}$ in a product mixture. The $\%V^{V}$ is readily calculated: $\%V^{IV} + \%V^{III} + \%V^{V} = 100\%$ total vanadium.

Reactions between tunichrome Mm-1 and V^{V} or V^{IV} were conducted under argon in pH 7 buffer, acidified with excess pH 2 buffer, and transferred to EPR tubes via cannula. The amount of vanadium starting material was varied as follows: V:Mm-1 = 1:1, 2:1, and 4:1 mol eq. All V^{IV} signals measured are superimposable on pentaaquavanadyl spectra in pH 2 buffer (intensities vary); no evidence of complexation was found in pH 2 buffer, as expected. The calculated percentages (not shown) of V^{V}, V^{IV}, and V^{III} suggested a large error (estimated at 20%) in comparing a product mixture to its control. More accurate calculations were made by comparing EPR measurements from the same EPR tube. Therefore, the ratios of V^{IV} to V^{III} in product mixtures were calculated. The ratios are shown in Table 2.

We tested whether redox reactions go to completion at pH 7, before acidification with pH 2 buffer which shifts redox potentials for V^{V}, V^{IV}, and catechols. The addition of excess EDTA instead of pH 2 buffer dissociated EPR-silent V^{IV} dimers and hydroxides in neutral aqueous media. For product mixtures treated with excess EDTA, the EPR measurements of $\%V^{IV}$ were the same as, or similar to ($\pm$ 15%), the measurements using pH 2 buffer, indicating that redox reactions proceed at pH 7. Measurements of V^{III} are not possible for the EDTA reactions because the oxygenation procedure at neutral pH converts V^{III} (and also V^{IV}) to EPR-silent V^{V}.

The EPR studies provide indirect evidence that tunichrome reduces V^{V} and V^{IV} to appreciable levels of V^{III}. The results suggest that V^{III} is produced under different conditions for V^{V} and V^{IV}. This differ-

TABLE 2

Ratio of V^{IV} vs. V^{III} Indicated by EPR Analysis of Mm-1 (see Fig. 1) Treated with V^{V} or V^{IV} (1, 2, and 4 mol eq) at pH 7[a]

Entry	Starting metal	V^{IV} found vs. V^{III} found
1	1 mol eq of V^{IV}	85:15
2	2 mol eq of V^{IV}	90:10
3	4 mol eq of V^{IV}	60:40
4	1 mol eq of V^{V}	Trace:$<$100
5	2 mol eq of V^{V}	60:40
6	4 mol eq of V^{V}	93:7

[a]Samples were adjusted to pH 2 before EPR measurement.

ence is evident by comparing 1 mol eq of each metal (entries 1 and 4). Whereas V^{IV} reactant was mostly not reduced (85%), V^{V} reactant appeared to be fully reduced to V^{III}. Reduction of V^{V} occurred at all mole equivalents of Mm-1 tested. In contrast, V^{IV} appeared to be more fully reduced when larger excesses of V^{IV} were added. This suggests that the redox equilibrium for V^{IV} depends on a complexation equilibrium with tunichrome or with itself, e.g., formation of V^{IV} pairs. The studies indicate that Mm-1 can donate at least four electrons to a suitably strong oxidant, such as V^{V} (entry 6).

Clues to the types of reactions observed spectrophotometrically for Mm-1 were obtained from spectrophotometric studies of catechol treated with up to 4 mol eq of V^{V} or V^{IV} in deionized water and aqueous buffers at pH 2 and 7. The concentration of these reactions was ~0.25 mM (10-fold more concentrated than the tunichrome reactions). These concentrations are significantly more dilute than those of other studies of vanadium-catechol reactions in aqueous media. For V^{V} with catechol, a reaction was not apparent in water, pH 4 buffer, or pH 7 buffer. In these media, a small, red-shifted shoulder may be due to a minor reaction between catechol and V^{V}. For V^{V} in pH 2 buffer, however, the oxidation of catechol (λ_{max} 275 nm) to *ortho*-benzoquinone was evident from the peak at 385–392 nm that developed over ~1 hr and then diminished with broadening [118]. The same peak was generated instantaneously by the addition of 1 mol eq of $NaIO_4$ to catechol in pH 2 buffer. Periodate is known to oxidize catechol to *ortho*-benzoquinone [130]. For V^{IV} in water or pH 2 buffer, no reaction with catechol was observed. In pH 7 buffer, however, the addition of 1 mol eq (or more) of V^{IV} produced a new peak at 292–3 nm that was relatively stable over time. This product was demonstrated to be a complex when EDTA (10 mol eq) was added to the product of catechol and V^{IV} (2 mol eq), and the original peak for free catechol was immediately restored.

The complexation of catechol by V^{IV} in pH 7 buffer suggests that tunichrome may similarly form complexes with V^{IV} in pH 7 buffer. Such complexes were indicated when the product of Mm-1 and V^{IV} (4 mol eq) in pH 7 buffer was treated with EDTA (40 mol eq). The original peak for free Mm-1 was restored after 1–1.5 hr, a significantly longer time than for the comparable complex of catechol treated with EDTA. A parallel study for V^V demonstrated that tunichrome forms complexes with V^V in pH 7 buffer. (A complex was clearly produced from V^V and Mm-1 under similar conditions in our X-ray absorption studies discussed below.) For V^V with Mm-1 in pH 7 buffer, EDTA addition to the product generates a spectrum after 1–1.5 hr that appears to be a mixture of free Mm-1 and free, oxidized Mm-1 (the product in pH 2 buffer; discussion follows). The catechol studies also suggest that V^V oxidizes tunichrome at pH 2. This was demonstrated as follows by comparing V^V and periodate treatments. The absorption curve for Mm-1 in pH 2 buffer treated with V^V (4 mol eq) shows two peaks of similar intensity at 285 and 318 nm. Under parallel conditions, treating Mm-1 in pH 2 buffer with $NaIO_4$ (2 mol eq) generated a strikingly similar curve with two peaks at 286 and 321 nm. For both reactions, a transient shoulder was initially apparent at 370–3 nm in the region of *ortho*-benzoquinone absorption. The lifetime of this transient was ~45 min for a tunichrome concentration of 0.021 mM. The final oxidation product has absorption features similar to those of Mm-1 itself. Studies of the tyrosinase oxidation of α,β-unsaturated *N*-acetyldopamine, a compound that is directly analogous to a portion of tunichrome, showed a transient quinonoid product (λ_{max} 500 nm) [131]. The absorption curve of the final product (λ_{max} 280 nm, shoulder at 310 nm) was quite similar to that of the starting material, but attenuated. A single product was isolated, characterized, and found to be a dimer. In the oxidation of Mm-1, the final product has not yet been determined.

The EPR method for analyzing vanadium oxidation states provides *indirect* evidence for the V^{III} state. Direct evidence was sought using XAS in collaboration with Patrick Frank, Britt Hedman, Keith Hodgson and group at Stanford University and the Stanford Synchrotron Radiation Laboratory. Tunichrome complexes of vanadium were prepared by treating Mm-1 and An-1,2 (see Fig. 1) with 1 mol eq of V^V. Therefore, we expected the metal to be reduced to V^{III} according to the EPR studies (Table 2, entry 4). Specifically, a solution of each tunichrome in methanol at 0°C was treated with 1 mol eq of V^V in aqueous NaOH (pH 11) under argon [Eq. (2)]. (Phosphate buffer at pH 7 was not used to avoid complicating the ligand environments for the EXAFS studies.) Greenish black precipitates formed from the Mm-1 and An-1,2 reactions. The precipitates were centrifuged and the pellets lyophilized. Dry pellets were ground with boron nitride in a glove box and loaded into sample holders for X-ray absorption studies. Vanadium catecholate complexes were prepared according to literature methods [81] to provide model complexes for comparison to the tunichrome complexes.

$$\text{An-1,2} \xrightarrow[\text{MeOH}]{+\ V^V \text{ in NaOH (aq), pH 11}} \text{green ppt.} \tag{2}$$

The K-edge X-ray absorption spectrum for the vanadium-An-1,2 complex is shown in Fig. 9. The most striking feature of the curve is the lack of a large pre-edge spike. Such a spike is characteristic of the V=O group(s) of vanadyl and vanadate complexes (e.g., vanadyl biscatecholate in Fig. 10). Vanadium complexes which lack oxo ligands, e.g., *tris*-catecholate complexes, do not show the large pre-edge spike. (For one oxo ligand, the intensity of the spike is generally ~50% of the height of the major edge jump.) The bulk of the An-1,2 complex appears to lack oxo ligands and could be a *tris*-catecholate-type complex.

The energy of the edge jump is useful for determining the formal oxidation state of the vanadium. The An-1,2 complex falls between V^{III} *tris*-catecholate and V^{IV} *tris*-catecholate but lies much closer to the V^{IV} complex (Fig. 10). This implies that the An-1,2 complex is predominantly in the V^{IV} oxidation state. In aqueous media, it is unusual to generate a V^{IV} complex that lacks an oxo ligand [132,133]. It is possible that a V^{III} complex generated initially was oxidized by accidental air exposure to such a V^{IV} complex.

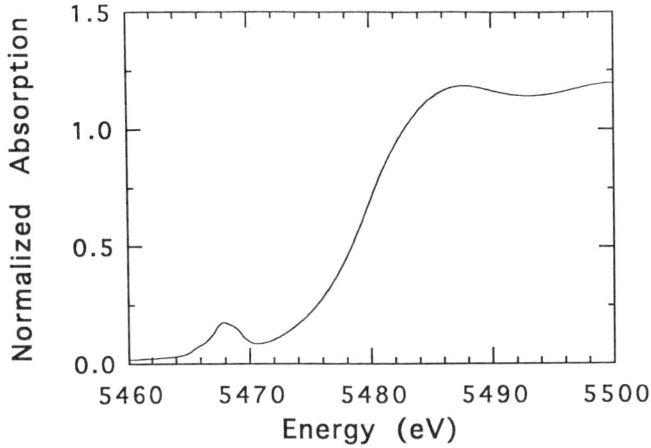

FIG. 9. Vanadium K-edge X-ray absorption spectrum of vanadium-An-1,2 complex.

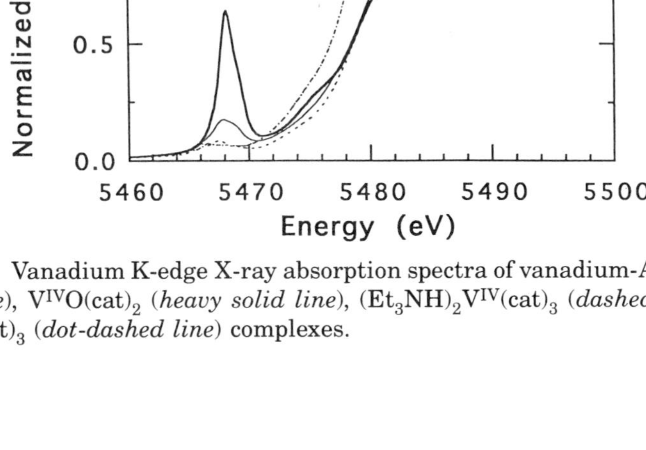

FIG. 10. Vanadium K-edge X-ray absorption spectra of vanadium-An-1,2 (*thin solid line*), $V^{IV}O(cat)_2$ (*heavy solid line*), $(Et_3NH)_2V^{IV}(cat)_3$ (*dashed line*), and $K_3V^{III}(cat)_3$ (*dot-dashed line*) complexes.

Comparison of the X-ray absorption spectra for the An-1,2 complex and the catecholate model complexes (Fig. 10) suggests that the An-1,2 complex may be composed of a mixture of V^{IV} and V^{III} states. The spectra for the catecholate models show similarities to the vanadium-An-1,2 spectrum; however, some features for the An-1,2 complex are not well modeled. We would like to determine the structure of the tunichrome ligand in the precipitate as it may be derived from tunichrome that was oxidized by V^V reagent. With this knowledge, pure V^{III} and V^{IV} complexes of the actual ligand can be prepared and analyzed by XAS. These spectra will allow an accurate determination of the vanadium oxidation state(s) in precipitates generated from tunichrome and V^V. Pure complexes will also allow interpretation of the EXAFS region. From this we hope to obtain coordination number, bond distances, and types of coordinated atoms.

3.3. Conclusions

Spectrophotometric studies of tunichrome (Mm-1) have demonstrated that (1) V^V oxidizes Mm-1 in pH 2 buffer and (2) V^V and V^{IV} form complexes with Mm-1 in pH 7 buffer. For the V^V reaction in pH 7 buffer, Mm-1 reduces V^V to V^{IV} as directly observed by EPR spectroscopy. EPR evidence suggests that tunichrome reduces V^V, and also V^{IV}, to the +3 oxidation state, V^{III}. We are currently screening reactions of Mm-1 with V^V or V^{IV} under various conditions for direct evidence of V^{III} formation.

4. ASCIDIAN ECOPHYSIOLOGY

4.1. Global Overview

It is somewhat peculiar and portentous that over seven dozen species of ascidians, comprising all three suborders, i.e., stolidobranchs, phlebobranchs, and aplousobranchs, are denizens of deep water, such as the Kuril-Kamchatka Trench at −8430 m [38]. Several related accounts on the ecological distribution of (epi)benthic macrofauna from deep, oxygen-limited basins off southern California (i.e., dissolved O_2 0.3–0.4 mL/L) [134], the Baltic Sea (0–4 mL O_2/L) [135], the Black Sea ($\ll$3 mL

O_2/L) [136], a fjord on Vancouver Island (<1 mL O_2/L) [137], and two Norwegian polls, i.e., smaller, more enclosed fjords (0–3 mL O_2/L) [138], collectively demonstrate that dissimilar species of ascidians tolerate prolonged periods of O_2 limitation to extraordinary extents. Within these remote basins, natural physical and biological discontinuities partition many nutrients into discrete zones, according to their relative utility, solubility, and redox potential, whereby vanishingly low O_2 levels effectively predetermine the lifestyles of all native organisms [139–142].

However, virtually all forms of life are inextricably linked to the catabolism (oxidation) of food as a ready source of nourishment (energy and cellular metabolites). To this end and because of a pervasive evolutionary trend, O_2 became the respiratory oxidant par excellence, owing to its high efficiency as a terminal electron acceptor (respirant), i.e., as many as 38 mol ATP generated/mol glucosylglycogen vs. 3 mol ATP/mol glucosylglycogen fermented to lactates, despite the fact that O_2 is also toxic [143]. In stark contrast to ascidians (vide infra), aquatic organisms with more modern respiratory systems thus tend to rely on aerobic catabolism to the fullest extent possible, provided oxygen tensions do not drop from as high as 12+ to below 1–3 mL O_2/L [138,139,144]. Nevertheless, most organisms experience bouts of O_2 limitation and have retained a limited capacity for anaerobic energy transduction. Indeed, many marine animals are demonstrably "good anaerobes," owing to the presence of multiple, compensatory mechanisms to cope with O_2 deprivation, e.g., *O_2 conformity*, a response whereby O_2 uptake rates drop abruptly; *electron transport system (ETS) arrest*, once O_2 or adenylate levels within mitochondria become rate limiting; and *glycolytic arrest*, such that carbohydrate stores are spared from the "Pasteur effect," i.e., overinduction of fermentative glycolysis [145–147]. Correspondingly, the maintenance of organismal (ion) homeostasis is often achieved through a reiterative modulation of ventilation, respiration, thermogenesis, osmoregulation, and dormancy, via coordinate control of membrane-mediated electron transport chains and ion channels, especially those which couple the oxidation of catabolically generated reducing equivalents, like NADH, to the regeneration of energy storage compounds. Yet despite such sophistication many "good anaerobes" must still cope with a potentially noxious collection of acidic end products, such as acetate, lactate, propionate, succinate, or volatile fatty acids, and have devised stop-gap measures to do so, depending on the duration and severity of O_2 limitation [145,148]. In effect, the buildup of such organic acids usually entails repaying an "oxygen debt" once O_2 becomes more available. According to one prominent investigator, "the picture emerging is that several processes contribute to the hypoxia tolerance of good anaerobes, but of these, metabolic arrest mechanisms yield by far the most effective protection against O_2 lack," whereby anoxia tolerance in clams (*Mytilus* spp.), diving turtles and brine shrimp can be increased by factors of 20, 60, and several orders of magnitude, respectively [149].

Analogous ecophysiological responses surely pertain to ascidians, for many varieties proficiently colonize biotopes with fluctuating O_2 levels, temperatures, salinities, illumination, and sulfur content [38, 138]. By comparison, certain facultative anaerobic microbes utilize one or more alternative respiratory oxidants during O_2 limitation, such as V^{V}, Fe^{III}, Cr^{VI}, Se^{VI}, Mn^{IV}, NO_3^-, or SO_4^{2-}, if not acetate, pyruvate, or fumarate, etc., as well [150,151]. Accordingly, even though perpetually stagnant, anoxic waters ordinarily possess less than 0.1 mL O_2/L and high levels of (noxious) hydrogen sulfide, they are seldomly devoid of life, largely because of the unique bioenergetics of such "sulfur-sulfide" biomes [152–154] (vide infra). Along these very hydrological lines, Dybern noted that the archetypical cosmopolitan ascidian [42], *Ciona intestinalis*, can live just above the H_2S zone in Vågsböpollen: "Dead *Ciona* specimens were found below the upper surface of the oxygen-free bottom water, while living individuals were settled just above it. The lowest ones had been successfully killed by the slowly rising oxygen-free water and were for a time preserved in it" [138]. Correspondingly, as Tunnicliffe observed in a fjord on Vancouver Island, "At −85 m . . . the epifauna is *dominated by ascidians* . . . despite less than 1.0 mL/L of oxygen over half the monitored time," i.e., 5 of 10 months [137]. Furthermore, in light of the nearly frozen biotopes that particular ascidians occupy, e.g., polar waters [38,155,156], it is also pertinent that they are poikilotherms ($\approx$ variable body temperatures) [157,158] whose longevity within and between species tends to increase with decreasing ambient (geographic) temperature [38,157,159–161]. The longevity of different species also appears to be related to the structure of the cuticle (tunic), in that a positive correlation exists between maximal lifespans and higher cuticle/body size ratios [160]. Along these lines, the juvenile mortality rates of (tropical) *A. nigra* drop abruptly at 19 days, once their black pigmentation (melanin) becomes visible [162–164], which may be derived from the oxidative polymerization of the tunichromes, because many blood cells, including several "polyphenol" varieties [165], are destined to senesce/lyse before and/or in the tissues that form the tunic (vide infra) [15,166]. By comparison, the settlement patterns of different species within two Norwegian polls were primarily influenced by relative salinity $>$ O_2 tension $\gg$ illumination [138]. Moreover, periodic shifts in salinity and O_2 availability within such oxic/anoxic haloclines elicit ecophysiological responses to maintain ion homeostasis that can be rather energy-expensive [158,167–169].

Somewhat similarly, sandwiched between other, deeper oxic/anoxic gradations lies a primeval dysaerobic [Greek *dys* (abnormal) + *aero* (air) + *bios* (life)] zone together with its own characteristic biome, where O_2 tensions persist at ~0.1–0.5 mL O_2/L [139,141]. The term "dysaerobic" originally denoted the microaerobic zone where magnetotactic bacteria [170,171] and algae [172] utilized iron or manganese oxides, instead of O_2, as terminal electron acceptors [171] and, in doing so, produced/secreted crystalline magnetite from their membrane-delimited magnetosomes [141,173]. Such *assimilatory* "iron respiration-driven proton translocation" is reversible once O_2 levels are no longer limiting [171,174]. By comparison, *dissimilatory* metal reduction by *Geobacter metallireducens* appears to involve the hydroquinone menaquinone, *c*-type cytochromes, and ferredoxin [175]. Both microbial processes bear a striking resemblance to a recent formulation that ascidian (blood) cells laden with hydroquinonoid tunichromes and/or metals like vanadium participate in analogous redox chemistries (vide infra) [27]. The biomineralization of magnetite carries considerable geochemical and ecological significance [151,176–178] as well as further topical intrigue, since four of the eight sampling sites where ascidians were retrieved from O_2-limited zones within the Black Sea displayed analogous ferromanganese concretions [136] (a.k.a. nodules or reduction spheroids) [178]. As noted above, the transition between dysaerobic and aerobic zones displays prominent microbial activity, sustained by substantial levels of nutrients in the form of detrital carbon and inorganic nitrogen [140,141,167,179]. It would thus afford a suitable habitat for invertebrate filter feeders, especially "good anaerobes" like ascidians (vide infra), whose basic diet consists of plankton and detritus, plus a few special metals. Consequently, a promising clue to the elusive purpose of vanadium for ascidians may reside within a related proposal that reduction spheroids are formed through microaerobic (respiratory?) redox processes of microbes because the most common authigenic mineral in their cores is roscoelite (~30%), which contains about 25% V^{III} in the form of air-sensitive V_2O_3 [178]. More to the point, Yurkova and Lyalikova [180] recently demonstrated that vanadate does indeed act as a terminal electron acceptor during anaerobic respiration, for two facultative chemolithotrophic bacteria isolated from a vanadium slag processing plant and from *Ascidia phalosia*. Both strains are reported to assimilate and reduce V^V to V^{III}, while a crystalline mineral that closely resembled sherwoodite $[Ca_3(V^{IV}O)_2V^V{}_6O_{20}\cdot 15H_2O]$ was obtained once the ultracentrifuged culture supernatant fluid was stored for several months [181], thereby raising intriguing implications for (global) vanadium geochemistry.

4.2. Evolutionary Perspectives

With respect to geological timeframes, reduction spheroids are found in continental and marine (iron) red-beds of Proterozoic and Phanerozoic Ages, from ca. 2 and 0.7 billion years ago, respectively, although most occur in more recent Permian continental red-beds from ca. 250 million years ago [178]. Within this broad timeframe, several pervasive biotic and abiotic trends in planetary metal ion metabolism seem to mirror the evolution of the earth's atmosphere, i.e., fluctuating increases in ambient O_2 levels [182–185]. Present day marine invertebrates from O_2-limited basins thereby offer an evolutionary approximation of a foregone but unforgotten era of metazoan species profusion [139,140,142, 148,185]. Current guesstimates as to the origin of ascidians place the original stock on the planet somewhere between 0.5 and 1 billion years ago [37,42]. By comparison, the only known record of a fossil ascidian is of a *Chelysoma*-like species in a Permian paleosurface from ca. 250 million years ago [155,186]. Along these lines and with respect to the origins of mineral nutrient requirements [187], the validity of an evident correlation between ascidian phylogeny and specificity of metal ion accumulation [52,55,64,188] has been revisited and questioned by Michibata et al. [189], i.e., the relative distribution of vanadium, iron, and manganese within diverse, solitary species. In any event, the novel roles of vanadium within dissimilar ascidians almost surely overlaps several of those performed by iron in other species.

It is also noteworthy that symbiotic associations between microbes and invertebrates are very common within nutrient-deficient marine habitats [154,179,190–192]; such associations are frequently thought to have facilitated the evolutionary advancement of particular organisms [193], including ascidians [194]. Three intriguing instances of modern ascidian symbioses thus merit brief mention, although such associations *appear* to be restricted to a few specific genera [38]: (1) the tripartite (bacterial-fungal-ascidian) symbiosis of *Molgula* (*manhattensis*) sp. [195,196] (*M. manhattensis* sequesters tunichrome Mms and iron); (2) shallow water didemnids [which produce tunichrome-like lamellarins (Fig. 1) and supplement their O_2 requirements via oxygenic photosynthetic bacteria] [7,197,198]; and (3) *C. intestinalis*, which can maintain blue-green alga-like cells in its tunic [159,199] (and assimilates vanadium, iron, and lesser amounts of manganese). Interestingly, *C. intestinalis* from the (shallow) harbor of Palermo, Italy, displays a distinct preference for well-illuminated habitats, to the point where it "certainly dies in the dark" [199,200]. Yet the opposite trend applies to (deeper) Norwegian polls, where "*Ciona [intestinalis]* prefer shady places . . . [although] survives fairly well also in light positions. It then sometimes adopts the reddish colour [e.g., melanin] known from other populations, mostly in shallow waters" [138]. A companion report notes that *C. intestinalis* in a Swedish fjord "of the generation which has lived through the summer (gen. I) are covered with a 'skin' consisting of bluish-green [symbiotic?] algae and other organisms. The individuals of gen. II, which have their first period of growth in the late summer and autumn appear rather naked and transparent" [159]. Evidently, varieties like this cosmopolitan species are versatile and poised to adapt divergent ecophysiological habits, depending on which nutrients become truly limiting. The relatively sparse literature thus conveys a tentative indication that microbial associations are not obligatory to all (present day) ascidians.

More to the point, ancestral ascidians are frequently placed at the very origin of vertebrate evolution, for they possess a perforated pharynx, a dorsal tubular nervous system, a notocord, and the early developmental patterns of primitive vertebrates, yet they lack several crucial specializations that true vertebrates embody [37,42]. Apparently, ascidians per se arose before the shift from passive (filter-feeding) to more active forms of predation occurred, e.g., efficient locomotion. In particular, ascidians lack ancillary improvements in cardiovascular design and respiratory ventilation, as do their many, rudimentary, physiological traits [37,39,148,201–203]. For instance, the respiratory gas carrying capacity of ascidian blood is marginal and equal to seawater, nor does it bind O_2 reversibly [9,204–206]. Accordingly, the purported increase in ascidian body size over evolutionary time, from less than 1 mm to over 10 cm, has been postulated to be a reflection of decreasing respiratory constraints, owing to an increase in O_2 availability, insofar as ascidians appear to be "almost [O_2] diffusion-limited" animals [205,206]. However, this interpretation overlooks the chronic bouts of hypoxia experienced by the very same (intertidal) species [66,207,208], as well as the presence of modern sizes within deep basins, where vanishingly low O_2 levels persist, ~0.3–0.4 mL/L [134]. In other words, the purported increase in ascidian body size probably reflects evolutionarily derived innovations in anaerobic metabolism much more accurately.

As with vertebrates, the embryonic loci for many evolutionary improvements in ascidian metabolism reside within neural crest cells, which differentiate into a wide variety of cells and organs during embryogenesis [202,203,209]. According to Barrington [209], neural crest cells have "retained a primitive capacity for tyrosine metabolism and the handling of amines, leading either inside or outside the nervous system, to the formation of melanin and catecholamines." In other marine invertebrates, the respiratory pigment haemocyanin is likewise thought to be evolutionarily derived from a tyrosinase-like ancestor [210]. Somewhat similarly, tyrosinase inhibitors in *C. intestinalis*, *Ascidiella aspersa*, and *Phallusia mammillata* exert the same effects as inhibitors of aerobic respiration (vide infra) [162,200,211,212].

Likewise, the hemopoietic (pharyngeal, digestive, and body wall) tissues that give rise to the circulating blood cells of ascidians are reported to stem from neural crest cells [16,19,201,213]. Correspondingly, the morula (blood) cells and other "chromaffin" cells, i.e., those which stain deeply with chromium salts, of ascidians [214], display a striking capacity to sequester particular ions and catecholamines, e.g., vanadium and the (tyrosine-based) tunichromes (Fig. 1) [98]. Another relevant differentiation product of neural crest cells that coordinates metabolic shifts in (ion) homeostasis are iodinated (tyrosine-like) catecholamines from the thyroid gland [215]; endogenous, iodinated catecholamines also stimulate polyphenol oxidase activity during ascidian tunic biogenesis [162,164,201,209,216–220].

In view of the migration of diverse blood cell subpopulations (several of which almost surely contain vanadium and/or some form of tunichromes, i.e., "polyphenol cells") to the site of tunic biogenesis [14,15,39,165,166,221–225], an early (pretunichrome) histochemical observation of Kalk [14] merits reiteration: "One feature seems common to all the cell activities invoked in the formation and activity of the pigment, haemovanadin [i.e., an erroneous, haemoglobin-like inference concerning the vanadium complex], in *Ascidia pygmaea*: in every kind of cell described the sphere of activity, at any one time, is cut off from the rest of the cell by a vacuolar membrane. The membrane may itself participate enzymatically in the activity directly or indirectly." Another promising clue for vanadium's bioactivity may thus stem from the role of migrating blood cells in the production of cellulose-like polymers within the tunic [224,226], i.e., cellulose biosynthesis seemingly requires a membrane (electrical) potential for full activity of the plasma membrane-associated multienzyme complex, at least in certain plants and microbes [227–229]. Vanadium in green algae likewise appears to participate in photosynthesis and chlorophyll formation [230]. In view of this and the associated bioenergetics of ascidian tunic biogenesis, it is noteworthy that preparations of chromaffin cell ghosts (from vertebrates) possess the capacity to synthesize ATP, if an electrochemical proton gradient is imposed, owing to the presence of membrane-bound proton-translocating ATPases and monoamine/H^+ antiporters [231,232]. The same applies to the vanadate-sensitive, reversible proton-translocating ATPase of turtle urinary bladders, vis-à-vis its coupling to anaerobic glycolysis [233–235]. In general, as elaborated by Mitchell [236,237], many bioenergetic processes within cells involve a relatively impermeable (intact) lipid bilayer to maintain an electrochemical (chemiosmotic) ion gradient, wherein membrane-bound "translocators" drive the vectorial shuttling of electrons and protons or other ions through channels to regenerate energy storage compounds or perform useful work. More often than not, such translocators are proteins, e.g., ATPases, although lipophilic hydroquinones can also be instrumental [236,238–244], e.g., ubiquinone (coenzyme Q) in the membrane-based,

ETS-driven, oxidative phosphorylations of mitochondria. Primeval hydroquinone-laden vesicles have even been proposed to have facilitated energy transduction in prebiotic and primordial cells [185,245,246], whereas a more proximal analogy to tunichrome's predicted involvement in tunic biogenesis, vis-à-vis hemovanadin [226], and anaerobic energy transduction [27] would seemingly reside within the evident blood-clotting [247] and energy-transducing [239] activities of the hydroquinone, vitamin K (vide infra).

4.3. Comparative Biochemistry

For several decades, it has been implicitly assumed that ascidians require only modest amounts of catabolic oxidants, owing to their very low rates of O_2 uptake. However, the decisive compensatory bioenergetics which underlie ascidian (ion) homeostasis would seemingly argue otherwise, i.e., microaerobic, haloclinal, or thermogenic shifts that can be energy-expensive, not to mention additional allocations toward metal ion assimilation/reduction and tunic formation. A similar misapprehension prevailed with helminths, for Malcolm Smith [248] observed "a continuous rearguard action to bring all such data into line with the pre-existing belief that such parasites require little or no oxygen in vivo." A comparable interpretation was offered by Webb [54] in 1939: "There seems to be no reason why ascidians should require any respiratory pigment at all; they are completely sedentary, and their structure is such that they offer a wide surface for respiratory exchange by diffusion. The data tabulated by Krogh (1916) show that the only forms whose rate of respiratory exchange approaches the low figures of the ascidians are echinoderms, coelenterates, lamellibranchs [which produce tunichrome-like lamellarins], and opisthobranchs [which produce bursatellin (Fig. 1)], none of which are provided with a respiratory pigment." On the other hand, as early as 1847 a German physiologist had already reported that upon exposure to air for a few minutes, ascidian blood turned from yellow-green to deep blue [249], in spite of its more recently documented inability to bind O_2 reversibly (vide supra). Correspondingly, the fact that *P. mammillata* turns blue before its death is generally attributed to oxidation of the greenish vanadium chromogen [1,54,56]. It has even been suggested that a respiratory role of blood from marine invertebrates that lack oxygen carriers may be so small that anaerobic pathways exist in deep tissues even under aerobic conditions [250]. The O_2 tension of ascidian blood plasma has since been reported to be very low and, if raised artificially, cellular V^{IV} levels rise concomitantly, via V^{III} oxidation [64]. Indeed, numerous studies have now confirmed that a variety of ascidians display the hallmarks of good anaerobes, including the O_2 uptake, respiratory, and ventilatory responses of true oxyconformers [39,66,169,251–255], a common trait for many invertebrate filter feeders (noted above) within analogous biotopes. To place this in perspective, the low O_2 uptake rates of *A. nigra*, which stores copious quantities of tunichrome Ans and vanadium, is comparable to the clam *Mytilus edulis* [39], a "good anaerobe" that has been estimated to be able to endure O_2 deprivation for 150 days, based on the presence of the succinic dehydrogenase (SDH) pathway of anaerobic energy transduction and its sizable glycogen stores [147]. Correspondingly, Fisher [252] found that it was "possible to maintain *Styela* [*plicata*] in the laboratory for several weeks at temperatures as low as 2–3°C, but no pumping or squirting behavior was observed," whereas Kajihara has reported that O_2 consumption by *M. manhattensis* (which accumulates tunichrome Mms and iron) dropped from the already low rates of $\sim$0.08 to $<$0.01 mL/g/hr upon exposure to hyposaline conditions [169,254]. In contrast, as noted above, shallow water didemnid ascidians supplement their respiratory requirements for catabolic oxidants via oxygenic, photosynthetic, symbiotic bacteria, whereby "discontinuous respiration" may prevail [197,200].

At this juncture, despite an acute shortage of physiological investigations on the oxidative metabolism of ascidians, several preliminary findings and (correspondingly) tentative indications merit comment and further investigation. Compared to most other marine animals [256], ascidian tissues appear to possess surprisingly low levels of classical energy storage compounds, i.e., the adenine nucleotides (and their derivatives) as well as arginine and creatine phosphagens [257–259]. Like many marine invertebrates, however, ascidians possess prodigious levels of taurine (2-aminoethanesulfonic acid) and proline, as well as glutamic acid, glycine, and alanine [258–260]—amino acids that can be instrumental in homeostatic responses to shifts in salinity and O_2 availability [147,158,261–263]. The presence of large glycogen stores [166,257,264–266] and a functional Krebs cycle [207], as with other animals, could be interpreted as an adaptation to chronic O_2 limitation [147]. Even so, Schofield and Stewart [207] found that the operation of the Krebs' cycle in the intertidal ascidian *Pyura stolonifera* "may be minimal" and that "the rate of glucose utilization . . . was found to be extremely low"; these already low glycolytic rates were largely unaffected by O_2 limitation [267], thereby suggesting that such ascidians exhibit a *reversed Pasteur effect* and, hence, *metabolic arrest* [146,147,149], i.e., such animals might depend on anabolic pathways that closely modulate carbohydrate and lipid metabolism. Moreover, high levels of lactate [257], lactate dehydrogenase [207], and mitochondrial SDH [200,207,212] offer clear indications that numerous species probably utilize one of a variety of anaerobic pathways, depending on the duration and severity of metabolic stress (vide infra). Along these lines, the incisive work of Hatcher [255] on the use of respiratory quotients (RQs) for determining catabolic substrates found that the ascidian *Herdmania momus* predominantly catabolizes phospholipids during several months of progressive nutritional stress (starvation); phospholipid catabolism also seems to prevail in developing eggs of *A. ceratodes* and *C. intestinalis* [268,269]; moreover, *C. intestinalis, P. mammillata*, and *A. aspersa* eggs and embryos appear to possess a somewhat related yet unconventional pathway of respiratory oxidation (vide infra) [162,200,211,212]. Correspondingly, the O/PO_4^{3-} RQ ratios of *H. momus* maintained in ambient (*oxic*) seawater were actually *lower* than the theoretical limits, which are "based on the amount of oxygen required," thereby indicating that *catabolic oxidants besides O_2 are utilized routinely*. The relevance that "anoxic endogenous oxidation" of unsaturated phospholipids is one of many anaerobic responses utilized by other marine invertebrates, such as *Mytilus galloprovincialis*, shall be discussed in detail below [270–274]. The latter mechanism also appears to be instrumental to the "detoxification" of heavy metals [275]. It is thus pertinent that vanadate exhibits prooxidant properties toward lipids [276].

To date, of the few (known) conventional mechanisms that ascidians might activate in order to sustain prolonged periods of O_2 limitation [200,207,212], mitochondrial SDH looks exceptionally promising. Many good anaerobes rely on it, i.e., up to 4–6 mol ATP/mol glucose catabolized via reductive hydrogenation of fumarate by SDH (in reverse) [147,148,250,277]. Moreover, several species of ascidians convey distinct indications of the aforementioned unconventional metal-mediated "alternative pathway of oxidation" [27,200] that may also involve SDH [212]. By comparison, the lactate pathway could (in theory) generate ~3 mol ATP per mol glucosylglycogen fermented. An alternative that is often utilized for anaerobic work that entails either a high rate of energy output for a short duration, e.g., diving seals (<1.2 hr), anoxic goldfish (<4 days), or a metabolically arrested output over a prolonged period, e.g., hibernating turtles and frogs (several months of torpor) [147,148,278]. In light of the above, one might question whether any combination of *conventional* anaerobic pathways and responses could adequately fulfill the *elemental* requirements of ascidians (vide infra), especially those encumbered by perpetual O_2 limitation. Under the most adverse conditions of nutrient deficiency (starvation?), species such as the colonial *Perophora viridis* (which accumulate tunichrome Ans and vanadium) undergo a profound morphological regression whereby the entire zooid (body) is resorbed into a vegetative stolon, only to be regenerated once growth conditions improve [201,279–282]. By comparison, several other species of ascidians endure adverse growth conditions via hibernation or aestivation [38], i.e., metabolic arrest.

4.4. Implications for the Tunichromes

Section 2 above has persuasively outlined how vanadium in the blood cells of *A. ceratodes* appears to be maintained in a single major state, as V^{III}_{aq}. As Frank and Hodgson noted, such findings do not rule out the potential presence of as much as 10% V^{IV} as *tris*-chelates, 10% V^{III} as *tris*-chelates, and 5% vanadyl/vanadate. Therefore, their results, together with several histochemical findings noted below, do not detract from the inference that endogenous organic ligands drive earlier stages of vanadium assimilation. Indeed, in light of the aforementioned ecophysiological constraints imposed on ascidians in general, one is hard-pressed not to envision that organic ligands mediate transition metal assimilation/reduction, namely, water ligands alone, as in V^{III}_{aq}, are unlikely to facilitate either metal assimilation or reduction. With regard to the prospective ecophysiological involvement of the tunichromes, two recent compendia on marine natural products disclose an intriguing array of related secondary metabolites from ascidians and diverse marine invertebrates (Fig. 1) [36,283]. The striking structural and ecophysiological commonalities disclosed by such representatives merit close scrutiny, i.e., halocyamines [4], (dihydro)tunichrome Pms [3], (iron-binding) ferreascidin [5], celenamides and clionamides [6], lamellarins [7], bursatellin [8], and Fe^{III}-adenochromines [10]. As noted above and previously [2], these and numerous other more distantly related redox agents, i.e., the sulfated siphonodictals [11] and sulfated polyclinal [12], are *air-sensitive* and potent *metal chelators* and all occur in animals prone to periods of O_2 limitation. Therefore, the prospect that a unifying purpose underlies the common chemistries of this entire class of natural products does indeed bear fantastic implications [35]. To this end, the implementation of a universal assay for potent metal ligands might disclose novel variants [284]. Along these lines, brief mention of yet another vanadium-binding substance, "vanadobin," from diverse ascidians is warranted in light of its distribution and capacity to complex/reduce V^{V} to V^{IV}; however, its proposed role in ascidian vanadium assimilation is difficult to scrutinize because its structure remains unknown [22].

Parenthetically, a misapprehension [3,35,40,41] that the tunichromes Ans and Mms (and, by implication, celenamide D; Fig. 1) from *A. nigra*, *A. ceratodes*, *M. manhattensis*, and *P. viridis* [2] represent (vinylogous) artifacts of the isolation procedure overlook several critical details. (**1**) The tunichrome Ans and Mms and their peracetates are *vivid yellow*, e.g., An-1 UV (MeOH) 210 nm (ϵ = 68 000), 245 (sh), 285 (sh), 340 (19 600) [120]; An-1 Ac UV (MeOH) 280 (sh), 319 (26 000) and *highly fluorescent*, as are the tunichrome-laden (morula) cells from such species, i.e., 20–50% tunichromes on a dry weight basis [2,98]. (**2**) In stark contrast, the (dihydro)tunichrome Pms have been isolated from *P. mammillata* (a.k.a. *Ascidia mammillata*) [3], which possess "nonfluorescent" [101,285] morula cells (a.k.a. *vanadocytes*), whose pigmentation ranges from "*faint yellow*" to "*faint yellow-green*" [56] to "*greenish*," but this is sometimes so pale as to be scarcely perceptible [54]. Such cells afford organic ligands that are "*faint yellow*," whereas other species, like *Phallusia fumigata*, possess both "*colourless*" and "*bright lemon yellow*" morula cells [54,64]. (**3**) Accordingly, closer structural and spectrophotometric approximations to the (dihydro)tunichrome Pms (λ_{max} 301 nm [3]) reside with "*colorless*" clionamide (peracetate) [286,287] and "*white*" celenamides A and B (peracetates) [288], e.g., UV (MeOH) 227 nm (ϵ = 59 000), 290 (37 000) and 227 nm (72 000), 289 (51 000), which possess a diminished potential for extended conjugation through the backbone, compared to the tunichromes Ans, Mms, and celenamide D (Fig. 1). Therefore, in light of celenamides A–D, it is likely that some ascidians possess a *natural* mixture of tunichromes Ans and Pms, e.g., *P. mammillata* [3]. (**4**) It is equally implausible that species like "*A. ceratodes* . . . generate metal-free [divinyl] tunichrome on cell lysis . . . [from a] . . . precursor compound, which is thought to be a 3,4,5-trihydroxyphenylalanine-containing protein" [40,41], for the introduction of two vinyl groups within *rapidly freezed-dried* morula cells [2] would have to be invoked (within an organism whose metabolism is extraordinarily reducing [56]), not to mention that fresh morula cells from *A. ceratodes* [UV ("plasma-saline solution") 245 nm (sh), 315 (br sh), 410 (br sh)] [21,55,98] and *A. nigra* [98] exhibit similar fluorescence and *vivid yellow* pigmentation.

A related misapprehension that "if tunichrome is involved in the reduction and accumulation of vanadium ions in the ascidian blood cells, it would be necessary that the tunichrome should be contained in the signet ring cell. . . . However, no fluorescence due to the tunichrome was detected in this cell" [20,40,289] overlooks several key details, all of which have been covered in previous reports and rebuttals. (**5**) Specifically, paramagnetic metals effectively quench autonomous fluorescence [290] and (V^{III}-laden) signet ring cells emit negligible to no fluorescence *compared to virtually any other kind of cell* [2,34,291]. (**6**) Microelemental findings on vanadium uptake by *Ascidia gemmata* blood cells comport with the inference that vanadium-laden signet ring cells represent the (metabolically driven?) V^{III} terminus of vacuolated blood cell differentiation, even though "compartment cells . . . appeared to incorporate the highest amount of ^{48}V. Signet ring cells and morula cells accumulated smaller amounts of ^{48}V" [292]. (**7**) Additional and mutually compatible findings on the distribution of *free* tunichrome and total vanadium in *A. nigra* and *A. ceratodes* blood cell subpopulations were obtained via fluorescence-activated cell sorting (FACS), together with acetylation assays and atomic absorption elemental analyses [98]. The FACS technique minimizes errors due to cell lysis or extracellular contamination because it selects only intact cells, the integrity of which is reflected by the optical parameters monitored. In complete accord with point **6**, the yellow to yellow-green morula cells (characterized by multiple vacuoles) held nearly all of the *free* tunichrome and up to 30% of the vanadium found (in *fluorescently dimmer* subpopulations), whereas the green-grey signet ring cells (which contain one large vacuole) yielded only a trace of *free* tunichrome and the bulk of vanadium [98]. These data thus offer persuasive evidence that *tunichrome and vanadium reside within the same cell* and, hence, the same (morula cell) vacuole, i.e., predictably as somewhat insoluble, greenish vanadium chelates (vide infra). Points **6** and **7** thus harmonize with the contentious inference, e.g., [18], that upon accumulating vanadium, morula cells differentiate into signet ring cells. The above distribution pattern for vanadium also agrees with earlier X-ray microanalyses [84,93,293–295] and more recent staining techniques using 2,2′-bipyridine and 1,10-phenanthroline. The latter approach further indicated that some morula cells appear to possess V^{III} [64,296]. (**8**) Correspondingly, Anderson and Swinehart [21] noted in *A. ceratodes* that "some cells having the vacuole size of pigment cells [their term for the vanadium-laden (signet ring) cells] but the colour of morula cells or a colour intermediate between that of pigment and morula cells were observed . . . those cells which appeared similar to pigment cells but had the colour of morula cells did not show the [sulfur/vanadium] ratio characteristic of the pigment cell but that of morula cells." (**9**) Transition metal assimilation/reduction in ascidian blood cells is massive (vide supra), which may thus entail stoichiometric (quasi-reversible?) ligand oxidation such that if the tunichromes were true vanadium ligands in vivo, they would eventually be polymerized (oxidatively) into a product that defies ready detection [34]. The above interpretations would thus reconcile the predicted involvement of the tunichromes in an earlier stage of vanadium assimilation with the virtual absence of *free* tunichromes in signet ring cells, where V^{III}_{aq} appears to accumulate as the quasi-terminal complex. Indeed, exposure of morula cells to 0.1 M HCl transforms them into a variety that "resembles a signet ring cell" visually [41] and spectrophotometrically [2]. Once again, an analogous, natural transformation has been predicted [18,34], which could also reconcile the acidic vs. neutral pH blood cell controversy concomitantly, namely, vanadium uptake by morula cells, at a neutral [75] pH, could lead to their differentiation into (acidic [70]) signet ring cells, owing to the extraordinarily reducing metabolism of ascidians (vide supra). The microelemental analyses of *A. ceratodes*, *A. nigra*, and *A. gemmata* blood cells [21,98,292] thus confer considerable support to the inference that upon accumulating vanadium morula cells mature (differentiate) into signet ring cells, possibly via compartment cell subpopulations. If true, the natural gradations observed in cell coloration, fluorescence, degree of vacuolation, chemical composition, and so forth, would readily account for the ill-defined boundaries between these three subpopulations of vacuolated blood cells.

With regard to potential structure-activity relationships for ascidian vanadium complexes, the tunichrome-vanadium precipitates noted in Sec. 3.2 recall high molecular weight vanadium complexes isolated from ascidian blood cell lysates [2], i.e., apparent multinuclear network polymers and, hence, "dendritic" tree-like supermolecules [297]. Along these lines, according to the histological analyses of Michael Smith [15], "A cell is considered a *signet ring cell* when the vacuole is larger than the nucleus and when the vacuole *does not contain granular inclusions*." Likewise, Barrington and Thorpe [165] noted several subpopulations of chromaffin-positive, "polyphenol" blood cells that *degranulated naturally*, when they arrived at the mantle epithelium, i.e., site of tunic biogenesis, and also during histochemical fixation. These observations harmonize with the predictable dissociation of putative tunichrome(quinone?)-vanadium complexes in vivo, vis-à-vis endogenous degranulation. Therefore, by definition, it is unwise to expect that highly differentiated [15,21], vanadium-laden [98], seemingly acidic [22,70] signet ring cells should readily avail evidence of *free* tunichromes or *insoluble* vanadium complexes [14,15,18,64,93,223,293, 298], as opposed to V^{III}_{aq}, despite compelling, indirect, microanalytical [21,98], and spectrophotometric [2] evidence that insoluble tunichrome-vanadium complexes probably occur in an earlier stage [34] of the very same cell, i.e., its *putative* morula cell precursor. Even so, one additional, mutually compatible alternative to the (exclusive) uptake of vanadium by circulating blood cells is that blood cell differentiation (hemopoiesis) may involve some degree of vanadium prepackaging, prior to hemocyte release into the blood plasma. If this is actually true, one inference would be that signet ring cells are not derived from differentiated morula cells at all. However, little if any credible evidence supports this unphysiological inference [14–22,64,165,298–301].

In sum, ascidian hematology is confounded by a tenuous consensus concerning cell terminology and cellular trends within and between species. The accurate recognition and classification of blood cell subpopulations here are fraught with deeply entwined alterations in distribution, coloration, morphology, chemical content, pH, fragility, vitality, and so forth, which occur during differentiation, migration, and eventual senescence [18,22,27,34,98]. In any event, an issue of paramount importance still looms above and beyond such intricacies, i.e., *What (organic) ligands and underlying bioenergetic mechanisms drive vanadate up a steep gradient such that* V^{III}_{aq} *appears to be produced eventually?* One plausible model outlines how tunichromes (and homologous hydroquinonoid ligands) might mediate an energetically favorable uptake of particular oxoanions (phosphate, vanadate, etc.), followed by the reductive trapping of certain subvarieties [27,34,119,299], a mechanism whose farthest ancestors may actually antedate the origin of life [246]. Simply put, if ascidian ecophysiology is truly encumbered by an endemic microaerobic existence, such animals can hardly afford to squander what little classical energy reserves they seemingly maintain to accumulate massive levels of low-valence transition metals and reducing ligands—unless, of course, they gain a distinct bioenergetic edge in doing so [27].

4.5. Potential Roles for Vanadium

The recent disclosure of certain vanadium-accumulating varieties of fan worms (classified into Annelida, Polychaeta) [83] should provide new clues concerning the purpose of this metal for animals. Like ascidians, polychaetes are evolutionarily ancient marine invertebrates and good anaerobes, and often reside in or near the same sulfur-sulfide biomes [134,136,139]. Along these lines, and in light of the aforementioned sulfated natural products from ascidians and nonascidians alike (Fig. 1), the potential that vanadium—in conjunction with tunichrome-like hydroquinones—might somehow mediate the assimilation of sulfur oxides [119,298,299,302], vis-à-vis taurine biosynthesis by other marine invertebrates and/or their endosymbiotic chemoautotrophic bacteria [147,191,192,303], thus merits careful consideration. Along similar lines, sulfate reduction in the bacterium *Desulfovibrio* has been proposed to occur via ATP adenyl sulfate and cytochrome c_3 and seems to involve abundant endogenous menaquinone (vitamin K_2) [239,304]; the latter compound bears a striking structural similarity to the unsulfated form of siphonodictal E (Fig. 1). Furthermore, ascidians possess copious quantities of taurine [258–260], (lysosomal) arylsulfatases [305], sulfates, and sulfonates associated with V^{III} [89], alkyl sulfates [306], and sulfated saccharides [307–311].

Other (published) proposals regarding the role of vanadium in ascidians include protection from microbes [312] or other predators [25,313,314], histoincompatibility responses [315], peroxide production/sterilization reactions [53,315], and tunic biogenesis [226], any of which may be compatible with a more fundamental purpose. Specifically, by virtue of their facility to inhabit nearly frozen, anoxic haloclines, as well as brackish water [254], diverse ascidians are demonstrably *good oxyconformers*, *osmoconformers*, and *poikilotherms*. Therefore, two additional corollaries to the formulation concerning ascidian anaerobic energy transduction [27] are that thermogenesis and osmoregulation should now be included as potential homeostatic factors (purposes) for vanadium accumulation/tunichrome production as well. For instance, many biothermogenic reactions involve hydroquinones [239,316,317], whereas several related model systems incorporate ETS shunts and vanadate-catalyzed NAD(P)H oxidation, and are stimulated by high concentrations of phosphate, sulfate, and NADH [25,318–323]. Furthermore, mitochondrial ubiquinone and succinic dehydrogenase are proposed to couple the shuttling of electrons toward specific (thermogenic) microsomes [318]. Ironically, according to Hochachka [149], "Metabolic arrest mechanisms yield by far the most effective protection against O_2 lack. . . . Simultaneous resistance to lack of O_2 and to low temperature arises because some of the fundamental mechanisms used by hypoxia-tolerant animals to protect tissues and organs against hypoxia are the same as, or at least remarkably similar to, those used by cold-tolerant organisms. . . . It is proposed that the *missing element* is to be found at the interface between cell membrane functions and cell metabolism" (emphasis added). Might this rhetorical "element" thus be vanadium on occasion? If so, multiple ETS, metabolic, and/or channel arrest mechanisms of good anaerobes would be readily activatable, i.e., vanadate acts as an inhibitor/uncoupler of mitochondrial and glycolytic energy metabolism [25,324–326].

Astonishingly enough, the classic observations of Mingati [162,211] on melanogenesis in *A. aspersa* and *P. mammillata*, the action of tyrosinase inhibitors on the respiratory systems of eggs and embryos of *P. mammillata* (which accumulate tunichrome Pms and vanadium), and follow-up studies [164,200,212] shed considerable light on this absolutely decisive issue and thus merit revisitation: "The experimental results give evidence that the substances used as tyrosinase inhibitors [i.e., metal complexing agents] are not strictly specific, and have two distinct effects on ascidian development: (1) they suppress melanogenesis in the sensory organs, and (2) they bring about characteristic abnormalities in the development and morphology of the tail of the tadpoles" [162]. "On the other hand, the abnormalities, both morphological and functional, thus caused, are identical to those brought about by other substances, which are assumed to be specific inhibitors of respiratory enzymes, accumulated in the muscle cells" [327]. "As these enzymes are in no way affected by tyrosinase inhibitors, as demonstrated below, *we suppose that a respiratory mechanism, other than cytochrome system, but with the same localization and significance for the morphogenesis, may be present in the muscle territory of the embryo* (emphasis added). . . . We can assume that this system is particularly active in the same cells, and has the same significance for the morphogenesis, as the cytochrome system, thus explaining the fact that the caudal abnormalities are identical to those brought about by inhibitors of cytochrome-oxidase and succino-dehydrogenase, as sodium azide, malonate and selenite. . . . The nature of the respiratory system postulated above is a matter of speculation. A suggestion is given by the mode of action of the substances inhibiting it, which have the common property to form complexes with heavy metal ions. So, probably a metal-containing enzyme (or enzymes) is involved" [211]. "In the suggested hypothesis the release of a blocked carrier would render cytochrome oxidase capable of higher activity" [200]. Correspondingly, nontoxic concentrations of copper, iron, and aluminum salts [37] accelerate the metamorphosis of embryonic ascidians, as do several reducing substances [76]. As noted above, vanadate is a general uncoupler of all three sites of ETS-driven oxidative phosphorylation in mitochondria [25,324–326] which, significantly, involve several membrane-based cytochromes.

An unorthodox formulation along these very lines has recently disclosed how energetic dividends could redound if circulating (respiratory) blood pigments, such as the tunichromes, were to mediate the uptake and reduction of vanadium as an ecophysiological response to ascidian anaerobiosis [27,34,328]. The air sensitivity of ascidian blood cells, the tunichromes, and V^{III} comports well with the independent formulations of Mingati and Mitchell Smith; in essence, the former inferred the presence of a cognate respiratory system in eggs and embryos, whereas the latter invoked vanadium and the hydroquinonoid tunichromes to be the key (anaerobic) redox agents therein. As noted above, vanadate can be "respired" (anaerobically) by facultative chemolithotrophic pseudomonads isolated from *A. phalosia* [180,181]. In contradistinction to this, Henze [329], the scientist who discovered vanadium in ascidians, suggested that the vanadium chromogen acted as an *aerobic* respiratory pigment ("respiratorische Farbstoffe"), viz., this chromogen [is] a catalyst and specifically an *oxygen carrier*: "Dieses Chromogen als einen Katalysator und zwar speziell als Sauerstoffüberträger. . ." (emphasis added). The first and second of three necessary preconditions to validate the formulation that ascidians "respire" vanadium have thus been fulfilled: (1) many, if not all, ascidians experience anaerobiosis; (2) vanadium can be respired anaerobically (by facultative chemolithotrophic bacteria from ascidians); and (3) vanadium-accumulating ascidians may prove to utilize anaerobic electron transport pathways akin to "iron respiration-driven proton translocation" of magnetotactic bacteria, which happen to reside within some of the very same biotopes (vide supra). If true, this formulation confers a genuine purpose to vanadium for an animal. With further respect to the potential thermodynamic advantages of vanadium-based redox processes for ascidians, Carlisle [66] noted that "this can only be true of the mechanisms occurring in the [blood] cell itself and not as a general redox mechanism for the whole body, for there would be no useful purpose served if metabolites had to diffuse into the blood from tissues before a reducing mechanism were called into play." However, one should recall that the above formulation by Mingati was based on studies of *whole* embryos and eggs. Moreover, a variety of blood cells are intimately associated with, and *actually diffuse through*, ascidian tissues displaying prominent metabolic activity, e.g., the tunic-forming epidermis [14,15,166, 209,221,222,224,330], such that this vanadium redox loop entails widespread dissemination, with an apparent time span of several weeks (vide supra). One might thus wonder what ligands and redox processes are associated with the sizable quantities of vanadium found in cells and tissues before and beyond the blood [18,55,188,189,295].

In general, the reduction of vanadate via NADH oxidation in microsomes and plasma membranes is well documented [320–323,331,

332]. Furthermore, certain molluscs and annelids offer biological precedents for unconventional, vacuole-based mechanisms of anaerobic energy transduction. The first process, anoxic endogenous oxidation, was formulated by Zsnagy with reference to molluscs that possess yellow, membrane-delimited, lipochrome (a.k.a., lipofuscin) granules or cytosomes in nervous and other tissue [271,274,333,334]. It appears to involve hydroquinones [272], succinic dehydrogenase [274,335], and NADH oxidation, via reductive hydrogenation of unsaturated fatty acids, phospholipids, and carotenoids [336], in order to produce ATP, absent O_2 [271,273]. It may also be instrumental to the detoxification of heavy metals in *M. galloprovincialis* [275]. Moreover, although many lamellibranchs display anoxic endogenous oxidation [274], some also possess tunichrome-like lamellarins (Fig. 1) [7]. A close histochemical relationship between melanin and lipofuscin has also been noted, which has led to several suggestions concerning their potential interconvertibility [337]. The formulation of Zsnagy seems to have garnered considerable respect [145,208,250,277,338–341], if not widespread acceptance. A similar process has been formulated by Fischer, who suggested that the lysosome-like chloragosomes of earthworms, which sequester metal ions (Ca, Zn, Fe, etc.), phosphate, sulfur, and various phenols (yellow coumarins, flavones, etc.) [338,342–345], act as specific electron acceptors during hypoxia and anoxia [338,346], evidently for the "energy dependent transport of particular ions over the chloragosome membrane" [338], via NADH oxidation [347].

In light of the last interpretation, as well as several important discrepancies concerning the pathways, kinetics, and energetics of vanadium accumulation by ascidian blood cells [20,21,292,299], further investigations, though nontrivial, are surely warranted. Specifically, the inference that vanadium uptake (via a phosphate translocator) "cannot be directly coupled to ATP hydrolysis" [299] was based on comparable (initial) rates of [^{48}V]vanadate uptake—in the presence or absence of glycolytic inhibitors and uncouplers of oxidative phosphorylation, e.g., the protonophore carbonylcyanide *p*-trifluoromethylphenylhydrazone (FCCP). However, these findings do not preclude unconventional sources of *anaerobic* energy transduction, apart from classical fermentative glycolysis. Correspondingly, the energy for high-affinity nickel transport by *Methanobacterium bryantii* stems from a transmembrane pH gradient, but for reasons not understood, FCCP did not inhibit uptake [348]; endogenous nickel complexes in such bacteria somehow catalyze anaerobic energy transduction (methane production) [349]. In addition, potassium uptake by *Mycoplasma mycoides* (which lack oxidative proton pumps) is also dependent on an energy source and is likewise resistant to FCCP, even though the concentration used abolished its proton motive force (PMF) [350]. Somewhat similarly, phosphate uptake by the anaerobe *Streptococcus faecalis* depends on concurrent energy metabolism, yet the protonophore carbonylcyanide *m*-chlorophenylhydrazone failed to block phosphate uptake, even though it abolished the PMF [351]. It may thus be relevant that provocative inverse correlations have been drawn between membrane leakiness and tolerance to anaerobiosis/poikilothermy [149].

Several comprehensive reviews on vanadium biochemistry offer considerable insights with respect to its coupling to electron transport, mechanisms of action, and potential roles in cellular regulation [24–26]. To this end, several predictable (proteinaceous) redox targets for vanadium in ascidians follow, simply to highlight particularly promising candidates. Direct interactions with vanadium are not necessarily assumed or implied, however; hydroquinones are often instrumental to membrane-based, metal-mediated electron transfer chains [236,240–244]. As noted above, the electrons transferred to vanadate might be derived from one of a variety of cellular metabolites known to be oxidized by vanadium: lipids (to peroxides), catecholamines, ascorbic acid, glutathione, and/or NAD(P)H [352]. (**1**) *Cytochrome* c: Califano and Boeri were the first to demonstrate that ascidian "haemovanadin" (the erroneous, hemoglobin-like inference concerning the vanadium complex) reduced beef heart cytochrome *c* both aerobically and anaerobically [353]. Recently, Stern et al. [352] found that a delicately poised redox balance exists between V^{IV} and V^{V} chelates and horse heart cytochrome *c*, whereby the latter was paradoxically reduced or oxidized, depending on the pH, O_2 availability, and structure of the chelate; the latter finding would, in theory, enable vanadium to be incorporated into a reversible redox loop. In light of vanadium's proclivity to esterify tyrosine residues in proteins [354], the "swinging" tyrosine residue within the active site of cytochrome *c* [239] is strongly implicated. Somewhat similarly, as noted above, *c*-type cytochromes and menaquinone (vitamin K_2) have been proposed to facilitate sulfate reduction [239,304] and dissimilatory metal reduction [175]. (**2**) *Cytochrome* c *oxidase*: Oxidizes (positive) vanadyl and (negative) vanadite ions [355]. (**3**) *Flavin reductases*: By analogy to flavin-containing ferrisiderophore reductases in microorganisms [356], metallic flavin reductases like succinic dehydrogenase are particularly intriguing in view of their potential to facilitate anaerobic energy transduction [147,148,200,250], thermogenesis [239,318–320], and their ancient relationships with hydroquinones like coenzyme Q [357]. (**4**) *Na^+,K^+-ATPases*: Contain significant NADH-vanadate reductase activity [358,359]. Vanadate is a potent inhibitor [24] that could conceivably diminish ecophysiological stresses arising from hypoxia, hypothermia, and hyposalinity via responses involving ion channel/glycolytic/metabolic arrest [147,149]. (**5**) *Vanadate-sensitive Proton Translocating ATPases*: Occur in Madagascar periwinkle cells [360], diving turtles (anaerobic glycolysis) [233–235], *Mycobacterium phlei* [361], and chromaffin cells (absent vanadium studies) [231,232]. In light of the sheer abundance and distribution of vanadium in ascidians, one can thus anticipate that its modes of action are pleiotropic, stoichiometric, and catalytic.

5. OUTLOOK

Several pervasive trends concerning the bioinorganic chemistry and ecophysiology of vanadium are now evident. To this end, the bioenergetics of ascidian ion homeostasis involves driving ions like vanadate (V^V) up million-fold concentration gradients, after which $V^{III}{}_{aq}$ appears to be formed, perhaps as a metabolic terminus, until it is recycled anew. However, the vanadium-laden (signet ring) blood cells, by definition, do not display insoluble vanadium granules. Accordingly, neither *free* tunichromes nor their insoluble vanadium complexes should be sought there; such species may exist in other subpopulations of vacuolated cells that have vanadium associated with intracellular membranes and granules. Therefore, the interpretation that $V^{III}{}_{aq}$ represents a major portion of blood cell vanadium is not incompatible with the predicted role of endogenous organic ligands in vanadium assimilation. Meanwhile, these unobtrusive creatures are hardly encumbered by their microaerobic metabolism, for they excel at generating a resilient tunic composed of melanin and cellulose-like polymers. These polymers appear to be derived from anomalous *air-sensitive* blood cells, some of which possess high levels of *air-sensitive* (aquo) V^{III} and/or *air-sensitive* hydroquinonoid ligands. The production of cellulose-like polymers and V^{III} by these primitive animals thus recall primeval microbial processes of carbon fixation and metal respiration, which evidently pervade some of the very same biotopes, as well as *infected* ascidians. A profound degree of intellectual satisfaction thus awaits any intrepid interdisciplinarian who completes the riddle around the assimilation, complexation, reduction, recycling, and purposes of vanadium. To this end, a rhetorical remark by George Wald offers heuristic encouragement: "I have often had cause to feel that my hands are cleverer than my head. That is a crude way of characterizing the dialectics of experimentation. When it is going well, it is like a quiet conversation with nature. One asks a question and gets the next answer. An experiment is a device to make nature speak intelligibly. After that one has only to listen."

ACKNOWLEDGMENTS

Many of the findings presented were derived from the investigations of our esteemed coworkers. M.J.S. would also like to express special gratitude to his associates at the FDA for their able technical editing and reference services, as well as J. B. Neilands, M. O. Walderhaug, R. H. Garrett, S. P. Maroney, and R. H. Albert for their helpful scientific guidance. The work in Sec. 2 was supported by NSF CHE 91-21576 and NIH RR-01209 (to K.O.H.). The work in Sec. 3 was supported by National Research Service Award NIH GM-14042 (to D.E.R.) and NIH AI-10187 (to K.N.). SSRL is supported by the Department of Energy, Office of Basic Energy Sciences, Divisions of Chemical and Materials Sciences. Further support is provided by the National Institutes of Health, Division of Research Resources, Biomedical Research Technology Program and by the Department of Energy, Office of Health and Environmental Research.

ABBREVIATIONS

acac	acetylacetonate
ATP	adenosine 5′-triphosphate

DBcatH	3,5-di-*tert*-butylcatechol
DBSQ	3,5-di-*tert*-butylsemiquinone
EDTA	ethylenediamine-*N*,*N*,*N'*,*N'*-tetraacetic acid
EPR	electron paramagnetic resonance
ETS	electron transport system
EXAFS	extended x-ray absorption fine structure
FACS	fluorescence-activated cell sorting
FCCP	carbonylcyanide *p*-trifluoromethylphenylhydrazone
HPLC	high-performance liquid chromatography
LUMO	lowest unoccupied molecular orbital
NAD(P)H	nicotinamide adenine dinucleotide phosphate, reduced form
NADH	nicotinamide adenine dinucleotide, reduced form
NHE	normal hydrogen electrode
NMR	nuclear magnetic resonance
PMF	proton motive force
RQ	respiratory quotient
salen	*N*,*N'*-ethylenebis(salicylideniminato)
SCE	saturated calomel electrode
SDH	succinic dehydrogenase
SQUID	superquantum interference device
THF	tetrahydrofuran
UV	ultraviolet
XAS	X-ray absorption spectroscopy

REFERENCES

1. M. Henze, *Hoppe-Seyler's Z. Physiol. Chem.*, *72*, 494 (1911).
2. E. M. Oltz, R. C. Bruening, M. J. Smith, K. Kustin, and K. Nakanishi, *J. Am. Chem. Soc.*, *110*, 6162 (1988).
3. E. Bayer, G. Schiefer, D. Waidelich, S. Scippa, and M. de Vincentiis, *Angew. Chem. Int. Ed. Engl.*, *31*, 52 (1992).
4. K. Azumi, H. Yokosawa, and S. Ishii, *Biochemistry*, *29*, 159 (1990).
5. L. C. Dorsett, C. J. Hawkins, J. A. Grice, M. F. Lavin, P. M. Merefield, D. L. Parry, and I. L. Ross, *Biochemistry*, *26*, 8078 (1987).

6. R. J. Stonard and R. J. Anderson, *Can. J. Chem.*, *58*, 2121 (1980).
7. R. J. Anderson, D. J. Faulkner, H. Cun-heng, G. D. Van Duyne, and J. Clardy, *J. Am. Chem. Soc.*, *107*, 5492 (1985).
8. M. J. Garson, *Chem. Rev.*, *93*, 1699 (1993).
9. I. G. Macara, G. C. McLeod, and K. Kustin, *Comp. Biochem. Physiol.*, *62A*, 821 (1979).
10. S. Ito, G. Nardi, A. Palumbo, and G. Prota, *J.C.S. Perkin I*, 2617 (1979).
11. B. W. Sullivan, D. J. Faulkner, G. K. Matsumoto, H. Cun-heng, and J. Clardy, *J. Org. Chem.*, *51*, 4568 (1986).
12. N. Lindquist, W. Fenical, L. Prákányl, and J. Clardy, *Experientia*, *47*, 503 (1991).
13. R. C. Bruening, E. M. Oltz, J. Furukawa, and K. Nakanishi, *J. Nat. Prod.*, *49*, 193 (1986).
14. M. Kalk, *Quart. J. Micr. Sci.*, *104*, 483 (1963).
15. M. J. Smith, *Biol. Bull.*, *138*, 354 (1970).
16. T. H. Ermak, *Experientia*, *31*, 837 (1975).
17. T. H. Ermak, *Am. Zool.*, *22*, 795 (1982).
18. A. F. Rowley, *Cell Tissue Res.*, *223*, 403 (1982).
19. D. A. Raftos, D. L. Stillman, and E. L. Cooper, *In vitro Cell. Div. Biol.*, *26*, 962 (1990).
20. D. H. Anderson, J. R. Berg, and J. H. Swinehart, *Comp. Biochem. Physiol.*, *99A*, 151 (1991).
21. D. H. Anderson and J. H. Swinehart, *Comp. Biochem. Physiol.*, *99A*, 585 (1991).
22. H. Michibata, *Adv. Biophys.*, *29*, 105 (1993).
23. W. R. Biggs and J. H. Swinehart, Vanadium in Selected Biological Systems, *Metal Ions in Biological Systems* Vol. 6 (H. Sigel, ed.), Marcel Dekker, New York, 1976, p. 141.
24. N. D. Chasteen, *Struct. Bonding*, *53*, 107 (1983).
25. T. Ramasarma and F. L. Crane, *Curr. Topics Cell. Reg.*, *20*, 247 (1981).
26. B. R. Nechay, *Ann. Rev. Pharmacol. Toxicol.*, *24*, 501 (1984).
27. M. J. Smith, *Experientia*, *45*, 452 (1989).
28. E. Kime-Hunt, K. Spartalian, S. Holmes, M. Mohan, and C. J. Carrano, *J. Inorg. Biochem.*, *41*, 125 (1991).

29. A. Butler and C. J. Carrano, *Coord. Chem. Rev.*, *109*, 61 (1991).
30. D. Rehder, *Biometals*, *5*, 3 (1992).
31. E. M. Oltz, Ph.D. dissert., Columbia Univ., 1987.
32. B. A. Horenstein, Ph.D. dissert., Columbia Univ., 1988.
33. S. Lee, Ph.D. dissert., Columbia Univ., 1989.
34. M. J. Smith, D. Kim, B. Horenstein, K. Nakanishi, and K. Kustin, *Acc. Chem. Res.*, *24*, 117 (1991).
35. J. P. Michael and G. Pattenden, *Angew. Chem. Int. Ed. Engl.*, *32*, 1 (1993).
36. B. S. Davidson, *Chem. Rev.*, *93*, 1771 (1993).
37. N. J. Berrill, *The Origin of Vertebrates*, Clarendon Press, Oxford, 1955.
38. R. H. Millar, *Adv. Mar. Biol.*, *9*, 1 (1971).
39. I. Goodbody, *Adv. Mar. Biol.*, *12*, 1 (1974).
40. C. J. Hawkins, *J. Inorg. Biochem.*, *43*, 402 (1991).
41. S. W. Taylor, D. L. Parry, C. J. Hawkins, and J. H. Swinehart, *Comp. Biochem. Physiol.*, *106A*, 531 (1993).
42. N. J. Berrill, *Encyclopedia Britannica*, Vol. *22*, William Benton, Chicago, 1964, p. 550.
43. S. P. Cramer, *Iron Proteins*, Part 3 (A.B. P. Lever and H. B. Gray, eds.), VCH, New York, 1989.
44. S. M. Heald and J. M. Tranquada, *Physical Methods in Chemistry*, Vol. 5 (B. W. Rossiter and J. F. Hamilton, eds.), Wiley, New York, 1990.
45. E. Norlander, S. C. Lee, W. Cen, Z. Y. Wu, C. R. Natoli, A. Di Cicco, A. Filipponi, B. Hedman, K. O. Hodgson, and R. H. Holm, *J. Am. Chem. Soc.*, *115*, 5549 (1993).
46. V. K. Yachandra, V. J. DeRose, M. J. Latimer, I. Mukerji, and K. Sauer, *Science*, *260*, 675 (1993).
47. R. S. Mulliken, *J. Am. Chem. Soc.*, *86*, 3183 (1964).
48. B. K. Agarwal, *X-ray Spectroscopy*, Springer-Verlag, New York, 1979.
49. A. P. Hitchcock, J. A. Horsley, and J. Stöhr, *J. Chem. Phys.*, *85*, 4835 (1986).
50. C. T. Chen, Y. Ma, and F. Sette, *Phys. Rev. A.*, *40*, 6737 (1989).
51. S. P. Cramer and K. O. Hodgson, *Prog. Inorg. Chem.*, *25*, 1 (1979).

52. C. J. Hawkins, P. Kott, D. L. Parry, and J. H. Swinehart, *Comp. Biochem. Physiol.*, *76B*, 555 (1983).
53. P. Frank, R. M. K. Carlson, and K. O. Hodgson, *Inorg. Chem.*, *25*, 470 (1986).
54. D. A. Webb, *J. Exp. Biol.*, *16*, 499 (1939).
55. J. H. Swinehart, W. R. Biggs, D. J. Halko, and N. C. Schroeder, *Biol. Bull.*, *146*, 302 (1974).
56. H. J. Bielig, E. Bayer, H. D. Dell, G. Rohns, H. Mollinger, and W. Rüdiger, *Protides Biol. Fluids*, *14*, 197 (1966).
57. E. Boeri and A. Ehrenberg, *Arch. Biochem. Biophys.*, *50*, 404 (1954).
58. L. T. Rezaeva, *Zh. Obsch. Biol.*, *25*, 347 (T836) (1964).
59. H. J. Bielig and E. Bayer, *Pubbl. Staz. Zool. Napoli*, *29*, 109 (1957).
60. R. M. K. Carlson, *Proc. Natl. Acad. Sci. USA*, *72*, 2217 (1975).
61. T. D. Tullius, W. O. Gillum, R. M. K. Carlson, and K. O. Hodgson, *J. Am. Chem. Soc.*, *102*, 5670 (1980).
62. S. Lee, K. Kustin, W. E. Robinson, R. B. Frankel, and K. Spartalian, *J. Inorg. Biochem.*, *33*, 183 (1988).
63. J. Hirata and H. Michibata, *J. Exp. Zool.*, *257*, 160 (1991).
64. S. G. Brand, C. J. Hawkins, A. T. Marshall, G. W. Nette, and D. L. Parry, *Comp. Biochem. Physiol.*, *93B*, 425 (1989).
65. L. S. Ciereszko, E. M. Ciereszko, E. R. Harris, and C. A. Lane, *Comp. Biochem. Physiol.*, *8*, 137 (1963).
66. D. B. Carlisle, *Proc. Roy. Soc. B*, *171*, 31 (1968).
67. M. V. Bell, B. J. S. Pirie, D. B. McPhail, B. A. Goodman, I. B. Falk-Peterson, and J. R. Sargent, *J. Mar. Biol. Assoc. U.K.*, *62*, 709 (1982).
68. P. Frank, R. M. K. Carlson, and K. O. Hodgson, *Inorg. Chem.*, *27*, 118 (1988).
69. P. Frank, B. Hedman, R. M. K. Carlson, and K. O. Hodgson, *Inorg. Chem.*, *33*, 3794 (1994).
70. H. Michibata, Y. Iwata, and J. Hirata, *J. Exp. Zool.*, *257*, 306 (1991).
71. J. Hirata and H. Michibata, *Zool. Sci.*, *9*, 207 (1992).
72. A. L. Dingley, K. Kustin, I. G. Macara, G. C. McLeod, and M. F. Roberts, *Biochim. Biophys. Acta*, *720*, 384 (1982).

73. C. J. Hawkins, G. A. James, D. L. Parry, J. H. Swinehart, and A. L. Wood, *Comp. Biochem. Physiol., 76B*, 559 (1983).

74. S. G. Brand, C. J. Hawkins, and D. L. Parry, *Inorg. Chem., 26*, 629 (1987).

75. S. Lee, K. Nakanishi, and K. Kustin, *Biochim. Biophys. Acta, 1033*, 311 (1990).

76. K. Kustin, W. E. Robinson, and M. J. Smith, *Invert. Rep. Dev., 17*, 129 (1990).

77. A. R. Bulls, G. C. Pippin, F. E. Hahn, and K. N. Raymond, *J. Am. Chem. Soc., 112*, 2627 (1990).

78. P. Frank, K. Kustin, W. Robinson, L. Linebaugh, and K. O. Hodgson, in preparation, 1994.

79. B. Hedman, P. Frank, S. F. Gheller, A. L. Roe, W. E. Newton, and K. O. Hodgson, *J. Am. Chem. Soc., 110*, 3798 (1988).

80. M. Benfatto, C. R. Natoli, A. Bianconi, J. Garcia, A. Marcelli, M. Fanfoni, and I. Davoli, *Phys. Rev. B, 34*, 5774 (1986).

81. S. R. Cooper, Y. B. Koh, and K. N. Raymond, *J. Am. Chem. Soc., 104*, 5092 (1982).

82. T. Miyanaga, I. Watanabe, and S. Ikeda, *Bull. Chem. Soc. Jpn., 63*, 3282 (1990).

83. T. Ishii, I. Nakai, C. Numako, K. Okoshi, and T. Otake, *Naturwissenschaften, 80*, 268 (1993).

84. B. J. S. Pirie and M. V. Bell, *J. Exp. Mar. Biol. Ecol., 74*, 187 (1984).

85. D. Stoecker, *J. Exp. Mar. Biol. Ecol., 48*, 277 (1980).

86. T. W. Newton and F. B. Baker, *Inorg. Chem., 3*, 569 (1964).

87. J. K. Beattie, S. P. Best, B. W. Skelton, and A. H. White, *J. Chem. Soc. Dalton Trans.*, 2105 (1981).

88. F. A. Cotton, C. K. Fair, G. E. Lewis, G. N. Mott, F. K. Ross, A. J. Schultz, and J. M. Williams, *J. Am. Chem. Soc., 106*, 5319 (1984).

89. P. Frank, B. Hedman, R. M. K. Carlson, T. A. Tyson, A. L. Roe, and K. O. Hodgson, *Biochemistry, 26*, 4975 (1987).

90. K.-I. Okamoto, J. Hidaka, M. Fukagawa, and K. Kanamori, *Acta Crystallogr., C48*, 1027 (1992).

91. R. E. Shepard, W. E. Hatfield, D. Ghosh, C. D. Stout, F. J. Kristine, and J. R. Ruble, *J. Am. Chem. Soc., 103*, 5511 (1981).

92. B. Hedman, P. Frank, J. E. Penner-Hahn, A. L. Roe, K. O.

Hodgson, R. M. K. Carlson, G. Brown, J. Cerino, R. Hettel, T. Troxel, H. Winick, and J. Yang, *Nucl. Instr. Meth. Phys. Res.*, *A246*, 797 (1986).

93. A. F. Rowley, *J. Mar. Biol. Assoc. UK*, *62*, 607 (1982).
94. S. Scippa, K. Zierold, and M. de Vincentiis, *J. Submicrosc. Cytol. Pathol.*, *20*, 719 (1988).
95. K. Kanamori, Y. Ookubo, K. Ino, K. Kawai, and H. Michibata, *Inorg. Chem.*, *30*, 3832 (1991).
96. P. Frank, PhD dissert., Stanford University, 1980.
97. T. Kimura, M. Morinaga, and K. Nakano, *Nippon Kagaku Kaishi*, 664 (1972).
98. E. M. Oltz, S. Pollack, T. Delohery, M. J. Smith, M. Ojika, S. Lee, K. Kustin, and K. Nakanishi, *Experientia*, *45*, 186 (1989).
99. H. Michibata, J. Hirata, M. Uesaka, T. Numakunai, and H. Sakurai, *J. Exp. Zool.*, *244*, 33 (1987).
100. H. Michibata and T. Uyama, *J. Exp. Zool.*, *254*, 132 (1990).
101. W. R. Biggs and J. H. Swinehart, *Experientia*, *35*, 1047 (1979).
102. W. C. George, *Quart. J. Microsc. Sci.*, *81*, 391 (1939).
103. D. R. Lide, *CRC Handbook of Chemistry and Physics*, CRC Press, Boca Raton, 1993.
104. L. Horner and E. Geyer, *Chem. Ber.*, *98*, 2016 (1965).
105. S. Y. Shnaiderman, A. N. Demidovskaya, and V. G. Zaletov, *Russ. J. Inorg. Chem.*, *17*, 348 (1972).
106. J. Zelinka, M. Bartusek, and A. Okac, *Collect. Czech. Chem. Commun.*, *39*, 83 (1974).
107. K. Kustin, S.-T Liu, C. Nicolini, and D. L. Toppen, *J. Am. Chem. Soc.*, *96*, 7410 (1974).
108. K. Kustin, C. Nicolini, and D. L. Toppen, *J. Am. Chem. Soc.*, *96*, 7416 (1974).
109. L. C. Cantley, J. H. Ferguson, and K. Kustin, *J. Am. Chem. Soc.*, *100*, 5211 (1978).
110. D. F. Ferguson and K. Kustin, *Inorg. Chem.*, *18*, 3349 (1979).
111. P. Wardman, *J. Phys. Chem. Ref. Data*, *18*, 1637 (1989).
112. T. Kitagawa, J. Toyoda, K. Nakasuji, H. Yamamoto, and I. Murata, *Chem. Lett.*, 897 (1990).

113. S. Lee, K. Nakanishi, M. Y. Chiang, R. G. Frankel, and K. Spartalian, *J. Chem. Soc. Chem. Commun.*, 785 (1988).

114. J. A. Bonadies, W. M. Butler, V. L. Pecoraro, and C. J. Carrano, *Inorg. Chem.*, *26*, 1218 (1987).

115. M. E. Cass, R. M. Greene, R. M. Buchanan, and C. G. Pierpont, *J. Am. Chem. Soc.*, *105*, 2680 (1983).

116. M. E. Cass, N. R. Gordon, and C. G. Pierpont, *Inorg. Chem.*, *25*, 3962 (1986).

117. Y. Tatsuno, C. Nakamura, and T. Saito, *J. Mol. Catal.*, *42*, 57 (1987).

118. H. S. Mason, *J. Biol. Chem.*, *181*, 803 (1949).

119. I. G. Macara, G. C. McLeod, and K. Kustin, *Biochem. J.*, *181*, 457 (1979).

120. R. C. Bruening, E. M. Oltz, J. Furukawa, K. Nakanishi, and K. Kustin, *J. Am. Chem. Soc.*, *107*, 5298 (1985).

121. I. G. Macara, G. C. McLeod, and K. Kustin, *Comp. Biochem. Physiol.*, *63B*, 299 (1979).

122. B. Horenstein and K. Nakanishi, *J. Am. Chem. Soc.*, *111*, 6242 (1989).

123. D. Kim, B. Li, and K. Nakanishi, *Tetrahedron Lett.*, *31*, 7119 (1990).

124. D. Kim, Y. Li, and K. Nakanishi, *J. Chem. Soc. Chem. Commun.*, 9 (1991).

125. D. E. Ryan, N. D. Ghatlia, A. E. McDermott, N. J. Turro, K. Nakanishi, and K. Kustin, *J. Am. Chem. Soc.*, *114*, 9659 (1992).

126. G. E. Pake and T. L. Estle, *The Physical Principles of Electron Paramagnetic Resonance*, W. A. Benjamin, Reading, MA, 1973.

127. J. E. Wertz and J. R. Bolton, *Electron Spin Resonance: Elementary Theory and Applications*, McGraw-Hill, New York, 1972, p. 329.

128. J. Francavilla and N. D. Chasteen, *Inorg. Chem.*, *14*, 2860 (1975).

129. N. D. Chasteen, *Biol. Magn. Reson.*, *3*, 53 (1981).

130. E. Alder, and R. Magnusson, *Acta Chem. Scand.*, *13*, 505 (1959).

131. M. Sugumaran, H. Dali, V. Semensi, and B. Hennigan, *J. Biol. Chem.*, *262*, 10546 (1987).

132. M. Branca, G. Micera, A. Dessi, D. Sanna, and K. N. Raymond, *Inorg. Chem.*, *29*, 1586 (1990).
133. T. A. Kabanos, A. J. P. White, D. J. Williams, and J. D. Woolins, *J. Chem. Soc. Chem. Commun.*, 17 (1992).
134. O. Hartman and J. L. Barnard, *The Benthic Fauna of the Deep Basins Off Southern California*, Vol. 22, Los Angeles, U.S.C. Press, 1958, p. 1.
135. B. I. Dybern, *Limnologica (Berlin)*, *7*, 27 (1969).
136. M. Baçesco, "Contribution à la Biocoenologie de la Mer Noire L'étage Périazoïque et le Faciès Dreissenifère Leurs Caractèristiques," *International Commission for the Scientific Exploration of the Mediterranean Sea. Rapports et proces-verbaux des réunions. [Paris]*, *17(2)*, 107 (1963).
137. V. Tunnicliffe, *Nature*, *294*, 354 (1981).
138. B. I. Dybern, *Sarsia*, *37*, 21 (1969).
139. D. C. Rhoads and J. W. Morse, *Lethaia*, *4*, 413 (1971).
140. J. B. Thompson, H. T. Mullins, C. R. Newton, and T. L. Verocoutere, *Lethaia*, *18*, 167 (1985).
141. S. Muslow and D. C. Rhoads, "Biomineralization Processes of Iron and Manganese: Modern and Ancient Environments," Vol. 21 (H. C. W. Skinner and R. W. Fitzpatrick, eds.), Catena Verlag, Cremlingen-Destedt, Germany, 1992, p. 303.
142. C. E. Savrda and D. J. Bottjer, *Nature*, *327*, 54 (1987).
143. I. Fridovich, *Bioscience*, *27*, 462 (1977).
144. T. von Brand, *Biodynamica*, *4*, 185 (1944).
145. A. DeZwaan, F. M. Kluytmans, and D. I. Zandee, *Biochem. Soc. Symp.*, *41*, 133 (1976).
146. P. W. Hochachka and M. Guppy, *Metabolic Arrest and the Control of Biological Time*, Harvard University Press, Cambridge, MA, 1987.
147. P. W. Hochachka and G. N. Somero, *Biochemical Adaptation*, Princeton Univ. Press, Princeton, NJ, 1984.
148. D. R. Livingstone, *J. Geol. Soc. London*, *140*, 27 (1983).
149. P. W. Hochachka, *Science*, *231*, 234 (1986).
150. A. D. Moodie and W. J. Ingledew, *Advances in Microbial Physiology*, Vol. 31, Academic Press, San Diego, 1990, p. 225.

151. D. R. Lovley, *Ann. Rev. Microbiol.*, *47*, 263 (1993).

152. H. Theede, A. Ponat, K. Hiroki, and C. Schlieper, *Mar. Biol.*, *2*, 325 (1969).

153. T. M. Fenchel and R. J. Riedl, *Mar. Biol.*, *7*, 225 (1970).

154. C. R. Fisher, *Rev. Aquatic Sci.*, *2*, 339 (1990).

155. P. Kott, Antarctic Ascidiacea. *Antarct. Res. Ser.*, *13*, 1, (1969).

156. J. B. McClintock, J. Heine, M. Slattery, and J. Weston, *J. Exp. Mar. Ecol.*, *147*, 163 (1991).

157. T. A. Nomaguchi, *Exp. Geront.*, *9*, 231 (1974).

158. L. L. Sims, *J. Exp. Mar. Biol. Ecol.*, *82*, 117 (1984).

159. B. I. Dybern, *Oikios*, *16*, 109 (1965).

160. I. Svane and T. Lundälv, *Sarsia*, *67*, 249 (1982).

161. B. Rinkevich, R. J. Lauzon, B. W. M. Brown, and I. L. Weissman, *Proc. Natl. Acad. Sci. USA*, *89*, 3546 (1992).

162. A. Mingati, *Acta Embryol. Morphol. Exp.*, *1*, 37 (1957).

163. I. Goodbody and J. Gibson, *Biol. Bull.*, *146*, 217 (1974).

164. J. R. Whittaker, *Histochemistry*, *71*, 349 (1981).

165. E. J. W. Barrington and F. R. S. Thorpe, *Proc. Roy. Soc. B*, *171*, 91 (1968).

166. J. D. Deck, E. D. Hay, and J.-P. Revel, *J. Morphol.*, *120*, 267 (1966).

167. R. Rosenberg, I. Olsson, and E. Ölundh, *Mar. Ecol.*, *42*, 99 (1977).

168. R. Rosenberg, in *Fjord Oceanography* (H. J. Freeland, D. M. Farmer, C. D. Levings, eds.), Plenum Press, New York, 1979, p. 499.

169. B. K. Evans and R. Huntington, *J. Exp. Zool.*, *262*, 462 (1992).

170. R. P. Blakemore, K. A. Short, D. A. Bazylinski, C. Rosenblatt, and R. B. Frankel, *Geomicrobiol. J.*, *4*, 53 (1985).

171. W. F. Guerin and R. P. Blakemore, *Appl. Environ. Microbiol.*, *58*, 1102 (1992).

172. F. F. TorresDeAraujo, M. A. Pires, R. B. Frankel, and C. E. M. Bicudo, *Biophys. J.*, *50*, 375 (1986).

173. Y. A. Gorby, T. J. Beveridge, and R. P. Blakemore, *J. Bacteriol.*, *170*, 834 (1988).

174. K. A. Short and R. P. Blakemore, *J. Bacteriol.*, *167*, 729 (1986).

175. D. R. Lovley, S. J. Giovannoni, D. C. White, J. E. Champine, and E. J. P. Phillips, *Arch. Microbiol.*, *159*, 336 (1993).
176. H. C. W. Skinner and R. W. Fitzpatrick, *Biomineralization Processes of Iron and Manganese: Modern and Ancient Environments* (H. C. W. Skinner and R. W. Fitzpatrick, eds.), Catena Verlag, Cremlingen-Destedt, Germany, 1992, p. 1.
177. D. R. Lovley, *Trends Ecol. Evol.*, *8*, 213 (1993).
178. B. A. Hoffman, *Chem. Geol.*, *81*, 55 (1990).
179. D. Siebers, *Am. Zool.*, *22*, 723 (1982).
180. N. A. Yurkova and N. N. Lyalikova, *Microbiology*, *59*, 672 (1991).
181. N. N. Lyalikova and N. A. Yurkova, *Geomicrobiol. J.*, *10*, 15 (1992).
182. L. V. Berkner and L. C. Marshall, *Proc. Natl. Acad. Sci. USA*, *53*, 1215 (1965).
183. P. Cloud, *Am. Sci.*, *62*, 54 (1974).
184. G. Wald, *Proc. Natl. Acad. Sci. USA*, *52*, 595 (1964).
185. W. Day, *Genesis on Planet Earth*, Talos, East Lansing, MI, 1979.
186. O. Jaekel, *Paläont. Z.*, *2*, 66 (1915).
187. J. H. McClendon, *J. Mol. Evol.*, *8*, 175 (1976).
188. D. A. Webb, *Pubbl. Staz. Zool. Napoli*, *28*, 273 (1956).
189. H. Michibata, T. Terada, N. Anada, K. Yamakawa, and T. Numakunai, *Biol. Bull.*, *171*, 672 (1986).
190. L. Muscatine and J. W. Porter, *Bioscience*, *27*, 454 (1977).
191. H. Felbeck, J. J. Childress, and G. N. Somero, *The Mollusca: Metabolic Biochemistry and Molecular Biomechanics*, Vol. 2 (P. W. Hochachka, ed.), Academic Press, New York, 1983, p. 331.
192. N. M. Conway and J. E. McDowell-Capuzzo, *Comp. Biochem. Physiol.*, *102B*, 175 (1992).
193. L. Margulis, *Chemical Evolution and the Origin of Life* (R. Buvet and C. Ponnamperuma, eds.), North-Holland, Amsterdam, 1971, p. 480.
194. M. B. Saffo, *Mar. Biol.*, *107*, 291 (1990).
195. M. B. Saffo, *Am. Zool.*, *32*, 557 (1992).
196. M. A. Miller and R. K. Packer, *Comp. Biochem. Physiol.*, *57A*, 424 (1977).
197. R. L. Pardy, *Comp. Biochem. Physiol.*, *79A*, 345 (1984).

198. R. A. Lewin, in *Bergey's Manual of Systematic Bacteriology*, Vol. 3 (J. T. Stanley, M. P. Bryant, N. Pfenning, and J. G. Holt, eds.), Williams & Wilkins, Baltimore, 1989, p. 1799.
199. G. DeLeo and E. Patricolo, *Cell Tissue Res.*, *212*, 91 (1980).
200. A. Mingati, *Acta Embryol. Morphol. Exp.*, *1*, 157 (1957).
201. G. Freeman, *J. Reticuloendothel. Soc.*, *7*, 183 (1970).
202. C. Gans and R. G. Northcutt, *Science*, *220*, 268 (1983).
203. R. G. Northcutt and C. Gans, *Quart. Rev. Biol.*, *58*, 1 (1983).
204. H. Winterstein, *Biochem. Z.*, *29*, 384 (1909).
205. D. Baldwin, M. McCabe, and F. Thomas, *Comp. Biochem. Physiol.*, *79A*, 479 (1984).
206. M. McCabe, *Adv. Exp. Med. Biol.*, *200*, 21 (1985).
207. P. J. Schofield and B. W. Stewart, *Comp. Biochem. Physiol.*, *45B*, 579 (1973).
208. I. Hardewig, A. D. F. Addink, M. K. Grieshaber, H. O. Pörtner, and G. Van den Thillart, *J. Exp. Biol.*, *157*, 143 (1991).
209. E. J. W. Barrington, *J. Exp. Zool.*, *194*, 271 (1975).
210. A. Toulmond, *Symp. Soc. Exp. Biol.*, *39*, 163 (1985).
211. A. Mingati, *Acta Embryol. Morphol. Exp.*, *1*, 71 (1957).
212. T. DAnna, *Acta Embryol. Exp. (Palermo)*, 165 (1973).
213. L. Cuénot, *Archs. Zool. Exp. Gén.*, *9*, 13 (1891).
214. D. A. Sakharov and L. T. Zhuravlyova, *Zh. Obsch. Biol.*, *49*, 218 (1988).
215. A. J. Hulbert and P. L. Else, *Am. J. Physiol.*, *241*, R350 (1981).
216. H. Fujita and H. Nanba, *Z. Zellforsch.*, *121*, 455 (1971).
217. G. Fredriksson, T. Öfverholm, and L. E. Ericson, *Cell Tissue Res.*, *253*, 403 (1988).
218. M. C. Thorndyke, *J. Endocr.*, *58*, 679 (1973).
219. M. Srivastava and S. P. Gupta, *Z. Parasitenkd.*, *55*, 55 (1978).
220. A. D. Dunn, *Gen. Comp. Endocr.*, *40*, 473 (1980).
221. R. A. Cloney and L. Grimm, *Z. Zellforsch.*, *107*, 157 (1970).
222. M. J. Smith, *Biol. Bull.*, *138*, 379 (1970).
223. R. Hori and H. Michibata, *Protoplasma*, *108*, 9 (1981).
224. H. Mukai, K. Hashimoto, and H. Watanabe, *J. Morphol.*, *206*, 197 (1990).

225. B. Lübbering, T. Nishikata, and G. Goffinet, *Tissue and Cell*, *24*, 121 (1992).

226. M. Henze, *Hoppe-Seyler's Z. Physiol. Chem.*, *213*, 125 (1932).

227. N. C. Corpita and D. F. Delmer, *Plant Physiol.*, *66*, 911 (1980).

228. D. P. Delmer, M. Benziman, and E. Padan, *Proc. Natl. Acad. Sci. USA*, *79*, 5282 (1982).

229. M. Futai and T. Tsuchiya, *Ion Transport in Prokaryotes* (B. P. Rosen and S. Silver, eds.), Academic Press, San Diego, 1987, p. 3.

230. H.-U. Meisch and H.-J Bielig, *Arch. Microbiol.*, *105*, 77 (1975).

231. D. Njus, M. Zallakian, and J. Knoth, *Chemiosmotic Proton Circuits in Biological Membranes* (V. P. Skulachev and P. C. Hinkle, eds.), Addison-Wesley, Reading, MA, 1981, p. 365.

232. M. P. Roisin, D. Scherman, and J. P. Henry, *FEBS Lett.*, *115*, 143 (1980).

233. T. E. Dixon and Q. Al-Awaqati, *Proc. Natl. Acad. Sci. USA*, *76*, 3135 (1979).

234. P. R. Steinmetz, R. F. Husted, and A. Mueller, *Trans. Assoc. Am. Physicians*, *93*, 289 (1980).

235. J. A. L. Arruda, S. Sabatini, and C. Westenfelder, *Kidney Int.*, *20*, 772 (1981).

236. P. Mitchell, *Nature*, *191*, 144 (1961).

237. P. Mitchell, *Biol. Rev.*, *41*, 445 (1966).

238. V. P. Skulachev, *Ion Transport in Prokaryotes* (B. P. Rosen and S. Silver, eds.), Academic Press, San Diego, 1987, p. 131.

239. D. E. Metzler, "Biochemistry: The Chemical Reactions of Living Cells." New York: Academic Press, 1977.

240. T. Ohnishi, *J. Bioenerg. Biomemb.*, *25*, 325 (1993).

241. A. L. Moore, G. Leach, and D. G. Whitehouse, *Biochem. Soc. Trans.*, *21*, 765 (1993).

242. T. Unemoto and M. Hayashi, *J. Bioenerg. Biomemb.*, *25*, 385 (1993).

243. D. M. Dooley, *FASEB J.*, *7*, A1054 (1993).

244. T. Yagi, *Biochim. Biophys. Acta*, *1141*, 1 (1993).

245. W. Stillwell, *J. Theor. Biol.*, *65*, 479 (1977).

246. W. Stillwell, *Origins of Life*, *10*, 277 (1980).

247. S. F. Dyke, *The Chemistry of Vitamins*, Vol. 6 (K. W. Bentley, ed.), Interscience, London, 1965.

248. M. H. Smith, *Nature*, *223*, 1129 (1969).

249. E. Harless, *Mueller's Arch.*, 148 (1847).

250. S. D. Long, G. E. Rodrick, and F. E. Friedl, *Comp. Biochem. Physiol.*, *78B*, 467 (1984).

251. C. B. Jørgensen, *Biol. Bull.*, *103*, 356 (1952).

252. T. R. Fisher, *Biol. Bull.*, *151*, 297 (1976).

253. A. Fiala-Médioni, *Mar. Ecol. Prog. Ser.*, *1*, 49 (1979).

254. M. Nakauchi and T. Kajihara, *Rep. Usa. Mar. Biol. Inst.*, *3*, 61 (1981).

255. A. Hatcher, *Mar. Biol.*, *108*, 433 (1991).

256. S. Konosu and K. Yamaguchi, *Chemistry and Biochemistry of Marine Food Products* (R. E. Martin, G. J. Flick, C. E. Hebard, and D. R. Ward, eds.), AVI, Westport, Conn., 1982, p. 367.

257. A. Nontratip, S. Wada, and H. Yamanaka, *Nippon Suisan Gakkaishi*, *57*, 761 (1991).

258. K. Watanabe, H. Uehara, M. Sato, and S. Konosu, *Nippon Suisan Gakkaishi*, *51*, 1293 (1985).

259. C.-K. Park, T. Matsui, K. Watanabe, K. Yamaguchi, and S. Konosu, *Nippon Suisan Gakkaishi*, *56*, 1319 (1990).

260. K. Watanabe, H. Maezawa, H. Nakamura, and S. Konosu, *Nippon Suisan Gakkaishi*, *49*, 1755 (1983).

261. B. M. Hitzig, M. P. Kneussel, V. Shih, R. D. Brandstetter, and H. Kazemi, *J. Appl. Physiol.*, *58*, 1751 (1985).

262. G. E. Nilsson and P. L. Lutz, *Am. J. Physiol.*, *261*, R32 (1991).

263. T. Nakada, K. Hida, and I. L. Kwee, *Neuroreport*, *2*, 325 (1991).

264. T. H. Ermak, *Cell Tissue Res.*, *176*, 47 (1977).

265. F. Gaill, *Cell Tissue Res.*, *208*, 197 (1980).

266. M. J. Cavey, *Am. Zool.*, *22*, 807 (1982).

267. P. J. Schofield and B. W. Stewart, *Comp. Biochem. Physiol.*, *45B*, 567 (1973).

268. T. DAnna, *Acta Embryol. Morphol. Exp. New Ser.*, *4*, 73 (1983).

269. G. Perry and C. Lambert, *Comp. Biochem. Physiol.*, *90B*, 785 (1988).

270. C. Bourcart, P. Lubet, and H. Ranc, *C. R. Séanc. Soc. Biol. Filiales Lyon*, *158*, 1638 (1964).
271. I. Zsnagy, *Acta Biochim. Biophys. Acad. Sci. Hung.*, *8*, 143 (1973).
272. I. Zsnagy and M. Ermini, *Comp. Biochem. Physiol.*, *43B*, 39 (1972).
273. I. Zsnagy and M. Ermini, *Comp. Biochem. Physiol.*, *43B*, 593 (1972).
274. I. Zsnagy, *Int. Rev. Cytol.*, *49*, 331 (1977).
275. M. C. Sarasquete, M. L. G. Canales, and S. Gimeno, *Eur. J. Histochem.*, *36*, 223 (1992).
276. J. Donaldson and F. LaBella, *J. Toxicol. Environ. Health*, *12*, 119 (1983).
277. D. A. Holwerda and A. DeZwaan, *Comp. Biochem. Physiol.*, *67B*, 447 (1980).
278. D. G. Penney, *Comp. Biochem. Physiol.*, *86A*, 609 (1987).
279. P. Burighel and A. Schiavinato, *Cell Tissue Res.*, *235*, 309 (1984).
280. N. J. Berrill, *Biol. Rev.*, *26*, 456 (1951).
281. L. G. Barth and L. J. Barth, *J. Morphol.*, *118*, 451 (1966).
282. B. Rinkevich and I. L. Weissman, *J. Exp. Zool.*, *253*, 189 (1990).
283. *CHCD Dictionary of Natural Products on CD-ROM*, Chapman & Hall, London, 1992.
284. B. Schwyn and J. B. Nielands, *Anal. Biochem.*, *160*, 47 (1987).
285. M. DeVincentiis and W. Rüdiger, *Experientia*, *23*, 245 (1967).
286. R. J. Anderson and R. J. Stonard, *Can. J. Chem.*, *57*, 2325 (1979).
287. U. Schmidt, A. Lieberknecht, H. Griesser, and H. Bökens, *Liebigs Ann. Chem.*, 785 (1985).
288. R. J. Stonard and R. J. Anderson, *J. Org. Chem.*, *45*, 3687 (1980).
289. H. Michibata, T. Uyama, and J. Hirata, *Zool. Sci.*, *7*, 55 (1990).
290. F. E. Lytle, *Appl. Spectrosc.*, *24*, 319 (1970).
291. T. Delohery. Personal communication, 1988.
292. H. Michibata, Y. Seki, J. Hirata, M. Kawamura, I. Kunihisa, Y. Iwata, and T. Ido, *Zool. Sci.*, *8*, 447 (1991).
293. L. Botte, S. Scippa, and M. de Vincentiis, *Experientia*, *35*, 1228 (1979).
294. D. J. W. Lane and S. L. Wilkes, *Acta Zool. (Stockh.)*, *69*, 135 (1988).

295. S. Scippa, M. de Vincentiis, and K. Zierold, *Invert. Rep. Dev.*, *17*, 141 (1990).

296. C. L. Simpson and C. G. Pierpont, *Inorg. Chem.*, *31*, 4308 (1992).

297. R. Dagani, *C&E News*, February, 28 (1993).

298. M. Kalk, *Nature*, 198, 1010 (1963).

299. A. L. Dingley, K. Kustin, I. G. Macara, and G. C. McLeod, *Biochim. Biophys. Acta*, *649*, 493 (1981).

300. D. A. Raftos, D. L. Stillman, and E. L. Cooper, *Immun. Cell Biol.*, *69*, 225 (1991).

301. D. A. Raftos and E. L. Cooper, *J. Exp. Zool.*, *260*, 391 (1991).

302. C. J. Hawkins, D. L. Parry, and B. J. Wood, *Inorg. Chim. Acta*, *78*, L29 (1983).

303. J. G. Jacobsen and L. H. Smith, *Physiol. Rev.*, *48*, 424 (1968).

304. G. C. Wagner, R. J. Kassner, and M. D. Kamen, *Proc. Natl. Acad. Sci. USA*, *71*, 253 (1974).

305. W. Mraz and H. Jatzkewitz, *Hoppe-Seyler's Z. Physiol. Chem.*, *355*, 33 (1974).

306. S. N. Federov, A. D. Chumak, V. A. Denisenko, V. A. Stonik, and V. V. Isakov, *Chem. Nat. Prod.*, *18*, 634 (1982).

307. G. Materazzi, *Boll. Soc. Ital. Biol. Sper.*, *43*, 1919 (1967).

308. G. R. Vasta, *Phylogenesis of Immune Functions* (G. W. Warr and N. Cohen, eds.), CRC Press, Boca Raton, 1991, p. 74.

309. R. M. Albano, M. S. G. Pavão, P. A. S. Mourão, and B. Mulloy, *Carbohydr. Res.*, *208*, 163 (1990).

310. M. S. G. Pavão, P. A. S. Mourão, and B. Mulloy, *Carbohydr. Res.*, 153 (1990).

311. J. Santos, B. Mulloy, and P. A. S. Mourão, *Eur. J. Biochem.*, *204*, 669 (1992).

312. A. F. Rowley, *J. Exp. Zool.*, *227*, 319 (1983).

313. D. Stoecker, *Ecology*, *61*, 1327 (1980).

314. D. Stoecker, *Mar. Ecol. Prog. Ser.*, *3*, 257 (1980).

315. V. L. Scofield and L. S. Nagashima, *Biol. Bull.*, *165*, 733 (1983).

316. G. D. Parry, *J. Theor. Biol.*, *101*, 453 (1983).

317. J. Himms-Hagen, *Ann. Rev. Physiol.*, *38*, 315 (1976).

318. R. Ramasarma and L. Susheela, *Biomembrane-Architecture, Biogenesis, Bioenergetics and Differentiation* (L. Packer, ed.), Academic Press, New York, 1973, p. 261.
319. T. Ramasarma and S. Sivaramakrishnan, *World Rev. Nutr. Diet.*, *31*, 226 (1978).
320. T. Ramasarma, W. C. MacKellar, and F. L. Crane, *Biochim. Biophys. Acta*, *646*, 88 (1981).
321. M. S. Patole, C. K. R. Kurup, and T. Ramasarma, *Biochem. Biophys. Res. Commun.*, *141*, 171 (1986).
322. M. S. Patole, C. K. R. Kurup, and T. Ramasarma, *Mol. Cell. Biochem.*, *75*, 161 (1987).
323. M. Rau, M. S. Patole, S. Vijaya, C. K. R. Kurup, and T. Ramasarma, *Mol. Cell. Biochem.*, *75*, 151 (1987).
324. J. N. Hathcock, C. H. Hill, and S. B. Tove, *Can. J. Biochem.*, *44*, 983 (1966).
325. E. G. DeMaster and R. A. Mitchell, *Biochemistry*, *12*, 3616 (1973).
326. J. Z. Byckowski, L. Byckowski, and J. Tluczkiewicz, *Int. J. Biochem.*, *10*, 1007 (1979).
327. G. Reverberi, *Acta Embryol. Morphol. Exper.*, *1*, 12 (1957).
328. M. J. Smith. In preparation, 1994.
329. M. Henze, *Hoppe-Seyler's Z. Physiol. Chem.*, *79*, 215 (1912).
330. B. Lübbering and G. Goffinet, *Belg. J. Zool.*, *121*, 39 (1991).
331. Z. Y. Kikvidze, D. V. Chikvashvili, and Z. P. Kometiani, *Biochemistry (Engl. Transl.)*, *47*, 1542 (1982).
332. D. W. Reif, R. A. Coulombe, Jr., and S. D. Aust, *Arch. Biochem. Biophys.*, *270*, 137 (1989).
333. I. Zsnagy, *Comp. Biochem. Physiol.*, *49A*, 399 (1974).
334. I. Zsnagy, D. A. Holwerda, and D. I. Zandee, *Basic Appl. Histochem.*, *26*, PR8 (1982).
335. I. Zsnagy and S. Kerpel-Fronius, *Acta Biol. Acad. Sci. Hung.*, *21*, 105 (1970).
336. I. Zsnagy and C. Galli, *Acta. Biol. Acad. Sci. Hung.*, *28*, 123 (1977).
337. K. J. Fan, J. Kovi, and S. D. Duhaney, *Acta Neuropath. (Berl.)*, *41*, 249 (1978).
338. P. Prento, *Cell Tissue Res.*, *196*, 123 (1979).

339. M. Damerval and G. Prunus, *C. R. Acad. Sci. Paris. Sér. III*, *309*, 105 (1989).

340. I. A. Skulskii, N. B. Pivovarova, S. A. Gapon, and S. M. Korotkov, *Neurophysiology*, *23*, 234 (1991).

341. D. A. Holwerda and P. R. Veenhof, *Comp. Biochem. Physiol.*, *78B*, 707 (1984).

342. B. I. Roots and P. V. Johnston, *Comp. Biochem. Physiol.*, *17*, 285 (1966).

343. E. Fischer, *Acta Histochem.*, *60*, 46 (1977).

344. E. Fischer, *Exp. Gerontol.*, *12*, 69 (1977).

345. A. J. Morgan and C. Winters, *Histochemistry*, *73*, 589 (1982).

346. E. Fischer, *Acta Biol. Acad. Sci. Hung.*, *26*, 135 (1975).

347. E. Fischer, *Acta Histochem.*, *46*, 1 (1973).

348. S. Silver and J. E. Lusk, *Ion Transport in Prokaryotes* (B. P. Rosen and S. Silver, eds.), Academic Press, London, 1987, p. 165.

349. L. Efros, PhD dissert., Yale University, 1994.

350. M. O. Walderhaug, D. C. Dosh, and W. Epstein, *Ion Transport in Prokaryotes* (B. P. Rosen and S. Silver, eds.), Academic Press, London, 1987, p. 85.

351. F. M. Harold and E. Spitz, *J. Bacteriol.*, *122*, 266 (1975).

352. A. Stern, A. J. Davison, Q. Wu, and J. Moon, *Free Radical Biol. Med.*, *12*, 373 (1992).

353. L. Califano and E. Boeri, *J. Exp. Biol.*, *27*, 253 (1950).

354. R. L. VanEtten, P. P. Waymack, and D. M. Rehkop, *J. Am. Chem. Soc.*, *96*, 6782 (1974).

355. F. L. Crane, *Biochem. Biophys. Res. Commun.*, *63*, 355 (1975).

356. M. Fontecave, J. Covès, and J.-L. Pierre, *Biometals*, *7*, 3 (1994).

357. E. A. Boichenko, *J. Evol. Biochem. Physiol. (Engl. Transl.)*, *9*, 10 (1973).

358. L. C. Cantley, Jr., L. G. Cantley, and L. Josephson, *J. Biol. Chem.*, *253*, 7361 (1978).

359. E. Erdmann, W. Krawietz, G. Philipp, I. Hackbarth, W. Schmitz, H. Scholz, and F. L. Crane, *Nature*, *282*, 335 (1979).

360. M. Belkoura and G. Marigo, *Biochimie*, *68*, 1299 (1986).

361. F. Yoshimura and A. F. Brodie, *J. Biol. Chem.*, *256*, 12239 (1981).

14
Biochemical Significance of Vanadium in a Polychaete Worm

Toshiaki Ishii,[1] *Izumi Nakai,*[2] *and Kenji Okoshi*[3]

[1]Division of Marine Radioecology, National Institute of Radiological Sciences, 3609 Isozaki, Nakaminato, Ibaraki 311-12, Japan

[2]Department of Chemistry, University of Tsukuba, Tsukuba 305, Japan

[3]Department of Biotechnology, Senshu University of Ishinomaki, Ishinomaki 986, Japan

1. INTRODUCTION

1.1. Vanadium in Marine Organisms

High concentrations of trace metals have often been found in various tissues or organs of marine organisms. Examples of these include manganese and zinc in the kidneys of clams [1–4], cobalt and uranium in the branchial heart of *Cephalopoda* molluscs [5,6], and iron in the radular teeth of limpet [7] and chiton [8].

It is also well known that there is a remarkably high level of vanadium in the blood of the ascidian *Ascidiacea, Protochordata* [9–11], but the specific accumulation of vanadium by other marine organisms has not been reported. We ourselves have analyzed over 500 species of marine organisms such as algae, echinoderms, molluscs, crustaceans, fishes, and so forth collected from the coast around Japan and found that vanadium concentrations in their whole bodies ranged from only 0.1 to 25 μg/g on a dry weight basis.

1.2. Discovery of a New Vanadium Accumulator

In spite of many efforts to find other vanadium accumulators which are indispensable to a comparative study, only ascidians in the animal kingdom have maintained the unique position of the high vanadium accumulation since the first report by Henze [9].

We were the first to report a high level of vanadium in the fan

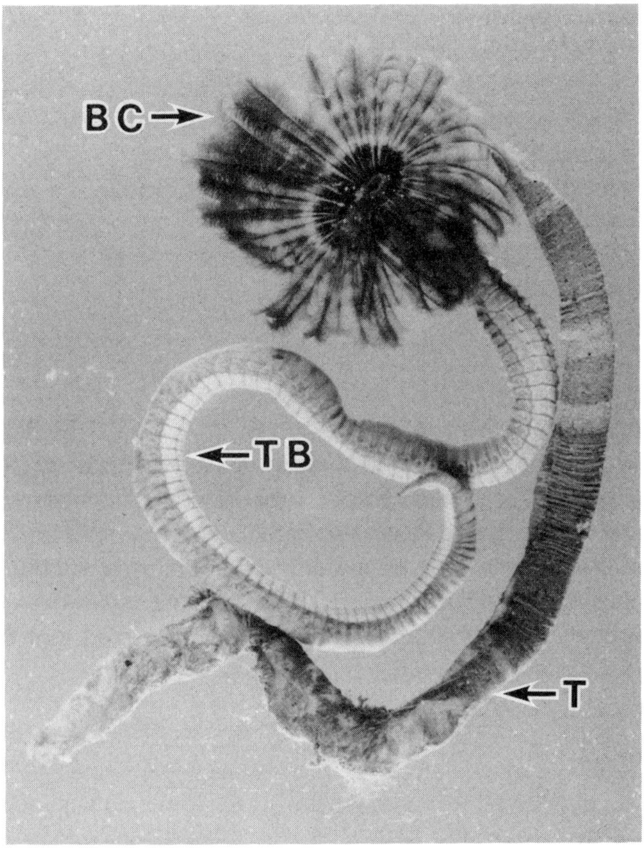

FIG. 1. Lateral view of *P. occelata*. This fan worm has a trunk body (TB) and a branchial crown (BC) which is composed of many bipinnate radioles. The fan worm constructs a tube (T) of sand grains embedded in mucus.

worm *Pseudopotamilla occelata* (Fig. 1) classified into *Polychaeta*, *Annelida* [12]. It is of great interest that the new vanadium accumulator is found in the teloblast series of the phylogenetic tree. This suggests that some vanadium accumulators are present in the animal kingdom, independently of their taxonomic ranks. We could not prove that the high accumulation of vanadium is a widespread phenomenon in the group of fan worms. Although a worldwide survey of fan worms is necessary, two species of fan worms occurring in Japan, *Sabellastarte japonica* (12 μg/g dry wt) and *Hydroides ezoensis* (17 μg/g dry wt), showed very low vanadium contents.

In this chapter, we report the concentration, distribution, and chemical states of vanadium in *P. occelata* and describe some of its physiological roles.

2. VANADIUM CONTENT IN FAN WORMS AND ASCIDIANS

2.1. Qualitative and Quantitative Multielement Analyses

We noticed a large amount of vanadium in *P. occelata* when qualitative analyses by inductively coupled plasma mass spectrometry (ICP-MS), neutron activation analysis (NAA), and wavelength-dispersive X-ray fluorescence (WLD-XRF) were applied to various species of marine organisms. Determinations of elements in fan worms and ascidians were carried out by means of atomic absorption spectrometry (AAS) and inductively coupled plasma atomic absorption spectrometry (ICP-AES).

2.2. Vanadium Concentrations in *Pseudopotamilla occelata*

The concentrations of 11 elements in the bipinnate radiole, the trunk body, and the whole soft body of *P. occelata* are listed in Table 1. Most elements showed little difference in concentration in the bipinnate radiole and the trunk body, whereas the concentrations of iron, zinc, and vanadium were higher in the bipinnate radiole than in the trunk body. In particular, the vanadium concentration (mean ± SD = 5600 ± 1900 μg/g dry wt, n = 25) in the bipinnate radiole was about 100 times higher than in the trunk body (mean ± SD = 66 ± 30 μg/g dry wt, n = 25). Approximately 90% of the total body burden was concentrated in the bipinnate radiole, which occupies only 7.5 ± 1.7% (n = 55) of the whole soft body weight.

The vanadium concentration in the whole body of *P. occelata* ranged from 350 to 1200 μg/g on a dry weight basis. This value was comparable to that in three species of ascidians *Ascidia ahodori* (1500

TABLE 1

Concentrations (μg/g dry wt) of 11 Elements in the Fan Worm *P. occelata*

Element	Bipinnate radiole (n = 25)	Trunk body (n = 25)	Whole body (n = 30)
Ca	960 ± 170	1200 ± 250	1170 ± 120
Cu	8.9 ± 3.4	4.9 ± 1.4	6.4 ± 2.7
Fe	330 ± 65	140 ± 16	190 ± 38
K	10900 ± 1520	9930 ± 970	10000 ± 2020
Mg	2090 ± 250	1540 ± 480	1800 ± 540
Mn	7.3 ± 1.9	5.5 ± 2.1	6.6 ± 3.3
Na	12300 ± 4500	6210 ± 1860	10300 ± 4320
P	6450 ± 490	4820 ± 680	5410 ± 1340
Sr	14 ± 5	23 ± 12	18 ± 10
V	5600 ± 1900	66 ± 30	550 ± 260
Zn	220 ± 95	67 ± 20	89 ± 26

μg/g dry wt), *Ascidia sydneiensis samae* (260 μg/g dry wt), and *Ciona savignyi* (220 μg/g dry wt).

3. DISTRIBUTION OF VANADIUM IN THE BRANCHIAL CROWN (RADIOLE)

3.1. Pretreatment for EPMA and AEM Measurements

In order to examine the distribution of vanadium in the radioles, specimens for electron probe X-ray microanalysis (EPMA) and analytical electron microscopy (AEM) analyses were prepared in accordance with conventional procedures (i.e., fixing with glutaraldehyde and osmic solutions and dehydrating several times in ethanol). However, no characteristic X rays of vanadium were detected from any part of the specimens. We presumed that vanadium was eluted from its localized sites to the medium. Therefore, freeze-cracking and cryosectioning methods as more preferable sample preparations were applied to *P. occelata*.

3.2. Vanadium Distribution in the Radiole

We have reported that most of the vanadium in *P. occelata* was concentrated only in the epidermis covering the radioles and little vanadium was contained in the muscular, connecting, and supporting tissues or in blood plasma [12].

Line analysis by EPMA for a transverse fracture surface of the radiole indicates that vanadium is not distributed uniformly in the

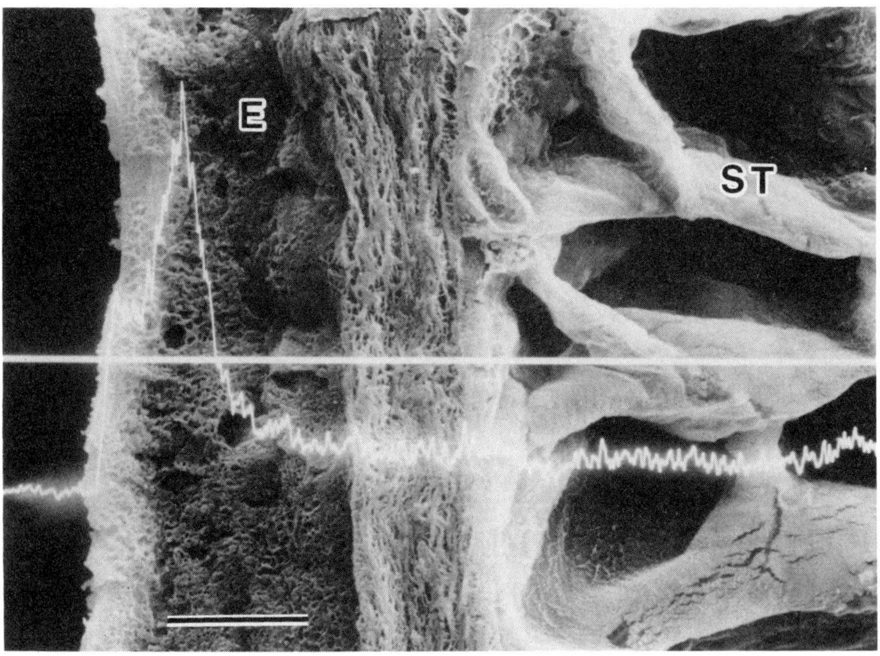

FIG. 2. Line analysis by electron probe X-ray microanalysis (EPMA) for a transverse fracture surface of the bipinnate radiole. A clear peak indicating the highest vanadium content is observed in the outer portion of the epidermis (E). ST, supporting tissue. Scale bar = 10 μm.

epidermis and that a clear peak indicating the highest vanadium content is observed in the outer position of the epidermis (Fig. 2). The same tendency is also observed for the longitudinal cryosections which were cut with a Leitz 1720 digital cryostat. On the other hand, among several types of blood cells of ascidians, the vanadium-containing blood cell of *Ascidia ahodori* [13], *A. sydneiensis samae* [14], and *A. gemmata* [15] has been identified as the signet ring cell. It is very interesting that the vanadocytes (epidermal cells and blood cells) are quite different between the two vanadium accumulators. The similarity and difference in vanadium distribution between *P. occelata* and ascidians are expected to provide important information on the mechanisms of bioconcentration and physiological roles of vanadium.

3.3. Subcellular Distribution in the Epidermal Cells

An electron micrograph of the cryosection of biological materials without osmic fixation and staining usually gives low contrast. However, as shown in Fig. 3a of the freeze-dried ultrathin section (100 nm), many electron-dense deposits suggesting the localization of certain elements are present in the apical portion of the cytoplasm. We found from the transmission electron micrograph of Fig. 3b that electron-dense deposits are situated at the same places where many vacuoles with low electron density are located.

As illustrated in Fig. 4, clear peaks of characteristic X rays of sulfur (2.31 keV) and vanadium (4.95 keV) appear from all of the electron-dense deposits. We concluded that most of the vanadium in *P. occelata* is localized in the vacuoles of the epidermal cells [16]. A small quantity of vanadium is also found in the cuticle, but no vanadium is detected from other cellular components such as the nucleus, cytoplasm, etc.

Scippa et al. applied X-ray microanalysis in scanning transmission electron microscopy to vacuolated and granular amebocyte, signet ring cell, and a new type of compartment cell of the ascidian *Phallusia mammillata* [17]. They found vanadium deposits on the vacuolar membranes and in intravacuolar and cytoplasmic electron-dense inclusions.

Most of the vanadium in ascidians is thought to be concentrated in the vacuole of the signet ring cell, based on results from the separation

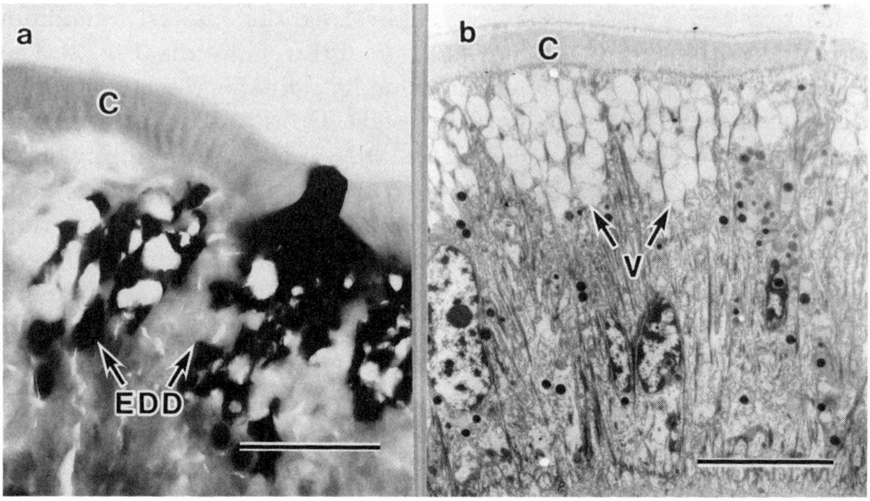

FIG. 3. (a) Transmission electron micrograph of the cryosection of the bipinnate radiole. Electron-dense deposits (EDDs) are observed in the apical portion of the epidermis. C, cuticle. Scale bar = 3 μm. (b) Transmission electron micrograph for an ultrathin section of the bipinnate radiole. Many vacuoles (V) are observed in the apical portion of the epidermal cells. Scale bar = 5 μm.

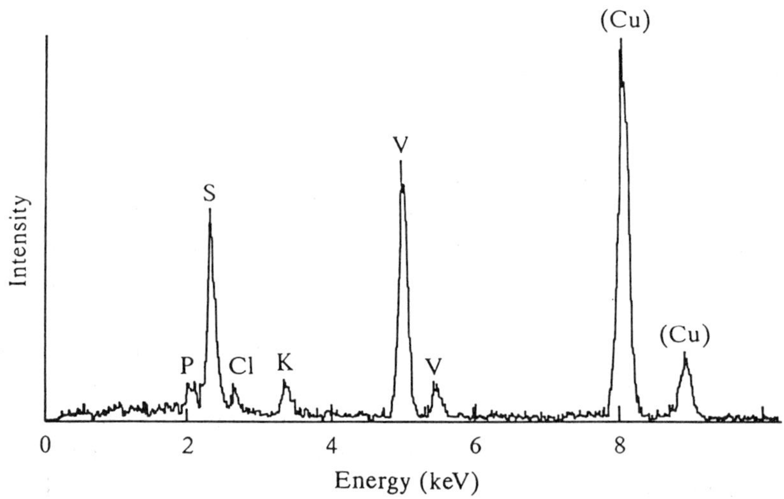

FIG. 4. X-ray spectrum of the electron-dense deposit obtained by analytical electron microscopy.

of the vanadocytes [13] and confirmation of the locality of vanadium deposits [17]. It is significant that the subcellular distribution of vanadium is the same, although the vanadocytes are quite different between the two vanadium accumulators.

4. CHEMICAL SPECIATION OF VANADIUM IN LIVING ORGANISMS

4.1. Application of XAFS to In Vivo Measurement

Knowledge about the oxidation state and coordination form of vanadium in *P. occelata* is important to study the physiological roles of vanadium in this animal. X-ray absorption fine structure (XAFS) is a powerful technique for investigation of the electronic structure and the local environment of specific atoms in biological samples [18]. This technique was successfully applied to the study of vanadium in living ascidian blood cells (vanadocytes) [19]. It was revealed that the vanadium in the living cells is present as V(III) ions surrounded by six oxygen atoms. We thought that XAFS is particularly useful for the study of the new vanadium accumulator because it allows literally nondestructive analysis of living organisms; the oxidation state of vanadium is very sensitive to aerooxidation and in vivo analysis is an often crucial requirement. Therefore, we utilized the XAFS technique to clarify the oxidation state and local structure of vanadium in *P. occelata* [12,20]. This is possibly the first successful application of the XAFS technique to the whole soft body of a living marine animal. Since the principle and theory of the XAFS technique have already been explained in the literature [18,21] and also in Chap. 19 of this volume, they are not repeated here.

XAFS spectral data of the living *P. occelata* were measured at the Photon Factory of the National Laboratory for High Energy Physics, Tsukuba, Japan with a ring energy of 2.5 GeV. Data were collected in the fluorescence mode using a two-crystal monochromator with Si(111) crystals and a Lytle-type fluorescence detector [22] at a bending magnet beam line.

The animal specimens were collected on the Sanriku coast of the Pacific Ocean in the northeastern region of Japan. They were kept in fresh seawater and transported to the Photon Factory within 24 hr and subjected to the XAFS measurement in the living condition. The whole bodies of the animals were pulled out from their tubes and three specimens were placed in a polyethylene bag. The radioles were exposed to the X-ray irradiation. The X-ray absorption spectra were measured from 4.96 to 6.43 keV with a step size of approximately 1.2 eV in the edge region and 4s/point (total 545 steps). It was remarkable that the animals were still alive even after the X-ray irradiation. For comparison, XAFS spectra of six vanadium compounds of known structures were also measured: $V_2(SO_4)_3$, $V(acac)_3$, $VO(acac)_2$ were prepared by the methods of Claunch and Jones [23], Grdenic and Korpar-Colig [24], Rowe and Jones [25], respectively; V_2O_5 and NH_4VO_3 were purchased from Kanto Chemical Co. and $VO(C_2O_4)_2 \cdot nH_2O$ from Wako Chemical Co., Japan.

4.2. Oxidation State of Vanadium in *P. occelata*

Figure 5 shows the vanadium K-edge spectra of the specimens and the reference compounds. It is found that the K-edge spectra of certain compounds exhibit a strong pre-edge peak at around 5.47 keV. This peak is ascribed to the 1s-to-3d transition, which is formally forbidden by dipole selection rules if the coordination sphere of vanadium has octahedral (O_h) symmetry with a center of inversion [26]. When the symmetry of the ligand is lowered from O_h, the inversion center is broken, and the pre-edge absorption becomes dipole allowed due to a combination of stronger 3d-4p mixing and overlap of the metal 3d orbitals with the 2p orbitals of the ligand. Therefore, if a vanadium compound contains a terminal oxo ligand (V=O) as in the vanadium(IV) compounds, the octahedral symmetry is broken and the compound exhibits intense pre-edge absorption. It was reported that the intensity of the pre-edge absorption is roughly proportional to the number of oxo groups present in the molecule [19].

As can be seen in Fig. 5, the intensity variation of the pre-edge peaks across the series of the reference compounds from V(III) to V(V) compounds is remarkable. On the other hand, the pre-edge absorption in the spectrum of *P. occelata* is almost negligible resembling that of the symmetrically coordinated $V^{III}(acac)_3$. This indicates that most of the

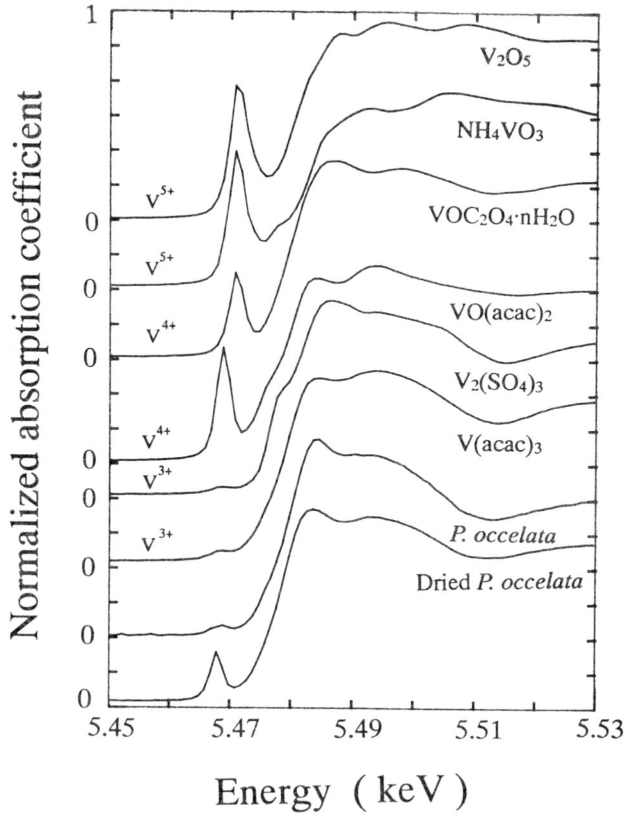

FIG. 5. V-K edge XANES spectra of the radioles of living *P. occelata* and reference vanadium compounds. Intense pre-edge peaks at around 5.47 keV were observed for compounds containing V(IV) or V(V) ions.

vanadium exists in the V(III) state in the living condition with lack of any significant quantity of a VO^{2+} component.

It is also noteworthy that a significant increase in the pre-edge intensity was observed for the dried sample (Fig. 5). A similar effect was obtained when the radioles were removed from the body and were ground in an agate mortar in air. This suggests that vanadium was partly oxidized by air to produce a VO^{2+} component. Therefore, in vivo measurement of the XAFS spectra is essential for the determination of the chemical state of vanadium in *P. occelata*.

4.3. Coordination Structure of Vanadium in *P. occelata*

The analysis of the data was made following a procedure reported in the literature [27]. The EXAFS oscillation $\chi(k)$, was extracted from the spectra by pre-edge and background subtraction, and energy conversion into the photoelectron wave vector k ($k = [8m(E - E_0)/h^2]^{1/2}$, where m is the mass of the electron, h is the Planck constant, E is the incident photon energy, and E_0 is the threshold energy, i.e., the energy at $k = 0$, of that particular absorption edge) [21]. Figure 6 shows the k^2-weighted $\chi(k)$. It can be seen that the difference in amplitude of the $\chi(k)$ function

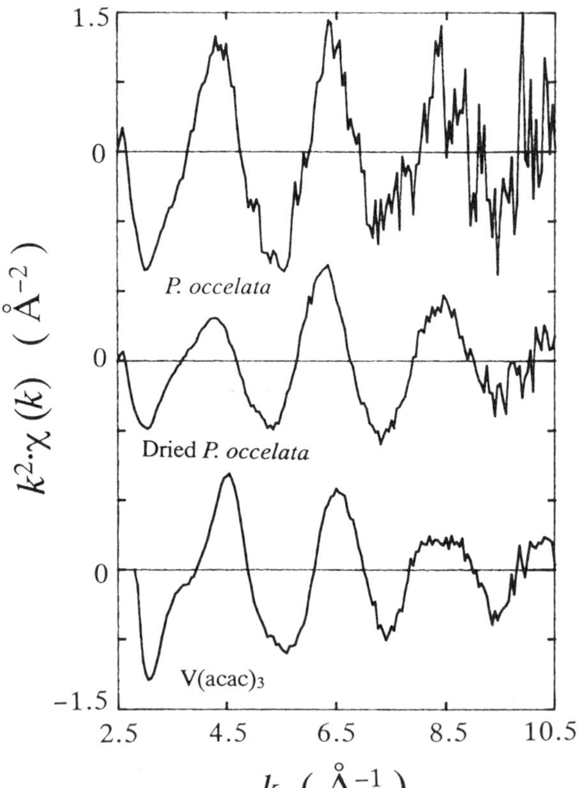

FIG. 6. k^2 weighted χ(k) for living *P. occelata*, dried *P. occelata*, and V(acac)$_3$.

between living *P. occelata* and the dried specimen is significant. The Fourier transforms of the $k^2\chi(k)$ extended X-ray absorption fine structure (EXAFS) oscillation yielded the radial structure functions given in Fig. 7, which are not corrected for phase shift. This gives a single main peak with the absence of any ordered second shell of scattering atoms at longer distance. Considering the result of the X-ray absorption near-edge structure (XANES) analysis, the lack of the second coordination atoms would indicate that vanadium is present in solution in the form of an aqua complex with symmetric octahedral coordination.

We have obtained the structural parameters of vanadium by the

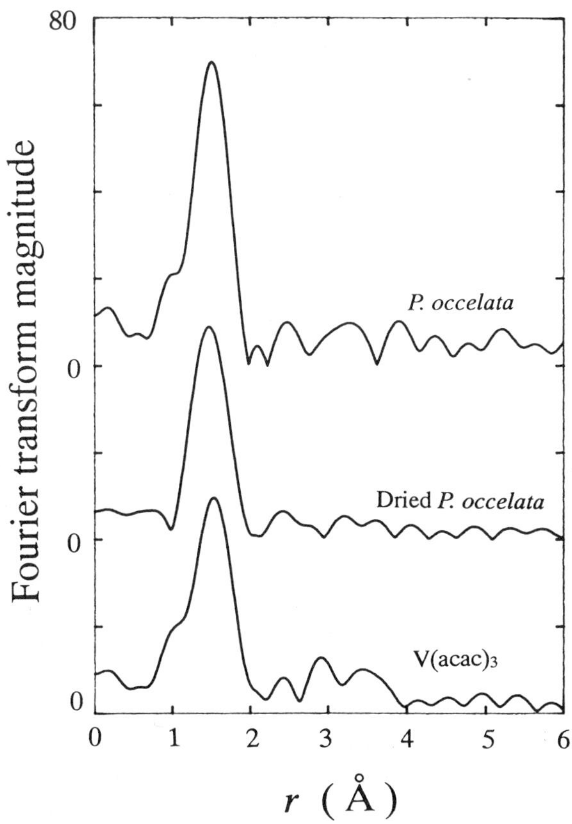

FIG. 7. Radial structure functions for the radioles of living *P. occelata*, dried *P. occelata*, and $V(acac)_3$ calculated without correcting for the phase shifts.

Fourier filtering technique. This was accomplished by an inverse Fourier transformation of the main peak in the radial structural function. This peak defines the contribution of the first coordination shell belonging to the vanadium(III) atom to the EXAFS oscillation. Then, the resulting "observed" $\chi(k)$ function was analyzed using iterative nonlinear least squares curve fitting with the $\chi(k)$ function calculated from a coordination model and backscattering amplitude and phase parameters given by Teo and Lee [28]. As a model, the single oxygen wave fit reported for the vanadocyte of ascidians was used in the fitting analysis and the coordination number was fixed to 6. The agreement between experimental and calculated data was fair ($R = 0.072$) considering that the data were obtained from the living animal. The V–O distance thus obtained was 2.00 Å with a Debye-Waller factor σ of 0.021.

There have been many studies concerning the chemical form of vanadium in ascidians. Electron paramagnetic resonance (EPR) analysis disclosed that the whole blood of *Ascidia nigra* contained V(III) as the major component with a minor amount of vanadyl ion [29,30]. A magnetic study of the vanadocyte using a superconducting quantum interference device (SQUID) susceptometer also supported the predominance of vanadium(III) in ascidians, i.e. the intact blood cells of the tunicate *A. nigra* contained 90% vanadium(III) and 10% vanadium(IV) [31]. XAFS examination of living *Ascidia ceratodes* vanadocytes revealed that the first coordination sphere of vanadium(III) consisted of six oxygen atoms at an average distance of 1.99 Å [19]. No evidence for a ligand second shell was found, and the possibility that the vanadium was confined either by a rigid low molecular weight chelate or a highly ordered protein binding site was denied. In addition, XANES analysis of sulfur within the plasma cells of *A. ceratodes* disclosed the presence of a large endogenous sulfate concentration, as well as smaller significant amounts of thiol or thioether sulfur [32]. EPR linewidth/pH correlation showed that the internal vanadophoric pH in *A. ceratodes* vanadocytes was 1.8 ± 0.1 [33]. This pH is sufficiently low to prevent hydrolysis of endogenous $V(H_2O)_6^{3+}$ or its autoxidation. In addition, nuclear magnetic resonance (NMR) analysis indicated the presence of four to five ligating waters [34]. From these observations it was concluded that the most reasonable form to be deduced for vanadophoric vanadium(III) in the presence of high $[SO_4^{2-}]$ is as the $V(SO_4)(H_2O)_{4-5}^{+}$ complex ion [31]. On the other hand, our preliminary study of the S-K edge XANES spectrum of the radiole of *P. occelata* has indicated that the predominant chemical form of sulfur is the sulfate ion. Moreover, our local analysis by AEM disclosed coexistence of vanadium and sulfur (Fig. 4). It was reported that the signal strengths of the two elements in the cells from *Ascidia mentula* and *Ascidiella aspersa* determined by their X-ray microanalysis were usually similar [35]. As can be seen in Fig. 4, this tendency was also observed in the case of *P. occelata*. These apparent similarities between the two animals suggest the possibility that vanadium in *P. occelata* exists in a similar chemical form as in ascidians, i.e., an aqua complex of V(III) in sulfuric acid media. As to the existence of a V(IV) component in ascidians, EPR analysis indicated the presence of a vanadyl ion in the whole-cell preparation of *A. ceratodes*, although there is a possibility that the observed EPR signal of the vanadyl ion arose from the signet ring cells or compartment cells rather than from the vanadocytes, which are the major constituents of the plasma cells [35]. The existence of a vanadyl ion is also supported by the magnetic study cited above [31]. The presence of a vanadyl ion as a minor component of *P. occelata* could not be ruled out because the V-K edge XANES spectrum of *P. occelata* exhibited a faint pre-edge peak assignable to a VO^{2+} component. In order to solve this problem, EPR analysis of this animal will be essential.

5. PHYSIOLOGICAL ROLES OF VANADIUM

Many biologists and chemists were interested in the biochemical significance of the high vanadium concentration in ascidians since the first report by Henze [9]. In particular, physiologists have studied the biological functions of vanadium in ascidians and presented hypotheses such as oxygen binding, tunicin synthesis, antimicrobial defenses, and antipredation mechanisms. But as yet there is no widespread agreement supporting anyone of them.

Reasonable data have not been obtained yet, although we are intensively studying the physiological roles of vanadium in *P. occelata*. However, results on the localization and chemical states of vanadium in the vacuoles might provide important suggestions toward solving the mystery of the high vanadium concentration in *P. occelata*.

We propose several ideas regarding the physiological roles of vanadium in *P. occelata*:

(1) Generally speaking, the localization of vanadium in the apical vacuoles of the epidermal cells and in the cuticle implies that vanadium itself or vanadium-containing substances are absorbed, secreted, or excreted at the surface of the radioles.

(2) Vanadium may be related to the regulation of oxidation-reduction reactions in the surface of the bipinnate radioles.

(3) Since the branchial crown serves as a site of gas exchange [36], it is probable that vanadium is related to the absorption of O_2 through many cytoplasmic processes in the cuticle, or the storage of oxygen gas while the worm is withdrawn into the tube.

(4) There is a possibility that vanadium in *P. occelata* does not play any important role in physiology after vanadium reaches the vacuoles. This is because the vacuoles are often thought to be used as an internal disposal site to isolate or confine toxic substances, as indicated in examples of metal-containing granules which appear intracellularly in the vacuoles of the kidney cells of clams [1,2,4] or the branchial heart of the Cephalopoda molluscs (unpublished data).

(5) The fact that vanadium in the vacuoles (probably in the cuticle too) exists as an aqua complex of V(III) ion in sulfuric acid solution leads us to think that a simple cation, only surrounded by solvents, has a great advantage for swiftly transporting signals which are caused by stimulation from the outer atmosphere, to reaction systems like the nervous system.

(6) Vanadium absorbed from the surrounding seawater is thought to be associated temporally with organic materials like proteins. When vanadium-containing substances pass through the vacuole membrane and change into an aqua complex ion, a certain amount of energy may be produced and stored in the vacuoles.

(7) *P. occelata*, which is used as a fishing bait, has several defense mechanisms against predators, like fishes. The most important defense is swift withdrawal into its tube when predators approach [36]. According to our observations, *P. occelata* in a colony could live with three species of starved carnivorous fishes in an aquarium for over 2 weeks. We thought that a certain chemical defense substance including vanadium may be secreted from the radioles.

(8) Vanadium is known to be an integral component of some enzymes such as bromoperoxidase of brown algae, nitrogenase of nitrogen-fixing bacteria, etc. Since free peroxide would oxidatively destroy many cellular components, peroxidase plays an important role

in the detoxification of hydrogen peroxide. Although the presence of vanadium enzymes in the vacuoles of the epidermal cells of *P. occelata* has not been examined, vanadium may have some biological functions in the process of catalysis of certain enzymes in relation to detoxification.

More than 80 years have passed since Henze first reported that a high level of vanadium was contained in the blood of ascidians. However, its biological function remains unknown in spite of many efforts by biologists.

We hope that detailed comparative studies between *P. occelata* and ascidians will be helpful in clarifying the mechanism of the specific accumulation or physiological roles of vanadium in the two vanadium accumulators.

ABBREVIATIONS

AAS	atomic absorption spectrometry
acac	acetylacetonate
AEM	analytical electron microscopy
EPMA	electron probe X-ray microanalysis
EPR	electron paramagnetic resonance
EXAFS	extended X-ray absorption fine structure
ICP-AES	inductively coupled plasma atomic emission
ICP-MS	inductively coupled plasma mass spectrometry
NAA	neutron activation analysis
NMR	nuclear magnetic resonance
SQUID	superconducting quantum interference device
WLD-XRF	wavelength-dispersive X-ray fluorescence
XAFS	X-ray absorption fine structure
XANES	X-ray absorption near-edge structure

REFERENCES

1. N. G. Carmichael, K. S. Squibb, and B. A. Fowler, *J. Fish. Res. Bd. Can.*, *36*, 1149 (1979).
2. S. G. George, B. J. S. Pirie, and T. L. Cooms, *J. Exp. Mar. Biol. Ecol.*, *42*, 143 (1980).

3. T. Ishii, T. Otake, M. Hara, M. Ishikawa, and T. Koyanagi, *Bull. Jpn. Soc. Sci. Fish.*, *52*, 147 (1986).
4. T. Ishii, K. Okoshi, T. Otake, and M. Nakahara, *Nippon Suisan Gakkaishi*, *58*, 1285–1290 (1992).
5. T. Ueda, M. Nakahara, T. Ishii, Y. Suzuki, and H. Suzuki, *J. Rad. Res.*, *20*, 338 (1979).
6. T. Ishii, M. Nakahara, M. Matsuba, and M. Ishikawa, *Nippon Suisan Gakkaishi*, *57*, 779 (1991).
7. S. Mann, C. C. Perry, J. Webb, B. Luke, and R. J. P. Williams, *Proc. Roy. Soc. Lond.*, *B227*, 179 (1986).
8. K. S. Kim, J. Webb, D. J. Macey, and D. D. Cohen, *J. Inorg. Biochemistry*, *28*, 337 (1986).
9. M. Henze, *Hoppe-Seyler's Z. Physiol. Chem.*, *72*, 494 (1911).
10. I. G. Macara, G. C. McLeod, and K. Kustin, *Comp. Biochem. Physiol.*, *63B*, 299 (1979).
11. H. Michibata, T. Terada, N. Anada, K. Yamakawa, and T. Numakunai, *Biol. Bull.*, *171*, 672 (1986).
12. T. Ishii, I. Nakai, C. Numako, K. Okoshi, and T. Otake, *Naturwissenschaften*, *80*, 268 (1993).
13. H. Michibata, J. Hirata, M. Uesaka, T. Numakunai, and H. Sakurai, *J. Exp. Zool.*, *244*, 33 (1987).
14. H. Michibata and T. Uyama, *J. Exp. Zool.*, *254*, 132 (1990).
15. J. Hirata and H. Michibata, *J. Exp. Zool.*, *257*, 160 (1991).
16. T. Ishii, T. Otake, K. Okoshi, M. Nakahara, and R. Nakamura, *Marine Biology*, *120* (1994).
17. S. Scippa, K. Zierold, and M. de Vincentiis, *J. Submicrosc. Cytol. Pathol.*, *20*, 719 (1988).
18. L. Powers, *Biochim. Biophys. Acta*, *683*, 1 (1982).
19. T. D. Tullius, W. O. Gillum, R. M. K. Carlson, and K. O. Hodgson, *J. Am. Chem. Soc.*, *102*, 5670 (1980).
20. I. Nakai, C. Numako, K. Okoshi, and T. Ishii, *J. Am. Chem. Soc.*, submitted.
21. B. K. Teo, *EXAFS: Basic Principles and Data Analysis*, Springer-Verlag, Berlin, 1986.
22. F. W. Lytle, R. B. Greegor, D. R. Sandstom, E. C. Marques, J. Wong, C. L. Spiro, G. R. Huffman, and F. E. Huggins, *Nucl. Instr. Meth.*, *226*, 542 (1984).

23. R. T. Claunch and M. M. Jones, *Inorganic Syntheses*, Vol. 7, McGraw-Hill, New York, 1963, p. 92.
24. D. Grdenic and B. Korpar-Colig, *Inorg. Chem.*, *3*, 1328 (1964).
25. R. A. Rowe and M. M. Jones, *Inorganic Syntheses*, Vol. 5, McGraw-Hill, New York, 1957, p. 114.
26. J. Wong, F. W. Lytle, R. P. Messmer, and D. H. Maylotte, *Phys. Rev.*, *30*, 5596 (1984).
27. K. Ozutumi and T. Kawashima, *Inorg. Chim. Acta*, *180*, 231 (1991).
28. B. K. Teo and P. A. Lee, *J. Am. Chem. Soc.*, *101*, 2815 (1979).
29. K. Kustin, D. S. Levine, G. C. McLeod, and W. A. Curby, *Biol. Bull.*, *150*, 426 (1976).
30. A. L. Dingley, K. Kustin, I. G. Macara, and G. C. McLeod, *Biochim. Biophys. Acta*, *649*, 493 (1981).
31. S. Lee, K. Kustin, W. E. Robinson, R. B. Frankel, and K. Spartman, *J. Inorg. Biochem.*, *33*, 183 (1988).
32. P. Frank, B. Hedman, R. M. K. Carlson, T. A. Tyson, A. L. Roe, and K. O. Hodgson, *Biochemistry*, *26*, 4975 (1986).
33. P. Frank, R. M. K. Carlson, and K. O. Hodgson, *Inorg. Chem.*, *25*, 470 (1986).
34. R. M. K. Carlson, *Proc. Natl. Acad. Sci. USA*, *72*, 2217 (1975).
35. M. V. Bell, B. J. S. Pirie, D. B. McPhail, B. A. Goodman, I. B. Falk-Peterson, and J. R. Sargent, *J. Mar. Biol. Assoc. U.K.*, *62*, 709 (1982).
36. R. D. Barnes, *Invertebrate Zoology*, Saunders College, Philadelphia, 1980, pp. 467–566.

15
Vanadium Transport in Animal Systems

Kenneth Kustin[1] *and William E. Robinson*[2]

[1]Department of Chemistry, Brandeis University, Waltham, Massachusetts 02254-9110, USA

[2]Environmental Sciences Program, University of Massachusetts—Boston, 100 Morrissey Blvd., Boston, Massachusetts 02125-3393, USA

1. INTRODUCTION

Transport is the shuttling of substances, such as vanadium in the present case, from one bodily compartment to another by the components of the circulatory system, either the blood plasma or the blood cells. In mammalian systems, the blood plasma is generally considered to play the dominant role in the transport of physiologically important elements, metabolic substrates, as well as a variety of endogenous compounds and environmental contaminants [1]. In the invertebrates, blood cells have traditionally been thought to play a predominant role in transport. However, recent studies have placed this predominance in question, and have proposed that invertebrate blood plasma may play an analogous role to that of the mammalian plasma in this process [2–4].

Transport may also be considered as one component of the cellular absorption process. Absorption of substances must occur across an epithelium, whether it be an epidermal layer in contact with the external aqueous environment in the case of aquatic species, the lining of the lung in the case of respiratory exposure, or the epithelium of the gastrointestinal tract in the case of oral exposure. In order for vanadium to find its way from the external environment to the blood, it generally must pass through these epithelial cells. Thus, we can consider transcellular transport (the transport through the cell's cytoplasm) as an additional type of transport.

As evidenced in the present volume, much of the recent work on vanadium in biological systems has dealt with vanadium's toxicological and pharmacological effects, its unique inorganic and organic biochemistry, and its possible function in various specific biological systems. Absorption and transport of this element have received much less attention, except as they relate to the nutritional uptake of vanadium [5]. Our emphasis, while including exposure by ingestion, will explore other avenues of possible exposure of vanadium to various animal species.

In order to present a complete picture of vanadium transport in animal systems, we will first outline what is currently known about the end result of vanadium transport—the target tissues. We will then summarize vanadium absorption mechanisms and, finally, the mechanisms of vanadium transport in blood plasma and blood cells.

The reader will quickly realize the most of what is known about vanadium transport relates to aquatic organisms rather than terrestrial mammals. Even within the aquatic invertebrates, there is a preponderance of data on only one group, the class Ascidiacea within the subphylum Urochordata. The ascidians have received considerable attention ever since Henze first discovered that the blood cells within some members of this group concentrated vanadium to an astounding extent [6] in the unusual form of a vanadium(III) compound [7]. Over the years these findings have been confirmed and extended [8,9]. This review of vanadium transport will attempt to summarize what is known about vanadium transport throughout the animal system but will of necessity rely heavily on data gathered for the ascidians.

2. TARGET TISSUES AND ORGANS

Vanadium still lacks a clearly defined role in animal systems. Since it is known to be a potent inhibitor of the membrane-bound Na^+,K^+-ATPase, it has been hypothesized to play a role in the regulation of cation exchange [10]. It is also known to inhibit a wide variety of other phosphatase and kinase enzymes and to interact with a multitude of intracellular compounds [11]. Vanadium has been touted as being an essential trace element [12,13], although evidence has still proven equivocal [10,14]. To date, the best evidence for a biological role of vanadium comes not from the animal kingdom but from bacteria where it has been shown to be a necessary component in an alternative nitrogen fixation enzyme in *Azotobacter chroococcum* [15,16] and from the plant kingdom where it is integral to bromoperoxidase function in many marine alga [17,18].

If vanadium has a functional role in animals, it is clear that only ultratrace levels are required to fulfill this role. It is not surprising, therefore, that no single target tissue for vanadium has been identified in mammals. Vanadium is present in most tissues of humans at concentrations generally between 0.5 and 140 ng/g wet tissue [19]. Of the approximately 100 μg of vanadium present in the entire human body, most is located in the lungs, hair, liver, and kidneys [10,19,20], suggesting that most of the body's vanadium content is not being purposely sequestered. Goldfish, *Carassius auratus*, exposed to dissolved vanadium for 4 days exhibited highest concentrations in the intestine and liver, followed by the gill, heart, and kidney [21]. In both unexposed and vanadate-exposed catfish, *Clarias batrachus*, the kidney contained the highest vanadium concentration, the liver second most, and both gill and intestinal tissues held considerably lower concentrations [22]. Mice which were fed mussels with elevated vanadium body burdens concentrated the vanadium in the liver and kidney, with lowest doses in the muscle and skin [23]. The elevation in gastrointestinal and excretory tissues of mammals probably has more to do with the uptake and excretion of vanadium than storage or site of action. However, the possibility that these organs may serve to regulate the distribution and body burden of vanadium cannot be ruled out.

Over 90% of intracellular vanadium is complexed to phosphates, with lesser amounts bound to ferritin and, although rather weakly, to glutathione and ascorbate [10]. Since less than 1% of the intracellular vanadium is uncomplexed, it is unclear as to whether the concentrations of free vanadium (predominantly intracellular VO^{2+}) are sufficiently high to elicit enzyme inhibition in vivo [10].

Unlike mammals, very high concentrations of vanadium have been measured in the body tissues of several aquatic (primarily marine) animals (Table 1). The highest concentrations have been measured in the blood cells of a number of ascidian species (described later) and in the branchial crown of the sabellid polycheate *Pseudopotamilla occelata* [24,25]. *P. occelata* concentrates V(III) within epidermal cells of the branchial crown radioles. Over 90% of the body's vanadium is sequestered in this organ, reaching concentrations as high as 3270–7150 μg/g dry tissue wt [24]. Previously, whole-body tissue analysis of another sabellid polychaete, *Eudistylia vancouveri*, revealed a vanadium concentration of 94–786 μg/g dry wt [26], a value comparable to the estimated whole-body concentration of 399 μg/g measured in *P. occelata*. Additional analyses will be needed to determine whether the vanadium concentration in the branchial crown of *E. vancouveri* will equal or surpass that in the crown of *P. occelata*.

TABLE 1

Vanadium Concentrations in Selected Species[a]

Organism	[V]	BCF	Ref.
Bivalves			
Mytilus galloprovincialis (mussel)			30
Total soft parts	0.4	35	
Shell	3.0	1500	
Byssus	16	1500	
Visceral mass	3.0	350	
Tapes japonica (clam)			83
Total soft parts	1.1	110	
Cyrtodaria kurriana (clam)			84
Total soft parts	4.3–25	430–2500[b]	
Crassostrea virginica (oyster)			85
Entire body (?)	0.8	80	
Crustaceans			
Lysmata seticaudata (shrimp)			31
Total parts (soft and hard)	2.9	410	
Exoskeleton	2.5	500	
Digestive tract	17	1750	
Penaeus setiferous (shrimp)			85
Total parts (soft and hard?)	1.9	190	
Pandalus sp. (prawn)			83
Total soft parts	0.07	7	
Carcinus maenas (crab)			31
Total parts (soft and hard)	3	450	
Exoskeleton	3	1000	
Digestive gland	4	600	
Callinectes sapidus (crab)			85
Total parts (soft and hard?)	1.5	150	
Echinoderms			
Marthasterias glacialis (seastar)			32
Total parts (soft and hard)	1.5	150	
Body wall	1.1	165	
Digestive tract	0.7	70	

TABLE 1

Continued

Organism	[V]	BCF	Ref.
Echinoderms (continued)			
Paracentrotus lividus (urchin)			32
Total parts (soft and hard)	3.7	600	
Body wall	4.2	1350	
Coelomic fluid	3.7	40	
Holothuria forskali (holothurian)			32
Total parts (soft and hard)	4.0	215	
Body wall	2.6	190	
Digestive tract	2.4	115	
Polychaetes			
Eudistylia vancouveri (sabellid)			26
Whole body	94–786	$(0.9–7.9) \times 10^{4}$ [b]	
Pseudopotamilla occelata (sabellid)			24
Whole body	399	4×10^{4}	
Trunk/body	56	5600	
Bipinnate radioles	3270–7150	$(3.3–7.2) \times 10^{5}$	
Ascidians			
Ascidia nigra (phlebobranch)			86
Body (without tunic)	2120	2.1×10^{5} [b]	
Tunic	320	3.2×10^{4} [b]	
Blood cells	26760	2.7×10^{6} [b]	
Ciona intestinalis (phlebobranch)			86
Body (without tunic)	2460	2.5×10^{5} [b]	
Tunic	50	500[b]	
Blood cells	700	7×10^{4} [b]	
Ascidia ahodori (phlebobranch)			29
Tunic	1230	1.2×10^{5} [b]	
Blood cells	21121	2.1×10^{6} [b]	

TABLE 1

Continued

Organism	[V]	BCF	Ref.
Ascidians (continued)			
Ascidia dispar (phlebobranch)			61
Total soft tissue	25.4	2540[b]	
Tunic	0.4	40[b]	
Blood cells	692.9	6.9×10^{4} [b]	
Styela plicata (stolidobranch)			29
Tunic	2.6	260[b]	
Blood cells	3.5	350[b]	
Molgula manhattensis (stolidobranch)			86
Body (without tunic)	14	1400[b]	
Tunic	40	4000[b]	
Blood cells	25	2500[b]	

[a]Whole-body values are presented along with exoskeletons, and tissues with the highest vanadium concentrations. Tissue vanadium concentrations [V] expressed as μg/g dry weight. Bioconcentration factors (BCFs) calculated from vanadium concentrations normalized for tissue wet weight and a mean seawater vanadium concentration of 2 μg/L [27].
[b]Vanadium concentrations converted from dry weight basis, assuming 80% water content.

Vanadium concentrations in other aquatic species never reach the levels found in ascidian blood cells or the tissues of these two sabellid polychaetes (Table 1). Vanadium concentrations in *P. occelata* branchial crowns are 10^5 higher than the concentrations present in natural seawater (1–3 μg/L; [27]). The degree of vanadium concentration in ascidians correlates with phylogeny, with high concentrations found in the orders Phlebobranchia and Aplousobranchia, and low vanadium concentrations in the Stolidobranchia [28,29]. The vanadium concentrations in the blood cells from several phlebobranch ascidians are over 10^6 times greater than the surrounding water. In marked contrast, those ascidians that primarily concentrate Fe instead of vanadium (e.g., the stolidobranchs *Molgula manhattensis* and *Styela plicata*) have vanadium concentrations in body tissues and blood cells that are only 10^2–10^3 times background seawater concentrations. This factor is in the same range of vanadium bioconcentration factors as those in various tissues and exoskeletons of the bivalves, crustaceans, and echinoderms (10^1–10^3; Table 1).

Adsorption of vanadium to the outside of the bivalve shell, periostracum and byssus, the crustacean cuticle, and the echinoderm body wall appears to play the prominent role in the "accumulation" of metal by these organisms (Table 1). Short-term uptake studies using ^{48}V indicate that the vast majority of newly introduced label simply sticks to the outside of these animals and is not bioavailable [21,30–32]. A similar distribution pattern is exhibited for the stable vanadium measured in these experiments, indicating that the long-term accumulation process may also be dominated by surface adsorption.

Vanadium uptake by internal organs of bivalves, crustaceans, and echinoderms is a relatively slow process [30–32]. Compared with the bioaccumulation of such contaminants as Cd, Se, Zn, Mn, and Fe, soft tissues of *M. galloprovincialis* exhibit an order of magnitude lower bioconcentration factor for vanadium [30]. Although digestive and excretory tissues generally contain a slightly higher vanadium concentration than muscle tissues in these species, these tissues contain a much lower vanadium concentration than external exoskeletons. If external coverings are excluded from consideration, specific vanadium target tissues are as difficult to discern in these aquatic organisms as in mammals. Thus, with the exception of the ascidians and the sabellid polychaetes, these aquatic species will probably not prove useful in helping to unravel the function of vanadium in biological systems.

3. ABSORPTION

In its strictest sense, absorption is the transfer of material from the outside environment into the blood system of an organism (Fig. 1) and may involve such processes as passive diffusion, active transport, facilitated transport, endocytosis, and filtration [33]. Absorption generally necessitates the crossing of at least two cellular membranes (one on the

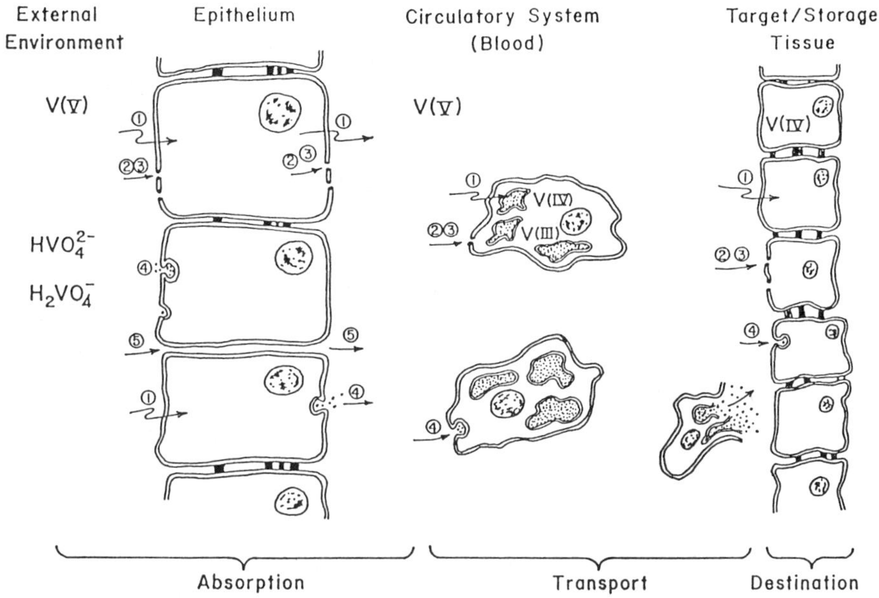

FIG. 1. Processes of vanadium absorption and transport in animal systems. Environmental vanadium, whether in water or gastrointestinal contents, is primarily V(V). Absorption may occur by one or more of five mechanisms: (1) passive diffusion; (2) active transport; (3) facilitated transport; (4) endocytosis; and (5) filtration. Mechanisms 1–4 necessitate the crossing of at least two cellular membranes. Transport occurs via either the plasma or following absorption by the blood cells (mechanisms 1–4). Vanadium is passed off to target/storage tissues, necessitating an additional absorptive process (mechanisms 1–4).

environmental side of an epithelial cell and the second on the blood side) with transcellular transport from one side of the cell's cytoplasm to the other. There is no reason, a priori, why the same mechanism used to cross the outer membrane of a cell is not used to cross the inner membrane. Absorption via the intracellular route (such as the specialized case of filtration) is rare since intercellular junctions generally restrict flow. Although endocytosis may play a role in the uptake of particulate-bound and soluble vanadium, it has not been specifically implicated to date. Given the charged nature of dissolved vanadium in aquatic systems (e.g., $H_2VO_4^-$, HVO_4^{2-}), one would not expect that the major mechanism of absorption would be diffusion. An anion transport system (either active or facilitated transport) would more likely be involved, such as used for the absorption of negatively charged organic acids [10]. Vanadate may be taken up via the same anion exchange mechanism used by phosphate [34]. Additional evidence for anionic uptake comes primarily from studies on the uptake of vanadate by mammalian erythrocytes and ascidian blood cells, respectively (as described later).

Absorption of vanadium can occur across many types of epithelia depending on whether exposure is from air, water, soil, or food, via the respiratory, gastrointestinal, or dermal routes [10]. Vanadium is almost exclusively present in its most oxidized form, V(V), in air, water, and soil. In several tissues, such as mammalian lung and heart, it is present as V(V) [35]. It may also be stored, either totally or in part, in a reduced form in some tissues, e.g., as V(IV) in mammalian kidney, liver, and erythrocytes [35,36], and as V(IV) or V(III) in ascidian blood cells [37–39]. This reduced vanadium would probably be oxidized to V(V) once the tissues are ingested as food. Thus, it appears that V(V) is the primary form of vanadium that is absorbed, regardless of exposure route. This is one area where additional research is needed.

In mammals, the primary route of vanadium exposure may be by the respiratory tract [10], although vanadium is known to be absorbed by the gastrointestinal tract as well. Most ingested vanadium is not absorbed, however, but is eliminated by the feces [10,19,40]. Subsequent elimination of absorbed vanadium from the body occurs primarily via the kidneys [41,42].

The main route of vanadium absorption by the common eel *Anguilla anguilla* was via the intestine, whereas uptake across gill epithelia was much slower. Orthovanadate [V(V)] absorption was linearly dependent on exposure concentration, was not saturable over the range tested (10^{-8}–10^{-4} M), was not competitively inhibited by Cl^-, or inhibited by the anion carrier blocker 4,4′-diisothiocyanostilbene-2,2′-disulfonic acid disodium salt (DIDS) or the active transport blocker ouabain [43]. Although diffusion was not ruled out, the charged nature of the vanadium ion would seem to preclude this pathway. Bell et al. [43] concluded that the eel gut was a leaky epithelium and that much of the absorption occurred through the loose junctions between intestinal cells. Vanadium uptake was therefore not controlled, except by exposure concentration.

While vanadium accumulation has been measured in various aquatic invertebrates (Table 1), the actual mechanism of uptake has not been worked out. Direct absorption of soluble vanadium seems to be the primary route in the echinoderms rather than gut absorption [32]. The best available information is on the ascidians.

Following exposure to ^{48}V, radiolabel is detected in ascidian branchial sac, branchial sac mucus, and the gastrointestinal tract [44–47]. Dissolved vanadium can be absorbed directly from seawater [46,48], although the uptake of particulate vanadium cannot also be ruled out. The branchial sac appears to be the primary site of direct absorption of dissolved vanadium [44,48]. The overlying mucus may be involved by complexing dissolved vanadium, which is then made available to the gut epithelium following ingestion of the mucus food string. In a similar manner, particulate vanadium (inorganic complexes, food particles, etc.) is captured, ingested, and introduced to the gut epithelium [45,49]. Similarly, the primary route for iron uptake in ascidians may be via the ingestion and assimilation of particulate iron from seawater [50].

A number of studies have examined absorption of vanadium by ascidians by monitoring the depletion of vanadium from the surrounding water. By itself this type of experiment provides no information on the mechanism of absorption. Moreover, if the rate at which water is pumped through the siphons and branchial sac is less than the absorption rate, vanadium depletion would simply measure the hydrodynamic transport rate and not the chemical absorption rate [51]. As shown for *Ciona intestinalis* [46,48], water transport is the faster of the two sequential processes, i.e., water transport and vanadium absorption. Tunicate seawater pumping, even when intermittent, appears to be sufficiently rapid to ensure that absorption of vanadium is rate limiting. Partitioning studies with *C. intestinalis* established that with this organism absorption through the outer tunic is negligible [52], a conclusion that has been confirmed in the case of iron uptake by the stolidobranch ascidian *Styela clava* [50] and has been extended to tunicates in general.

The following three examples will highlight our present knowledge of vanadium absorption in ascidians.

3.1. *Ciona intestinalis*

Because it identified and treated several distinct aspects of vanadium transport and accumulation, the pioneering study by Goldberg et al. [46] guided radioactive tracer studies that followed. Initially these investigators did not know what to expect from their whole-animal studies. For example, would measurements of radiovanadium activity taken from *Ciona*-containing seawater aquaria over a period of time as long as a week yield erratic or meaningful data? To their delight (we assume) they found meaningful kinetics curves, and reported that uptake could be divided into two first-order processes, with the faster process yielding to the slower process after 2–3 days. This and chemical analysis data established that *Ciona* accumulates vanadium. It was also demonstrated that the tunicate's seawater pumping rate assured an adequate supply of vanadium and that prior filtration of the seawater to remove particulates did not eliminate uptake. This study also presented data that indicated roles for the stomach, gut, and branchial sac in the absorption of vanadium, but did not distinguish between uptake from the soluble or particulate phases.

The uptake and distribution of vanadium in *C. intestinalis* were reexamined in a second study [45]. A repeat experiment to measure the kinetics curve of radiovanadium depletion from a *Ciona*-containing aquarium as a function of time was not carried out. Instead, these investigators measured the fraction of vanadium removed by single animals in a 6-hr period. They reported a saturation effect: with increasing concentration of vanadate in seawater, the fraction absorbed declines and appears to level off (but see below). Isostructural chromate ion was not absorbed by *C. intestinalis*. The most important absorption site was the branchial sac, followed by the gastrointestinal tract. Most of the accumulated vanadium resides in whole blood. Uptake of vanadium by blood cells was a gradual process requiring several days before appreciable quantities of radiovanadium were detected. From their uptake and distribution studies they concluded that uptake follows the sequence seawater/body interface → plasma → blood cells.

The branchial sac was the focus of a radiovanadium uptake study with *C. intestinalis* [49]. Uptake from seawater to the branchial sac was temperature-dependent, ouabain-inhibited, and inhibited by "mersalyl" an organomercury SH-reagent. Unlike the whole-animal study, in which no chromate uptake was found, the branchial sac absorbed chromate, niobate, and molybdate anions, and iron (form not specified), but in much lower amounts than vanadate. Uptake of phosphate and arsenate inhibited vanadate uptake, and phosphate uptake was ouabain-inhibited.

A later study addressing several uptake issues reported in the earlier studies extended or modified some of the previous conclusions [48]. *C. intestinalis* absorbed dissolved vanadium from seawater, since particulates were excluded by filtration through a 1-μm pore-sized filter. Kinetics curves of ^{48}V uptake were studied as a function of increasing vanadate concentration. The curves fit a one-step exponential decrease of seawater ^{48}V concentration: *two-step kinetics was not observed*; *saturation kinetics was not observed* (Fig. 2). While the former is an experimental observation, the latter statement is a difference in interpretation. Both earlier studies had treated uptake as a clearance process. However, uptake is an example of an *exchange* process. The ^{48}V tracer exchanges between two environments: seawater and tunicate. Decline in the seawater compartment's radioactivity is always a first-order process regardless of the details of the exchange mechanism [53]. Further experiments must be performed to determine the rate law from which the exchange mechanism may be deduced. Such experiments established that the rate of radiovanadium exchange was first order in vanadate concentration over a 1000-fold increase in vanadate concentration above the natural concentration of the element in seawater.

Taken together, these studies imply that for *Ciona intestinalis*, which stores vanadium primarily in the +4 oxidation state [54], extraction of V(V) from seawater is a monophasic regulated process. None of the *C. intestinalis* uptake studies extended beyond 7 days; therefore, it is possible that the monophasic approach to exchange equilibrium is only part of the story and that the total approach could be biphasic. Appreciable steady-state vanadium concentrations build up in the tissues of the branchial sac and gastrointestinal tract. Anionic uptake occurs at the primary absorption site, where there is evidence of selection in favor of vanadate. Coordination of vanadium by a plasma protein is not indicated in these studies. Whether accumulation of vanadium by blood cells occurs by transport from plasma or by production of blood cells was not clarified.

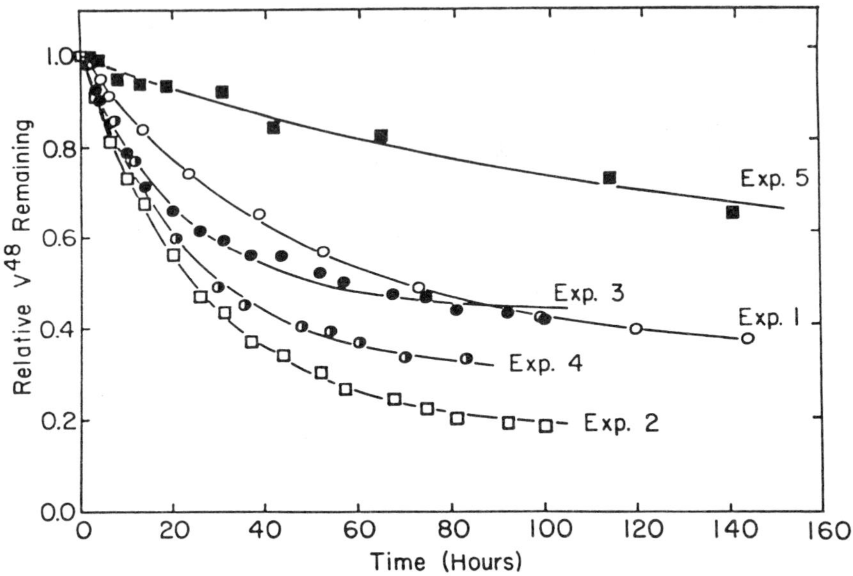

FIG. 2. Relative radioactivity remaining in solution vs. time for *Ciona intestinalis*. The smooth curves fit the expression $y = \beta_1 e^{-\beta_2 t} + \beta_3$ where y is the relative activity of the radiovanadium remaining in solution at time t. Experiment 1: 100-mL volume, four tunicates total wet mass 3.6 g, initial radioactive vanadium 0.835 μCi; initial nonradioactive vanadium 3.4 ± 0.5 ppb. Experiment 2: 100-mL volume, four tunicates total wet mass 6.1 g, initial radioactive vanadium 0.042 μCi; initial nonradioactive vanadium 4.1 ± 0.5 ppb. Experiment 3: 500-mL volume, four tunicates total wet mass 28.6 g, initial radioactive vanadium 0.170 μCi; initial nonradioactive vanadium 4.1 ± 0.5 ppb. Experiment 4: 500-mL volume, two tunicates total mass 31.0 g, initial radioactive vanadium 1.142 μCi; initial nonradioactive vanadium 3.4 ± 0.4 ppb. Experiment 5: 1500-mL volume; 10 tunicates total wet mass 17.0 g, initial radioactive vanadium 2.12 μCi; initial nonradioactive vanadium 2.8 ± 0.9 ppb. (Reproduced from the *Journal of General Physiology* [48] by copyright permission of the Rockefeller University Press.)

3.2. *Phallusia mamillata*

By the beginning of the 1960s *P. mamillata* (which stores +3 vanadium) was probably the best-studied ascidian in terms of characterization of blood. Transport studies with this animal are surprisingly few, however, perhaps due to difficulties in maintaining animals, as suggested in the only such study reported [45]. Examining ^{48}V uptake by whole animals, these investigators reported that younger, smaller animals absorbed more vanadium than older, heavier animals. After 48 hr most ^{48}V remained in the branchial sac and mantle. Most of the radiolabel in the blood after that time period was in the plasma. Uptake was measured by plotting residual seawater radioactivity against specimen weight. For reasons given above, this type of analysis should not be interpreted in the manner presented in the original publication. We conclude only that over a period of time no longer than 4 days, exchange kinetics was exhibited by the animals exposed to radiovanadium in their seawater aquaria.

3.3. *Ascidia ceratodes*

Forty years elapsed between the pioneering study of Goldberg et al. [46] and the next reported uptake study of *A. ceratodes* [47]. Not surprisingly, the more recent study is a much more sophisticated analysis of uptake dynamics and includes plasma/blood cell exchange studies. The earlier study [46] reported uptake curves that were first-order, showing either a one-step or a two-step exponential decrease in seawater radioactivity. Vanadium uptake occurred from unfiltered as well as filtered seawater, and with or without added phosphate, which retarded the rate of uptake. Autoradiography disclosed considerable radiolabel in the mantle, branchial sac, gut, ovarian tissues, and mature ova.

In the study by Anderson et al. [47] the amount of tracer in three separate environments—seawater, plasma, and blood—was studied as a function of time over a relatively long period of time (56 days). They report that the "seawater activity initially decreases rapidly, but the decrease slows and some activity is always found in the seawater." A smooth curve is drawn through the data, but no indication of curve-fitting parameters is given. It is worth considering the implications of their results.

Let there be three different and distinguishable compartments or environments containing the radioactive tracer, and let these be designated "sw" for seawater, "c1" for one vanadium-exchanging compartment within a tunicate, and "c2" for a second, different vanadium-exchanging compartment within a tunicate. Then any consecutive irreversible sequence of transport processes between compartments, such as $^{48}V_{sw} \rightarrow {}^{48}V_{c1} \rightarrow {}^{48}V_{c2}$, must show a single exponential decrease in the seawater activity regardless of which step is faster or whether the two steps have comparable half-lives. If the tracer disappears from the seawater by parallel reactions such as $^{48}V_{sw} \rightarrow {}^{48}V_{c1}$ and $^{48}V_{sw} \rightarrow {}^{48}V_{c2}$ a monophasic decline would also occur controlled by the sum of the two rate constants. Only if the back reactions contribute to the measured activity, i.e., if reversible exchange is being observed, can the approach to equilibrium show biphasic dynamics. Reversible exchange explains why a residual activity is always observed in seawater. We believe, however, that these data are not conclusive regarding the question of whether mono- or biphasic vanadium transport prevails in whole-animal tracer incorporation experiments.

In addition to incorporation experiments, a pulse experiment was performed. Five animals in a single aquarium were exposed to ^{48}V-dosed seawater for one day. Two animals remained in this environment, but three animals were removed and placed in an aquarium containing 3 L of tracer-free seawater. The activities of seawater in both aquaria were monitored; 1 and 9 days after the separation occurred, the ^{48}V activities of blood cells and plasma of one animal from each environment were measured. The key finding was that at day 9 postincubation the activity in blood cells of the animal placed in the unlabeled aquarium was approximately 8.5 times greater than the activity in blood cells at day 1 postincubation. Since the activity in plasma at day 9 was reduced almost to background, the expected increase in blood cell activity, if the plasma were a closed compartment and the only source of ^{48}V, would have been 4.8 times the day 1 postincubation activity. The additional increase must have originated from other tissues which continually released their vanadium stores into the plasma in an attempt to attain a new steady state. Thus, a relatively low and constant vanadium concentration was maintained in the plasma.

4. PLASMA TRANSPORT

Absorption culminates in heightened vanadium concentrations in blood plasma, which is then transported throughout the body and which may or may not also lead to vanadium accumulation in blood cells. Very low concentrations of vanadium are found in human whole blood (0.032–0.095 ng/mL [55]). In mammals, approximately 90–95% of the circulating vanadium is associated with the blood plasma rather than the blood cells [20,40,41]. Most of this plasma vanadium is bound to transferrin [20] as the V(IV) VO^{2+} ion. The remainder is present in the plasma as V(V). We could speculate that V(V) enters the plasma and is enzymatically reduced once it is bound to transferrin, in a manner similar to the reduction of Fe^{3+} to Fe^{2+} by ceruloplasmin [56]. Transferrin-bound vanadium could then be passed off to other tissues of the body via the transferrin receptor system [40,57]. A similar mechanism for plasma-bound vanadium may prove to be present in other chordates, since transferrin is ubiquitous within this group. Since invertebrates lack true transferrins [58], a homologous system is absent.

Almost nothing is known about vanadium transport in the plasma (or the blood cells) of invertebrates, with the exception of the ascidians. Because of the particularly high concentrations of vanadium found in the ascidians, vanadium transport has received considerable attention in this group. Vanadium may have either a transient or permanent appearance in blood plasma. In either case, vanadium concentrations in blood plasma are relatively low, although higher than what is found in the surrounding seawater (Table 2). Since the primary site of vanadium concentration in ascidians is the blood cells, plasma concentrations are far lower than that in the blood cells (Fig. 3). Phlebobranchs and aplousobranchs, species which concentrate vanadium, tend to have higher concentrations of vanadium in their blood plasma than do the Fe-concentrating stolidobranchs. Blood cell/plasma ratios range from 31 to 827 in those phlebobranch species (Fig. 3) where both blood cell and plasma data are available (this range is actually much higher since plasma concentrations were below detection limits for several of the species measured by Michibata et al. [29]). In contrast, blood cell/plasma vanadium ratios range from 0.3 to 110 for eight stolidobranch species (Table 2).

Most investigators have found that ascidian plasma is near neutral in pH [59–61]. Instances of low plasma pH have been attributed to blood cell breakage [62], although in at least one instance a low pH was found for the plasma of *Ascidiella aspersa* in which cell lysis or leakage was not evident [63]. A neutral pH would allow vanadium to be present

TABLE 2

Vanadium in Ascidian Blood Plasma[a]

Species	$[V]_{plasma}$	Blood cell/ plasma	Ref.
Phlebobranchia			
Ciona intestinalis	0.4	827	29
Ciona savignyi	12.9	122	29
Ascidia ahodori	522.2	40	29
Ascidia sydneiensis samea	27.6	169	29
Ascidia nigra	11	2433	86
Ascidia dispar	22	31	31
Ascidia ceratodes	335	94	87
Ascidia ceratodes	~340	~75	88
Aplousobranchia			
Podoclavella moluccensis	4.2	276	68
Stolidobranchia			
Polycarpa cryptocarpa	1.0	13.4	29
Styela plicata	1.3	2.7	29
Pyura sacciformis	2.0	0.3	29
Halocynthia roretzi	0.6	6	29
Halocynthia aurantium	0.02	110	29
Halocynthia papillosa	0.9	8.5	29
Molgula manhattensis	0.2	13	29

[a]Plasma concentrations $[V]_{plasma}$ presented as μg/g dry wt of plasma.

either as aquo anions or as coordination compounds with plasma proteins and other organic constituents.

Available evidence, while not extensive, indicates that vanadium is present in ascidian blood plasma as the oxidized V(V) form. In *Ascidia ceratodes*, 0.01–0.06 mM vanadium resides in the plasma; ^{51}V nuclear magnetic resonance (NMR) has shown that this vanadium is present as V(V), although it is complexed differently from the V(V) present in the surrounding seawater [54]. This vanadium could be bound to plasma constituents. Several investigators, using size exclusion chromatogra-

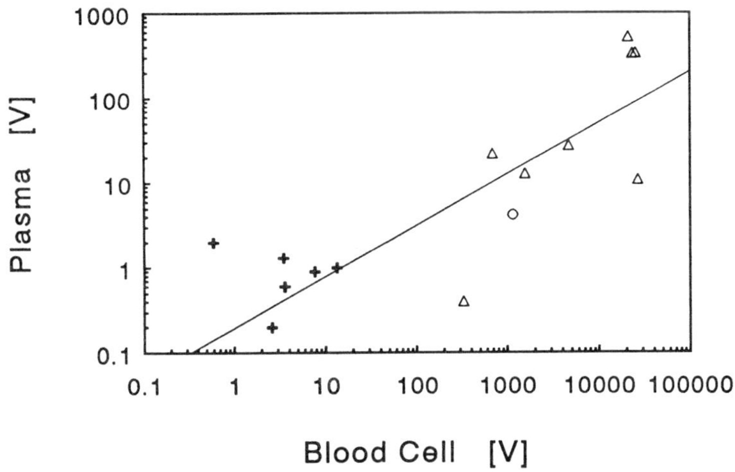

FIG. 3. Relationship between plasma and total blood cell vanadium concentrations within the ascidians. Best-fit regression of the data yields the function $y = 0.195\,e^{(0.605)x}$, $r = 0.832$, where y is plasma [V] and x is blood cell [V]. Data points presented for phlebobranchs (△), aplousobranchs (○), and stolidobranchs (+).

phy, have demonstrated that plasma vanadium associates with specific plasma constituents. Hawkins et al. [64] reported that some vanadium was associated with high molecular weight constituents of *Ascidia ceratodes* plasma but that the majority was associated with the low molecular weight fraction. Based on gel electrophoresis, they concluded that vanadium was primarily present in a free, noncomplexed form. Roman et al. [61] also demonstrated that most vanadium in *Ascidia dispar* plasma elutes with the column bed volume, indicating that the metal either is associated with low molecular weight constituents or is free.

Studies on plasma-associated vanadium contrast sharply with those on plasma iron in the Fe-concentrating ascidians. While these results seem to indicate that vanadium may be present in the free form or associated (possibly weakly) with low molecular weight components of ascidian plasma, studies on the binding of iron to plasma proteins of Fe-concentrating ascidians point to some specific plasma proteins that bind iron. A "transferrin-like" protein with a molecular weight of 40 kDa (half that of mammalian transferrin) was implicated in *Pyura stolonifera* [65,66]. Other investigators have characterized Fe-binding proteins in ascidian plasma but downplayed the possibility that a transferrin-like compound is present [60]. Rayner-Canham [67], while working with niobium, suggested that a plasma α-hydroxycarboxylic compound may be involved in transition metal transport. Roman et al. [61] presented evidence of a high molecular weight Fe-complexing protein in *Pyura chilensis*. In addition, Hawkins et al. [68] showed that iron binds to a high molecular weight fraction as well as a low molecular weight fraction which may contain an *n*-acetylamino sugar in *Pyura stolonifera*. Using gel electrophoresis, Agudelo et al. [69] identified a single high molecular weight protein in the plasma of *Boltenia ovifera*, which strongly bound iron.

In the ascidians, vanadium levels in the plasma fluctuate depending on seawater vanadium concentrations and the rates at which various compartments (interim storage tissues, long-term storage tissues, blood cells, etc.) take up the metal. Blood plasma cannot be considered either a static or a closed compartment. In *C. intestinalis*, for example, uptake from plasma by blood cells causes blood cell vanadium concentrations to rise while the concentration of vanadium in plasma falls as a function of time [49]. In the same study it was determined that plasma retains vanadium better than seawater, but does not bind vanadium as well as transferrin binds iron.

In a pulse exposure to ^{48}V, *Ascidia ceratodes* rapidly removed radiolabel from the surrounding water [47]. The activity in the plasma reached a maximum in 1–2 days and then decreased, whereas the blood cells reached a 10-fold higher maximum in approximately 20 days. The scatter in the blood cell data was very high, indicating that the maximum could have occurred anywhere between day 10 and day 20. Not all of the ^{48}V activity lost from seawater was accounted for in the blood plasma or blood cells [44,47]. Initially, vanadium is found in a variety of tissues. After the first 3 days, 67% of the activity removed from seawater was detected in the branchial sac, gastrointestinal tract, and blood. This temporary storage in branchial and digestive tissues was followed by the long-term accumulation of vanadium by the blood cells. The plasma played a role both in the short-term storage and the long-term transfer of vanadium to the blood cells.

A very thorough analysis of blood plasma has been carried out for two ascidians from Australian waters: vanadium-storing *Nephtheis fascicularis* and non-vanadium-storing *Polycarpa pigmentata* [70]. Four low molecular weight organic components and proteins, which might be capable of metal transport, were separated from the blood plasma of these ascidians. Among the organic components isolated was a substituted 18-crown-6 ether, which was found in both species. Although the function of such a compound has normally been associated with transport of (cationic) alkali metal ions via complexation in the annular space within the molecule, the substituents attached to the crown ether skeleton may facilitate conventional metal ion or metal-containing oxoanion chelation.

The transport of vanadium in ascidians should be reinterpreted in light of recent emphasis on the importance of invertebrate blood plasma in the transport of physiologically important elements and contaminant metals [4]. A specific vanadium-binding transport agent (protein or chelator) has not been identified in the ascidians. The question then arises as to why a specific transport protein is needed.

One reason cited for a specific transport agent is that certain transition metal ions catalyze the formation of highly reactive free radicals in oxic environments. The best known example is the redox cycling of the iron(II)/(III) couple which releases OH radical [71,72]. The vanadium(IV)/(V) couple functions in a similar manner in the presence of reducing sugars, forming superoxide radicals [73]. Nevertheless, a highly specific transport agent, capable of extremely tight binding to vanadate anion, may not be necessary since the production of superoxide anion in and of itself has minor toxicological significance. Lack of an obligatory transport protein for vanadium, such as transferrin for iron, may also reflect profound differences in the effects caused by cations (especially transition metal cations) and anions in living systems. In addition to being redox-active, transition metal cations exert a strong polarizing effect due to their high positive charge–to–ionic radius ratio. This effect allows transition metal cations to replace similarly charged biologically active ions such as Mg^{2+} and Ca^{2+} and to interact as well with normally metal-free proteins, distorting them and altering their tertiary structure. For these reasons free (uncomplexed) transition metal cations and other, nonessential metal cations are highly toxic to biological systems. Free anions, on the other hand, are far less toxic unless they are phosphate mimics and are in close proximity to phosphatase enzymes. Since much higher concentrations of free anions can be tolerated, there is less need to complex these anions with specific protective compounds.

As a result, we would expect that vanadium would be present in ascidian plasma either as a free ion (e.g., $H_2VO_4^-$), as a complex of a low molecular weight organic ligand (e.g., a *cis*-diol), or else weakly complexed to plasma proteins. The latter case would be analogous to calcium, cadmium, and nickel binding to human serum albumin [74], and to cadmium binding to plasma proteins in the marine clams *Mercenaria mercenaria* and *Mytilus edulis* [2,3]. In both cases, metal binding is weak (stability constants $\leq 10^4$) and nonspecific with respect to binding sites. This weak binding allows the metal to be readily passed off to target ligands which have higher affinities and specific binding sites.

5. CELLULAR TRANSPORT

The role of blood cell in vanadium transport appears to be minimal in mammalians. Over 90–95% of mammalian blood vanadium is found in the plasma [20,41]. Nevertheless, human red blood cells are capable of accumulating vanadium [as V(V)] by two separate mechanisms: one which is sensitive to anion carrier inhibitors and the second which is as yet unidentified [75]. Once inside, vanadium is stored as the V(IV) ion, VO^{2+} [10]. Macara et al. [76] identified a fast and a slow accumulation phase. The slow phase requires energy and correlates with the rate of internal V(IV) production within the erythrocytes. These investigators concluded that the rapidly accumulated V(V) was reduced by cytoplasmic glutathione, thereby protecting sensitive Na^+,K^+-ATPases. Alternatively, cellular phosphate stores may provide the protective function [10]. Nevertheless, the slow reduction of V(V) to V(IV) is probably the rate-limiting step in the overall uptake process. Although it is clear that vanadium is taken up by mammalian erythrocytes, it is not at all clear as to whether this vanadium is actually transported to other tissue compartments or whether it is simply stored until such time as the cells die and lyse.

The ascidians provide the only known data on blood cell transport in the invertebrates. Ascidians can have over eight different types of blood cells in the same species, several of which are known to accumulate vanadium [77]. These vanadium-containing blood cells are unusual in that they may be thought of as both the target site for the initial sequestration of vanadium in the animal (and therefore the sites that have the highest vanadium concentrations in the animals) and as a transport vehicle that conveys the vanadium to its final site of action. While the ultimate function of vanadium in ascidians remains a matter of controversy [78], the role of the blood cells in vanadium *transport* is generally accepted. Three examples will be used to summarize our current knowledge of ascidian blood cell transport.

5.1. *Ascidia nigra*

Transport of vanadate from external solution to unfractionated blood cells of the unusual velvety black ascidian *A. nigra* was investigated by tracer and electron paramagnetic resonance (EPR) methods [79]. In vitro exchange between external medium and blood cells was rapid (half-life 57 sec at 0°C), mainly monophasic, and exhibited Michaelis–Menten kinetics. The accumulated ^{48}V was not washed out by treatment with ethylenediamine-*N,N,N′,N′*-tetraacetic acid (EDTA) solution. Inhibitors of glycolysis and uncouplers of oxidative phosphorylation had no significant effect on the exchange rate. The specificity of the accumulation process was probed by studying exchange in the presence of the anions chromate, phosphate, and sulfate. Chromate and sulfate did not significantly affect the exchange dynamics and were not accumulated. Phosphate competitively inhibited vanadate uptake but was not accumulated. Potent competitive inhibitors of anion exchange across the plasma membrane of the human erythrocyte, such as 4,4′-dinitrostilbene-2-2′-disulfonic acid (DNDS), also blocked transmembrane vanadium exchange in this ascidian. Uptake studies in the presence of a fivefold excess of EDTA, which strongly complexes vanadate ions, were the same, within experimental error, as those in the absence of EDTA, indicating that nonspecific adsorption to either the microfuge tubes or the outside of the blood cells themselves was an insignificant component of radiovanadium exchange. Upon being exposed to external vanadate, the intracellular vanadium(IV) concentration, as determined by EPR spectroscopy, reaches a maximum within 1–2 min. The natural intracellular V(IV) abundance before and after exposure to exogenous vanadate measured in this experiment was about 4%, in good agreement with X-ray absorption spectroscopy (XAS) [37] and magnetic susceptibility [38] measurements.

These experiments show that for *A. nigra*, which stores vanadium primarily in the +3 oxidation state, the transport of vanadate into vanadium-accumulating blood cells is a highly specific process which can be saturated. Accumulation appears to be driven by a mechanism of reduction followed by trapping the reduced vanadium ion. The rate-determining step of the overall process is entry of vanadate into the blood cell. Since neither phosphate nor the stilbene derivatives completely blocked vanadate influx into the blood cells, more than a single process of transmembrane vanadate uptake may exist.

Although ascidian plasma is isosmotic with seawater, it is anomalously low in sulfates and carbonates, and is high in chloride [61,80]. This solution composition would support the hypothesis that an oxyanion porter system is responsible for the uptake of vanadate by the blood cells from the plasma. Chromium, as $HCrO_4^-$ or CrO_4^{2-}, would also be expected to be accumulated by this system. A number of electron probe studies have measured elevated sulfur levels in several ascidian blood cells, again lending credence to the oxyanion porter system. However, chromium has not been measured in any appreciable concentrations in ascidian blood cells. In addition, Dingley et al. [79] were unable to demonstrate in vivo uptake of sulfate by *Ascidia nigra* blood cells.

5.2. *Ascidia ceratodes*

As mentioned above [47], loss of ^{48}V label from seawater is not matched by gain of the label in plasma and blood following the first 6 days of exposure. To understand better this observation, additional experiments were performed to determine the distribution of tracer between (1) ^{48}V-labeled plasma and unlabeled blood cells and (2) unlabeled plasma and ^{48}V-labeled blood cells. As in the similar *A. nigra* experiments, in vitro exchange from plasma to cell is completed very quickly, in this case 3 min. On average 18% of the extracellular activity is absorbed by the blood cells. After a 1-day in vivo exposure to ^{48}V in the surrounding seawater, 33% of the blood cell activity was released to unlabeled plasma in vitro. After a 6-day in vivo exposure, only 2% was subsequently released. Adding external vanadate does not affect the percentage of exchange of ^{48}V activity from plasma and seawater to unlabeled cells.

The equilibration of vanadium between plasma and blood cells apparently does not involve all of the numerous vanadium-accumulating blood cells in *A. ceratodes*. Instead, Anderson et al. [47] speculate that exchange involves a cell type found at "low and/or variable levels in blood." A substantial difference occurs between seawater/whole-animal ^{48}V and solution/blood cell ^{48}V exchanges with regard to distribution. In vivo (labeled seawater/whole animal) 15% of the incorporated vanadium is associated with circulating blood cells, whereas in vitro (labeled plasma/unlabeled cells) approximately 75% associates with blood cells. In separate experiments [54], ^{51}V NMR data of V(V) in plasma show that the equilibration is between blood cells and a nontransferrin-like environment. Blood cells labeled with ^{48}V in vivo and then transferred to unlabeled plasma release less activity with increasing in vivo incubation time. Therefore, the percentage of V(V) in the blood cells available for equilibration decreases with increasing time.

Based on their activity imbalance and pulse experiment data, these investigators [47] argue that two paths are available for vanadium uptake by *A. ceratodes* blood cells. One path is direct absorption of V(V) from plasma to blood cells. The first stage in this path is reversible but is followed by a slower, irreversible incorporation step, either in the same or different blood cell types. A second path involves accumulation of vanadium in a tissue other than blood, followed either by slow loss into plasma and thence into blood cells through the direct path, or by production and proliferation of fresh vanadium-containing blood cells.

5.3. *Ascidia gemmata*

Measurements have been made of ^{48}V exchange between artificial seawater and blood cells of *A. gemmata* [81], a phlebobranch ascidian expected to store +3 vanadium. At least half the dose was incorporated into the blood cells over a 2-hr period, with uptake continuing over the next 22 hr. After a 2-hr exposure to label, cells were sorted by Percoll gradient density separation. Of the vacuolated blood cells, compartment cells accumulated more vanadium than morula or signet ring cells (375.3, 114.5, and 115.6 Bq/10^6 cells, respectively).

These investigators attempted to expose *A. gemmata* blood cells to both V(V) and V(IV) solutions, and interpreted their findings as showing that both V(V) and V(IV) could be absorbed by several of the blood cell types present. However, air oxidation of V(IV) to V(V) in the exchange medium would have been appreciable over the total time period of the experiment (including both solution preparation and exposure), resulting in a mixture of both +4 and +5 oxidation states [82]. This places their claim that V(IV) is accumulated in doubt. We conclude that exchange between external solution and blood cells seems to be slower in *A. gemmata* than in *A. nigra* or *A. ceratodes*. These differences cannot be attributed to nonspecific adsorption since EDTA had no effect on the apparent uptake rates in either the *A. gemmata* [82] or *A. nigra* [79] experiments. While all three studies document the blood cell accumulation of V(V), the possible uptake of V(IV) needs further investigation.

6. GENERAL CONCLUSIONS

Most of what is known about the transport of vanadium relates to either humans (including the human laboratory models: rats, mice, etc.) or to the unusual group of marine epibenthic urochordates, the ascidians. Very few data are available on birds, reptiles, amphibians, fish, and invertebrates. In both terrestrial and aquatic animals, the primary source of vanadium is in the +5 oxidation state, notwithstanding whether exposure is via air particulates, water, or food. In air-breathing vertebrates, absorption occurs most efficiently via the lungs, with gut absorption occurring at a considerably slower rate. Absorption in aquatic organisms is more difficult to generalize; gut absorption from solid or semisolid food sources predominates in such animals as the common eel; absorption of dissolved vanadium has been documented in laboratory studies using goldfish, catfish, and the ascidians.

Absorbed vanadium is quickly partitioned between the plasma

and blood cells. In mammals, the vast majority of the V(V) is retained in the blood plasma and is generally complexed to transferrin and reduced to V(IV). Erythrocytes absorb vanadium from the plasma, reducing it to V(IV). Target tissues probably receive vanadium primarily from the plasma, and generally retain it as V(IV) or reduce it to this oxidation state.

Tunicates (and possibly the sabellid worms) are quite different from other animals with regard to their ability to concentrate vanadium in their tissues. For both groups of animals the bioconcentration factor is 100–1000 times greater than in bivalves, crustaceans, and echinoderms. Tunicates can obtain vanadium directly from the aqueous phase of seawater. However, absorption of vanadium in nature may be more complicated, if it can be confirmed that soluble vanadium is chelated by branchial sac mucus and is subsequently ingested. Once absorbed, vanadium can be found in both the plasma and the blood cells. In the plasma, it is present either as weak protein complexes or as V(V) oxoanion.

Only in the tunicates and the sabellid *P. occelata* is the oxidation state of stored vanadium driven down to +3—most unusual for a living system. In tunicates it appears that vanadium, stored in blood cells and tissues in either of the two reduced oxidation states +3 or +4, is taken up as +5 vanadium by ascidian blood cells from the plasma, probably by an anion porter system which may transport phosphate, sulfate, and vanadate. Evidence for this is found in (1) in vitro experiments using ^{48}V(V), (2) inhibition of vanadium uptake by anion channel blockers, and (3) anomalously low sulfate and carbonate in ascidian plasma. While laboratory studies have determined that vanadium can be directly absorbed by one or more tunicate blood cell types from the plasma, they do not rule out the possibility that a second pathway of accumulation may also occur. This second pathway, the accumulation of vanadium during blood cell ontogeny, either prior to or subsequent to release into the circulation, must be further investigated.

While much has been learned concerning the uptake and transport of vanadium, much additional study is needed to determine the function of this element in animal systems. It is clear that mammals need only trace quantities of vanadium, if it is needed at all. The ultratrace levels involved make experimentation difficult. The extremely high vanadium concentrations in several of the ascidian blood cells may allow them to be used as a more easily experimented on model system to address this functionality issue. However, the ultimate fate of the vanadium in ascidians, just as in mammals, is unknown. This fate will be intricately tied to the mechanism of vanadium release from blood cells, a process that is totally unknown at the present time. The function of vanadium in ascidians and in animal systems in general will remain a fruitful area of inquiry for years to come.

ACKNOWLEDGMENT

We are indebted to MIT/SeaGrant GC-R-32189-7 for support of our current research on vanadium in ascidians.

ABBREVIATIONS AND DEFINITIONS

DIDS	4,4′-diisothiocyanostilbene-2,2′-disulfonic acid disodium salt
DNDS	4,4′-dinitrostilbene-2,2′-disulfonic acid
EDTA	ethylenediamine-*N,N,N′,N′*-tetraacetic acid
EPR	electron paramagnetic resonance
NMR	nuclear magnetic resonance
XAS	X-ray absorption spectroscopy

REFERENCES

1. F. W. Putnam, *The Plasma Proteins*, 2nd ed., Vol 1, Academic Press, Orlando, 1975.
2. W. E. Robinson and D. K. Ryan, *Mar. Biol., 97*, 101 (1988).
3. W. E. Robinson, D. K. Ryan, P. A. Sullivan, and C. C. Boggs, *Am. Zool., 28*, 68A (1988).
4. G. Roesijadi and W. E. Robinson, in *Aquatic Toxicology: Molecular, Biochemical, and Cellular Perspectives* (D. C. Malins and G. K. Ostrander, eds.), Lewis, Boca Raton, 1994, pp. 387–420.
5. F. H. Nielsen, in Chap. 16 of this volume.

6. M. Henze, *Hoppe-Seyler's Z. Physiol. Chem., 72*, 494 (1911).
7. M. Henze, *Hoppe-Seyler's Z. Physiol. Chem., 79*, 215 (1912).
8. D. A. Webb, *J. Exp. Biol., 16*, 499 (1939).
9. L. T. Rezaeva, *Fed. Proc. Transl. Suppl., 24*, T836 (1964).
10. B. R. Nechay, L.B. Nanninga, P. S. E. Nechay, R. L. Post, J. J. Grantham, I. G. Macara, L. F. Kubena, T. D. Phillips, and F. H. Nielsen, *Fed. Proc., 45*, 123 (1986).
11. P. J. Stankiewicz, D. C. Crans, and A. S. Tracey, in Chap. 9 of this volume.
12. L. L. Hopkins, Jr., and H. E. Mohr, *Fed. Proc., 33*, 1773 (1974).
13. K. Schwarz and D. B. Milne, *Science, 174*, 426 (1971).
14. F. H. Nielsen, *Ann. Rev. Nutr., 4*, 21 (1984).
15. R. L. Robson, R. R. Eady, T. H. Richardson, R. W. Miller, M. Hawkins, and J. R. Postgate, *Nature, 322*, 388 (1986).
16. R. R. Eady, in Chap. 11 of this volume.
17. R. Wever and K. Kustin, in *Advances in Inorganic Chemistry* (A. G. Sykes, ed.), Academic Press, New York, 1990, pp. 81–115.
18. H. Vilter, in Chap. 10 of this volume.
19. A. R. Byrne and L. Kosta, *Sci. Tot. Environ., 10*, 17 (1978).
20. E. Sabbioni, E. Marafante, L. Amantini, L. Ubertalli, and C. Birattari, *Bioinorg. Chem., 8*, 503 (1978).
21. J. Edel and E. Sabbioni, *Sci. Tot. Environ., 133*, 139 (1993).
22. D. Ray, S. K. Banerjee, and M. Chatterjee, *J. Inorg. Biochem., 38*, 169 (1990).
23. M. Ünsal, *Mar. Pollut. Bull., 13*, 139 (1982).
24. T. Ishii, I. Nakai, C. Numako, K. Okoshi, and T. Otake, *Naturwissenschaften, 80*, 268 (1993).
25. T. Ishii, I. Nakai, and K. Okoshi, in Chap. 14 of this volume.
26. J. D. Popham and J. M. D'Auria, *Mar. Pollut. Bull., 13*, 25 (1982).
27. H. V. Weiss, M. A. Guttman, J. Korkisch, and I. Steffan, *Talanta, 24*, 509 (1977).
28. C. J. Hawkins, P. Kott, D. L. Parry, and J. H. Swinehart, *Comp. Biochem. Physiol., 76B*, 555 (1983).
29. H. Michibata, T. Terada, N. Anada, K. Yamakawa, and T. Numakunai, *Biol. Bull., 171*, 672 (1986).

30. P. Miramand, J. C. Guary, and S. W. Fowler, *Mar. Biol., 56*, 281 (1980).
31. P. Miramand, J. C. Guary, and S. W. Fowler, *J. Exp. Mar. Biol. Ecol., 49*, 267 (1981).
32. P. Miramand, S. W. Fowler, and J. C. Guary, *Mar. Biol., 67*, 127 (1982).
33. J. Doull, C. D. Klassen, and M. O. Amdur (eds.), *Casarett and Doull's Toxicology: The Basic Science of Poisons*, 3rd ed., Macmillan, New York, 1986.
34. L. C. Cantley, Jr., M. D. Resh, and G. Guidotti, *Nature, 272*, 552 (1978).
35. J. L. Johnson, H. J. Cohen, and K. V. Rajagopalan, *Biochem. Biophys. Res. Commun., 56*, 940 (1974).
36. L. C. Cantley and P. Aisen, *J. Biol. Chem., 254*, 1781 (1979).
37. T. D. Tullius, W. O. Gillum, R. M. K. Carlson, and K. O. Hodgson, *J. Am. Chem. Soc., 102*, 5670 (1980).
38. S. Lee, K. Kustin, W. E. Robinson, R. B. Frankel, and K. Spartalian, *J. Inorg. Biochem., 33*, 183 (1988).
39. S. G. Brand, C. J. Hawkins, A. T. Marshall, G. W. Nette, and D. L. Parry, *Comp. Biochem. Physiol., 93B*, 425 (1989).
40. B. W. Patterson, S. L. Hansard II, C. B. Ammerman, P. R. Henry, L. A. Zech, and W. R. Fisher, *Am. J. Physiol., 251*, R325 (1986).
41. L. L. Hopkins and B. E. Tilton, *Am. J. Physiol., 211*, 169 (1966).
42. J. J. Grantham, *Am. J. Physiol., 239*, F97 (1980).
43. M. V. Bell, K. F. Kelly, and J. R. Sargent, *J. Exp. Biol., 102*, 295 (1983).
44. H. J. Bielig, E. Jost, K. Pfleger, W. Rummel, and E. Seifen, *Hoppe-Seyler's Z. Physiol. Chem., 325*, 122 (1961).
45. H.-J. Bielig, K. Pfleger, W. Rummel, and E. Seifen, *Hoppe-Seyler's Z. Physiol. Chem., 325*, 249 (1961).
46. E. D. Goldberg, W. McBlair, and K. M. Taylor, *Biol. Bull., 101*, 84 (1951).
47. D. H. Anderson, J. R. Berg, and J. H. Swinehart, *Comp. Biochem. Physiol., 99A*, 151 (1991).
48. K. Kustin, K. V. Ladd, and G. C. McLeod, *J. Gen. Physiol., 65*, 315 (1975).

49. W. Rummel, H.-J. Bielig, W. Forth, K. Pfleger, W. Rüdiger, and E. Seifen, in *Protides of the Biological Fluids*, Vol. 14, Elsevier, Amsterdam, 1966, p. 205 ff.
50. M. A. Curtin, K. Kustin, and W. E. Robinson, *Biol. Bull.*, *169*, 152 (1985).
51. K. Kustin and G. C. McLeod, *Topics Curr. Chem.*, *69*, 1 (1977).
52. K. Kustin, K. V. Ladd, D. L. Toppen, and G. C. McLeod, *Biol. Bull.*, *147*, 608 (1974).
53. J. W. Moore and R. G. Pearson, *Kinetics and Mechanism*, 3rd ed., John Wiley and Sons, New York, 1981, p. 311 ff.
54. D. H. Anderson and J. H. Swinehart, *Comp. Biochem. Physiol.*, *99A*, 585 (1991).
55. J. Kučera, A. R. Byrne, A. Mravcová, and J. Lener, *Sci. Tot. Environ.*, *15*, 191 (1992).
56. S. Osaki, J. A. McDermott, D. A. Johnson, and E. Frieden, in *The Biochemistry of Copper* (J. Peisach, P. Aisen, and W. E. Blumberg, eds.), Academic Press, New York, 1966, p. 559 ff.
57. N. D. Chasteen, J. K. Grady, and C. E. Holloway, *Inorg. Chem.*, *25*, 2754 (1986).
58. P. Aisen and I. Listowsky, *Ann. Rev. Biochem.*, *49*, 357 (1980).
59. K. Kustin, D. S. Levine, G. C. McLeod, and W. A. Curby, *Biol. Bull.*, *150*, 426 (1976).
60. J. Webb and P. Chrystal, *Mar. Biol.*, *63*, 107 (1981).
61. D. A. Roman, J. Molina, and L. Rivera, *Biol. Bull.*, *175*, 154 (1988).
62. W. R. Biggs and J. H. Swinehart, in *Metal Ions in Biological Systems*, Volume 6 (H. Sigel, ed.), Marcel Dekker, New York, 1976, pp. 141–196.
63. M. V. Bell, B. J. S. Pirie, D. B. McPhail, B. A. Goodman, I.-B. Falk-Petersen, and J. R. Sargent, *J. Mar. Biol. Assoc. U.K.*, *62*, 709 (1982).
64. C. J. Hawkins, P. M. Merefield, D. L. Parry, W. R. Biggs, and J. H. Swinehart, *Biol. Bull.*, *159*, 656 (1980).
65. A. W. Martin, E. Huebers, H. Huebers, J. Webb, and C. A. Finch, *Blood*, *64*, 1047 (1984).
66. C. A. Finch and H. A. Huebers, *Clin. Physiol. Biochem.*, *4*, 5 (1986).
67. G. B. Rayner-Canham, *Polyhedron*, *3*, 1029 (1984).
68. C. J. Hawkins, D. L. Parry, and C. Pierce, *Biol. Bull.*, *159*, 669 (1980).

69. M. I. Agudelo, K. Kustin, G. C. McLeod, W. E. Robinson, and R. T. Wang, *Biol. Bull.*, *165*, 100 (1983).

70. S. G. Brand, PhD dissert., Northern Territory University, Darwin, Australia, 1992.

71. W. G. Barb, J. H. Baxendale, P. George, and K. R. Hargrave, *Trans. Faraday Soc.,*, *47*, 462 (1951).

72. W. G. Barb, J. H. Baxendale, P. George, and K. R. Hargrave, *Trans. Faraday Soc.*, *47*, 591 (1951).

73. S. Liochev and I. Fridovich, *Biochim. Biophys. Acta*, *924*, 319 (1987).

74. B. J. Scott and A. R. Bradwell, *Clin. Chem.*, *29*, 629 (1983).

75. A. Heinz, K. A. Rubinson, and J. J. Grantham, *J. Lab. Clin. Med.*, *100*, 593 (1982).

76. I. G. Macara, K. Kustin, and L. C. Cantley, Jr., *Biochim. Biophys. Acta*, *629*, 95 (1980).

77. R. K. Wright, in *Invertebrate Blood Cells*, Vol. 2 (N. A. Ratcliffe and A. F. Rowley, eds.), Academic Press, London, 1981, pp. 565–626.

78. M. J. Smith, *Experientia*, *45*, 452 (1989).

79. A. L. Dingley, K. Kustin, I. G. Macara, and G. C. McLeod, *Biochim. Biophys. Acta*, *649*, 493 (1981).

80. I. Goodbody, *Adv. Mar. Biol.*, *12*, 1 (1974).

81. H. Michibata, Y. Seki, J. Hirata, M. Kawamura, K. Iwai, R. Iwata, and T. Ido, *Zool. Sci.*, *8*, 447 (1991).

82. P. North and R. L. Post, *J. Biol. Chem.*, *259*, 4971 (1984).

83. R. Fukai and W. W. Meinke, *Limnol. Oceanogr.*, *7*, 186 (1962).

84. B. P. Bourgoin and M. J. Risk, *Bull. Environ. Contam. Toxicol.*, *39*, 1063 (1987).

85. A. J. Blotcky, C. Falcone, V. A. Medina, and E. P. Rack, *Anal. Chem.*, *51*, 178 (1979).

86. I. G. Macara, G. C. McLeod, and K. Kustin, *Comp. Biochem. Physiol.*, *63B*, 299 (1979).

87. W. E. Robinson and K. Kustin, unpublished data.

88. P. Frank, R. M. K. Carlson, and K. O. Hodgson, *Inorg. Chem.*, *25*, 470 (1986).

16

Vanadium in Mammalian Physiology and Nutrition

Forrest H. Nielsen

USDA, ARS, Grand Forks Human Nutrition Research Center, 2420 2nd Avenue North, P.O. Box 9034, University Station, Grand Forks, ND 58202-9034, USA

1. BRIEF INTRODUCTION AND HISTORY

In 1876, Priestley and Gamgee [1] reported on the toxicity of sodium vanadate in frogs, pigeons, guinea pigs, rabbits, dogs, and cats. Two other reports appeared shortly thereafter in which the toxic action of vanadium on heart muscle contraction [2] and in liver resulting in fatty degeneration [3] were described. But following these studies, interest in the biological actions of vanadium essentially lay dormant until the turn of the nineteenth century, when various French physicians used vanadium as a panacea for a number of human disorders [4]. Since that time to the present, vanadium has been brought to the fore numerous times as an element of possible pharmacological or nutritional importance. Through the years, enthusiasm has risen and fallen for the possible use of vanadium as a pharmaceutical for treating syphilis [5], reducing serum cholesterol [6,7] and preventing caries [8]. Recently, vanadium has been receiving much attention because of its insulin-mimetic properties (see Chap. 17).

Research examining the hypothesis that vanadium plays a physiological role or is an essential nutrient in higher animals also can be characterized as having periods of convincement followed by periods of much skepticism. In 1931 it was found that vanadium as a supplement with an incompletely effective dose of iron was superior to copper in alleviating anemia caused by feeding a milk diet to rats [9]; vanadium also improved growth [10]. Apparently, because other elements in addition to vanadium had similar effects, these findings neither stimulated further research in nor resulted in the acceptance of vanadium as an essential element. In 1949, Rygh [11] suggested that vanadium may be nutritionally important for animals because it stimulated the mineralization of bones and teeth in rats and guinea pigs. However, almost 20 years later, Schroeder et al. [12] stated that although vanadium behaves like an essential trace metal, final proof of essentiality for mammals was still lacking. Findings reported between 1971 and 1974 by four different research groups led many to conclude that vanadium is an essential nutrient. In a 1974 review of those findings, Hopkins and Mohr [13] stated "we are secure in the concept that vanadium is an essential nutrient." However, 11 years later, a review appeared [14] that presented a convincing argument that the evidence for the nutritional essentiality of vanadium was still inconclusive. This review indicated that the findings accepted as evidence for vanadium essentially were mostly manifestations of high vanadium supplements which induced pharmacological changes in animals fed unbalanced diets. Some substantive evidence for vanadium essentiality has appeared since 1987 and is described in the following. Nonetheless, still today, because a defined biochemical function for vanadium in higher animals has not been described, vanadium is not unequivocally accepted as an essential nutrient for higher animals. It seems as if the statement made by Schroeder et al. [12] in 1963 still holds true today, i.e., "no other trace metal has so long had so many supposed biological activities without having been proved to be essential."

2. METABOLISM

2.1. Absorption

2.1.1. Amounts Absorbed

In 1917, Proescher et al. [5] fed a healthy young man a relatively high dose of vanadium, or 12.5 mg of sodium tetravanadate. They found, by using a potassium permanganate titration method, that 12.4% of the dose was voided in the urine and 87.6% passed through the feces. More recent research indicates that 8–10% of absorbed vanadium is excreted through the bile [15–17]. Thus, this first study of vanadium absorption suggested that about 15% of a large single bolus of vanadium is absorbed. Some studies in which rats were given relatively high doses of vanadium also indicated that vanadium absorption can exceed 10%. Wiegmann et al. [17] recovered in the feces only 69.1% of a 5-μmol dose of Na_3VO_4 containing the tracer ^{48}V gavaged into fasted rats; recovery from the feces increased to 85.7% if a suspension of $Al(OH)_3$ was administered simultaneously with the vanadium. Bogden et al. [18] used the balance method to determine that rats absorbed about 40%, and excreted about 59% in the feces, of vanadium ingested as sodium metavanadate supplemented at 5 or 25 μg/g to a casein-sucrose-dextrin–based diet.

Most studies indicate that the amount of ingested vanadium absorbed in higher animals is much lower than 10%, and probably is normally near 1–3% with usual intakes in the presence of natural-type foods. For example, based on the very low concentrations of vanadium, generally less than 0.8 μg/L, found in urine [19], in comparison with the estimated daily intake of 10–30 μg [19,20], and the high fecal content of vanadium [19], less than 5% of vanadium ingested is normally absorbed. Two research groups found that the administration of tartrate salts of vanadium with meals to humans resulted in quite variable amounts of vanadium in the urine, and thus variable absorption. Curran et al. [21] calculated that about 0.1–1.0% of vanadium in 100 mg of the very soluble diammonium oxytartratovanadate fed daily for 6 weeks to five normal, healthy men, aged 23–26 years and consuming identical diets, was absorbed. Dimond et al. [22] obtained similar findings with five females and one male, aged 30–55 years. When these patients were fed 25–125 mg of ammonium vanadyl tartrate daily for a minimum of 6 weeks, urinary vanadium ranged from 25 to 1514 μg/24 hr; these amounts suggest an absorption of less than 1%.

Some animal studies also support the concept that vanadium absorption by the gastrointestinal tract is quite low. Conklin et al. [23] found that, based on the uptake of ^{48}V in 40 μg of vanadium in the form of V_2O_5 administered by gavage to rats, 2.6% of the dose was absorbed. Sheep consuming a natural-type diet providing 1.9 mg V/day, or supplemented with ammonium metavanadate to supply 38 or 153 mg V/day, were estimated to have absorbed 0.13–0.75% (mean of 0.34%) of the ingested vanadium [24].

2.1.2. *Factors Affecting Absorption*

The variable urinary excretion findings described in the preceding section indicate that the absorption of vanadium can be modified by a number of different factors. Except for the $Al(OH)_3$ finding mentioned above, the direct determination of changes in vanadium absorption caused by various factors has received very little attention. However, indirect evidence that certain factors affect the absorption of vanadium has resulted from two other types of studies, i.e., a factor may be modifying vanadium absorption if it modifies vanadium toxicity, or tissue retention or accumulation. Table 1 lists a number of such factors.

Dietary vanadium probably occurs mainly as VO^{2+} (vanadyl, V^{4+}) or as HVO_4^{2-} (vanadate, V^{5+}). Most ingested vanadium probably is transformed in the acidic milieu of the stomach to VO^{2+} and remains in this form as it passes into the duodenum and when excreted in the feces [39]. However, vanadate is absorbed three to five times more effectively than vanadyl. Thus, the effect of other dietary components on the form of vanadium in the stomach and the speed at which it is transformed into vanadyl probably markedly affect the percentage of ingested vanadium absorbed. Moreover, the extent of binding, especially the cationic form of vanadium, to macromolecules such as complex carbohydrates and proteins apparently has a profound effect on vanadium absorption (see Table 1).

2.1.3. *Mechanisms Involved in Absorption*

Very little is known about the mechanisms involved in the absorption of vanadium in the cationic or VO^{2+} form, which apparently is the most prevalent form in the duodenum. In vitro studies suggest that anionic vanadium or HVO_4^{2-} can enter cells through phosphate or other anion transport systems [40]. Kinetic modeling of whole-body vanadium metabolism in sheep indicates that a significant amount of vanadium absorption occurs in the upper gastrointestinal tract [41].

TABLE 1

Factors Affecting Vanadium Toxicity and Retention

Factor	Effect	Ref.
Aluminum hydroxide	Retention of vanadium gavaged as ^{48}V increased in feces and decreased in urine and tissues when administered with $Al(OH)_3$.	17
Ascorbic acid	Ascorbic acid alleviated growth retardation and bone vanadium retention in growing chicks caused by high dietary vanadium.	25,26
	Ascorbic acid alleviated the reduction in egg quality and production, and in body weight caused by high vanadium in the diet of laying hens.	26
Carbohydrate form (simple vs. complex)	Replacing grain with sucrose in the diet of laying hens intensified the depression in egg quality and production, and body weight induced by high dietary vanadium.	27
	Lactose addition to the diet exacerbated vanadium toxicity signs of growth depression and mortality in growing chicks.	28
Chloride	Increasing the dietary supplement of NaCl from 0.5% to 2.0% ameliorated vanadate toxicity as assessed by growth rate and inconsistently decreased tissue accumulation of vanadium when its intake was high; the effect was found to reside in the chloride ion.	29,30
Chromium	Chromium reduced mortality, growth depression, and the uncoupling of oxidative phosphorylation caused by high dietary vanadium in growing chicks.	28,31
	Chromium reduced bone vanadium content.	32
Copper	High dietary copper alleviated the growth-retarding effect of vanadate for chicks but did not affect the femur concentration of vanadium.	33

TABLE 1

Continued

Factor	Effect	Ref.
EDTA	EDTA prevented vanadium toxicity in chicks assessed by growth depression and mortality, and decreased the amount of ^{48}V from a tracer dose found in tissues.	31,34
Form or oxidation state	The tissue retention of vanadium was higher in rats fed 50 μg V/g diet as sodium orthovanadate than in rats fed 50 μg V/g diet as vanadyl sulfate.	35
Iron	Vanadium toxicity as assessed by chick growth was more severe with iron deficiency; however, ^{48}V absorption was not affected by dietary iron. Iron-deficient animals retained more vanadium in the blood and liver, and less in bone.	36
Protein intake	Increasing dietary protein from 10% to 30% reduced mortality and growth retardation in chicks caused by vanadium toxicity.	37
Protein source	Replacement of soybean meal by cottonseed meal reduced the effects of high dietary vanadium in laying hens, including elevated tissue vanadium concentrations, reduced egg quality and production, and decreased body weight; replacement of soybean meal with herring fish meal intensified the effects of vanadium toxicity.	27,38

2.2. Transport and Retention

2.2.1. Form in Blood and Organs

About 90% of the vanadium in blood is associated with the plasma fraction. Although it was once thought that extracellular vanadium existed mainly as vanadate [42], current thought is that the vanadyl form is most prevalent in blood. If vanadate appears in blood, it apparently is quickly converted to vanadyl in the erythrocytes by glutathione [43,44], or in plasma by reductants such as ascorbate [45], catecholamines [46], and cysteine [47]. However, as a result of oxygen tension, vanadate still exists in blood. Vanadyl is bound and transported by transferrin and albumin [48]. Vanadate apparently is transported by transferrin only [48]. Chasteen et al. [39] estimated that at least 50% of vanadium in plasma is associated with transferrin [39]. Harris et al. [49] found that radiolabeled vanadium as either vanadyl or vanadate injected intravenously into adult beagle dogs required about 30 hr to reach the maximum degree of transferrin binding, which was about 77%. Low molecular weight species account for only 3% of the vanadium present in plasma [48]. Chasteen et al. [48] stated that "the actual distribution of vanadium between oxidation states and between transferrin and albumin will depend on kinetic factors such as concentration and identity of reductants in plasma, the fluctuation of oxygen tension between 40 and 100 torr for venous and arterial blood, respectively, and the flux of the metal in and out of cells."

According to Chasteen et al. [50], vanadium is targeted to cells that are rich in iron where a significant amount of it as vanadyl is bound to ferritin, the iron storage protein. Evidence for this statement is that radiolabeled vanadium orally administered, injected intravenously (IV) or instilled intratracheally (IT) is retained in high amounts in liver and spleen [23,51–54]. Sabbioni and Marafante [55] found that one day after IV administration of $^{48}VO^{2+}$, 29% of ^{48}V incorporated in rat liver cytosol existed as a vanadium–low molecular weight complex (<5000 MW). By day 9, however, they found that the low molecular weight complex had disappeared and vanadium was present only as vanadium-ferritin (15%) and vanadium-transferrin (85%) in rat liver cytosol. Nine days after the administration of vanadium, partial purification of heart myoglobin, liver mitochondrial and microsomal cytochromes *b*, b_5, and *c*, and ferriporphyrin, and red blood cell hemoglobin showed no significant incorporation of ^{48}V into these proteins [56]. Although these findings indicate that only nonheme iron metalloproteins are involved in the metabolism of vanadium in vivo, it remains to be determined as to whether ferritin is a storage vehicle for vanadium, and whether vanadyl-transferrin can transfer vanadium to cells through the transferrin receptor.

Vanadium in milk of lactating rats injected with ^{48}V was found mainly in the protein fraction associated with a transferrin-like protein, apparently lactoferrin [57]. Nursing pups absorbed a significant amount of the ^{48}V in the milk; this suggests that a lactoferrin-vanadium complex is important in vanadium metabolism in the suckling rat pup.

Other organs that retain relatively high amounts of orally, IV, or IT administered vanadium are kidney, bone, and testis [23,51–54]. The high amount in kidney may reflect its being the site of vanadium excretion. Part of the retention of vanadium in bone is thought to occur because vanadate substitutes for phosphate in the hydroxyapatite crystal [58]. Bone also contains the iron-protein-rich marrow. Why testis retains relatively high amounts of vanadium and how it is bound are unclear.

2.2.2. *Amounts in Organs and Fluids*

Heydorn [59] reviewed analytical methods for the determination of vanadium in the low amounts found in tissues, blood, and urine. For this task, especially for human plasma and serum, methods using atomic emission spectrometry, particle-induced X-ray emission, flame atomic absorption spectrometry, and catalysis were found to be inadequate. Methods that apparently can determine vanadium accurately in low amounts are electrothermal atomic absorption spectrometry, neutron activation analysis with radiochemical separation, and neutron activation with preirradiation separation. Thus, early reports of the vanadium content of organs and fluids determined by unreliable and insensitive techniques should be disregarded, including that by Schroeder et al. [12]. Recent analyses using reliable techniques indicate that very little vanadium is retained under normal conditions in the body; most tissues contain less than 10 ng V/g wet weight (Table 2). The total body burden of vanadium is about 100 μg.

2.2.3. *Factors Affecting Retention*

There are reports that occupational exposure and several pathological conditions can affect the tissue retention of vanadium. Most of these reports must be accepted with wariness because the vanadium concentration values given, even for healthy, normal individuals, do not con-

TABLE 2

Reported Mean Vanadium Concentrations in Human Fluids and Organs on a Fresh Weight Basis

Fluid/organ	Vanadium (ng/g or mL)	Ref.	Fluid/organ	Vanadium (ng/g or mL)	Ref.
Amniotic fluid	12	60	Milk		
Bile	1.0	19	Guatemala	0.21	67
Blood	0.058	61	Hungary	0.11	67
Bone	3.5	19	Nigeria	0.46	67
Brain	0.75	19	Philippines	0.69	67
Bronchus	60	62	Sweden	0.13	67
Fat, subcutaneous	0.72	19	Zaire	0.27	67
Fingernails			Placenta (dry wt)	8.3	68
Canada, India, USA	80	63		4.8	62
Japan	150	63	Red blood cells	0.026	69
Poland	110	63	Serum	0.031	69
Hair, scalp	40	19		0.071	61
	29	64	Subpleura	58	62
Heart	1.1	19	Teeth	3.6	19
Kidney	3.0	19	Primary	4.3	70
Liver	7.0	65	Permanent	3.9	70
Lung	30 (median)	19	Thyroid	3.1	19
	30	66	Trachea	38	62
Muscle	0.54	19	Urine	0.24	71

form with those shown in Table 2 (usually greater by a factor of at least 10). Nonetheless comparisons are made between healthy normals and targeted groups. Thus, a relative difference probably can be accepted as evidence that vanadium retention has been modified. When compared to healthy or normal controls, the vanadium concentration was higher in the hair of welders [72], patients with chronic renal failure [73], and patients with manic-depressive illness [64,74,75] and lower in the hair of multiple sclerosis patients [76]. Elevated blood, serum, or plasma vanadium concentrations have been found in patients with chronic renal disease [77–79] and manic-depressive illness [74,75,80–83]. Kidneys taken at autopsy from patients in which the cause of death was related to pulmonary pathology were found to contain increased vanadium concentrations [84].

Based on animal studies, high dietary vanadium also increases the retention of vanadium by the body. In rats, on a wet weight basis, vanadium increased from 10 to 554 ng/g in liver, from 2 to 210 ng/mL in plasma, from 2 to 110 ng/mL in whole blood, and from 10 to 1812 ng/g in kidney cortex when dietary vanadium was increased from 0.1 to 25 μg/g [18]. In sheep, bone vanadium increased from 220 to 3320 ng/g ash weight when dietary vanadium was increased from 10 to 220 μg/g [85]. Bone apparently is a major sink for retained vanadium.

Vanadium retention may also be affected by age. Hair vanadium concentrations were found to increase in infants from birth to 6 months of age [86]. In rats between ages 21 and 115 days, the vanadium concentration decreased in kidney, liver, lung, and spleen, and increased in fat and bone [87].

2.3. Excretion

2.3.1. Sites and Mechanisms of Excretion

Most ingested vanadium is excreted via the feces; most of the vanadium is that which was unabsorbed. However, based on studies in which vanadium was administered parenterally, a significant amount of absorbed vanadium is reexcreted via the intestinal tract. Kent and McCance [16] recovered in 2 weeks about 10% of an IV dose of sodium tetravanadate in the feces of two human subjects. Hopkins and Tilton [15], using the radiotracer ^{48}V, found that rats excreted 8.6% of the dose in the feces within 96 hr after being IV injected. Wiegmann et al. [17] obtained similar results with an intraperitoneal (IP) injection of vanadium containing ^{48}V. Sabbioni and Marafante [51] found that about 10% of IV-injected 10 μg/g of ^{48}V-labeled vanadium as ammonium metavanadate appeared in the feces. It was suggested, based on experiments with cannulated rats, that the bile contributes to the intestinal elimination of vanadium during the initial hours after absorption when high amounts of circulating vanadium are present, but upon binding to hepatic constituents, biliary excretion is limited [88]. However, Hansard et al. [54] found that the appearance of an IV dose of carrier-free ^{48}V as dioxovanadium chloride in feces progressively increased over 6 days at which time 5.3% of the dose had appeared. This suggests that even when circulating vanadium is low a significant amount is excreted via the bile. Further evidence that bile is an important excretory route is the finding of about 1.0 ng V/g of human bile [19].

Based on studies in which vanadium is administered parenterally, urine is the major excretory route for absorbed vanadium. Kent and McCance [16] gave two men six IV injections of sodium tetravanadate totaling 18 and 24 mg of vanadium in one week; at the end of 2 weeks, 86% and 83%, respectively, of the vanadium was found in the urine. Hansard et al. [54] found that 57–70% of an IV administered tracer dose of ^{48}V appeared in the urine of sheep in 6 days. IV or IP vanadium appeared somewhat slower in rats. At 96 hr after an IV injection of ^{48}V in rats, 30–46% of the dose had been excreted in the urine [15,51]. Within five days after IP administration of ^{48}V-labeled vanadate to rats, only 41% of the dose had appeared in the urine. However, Roshchin et al. [52] found that 66% of ^{48}V injected intramuscularly (IM) as $VOCl_2$ was eliminated in the urine within 24 hr.

2.3.2. Forms Excreted

Both high and low molecular weight complexes have been found in urine. Sabbioni and Marafante [51] found that ^{48}V in the urine was mainly associated with high molecular weight components 3 days after an IV injection of ^{48}V. They suggested that one high molecular weight component binding vanadium was transferrin. On the other hand, Chasteen et al. [39] found that the majority of vanadium excreted in the urine was as a low molecular weight VO^{2+} complex; this complex was not identified.

3. ESSENTIALITY

3.1. Signs of Deprivation

Between 1971 and 1974, four research groups described possible signs of vanadium deficiency. However, there was much inconsistency in the findings. Strasia [89] reported that rats fed less than 100 ng V/g diet exhibited slower growth, higher plasma and bone iron, and higher hematocrits than controls fed 500 ng V/g diet. Williams [90], using a diet lower in casein and higher in iron and ascorbic acid, was not able to produce those deficiency signs in the same laboratory in which Strasia worked. Schwarz and Milne [91] reported that a vanadium supplement of 0.25–0.50 μg/g diet gave a positive growth response in suboptimally growing rats fed an amino acid–based diet probably deficient in riboflavin and with an unknown vanadium content. On the other hand, Hopkins and Mohr [13] reported that the only effect of vanadium deprivation in rats was impaired reproductive performance (decreased fertility and increased perinatal mortality) that became apparent only in the fourth generation; the diet used contained luxuriant amounts of methionine and was probably deficient in cysteine.

Deprivation studies with chicks also gave inconsistent findings. Hopkins and Mohr [13,92] found that vanadium-deprived chicks exhibited significantly depressed wing and tail feather development, depressed plasma cholesterol and triglycerides at age 28 days, and elevated plasma cholesterol at age 49 days. Nielsen and Ollerich [93,94] reported that vanadium deprivation depressed growth, elevated hematocrits and plasma cholesterol, and adversely affected bone development.

In the preceding vanadium deprivation studies, "controls" or supplemented animals were fed 0.5–3.0 μg V/g diet. Although these amounts are near those in natural diets which apparently contain about 1.0 μg V/g [95], those doses of highly available vanadium apparently are 10–50 times those normally found in purified or semipurified diets [95]. As indicated elsewhere in this book, vanadium is a relatively toxic and pharmacologically active element. As a result, it is difficult to determine whether the "deficiency signs" in early experiments with questionable diets were true deficiency signs, indirect changes caused by an enhanced need for vanadium in some metabolic function, or manifestations of a pharmacological action of vanadium. However, some recent findings suggest that the early findings of impaired reproductive performance, bone structure changes, and plasma cholesterol changes may have been true vanadium deficiency signs.

The uncertainty about vanadium deficiency signs stimulated new efforts in the 1980s to produce deficiency signs in animals fed diets apparently containing adequate and balanced amounts of all known nutrients; findings from these studies are described below. Vanadium deficiency signs for humans have not been described.

3.1.1. Gross Signs

In nine experiments, Anke and coworkers [96–99] found that, when compared to controls fed in the first experiments 2 μg V/g diet and in the latest experiments 0.5 μg V/g diet, goats fed less than 10 ng V/g diet had more difficulty in conceiving, exhibited a higher rate of spontaneous abortion, increased ratio of female to male kids born, and those animals that delivered offspring produced less milk. Forty-two percent of kids from vanadium-deprived goats died between days 7 and 91 of life with some deaths preceded by convulsions; only 8% of kids from vanadium-supplemented goats died during this time. Vanadium-deficient goats had only 50% the life-span of control goats. Also, deficient goats exhibited pain in the extremities, swollen forefoot tarsal joints, and skeletal deformations in the forelegs; some of these changes have been illustrated [96,99]. Uthus and Nielsen [100] reported that, when compared to controls fed 1 μg V/g diet, vanadium deprivation (2 ng V/g diet) increased thyroid weight and thyroid weight/body weight ratio and tended to decrease growth of rats.

3.1.2. Biochemical Signs

Uthus and Nielsen [100] found that as dietary iodine increased from 0.05 to 0.33 to 25 μg/g, thyroid peroxidase activity decreased, and the decrease was more marked (38.1 to 12.3 to 3.5 mGU/mg protein) in vanadium-supplemented (1 μg V/g diet) than in vanadium-deprived (2 ng V/g diet) rats (18.7 to 10.2 to 6.8 mGU/mg protein). Also, as dietary iodine increased, plasma glucose increased in the vanadium-deprived rats but decreased in the vanadium-supplemented rats. Vanadium deprivation also depressed red blood cell glucose-6-phosphate dehydrogenase and cecal total carbonic anhydrase [101].

Seaborn et al. [102] found that, when compared to guinea pigs fed less than 10 ng V/g diet, guinea pigs supplemented with 0.5 μg V/g diet exhibited decreased hepatic glycogen and bone copper, and increased hepatic lipids, fecal bile acids, plasma cortisol, and bone calcium and magnesium. Vanadium supplementation also increased HDL cholesterol [103]. An interaction between dietary vanadium and ascorbate affected several variables in guinea pigs. Vanadium supplementation increased plasma 3-hydroxy-3-methylglutaryl coenzyme A reductase activity, fecal cholesterol, and hepatic copper and iron in ascorbate-deficient, but decreased these variables in ascorbate-adequate, guinea pigs; just the opposite occurred with plasma cholesterol and bile acids in bile [102]. Anke et al. [96] reported that, when compared to goats fed less than 10 ng V/g diet, goats fed 2 μg V/g diet exhibited decreased serum β-lipoproteins, creatinine, isocitrate dehydrogenase, and lactate dehydrogenase, and increased serum glucose.

Unfortunately, the conclusiveness of the findings from the recent studies is somewhat diminished by the fact that the control diets, like those in the early vanadium essentiality studies, contained high amounts of highly available vanadium (0.5–2 μg/g); these amounts may have had unknown pharmacological effects in these studies. For example, findings similar to the biochemical findings with control goats fed 2 μg V/g diet reported by Anke et al. [96] were not mentioned in later reports in which control goats were fed 0.5 μg V/g diet [98,99]. Seaborn et al. [102] suggested that the increased plasma cortisol in guinea pigs supplemented with 0.5 μg V/g diet was evidence that this amount of dietary vanadium was a stressor, and thus it may have been exerting pharmacological or toxicological effects. Nonetheless, because the diets used apparently were complete and well balanced, the recent vanadium deprivation studies probably have found some true deficiency signs. It seems unlikely that the diets lacked any nutrient that would cause some of the marked deficiency signs (especially the bone changes) that would be prevented by a pharmacological action of the small vanadium supplements used.

3.2. Possible Biochemical Functions

A defined biochemical function for vanadium in higher animals, and thus humans, has not been described. Recently, however, functional roles for vanadium have been defined for some algae, lichens, fungi, and bacteria; these include roles in haloperoxidase [104] and nitrogenase [105] enzymes, a compound called amavadin found in mushrooms [106], and a substance isolated from ascidian blood cells called vandobin [107]. These findings, along with the animal findings above, suggest that vanadium has an essential functional role in higher animals, including humans.

Numerous biochemical and physiological functions for vanadium have been suggested based upon its in vitro and pharmacological actions; these have been discussed in several reviews [108–112] and are too extensive to discuss in detail here. In vitro studies with cells and pharmacological studies with animals have shown that vanadium has insulin-mimetic properties; numerous stimulatory effects on cell proliferation and differentiation; effects on cell phosphorylation/dephosphorylation; inhibitory effects on the motility of sperm, cilia, and chromosomes; effects on glucose and ion transport across the plasma membrane; interfering effects on intracellular ionized calcium movement; and effects on oxidation-reduction processes. In vitro cell-free systems have shown that vanadium inhibits numerous ATPases, phosphatases, and phosphoryl transfer enzymes. Vanadium also stimulates the activity of numerous enyzmes in vitro including NADPH/NADH oxidase. However, vanadium has not been found to be the specific activator or inhibitor of any enzyme in higher forms of life.

With the entanglement of pharmacological and nutritional observations, it is difficult to speculate on a possible function for vanadium. However, two promising sites for a functional role of vanadium is in thyroid and in bone (and/or connective tissue) metabolism.

The finding that some haloperoxidases in lower forms of life require vanadium for activity suggests a similar function in higher animals. The best known haloperoxidase in animals is thyroid peroxidase. As described previously, an apparent vanadium deprivation affected the response of thyroid peroxidase activity to changing dietary iodine in addition to elevating thyroid weight. Other findings pointing toward a role for vanadium in thyroid hormone metabolism are that the thyroid apparently contains relatively high amounts of vanadium [113], thyroidectomy-parathyroidectomy increases tissue retention of vanadium [114], and thyroxine administration alleviates the retention of vanadium in pancreas, kidney, and bone caused by hypophysectomy [115].

The changes in bone structure in the vanadium deprivation studies found by Anke et al. [96–99] and Nielsen and Ollerich [93,94] suggest a role in bone or connective tissue metabolism. Also supporting such a role is the finding that the uptake of an IP-injected dose of ^{48}V as vanadyl chloride by chick tibia occurred only in areas of calcification, i.e., secondary center of ossification, and calcifying cartilage and bone; the highest uptake occurred in newly calcified cartilage [116]; similar findings were obtained with fish [117]. Additionally, vanadium has been found to stimulate the mineralization of bones and teeth [11] and the repair of bones [118]. The possibility that vanadium affects bone through an effect on thyroid metabolism has been discussed [119]. Another possible mechanism is through vanadium affecting or acting like a growth factor. Vanadate can mimic growth factors such as epidermal growth factor, fibroblast growth factor, and insulin [120,121]. For example, orthovanadate stimulates bone cell proliferation and collagen synthesis in vitro [120,122]. Also, vanadate was shown to increase proteoglycan synthesis by stimulating the conversion of poorly differentiated chondrocyte cultures from a "fibroblastic" expression to a "chondrocytic" expression [123]. Low concentrations of vanadium were found to be required for optimal growth of fibroblasts in tissue culture [124]. Finally, the oral administration of 1.0–10.0 μmol V/100 g to rats as vanadium pentoxide was found to increase bone DNA content and alkaline phosphatase activity [125].

Recently, the peroxo form of vanadium [126–128] has been discussed as being responsible for many biological actions of vanadium including its insulin-mimetic action and haloperoxidase roles. Vanadate can interact with $O_2^{\cdot-}$ formed by NADPH oxidase to generate peroxovanadyl [V(IV)–OO]. Peroxovanadyl can in turn remove hydrogen from NADPH to yield vanadyl hydroperoxide [V(IV)–OOH] [126]. Vanadium(III)- and (V)-superoxide complexes also have been postulated

[129]. Perhaps a peroxovanadyl form of vanadium is important at the physiological or nutritional level.

4. TOXICITY

4.1. Signs of Toxicity in Animals

Reports describing toxicological aspects of vanadium are too extensive to discuss in detail here; several comprehensive reviews of vanadium toxicity have been published [1,4,5,130–133]. From their in-depth study of vanadium toxicity, Proescher et al. [5] concluded that vanadium is neurotoxic and a hemorrhagic-endotheliotoxic poison with nephrotoxic, hepatotoxic, and probably leukocytotoxic components. Thus, it is not surprising that a variety of toxicity signs have been described for animals and that they can vary among species and with dosage. Some of the more consistent signs include depressed growth, diarrhea, depressed food intake, and death.

4.2. Signs of Toxicity in Humans

4.2.1. Chronic

Vanadium is a relatively toxic element for humans. Signs of toxicity through occupational sources have been well described [130,131,133]. The threshold level for toxicity through ingestion apparently is near 10–20 mg/day or 10–20 μg/g of diet; this is supported by animal findings [110] and the following human findings. Schroeder et al. [12] fed 15 patients 4.5 and 9 mg V/day as diammonium oxytartratovanadate for 6–16 months without apparent detrimental effect. Curran et al. [21] fed each of five subjects 13.5 mg/day in three divided doses as diammonium oxytartratovanadate for 6 weeks; no sign of intolerance or toxicity was found. On the other hand, Somerville and Davis [134] gave each of 12 patients 13.5 mg V/day for 2 weeks and then 22.5 mg V/day for 5 months as diammonium vanadotartrate; five patients exhibited persistent upper abdominal pain, anorexia, nausea, and weight loss, and five patients exhibited green tongue. It should be noted that, according to Curran and Burch [7], "the infamous green tongue which has been cited as an example of vanadium toxicity is merely a staining from mucosal contact with vanadium salts." Dimond et al. [22] gave ammonium vanadyl tartrate orally to six subjects for 6–10 weeks in amounts ranging from 4.5 to 18 mg V/day; green tongue, cramps, and diarrhea were observed at the larger doses.

Excessive in vivo amounts of vanadium have been suggested as a factor in manic-depressive illness [64,74–76]. In patients with this illness, elevated concentrations of vanadium have been found in plasma and hair. Improvement in the illness occurred with treatments that alleviate vanadium toxicity [74]. Recently, it has been suggested that chronic exposure to vanadium may result in hypertension [135,136].

4.2.2. Acute

Acute signs of vanadium toxicity in humans have been described by Proescher et al. [5]. They reported that in man, after an IV injection of 20 mg V_2O_5 as sodium tetravanadate, constriction of the throat, salivation, lacrimation, disappearance of pulse, free discharge of feces, vomiting, cessation of respiration, and a drop of 3° in temperature occurred.

5. ASSESSMENT OF STATUS

Because no function for vanadium in higher animals has been defined, a biochemical indicator of vanadium status does not exist. Blood, serum, and hair vanadium concentrations are possible indicators of vanadium status. Hair vanadium concentrations have been found to be elevated in patients with manic-depressive illness [64,74,75] and renal failure [73]. Elevated blood, serum, or plasma vanadium concentrations also were found in these patients [77–83]. Human whole-blood vanadium concentrations ranged from 0.024 to 0.226 (mean 0.056) ng/mL in nonexposed children near Prague, Czechoslovakia [61]. The range of concentrations in potentially exposed children near Prague were similar (0.018–0.239 ng/mL), but their mean value was higher (0.099 ng/mL) [61]. These findings suggest that vanadium concentrations in blood, serum, or hair significantly above or below those given in Table 2 may indicate an abnormal vanadium status.

6. SAFE AND ADEQUATE INTAKES

Any human requirement for vanadium would likely be very small. The diets used in animal deprivation studies contained only 1–25 ng V/g; often these did not markedly affect the animals. Vanadium deficiency has not been identified in humans; yet, as indicated in Table 3, most diets supply less than 30 μg daily, with many near 15 μg daily. These observations suggest that a dietary intake of 10 μg daily probably meets any postulated vanadium requirement.

The preceding discussion of toxicity indicated that a daily intake of over 10 mg of vanadium produced overt signs of toxicity. However, much lower amounts of vanadium were found to pharmacologically affect animals and humans, which suggests that they may have toxic manifestations under certain situations. For example, Schroeder et al. [12] found that 4.5 mg V/day decreased serum cholesterol in humans. In the preceding discussion of essentiality it was suggested that doses of highly available vanadium 10–50 times those normally found in purified or semipurified diets were having pharmacological effects in animals. Extrapolating these factors to humans indicates that the daily intake of vanadium probably should not exceed 100 μg.

Suggesting a safe and adequate intake of vanadium with the above limited information is difficult. However, based on the preceding, the safe and adequate intake of vanadium probably will be found to be near 10–50 μg/day.

7. DIETARY INTAKES

As indicated by Table 3, the daily intake of vanadium is relatively low in comparison with other essential trace elements. Foods rich in vanadium (greater than 40 ng/g) include shellfish, mushrooms, parsley, dill seed, black pepper, and some prepared foods [142,143]. Cereals, liver, and fish tend to have intermediate amounts (5–40 ng V/g) [19,137,140,142].

TABLE 3

Reported Dietary Intakes of Humans

Country	Daily intake (μg)	Ref.	Country	Daily intake (μg)	Ref.
Brazil	16.7[a,c]	61	United States, age/sex		
Britain	13[b]	137	6–11 mo	6.7[b]	140
	30[b]	138	2 yr	6.5[b]	140
Iran	34.7[a,c]	61	14–16 yr, F	7.1[b]	140
Italy	17.7[a,c]	61	14–16 yr, M	11.0[b]	140
Japan	2.5[b]	139	25–30 yr, F	8.1[b]	140
Spain	11.8[a,c]	61	25–30 yr, M	18.3[b]	140
Thailand	10.0[a,c]	61	60–65 yr, F	6.2[b]	140
Turkey	21.9[a,c]	61	60–65 yr, M	10.6[b]	140
United States			United States,		
Southeastern	15.5[b,c]	61	TPN solutions	29[b]	141
North-central	22.8[b,c]	61			
Northeastern	20.9[b,c]	61			
Southern	16.4[b,c]	61			
Western	25.3[b,c]	61			
North Dakota	20.4[b]	20			

[a]Median.
[b]Mean.
[c]Daily intake calculated from dry weight concentration and based on a consumption of about 0.5 kg of food/day dry weight.

Beverages, fats and oils, fresh fruits, and fresh vegetables generally contain less than 5.0 ng V/g, and often less than 1.0 ng/g [19,137,140,142].

8. CLINICAL SIGNIFICANCE OF VANADIUM

The clinical importance of vanadium is uncertain. Because neither a defined biochemical function nor signs of vanadium deficiency have been described for humans, any discussion of its possible nutritional importance seems inappropriate at this time.

Toxic effects from the intake of high amounts of vanadium in the diet are unlikely. Except for a few vanadium accumulator plants, only small amounts of vanadium are taken up by plants from the soil and deposited on the edible portion. Low plant uptake prevents the production of foods with high vanadium content even if local soil concentrations are high and soil contamination occurs because of combustion of petroleum and other naturally occurring hydrocarbons, such as asphaltite, which contain appreciable amounts of vanadium [144]. Although tissue vanadium can be elevated in livestock fed high-vanadium phosphates [110], vanadium toxicity occurs in the animal long before the vanadium content of the food products from those animals are of concern. Toxicity of vanadium usually occurs only as a result of industrial exposure to high amounts of airborne vanadium. However, because vanadium is so active pharmacologically, a beneficial pharmaceutical role for this element may be found. Also, nutritional supplements containing vanadium are now being marketed. Thus, the possibility of oral toxicity of vanadium in the future may not be remote.

The primary clinical research need for vanadium is the identification of an essential biochemical function in higher animals. This is necessary to disentangle pharmacological from nutritional or physiological observations in order to assess the clinical importance of vanadium. Determination of a defined function also will facilitate other research needs of vanadium in nutrition including the determination of status assessment indicators as well as safe and adequate intakes. Furthermore, until a defined function is described, it is unlikely that vanadium will be unequivocally accepted as an essential nutrient for higher animals.

ABBREVIATIONS AND DEFINITIONS

ATP	adenosine 5′-triphosphate
EDTA	ethylenediamine-*N,N,N′,N′*-tetraacetate
HDL cholesterol	high-density lipoprotein–cholesterol
IM	intramuscularly
IP	intraperitoneally
IT	intratracheally
IV	intravenously
mGU	milliguaiacol unit (amount of enzyme that gives a change of 1.0 absorbance unit per second with guaiacol as substrate divided by 1000)
NADH	nicotinamide adenine dinucleotide (reduced)
NADPH	nicotinamide adenine dinucleotide phosphate (reduced)
TPN	total parenteral nutrition

Pharmacological action: The effect of a relatively high dietary intake of a substance that either alleviates an abnormality caused by something other than a nutritional deficiency of that substance or alters some biochemical function or biological structure in a manner that may be construed as beneficial.

REFERENCES

1. J. Priestley and A. Gamgee, *Phil. Trans. Roy. Soc. London, 166,* 495 (1876)
2. A. Gamgee and L. Larmuth, *J. Anat. Physiol., 11,* 235 (1876).
3. G. F. Dowdeswell, *J. Physiol. (London), 1,* 257 (1878)
4. D. E. Jackson, *J. Pharmacol., 3,* 477 (1912).
5. F. Proescher, H. A. Seil, and A. W. Stillians, *Am. J. Syph., 1,* 347 (1917).
6. C. E. Lewis, *AMA Arch. Ind. Health, 19,* 419 (1959).
7. G. L. Curran and R. E. Burch, *Trace Sub. Environ. Health, 1,* 96 (1968).
8. G. Tank and C. A. Storvick, *J. Dent. Res., 39,* 473 (1960).

9. V. C. Myers and H. H. Beard, *J. Biol. Chem.*, *94*, 89 (1931).
10. H. H. Beard, *J. Biol. Chem.*, *94*, 135 (1931).
11. O. Rygh, *Bull. Soc. Chim. Biol.*, *31*, 1052, 1403, 1408 (1949).
12. H. A. Schroeder, J. J. Balassa, and I. H. Tipton, *J. Chron. Dis.*, *16*, 1047 (1963).
13. L. L. Hopkins, Jr. and H. E. Mohr, *Fed. Proc.*, *33*, 1773 (1974).
14. F. H. Nielsen, *J. Nutr.*, *115*, 1239 (1985).
15. L. L. Hopkins, Jr. and B. E. Tilton, *Am. J. Physiol.*, *211*, 169 (1966).
16. N. L. Kent and R. A. McCance, *Biochem. J.*, *35*, 837 (1941).
17. T. B. Wiegmann, H. D. Day, and R. V. Patak, *J. Toxicol. Environ. Health*, *10*, 233 (1982).
18. J. D. Bogden, H. Higashino, M. A. Lavenhar, J. W. Bauman, Jr., F. W. Kemp, and A. Aviv, *J. Nutr.*, *112*, 2279 (1982).
19. A. R. Byrne and L. Kosta, *Sci. Tot. Environ.*, *10*, 17 (1978).
20. D. R. Myron, T. J. Zimmerman, T. R. Shuler, L. M. Klevay, D. E. Lee, and F. H. Nielsen, *Am. J. Clin. Nutr.*, *31*, 527 (1978).
21. G. L. Curran, D. L. Azarnoff, and R. E. Bolinger, *J. Clin. Invest.*, *38*, 1251 (1959).
22. E. G. Dimond, J. Caravaca, and A. Benchimol, *Am. J. Clin. Nutr.*, *12*, 49 (1963).
23. A. W. Conklin, C. S. Skinner, T. L. Felten, and C. L. Sanders, *Toxicol. Lett.*, *11*, 199 (1982).
24. S. L. Hansard, II, C. B. Ammerman, and P. R. Henry, *J. Anim. Sci.*, *55*, 350 (1982).
25. L. R. Berg and W. W. Lawrence, *Poult. Sci.*, *50*, 1399 (1971).
26. C. H. Hill, *J. Nutr.*, *109*, 84 (1979).
27. L. E. Ousterhout and L. R. Berg, *Poult. Sci.*, *60*, 1152 (1981).
28. Y. S. M. Hafez and F. H. Kratzer, *Poult. Sci.*, *55*, 918 (1976).
29. C. H. Hill, *Nutr. Res.*, *Suppl. I*, 555 (1985).
30. C. H. Hill, *Biol. Trace Elem. Res.*, *23*, 1 (1990).
31. C. H. Hill, in *Chromium in Nutrition and Metabolism* (D. Shapcott and J. Huber, eds.), Elsevier, Amsterdam, 1979, p. 229 ff.
32. B. J. Stoecker and Y. C. Li, in *Trace Elements in Man and*

Animals 6 (L. S. Hurley, C. L. Keen, B. Lönnerdal, and R. B. Rucker, eds.), Plenum Press, New York, 1988, p. 553 ff.

33. C. H. Hill, *Biol. Trace Elem. Res.*, *23*, 17 (1990).
34. J. N. Hathcock, C. H. Hill, and G. Matrone, *J. Nutr.*, *82*, 106 (1964).
35. R. D. R. Parker and R. P. Sharma, *J. Environ. Pathol. Toxicol.*, *2*, 235 (1978).
36. T. L. Blalock and C. H. Hill, *Biol. Trace Elem. Res.*, *14*, 225 (1987).
37. C. H. Hill, *J. Nutr.*, *109*, 501 (1979).
38. J. L. Sell, C. Y. Davis, and S. E. Scheideler, *Poult. Sci.*, *65*, 138 (1986).
39. N. D. Chasteen, E. M. Lord, and H. J. Thompson, in *Frontiers in Bioinorganic Chemistry* (A. V. Xavier, ed.), VCH, Weinheim, 1986, p. 133 ff.
40. L. C. Cantley, Jr., M. D. Resh, and G. Guidotti, *Nature*, *272*, 552 (1978).
41. B. W. Patterson, S. L. Hansard, II, C. B. Ammerman, P. R. Henry, L. A. Zech, and W. R. Fisher, *Am. J. Physiol.*, *251*, R325 (1986).
42. B. R. Nechay, L. B. Nanninga, and P. S. E. Nechay, *Arch. Biochem. Biophys.*, *251*, 128 (1986).
43. I. G. Macara, K. Kustin, and L. C. Cantley, Jr., *Biochem. Biophys. Acta*, *629*, 95 (1980).
44. T. V. Hansen, J. Aaseth, and J. Alexander, *Arch. Toxicol.*, *50*, 195 (1982).
45. K. Kustin and D. L. Toppen, *Inorg. Chem.*, *12*, 1404 (1973).
46. L. C. Cantley, Jr., J. H. Ferguson, and K. Kustin, *J. Am. Chem. Soc.*, *100*, 5210 (1978).
47. H. Sakurai, S. Shimomura, and K. Ishizu, *Inorg. Chim. Acta*, *55*, 67L (1981).
48. N. D. Chasteen, J. K. Grady, and C. E. Holloway, *Inorg. Chem.*, *25*, 2754 (1986).
49. W. R. Harris, S. B. Friedman, and D. Silberman, *J. Inorg. Biochem.*, *20*, 157 (1984).
50. N. D. Chasteen, E. M. Lord, H. J. Thompson, and J. K. Grady, *Biochim. Biophys. Acta*, *884*, 84 (1986).
51. E. Sabbioni and E. Marafante, *Bioinorg. Chem.*, *9*, 389 (1978).

52. A. V. Roshchin, E. K. Ordzhonikidze, and I. V. Shalganova, *J. Hyg. Epidemiol. Microbiol. Immunol.*, *24*, 377 (1980).
53. J. Edel and E. Sabbioni, *J. Trace Elem. Electrolytes Health Dis.*, *2*, 23 (1988).
54. S. L. Hansard, II, C. B. Ammerman, P. R. Henry, and B. W. Patterson, *J. Anim. Sci.*, *62*, 804 (1986).
55. E. Sabbioni and E. Marafante, in *Trace Element Metabolism in Man and Animals* (J. McC. Howell, J. M. Gawthorne, and C. L. White, eds.), Australian Academy of Science, Canberra, 1981, p. 629 ff.
56. E. Sabbioni and E. Marafante, *J. Toxicol. Environ. Health*, *8*, 419 (1981).
57. J. Edel and E. Sabbioni, *Biol. Trace Elem. Res.*, *22*, 265 (1989).
58. S. Wallach, in *6th International Trace Element Symposium, Vol. 1, Molybdenum, Vanadium* (M. Anke, W. Baumann, H. Bräunlich, C. Brückner, B. Groppel, and M. Grün, eds.), Karl-Marx-Universität, Leipzig, 1989, p. 28 ff.
59. K. Heydorn, *Biol. Trace Elem. Res.*, *26/27*, 541 (1990).
60. N. Ward, D. Bryce-Smith, M. Minski, J. T. Zaaijman, and B. Pim, in *Trace Element Analytical Chemistry in Medicine and Biology*, Vol. 2 (P. Brätter and P. Schramel, eds.), Walter de Gruyter, Berlin, 1983, p. 483 ff.
61. A. R. Byrne and J. Kučera, in *Trace Elements in Man and Animals*, Vol. 7 (B. Momčilović, ed.), IMI, Zagreb, 1991, p. 25-18 ff.
62. E. Sabbioni, G. R. Nicolaou, R. Pietra, E. Beccaloni, E. Coni, A. Alimonti, and S. Caroli, *Biol. Trace Elem. Res.*, *26/27*, 757 (1990).
63. Y. Takagi, S. Matsuda, S. Imai, Y. Ohmori, T. Masuda, J. A. Vinson, M. C. Mehra, B. K. Puri, and A. Kaniewski, *Bull. Environ. Contam. Toxicol.*, *41*, 690 (1988).
64. G. J. Naylor, A. H. W. Smith, D. Bryce-Smith, and N. I. Ward, *Biol. Psychiat.*, *19*, 759 (1984).
65. R. Cornelis, L. Mees, J. Hoste, J. Ryckebusch, J. Versieck, and F. Barbier, in *Nuclear Activation Techniques in the Life Sciences 1978*, International Atomic Energy Agency, Vienna, 1979, p. 165 ff.
66. A. Alimonti, L. Fonarelli, E. Coni, G. B. Bolis, E. Cristallini, and S. Caroli, *Microchem. J.*, *46*, 191 (1992).

67. R. M. Parr, E. M. DeMaeyer, V. G. Iyengar, A. R. Byrne, G. F. Kirkbright, G. Schöch, L. Ninistö, O. Pineda, H. L. Vis, Y. Hofvander, and A. Omololu, *Biol. Trace. Elem. Res.*, *29*, 51 (1991).

68. N. I. Ward, T. D. Macmahon, and J. A. Mason, *J. Radiochem. Nucl. Chem.*, *113*, 501 (1987).

69. R. Cornelis, J. Versieck, L. Mees, J. Hoste, and F. Barbier, *Biol. Trace Elem. Res.*, *3*, 257 (1981).

70. V. Vrbič, J. Štupar, and A. R. Byrne, *Caries Res.*, *21*, 37 (1987).

71. O. Ishida, K. Kihira, Y. Tsukamoto, and F. Marumo, *Clin. Chem.*, *35*, 127 (1989).

72. R. E. Jervis, G. J. Evan, and P. J. Hewitt, *Nutr. Res.*, *Suppl. I*, 627 (1985).

73. F. Marumo, Y. Tsukamoto, S. Iwanami, T. Kishimoto, and S. Yamagami, *Nephron*, *38*, 267 (1984).

74. G. J. Naylor, in *Metal Ions in Neurology and Psychiatry* (S. Gabay, J. Harris, and B. T. Ho, eds.), Alan R. Liss, New York, 1985, p. 91 ff.

75. G. J. Naylor, F. M. Corrigan, A. H. W. Smith, P. Connelly, and N. I. Ward, *Br. J. Psychiat.*, *150*, 656 (1987).

76. N. I. Ward, D. Bryce-Smith, M. Minski, and W. B. Matthews, *Biol. Trace Elem. Res.*, *7*, 153 (1985).

77. E. N. Bello-Reuss, T. P. Grady, and D. C. Mazumdar, *Ann. Intern. Med.*, *91*, 743 (1979).

78. J. A. Navarro, V. A. Granadillo, O. Salgado, R. Garcia, B. Rodriguez-Iturbe, and R. A. Romero, in *Trace Elements in Man and Animals*, Vol. 7 (B. Momčilović, ed.), IMI, Zagreb, 1991, p. 29-9 ff.

79. S. Hosokawa and O. Yoshida, *J. Trace. Elem. Exp. Med.*, *2*, 140 (1989).

80. D. A. T. Dick, E. G. Dick, and G. J. Naylor, *J. Physiol. (London)*, *310*, 24P (1981).

81. D. A. T. Dick, G. J. Naylor, and E. G. Dick, *Psychol. Med.*, *12*, 533 (1982).

82. M. Simonoff, Y. Llabador, C. Hamon, B. Berdeu, G. Simonoff, C. Conri, B. Fleury, P. Couzigou, and A. Lucena, *J. Radioanal. Nucl. Chem.*, *113*, 107 (1987).

83. N. S. Harvey and N. I. Ward, *Lithium*, *2*, 91 (1991).

84. C. N. Corder, *J. Environ. Pathol. Toxicol.*, *6*, 61 (1986).

85. S. L. Hansard, II, C. B. Ammerman, K. R. Fick, and S. M. Miller, *J. Anim. Sci.*, *46*, 1091 (1978).

86. R. S. Gibson and M. S. Dewolfe, *Nutr. Rep. Int.*, *21*, 341 (1980).

87. J. Edel, R. Pietra, E. Sabbioni, E. Marafante, A. Springer, and L. Ubertalli, *Chemosphere*, *13*, 87 (1984).

88. E. Marafante, R. Pietra, J. Rade-Edel, and E. Sabbioni, in *Heavy Metals in the Environment*, CEP Consult Ltd., Edinburgh, 1981, p. 498 ff.

89. C. A. Strasia, in Vanadium: Essentiality and Toxicity in the Laboratory Rat, PhD thesis, Purdue University, Lafayette, IN, 1971.

90. D. L. Williams, Biological Value of Vanadium for Rats, Chickens, and Sheep, PhD thesis, Purdue University, Lafayette, IN, 1973.

91. K. Schwarz and D. B. Milne, *Science*, *174*, 426 (1971).

92. L. L. Hopkins, Jr. and H. E. Mohr, in *Newer Trace Elements in Nutrition* (W. Mertz and W. E. Cornatzer, eds.), Marcel Dekker, New York, 1971, p. 195 ff.

93. F. H. Nielsen and D. A. Ollerich, *Fed. Proc.*, *32*, 929 (1973).

94. F. H. Nielsen, *Food Technol.*, *28*, 38 (1974).

95. P. G. Reeves, F. H. Nielsen, and G. C. Fahey, Jr., *J. Nutr.*, *123*, 1939 (1993).

96. M. Anke, B. Groppel, K. Gruhn, T. Kośla, and M. Szilágyi, in *5. Spurenelement-Symposium New Trace Elements* (M. Anke, W. Baumann, H. Bräunlich, C. Brückner, and B. Groppel, eds.), Friedrich-Schiller-Universität, Jena, 1986, p. 1266 ff.

97. M. Anke, B. Groppel, K. Gruhn, M. Langer, and W. Arnhold, in *6th International Trace Element Symposium, Vol. 1, Molybdenum, Vanadium* (M. Anke, W. Baumann, H. Bräunlich, C. Brückner, B. Groppel, and M. Grün, eds.), Karl-Marx-Universität, Leipzig, 1989, p. 17 ff.

98. M. Anke, B. Groppel, and U. Krause, in *Trace Elements in Man and Animals*, Vol. 7 (B. Momčilović, ed.), IMI, Zagreb, 1991, p. 11-9 ff.

99. M. Anke, B. Groppel, W. Arnhold, M. Langer, and U. Krause, in *Trace Elements in Clinical Medicine* (H. Tomita, ed.), Springer-Verlag, Tokyo, 1990, p. 361 ff.

100. E. O. Uthus and F. H. Nielsen, *Magnesium Trace Elem.*, *9*, 219 (1990).
101. R. A. Poellot, C. D. Seaborn, and E. O. Uthus, *Proc. ND Acad. Sci.*, *46*, 75 (1992).
102. C. D. Seaborn, E. D. Mitchell, and B. J. Stoecker, *Magnesium Trace Elem.*, *10*, 327 (1991–92).
103. Y.-C. Li, M. M. Wang, W. K. Oladut, and B. J. Stoecker, *Fed. Proc.*, *44*, 1848 (1985).
104. R. Wever and B. E. Krenn, in *Vanadium in Biological Systems* (N. D. Chasteen, ed.), Kluwer, Dordrecht, 1990, p. 81 ff.
105. R. R. Eady, in *Vanadium in Biological Systems* (N. D. Chasteen, ed.), Kluwer, Dordrecht, 1990, p. 99 ff.
106. J. J. R. Fraústo da Silva, *Chem. Spec. Bioavail.*, *1*, 139 (1989).
107. H. Michibata and H. Sakurai, in *Vanadium in Biological Systems* (N. D. Chasteen, ed.), Kluwer, Dordrecht, 1990, p. 153 ff.
108. D. W. Boyd and K. Kustin, *Adv. Inorg. Biochem.*, *6*, 311 (1984).
109. B. R. Nechay, *Ann. Rev. Pharmacol. Toxicol*, *24*, 501 (1984).
110. F. H. Nielsen, in *Trace Elements in Human and Animal Nutrition*, Vol. 1 (W. Mertz, ed.), Academic Press, San Diego, 1987, p. 275 ff.
111. G. R. Willsky, in *Vanadium in Biological Systems* (N. D. Chasteen, ed.), Kluwer, Dordrecht, 1990, p. 1 ff.
112. F. H. Nielsen, in *Handbook on Metal-Ligand Interactions in Biological Fluids* (G. Berthon, ed.), Marcel Dekker, New York and Basel, in press.
113. H. A. Schroeder and J. J. Balassa, *J. Nutr.*, *92*, 245 (1967).
114. R. A. Peabody, S. Wallach, R. L. Verch, and J. Kraszeski, *Trace Subst. Environ. Health*, *10*, 441 (1976).
115. R. A. Peabody, S. Wallach, R. L. Verch, and M. L. Lifschitz, *Trace Subst. Environ. Health*, *11*, 297 (1977).
116. C. D. Hunt, *Anat. Rec.*, *190*, 427 (1978).
117. R. Söremark, *J. Nutr.*, *92*, 183 (1967).
118. O. D. Nemsadze, *Stomatologiya* (Moscow), *56*, 1 (1977).
119. F. H. Nielsen and E. O. Uthus, in *Vanadium in Biological Systems* (N. D. Chasteen, ed.), Kluwer, Dordrecht, 1990, p. 51 ff.

120. K.-H. W. Lau, H. Tanimoto, and D. J. Baylink, *Endocrinology, 123*, 2858 (1988).

121. B. R. Nechay, K. Norcross-Nechay, and P. S. E. Nechay, in *Spurenelement-Symposium, Vol. 1, Molybdenum, Vanadium* (M. Anke, W. Baumann, H. Bräunlich, C. Brückner, B. Groppel, and M. Grün, eds.), Karl-Marx-Universität, Leipzig, 1989, p. 10 ff.

122. E. Canalis, *Endocrinology, 116*, 855 (1985).

123. Y. Kato, M. Iwamoto, T. Koike, and F. Suzuki, *J. Cell Biol., 104*, 311 (1987).

124. W. L. McKeehan, K. A. McKeehan, S. L. Hammond, and R. G. Ham, *In Vitro, 3*, 399 (1977).

125. M. Yamaguchi, H. Oishi, and Y. Suketa, *Res. Exp. Med., 189*, 47 (1989).

126. S. I. Liochev and I. Fridovich, *Arch. Biochem. Biophys., 279*, 1 (1990).

127. I. G. Fantus, S. Kadota, G. Deragon, B. Foster, and B. I. Posner, *Biochemistry, 28*, 8864 (1989).

128. R. Wever and K. Kustin, *Adv. Inorg. Chem., 35*, 81 (1990).

129. P. J. Stankiewicz, A. Stern, and A. J. Davison, *Arch. Biochem. Biophys., 287*, 8 (1991).

130. T. G. Faulkner Hudson, in *Vanadium: Toxicology and Biological Significance*, Elsevier, Amsterdam, 1964.

131. National Research Council, *Vanadium*, National Academy of Sciences, Washington, DC, 1974.

132. National Research Council, 'Vanadium' in *Mineral Tolerance of Domestic Animals*, National Academy of Sciences, Washington, DC, 1980, p. 534 ff.

133. B. Lagerkvist, G. F. Nordberg, and V. Vouk, in *Handbook on the Toxicology of Metals*, 2nd ed. (L. Friberg, G. F. Nordberg, and V. Vouk, eds.), Elsevier, Amsterdam, 1986, p. 638 ff.

134. J. Somerville and B. Davies, *Am. Heart J., 64*, 54 (1962).

135. F. M. Corrigan, G. Ijomah, J. Holliday, D. F. Horrobin, E. R. Skinner, and N. I. Ward, *Trace Elem. Med., 9*, 90 (1992).

136. P. Boscolo, M. Carmignani, E. Sabbioni, A. R. Volpe, G. Giuliano, and P. Preziosi, *Int. J. Toxicol. Occup. Environ. Health, 2*, 21 (1993).

137. W. H. Evans, J. I. Read, and D. Caughlin, *Analyst, 110*, 873 (1985).
138. D. H. Buss and H. J. Rose, *Food Chem., 43*, 209 (1992).
139. K. Ikebe, *Bull. Environ. Contam. Toxicol., 49*, 509 (1992).
140. J. A. T. Pennington and J. W. Jones, *J. Am. Diet. Assoc., 87*, 1644 (1987).
141. Y. N. Berner, T. R. Shuler, F. H. Nielsen, C. Flombaum, S. A. Farkouh, and M. Shike, *Am. J. Clin. Nutr., 50*, 1079 (1989).
142. D. R. Myron, S. H. Givand, and F. H. Nielsen, *Agric. Food Chem., 25*, 297 (1977).
143. K. Ikebe and R. Tanaka, *Bull. Environ. Contam. Toxicol., 21*, 526 (1979).
144. L. L. Hopkins, Jr., H. L. Cannon, A. T. Miesch, R. M. Welch, and F. H. Nielsen, in *Geochemistry and the Environment, Vol. 2, The Relation of Other Trace Elements to Health and Disease*, National Academy of Sciences, Washington, DC, 1977, p. 93 ff.

17
Vanadium Compounds as Insulin Mimics

Chris Orvig,[1] *Katherine H. Thompson,*[2] *Mary Battell,*[2] *and John H. McNeill*[2]

[1]Medicinal Inorganic Chemistry Group, Department of Chemistry, University of British Columbia, 2036 Main Mall, Vancouver, BC, V6T 1Z1, Canada

[2]Faculty of Pharmaceutical Sciences, University of British Columbia, 2146 East Mall, Vancouver, BC, V6T 1Z3, Canada

1. INTRODUCTION

The appearance of this volume and its antecedent [1] as well as reviews of biological vanadium chemistry [2–7] attest to the recent interest in vanadium from a distinctly biological perspective. This has largely come about from the use of vanadate (VO_4^{3-}) as a probe in biochemistry over the last 30 years, with increasing interest in its pharmacological potential, as well as from the slow unfolding of vanadium's role in certain metalloproteins and its ubiquity in nature.

Vanadium exists in several valence states. In aerobic water at ambient pH and temperature, two oxidation states predominate: V(V) and V(IV), the former as orthovanadate, a mixture of HVO_4^{2-} and $H_2VO_4^-$ at pH 7; and the latter usually as the vanadyl ion, VO^{2+}, which undergoes a pH- and concentration-dependent oxidation by dissolved oxygen in pH >3 solutions. Vanadium is omnipresent in both plants and animals; however, its biological role is unclear [8]. In living systems, vanadium is an ultratrace element and is found in both the vanadate and vanadyl forms [9]. In man, the total body pool of vanadium is about 100 μg, present in virtually all cells at 0.1–1.0 μM concentrations (see Chap. 16). Approximately 90% of circulating vanadium is bound to proteins, with less than 10% in the free state [2,3]. Vanadium in plasma exists in both oxidation states as a result of oxygen tension and the presence of endogenous reducing agents (ascorbate, catecholamines), and forms complexes with transferrin preferentially [10]. Free vanadate may enter the cell via an anion transport system. There it is rapidly reduced to vanadyl by glutathione [11,12] and possibly by other low molecular weight peptides. Less than 1% of intracellular vanadium remains unbound [10].

Vanadium is not generally classified as an essential trace element. There has been no demonstration to date that an induced vanadium deficiency reproducibly and consistently impairs a discrete biological function in any animal; however, vanadium is essential for development and growth of whole animals and cells [2,3,13,14]. The fact that vanadium is a very active substance pharmacologically makes it difficult to establish its essentiality (see Chap. 16).

One of the unexpected byproducts of the intense investigation of vanadium over the past two decades has been the discovery that vanadium compounds have insulin-mimetic properties (reviewed in [5,15–17]), which will be the main topic of this chapter.

2. DIABETES, GLUCOSE, AND INSULIN ACTION

Insulin is a signaling hormone (a 6- to 7-kDa protein) which has numerous regulatory roles (for review, see [18]). The most important for the purposes of this review are the uptake from the circulatory system of amino acids, fatty acids, and glucose for storage as proteins in muscle, as triglycerides in adipose tissue, and as glycogen in muscle and liver, respectively. Insulin also serves to counteract other catabolic hormones responsible for the reversal of these storage reactions. The action of insulin is initiated by its specific and high-affinity binding to the surface receptors which are prevalent in mammalian cells [19]. Insulin receptors (IRs) are membrane-spanning tyrosine-specific protein kinases activated by insulin on the extracellular side to catalyze intracellular protein tyrosine phosphorylation, which is a key early step in the cascade of reactions leading to the uptake and metabolism of glucose [20]. Serine and threonine residues on the insulin receptor are also phosphorylated, but by other intracellular kinases [21]. Much remains to be clarified with respect to later steps in the insulin-signaling cascade.

Diabetes is characterized by an absolute or relative lack of insulin, and/or insulin resistance. Glucose uptake into peripheral tissues such as skeletal muscle and fat is impaired, as it is an insulin-dependent process [22]. Glucose tolerance, generally defined in terms of response to an orally administered glucose load, is also impaired. Glucose utilization in the energy-dependent processes within cells may also be abnormal.

Generally, diabetes mellitus in humans can be categorized as either type I or type II. Type I is insulin-dependent, while type II (in which insulin is present but the peripheral tissues are resistant to its actions) is classified as non-insulin-dependent [23]. In fact, some diabetic individuals fit into neither category, and treatment regimens may include exogenous insulin administration for both types. Like many proteins, insulin is not absorbed intact following oral administration, hence in insulin-deficient or resistant states, such as diabetes, it must be administered parenterally. The potential of vanadium compounds to mimic insulin while being administered orally would represent a significant advance over currently available therapies.

3. INSULIN-MIMETIC EFFECTS OF VANADIUM COMPOUNDS IN VITRO

Vanadium has many effects on enzyme action which have been demonstrated in both homogenized systems and in isolated cell preparations (for a complete survey of in vitro actions of vanadium, see Chap. 8). The first indications that addition of vanadium to an in vitro system resulted in insulin-like effects were reported in 1979 [24]. Vanadate was shown to stimulate glucose uptake and oxidation in rat adipocytes, glycogen synthesis in the rat diaphragm and liver, and to inhibit hepatic gluconeogenesis [24]. Subsequently, two independent investigative groups [25,26] recognized that the effects of vanadium on glucose transport and metabolism in intact adipocyte preparations were not linked to its well-documented [27] inhibitory effects on Na^+,K^+-ATPase, but the exact mechanism remained (and remains) unclear.

Vanadate inhibited glucose output from perfused rat liver [28] and increased glycolysis by affecting a variety of enzymes involved in the glycolytic pathway toward increased glucose breakdown [29]. For instance, vanadate inhibited glucose-6-phosphatase [30], stimulated 2,3-bisphosphoglycerate phosphatase [31], and counteracted glucagon effects on 6-phosphofructo-2-kinase activity [32] in rat adipocytes and hepatocytes. The insulin-like effects of vanadium extend to lipid metabolic pathways as well; vanadium inhibited lipolysis [28,33,34] and stimulated lipogenesis [35,36] in adipocytes.

Some in vitro effects of vanadium are not comparable to insulin's metabolic actions and may be tissue-dependent. In rat epitrochlearis muscle, vanadate mimicked insulin in increasing glucose uptake, glycogen synthesis and glucose oxidation but, unlike insulin, had no effect on protein synthesis or degradation [37]. In erythrocytes, vanadate reduced the rate of glycolysis by inhibiting glyceraldehyde-3-phosphate dehydrogenase [38]. In hepatocytes, vanadate activated glycogen synthase and inactivated phosphorylase [39], just the opposite of the effects observed in adipocytes [40]. Both increases [41,42] and decreases [40] in glycogen phosphorylase activity upon addition of vanadate have been reported. Similarly, amino acid transport has been reported to be inhibited by vanadate in the rat intestine [43], but stimulated in rat skeletal muscle [44,45].

A major effect of vanadate may be improved glucose transport in peripheral tissues. Vanadate's enhancement of glucose transport has been demonstrated in rat adipocytes [26], mouse brain [46], and rat skeletal muscle [47], an effect which could be attributed to an enhanced translocation of the insulin-regulatable transporter (GLUT-4) to the plasma membrane [48]. Vanadate increased glucose transporter expression in vitro in NIH 3T3 mouse fibroblasts [49] and in vivo in rat skeletal muscle [50]. Glucose uptake by isolated villus cells appeared to be stimulated by addition of 1 mM sodium orthovanadate [51]. However, vanadate inhibited glucose transport in rat intestine when added to the perfusate of an everted gut sac preparation [52], or when added as a supplement to the drinking water of diabetic and nondiabetic rats at concentrations up to 0.08 mM for 2–4 weeks [51,53].

4. VANADIUM COMPOUNDS AS INSULIN MIMICS IN VIVO

Orally administered vanadium has been shown to lower blood glucose in both chemically induced and spontaneously diabetic rats and mice [15,16,54–57]. Vanadium compounds mimic insulin in that they not only lower the average blood glucose levels of diabetic animals but also reverse some or all of the following symptoms in treated diabetic animals: hyperphagia, polydipsia [58,59], hyperlipidemia [60], and hypothyroidemia [61]. Plasma insulin levels are either not affected [62] or are lower than controls in treated diabetic and nondiabetic rats

[63,64]. As well, early studies showing improved cardiac function in vanadate-treated streptozotocin (STZ)–diabetic rats [54] have been corroborated by subsequent studies employing other vanadium compounds [65].

4.1. Chemically Induced Diabetes

A reproducible model of diabetes is obtained by administration of STZ to rats at doses of 45–75 mg/kg body weight. STZ-treated rats are insulinopenic, hyperphagic, and catabolic. The model does not completely parallel type I diabetes in humans in that STZ-diabetic rats can survive without administration of exogenous insulin; however, it is relatively simple, inexpensive, and reliable.

Moderate-to-good diabetic control has been obtained with several vanadium compounds in the STZ-diabetic rat, at oral doses of between 0.1 and 0.7 mmol/kg/day [62,66–70]. The dose of vanadium required to achieve good diabetic control varies with the initial diabetic state of the animal [70], the particular vanadium compound given [63,64], addition of other trace elements [71], and as yet undetermined individual factors [64,72]. Bendayan and Gingras [72] proposed a "dual sensitivity" to oral vanadium administration to explain the apparent biphasic nature of the glucose-lowering response to chronic vanadium treatment. This hypothesis has been borne out in subsequent studies [70]. More severely diabetic rats (as determined by blood glucose levels) required much higher doses of vanadyl sulfate (McNeill and coworkers, unpublished). Some diabetic rats which were not rendered normoglycemic by low-dose treatment with vanadium (<0.5 mmol/kg/day) nonetheless had significantly lower plasma triglyceride and cholesterol levels and improved glucose tolerance compared to untreated diabetic animals [63].

Oral vanadyl sulfate, 0.5 g/L in the drinking water (approximately 0.4 mmol/kg/day), restored liver enzymes (e.g., hexokinase, aldolase) to normal activity levels in diabetic rats [73]. Sorbitol accumulation was also reduced at this dose [74]. Cataract development, common in STZ-diabetic rats by 8 weeks following diabetic induction, was effectively prevented by oral vanadyl sulfate at a concentration of 1.0 g/L in the drinking water [75]. The time of the start of vanadium treatment after

STZ induction of diabetes (3, 7, and 17 days) did not affect the glucose-lowering potential of vanadyl sulfate [64]. Surprisingly, withdrawal of vanadyl sulfate for up to 13 weeks did not result in reversion to hyperglycemia in most animals [61,64].

Vanadyl sulfate treatment, up to 1.25 g/L in the drinking water, resulted in near-normal organ/body weight ratios of lung, heart, liver, kidney, and adrenal glands (as opposed to the significantly elevated ratios determined in untreated diabetic rats); histopathological tests showed no acceleration in morphological abnormalities in a variety of tissues, while overall mortality was 19% in the treated diabetic rats compared to 60% in the untreated diabetic animals [76]. Tissue vanadium levels varied from 6.5 to 15.1 μg/g in bone, from 3.6 to 7.3 μg/g in kidney, and from 0.2 to 0.5 μg/mL in plasma [76].

Reports of toxicity accompanying prolonged treatment with a variety of vanadium compounds [77–81] may greatly exaggerate the inherent danger of administering these compounds to diabetic animals. Mortality among vanadium-treated rats is often the result of a treatment protocol which does not allow for a several week period of adaptation to vanadium treatment in order to avoid a potentially lethal reduction in fluid intake [58,72]. Development of appropriately liganded vanadium and/or appropriately constituted vanadium coordination complexes may considerably improve the pharmacological potential of vanadium compounds as insulin mimetics by minimizing the dose required for effective diabetic control.

4.2. Orally Administered Vanadium Compounds in Spontaneously Diabetic Rats

The BioBreeding (BB) Wistar rat is a spontaneous model of diabetes which closely resembles type I diabetes. It is characterized by rapid onset of hyperglycemia between 60 and 120 days of age and by loss of pancreatic insulin production. Death due to ketoacidosis results unless parenteral insulin is administered [82]. Vanadyl sulfate reduced the dose of insulin required by diabetic BB rats but could not totally replace insulin in this model, thus demonstrating an insulin-sparing effect of vanadium compounds [83].

One of the most consistent findings in vanadium supplementation studies is the insulin-lowering effect of vanadium treatment [54,84]. This has particular relevance to type II diabetes, in which hyperinsulinemia is a frequent concomitant and may be a contributing factor in the development of secondary complications [85]. No animal model currently exists which completely mimics type II diabetes; the *ob/ob* and *db/db* mouse models are obese, hyperlipidemic, hyperinsulinemic, and eventually hyperglycemic [86], while the *fa/fa* Zucker rat is obese, hyperinsulinemic, and glucose-intolerant, but not usually hyperglycemic [82,87]. Vanadate decreased food and fluid intake, reduced weight gain, and improved hyperinsulinemia and glucose tolerance, while pancreatic insulin content remained unchanged, in female Zucker rats [88] over a treatment period of 3 months. Sodium orthovanadate at a dose of 0.1–0.2 mmol/kg/day was also shown to improve insulin-mediated glucose utilization, without an increase in glucose transporter levels [89]. In the *ob/ob* mouse [90], plasma glucose and insulin were halved within a week of beginning treatment with 0.3 g/L sodium orthovanadate in drinking water. Glucose tolerance was also shown to be improved in treated diabetic mice. In the *db/db* mouse [91], sodium orthovanadate, <0.6 g/L, also decreased food and fluid consumption and reduced weight gain compared to controls, and restored normoglycemia compared to diabetic animals. On the other hand, plasma triglycerides were not normalized in any of these models.

4.3. Vanadium Coordination Complexes In Vivo

Coordination complexes of vanadium which have been tested to date are summarized in Fig. 1. Ligands have been chosen to impart specific features to the resulting V complexes: improved lipophilicity (vanadyl cysteine methyl ester [92]; Naglivan [93]), improved oral absorption by passive diffusion (BMOV [69]), potentiation of in vitro insulin-mimetic effect (peroxovanadates [94–96]), or facilitation of transmembranal ion uptake (RL-252 and RL-262 [97]).

Naglivan, bis(*N*-octylcysteineamide)oxovanadium(IV), is insoluble in water. When administered in a suspension of 3% acacia gum by oral gavage at doses of 0.1–0.3 mmol/kg/day, Naglivan was effective in lowering blood glucose of STZ-diabetic rats to near-normal levels but had a very slow onset of action compared to treatment with inorganic

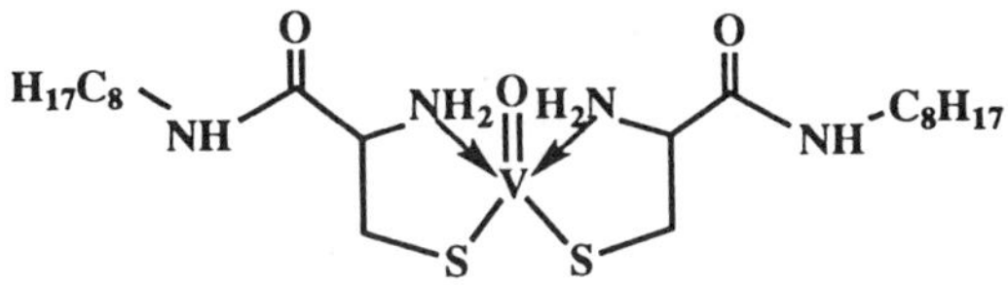

Naglivan

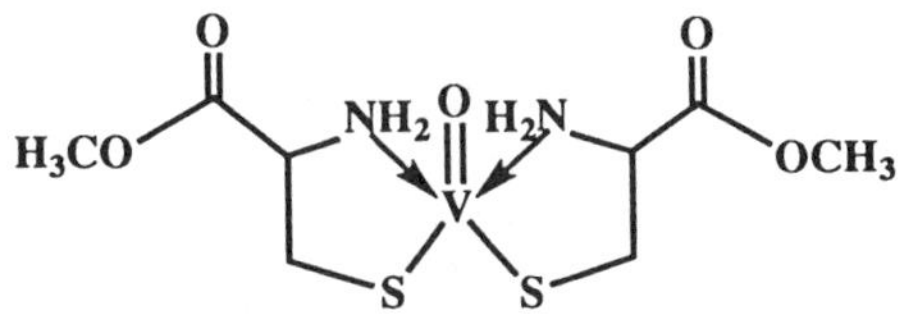

Bis(cysteine methyl ester)oxovanadium(IV)

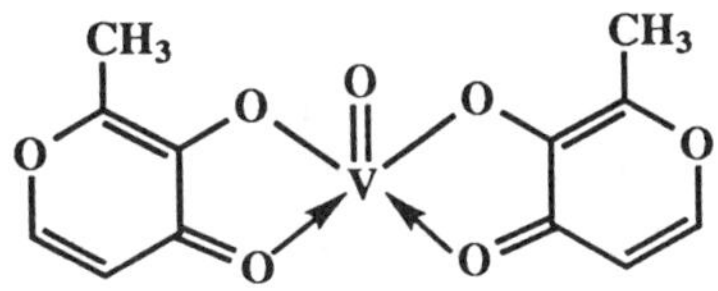

BMOV

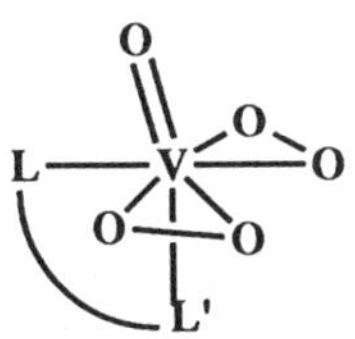

$[VO(O_2)_2(L\text{–}L')]^{n-}$
Ligandoxobis(peroxo)vanadate(V)

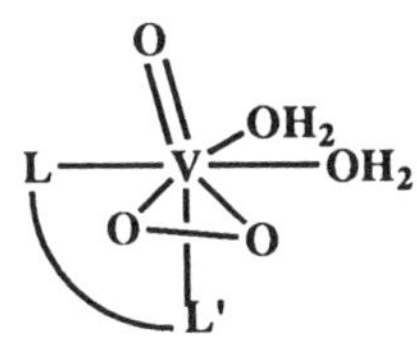

$[VO(O_2)(H_2O)_2(L\text{–}L')]^{n-}$
Ligandoxoperoxovanadium(V)

FIG. 1. Coordination complexes of vanadium(IV) and (V) recently reported to be insulin mimics.

vanadium. The advantages over vanadate or vanadyl treatment included lack of weight loss or reduction in food and fluid intake in control animals administered Naglivan [93].

A series of bis(ligand)oxovanadium(IV) complexes were studied in STZ-diabetic rats [92]. The ligands were salicylate, oxalate, malonate, tartrate, and cysteine methyl ester. The bis(cysteine methyl ester)vanadyl complex, at a dose of 10 mg V/kg body weight (0.2 mmol/kg), was slightly more effective than the others in lowering plasma glucose levels within 24 hr of administration by gavage. Vanadyl cysteine methyl ester was also effective in lowering blood glucose in diabetic rats when administered intraperitoneally at doses up to 0.2 mmol/kg/day, approximately the same as for intraperitoneal (ip) treatment with vanadyl sulfate [98]. Oral toxicity tests demonstrated no toxicity at 10 mg/kg/day vanadium (0.2 mmol/kg/day); however, at 100 mg/kg/day vanadium (2 mmol/kg/day), all test animals died of diarrhea within 4 days [92].

BMOV, bis(maltolato)oxovanadium(IV), was prepared as a vanadyl complex which would be water-soluble, charge-neutral, and readily available for gastrointestinal absorption by passive diffusion from drinking water solutions [69]. BMOV, at a dose of 0.4 mmol/kg/day, was effective in reducing blood glucose and lipid levels to near normal with no diarrhea and no mortality during a 6-month study [63,99]. The increased absorption of vanadium from BMOV administration was reflected in higher tissue concentrations of vanadium for a similar course of treatment with vanadyl sulfate [65]. An advantage of the apparent increased absorption is the more rapid onset of action of BMOV compared to other vanadium compounds tested so far.

A series of dihydroxamic acid chelators has been designed as hydrophobic carriers for the vanadyl ion [97]. In assays of lipogenic stimulation in rat fat cells, RL-252 and RL-262 were maximally effective at molar ratios of 10:1 vanadyl sulfate/chelator, suggesting a shuttle effect. Compounds were electrically neutral, lipid-soluble, and optionally chiral, and released the bound metal ion when treated with aqueous glutathione solutions.

Potentiation of vanadate complexes by addition of hydrogen peroxide was a serendipitous discovery [100,101]. New complexes, termed pervanadates, were found to stimulate lipogenesis, inhibit lipolysis, and promote protein synthesis in rat adipocytes at micromolar concentrations [102]. Addition of catalase to vanadate solution prior to the addition of peroxide abolished the synergism of vanadate with hydrogen peroxide; however, addition of catalase 15 min after the mixing of vanadate and peroxide preserved the synergism [94], suggesting that complexation of the peroxide to the vanadium had taken place [5]. Subsequently, a variety of new monoperoxovanadate and diperoxovanadate compounds have been synthesized as insulin mimetics [96,103], including monoperoxopicolinatovanadate and monoperoxo(pyridine-2,6-dicarboxylato)vanadate [mpV(pic) and mpV(2,6-pdc), respectively], which activated insulin receptor kinase activity and inhibited protein phosphotyrosine phosphatase (PPTPase) activity in rat liver endosomes at 5–80 μM concentrations, and also lowered plasma glucose at doses in the μmol/kg body weight range. The stimulation of insulin receptor kinase by pervanadate complexes presumably represents a different mechanism of insulin mimesis from vanadate or vanadyl, which appear to have post-IR (or cytoplasmic receptor kinase) stimulatory effects (vide infra).

5. ACUTE STUDIES AND GLUCOSE CLAMP INVESTIGATIONS

Vanadium's insulin-mimetic effects in vivo have been confirmed by acute and short-term studies demonstrating increasing percentage achievement of normoglycemia with increasing doses of vanadium for a variety of compounds [63]. At doses ranging from 16 to 280 μmol/kg body weight by intraperitoneal injection, plasma glucose was lowered to <8.3 mM within 8 hr in progressively more rats (all initially diabetic) as the dose was increased [63]. Tissue levels of vanadium ranged from 0.3 to 59 μg/g wet weight for kidney and 0.1 to 4.8 μg/mL for plasma (unpublished).

In short-term euglycemic clamp studies, indwelling venous and arterial catheters permit maintenance of plasma glucose levels in response to exogenous insulin by infusion of tritiated glucose. Using this technique, sodium metavanadate, 0.2 g/L [104], as well as vanadyl sulfate, 0.5 g/L [59], has been shown to suppress hepatic glucose production and improve glucose utilization induced by submaximal or maximal insulin levels, compared to control rats. In partially pancreatectomized Sprague-Dawley rats, sodium metavanadate, 0.2 g/L (36–39 μmol/kg/day), lowered fed and fasted plasma glucose levels without altering plasma insulin levels or affecting body weight gain [105].

Vanadium has also been combined with lithium [71,106], and zinc, lithium as well as magnesium in combination [106], to potentiate the effect of vanadium alone in improving glucose tolerance. Pancreatectomized rats given sodium vanadate, 0.05 g/L, and lithium carbonate, 0.3 g/L, in the drinking water for 2–3 weeks prior to euglycemic clamp investigation showed improved glucose uptake and skeletal muscle glycogenic rate, compared to untreated pancreatectomized rats [106]. Addition of magnesium sulfate, 0.3 g/L, and zinc sulfate, 0.3 g/L, to some of the treated rats resulted in further slight but significant improvement in tissue glucose uptake [106]. In a separate experiment, diabetic rats given sodium vanadate, 0.05 g/L, and lithium carbonate, 0.3 g/L, in the drinking water were rendered normoglycemic within 4 days and had partially restored liver and kidney superoxide dismutase activities following 16 days of treatment [71], demonstrating an ameliorative effect on antioxidant status in lithium- and vanadate-treated diabetic rats.

6. MECHANISM OF ACTION

A postulated mechanism of vanadium's in vitro effects is that vanadate behaves as a phosphate analog, stimulating protein-tyrosine phosphorylation by virtue of its inhibitory actions on PPTPases [107–109]. Routine addition of the vanadate ion to cell lysates and tyrosine kinase assays is testimony to the ability of vanadium to preserve the phosphotyrosine content of cells [109].

In early studies, vanadate activated autophosphorylation of solubilized IRs, but not serine or threonine residues of the receptors, in a fashion analogous to insulin [40,110,111]. Vanadate also stimulated tyrosine kinase activity of the IR β subunit [110,112]. Subsequently, it was found that vanadate was equally effective in stimulating glucose metabolism in rat fat cells when half of the IRs had been inactivated by insulin overstimulation [113]. In addition, oral vanadate failed to change IR kinase activity while exerting an insulin-like glucose-lowering effect in STZ-diabetic rats [114]. In rat adipocytes, vanadate appeared to bypass IR kinase inhibition, as shown by insulin-independent stimulation of glucose transport and metabolism [115]. Thus, current evidence suggests a postreceptor mechanism of action for vanadium in stimulating glucose utilization, perhaps involving an alternative, cytosolic (i.e., nonreceptor) protein tyrosine kinase which is stimulated preferentially by vanadium and is insulin-independent [17,116,117].

As intracellular vanadium appears to be largely vanadyl [118,119], which is not a potent phosphotyrosine phosphatase inhibitor [109,120], there is additional reason to believe that some alternative mechanism of action, at a later point in the insulin-signaling cascade of reactions, is most relevant to the observed insulin-mimetic effects of vanadate and vanadyl ions. In the case of pervanadate ions [95,96], or vanadate plus superoxide radicals [121], or vanadate with added hydrogen peroxide [122,123], increased autophosphorylation and activation of the insulin receptor, with consequent stimulation of glucose uptake, appears to be the mechanism. However, effects of intracellular vanadium on calcium influx [124,125] as well as intracellular and intravesicular pH modification [15,45,126] have not been ruled out as important factors in the mechanism of action of any of the vanadium ions as insulin-mimetic agents.

7. SUMMARY AND CONCLUSIONS

That vanadium compounds act in an insulin-mimetic fashion both in vitro and in vivo has been well established. Both inorganic and organic vanadium compounds have been shown to lower plasma glucose levels, increase peripheral glucose uptake, improve insulin sensitivity, decrease plasma lipid levels, and normalize liver enzyme activities in a variety of animal models of both type I and type II diabetes. Vanadium treatment of diabetic animals does not restore plasma insulin levels but may spare pancreatic insulin. Elucidation of the mechanism(s) of action and potentiation of vanadium's insulin-mimetic effect by appropriate ligand binding would seem to be the highest priorities for future investigation.

ABBREVIATIONS AND DEFINITIONS

BB	BioBreeding
BMOV	bis(maltolato)oxovanadium(IV)
GLUT-4	insulin-activated transporter protein which is encoded by a distinct gene
ip	intraperitoneal

IR	insulin receptor
iv	intravenous
Naglivan	bis(*N*-octylcysteineamide)oxovanadium(IV)
NIH	U.S. National Institutes of Health
PPTPase	protein phosphotyrosine phosphatase
RL-252/RL-262	dihydroxamic acid derivatives
STZ	streptozotocin

The following codes are genetic designations for homozygous mutants:

db/db	diabetic mice
fa/fa	fatty Zucker rats
ob/ob	obese mice

REFERENCES

1. N. D.Chasteen (ed.), *Vanadium in Biological Systems*, Kluwer, Dordrecht, 1990.
2. B. R. Nechay, *Ann. Rev. Pharmacol. Toxicol., 24*, 501 (1984).
3. B. R. Nechay, L. B. Nanninga, P. S. E. Nechay, R. L. Post, J. J. Grantham, I. G. Macara, L. F. Kubena, T. D. Phillips, and F. H. Nielsen, *Fed. Proc., 45*, 123 (1986).
4. R. Wever and K. Kustin, *Adv. Inorg. Chem., 35*, 81 (1990).
5. B. I. Posner, A. Shaver, and I. G. Fantus, in *New Antidiabetic Drugs* (C. J. Bailey and P. R. Flatt, eds.), Smith Gordon, 1990, p. 107.
6. A. R. Butler and C. J. Carrano, *Coord. Chem. Rev., 109*, 61 (1991).
7. D. Rehder, *Angew. Chem., Int. Ed. Engl., 103*, 148 (1991).
8. I. G. Macara, *Trends Biochem. Sci., 5*, 92 (1980).
9. F. H. Nielsen and E. O. Uthus, in *Vanadium in Biological Systems* (N. D. Chasteen, ed.), Kluwer, Dordrecht, 1990, p. 51.
10. N. D. Chasteen, J. K. Grady, and C. E. Holloway, *Inorg. Chem., 25*, 2754 (1986).
11. H. Sakurai, S. Shimomura, K. Fukuzawa, and K. Ishizu, *Biochem. Biophys. Res. Commun., 90*, 293 (1980).
12. K. A. Rubinson, *Proc. Roy. Soc. London. Ser. B, 212*, 65 (1981).

13. N. D. Chasteen, *Struct. Bonding (Berlin)*, *53*, 105 (1983).
14. R. J. French and P. J. H. Jones, *Life Sci.*, *52*, 339 (1992).
15. Y. Shechter, J. Meyerovitch, Z. Farfel, J. Sack, R. Bruck, S. Bar-Meir, S. Amir, H. Degani, and S. J. D. Karlish, in *Vanadium in Biological Systems* (N. D. Chasteen, ed.), Kluwer, Dordrecht, 1990, p. 129.
16. Y. Shechter, *Diabetes*, *39*, 1 (1990).
17. Y. Shechter and A. Shisheva, *Endeavour New Series*, *17*, 27 (1993).
18. M. P. Czech, *Ann. Rev. Biochim.*, *46*, 359 (1977).
19. C. R. Kahn and M. F. White, *J. Clin. Invest.*, *82*, 1151 (1988).
20. M. F. White, S. E. Shaleson, H. Keutmann, and C. R. Kahn, *J. Biol. Chem.*, *263*, 2969 (1988).
21. S. Takayama, M. F. White, and C. R. Kahn, *J. Biol. Chem.*, *263*, 3440 (1988).
22. G. E. Lienhard, J. W. Slot, D. E. James, and M. M. Mueckler, *Sci. Am.*, *267*, 86 (1992).
23. W. T. Garvey, *Diab. Care*, *15*, 396 (1992).
24. E. L. Tolman, E. Barris, M. Burns, A. Pansisni, and R. Partridge, *Life Sci.*, *25*, 1159 (1979).
25. Y. Shechter and S. J. D. Karlish, *Nature*, *284*, 556 (1980).
26. G. R. Dubyak and A. Kleinzeller, *J. Biol. Chem.*, *255*, 5306 (1980).
27. L. C. Cantley, Jr., L. Josephson, R. Warner, M. Yanagisawa, C. Lechene, and G. Giudotti, *J. Biol. Chem.*, *252*, 7421 (1977).
28. R. Bruck, H. Prigozin, Z. Krepel, P. Rotenberg, Y. Shechter, and S. Bar-Meir, *Hepatology*, *14*, 540 (1991).
29. J. E. Rodriguez-Gil, A. M. Gomez-Foix, C. Fillat, F. Bosch, and J. J. Guinovart, *Diabetes*, *40*, 1355 (1991).
30. J. Singh, R. C. Nordlie, and R. A. Jorgenson, *Biochim. Biophys. Acta*, *678*, 477 (1981).
31. G. L. Mendz, S. J. Hyslop, and P. W. Kuchel, *Arch. Biochem. Biophys.*, *276*, 160 (1990).
32. A. M. Gomez-Foix, J. E. Rodriguez-Gil, C. Fillat, J. J. Guinovart, and F. Bosch, *Biochem. J.*, *255*, 507 (1988).
33. H. Degani, M. Gochin, S. J. D. Karlish, and Y. Schechter, *Biochemistry*, *20*, 5795 (1981).

34. W. C. Duckworth, S. S. Solomon, J. Liepnieks, F. G. Hamel, S. Hand, and D. E. Peavy, *Endocrinology*, *122*, 2285 (1988).
35. L. Aguis and W. J. Vaartjes, *Biochem. J.*, *202*, 791 (1982).
36. J. Castro, A. Maquedano, and M. Olive, *Biochem. Inter.*, *9*, 413 (1984).
37. A. S. Clark, J. M. Fagan, and W. E. Mitch, *Biochem. J.*, *232*, 273 (1985).
38. J. E. Benabe, L. A. Echegoyen, B. Pastrana, and M. Martinez-Maldonado, *J. Biol. Chem.*, *262*, 9555 (1987).
39. F. Bosch, J. Arino, A. M. Gomez-Foix, and J. J. Giunovart, *J. Biol. Chem.*, *262*, 218 (1987).
40. S. Tamura, T. A. Brown, R. E. Dubler, and J. Larner, *Biochem. Biophys. Res. Commun.*, *113*, 80 (1983).
41. T. K. Jackson, A. I. Salhanick, J. D. Sparks, C. E. Sparks, M. Bolognino, and J. M. Amatruda, *Diabetes*, *37*, 1234 (1988).
42. H. V. Strout, P. P. Vicario, R. Saperstein, and E. E. Slater, *Endocrinology*, *124*, 1918 (1989).
43. J. J. Hajjar, M. P. Dobish, and T. K. Tomicic, *Arch. Toxicol.*, *63*, 29 (1989).
44. E. J. Henriksen, *Am. J. Physiol.*, *261*, C608 (1991).
45. P. Munoz, A. Guma, M. Camps, M. Furriols, X. Testar, M. Palacin, and A. Zorzano, *J. Biol. Chem.*, *267*, 10381 (1992).
46. S. Amir, J. Meyerovitch, and Y. Shechter, *Brain Res.*, *419*, 392 (1987).
47. N. Okumura and T. Shimazu, *J. Biochem. (Tokyo)*, *112*, 107 (1992).
48. M. R. Paquet, R. J. Romanek, and R. J. Sargeant, *Mol. Cell. Biochem.*, *109*, 149 (1992).
49. K. G. Mountjoy and J. S. Flier, *Endocrinology*, *127*, 2025 (1990).
50. H. V. Strout, P. P. Vicario, C. Biswis, R. Saperstein, E. J. Brady, P. F. Pilch, and J. Berger, *Endocrinology*, *126*, 2728 (1990).
51. J. J. Hajjar, M. P. Dobish, and T. K. Tomicic, *Proc. Soc. Exp. Biol. Med.*, *190*, 35 (1989).
52. G. L. Kellett and E. D. Barker, *Biochim. Biophys. Acta*, *979*, 311 (1989).

53. K. L. Madsen, V. M. Porter, and R. N. Fedorak, *Diabetes*, *42*, 1126 (1993).
54. C. E. Heyliger, A. G. Tahiliani, and J. H. McNeill, *Science*, *227*, 1474 (1985).
55. S. Ramanadham, J. J. Mongold, R. W. Brownsey, G. H. Cros, and J. H. McNeill, *Am. J. Physiol.*, *257*, H904 (1989).
56. S. Ramanadham, G. H. Cros, J. J. Mongold, J. J. Serrano, and J. H. McNeill, *Can. J. Physiol. Pharmacol.*, *68*, 486 (1990).
57. D. Heffetz, I. Bushkin, R. Dror, and Y. Zick, *J. Biol. Chem.*, *265*, 2896 (1990).
58. S. Pugazhenthi and R. L. Khandelwal, *Diabetes*, *39*, 821 (1990).
59. N. Venkatesan, A. Avidan, and M. B. Davidson, *Diabetes*, *40*, 492 (1991).
60. J. J. Mongold, G. H. Cros, L. Vian, A. Tep, S. Ramanadham, G. Siou, J. Diaz, J. H. McNeill, and J. J. Serrano, *Pharmacol. Toxicol.*, *67*, 192 (1990).
61. S. Ramanadham, R. W. Brownsey, G. H. Cros, J. J. Mangold, and J. H. McNeill, *Metabolism*, *38*, 1022 (1989).
62. N. Sekar, A. Kanthasamy, S. Williams, S. Subramanian, and S. Govinsasamy, *Pharmacol. Res.*, *22*, 207 (1990).
63. V. G. Yuen, C. Orvig, and J. H. McNeill, *Can. J. Physiol. Pharmacol.*, *71*, 263 (1993).
64. M. C. Cam, R. A. Pederson, R. W. Brownsey, and J. H. McNeill, *Diabetologia*, *36*, 218 (1993).
65. V. G. Yuen, C. Orvig, K. H. Thompson, and J. H. McNeill, *Can. J. Physiol. Pharmacol.*, *71*, 270 (1993).
66. J. Meyerovitch, Z. Farfel, J. Sack, and Y. Shechter, *J. Biol. Chem.*, *262*, 6658 (1987).
67. R. A. Pederson, S. Ramanadham, A. M. M. Buchan, and J. H. McNeill, *Diabetes*, *38*, 1390 (1989).
68. O. Blondel, J. Simon, B. Chevalier, and B. Portha, *Am. J. Physiol.*, *258*, E459 (1990).
69. J. H. McNeill, V. G. Yuen, H. R. Hoveyda, and C. Orvig, *J. Med. Chem.*, *35*, 1489 (1992).
70. K. H. Thompson, J. Leichter, and J. H. McNeill, *Biochem. Biophys. Res. Commun.*, *197*, 1549 (1993).

71. P. Srivastava, A. K. Saxena, R. K. Kale, and N. Z. Baquer, *Res. Commun. Chem. Path. Pharmacol.*, *80*, 283 (1993).

72. M. Bendayan and D. Gingras, *Diabetologia*, *32*, 561 (1989).

73. A. K. Saxena, P. Srivastava, and N. Z. Baquer, *Eur. J. Pharmacol.*, *216*, 123 (1992).

74. A. K. Saxena, P. Srivastava, R. K. Kale, and N. Z. Baquer, *Biochem. Int.*, *26*, 59 (1992).

75. K. H. Thompson and J. H. McNeill, *Res. Commun. Chem. Path. Pharmacol.*, *80*, 187 (1993).

76. S. Dai, K. H. Thompson, E. Vera, and J. H. McNeill, *Pharmacol. Toxicol.*, in press.

77. J. L. Domingo, M. Gomez, J. M. Llobet, J. Corbella, and C. L. Keen, *Pharmacol. Toxicol.*, *68*, 249 (1991).

78. J. L. Domingo, A. Ortega, J. M. Llobet, and C. L. Keen, *Trace Elements Med.*, *8*, 181 (1991).

79. J. L. Domingo, M. Gomez, J. M. Llobet, J. Corbella, and C. L. Keen, *Toxicology*, *66*, 279 (1991).

80. J. L. Domingo, D. J. Sanchez, M. Gomez, J. M. Llobet, and J. Corbella, *Res. Commun. Chem. Path. Pharmacol.*, *75*, 369 (1992).

81. J. L. Domingo, M. Gomez, D. J. Sanchez, J. M. Llobet, and C. L. Keen, *Life Sci.*, *50*, 1311 (1992).

82. R. H. J. Bell and R. J. Hye, *J. Surg. Res.*, *35*, 433 (1983).

83. M. L. Battell, V. G. Yuen, and J. H. McNeill, *Pharmacol. Commun.*, *1*, 291 (1992).

84. A. K. Saxena, P. Srivastava, R. K. Kale, and N. Z. Baquer, *Biochem. Pharmacol.*, *45*, 539 (1993).

85. G. M. Reaven, *Annu. Rev. Med.*, *44*, 121 (1993).

86. D. L. Coleman, *Diabetes*, *31 (Suppl. 1)*, 1 (1982).

87. C. J. Bailey and P. R. Flatt, *Diab./Metab. Rev.*, *9 (Suppl. 1)*, 43 (1993).

88. S. M. Brichard, A. M. Pottier, and J. C. Henquin. *Endocrinology*, *125*, 2510 (1989).

89. S. M. Brichard, F. Assimacopoulos-Jeannet, and B. Jeanrenaud, *Endocrinology*, *131*, 311 (1992).

90. S. M. Brichard, C. J. Bailey, and J.-C. Henquin, *Diabetes*, *39*, 1326 (1990).

91. S. Pugazhenthi, J. F. Angel, and R. L. Khandelwal, *Metabolism*, *40*, 941 (1991).
92. H. Sakurai, K. Tsuchiya, M. Nukatsuka, J. Kawada, S. Ishikawa, H. Yoshida, and M. Komatsu, *J. Clin. Biochem. Nutrit.*, *8*, 193 (1990).
93. M. C. Cam, G. H. Cros, J.-J. Serrano, R. Lazaro, and J. H. McNeill, *Diab. Res. Clin. Prac.*, *20*, 111 (1993).
94. S. Kadota, G. Fantus, G. Deragon, H. J. Guyda, B. Hersh, and B. I. Posner, *Biochem. Biophys. Res. Commun.*, *147*, 259 (1987).
95. A. Shisheva and Y. Shechter, *Endocrinology*, *133*, 1562 (1993).
96. B. I. Posner, R. Faure, J. W. Burgess, A. P. Bevan, D. Lachance, G. Zhang-Sun, I. G. Fantus, J. B. Ng, D. A. Hall, B. Soo Lum, and A. Shaver, *J. Biol. Chem.*, *269*, 4596 (1994).
97. Y. Shechter, A. Shisheva, R. Lazar, J. Libman, and A. Shanzer, *Biochemistry*, *31*, 2063 (1992).
98. H. Sakurai, K. Tsuchiya, M. Nukatsuka, M. Sofue, and J. Kawada, *J. Endocrinology*, *126*, 451 (1990).
99. S. Dai, V. G. Yuen, C. Orvig, and J. H. McNeill, *Pharmacol. Commun.*, *3*, 311 (1993).
100. S. Kadota, I. G. Fantus, B. Hersh, and B. I. Posner, *Biochem. Biophys. Res. Commun.*, *138*, 174 (1986).
101. S. Kadota, I. G. Fantus, G. Deragon, H. J. Guyda, and B. I. Posner, *J. Biol. Chem.*, *262*, 8252 (1987).
102. I. G. Fantus, S. Kadota, G. Deragon, B. Foster, and B. I. Posner, *Biochemistry*, *28*, 8864 (1989).
103. A. Shaver, J. B. Ng, D. A. Hall, B. Soo Lum, and B. I. Posner, *Inorg. Chem.*, *32*, 3109 (1993).
104. O. Blondel, D. Bailbe, and B. Portha, *Diabetologia*, *32*, 185 (1989).
105. R. Cordera, F. Andraghetti, R. A. DeFronzo, and L. Rossetti, *Endocrinology*, *126*, 2177 (1990).
106. L. Rossetti, A. Giaccari, E. Klein-Robbenhaar, and L. R. Vogel, *Diabetes*, *39*, 1243 (1990).
107. G. Swarup, K. V. J. Speeg, S. Cohen, and D. I. Garbers, *J. Biol. Chem.*, *257*, 7298 (1982).
108. A. S. Tracey and M. J. Gresser, *Proc. Natl. Acad. Sci. USA*, *83*, 609 (1986).
109. J. A. Gordon, *Meth. Enzmol.*, *201*, 477 (1991).

110. A. Ueno, N. Arakaki, Y. Takeda, and H. Fujio, *Biochem. Biophys. Res. Commun., 144,* 11 (1987).

111. R. Gherzi, C. Caratti, G. Andraghetti, S. Bertolini, A. Montemurro, G. Sesti, and R. Cordera, *Biochem. Biophys. Res. Commun., 152,* 1474 (1988).

112. D. M. Smith and G. J. Sale, *Biochem. J., 256,* 903 (1988).

113. A. Green, *Biochem. J., 238,* 663 (1986).

114. R. A. Mooney, K. L. Bordwell, S. Luhowskyj, and J. E. Casnelle, *Endocrinology, 124,* 422 (1989).

115. A. Shisheva and Y. Shechter, *Biochemistry, 31,* 8059 (1992).

116. A. Srivastava, *Int. J. Biochem., 11,* 1229 (1990).

117. A. Shisheva and Y. Shechter, *J. Biol. Chem., 268,* 6463 (1993).

118. I. G. Macara, K. Kustin, and L. C. Cantley, Jr. *Biochim. Biophys. Acta, 629,* 95 (1980).

119. M. Bruech, M. E. Quintanilla, W. Legrum, J. Koch, K. J. Netter, and G. F. Fuhrmann, *Toxicology, 31,* 283 (1984).

120. L. C. Cantley, Jr. and P. Aisen, *J. Biol. Chem., 254,* 1781 (1979).

121. S. Trudel, M. R. Paquet, and S. Grinstein, *Biochem. J., 276,* 611 (1991).

122. Y. Zick and R. Sagi-Eisenberg, *Biochemistry, 29,* 10240 (1990).

123. D. Hecht and Y. Zick, *Biochem. Biophys. Res. Commun., 188,* 773 (1992).

124. D. M. Delfert and J. M. McDonald, *Arch. Biochem. Biophys., 241,* 665 (1985).

125. A. Zhang, Z.-Y. Gao, P. Gilon, M. Nenquin, G. Drews, and J.-C. Henquin, *J. Biol. Chem., 266,* 21649 (1991).

126. D. Cassel, Y.-X. Zhuang, and L. Glaser, *Biochem. Biophys. Res. Commun., 118,* 675 (1984).

18

Antitumor Activity of Vanadium Compounds

Cirila Djordjevic

Department of Chemistry, The College of William and Mary, P.O. Box 8795, Williamsburg, Virginia 23185-8795, USA

1. INTRODUCTION

Only a small number of vanadium compounds have been tested for antitumor activity, and few of these have been found antitumor-active. In decades to come this may change because the number of studies of the chemistry, biochemistry, and biology of vanadium has recently grown immensely. Several reviews and books have appeared in the literature [1–11]. At the center of attention are biological and biochemical aspects of vanadium chemistry. However, the solid state properties of the compounds of this versatile light transition metal continue to fascinate chemists, biochemists, and physicists [12].

Vanadium metal and its derivatives have long had important applications in industry [13]. Alloy steels, heterogeneous and homogeneous catalysis, and environmental concerns [14] all request background research in synthetic, mechanistic, and analytical chemistry, and they have stimulated many investigations. Greatly improved techniques of vanadium analysis have contributed to a better assessment of vanadium concentrations, which leads to a better understanding of vanadium interactions in bioorganisms. Vanadium is not new to medicinal chemistry [5,9]. In the early 1900s, for example, vanadium compounds were used for treating tuberculosis, anemia, and diabetes, and were present in antiseptics or tonics. In 1950s and 1960s vanadium compounds were found to lower blood cholesterol levels, impair growth and reproduction, and disturb the metabolism of lipids. Interest in vanadium compounds increased in the 1980s because some vanadium complexes can act as insulin-mimetic agents [9,15] (see also Chap. 17). Antitumor activity of a few vanadium compounds was discovered but has so far not aroused a wide interest. However, scarce data on a small number of active complexes reveal some relations between chemical composition, antitumor activity, and toxicity dose, which deserve attention and require further investigations.

2. CHEMICAL FEATURES OF VANADIUM RELEVANT TO ANTITUMOR ACTIVITY

2.1. Redox Potentials and Radical Generation Ability

Maximum oxidation state of vanadium is V(V), with a d^0 electron configuration. In aqueous solutions V(V) exists in a pH-dependent equilibrium with V(IV), d^1 [2,9,16]. A number of monomeric and polymeric oxovanadium species of V(V) and V(IV) can be present in aqueous media (Table 1), and their composition depends on pH and vanadium concentration [9,16]. The thermodynamic standard electrode potentials of V(V) and V(IV) ions are affected by the pH and by the ligands that enter oxovanadium ligand spheres in a given environment. Thermodynamic potentials, measured at pH 0 in pure 1.0 M water solutions at 25°C (Table 2), numerically hardly apply to processes in the living matter at physiological pH, where metal ions are surrounded by a sea of potential donor atoms attached to small or large bio-ions, biomolecules, or biopolymers. Ligand spheres created under such conditions tune the potential of vanadium ions and govern the energy of electron transfer (e.g., see Fe(III)/Fe(II) potentials in Table 2).

Thermodynamic potentials in Table 2 indicate that pH-dependent V(V)/V(IV) electron transfer occurs in the vicinity of biologically rele-

TABLE 1

Predominant Species at Low Vanadium Concentration ($\leqslant 10^{-6}$ M) [9,16]

	Aqueous solutions		Aqueous solutions with peroxides
pH	V(V)	V(IV)	V(V)
<3	VO_2^{2+}	VO^{2+}	$VO(O_2)^+$
~3–7	$[VO_2(OH)_2]^- \equiv H_2VO_4^-$	$VO(OH)^+$ $[VO(OH)]_2^{2+}$	$H_2VO_2(O_2)_2^-$
~7–13	$[VO_3(OH)]^{2-} \equiv HVO_4^{2-}$ $[O(VO_3)_2]^{4-} \equiv V_2O_7^{4-}$	$VO(OH)_2(s)$ + (?)	$HVO_2(O_2)_2^{2-}$ $[O\{VO(O_2)_2\}_2]^{4-}$
>13	VO_4^{3-}	$VO(OH)_3^-$	$V(O_2)_4^{3-}$

TABLE 2

Selected Standard Thermodynamic Potentials (*V*)

Reaction	E^0 (pH = 0)	$E^{0\prime}$ (pH = 7)
$VO_2^+ + 2H^+ + e^- \rightarrow VO^{2+} + H_2O$	1.00	0.17
$VO^{2+} + 2H^+ + e^- \rightarrow V^{3+} + H_2O$	0.36	−0.49
$V^{3+} + e^- \rightarrow V^{2+}$	−0.26	
$O_2 + 2H^+ + 2e^- \rightarrow H_2O_2$	0.68	0.27
$O_2 + 4H^+ + 4e^- \rightarrow 2H_2O$	1.23	0.82
$H_2O_2 + 2H^+ + 2e^- \rightarrow 2H_2O$	1.76	1.35
$HO_2 + H^+ + e^- \rightarrow H_2O_2$	1.44	1.01
$Fe^{3+} + e^- \rightarrow Fe^{2+}$	0.77	
$Fe(CN)_6^{3-} + e^- \rightarrow Fe(CN)_6^{4-}$	0.36	
$Fe\ (o\text{-}phen)_3^{3+} + e^- \rightarrow Fe\ (o\text{-}phen)_3^{2+}$	1.1	

o-phen = 1,10-phenanthroline.

vant potentials of Fe(III)/Fe(II) and O_2/H_2O_2 reduction. Figure 1 is a simplified illustration of the pH-dependent decrease of relevant thermodynamic potentials, and the pH "independent" Fe(III)/Fe(II) line, which is strictly seen not "independent", as hydrolysis of Fe(III) occurs, forming hydroxo species with increasing pH [16]. Numerical values of these potentials shift in the bioenvironment, depending on the composition of complex metal species formed, but the general trends remain. At a certain pH the V(V)/V(IV) line overlaps with Fe(III)/Fe(II) and crosses over the O_2/H_2O_2 line. Accordingly, in acid solutions V(IV) is stable, and V(V) is reduced by H_2O_2. In alkaline solutions V(V) is stable. In weakly acid solutions and in the vicinity of pH 7, where V(V)/V(IV) and O_2/H_2O_2 lines are crossing, the electron transfer becomes complicated. The possibility exists for the generation of radicals, according to reactions (1) and (2).

$$VO^{2+} + H_2O_2 \rightarrow VO_2^+ + OH^{\bullet} + H^+ \quad (1)$$

$$VO^{2+} + O_2^{2-} + H_2O \rightarrow VO_2^+ + OH^- + OH^{\bullet} \quad (2)$$

Complicated equilibria, dependent on environment, govern the presence of existing complex vanadium species and the capability of vana-

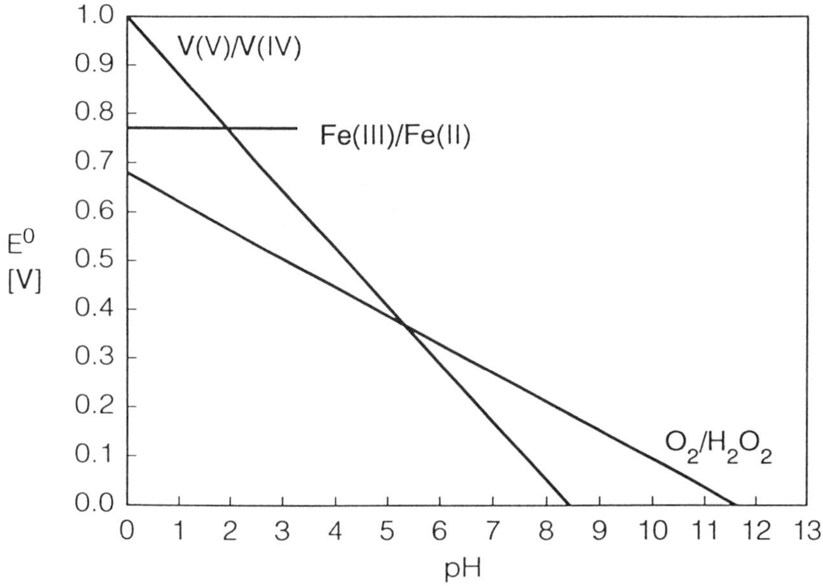

FIG. 1. Dependence of thermodynamic potentials upon pH.

dium to generate OH• radicals. The latter can trigger further radical reactions [17] according to reactions (3) and (4).

$$OH^{\bullet} + H_2O_2 \rightarrow H_2O + HO_2^{\bullet} \qquad (3)$$

$$^{\bullet}O_2^- + H_2O_2 \rightarrow O_2 + OH^- + OH^{\bullet} \qquad (4)$$

$$VO^{2+} + {}^{\bullet}O_2^- + H_2O \rightarrow VO_2^+ + H_2O_2 \qquad (5)$$

Superoxide, a product of the above reactions, although benign in comparison to other free radicals can spontaneously dismutate to oxygen and H_2O_2 [17], or it can further react with vanadium according to reaction (5). Singlet oxygen may be another product of this sequence. Selected vanadium ligand spheres are able to fine-tune the electron transfer energy of vanadium complex species present. As a consequence, reactions (1), (2), and (5) can proceed in either direction: the radical generation or the radical scavenging route. In a complex way, the direction of these reactions is obviously dependent on pH, vanadium concentration, and the availability of bioligands in the environment.

Oxidation of V(IV) by H_2O_2 recalls the similarity between the two redox couples V(V)/V(IV) (d^0/d^1) and Fe(III)/Fe(II) (d^5/d^6). The latter is well known for radical generation as proposed by the Haber–Weiss [18] reactions, and for the Fenton reagent reaction. "Fenton mechanism" reactivity is known to occur for a number of transition metal ions [14–22]. However, in the case of Fe(III)/Fe(II) and V(V)/V(IV), analogy between the (d^5/d^6) and (d^0/d^1) electron configuration and the proximity of redox potentials is unique, and must not be overlooked in deliberations on vanadium biochemistry and biological activity.

The lower oxidation states of vanadium, V(III), d^2, and V(II), d^3, are both far removed from the redox potentials of V(V)/V(IV) (Table 2).

2.2. Oxidation State and pH-Dependent Equilibria in Aqueous Solution

V(V) and V(IV) are the most common oxidation states of vanadium in aqueous solutions and in biochemistry. They have been thoroughly studied in aqueous solutions [9,16]. V(V)/V(IV), the (d^0/d^1) electron transfer interplay, depends on pH and vanadium concentration. Molar concentration of vanadium in human serum is of the order 10^{-7}–10^{-9} M [9,23] and is considered to be overall within 10^{-6}–10^{-8} M limits [2,26]. Selected animal or plant organisms concentrate vanadium to a much larger concentration [2,25,26]. In mammals variations occur from tissue to tissue, and within the cell locations, much as for most other trace metals [5,9,27]. In spite of local higher concentrations, molar concentration of vanadium remains very low. Excess of surrounding "ligands" amounts to a factor of about 10^3–10^5. Predominant V(V) and V(IV) species at low vanadium concentrations in water shown in Table 1 interact with bioligands, as dependent on pH variations in body tissue or body fluids, and pH gradients within the cell. The presence of V(IV) can be determined by instrumental methods based on d^1 paramagnetism. V(V) is a "silent" diamagnetic d^0 ion with no color, which is not easy to detect at low concentrations. It is also important to emphasize that the vanadium chemistry of any oxidation state in the presence and the absence of water, respectively, presents two different stories. In addition, no matter how non-alike biomedia are in comparison to pure aqueous solutions, in biomatter water molecules are always present.

Biochemical and biological investigations and deliberations of vanadium interactions must take these facts into account.

2.3. Stereochemical Nonrigidity and Lability of Ligands

Stereochemical nonrigidity implies a flexibility and ease in changing coordination polyhedra while retaining the coordination number, or changing this number in the process. Such interconvertibility occurs because of easy deformability and small energy difference between the two respective states. The nonrigidity is common for the coordination numbers 5 and 7 [28], and the V(V) and V(IV) complexes are an example for stereochemical nonrigidity. The trigonal bipyramid and square pyramid of coordination number five differ very little in energy; octahedral coordination can be decreased to five or increased to seven. The seven-coordinated pentagonal bipyramid, capped octahedron or capped trigonal prism are sometimes difficult to distinguish in distorted heteroligand spheres [29].

Lability refers to a rapid exchange of donor atoms in a coordination sphere. Vanadium polyhedra in aqueous media are known to interchange ligands with the environment and fast hydrolysis usually occurs in pure water [16]. The rearrangements related to stereochemical nonrigidity and lability of ligands are of importance in homogeneous catalysis, where very fast processes involve vanadium complexes as short-lived catalytic intermediates [1,30]. Many vanadium interactions in biomatter are expected to be of a similar nature.

3. ANTITUMOR ACTIVITY OF DIFFERENT TYPES OF VANADIUM COMPOUNDS

Speculation that metal complexes may possess antitumor activity has existed for a long time [31–35]. A survey of antitumor action of inorganic compounds appeared in 1931 [34], but vanadium compounds were not included. The activity of some elemental metals was tested later [36] (against spontaneous mice tumors), and vanadium was found to be inactive. Most cancer research interests prior to 1969 dealt with potential carcinogenicity of metals and their compounds [37–40], but vanadium and its compounds were not included. Metal antitumor compounds received serious attention after the discovery of the anticancer activity of "cisplatin", *cis*-$PtCl_2(NH_3)_2$ [41], and some other platinum coordination compounds [42]. Investigations of metal antitumor complexes have ever since been dominated by platinum compounds [33,42–44]. However, antitumor metal compounds of a few main group metals (Ga, Ge, Sn, Zn), and a number of transition metals were reported [33,43,44]. Among early transition metals are Ti, V, Nb, Ta, Cr, Mo, W, and others include Mn, Fe, Ru, Os, Co, Rh, Ir, Ni, Pd, Pt, Cu, and Au. Only the platinum compounds have been investigated more systematically.

The most active and thoroughly investigated vanadium antitumor active compound so far is the biscyclopentadienyldichloro-V(IV), of the formula $(C_5H_5)_2VCl_2$. This vanadium metallocene dichloride was prepared in 1954 [45] and reported active in 1979 [48]. A different type of vanadium antitumor-active compounds are the peroxovanadates(V), with or without a heteroligand [47], which were tested in one assay. The mechanism of their antitumor activity is expected to differ significantly from the mechanism proposed for vanadocene dichloride [43].

3.1. Metallocene Complexes

Metallocene and metallocenium compounds are among the most important antitumor-active complexes in the early transition metals family [43,44]. The most prominent are titanium compounds; the only other metal compound competing in the activity with them is vanadocene dichloride. This pale green complex, soluble in water and some organic solvents [45], has a distorted tetrahedral structure [49].

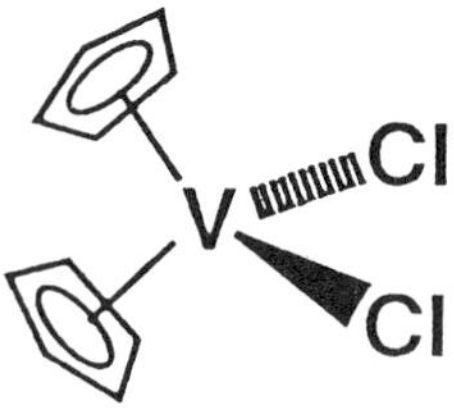

Vanadocene dichloride was screened for antitumor activity, along with analogous M(IV) = Ti, Nb, Ta, Mo, and W complexes, using several assays [43]. All six metallocene dichlorides screened exhibited antitumor activity, which was more pronounced for M(IV) = Ti, V, Nb, and Mo, than for M(IV) = Ta, W. In vitro vanadocene dichloride showed strong antiproliferative activity against human epidermoid (HEp-2) tumor cells, against human HeLa and KB tumor cells, against Ehrlich ascites tumor cells and human embrionic fibroblasts. In vivo, marked activity was observed against fluid and solid Ehrlich ascites tumors, as well as against other experimental tumor systems such as intraperitoneally growing mouse mammary tumor (TA3Ha). Vanadocene dichloride displayed the highest activity of all metals in cultured animal Ehrlich ascites tumor cells. However, the activity dose range was narrower than that of other metals. A distinct narrow dose-activity relationship was observed in in vivo tests as well [43,50–53]. A marginal activity of vanadocene dichloride was found only against the lymphoid leukemia L1210 and the lymphocytic leukemia P388 in vivo [50]. The characteristic narrow dose-activity relationship was observed in both, with a fast drop to toxicity past the optimum dose. Similar discrepancy in the antitumor activity against leukemias on the one hand, and against other experimental tumor systems on the other hand, has also been found for titanium, copper, gold, and germanium metal complexes [43].

After treatment with vanadocene dichloride no changes of urinary composition and no evidence for nephrotoxicity were found. The limiting toxicity seemed to be the injury of liver cells. An inhibition of nucleic acid synthesis, especially DNA, was observed, and vanadium accumulated in cellular regions which are rich in nucleic acids [43,53]. Organ distributions of vanadium analyzed in dry organ specimens found the main accumulation of vanadium in kidneys, whereas liver and intestine contained smaller concentrations. No transfer of vanadium across the intact blood-brain barrier occurred, and vanadium was not detectable in brain [43].

Hydrolysis and the aqueous chemistry of vanadocene dichloride was investigated [54], followed by a study of aqueous coordination chemistry with nucleotides and phosphoesters [55]. At physiological pH the compound was found to retain a stable V-(η^5-C_5H_5) bond, but chloride ions dissociated rapidly upon dissolution in water. This behavior is different from the relatively slow chloride hydrolysis observed for the reference compound, cisplatin. The acidity of water molecules bound to $(C_5H_5)_2V^{2+}$ is greater than that of $[Pt(NH_2)(H_2O)_2]^{2+}$. Investigations of aqueous vanadocene dichloride binding to DNA constituents, and the structure of a model compound, indicated that aqueous $(C_5H_5)_2V^{2+}$ species interact preferentially with the phosphate group of nucleotides. A labile outer sphere complex via a water bridge was proposed. Nucleotide-nucleotide base pairing was not disrupted by vanadocene dichloride. These studies indicated that the character of the vanadocene dichloride interaction with DNA differs very much from the interaction of cisplatin, although the proposed target for both is DNA [43,55].

3.2. Peroxo Complexes

Vanadium(V) peroxo complexes of the formula $M^I_x[VO(O_2)_2(L)]$ and $M^I_x[VO(O_2)(L)]$, where x depends on the negative charge of the heteroligand (L), have recently attracted the attention of biochemists because of their strong insulin-mimetic activity in vitro and in vivo [15]. They are also known in homogeneous catalysis [30,56], and some suggestions in the literature consider peroxo heteroligand vanadates(V) as a possible model for the mechanism of the vanadium bromoperoxidase activity [1,57]. Complexes of this type are of interest because of their stereochemistry, and because of the presence of a [V(V)-peroxo] moiety in a coordination sphere, where a heteroligand can modify the redox potential of V(V)/V(IV), and intramolecular one-electron transfer can then trigger radical generation. The reactivity of the peroxo group and the lability of the vanadium-peroxo bonds are strongly dependent on the character of the heteroligands. Examples are peroxovanadates with polycarboxylates [58,59], aminopolycarboxylates [60,61], 2,2′-bipyridyl and *o*-phenanthroline [62], and a variety of others. Bioheteroligands such as α-amino acids [63], α-hydroxypolycarboxylic acids [59], citrate [58], nicotinic acid [64], adenine and adenosine [65] are of special interest because short-lived intermediates of the peroxo-heteroligand type are expected to play an important role in bioprocesses. Tuning of the V(V)/V(IV) potential by heteroligands combined with sterical nonrigidity of coordination polyhedra and lability of some ligand sphere sites can combine to set off reversible rearrangements involving electron transfers. [V(V)/V(IV)-peroxo] interplay becomes capable of generating or scavenging oxygen radicals, leading to toxic or beneficial biological effects [17,66], including antitumor activity.

Marginal antitumor activity [47] against L1210 murine leukemia was found for four compounds shown in Table 3. Vanadocene dichloride and an inactive peroxo complex are added for comparison. A set of 14 vanadate(V) complexes were tested [47] to assess the relationship between analogous nonperoxo and peroxo compounds, and the role of heteroligands. Eleven of the compounds screened were peroxovanadates and three were analogous oxo complexes [47]. The X-ray structure is known for six of these compounds [58–61,67–70]. Eight of 11 peroxo complexes tested showed an increased life span (ILS) ⩾20%, and four ⩾25% (see Table 3). Antitumor activity of all peroxovanadates(V) was characterized by a narrow dose-response range and a steep drop to toxicity after the optimal dose. Screening of such compounds may miss the optimal dose, and complexes which may well be active in a dose range not used in testing would show an activity below the accepted number for a particular assay.

Sketches of two dimeric structures for the active compounds (1) and (4) are shown below, compared to the inactive monomeric $K_2[VO(O_2)NTA]$ (6), NTA = nitrilotriacetate, characterized by low toxicity and a remarkable stability toward decomposition [47,61]. Common to all three structures is the seven-coordinated V(V) in a distorted pentagonal bi-

TABLE 3

Antitumor Activity of Vanadium Compounds Against L1210 Murine Leukemia [47]

Compound	ILS (%)[a]	Ref.[b]
(1) $(NH_4)_4[O\{VO(O_2)_2\}_2]$	39	67,68
(2) $(NH_4)_3[VO(O_2)_2(C_2O_4)]$	32	62
(3) $K_3[VO(O_2)_2(C_2O_4)]$	25	62,69
(4) $(NH_4)[VO(O_2)(C_4H_4O_5)]$	29	59
(5) $(C_5H_5)_2VCl_2$	29	43
(6) $K_2[VO(O_2)NTA]2H_2O$	6	61

[a]Activity in the L1210 system is defined as ⩾25% increase in life span [ILS].
[b]References for synthesis and structure.
NTA^{3-} = nitrilotriacetate.

pyramid, and the location of the peroxo group in the pentagonal plane, cis to the V=O group. Antitumor active dimers have an elongated V–O bond trans to the oxo group. This V–O bond is longer than the bond in the monomer (6), where all three V–O(carboxylato) bond distances are comparable, and the peroxo group is nonsymmetrically coordinated. Two V–O(peroxo) bonds of different lengths were proposed to increase the V–O(peroxo) bond strength, due to the oxygen 2s atomic orbital's and the metal p_z atomic orbital's interaction [71].

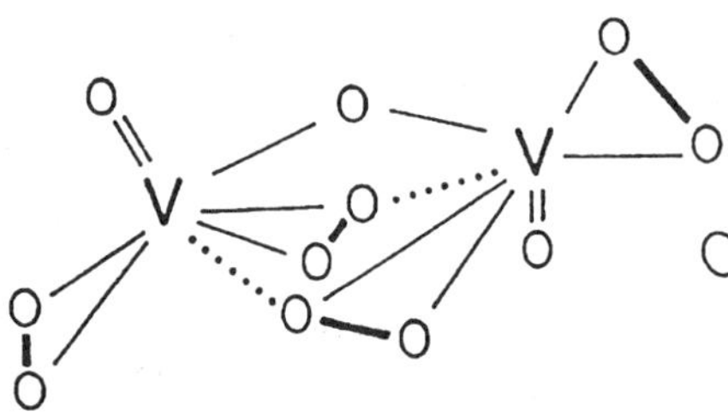

(1)

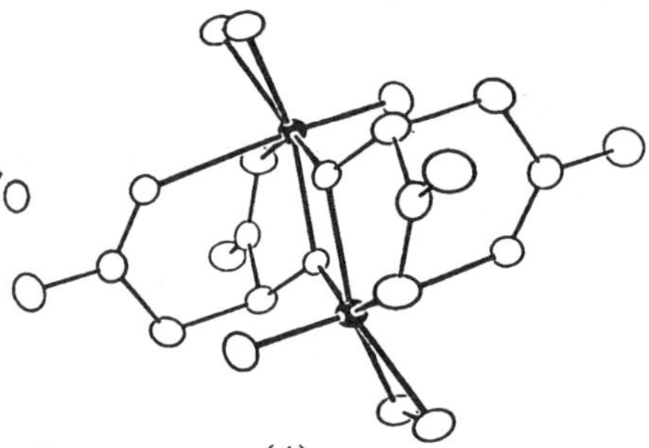

(4)

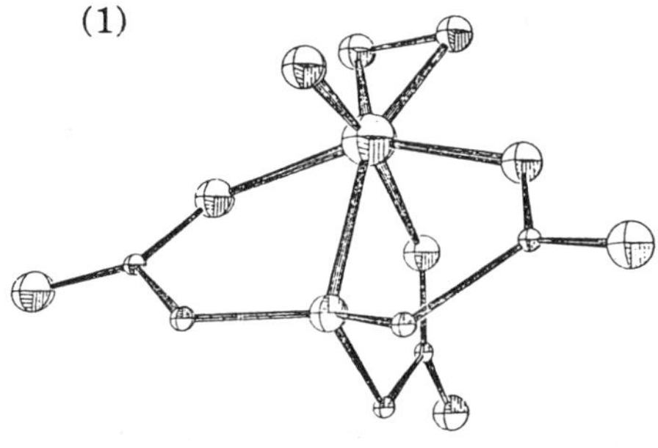

(6)

Vanadocene dichloride (5) showed strong antitumor activity in several in vivo and in vitro screening tests, but only 29% ILS against L1210 murine leukemia (see Table 3). In several screening tests, vanadocene dichloride also displayed a narrow dose-response range. Such behavior, remarkably similar to peroxovanadates(V), is interesting because of the very different chemical nature of (5).

Screening of the peroxo- and nonperoxovanadates(V) has provided some interesting observations. Both antitumor activity and toxicity are strongly dependent on the heteroligands, and analogous nonperoxovanadates(V) are not active. In addition, peroxoheteroligand complexes of Nb(V), Ta(V), and La(III) tested previously were inactive [72]. All this evidence supports the idea that the V(V)-peroxo moiety is unique, indicating that V(V)/V(IV) intramolecular transfer does play a role in some mechanisms of vanadium biological activity.

3.3. Other

One of the reasons for "no other" complexes described here is quite likely the general behavior of vanadium antitumor-active complexes in terms of the very narrow dose-response range and the sharp drop to toxicity after the optimal dose. Chances are small to "hit" a successful combination of three variables: (1) a well-chosen compound, (2) the correct dose, and (3) the appropriate assay for (1). Such interdependent investigation are risky, unpredictable, and expensive.

Among 84 nondiscreet vanadium compounds evaluated for anticancer activity by the National Cancer Institute in Bethesda, Maryland [73], only 10 exhibited activity. None of the 10 active by in vitro KB assay showed any activity in follow-up in vivo mouse testing, and none of vanadium complexes tested in the new in vitro human tumor cell line assay have been referred for possible in vivo testing [73]. Screening of some vanadium compounds is certainly going on in other research institutions, but none are described in the chemical literature.

4. RELATIONSHIP BETWEEN ANTITUMOR ACTIVITY AND SOME VANADIUM PROPERTIES

Correlations drawn between vanadium properties regarding oxidation state, stereochemistry, and ligand preference in solid state and aqueous media cannot be directly correlated to vanadium interaction with bioligands in biomatter. Metal transfer, metal concentration gradient in a specific body location, selected tissue or within a cell, as well as metal toxicity, have long been known to depend strongly on the chemical composition of the substances administered [74]. Furst also proposed a long time ago that metal species react in situ with a number of chelating agents available in the bioenvironment [34,35]. The search for metal antitumor activity is therefore a multidimensional and unpredictable task.

4.1. Oxidation State and Coordination Spheres

The best known vanadium antitumor active complex, the V(IV) vanadocene dichloride, has a distorted tetrahedral stereochemistry [49] in the solid state. After hydrolysis and chloride dissociation in aqueous solutions the $(C_5H_5)_2V^{2+}$ moiety is retained in solution [54] and water molecules enter the ligand sphere. They are believed to interact with the phosphate part of DNA via hydrogen bonding [55]. The number of water molecules in the complex which is interacting with DNA cannot be experimentally determined and most likely varies. The pH-dependent V(V)/V(IV) equilibria require the presence of V(V) and V(IV) species. Both oxidation states form stereochemically nonrigid polyhedra, which can undergo rapid exchange and rearrangement involving ligands in the bioenvironment. Under these circumstances, the coordination number of a complex vanadium species is expected to fluctuate, and V(V) and V(IV) should both be present, as observed in the bloodstream [24].

Antitumor-active peroxoheteroligand vanadates, starting with dimers of seven-coordinated V(V), undergo hydrolysis in water. New species are formed which are highly dependent on pH and the heteroligand. Aqueous solutions of the peroxoheteroligands 2–4 listed in Table 3 were not studied, and the moiety retained in water is not known. However, it is known that in aqueous solutions vanadium peroxo species depend on pH (Table 1), and vanadium-monoperoxo bonds are stronger than the bonds to the second peroxide in a diperoxo species [75]. In addition, the stabilization of a coordinated peroxide depends on the character of the heteroligand. The presence of a coordinated peroxo group stabilizes the V(V), and peroxide dissociation can lead to reduction of V(V). In biomedia, rearrangements of the polyhedra present and equilibria between V(V) and V(IV) species formed in the process are complicated by [V(V)/V(IV)-peroxo] interplay. The possibility of radical generation or radical scavenging exists, as discussed before, including singlet oxygen formation [76,77]. Speculations on mechanisms of antitumor activity of this type are endless. Interactions with proteins [78], DNA [65], and other biomolecules and biopolymers could occur via a direct bond of vanadium to donor atom(s) of biospecies, or by hydrogen bonding between biospecies and the peroxo group, or both. In all such interactions the coordination number remains flexible, and both oxidation states can be present. A safe assumption is only that no polymeric vanadium species participate at the existing low vanadium concentration [79].

4.2. Toxicity and Carcinogenesis

Toxicity of vanadium is currently in the limelight because of the environmental concern regarding human exposure to combustion products of vanadium-bearing fuel oils as well as fumes and dust generated by metallurgical refining [14,80]. Vanadium atmospheric emission is due mainly to burning of oils, and in most industrial situations vanadium enters the body through the respiratory tract.

Vanadium compounds exhibit highest toxicity if injected, an intermediate toxicity via the respiration route, and the lowest toxicity if administered orally [5,13]. Mechanisms of toxic action and interaction are not known [5], and under physiological conditions V(V) and V(IV) are present [24]. Numerous studies exist on enzyme inhibition or stimulation of activity [2,5] by vanadium compounds. However, except at an extremely high concentration, vanadium does not cause systemic toxicity, which varies considerably with the nature of compounds [5,13]. The conclusion that vanadium is not very toxic for humans when given orally is based on the data from high levels of vanadium prescribed medicinally at the turn of the century [5,13].

According to a 1979 review on "unusual" metals as carcinogens [40], vanadium was not found to be carcinogenic. Later literature [5,14] does not list any carcinogenic vanadium compounds. Although a number of metabolic alternations occur upon administration of vanadium complexes, no concrete data were found on carcinogenic effects of vanadium exposure in humans or experimental animals [14]. Lifetime experiments with rats and mice allowed to drink water containing VO^{2+} showed no evidence of carcinogenic activity [5].

Antitumor-active vanadium compounds showed a very interesting toxicity pattern, regardless of the type of the compound. Toxicity occurred sharply after the optimal dose for antitumor activity, and depended, as expected, on the type of the compound. Dose-response for vanadocene chloride ranged between 0.20 and 0.24 mmol V/kg (50–60 mg/kg), and toxicity was indicated at 0.28 mmol of V/kg (about 70 mg/kg). The range of the antitumor activity dose for peroxovanadate compounds (Table 3) was 0.025–0.040 mmol V/kg (5–12 mg/kg) followed by toxicity at about 0.05 mmol V/kg (6–16 mg/kg) [47]. The nonactive peroxo complex, $K_2[VO(O)_2NTA]$,* however, was toxic at $\geqslant$0.30 mmol V/kg (about 120 mg/kg).

Organ toxicity was determined for $(C_5H_5)_2VCl_2$ [50]. The toxic pattern was found to be very different from the toxicological features of cisplatin. Cis-$PtCl_2(NH_3)_2$ displays severe nephrotoxicity, and vanadocene dichloride did not disturb renal function or damage the renal cells. A short-lasting injury of liver cells was indicated, which was reversible 2–4 days later.

5. CONCLUSIONS AND PERSPECTIVES

Studies of antitumor activity of metal complexes are presently overshadowed by fascinating new organic compounds, such as, for example, taxol [81]. These compounds are active against common human cancers and have less severe side effects [82]. A systematic investigation of metal antitumor agents has been done only with platinum compounds, which are used widely in the treatment of ovarian, bladder, and testicular cancers. Considerable information is available on the molecular basis for the anticancer activity of platinum compounds where DNA is the primary target [83]. The search is already on for orally active platinum anticancer drugs [84] in order to avoid expensive intravenous administration, which involves very serious side effects such as renal toxicity and neurotoxicity.

Vanadium compounds are not anticipated to participate in cancer therapy in the foreseeable future. They are too exotic and alien to most researchers to be taken seriously even for screening for antitumor activity. Studies summarized in this chapter, however, revealed two apparently general properties: low organ toxicity and a narrow dose-response range. The molecular basis for antitumor activity as well as

*NTA^{3-} = nitrilotriacetate.

toxicity of vanadium complexes is expected to be very different from cisplatin and its analogs. DNA may remain the target, but via other types of interaction. Heteroligand peroxovanadates offer the possibility for a distinct antitumor activity: the radical generation and the consequent radical interactions. Normal and cancer cells differ in a number of ways [85]. Of interest in this case are three parameters: the lower pH, a larger free radical character, and a higher biowater content in cancer cells. These properties of cancer cells could serve as excellent sensors for interaction with specific vanadium complexes. Unique heteroligand vanadium coordination spheres could form in the cancer cells, distinguished from vanadium polyhedra in the normal cells, where pH, biowater content, and free radical character are not identical. Speculations on the possible anticancer mechanism abound and, combined with the lower toxicity of vanadium, a call for testing more vanadium complexes by using more screening assays becomes reasonable.

Deliberations of biochemical interactions of vanadium implicate comparisons to biospecies involving some other element. The "phosphate analogy," extended to oligomers of vanadates(V), is common in investigations involving protein or enzyme interactions of vanadium [86–88]. A vanadium-molybdenum analogy is relevant in the catalysis [30] involving d^0 V(V) and Mo(VI) derivatives, and in lower oxidation states for nitrogenases [2]. Radical generation capability in biological systems and the one-electron transfer processes recall the interesting correlation between the (d^0/d^1) V(V)/V(IV) and (d^5/d^6) Fe(III)/Fe(II) systems, described in Sec. 2.1. Thermodynamic potentials and electron configurations would also suggest some parallels between (d^0/d^1) V(V)/V(IV) and (d^9/d^{10}) Cu(II)/Cu(I) pairs. Consequently, the heterobimetallic complexes of the V-Fe and V-Cu type should possess some intriguing properties and are likely to be studied in the near future. Recently, the bimetallic complexes of Pt-Ti were synthesized [89].

In the fascinating field of bioinorganic chemistry a lot of work lies ahead in the biological chemistry and biochemistry of vanadium. An essential element of an unknown function, displaying a variety of interesting activities, vanadium will undoubtedly attract many scientists. In the process, new vanadium compounds will be synthesized and characterized, and screening for antitumor activity of these should become a routine. Vanadium chemistry is never short on surprises.

REFERENCES

1. R. Wever and K. Kustin, *Adv. Inorg. Chem.*, *35*, 81 (1990).
2. N. D. Chasteen (ed.), *Vanadium in Biological Systems*, Kluwer, Dordrecht, 1990.
3. A. Butler and J. V. Walker, *Chem. Rev.*, *93*, 1937 (1993).
4. A. Butler and C. J. Carrano, *Coord. Chem. Rev.*, *109*, 61 (1991).
5. B. R. Nechay, L. B. Nanninga, P. S. E. Nechay, R. L. Post, J. J. Grantham, I. G. Macara, L. F. Kubena, T. D. Phillips, and F. H. Nielsen, *Fed. Proc.*, *45*, 123 (1986).
6. R. Wever, H. Plat, and E. deBoer, *Biochim. Biophys. Acta*, *830*, 181 (1985).
7. D. W. Boyd and K. Kustin, in *Advances in Inorganic Biochemistry*, Vol. 6 (G. L. Eichhorn and L. G. Marzilli, eds.), Elsevier, New York, 1985.
8. H. Vilter, *Phytochemistry*, *23*, 1387 (1984).
9. N. D. Chasteen, *Struct. Bond.*, *53*, 105 (1983).
10. K. Kustin, G. C. McLeod, T. R. Gilbert, and L. B. R. Briggs, *Struct. Bond.*, *53*, 139 (1983).
11. K. Kustin and I. G. Macara, *Comm. Inorg. Chem.*, *2*, 1 (1982).
12. V. Soghomonian, Q. Chen, R. C. Hausehalter, J. Zubieta, and C. J. O'Connor, *Science*, *259*, 1596 (1993).
13. M. D. Waters, in *Advances in Modern Toxicology*, Vol. 2 (R. A. Goyer and M. A. Mohlman, eds.), Hemisphere, Washington, DC, 1977, p. 47.
14. R. Wennig and N. Kirsch, in *Handbook on Toxicity of Inorganic Compounds* (H. G. Seiler, H. Sigel, and A. Sigel, eds.), Marcel Dekker, New York, 1988, p. 749.
15. A. Shaver, J. B. Ng, D. A. Hall, B. S. Loon, and B. I. Posner, *Inorg. Chem.*, *32*, 3109 (1993).
16. C. F. Baes, Jr. and R. E. Mesmer, *The Hydrolysis of Cations*, Wiley-Interscience, New York, 1976, p. 197.
17. I. Fridovich, in *Advances in Inorganic Biochemistry*, Vol. 1 (G. L. Eichhorn and L. G. Marzilli, eds.), Elsevier, New York, 1979, p. 67.
18. F. Haber and J. Weiss, *Proc. Roy. Soc. London*, *A147*, 332 (1934).
19. G. Czapski, *Ann. Rev. Phys. Chem.*, *22*, 171 (1971).

20. B. H. J. Bielski and A. O. Allen, *J. Phys. Chem.*, *81*, 1048 (1977).
21. A. Sobkowiak, H.-C. Tung, and D. T. Sawyer, in *Progress in Inorganic Chemistry*, Vol. 40 (S. J. Lippard, ed.), Wiley, New York, 1992, p. 291.
22. W.-D. Wang, A. Bakac, and J. H. Espenson, *Inorg. Chem.*, *32*, 2005 (1993).
23. W. R. Harris, S. B. Friedman, and D. Silberman, *J. Inorg. Biochem.*, *20*, 157 (1984).
24. N. D. Chasteen, J. K. Grady, and C. E. Holloway, *Inorg. Chem.*, *25*, 2754 (1986).
25. G. Micera and A. Dessi, *J. Inorg. Biochem.*, *35*, 71 (1989).
26. E. M. Armstrong, R. L. Beddoes, L. J. Calviou, J. M. Charnock, D. Collison, N. Ertok, J. H. Naismith, and C. D. Garner, *J. Am. Chem. Soc.*, *115*, 807 (1993).
27. S. E. Bryan, in *Advances in Inorganic Biochemistry*, Vol. 3 (G. L. Eichhorn and L. G. Marzilli, eds.), Elsevier, New York, 1981, p. 87.
28. R. B. King, *Inorg. Chem.*, *24*, 1716 (1985).
29. D. L. Kepert, *Progr. Inorg. Chem.*, *25*, 42 (1979).
30. F. di Furia and G. Modena, *Rev. Chem. Intermed.*, *6*, 51 (1985).
31. W. A. Collier and F. Krauss, *K. Krebsforsch.*, *34*, 526 (1931).
32. M. J. Cleare, *Coord. Chem. Rev.*, *12*, 349 (1974).
33. M. J. Cleare and P. C. Hydes, in *Metal Ions in Biological Systems*, Vol. 11 (H. Sigel, ed.), Marcel Dekker, New York, 1980, p. 1.
34. A. Furst, in *Metal Binding in Medicine* (M. J. Seven and L. A. Johnson, eds.), J. B. Lippincott, Philadelphia, 1960, Chap. 14.
35. A. Furst, *The Chemistry of Chelation and Cancer*, Charles C Thomas, Springfield, IL, 1963.
36. M. Kanisawa and H. A. Schroeder, *Cancer Res.*, *27*, 1192 (1967).
37. C. P. Flessel, A. Furst, and S. B. Radding, in *Metal Ions in Biological Systems*, Vol. 10 (H. Sigel, ed.), Marcel Dekker, New York, 1980, p. 23.
38. C. E. Searle (ed.), *Chemical Carcinogens*, ACS Mograph 173, Washington, DC, 1976.
39. F. W. Sunderman, Jr., *Biol. Trace Elem. Res.*, *1*, 63 (1979).
40. A. Furst and S. B. Radding, *Biol. Trace Elem. Res.*, *1*, 169 (1979).

41. B. Rosenberg, L. van Camp, J. E. Trosco, and V. H. Mansour, *Nature*, *222*, 385 (1969).
42. B. Rosenberg, in *Metal Ions in Biological Systems*, Vol. 11 (H. Sigel, ed.), 1980, p. 127.
43. P. Köpf-Maier and H. Köpf, *Struct. Bond.*, *70*, 105–181 (1988).
44. P. Köpf-Maier and H. Köpf, *Chem. Rev.*, *87*, 1137 (1987).
45. G. Wilkinson and J. M. Birmingham, *J. Am. Chem. Soc.*, *76*, 4281 (1954).
46. P. Köpf-Maier and H. Köpf, *Naturforsch*, *34b*, 805–807 (1979).
47. C. Djordjevic and G. L. Wampler, *J. Inorg. Biochem.*, *25*, 51 (1985).
48. B. K. Keppler, C. Friesen, H. G. Moritz, H. Vongerichten, and E. Vogel, *Struct. Bond.*, *78*, 97 (1991).
49. J. L. Petersen and L. F. Dahl, *J. Am. Chem. Soc.*, *97*, 6422 (1975).
50. P. Köpf-Maier, W. Wagner, B. Hesse, and H. Köpf, *Eur. J. Cancer*, *17*, 665 (1981).
51. M. S. Murthy, J. H. Toney, L. N. Rao, L. Y. Kuo, and T. J. Marks, *Proc. Am. Assoc. Cancer Res.*, *27*, 279 (1986).
52. J. H. Toney, L. N. Rao, M. S. Murthy, and T. J. Marks, *Breast Cancer Res. Treat.*, *6*, 185 (1985).
53. P. Köpf-Maier and P. Funke-Kaiser, *Toxicology*, *38*, 81 (1986).
54. J. H. Toney and T. J. Marks, *J. Am. Chem. Soc.*, *107*, 947 (1985).
55. J. H. Toney, C. P. Broch, and T. J. Marks, *J. Am. Chem. Soc.*, *108*, 7263 (1986).
56. S. Campestrini, V. Conte, F. di Furia, and G. Modena, *J. Org. Chem.*, *53*, 5721 (1988).
57. C. Djordjevic, M. Lee, B. A. Jacobs, and B. C. Puryear, *J. Inorg. Biochem.*, *36*, 239 (1989).
58. C. Djordjevic, M. Lee, and E. Sinn. *Inorg. Chem.*, *28*, 719 (1989).
59. C. Djordjevic, M. Lee, and E. Sinn, to be published.
60. C. Djordjevic, S. A. Craig, and E. Sinn, *Inorg. Chem.*, *24*, 1281 (1985).
61. C. Djordjevic, P. L. Wilkins, E. Sinn, and R. J. Butcher, *Inorg. Chim. Acta*, in press.
62. N. Vuletic and C. Djordjevic, *J. Chem. Soc. Dalton Trans.*, 1137 (1973).
63. A. S. Tracey and J. S. Jaswal, *Inorg. Chem.*, *32*, 4235 (1993).

64. C. Djordjevic, B. C. Puryear, N. Vuletic, C. J. Abelt, and S. J. Sheffield, *Inorg. Chem.*, *27*, 2926 (1988).
65. C. Djordjevic, N. Vuletic, and B. C. Puryear, *J. Inorg. Biochem.*, *51*, 308 (1993).
66. R. A. Floyd, *Free Radicals and Cancer*, Marcel Dekker, New York, 1982.
67. T. B. Swenson and R. Stomberg, *Acta Chem. Scand.*, *25*, 898 (1971).
68. R. Stomberg, S. Olson, and I.-B. Swenson, *Acta Chem. Scand.*, *A38*, 653 (1984).
69. D. Begin, F. W. B. Einstein, and J. Field, *Inorg. Chem.*, *14*, 1785 (1975).
70. W. R. Scheidt, Chun-che Tsai, and J. L. Hoard, *J. Am. Chem. Soc.*, *93*, 3867 (1971).
71. T. R. Cundari, M. C. Zerner, and R. S. Drago, *Inorg. Chem.*, *27*, 4239 (1988).
72. C. Djordjevic, L. G. Gonshor, and M. A. Bynum, *Inorg. Chim. Acta*, *79*, 301 (1983).
73. J. Johnson, personal communications, 1994.
74. C. Carruthers and W. Regelson, *Oncologia*, *16*, 101 (1963).
75. M. Orhanovic and R. G. Wilkins, *J. Am. Chem. Soc.*, *89*, 278 (1967).
76. Q. J. Niu and C. S. Foote, *Inorg. Chem.*, *31*, 3472 (1992).
77. K. Böhme and H.-D. Brauer, *Inorg. Chem.*, *31*, 3468 (1992).
78. J. S. Jaswal and A. S. Tracey, *J. Am. Chem. Soc.*, *115*, 5600 (1993).
79. S. E. Harnung, E. Larsen, and J. Pedersen, *Acta Chem. Scand.*, *47*, 674 (1993).
80. P. B. Hammond and E. C. Foulkes, in *Metal Ions in Biological Systems*, Vol. 20 (H. Sigel and A. Sigel, eds.), Marcel Dekker, New York, 1989, p. 157.
81. A. Stierle, G. Strobel, and D. Stierle, *Science*, *260*, 214 (1993).
82. E. W. Rowensky, L. A. Cazenave, and F. C. Donehower, *J. Natl. Cancer Inst.*, *82*, 1247 (1990).
83. S. L. Bruhn, P. M. Pil, J. M. Essigmann, D. E. Housman, and S. J. Lippard, *Proc. Natl. Acad. Sci. USA*, *89*, 2307 (1992).
84. J. F. Hartwig and S. J. Lippard, *J. Am. Chem. Soc.*, *114*, 5646 (1992).
85. L. N. Ferguson, *Chem. Soc. Rev.*, *4*, 289 (1975).

86. D. G. Drueckhammer, J. R. Durrwachter, R. L. Pederson, D. C. Crans, L. Daniels, and C.-H. Wong, *J. Org. Chem.*, *54*, 70 (1989).

87. D. C. Crans, E. M. Willging, and S. R. Butler, *J. Am. Chem. Soc.*, *112*, 427 (1990).

88. D. C. Crans, R. W. Marshman, M. S. Gottlieb, O. P. Anderson, and M. M. Miller, *Inorg. Chem.*, *31*, 4939 (1992).

89. M. Berardini, T. J. Emge, and J. G. Brennan, *Inorg. Chem.*, *32*, 2724 (1993).

19

Methods for the Spectroscopic Characterization of Vanadium Centers in Biological and Related Chemical Systems

C. David Garner, David Collison, and Frank E. Mabbs

Department of Chemistry, University of Manchester, Oxford Road, Manchester M13 9PL, UK

1. INTRODUCTION

Vanadium is essential for a variety of biological processes [1–5]. Virtually all of the available information concerning the oxidation state of the vanadium, its coordination environment, and how these are modulated by the oxidation state of the system, the presence of substrates or inhibitors of enzymic activity and/or other ligands has been obtained from spectroscopic studies. Fortunately, vanadium is rich in the spectroscopic probes available. Also, a comprehensive coordination chemistry of vanadium has been developed which can provide compounds to serve as suitable calibrants for the various spectroscopic methods.

This chapter will present a consideration of the spectroscopic methods which have been used successfully to probe the nature of the vanadium sites in a variety of biological systems. We have considered X-ray absorption, vibrational, optical, electron paramagnetic resonance, and ^{51}V nuclear magnetic resonance spectroscopies. Space does not permit a comprehensive account of any of these techniques and the reader is referred to specialist texts for further information [6–10].

2. X-RAY ABSORPTION SPECTROSCOPY

2.1. Introduction

In the mid-1970s, with the availability of intense X-ray synchrotron sources, a powerful new technique, X-ray absorption spectroscopy (XAS), emerged [6]. This is a local structural probe, the information content of which derives from electron diffraction. For a metalloprotein, the electron source and detector is the metal atom which is probed, since selective excitation is achieved by scanning a range of X-ray wavelengths appropriate to the element of central interest. The selectivity and the local nature of the diffraction process give the technique its major strength. For example, metal-ligand distances can be determined to an accuracy of about ±0.03 Å. In addition, XAS does not require crystalline samples; thus, aqueous protein samples are readily investigated under a variety of conditions.

Since 1975, when the first XAS experiments using a synchrotron radiation source were accomplished [11], XAS has become established as an important technique for probing the local structure of particular atoms in a wide variety of systems. Many significant advances have been made, not least for the characterization of metal centers of isolated proteins [12,13].

2.2. Theoretical Basis

2.2.1. Overview

The absorption of X rays by a material may be expressed as:

$$I = I_0 \, e^{-\mu x} \tag{1}$$

where I_0 = incident X-ray intensity, I = transmitted X-ray intensity, μ = absorption coefficient, and x = thickness of absorber. Figure 1 provides an illustration of a typical X-ray absorption spectrum [14]. For all materials, the generally smooth variation of absorption with increasing photon energy is punctuated by sharp increases in absorption, called absorption edges. These occur at positions where the incident photons have an energy which is just sufficient to promote an electron from a core level of a particular atom to a valence level and beyond to an

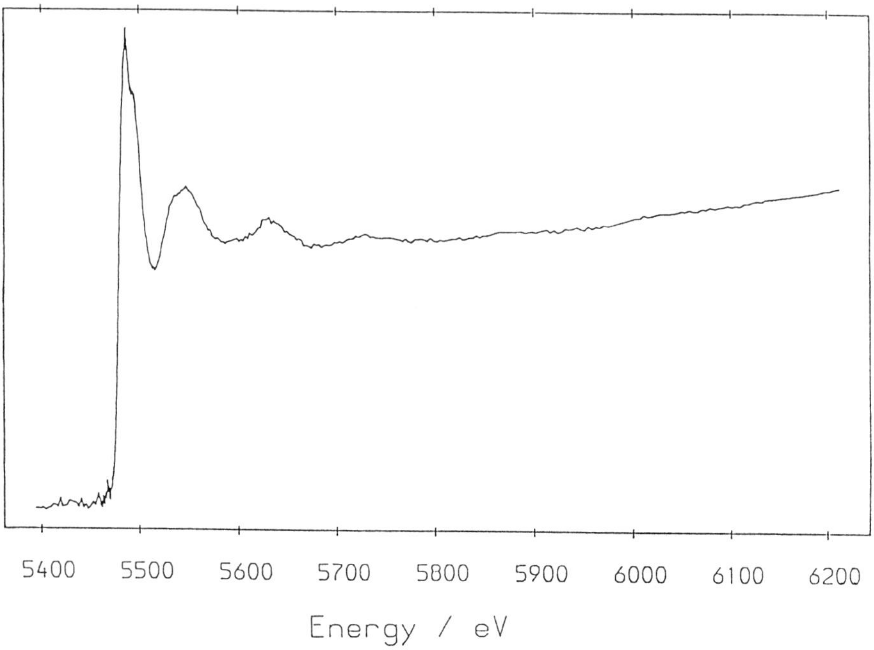

FIG. 1. Spectrum of the vanadium K-absorption edge for freeze-dried *Amanita muscaria* recorded in fluorescence [14].

unbound state. For a 1s (or K-shell) electron, a K-absorption edge results.

Above the absorption edge, low-amplitude oscillations are usually observed for several hundred eV. It is customary to classify the oscillations within ~50 eV of the edge as the X-ray absorption near-edge structure (XANES) and those which extend beyond this region as the extended X-ray absorption fine structure (EXAFS). The theoretical basis of the latter is considered mature and interpretations of EXAFS data have been reported confidently for some 15 years. In contrast, progress in the theoretical treatment and, therefore, the interpretation of XANES has been relatively slow and data in this spectral region are generally used qualitatively to "fingerprint" a metal site.

X-ray absorption spectra generally contain two other types of information:

1. The energy of a particular absorption edge depends on the oxidation state and the nature of the immediate chemical environment of the element. Typically, one unit increase of oxidation state increases the energy of an absorption edge by 1–3 eV.
2. In addition to the excitation of a core electron into the continuum, there may also be transitions to the valence levels. These promotions give rise to features on and/or before the absorption edge which can provide information concerning the environment and electronic structure of the atom of interest.

2.2.2. *EXAFS Interpretation*

The EXAFS (or XANES) amplitude $\chi(k)$ is defined as the modulation of the absorption coefficient, μ, of a particular atom relative to the smooth background absorption coefficient μ_s, normalized by the absorption coefficient that would be observed for a free atom:

$$\chi(k) = \frac{\mu - \mu_s}{\mu_0} \tag{2}$$

Although EXAFS and XANES are recorded as a function of energy, it is conventional to plot the data as a function of k, the photoelectron wave vector (3): where E is the energy of the X-ray photon, E_0 is the energy of the absorption threshold, m is the mass of the electron, and h is Planck's constant.

$$k = \frac{2\pi}{h}[2m(E - E_0)]^{1/2} \tag{3}$$

Except for the energies very close to the absorption threshold, a single scattering formalism is usually sufficient to describe the observed EXAFS [15,16]. As the photon energy increases beyond the ionization threshold, the wavelength of the photoelectron decreases and there will be a smooth and continuous movement through alternating constructive and destructive interference of the outgoing and backscattered waves. Constructive and destructive interference, respectively, increase and lower the absorption coefficient, as compared to the free atom value.

When the energy of the photoelectron is sufficiently high, the curvature of the photoelectron wave can be neglected and the theory can be greatly simplified as the plane-wave approximation:

$$\chi(k) = -\sum 3\cos^2\theta_j \frac{N_j}{kR_j^2}|f_j(\pi)| \sin(2kR_j + 2\delta_i + \phi_j) \cdot \exp(-2\sigma_j^2 k^2)\exp(-2R_j/\lambda) \tag{4}$$

This equation shows the structural basis of EXAFS, in that $\chi(k)$ is dependent on:

(i) the number of scattering atoms N_j;
(ii) the distance of the scattering atoms R_j from the primary absorber;
(iii) the type of scattering atom, through the characteristic energy dependence of the backscattering amplitude $|f_j(\pi)|$, the magnitude of which generally increases with increasing Z;
(iv) δ_i, the phase shift due to the potential of the emitting atom;
(v) ϕ_j, the phase shift induced by backscattering.

The mean square variation in R_j is represented by the Debye-Waller factor, σ_j^2; λ is the elastic mean-free path of the photoelectron and it is the damping term $[\exp(-2R_j/\lambda)]$ which invariably limits backscattering contributions to <4 Å from a metal atom in a biological system. The *j*th neighbor makes an angle θ_j with the polarization vector of the incident X ray and the term $3\cos^2\theta_j$ averages to 1 for solution and polycrystalline samples. At low photoelectron energies, the plane wave approximation (4) breaks down and leads to errors in the calculated phase which can result in an incorrect determination of the interatomic distances. The low energy part of the EXAFS spectrum can be analyzed by taking account of the curvature of the photoelectron wave [15]. This exact theory has considerable mathematical complexity and a requirement for large computational time. However, the calculations may be simplified by directional averaging [17] and this has been consolidated in the Daresbury EXCURV analysis programs. This simplification is well suited for virtually all studies of transition metals in biological systems.

The first stage of EXAFS data analysis is the removal of the background absorption, the extraction of the EXAFS, and its normalization to that for a unit metal atom. Often it is convenient to weight the data by k^n (usually $n = 2$ or 3) to enhance the profile at high k. Figure 2 shows the vanadium K-edge EXAFS × k^3 recorded for *Amanita muscaria* together with its Fourier transform.

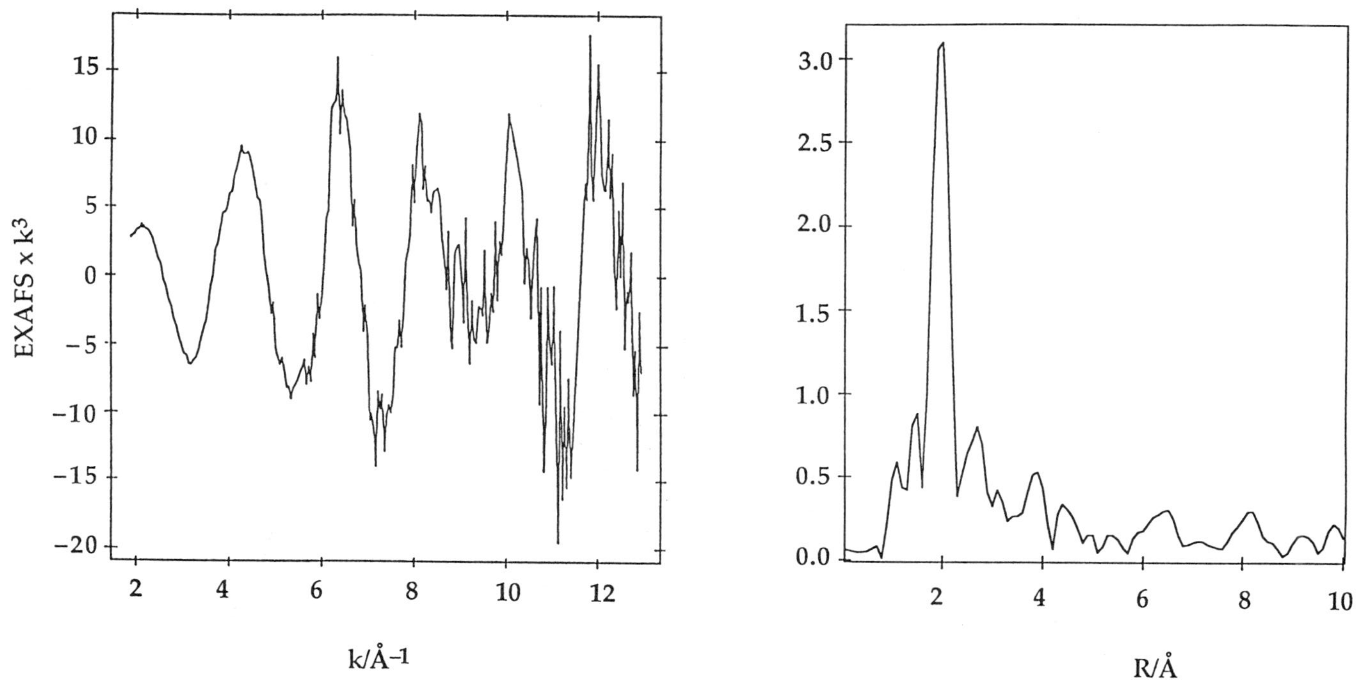

FIG. 2. Vanadium K-edge EXAFS × k^3 recorded for *Amanita muscaria* and the Fourier transform.

2.2.3. *Fourier Transform*

Sayers et al. [18] recognized that the Fourier transform $\chi(k)$ would yield a radial distribution function of the atomic arrangement about the primary absorber. This function peaks at distances close to the corresponding R_j values; the area of the peak depending on the number of backscattering atoms (N_j) within the shell and their backscattering amplitude ($|f_j(\pi)|$).

2.2.4. *EXAFS Analysis and Accuracy*

EXAFS data analysis consists of varying E_0 and the value of R_j, N_j, and σ_j, for each shell of atoms adjacent to the primary absorber, to produce optimum agreement between the experimental and simulated data and their Fourier transforms. The level of agreement may be assessed by a fit index and the improvement achieved by addition of an extra shell should be carefully monitored [19].

The major source of uncertainty in EXAFS analysis arises from the correlation between the coordination number (N_j) and the Debye-Waller factor (σ_j^2) for each shell, which results in the uncertainty in N_j being around $\pm 25\%$. The primary manifestation of the magnitude of R_j is the relative *phases* of the outgoing and backscattered waves and, in general, R_j values can be determined to an accuracy of around ± 0.03Å. The atomic number of the backscattering atoms cannot usually be determined to a precision better than ± 1 since $|f_j(\pi)|$ and ϕ_j do not show a marked dependence on Z. The present limitations in the accuracy with which N_j and Z values can be determined are especially frustrating for the characterization of metal centers in proteins. Therefore, wherever possible, the information obtained from other spectroscopic and structural techniques should be incorporated into the EXAFS analysis to help remove ambiguity.

2.2.5. *Multiple Scattering*

Multiple scattering contributions are generally not significant in the EXAFS region, except for systems which have exact or approximate colinear arrangements of three or more atoms. For metal centers, multiple scattering effects must be included for coordination by carbonyl, cyanide, imidazole, or porphyrin ligands *inter alia*. Theoretical developments [20] have allowed satisfactory data analyses of multiple scattering contributions [21]. Also, constrained and restrained refinements have been developed, in which information from known structures is used to restrict the number of parameters which can be adjusted independently in the simulation [22].

2.3. Experimental Considerations

2.3.1. *Source*

X-ray absorption spectra are recorded in single beam mode [6].

Within an X-ray absorption spectrum, the signal is generally small compared with the atomic absorption resulting from the excitation of the core electron of the atom of interest. Thus, good-quality EXAFS and XANES data require an intense and stable X-ray source. To date, all XAS studies of metalloproteins have required a synchrotron radiation source. Synchrotron radiation sources produce continuous and intense electromagnetic radiation, which allows the study of K-edge XAS for elements with $Z < 65$. However, studies of elements with $Z < 20$ are restricted by the beryllium windows which separate the high vacuum of the source from the experiment. Also, as Z increases, the excitation energy increases and the lifetime of the resultant hole in the 1s shell decreases. This leads to a further damping of the EXAFS oscillations and limits the K edges which can be studied. The heaviest element investigated by K-edge EXAFS is iodine ($Z = 53$), which has its K edge at 0.37 Å (33 500 eV) and a lifetime broadening >7 eV. For the higher Z elements, L-edge XAS can be recorded, concomitant with the excitation of a 2s or 2p electron; typically, L-edge data can be recorded for molybdenum ($Z = 42$) to uranium ($Z = 92$).

2.3.2. *Monochromator*

The "channel cut" monochromator is the simplest type employed. A channel is cut in a perfect crystal to provide two parallel reflecting surfaces which have a particular crystal plane parallel to the surface. The Bragg condition is used to select a particular wavelength and the reflected beam emerges parallel to the incident beam. The accuracy of data collected using channel cut crystal monochromators may be limited due to harmonic contamination of the reflected beam. Harmonic suppression can be achieved by use of a double-crystal monochromator which has the two crystal faces slightly offset.

2.3.3. Detectors

The conventional XAS experiment involves the direct measurement of the incident and transmitted beam intensity using ionization chambers. The first chamber contains a weakly absorbing gas, which transmits ~70% of the incident radiation, and the second chamber contains a mixture of inert gases which absorb virtually all the transmitted intensity. The absorption coefficient includes contributions from the matrix (μ_M) and the atom of interest (μ_A).

Transmission XAS is ultimately limited by the intensity of the beam and the ratio of μ_M to μ_A. In cases where $\mu_M/\mu_a = 1$ it is difficult to use the transmission method; for ratios greater than 10 it is almost impossible. The detection sensitivity can be enhanced if a discrimination can be made between the matrix and host absorption. X-ray fluorescence offers just this possibility. As the fluorescence yield is practically independent of the excitation energy over an EXAFS spectrum (~1000 eV above the edge), a change in the absorption cross-section is directly reflected by a change in the fluorescence yield (Fig. 1). This increased contrast arises since the fluorescence of the element of interest, in the region of one of its absorption edges, is considerably greater than that of the lighter, matrix, atoms. Fluorescence detection is now a standard procedure for recording XAS for metal atoms in biological systems and solid state detectors allow data to be collected at mM concentrations of the element of interest.

2.3.4. Sample

The basic requirement for any sample to be investigated by XAS is that the element of interest is present at a concentration of at least 1 mM in a volume of ~0.5 mL. Multiple scans (say up to 16) can be essential to achieve a reasonable signal-to-noise ratio. Recording data at low ($\leq$77 K) temperature reduces the Debye-Waller parameter and this usually permits extension of the EXAFS data range with a consequent improvement in resolution of the individual backscattering contributions. Therefore, it is advantageous to study metalloproteins in a frozen glass contained between thin plastic (e.g., Mylar) windows glued to a robust (e.g., aluminum or perspex) frame. The solution, containing the necessary buffers, salts, and/or other mediators, is prepared at room temperature, injected into the cell, and rapidly frozen. An "antifreeze" agent (e.g., glycerol or ethylene glycol) should be added to prevent crystallization. The study of metalloproteins in solution allows the introduction of redox partners, substrates or substrate analogs, inhibitors, or exogenous ligands.

The polarization inherent in synchrotron radiation readily permits the measurement of anisotropy in the X-ray absorption spectra for oriented samples. This has been used to good effect for single crystals of the iron-molybdenum protein of nitrogenase [23] and the manganese reaction center of photosystem II in oriented chloroplasts [24].

2.4. Applications to Vanadium in Biological Systems

2.4.1. Bromoperoxidase

Bromoperoxidase (BPO) from the marine brown alga *Ascophyllum nodosum* was the first enzyme isolated in which vanadium was identified at the active site [25]. The coordination of the vanadium center has been investigated using X-ray absorption spectra collected from native, native plus bromide, native plus hydrogen peroxide, and dithionite-reduced forms of the enzyme [26]. The vanadium K-edge and XANES spectra showed that it is only on reduction of the native enzyme that the metal site is substantially altered. All four spectra exhibit the large pre-edge feature characteristic of a V=O group, though in the reduced form of the enzyme the peak is smaller and broader. The positions of the edge and pre-edge peaks measured relative to a vanadium foil are given in Table 1. These values are consistent with the view that the reduced enzyme is a vanadium(IV) system and that the native enzyme samples all involve vanadium(V) [27]. The EXAFS of the native and reduced BPO reveal significant differences in the vanadium environments [26].

1. For the reduced sample, the dominant EXAFS contribution arises from light (C, N, O) atom backscattering at ~2 Å, with smaller backscattering contributions from light atoms at ~1.6, 3.0, and 4.3 Å.

TABLE 1

Positions of Edge and Pre-edge Peaks of Bromoperoxidase Relative to Vanadium Foil

Sample	ΔPEF (eV)[a]	ΔE(eV)[b]	I[c]	$\nu_{1/2}$(eV)[d]
Native BPO[e]	4.0	8.0	63	1.8
Native BPO + H_2O_2	4.0	8.1	59	1.7
Native BPO + Br^-	3.9	7.8	75	1.8
Reduced BPO	3.1	5.7	34	2.4

[a]Shift in energy of the pre-edge feature, relative to the energy of the pre-edge feature at 5466 eV of vanadium foil.
[b]Shift in energy of the edge position measured as the energy at half the edge height after normalization, relative to the energy of the edge position 5473.5 eV of vanadium foil.
[c]Intensity of the pre-edge feature expressed as the percentage height of the pre-edge feature relative to the normalized edge height.
[d]Width of the pre-edge feature at half maximum height.
[e]BPO is the vanadium-containing bromoperoxidase from *Ascophyllum nodosum.*

The best fit of the EXAFS, for the inner coordination sphere of the vanadium, is with one oxygen at 1.63 Å, three oxygens at 1.91 Å, and two nitrogens at 2.11 Å. The backscattering from light atoms at ~3 and 4.3 Å, together with the two nitrogens at 2.11 Å, strongly suggests that histidine is coordinated to the vanadium(IV) and, in order to fit the outer shells, two imidazole rings, coordinated via nitrogen to the vanadium, were used in the simulation. Inclusion of multiple scattering effects from the imidazoles significantly improved the fit of the EXAFS, particularly in the low-k region.

2. The EXAFS of the vanadium K edge in the native enzyme is well reproduced by a simulation involving backscattering from one oxygen at 1.61 Å, three oxygens at 1.72 Å, and two nitrogens at 2.11 Å. The major change from the reduced enzyme is that the dominant backscattering contribution occurs at a shorter distance (1.72 Å) in the native enzyme compared with the reduced form (1.91 Å). The evidence for imidazole coordination is less distinct for the native enzyme than for the reduced form. Possibly the vanadium changes its binding site upon

reduction. The vanadium K-edge EXAFS for the native enzyme is compatible with a distorted form of vanadate(V), $[VO_4]^{3-}$, with four tightly bound oxygen atoms, one at 1.61 Å and approximately three at 1.72 Å, plus approximately two longer bonds of 2.11 Å to O- or N-donor ligands.

The nature of the active site of vanadium in the haloperoxidases still poses interesting challenges (see Sec. 6) and the XAS data [26] provide valuable information but do not fully define the nature of the center. This study has demonstrated how information can be gleaned from X-ray absorption spectra. Thus:

1. The position of the absorption edge can provide information concerning the oxidation state of a transition metal in a biological molecule.
2. The pre-edge and/or edge features contain information relating to the electronic structure of the center.
3. Comparison of XANES profiles, since they arise from multiple scattering, can suggest similarity (or differences) in both radial and angular distribution of atoms about the primary absorber.
4. EXAFS data are readily interpreted to give a radial distribution function. The discrimination between backscattering from C, N, and O atoms and (say) P, S, and Cl is readily achieved. However, distinction between O- and N-donor ligands only arises when multiple scattering contributions enable, for example, imidazole groups to be suggested on the basis of the particular outer shell distances.

Given the bond lengths determined by EXAFS for the vanadium center of BPO in both its (IV) and (V) states, bond valence sum analysis (BVSA) [28] can be used to assess the coordination numbers [29]. This approach favors five coordination for both of these centers [30]. Bonds which are consistent with the BVSA sum are as follows: for vanadium(IV), one V=O of 1.63 Å, two V–N of 2.11 Å and two V–O of 1.91 Å; for vanadium(V), one V=O of 1.61 Å, two V–O of 1.72 Å, and two V–O (or one V–O and one V–N) of 2.11 Å.

2.4.2. *Amavadin*

Mushrooms of the genus *Amanita* are notable for their ability to accumulate vanadium up to concentrations of 200 ppm dry weight and the vanadium can be extracted as a low molecular weight complex, known as Amavadin [31]. Degradative organic chemistry has identified the anion of *S,S*-2-hydroxyiminodipropionic acid, $HON\{CH(CH_3)CO_2H\}_2$ (hidpaH_3), as the ligand which forms a 2:1 complex with vanadium [32]. Electron paramagnetic resonance (EPR) measurements show that vanadium(IV) is present [31] and electrochemical studies have established that this is capable of reversible oxidation to the vanadium(V) state [33]. Initially [32], a VO^{2+} center was postulated coordinated to two $(hidpaH_2)^-$ ligands, each acting as an N,O-chelate. However, this structure does not explain the high formation constants for the vanadium(IV) complexes with hidpa ($\log \beta_2 = 23$) and the related ligand from 2-hydroxyiminodiacetic acid, $HON\{CH_2CO_2H\}_2$ (hidaH_3) ($\log \beta_2 = 22$) when compared with values for similar chelating ligands. Therefore, an alternative structure was proposed, involving two hidpa^{3-} ligands each bonding to a vanadium(IV) center via an η^2-N,O group and two unidentate carboxylato groups, based on the structure of $[NH_4][N(CH_3)_4][V(hida)_2]$ [34]. The availability of structurally characterized chemical analogs of Amavadin (e.g., Fig. 3) allowed the calibration of the XAS data of the natural material (Figs. 1 and 2) [14,35]. The energy of the vanadium K edge is 5480.8 and 5482.4 ± 0.3 eV for these vanadium centers in their (IV) and (V) oxidation states, respectively. These details are consistent with the results of EPR spectroscopy and favor vanadium(IV) as the normal oxidation state of Amavadin.

The pre-edge feature, seen so clearly for the vanadium K-edge spectra of BPO [26] and typical of systems containing a V=O group, is absent from the XAS profile (Figure 1) of the *Amanita muscaria* extract (and those recorded for $[V(hidpa)_2]^{1-/2-}$ and $[V(hida)_2]^{1-/2-}$). The EXAFS for the *Amanita muscaria* extract (Figure 2) is simulated successfully by backscattering contributions from 6 oxygens at 2.02 Å, 2 nitrogens at 2.03 Å, 8 carbons at 2.94 Å and 4 oxygens at 3.89 Å, distances corresponding closely to those found for $[V(hidpa)_2]^{2-}$. Thus, there is no indication of a short V=O contribution to the EXAFS data for Amavadin.

The correspondence of the vanadium coordination sphere in Amavadin to that in $[V(hidpa)_2]^{2-}$ and $[V(hida)_2]^{2-}$, and in its oxidized form to that of $[V(hidpa)_2]^-$ and $[V(hida)_2]^-$, has also been confirmed by EPR, ^{51}V, ^{1}H and ^{13}C nuclear magnetic resonance (NMR) and circular dichroism (CD) spectroscopic studies [35,36]. Thus, Amavadin is proposed to have the eight-coordinate geometry shown in Fig. 3 but with the carbon

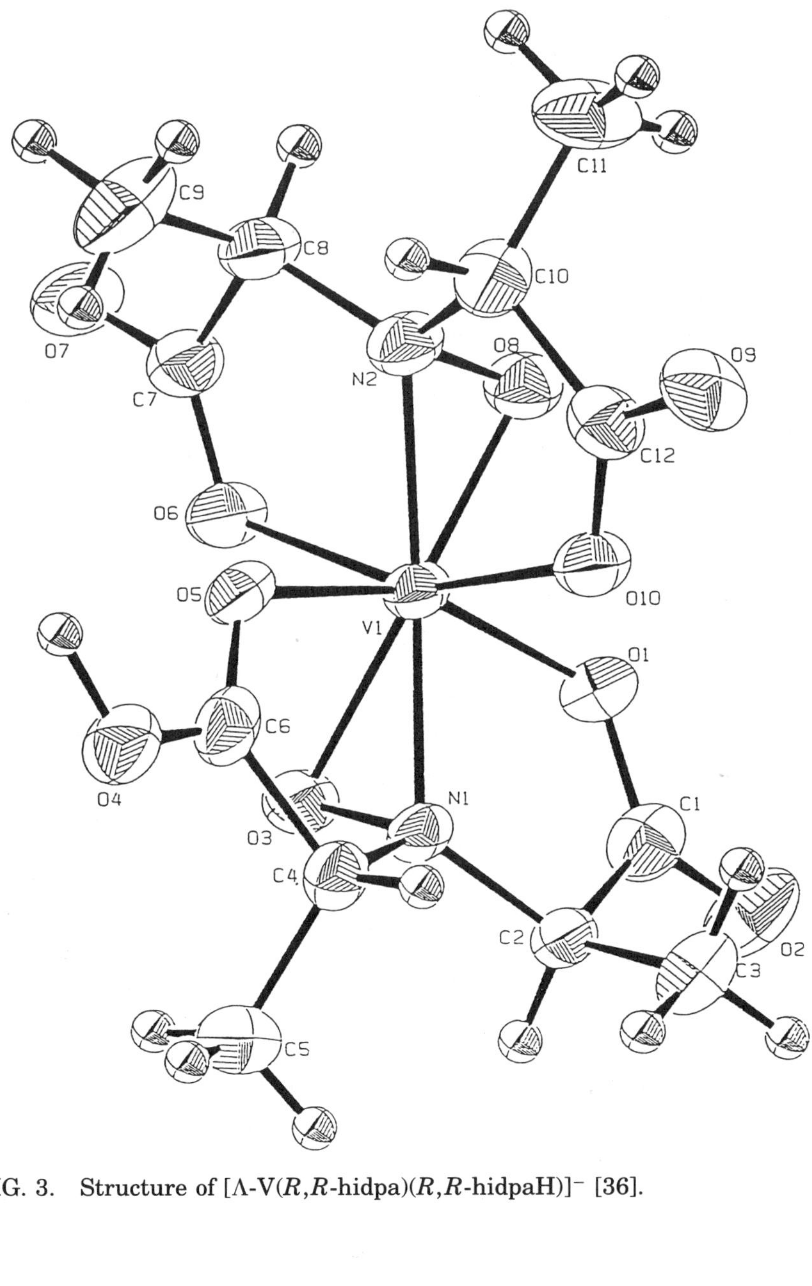

FIG. 3. Structure of [Λ-V(*R*,*R*-hidpa)(*R*,*R*-hidpaH)]$^-$ [36].

centers all in the S configuration and the vanadium being present as an approximately equal mixture of the Δ and Λ isomers [35].

2.4.3. *Vanadium in Ascidians*

The immediate coordination environment of vanadium in ascidian blood cells (vanadocytes) has been probed by vanadium [37] and sulfur [38] K-edge XAS. These results, together with EPR and ^{1}H NMR studies [39], have been interpreted in favor of the presence of vanadium(III), chiefly as a $\{V(SO_4)(H_2O)_{4-5}\}^+$ complex; the vanadium being bound to a shell of about six oxygen atoms at a distance of 1.99 Å. These conclusions have been disputed [40] and further spectroscopic investigations on these systems are in progress [41].

2.4.4. *Vanadium Nitrogenases*

One of the earliest applications of EXAFS to probe a metalloenzyme was the study of the molybdenum site of nitrogenase in the *Clostridium pasteurianum* and *Azotobacter vinelandii* MoFe proteins and the isolated iron-molybdenum cofactor (FeMoco) [42–45]. These studies showed definitively that molybdenum is present as part of a polynuclear cluster containing sulfur and iron, with Mo-S and Mo-Fe distances of ~2.35 and 2.72 Å, respectively. This work inspired the successful development of many chemical systems containing Mo-Fe-S clusters and XAS studies of these systems strengthened the interpretation of corresponding data for the natural system. The environment proposed is compatible with the results of the protein crystallographic investigations of Kim and Rees [46,47]. Thus, the molybdenum is located at one end of an $MoFe_7S_9$ cluster, ligated to three sulfurs of this framework (the imidazole group of His 442 and two oxygens from homocitrate) (Fig. 4). The EXAFS data provide accurate interatomic distances for the molybdenum environment in isolated FeMoco as: three O/N atoms at 2.19 Å, three sulfurs at 2.37 Å, and three irons at 2.70 Å with similar values being obtained for the MoFe protein [45]. These results identified a Mo-Fe separation of ~5.1 Å, corresponding to the three equivalent iron atoms in the other "half" of the cluster.

Genetic suppression of the "normal", molybdenum-dependent nitrogenase of certain classes of *Azotobacter* allows expression of the

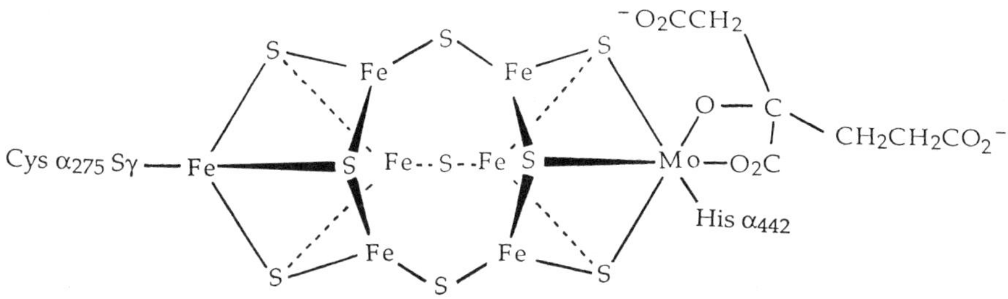

FIG. 4. Structure of the iron-molybdenum cofactor (FeMoco), modified from that in [46].

vanadium-dependent enzyme. The vanadium and molybdenum nitrogenase systems show many similarities and an iron-vanadium cofactor (FeVaco), analogous to FeMoco, has been isolated [48]. Clear evidence of a strong similarity between active sites in the MoFe and VFe proteins has been provided by vanadium K-edge XAS studies. VFe proteins have been investigated from *Azotobacter chroococcum* [49,50] and *Azotobacter vinelandii* [51]. The edge structure shows a weak, single 1s→3d transition, the intensity of which precludes the presence of terminal V=O bonds and implies an octahedral coordination around vanadium. The edge and XANES structure is very similar to that recorded for the VFe_3S_4 cubane-like cluster of $[N(CH_3)_4][VFe_3S_4Cl_3(DMF)_3]$ [52] and the EXAFS results [44–46] are consistent with vanadium in the VFe protein of *A. chroococcum* being ligated by approximately three oxygen (or nitrogen), sulfur, and iron atoms at ~2.13, 2.33, and 2.75 Å, respectively.

The average environment of iron in FeMoco [53] and FeVaco [54] has been investigated by XAS, showing that the two cofactors possess the same molecular topology, especially as the Fe-Fe distances across the central portion of the structure (Fig. 4) at 3.6–3.7 Å are clearly manifest in the EXAFS. *Vanadium would thus appear to substitute directly for molybdenum in FeMoco to form FeVaco.* The role of these metals in the catalysis of N_2 reduction remains to be elucidated. Possibly they facilitate electron transfer to and/or redox reactions within the cluster. Therefore, it is noteworthy that XAS studies indicate that neither the molybdenum nor the vanadium center experiences any significant structural change as the oxidation level of the cofactor is varied.

2.4.5. Vanadium in Oil and Coal

Vanadium is an important "impurity" in many oils and coals, especially as its presence can poison the catalysts used for dehydrosulfurization. Contrasting vanadium environments have been identified by two XAS studies. The vanadium in asphaltenic fractions of Boscan crude oil appears to be present primarily as VO^{2+}-porphyrin complexes with V–O and V–N distances of ~1.61 and 2.08 Å, respectively, and carbon atoms being apparent at ~3.12, 3.49, and 4.20 Å [55]. Coal from the Kentucky No. 9 seam involves vanadium in at least two environments, for both of which coordination to oxygen is suggested with no evidence for porphyrin coordination being obtained [56]. Both environments involve one V–O bond of ~1.6 Å and a group of longer contacts at ~2.0 Å, but the relative backscattering contributions vary within different coal fractions.

3. OPTICAL SPECTROSCOPIES

3.1. Introduction

The electronic absorption spectra of vanadium compounds in the three biologically relevant oxidation states (III), (IV), and (V) may derive from ligand-to-metal, or metal-to-ligand, charge transfer (LMCT or MLCT) processes and/or d-d transitions. The information content of these spectra gives a measure of the electronic excited states and this can be used in conjunction with EPR spectroscopy for an overall description of the electronic structure of the metal site. Electronic absorption spectra can also afford a sensitive measure of the effect of endogenous ligands, substrates, or redox agents. The following sections will consider the use of conventional spectrophotometry as well as plane and circularly polarized light.

3.2. Ultraviolet/Visible/Near-Infrared

3.2.1. Vanadium(V), $3d^0$

The absence of d electrons means that any absorptions in the UV/Vis/NIR region of the spectrum arise from either LMCT or intraligand processes. The solutions of many vanadate(V) species and vanadium(V) bound to oxygen donor ligands give a yellow color which is due to intense LMCT bands tailing from the UV region. The spectrum of thiovanadate, $[VS_4]^{3-}$, is dominated by an intense absorption in the visible region, e.g., $\lambda_{max} = 523$ nm for the Li^+ salt [57]. The binding of metals to this trianion can be readily monitored by band shifts and changes in ϵ values. In heterometallic clusters with iron or copper, the absorption spectrum of the VS_4 aggregate again has a characteristic fingerprint, e.g., $(PPh_3)_4Cu_3VS_4$ has its lowest energy band at 565 nm [58].

Heteroleptic vanadium(V) complexes are often colored. For instance, hydroxamates can be used to determine vanadium(V) colorimetrically through the intense blue (λ_{max} 530 nm, ϵ 4000 Lmol^{-1} cm^{-1}) of the bis(hydroxamate) complex [59]. By far the most abundant class of vanadium(V) complexes are those which contain a single oxo or a *cis*-dioxo group, and these can be found in the presence of many oxygen and/or nitrogen donor ligands and can give rise to colored complexes. Peroxo ligation gives rise to an LMCT process leading to a range of absorptions in the 300- to 460-nm range with ϵ values of a few hundred Lmol^{-1} cm^{-1} [60]. Alcohols, particularly aromatic ones, form vanadate esters which are intensely colored, again due to LMCT bands. Similarly, carboxylate ligands form colored complexes with vanadium(V).

3.2.2. *Vanadium(IV), 3d^1*

The Laporte-forbidden, spin-allowed, d-d transitions form the lowest energy electronic transitions of vanadium(IV) centers. There can be up to four absorption bands, ranging in energy from the NIR to the UV and possessing ϵ values in the range ~20 to ~200 Lmol^{-1} cm^{-1}. However, the highest energy band may be hidden under the much more intense CT features or it may appear as a shoulder on the CT tail.

The vast majority of vanadium(IV) centers are found as the VO^{2+} moiety and the description of the d-orbital splitting pattern derives readily from crystal field theory by considering a distortion from octahedral geometry (Fig. 5). The model for the ordering of the d orbitals for a VO^{2+} center was first treated by Jørgensen [61] and followed by the definitive work of Ballhausen and Gray [62]. Subsequently, there have been many detailed studies of the optical spectra of VO^{2+} complexes and the situation presented in Fig. 5 is generally accepted. The oxo group is

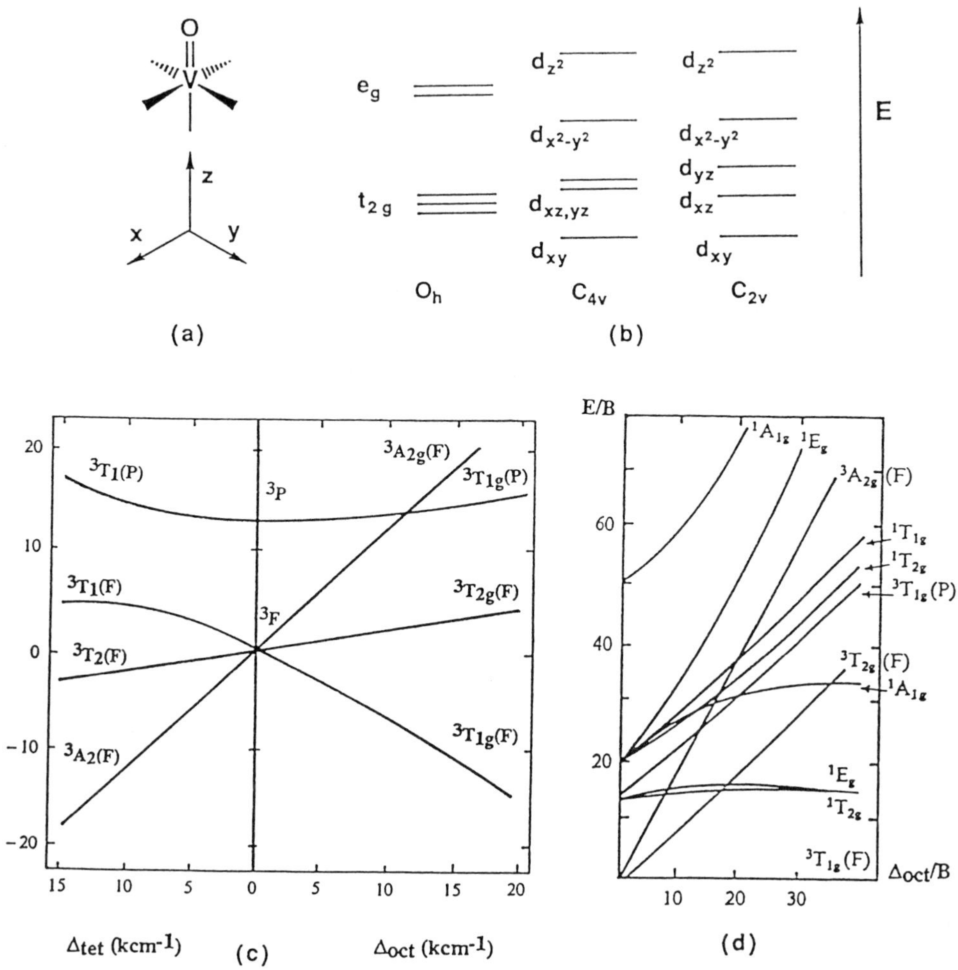

FIG. 5. (a) Representation of the structure of an oxovanadium center with molecular axes defined; (b) d-orbital splitting diagram for a metal center in octahedral (O_h), tetragonally distorted (C_{4v}), and rhombically distorted (C_{2v}) geometry; (c) energy level splitting diagram for the spin triplet states of the d^2-configuration in octahedral symmetry, modified from [64]; (d) Tanabe-Sugano diagram for V(III), also from [64].

both a strong σ and π donor and the distortion from octahedral (O_h) to square-bipyramidal (C_{4v}) geometry proceeds by an increase in the destabilization of the d_{z^2} orbital from the O^{2-} as a σ donor. The filled p orbitals on the O^{2-} interact with the d_{xz} and d_{yz} orbitals to form a degenerate pair of π bonds, shifting these d orbitals to higher energy. With the axis definition of Fig. 5, the d_{xy} orbital forms the ground state and is, to a first approximation, nonbonding. Thus, the orbital ordering $d_{xy} < d_{xz},d_{yz} \approx d_{x^2-y^2} < d_{z^2}$ is defined, but the relative ordering of the d_{xz}, d_{yz}, and $d_{x^2-y^2}$ orbitals depends on the strength of the "in-plane" ligands. For example, four N-bonded NCS^- ligands lead to the order $d_{x^2-y^2} > d_{xz},d_{yz}$, whereas for four Cl^- ligands the order is reversed [63]. If the ligand trans to the oxo group is removed the symmetry remains C_{4v} and the effect is to decrease the energy of the d_{z^2} orbital such that the transition $d_{xy} \rightarrow d_{z^2}$ may move from the UV to the visible region.

Further distortion of the metal site by incorporating mixed ligands or chelate rings, which is the likely situation for biological systems, leads to a lowering of the symmetry (seen in Fig. 5 for C_{2v} point symmetry), but the general pattern of the d-d transitions will remain. The axial distortion by the oxo group is the dominant perturbation of the ligand field. However, the "allowedness" of the transitions changes with the change in point symmetry, although values of ϵ do not rise above ~200 $L mol^{-1} cm^{-1}$ unless there is a substantial CT contribution to the transition.

The presence of the single electron in the d_{xy} orbital is responsible for the LMCT transitions being to higher energy than (most of) the d-d transitions. The considerations in Sec. 3.2.1. still apply to these higher energy transitions but now MLCT processes are also possible.

3.2.3. *Vanadium(III), 3d^2*

The d^2 electronic configuration introduces the added complication of interelectronic repulsions. The origin of the three transitions for the d^2 configuration can be rationalized on the basis of the strong field approximation used in Sec. 3.2.2. for vanadium(IV). In octahedral symmetry the d orbitals are split into t_{2g} and e_g sets. Electron occupation of these orbitals gives a $(t_{2g})^2$ ground state configuration and two excited state configurations, $(t_{2g})^1$ $(e_g)^1$ and $(e_g)^2$. These configurations produce various ligand field terms under the action of interelectronic repulsion.

Because of the spin selection rule for d-d transitions, the terms of importance are the $^3T_{1g}$ ground term from $(t_{2g})^2$, $^3T_{1g}$ and $^3T_{2g}$ from $(t_{2g})^1$ $(e_g)^1$, and $^3A_{2g}$ from $(e_g)^2$. The relative energies of these terms are shown in the Tanabe-Sugano diagram (Fig. 5) [64]. Transitions from the ground state to the excited spin-triplet states give rise to three spin-allowed transitions: $^3T_{1g}(F)$ to $^3T_{2g}$, $^3T_{1g}(P)$ and $^3A_{2g}$; for $[V(H_2O)_6]^{3+}$, these occur at 17 200, 25 600 and 38 000 cm^{-1}, respectively [65]. The ratios of the experimental band energies, when matched to ratios calculated from the Tanabe-Sugano diagram, allow values of Δ_0 and B (the Racah parameter for interelectron repulsion) to be obtained as 18 600 and 650 cm^{-1}, respectively, for $[V(H_2O)_6]^{3+}$. The values of Δ_0 and B follow the spectrochemical and nephelauxetic series, respectively. A descent in symmetry from octahedral increases the number of absorption bands and also increases their intensity as the inversion center at the metal center is destroyed.

The model compounds for vanadium tunichromes show characteristic spectra for the d^2-configuration. $[V(trencam)]^{3-}$, where trencam is a catecholamide derivative of triaminotriethylamine, has a spectrum dominated by a ligand-based $\pi \rightarrow \pi^*$ absorption at 330 nm and d-d absorptions between 550 and 700 nm [66]. On oxidation to $[V(trencam)]^{2-}$ only LMCT bands are observed in the visible region.

3.2.4. *Nonoxo Vanadium Complexes, d^0, d^1, and d^2*

One of the important properties of vanadium(V), (IV), and (III) complexes where no terminal (oxo or sulfido) groups are present is the facile redox activity between the states accompanied by characteristic changes in the optical spectra. Thus, for each of the complexes $[LV]^{n+}$, where $n = 0$, 1, or 2, and LH_3 = 1,4,7-tris(5-*t*-butyl-2-hydroxybenzyl)-1,4,7-triazacyclononane, there is a very intense, low-energy transition assigned to phenolate-to-metal-charge transfer [67]. The band shifts to longer wavelengths on increasing the formal oxidation state of the vanadium and there is a concomitant increase in ϵ value.

$[V(hidpa)_2]^{n-}$, for $n = 1$ and 2, are purple (λ_{max} = 506 nm, ϵ = 241 $L mol^{-1} cm^{-1}$) and blue (λ = 759, ϵ = 22; λ = 563, ϵ = 23), respectively. The latter spectrum accounts for the initial, erroneous assignment of Amavadin as an oxovanadium(IV) compound [35]. In this case, an orbital stacking diagram analogous to that for VO^{2+} (Fig. 5b) may be appropriate for both $n = 1$ and 2, with the latter having a $(d_{xy})^1$ ground state.

3.3. Circular Dichroism

The occurrence of both organic ligands of particular chirality (i.e., amino acids) and chelate rings at the vanadium site in biological systems gives rise to the possibility of optical activity associated with the metal. Circularly polarized light provides a method for distinguishing between the possible stereoisomers [68]. By definition, molecules which lack any element of symmetry or lack a rotation-reflection axis (S_n) are chiral.

Plane-polarized light can be thought of as the sum of two, interfering circularly polarized waves of equal amplitude and opposite rotation along the direction of propagation. The instantaneous sum gives an oscillatory electric field in the plane. A chiral molecule can rotate the plane of polarized light, which has the effect that the velocity of propagation of one of the circularly polarized components is not equal to the velocity of the other. The angle of rotation of plane-polarized light, α, is characteristic of the chiral molecule when measured at a specific wavelength and the change in α as a function of wavelength is called optical rotatory dispersion (ORD). The effect on α in the region of an electronic absorption band is part of the Cotton effect. Enantiomeric molecules can have different ϵ_M toward left and right circularly polarized light; the difference $\Delta = \epsilon_{left} - \epsilon_{right}$ may be positive or negative and is called circular dichroism (CD). Experimentally, plane-polarized light will be converted to elliptically polarized light in the region of an absorption band because one circularly polarized component is absorbed more than the other.

ORD and CD can therefore be used to assist in the determination of absolute configurations around a metal center, since enantiomers possess mirror image Cotton effects and hence mirror image spectra. However, a point of reference is needed and this can often only be obtained from single-crystal X-ray diffraction studies. In practice the usefulness of CD spectroscopy is to compare isomers (Fig. 6) and to study isomerization processes [36].

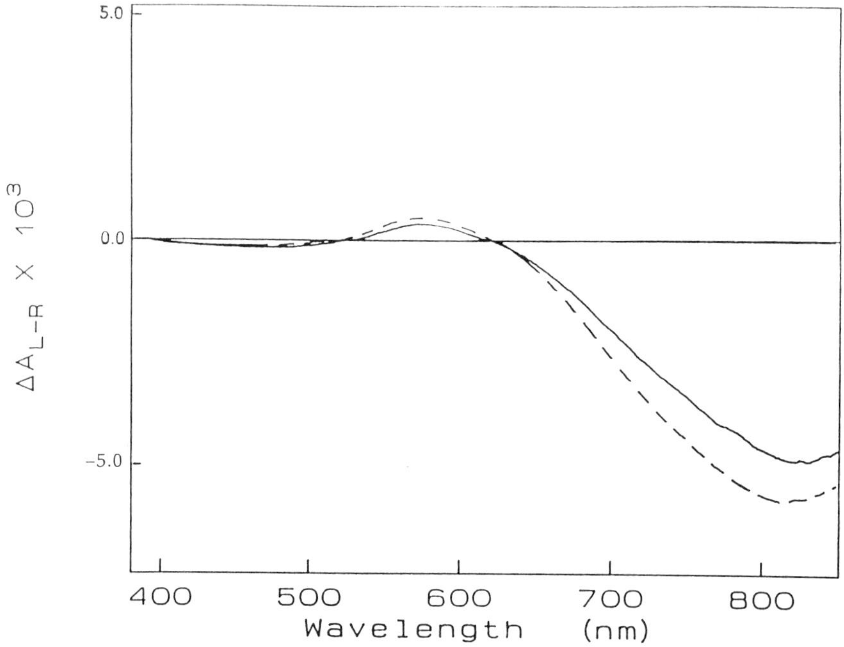

FIG. 6. CD spectra of isolated (——) and synthetic (----) Amavadin in H_2O (76 mM) [36].

3.4. Magnetic Circular Dichroism

In common with the CD spectra of the previous section, magnetic circular dichroism (MCD) spectra are determined from the difference between the absorption of left and right circularly polarized light as a function of wavelength of the incident light. However, in this case the sample is placed in a magnetic field parallel to the direction of propagation of the polarized light. The MCD results from the direct effect of the magnetic field on the degenerate electronic energy levels and does not need a chiral molecule for the effect to be observed. However, the molecule must be paramagnetic. MCD studies can be performed on fluid or frozen solutions of samples with concentrations similar to those used for UV/Vis/NIR spectroscopy. The wavelength range for MCD spans this latter energy region although often different spectrometers are required for the visible and NIR regions. The application of very low temperatures and high applied magnetic fields can give rise to useful information [69].

The application of a magnetic field to a paramagnetic system removes some of the orbital and spin degeneracies, and mixes the wavefunctions of the electronic states. The splitting of the electronic spin state is called the Zeeman effect and is the basis of the EPR method (Sec. 5). The Zeeman splitting ($\sim$1 cm^{-1}) is very much smaller than the bandwidth of a visible electronic transition ($\sim$1000 cm^{-1}) but MCD spectroscopy enables these optical transitions between the Zeeman split sublevels in the ground and excited electronic states to be detected using circularly polarized light. Experimentally, MCD lineshapes are composed of A,B, and C terms, where the A and C terms arise from the splitting of levels by the magnetic field and the B term from mixing of levels by the field. These lineshapes derive from the selection rules for circularly polarized light in electronic absorption (see Fig. 7). The A term is produced by transition to a degenerate excited state, has a derivative shape, and can be positive or negative at short wavelengths. The derivative form arises since the transitions using left and right circularly polarized light are at different energies, but may be of approximately equal intensity. C terms arise if the ground state is degenerate. Here the splitting by the magnetic field will give a differential population of the ground (electronic) state levels such that the intensities of the left and right circularly polarized transitions are unequal. The transition from the lowest energy state will dominate and the MCD band can occur entirely below the spectral baseline. The contribution of the C term is temperature-dependent. The B term is analogous to the term for temperature-independent paramagnetism in magnetic susceptibility and has a lineshape identical to that of the C term, but it is temperature-independent.

In practice, the greater number of features seen in an MCD spectrum, compared with the equivalent electronic absorption (UV/Vis/NIR) spectrum, makes the technique very useful for fingerprinting and can allow the deconvolution of electronic absorption bands. However, by varying the sample temperature or the applied magnetic field more information can be obtained. If two nondegenerate excited states are separated by an energy less than the spectral bandwidth and the ground state is degenerate, then if the C terms for the two transitions

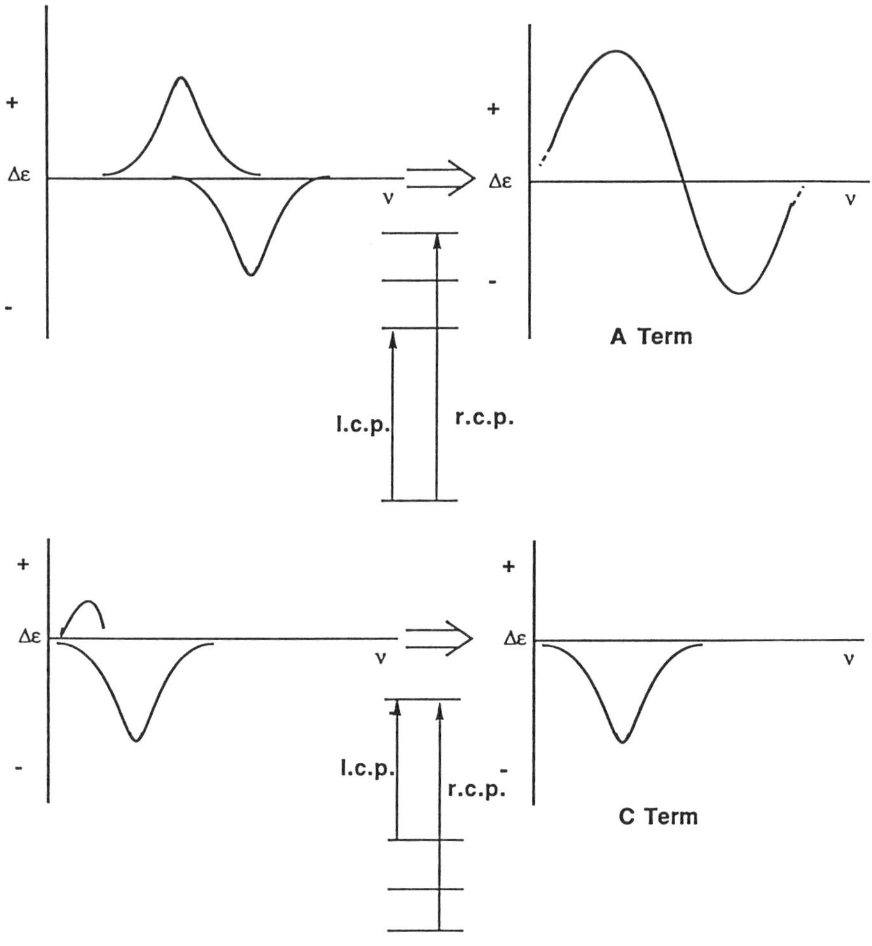

FIG. 7. Origin of A and C terms in MCD using an idealized energy level scheme; r.c.p. and l.c.p. are right and left circularly polarized transitions, respectively.

are of opposite sign, a derivative-shaped line can occur and this is called a pseudo-A term. Similarly, pseudo-A terms can arise from overlapping transitions with B terms of opposite sign. Variable temperature measurements could then distinguish C terms from combinations of A and B terms, and thus assign degeneracies to electronic states.

For an $S = 1/2$ system, the MCD intensity is proportional to the applied magnetic field and inversely proportional to the absolute temperature, as long as kT is greater than the Zeeman splitting in the ground state. At very low temperatures and high fields the sample becomes saturated and the MCD intensity becomes independent of temperature and magnetic field. A plot of MCD intensity as a function of magnetic field divided by temperature represents the MCD magnetization curve. From this sort of curve it is possible to derive ground state g values and zero field splittings when $S > 1/2$.

MCD spectroscopy has been applied widely to iron and copper proteins but only rarely to vanadium in coordination complexes [70] or biological systems.

4. VIBRATIONAL SPECTROSCOPIES

The appearance of characteristic frequencies of groups of atoms makes vibrational spectroscopies valuable both as a fingerprint and for use in structural elucidation [71]. For vanadium(V) and (IV) centers, modes associated with terminal or bridging oxo groups are observed in the range 700–1050 cm^{-1} and are indicative of V–O bond orders greater than unity. Bond-stretching modes associated with vanadium-ligand single bonds are usually found at <450 cm^{-1}. Frequencies characteristic of ligand-based modes are also diagnostic, e.g., ν(O–O) in peroxo complexes and ν(N–O) in η^2-hydroxy- or oxy-oximes. However, care is required in interpretation, e.g., the band at 980 cm^{-1} in the IR spectrum of Amavadin and in complexes of various isomeric forms of $[V(hidpa)_2]^{2-}$ (Sec. 2.4.2) derives from a CH_3-C mode and not a V=O stretching vibration.

4.1. Infrared

Experimentally, the measurement of IR spectra for low molecular weight coordination complexes is routine and reproducible data with a high signal-to-noise ratio can be obtained. Solid state spectra can be measured as mulls (nujol, hexachlorobutadiene), pressed disks (KBr,

CsI, polyethene), or evaporated films using Fourier transform infrared (FTIR) spectrometers.

The use of solution cells of small pathlength (~1 mm) allows solution phase measurements to be made, provided that the solvent has a "window" (i.e., no absorption) in the region of interest. This is the factor which causes the major limitation in IR measurements on biological systems. These are almost invariably obtained as high molecular weight species in aqueous media, and the broad background absorptions of water molecules dominate the spectrum. This problem can be obviated by application of Fourier transform attenuated total reflection infrared (FTATR IR) methods. Here, a boat containing, say, a zinc selenide crystal is loaded with the aqueous solution. The incident IR beam passes along the crystal, being reflected backward and forward between the crystal and solution interface, thus forming a multiple-pathlength experiment. The limiting factor is the recording of a relevant reference (background) and sample spectra under the same conditions; each spectrum is very sensitive to small movements in the sample with respect to the incident beam.

4.2. Raman

Raman spectroscopy is complementary to IR but the process of energy transfer arises due to a change in polarizability between the ground and excited states, rather than to a change in dipole moment, during the vibrational process [7,71]. Although the vibrational energies are within the IR region of the electromagnetic spectrum, a Raman spectrometer operates by using visible light, usually from a laser source of high intensity. The photons of interest are not absorbed but undergo inelastic scattering in which some of the photon energy is given up to vibrational transitions. Thus, the scattered photon is at a lower energy than the incident photon. This is called Stokes scattering and the unchanged frequency, ν_0, is termed the Rayleigh line. Because some of the molecules may initially be in an excited vibrational state, the incident photon may be scattered with an enhanced energy and appear on the high-energy side of the ν_0 line. Such a feature is called an anti-Stokes line but, since the majority of molecules will be in a ground vibrational state at room temperature, the Stokes lines will be much stronger than the anti-Stokes lines and are the ones usually detected in the Raman experiment. The Raman effect is weak with only one in $\sim 10^{12}$ photons scattered with a shift in energy, which is why a powerful laser source, sensitive photomultiplier detectors, and spectral accumulation are necessary to obtain good-quality spectra. A Raman spectrometer uses a monochromatic laser source and the scattered photons are detected at right angles to the incident beam, and the difference in frequency between the incident and scattered photons corresponds to the frequency of the vibrational modes.

Because IR-active and Raman-active modes derive from different selection rules, a combination of IR and Raman spectra can allow a more complete assignment of the vibrational processes. If a molecule has a center of symmetry, none of the modes can be both IR- and Raman-active and, under some circumstances, a vibrational mode may be inactive in both forms of spectroscopy. The light scattered in the Raman effect is polarized and this can afford additional information regarding the symmetry of the vibrational process involved. If two experiments are performed, in which the incident light is plane-polarized along the direction of propagation in two mutually perpendicular planes, some of the Raman lines will have the same intensity in the two experiments, while others may be of very different intensity. The lines with different intensity are termed *polarized* and correspond to totally symmetric vibrations. The lines with comparable intensity in the two experiments are termed *depolarized* and correspond to nontotally symmetric vibrations. Experimentally, it is often easier to perform Raman spectroscopy on aqueous solutions or on single crystals than to make equivalent measurements by IR spectroscopy.

4.3. Resonance Raman

Where relatively intense CT bands are present in the electronic absorption spectrum, the system may be amenable to study by resonance Raman spectroscopy. In contrast to conventional Raman spectroscopy, here an incident laser line frequency is used which falls within the envelope of an electronic absorption bond of the molecule. There can be an enhancement of the Raman effect for those vibrational modes which are associated with the electronic transition. Resonance Raman spectra often contain progressions in vibrational frequencies which are multiples of the fundamental frequency. An intense effect is seen if there is a change in geometry between the ground and excited electronic states of the molecule. Thus, the resonance Raman effect can aid assignment of electronic absorption bands and allow some structural information to be obtained. Also, since CT processes underlie this form of vibrational spectroscopy, the enhancement of the Raman effect may be correlated with the solvent dependence of the CT transition, i.e., through solvatochromism. Resonance Raman spectroscopy has been used to study the excited state transient species of vanadyl uroporphyrin I, [VO(uroP)], in basic solution, where photolability of OH^- and complex dimerization were detected [72].

5. ELECTRON PARAMAGNETIC RESONANCE SPECTROSCOPY

5.1. Introduction

Electron paramagnetic resonance (EPR) spectroscopy embraces techniques which provide extremely powerful probes of the electronic structures of materials with unpaired electrons. There is a range of EPR techniques: continuous wave (c.w.), electron nuclear double resonance (ENDOR), and pulsed techniques including electron spin echo envelope modulation (ESEEM). Each of these probes the system in a slightly different way and consequently provides information on electronic interactions of differing magnitudes. We consider c.w. EPR, ENDOR, and ESEEM separately. Although there can be some overlap in the information provided by each technique, each has its own special strengths and c.w. EPR is the technique of choice for obtaining information on what is normally the strong interaction of the unpaired electron(s) with the metal ion. Although c.w. EPR can give information when the unpaired electrons are delocalized onto ligand atoms which have nonzero nuclear spin, it has some serious limitations; therefore ENDOR and ESEEM are the preferred techniques.

The interactions which may be detected via EPR spectroscopy can be summarized in the general spin Hamiltonian:

$$\begin{aligned}\mathbf{H} = {} & \beta_e B \cdot g \cdot \mathbf{S} + \mathbf{S} \cdot D \cdot \mathbf{S} + \mathbf{S} \cdot A \cdot \mathbf{I} + \beta_n B \cdot g_n \cdot \mathbf{I} + \mathbf{I} \cdot P \cdot \mathbf{I} + \\ & \Sigma(\mathbf{S} \cdot a(i) \cdot \mathbf{I}(i) + \mathbf{I}(i) \cdot P(i) \cdot \mathbf{I}(i) + \beta_n B \cdot g(i) \cdot \mathbf{I}(i))\end{aligned} \tag{5}$$

where g is a tensor* representing the spin and orbital angular momenta of the electrons; D is the zero-field splitting tensor; A is a tensor* representing the interaction between the electron spin and the nuclear spin of the metal; g_n is a tensor* representing the interaction of the nuclear spin with the applied magnetic field, B; P is the nuclear quadrupole tensor of the metal; $a(i)$ is a tensor* representing the interaction between the electron spin and the nuclear spin of the ith ligand; $P(i)$ is the nuclear quadrupole tensor of the ith ligand; $g(i)$ is a tensor* representing the interaction of the nuclear spin of the ith ligand with the applied magnetic field; $\mathbf{S}$ and $\mathbf{I}$ are the electron spin and nuclear spin operators for the metal, and $\mathbf{I}(i)$ is the nuclear spin operator for the ith ligand.

Each of the above tensors is characterized by three mutually orthogonal principal values and their corresponding directions with respect to the atomic framework of the paramagnetic center. The directions of the principal values for any one tensor need not be coincident or coparallel with the directions of any of the other tensors. As we will see later, the magnitudes and relative directions of these principal values can provide valuable information concerning the electronic and geometric structure of a paramagnet.

The terms in (5) which are required to interpret an EPR spectrum depend on the number of unpaired electrons and on whether or not the various nuclei in the paramagnet have nonzero nuclear spins and nuclear quadrupole moments. The relevant oxidation states of vanadium are (IV) and (III), with one and two unpaired electrons, respectively. However, because of zero-field splitting effects, EPR spectra from vanadium(III) systems are difficult to observe.

5.2. The Effects of Point Symmetry

It is important to make the distinction between point symmetry and the more commonly used descriptions of the shapes of complexes. The point symmetry at the metal determines whether or not any of the principal *values* of g or of A are *required* to be equal to each other [8]. Also, it

---

*These are not strictly tensors, although experimentally they often have the form of a tensor and are usually referred to as such.

determines whether or not any of the principal *axes* of g and A are *required* to be coincident. These criteria are summarized in Table 2 with the accepted nomenclature for EPR behavior and their associated point symmetries [73,74]. Thus, each type of EPR behavior is associated with a restricted number of point symmetries. This in turn places constraints on the geometric structures of a paramagnet exhibiting a particular EPR symmetry. For example, if we can demonstrate experimentally that a paramagnet really has, say, rhombic EPR symmetry, then the associated geometry must belong to one of the point groups D_{2h}, C_{2v}, or D_2. It would be incorrect to assign a structure which belongs to a more symmetric arrangement, e.g., D_{4h} which is strictly axial.

For a system of unknown structure, if the EPR symmetry can be

TABLE 2

Relationships Between g and A Tensors, EPR Symmetry, and the Point Symmetry of Paramagnets [73,74]

EPR symmetry	g and A tensors	Coincidence of tensor axes	Molecular point symmetry
Isotropic	$g_{xx} = g_{yy} = g_{zz}$ $A_{xx} = A_{yy} = A_{zz}$	All coincident	O_h, T_d, O T_h, T
Axial	$g_{xx} = g_{yy} \neq g_{zz}$ $A_{xx} = A_{yy} \neq A_{zz}$	All coincident	D_{4h}, C_{4v}, D_4, D_{2d}, D_{6h}, C_{6v}, D_6, D_{3h}, D_{3d}, C_{3v}, D_3
Rhombic	$g_{xx} \neq g_{yy} \neq g_{zz}$ $A_{xx} \neq A_{yy} \neq A_{zz}$	All coincident	D_{2h}, C_{2v}, D_2
Monoclinic	$g_{xx} \neq g_{yy} \neq g_{zz}$ $A_{xx} \neq A_{yy} \neq A_{zz}$	One axis of g and A coincident	C_{2h}, C_s, C_2
Triclinic	$g_{xx} \neq g_{yy} \neq g_{zz}$ $A_{xx} \neq A_{yy} \neq A_{zz}$	Complete noncoincidence	C_1, C_i
Axial noncolinear	$g_{xx} = g_{yy} \neq g_{zz}$ $A_{xx} = A_{yy} \neq A_{zz}$	Only g_{zz} and A_{zz} coincident	C_3, S_6, C_4, S_4, C_{4h}, C_6, C_{3h}, C_{6h}

determined, we have invaluable structural information since the paramagnet can only belong to a restricted range of point symmetries. The problem is a practical one of how to obtain reliably the above type of information. As in all spectroscopic techniques, there is often a lack of uniqueness in the interpretation of the experimental data. This may arise either from the inadequacies of the data or from different combinations of interactions leading to the same spectrum.

5.3. Continuous Wave EPR Spectra from Vanadium(IV)

Spectra may be obtained on fluid solutions, frozen solutions, powdered solids, or single crystals. In the context of vanadium in biological systems, it is most likely that fluid and frozen solutions will be involved.

5.3.1. Fluid Solutions

In fluid solutions we would expect the paramagnets to be tumbling. If the tumbling rate is infinitely faster than the time for an EPR transition ($\sim 10^{-8}$ s^{-1}) then any anisotropy in the g and A tensors will be averaged and the system will appear to be isotropic. The appropriate spin Hamiltonian is:

$$\mathbf{H} = \beta_e B g_{av} \mathbf{S}_z + A_{av}(\mathbf{S}_x\mathbf{I}_x + \mathbf{S}_y\mathbf{I}_y + \mathbf{S}_z\mathbf{I}_z) \tag{6}$$

where $g_{av} = (g_{xx} + g_{yy} + g_{zz})/3$ and $A_{av} = (A_{xx} + A_{yy} + A_{zz})/3$. The application of (6) to the spin functions of vanadium(IV), $S = ½$, ^{51}V $I = 7/2$, leads to the energies of each m_s level being split into eight components. The selection rules $\Delta m_s = \pm 1$, $\Delta m_I = 0$ result in eight ($2I + 1$) lines of equal relative intensities. Tumbling rates in solution, although faster than the time for an EPR transition, are not usually infinitely faster with the result that the averaging of g and A is incomplete. The consequence of this incomplete averaging is to give m_I dependent linewidths, $\Delta B(m_I)$, which leads to variations in the amplitudes of the lines in the spectrum. The linewidths are given by (7) [75]:

$$\Delta B(m_I) = a' + a'' + bm_I + c(m_I)^2 + d(m_I)^3 \tag{7}$$

where: a', b, c, and d are related to the g- and A-tensor elements and the rotational correlation time; a'' is a residual linewidth which embraces other line-broadening mechanisms. The coefficients in (7) can be analyzed to give rotational correlation times [75–79].

5.3.2. Spectra from Powders or Frozen Solutions; Metal g and A Values

These types of spectra are the ones most commonly encountered when investigating materials of unknown structure. In this chapter the term powder spectrum will be used to cover spectra from either powders or frozen solutions. For these types of measurement to be useful, three essential criteria must be satisfied [80].

1. The paramagnets *must* be diluted at the *molecular* level. Thus for powders, no matter how finely they are powdered, *mechanical mixing* is unsatisfactory.
2. The paramagnets *must* be completely randomly oriented.
3. The randomness of the paramagnets in the sample should always be checked by repeating the spectrum after the sample has been turned through an angle, e.g., ~20°. If the two spectra are not *identical*, then the paramagnets in the sample are not completely randomly oriented.

Determination of the EPR symmetry and the spin Hamiltonian parameters from spectra is the next important step. This step is often inadequately performed, leading to possible incorrect assignments of the EPR symmetry and hence erroneous assignment of the electronic and geometric structures. With the aid of some selected examples, both the power and the limitations of methods of determining the EPR symmetry from powder spectra will be shown.

The simplest situations arise when there are no hyperfine interactions. In these cases it is usually a question of determining whether or not the paramagnet has axial or rhombic g values. If the anisotropy, in for example two of the g values, is such that at a particular microwave frequency the resonance magnetic fields are not well separated compared with the linewidths, then the spectra could be (mis-)assigned to an EPR symmetry which is too high. When spectra appear to correspond to axial EPR symmetry at, say, X band, it is important to do at least one but preferably two checks: (1) a full simulation of the spectrum, including linewidths and lineshapes as variables, and (2) a recording of the spectrum at a higher microwave frequency. Note that when only the g-value anisotropy is the experimental observable, there is a much greater choice of associated point symmetries. For example, for rhombic g values the point symmetry of the paramagnet could be any of those corresponding to the EPR symmetries rhombic, monoclinic, or triclinic (see Table 2).

The presence of a metal hyperfine interaction can be extremely useful in two particular ways. It provides an additional means of detecting anisotropy in the system and the possibility of determining the coincidence or otherwise of the principal axes of the g and A tensors. Let us first consider a typical visual difference between the spectra of an axial and a rhombic paramagnet. Figure 8 shows X-band spectra for two

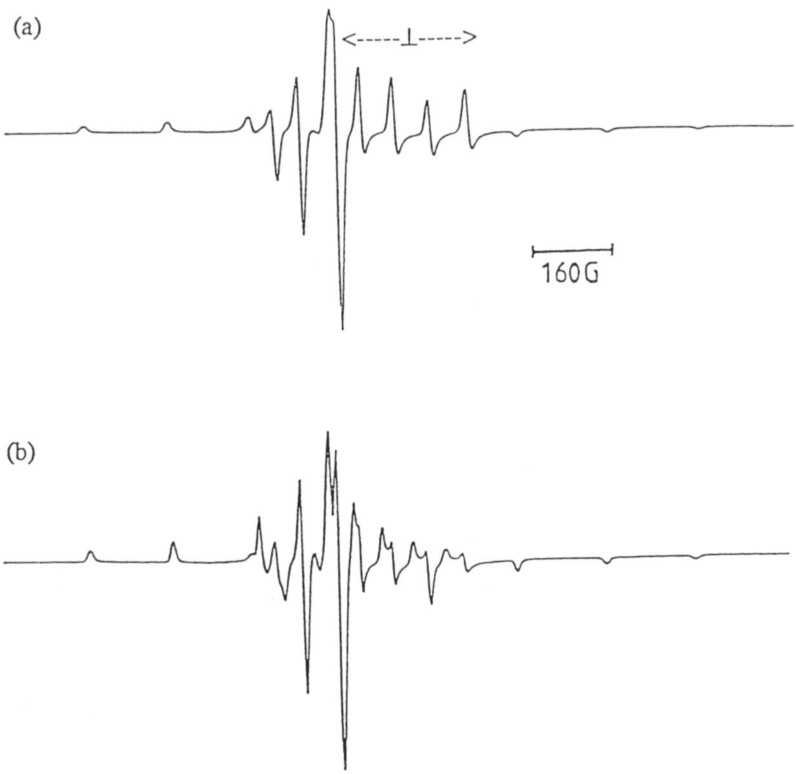

FIG. 8. Comparison of the experimental first-derivative powder c.w. EPR spectra of (a) an axial system represented by $[VO(mim)_4Cl]^+$, where mim = 1-methylimidazole, and (b) a rhombic system represented by $[VO(meox)_2]$, where meox = 2-methylquinolin-8-olato.

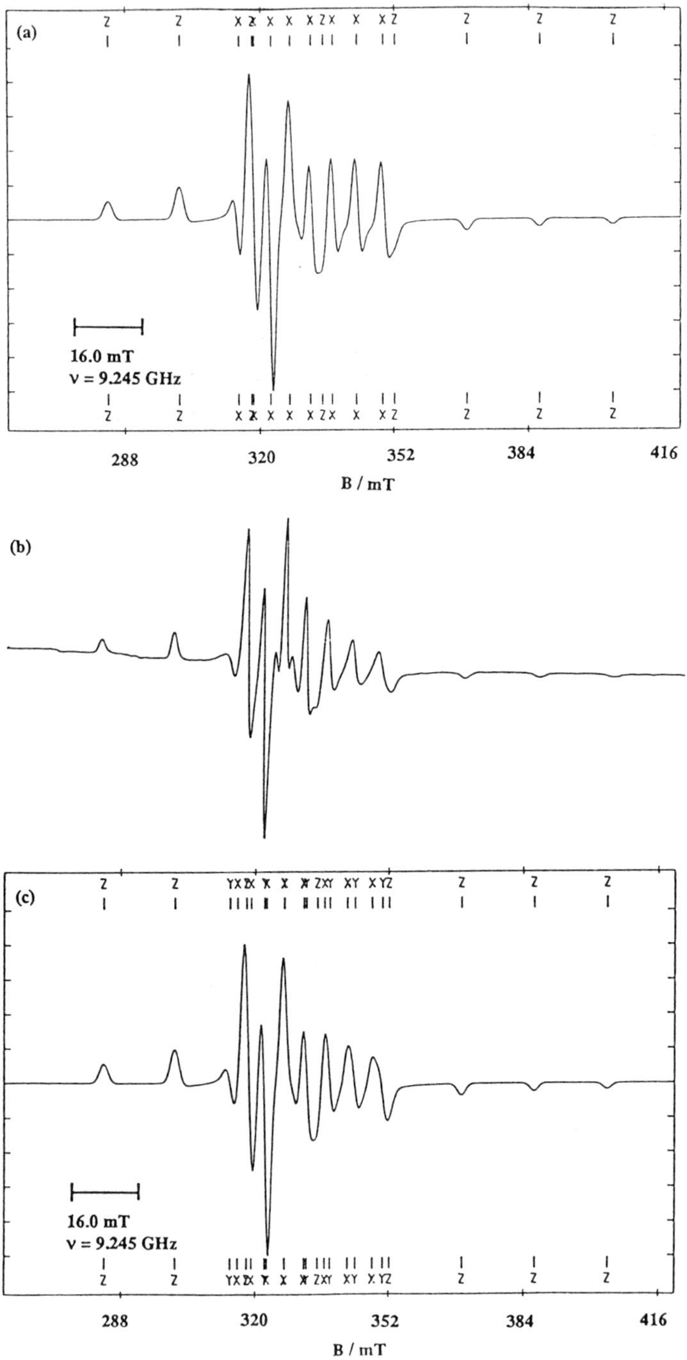
(a)
16.0 mT
ν = 9.245 GHz
288
320
352
384
416
B / mT
(b)
(c)
16.0 mT
ν = 9.245 GHz
288
320
352
384
416
B / mT

vanadium(IV) species, one of which has axial and one of which has rhombic, or lower, EPR symmetry. In these cases the major differences between the spectra are the splittings, which occur in the rhombic spectrum, of those features corresponding to those labeled "⊥" in the axial spectrum.

An example which represents a severe test of the combined use of more than one microwave frequency and of spectrum simulation is provided by $H_2[V(R,R\text{-hidpa})_2]$ either in frozen solution or diluted in the corresponding zirconium compound [81] (Figs. 9 and 10). Initial inspection of *both* the X-and Q-band spectra suggests that the system has axial EPR symmetry, since the features in the region of the spectra associated with the x and y directions (equivalent to those labeled "⊥" in Fig. 8) of the system are not split. However, all attempts to simulate both the X- and Q-band spectra, based on axial EPR symmetry, are not entirely satisfactory; cf. Figs. 9a and b, and 10a and b. The introduction of a small anisotropy into both g_{xx}, g_{yy} and A_{xx}, A_{yy}, produces an improved simulation at both frequencies; see Figs 9c and 10c. Although the rhombicity is small it is consistent with the known structure of the complex [35,36] for which the highest possible point symmetry at the vanadium is C_2 (Fig. 3).

The following example also provides additional reasons for performing detailed spectrum simulations. The features in a powder spectrum arise from turning points in the angular variation of the resonance fields within the system [82]. Turning points in $S = ½$ systems occur when the magnetic field is parallel to the principal axes of the g and A tensors. These features are marked X, Y, or Z on the simulated spectra in Fig. 11. However, for some combinations of the principal g and A values there may be turning points at other orientations, i.e., "off-axis" turning points. This gives rise to what are termed "overshoot" features, which are often observed in the X-band spectra of copper(II)

FIG. 9. Comparison of the experimental (b) and simulated (a) and (c) X-band powder spectra of $H_2[V(R,R\text{-hidpa})_2]$ diluted in the corresponding zirconium compound. (a) axial with $g_{\parallel} = 1.9195$, $g_{\perp} = 1.9839$, $A_{\parallel} = 17.15$ mT, $A_{\perp} = 4.90$ mT $\Delta B(p-p)_{\parallel} = \Delta B(p-p)_{\perp} = 2.0$ mT, (c) rhombic with $g_{zz} = 1.9195$, $g_{xx} = 1.9848$, $g_{yy} = 1.9829$, $A_{zz} = 17.15$ mT, $A_{xx} = 4.60$ mT, $A_{yy} = 5.2$ mT, $\Delta B(p-p)_{zz} = 2.0$ mT, $\Delta B(p-p)_{xx} = \Delta B(p-p)_{yy} = 1.6$ mT. A Gaussian lineshape was used in the simulations.

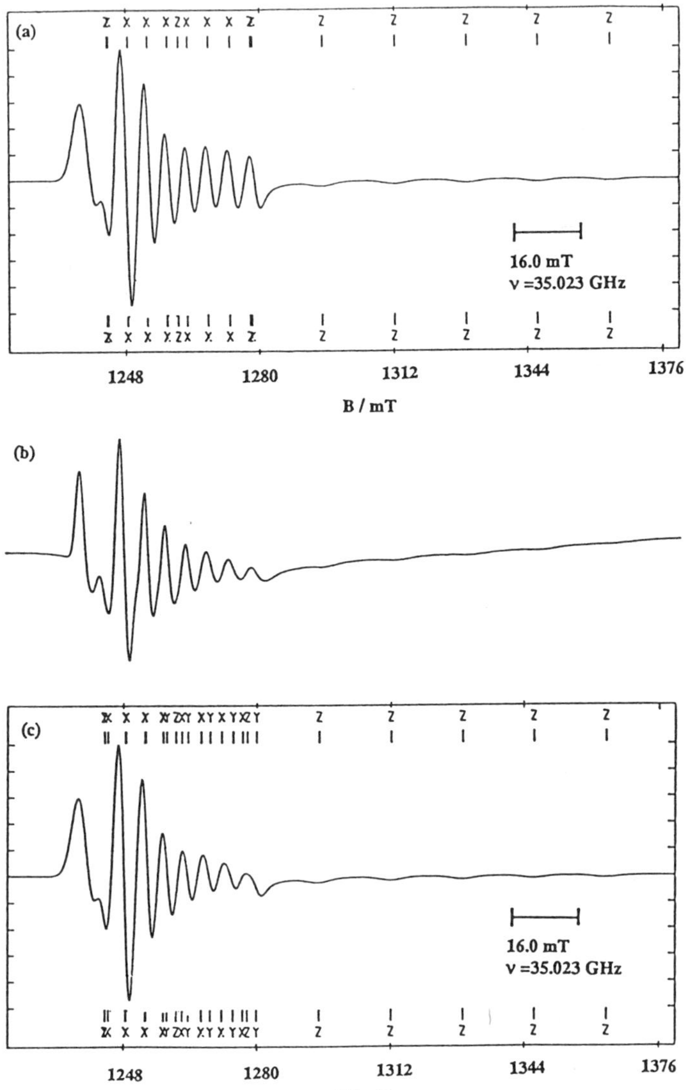

FIG. 10. Comparison of the experimental (b) and simulated (a) and (c) Q-band powder spectra of $H_2[V(R,R\text{-hidpa})_2]$ diluted in the corresponding zirconium compound. (a) Axial with $g_{\parallel} = 1.9195$, $g_{\perp} = 1.9839$, $A_{\parallel} = 17.15$ mT, $A_{\perp} = 4.90$ mT, $\Delta B(p-p)_{\parallel} = 5.4$ mT, $\Delta B(p-p)_{\perp} = 2.4$ mT, (c) rhombic with $g_{zz} = 1.9195$, $g_{xx} = 1.9848$, $g_{yy} = 1.9829$, $A_{zz} = 17.15$ mT, $A_{xx} = 4.60$ mT, $A_{yy} = 5.2$ mT, $\Delta B(p-p)_{zz} = 5.4$ mT, $\Delta B(p-p)_{xx} = \Delta B(p-p)_{yy} = 2.4$ mT. A Gaussian lineshape was used in the simulations.

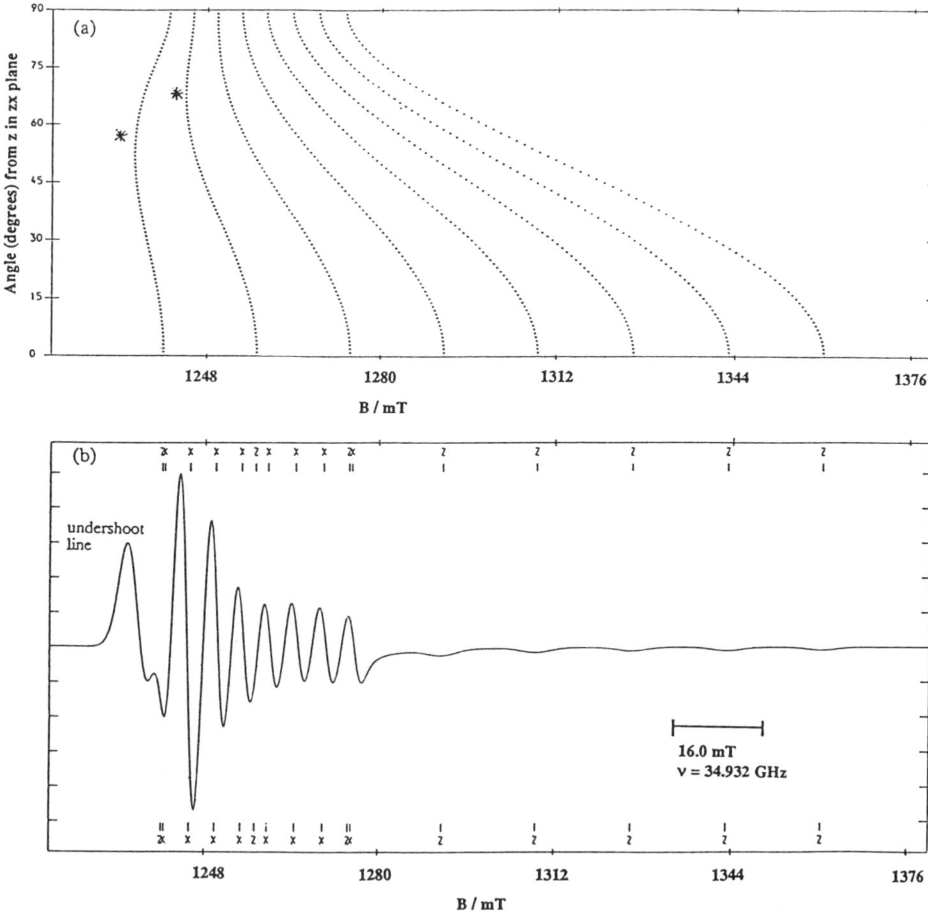

FIG. 11. Simulation of a Q-band axial powder spectrum of a vanadium(IV) species showing "undershoot" features. (a) Angular variation of the resonance fields with θ, showing "off-axis" turning points *. (b) Simulated spectrum for $g_{\|} = 1.9195$, $g_{\perp} = 1.9839$, $A_{\|} = 17.15$ mT, $A_{\perp} = 4.90$ mT, $\Delta B(p-p)_{\|} = 5.2$ mT, $\Delta B(p-p)_{\perp} = 2.4$ mT. Resonance fields at the axes are marked Z and X for the parallel and perpendicular directions, respectively. Any features not so marked belong to off-axis turning points.

species [83]. In Fig. 11 there are also off-axis turning points, but these now occur at the low-field end of the spectrum and can thus be termed "undershoot" features [81]. The features attributed to the off-axis turning points can have greater amplitudes in the powder spectra than some of those which belong to the principal directions. At first sight it would be tempting to assign the overshoot or undershoot features to the principal directions. Such an assignment would lead either to incorrect principal values of g and A, if the spectra were interpreted by simply "reading off" magnetic field positions, or to an inability to interpret all of the features in the spectra.

Trying to determine whether or not the g and A tensors are coincident is a much more difficult task. It is important that any putative noncoincidence be confirmed by *simulation* of spectra obtained at *more than one microwave frequency*. The first indication of noncoincidence is usually that there appear to be too many features in the spectrum or that features are in the wrong place compared with those predicted from simple considerations with the g and A tensor axes coincident. Some caution is advised with this approach, as one must be sure that any extra features are not due to impurities, or that failure of the extrapolated features to match those in the spectrum is not due to second-order effects. An additional indication that there is noncoincidence is the inability to simulate spectra assuming coincidence.

The considerations in the previous paragraph require suitable simulation programs and also considerable effort. However, if the information concerning noncoincidence of the g and A tensor axes can be obtained it restricts further the point groups to which the paramagnet can belong. The angles of noncoincidence between the directions of the principal values of g and A depend on the spin-orbit coupling of the metal ion and inversely on the energy separations to the excited states which are mixed into the ground state by spin-orbit coupling. The spin-orbit coupling constant for vanadium(IV) is small. This usually results in small angles of noncoincidence. If the angle of noncoincidence is small, say $<10°$, then it may be difficult to detect from powder spectra. The simulated spectra are often insensitive to such small angles of noncoincidence, particularly when the anisotropy in both the g and A tensor elements being rotated is small. Such small anisotropies are usual for vanadium(IV) systems. Sometimes the lower the experimental microwave frequency the easier it is to detect the presence of noncoincidence [84,85].

5.3.3. *Ligand Hyperfine Interactions*

The unpaired electron in the ground state of vanadium(IV) is usually in an essentially nonbonding orbital. This often leads to ligand hyperfine couplings which are smaller than the peak-peak linewidths in the c.w. EPR spectrum, with the consequence that the splittings are not often observed at high microwave frequencies, e.g., X or Q band. Lower frequency c.w. EPR offers some possibility of improved resolution. When coupling to a proton is suspected but not resolved in the spectrum, a narrowing of the linewidths upon deuteration can sometimes be used to indicate the presence of the interaction, as for example in the reduced form of BPO [86–88]. An additional complication, particularly important for low-symmetry systems, is that the principal axes of the ligand hyperfine tensor will not be coparallel with the principal axes of the metal g and A tensors. Thus, the maximal effect of the ligand hyperfine interaction does not occur at the turning points in a powder spectrum. Such a situation is observed for $[InCl\{VO\}Cl_2(tmu)_2]$, where tmu = *N,N,N′,N′*-tetramethylurea [89]. The c.w. EPR powder spectra at both X- and Q-band frequencies show no hyperfine splittings attributable to ligand nuclei, but single-crystal studies show splittings from two equivalent chlorine atoms. Analysis of these latter data show that the principal axes of the chlorine hyperfine and quadrupole tensors do not coincide with those of the vanadium g and A tensors.

The measurement of ligand hyperfine and quadrupole coupling requires other EPR methods which detect these interactions in a different way. The methods most appropriate are ENDOR (Sec. 5.4) and ESEEM (Sec. 5.5) spectroscopies.

5.4. ENDOR Spectroscopy

The basic ENDOR experiment consists of setting the magnetic field at the value for an EPR transition. If the microwave power is high enough to give significant saturation of the EPR signal, and the system is swept with radiofrequency radiation, there will be an increase in the EPR absorption when the radiofrequency equals the frequencies for transitions *between* the energy levels *within each* m_s manifold. Sharp lines result even when the associated c.w. EPR transition is too broad to permit resolution. Since the energy separations of the levels within an

m_s manifold are dependent on the ligand hyperfine and quadrupole interactions, ENDOR spectroscopy provides a means of measuring these two interactions. ENDOR sensitivity decreases with decreasing magnitude of the hyperfine interaction; thus small interactions, less than approximately 3–4 MHz, may be difficult to detect. In addition, c.w. ENDOR sensitivity is only ~0.1% of that of c.w. EPR, although this low sensitivity may be improved by nearly two orders of magnitude by using pulsed ENDOR. Pulsed ENDOR methodology was recently reviewed [90].

In a first-derivative powder c.w. EPR spectrum there are a number of features which are attributable to EPR transitions associated with the principal axes of the g and A tensor. These features arise from turning points in the EPR absorption. By setting the magnetic field successively at each of these positions, we may expect to obtain an ENDOR spectrum with respect to these directions, i.e., an angle-selected ENDOR spectrum [91–94]. Because the powder spectrum consists of the summed spectra from a random collection of paramagnets, any particular magnetic field within the spectrum may have contributions from a range of "molecular" orientations. It is important to recognize this when performing angle-selected experiments. Even the "obvious" strong features in the first-derivative c.w. EPR spectrum which are mainly associated with principal directions may contain significant contributions from other orientations.

5.5. ESEEM Spectroscopy

ESEEM spectroscopy, one of the methodologies within pulsed EPR, also measures the frequencies for transitions *between* the energy levels *within each* m_s manifold. However, in contrast to ENDOR spectroscopy, this is achieved by the application of a sequence of pulses of microwave radiation. There are a number of excellent reviews of pulsed methodologies by Schweiger et al. [95–98]. The response of the system to these pulses is to modulate the free-induction decay by the NMR frequencies in the m_s levels. ESEEM spectroscopy is more sensitive than ENDOR for detecting low-frequency transitions.

The magnitude of the interaction between an unpaired electron and a nuclear spin usually correlates with the distance of the interact-

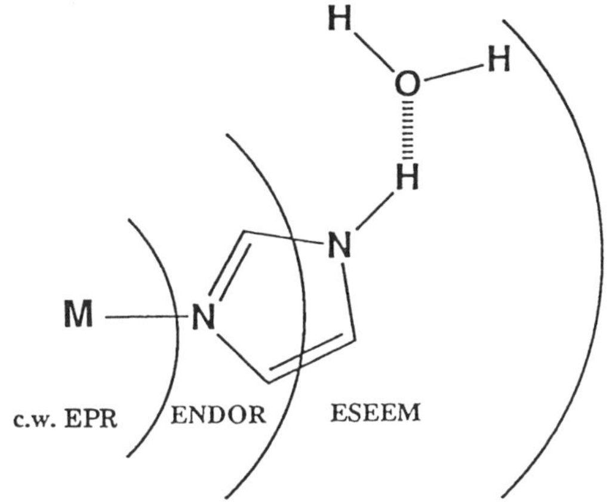

FIG. 12. Schematic diagram showing the areas with respect to the metal atom for which the various EPR techniques are usually most suited.

ing nucleus from the unpaired electron. On this basis, the usefulness of various EPR techniques can be summarized as in Fig. 12, where, by way of illustration, the ligands are taken as a coordinated imidazole and a more distant water molecule. The various regions in Fig. 12 should not be taken as mutually exclusive.

5.6. Selected Applications of ENDOR and ESEEM Spectroscopies for Vanadium(IV) Systems

One of several uses of ENDOR and ESEEM spectroscopies is the detection of the interaction between the unpaired electron and the nuclear spins of nitrogen and hydrogen atoms. The magnitudes and the orientations of the ^{14}N hyperfine and quadrupole tensors, relative to the metal g tensor, can provide structural information. This seems to be particularly the case when the nitrogen atoms are in ring systems such as the imidazole group of histidine.

An early angle-selected ENDOR study on some model complexes which contained imidazole groups bound to VO^{2+} gave values of the ^{14}N hyperfine coupling constant parallel to V=O, ($a_{\|}$), in the range 6.0–7.4 MHz [99]. In the case of imidazole and carnosine as ligands two distinct coupling constants (7.40, 6.64 MHz for imidazole and 7.10, 6.67 MHz for carnosine) were reported which were assigned to the two different nitrogen atoms of the imidazole ring. This type of assignment was applied in the case of the ENDOR spectrum of VO^{2+} added to horse spleen apoferritin, where two similar ^{14}N coupling constants were assigned to the same imidazole ring [100]. In contrast to the observation of these two coupling constants from ENDOR measurements, angle-selected ESEEM spectra give only one coupling constant [100–105]. In some cases this coupling constant must be due to coupling to the nitrogen atom directly coordinated to the vanadium, since the ligands used did not have the possibility of a distant nitrogen [101–104]. The assignment of two similar coupling constants to the different nitrogen atoms in the same imidazole ring was recently questioned by Dikanov et al. [105] on the grounds that the more distant nitrogen atom is two bonds further away and that in similar copper(II)-imidazole complexes the more distant nitrogen has a coupling constant of ~0.5% of that of the coordinated nitrogen. With this in mind, these authors studied the angle-selected ESEEM of $[VO(^{15}N\text{-imidazole})_4]^{2+}$ to demonstrate that the ^{14}N splitting constants in the range 6–7 MHz must belong to the coordinated nitrogen atom and that the remote ^{14}N atom in imidazole would not be observed in X-band ESEEM experiments. This latter study emphasizes the importance of applying more than one EPR technique, along with isotopic substitution wherever possible.

A combination of angle-selected three-pulse ESEEM and four-pulse two-dimensional ESEEM of the crystallographically characterized complexes $[VO(meox)_2]$ and $[VO(mim)_4Cl]^+$ (meox = 2-methylquinolin-8-olato and mim = 1-methylimidazole) defined the orientations of the ^{14}N hyperfine and quadrupole tensors relative to the g tensor [104]. For $[VO(meox)_2]$ the largest hyperfine and quadrupole values are found to coincide with the g-tensor element, which is known [106] to be parallel to the V=O direction in this molecule. In $[VO(mim)_4Cl]^+$ the largest values of the ^{14}N tensors are perpendicular to V=O. The crystal structures [107,108] of these two compounds show that the ring systems which contain the nitrogen atoms are perpendicular and parallel to the V=O direction in $[VO(meox)_2]$ and $[VO(mim)_4Cl]^+$, respectively. Hence, the orientation of the ^{14}N tensors can be correlated with the orientations of the rings.

Proton ENDOR and ESEEM, particularly when coupled to deuteration experiments, are extremely useful for obtaining the possible positions of hydrogen atoms relative to the metal ion. Early ENDOR work on VO^{2+} in frozen aqueous solution [109] demonstrated that the water molecules coordinated cis to the V=O group gave large proton hyperfine coupling constants because of covalency effects. In contrast, the protons from the water molecule trans to V=O have much smaller coupling constants, since they interact via a dipolar coupling mechanism. When the interaction between the unpaired electron and a nuclear spin is purely dipolar in nature, the coupling constants can be used to provide distance information using a point dipole coupling formalism.

The sensitivity of ESEEM to small coupling can be very useful for detecting the mode of coordination and proximity of species. For example, ESEEM spectra [110] on VO^{2+} in the divalent metal sites of rabbit muscle pyruvate kinase have demonstrated that (1) there is binding to lysine; (2) the substrate pyruvate acts as a bidentate ligand (using ^{13}C-labeled pyruvate); and (3) from the observation of weak coupling to ^{23}Na, ^{39}K, and ^{133}Cs, the added alkali metal ions are in close proximity to the vanadium site.

6. ^{51}V NUCLEAR MAGNETIC RESONANCE SPECTROSCOPY

The ^{51}V nucleus possesses excellent magnetic resonance properties [10]. NMR spectroscopy has proved to be a valuable means of detecting diamagnetic vanadium systems, especially in monitoring species formed in the series of equilibria which can occur in aqueous media [4, 111–113]. This has included the use of homonuclear two-dimensional ^{51}V NMR to determine the effect of temperature and ionic strength on chemical exchange of vanadate(V) species [114]. The ^{51}V chemical shift scale (vs. $VOCl_3$) for vanadium(V) systems ranges from +1457 ($[(VS_3)_2\mu\text{-}S]^{4-}$) and +1395 ($[VS_4]^{3-}$) ppm to −746 ($[\{VO(O_2)_2\}_2\mu\text{-}O]^{4-}$) ppm [115]. As a basis for the interpretation of ^{51}V NMR data recorded for biological systems, Rehder et al. [115,116] correlated ^{51}V chemical shifts for four-, five-, and six-coordinate vanadium(V) complexes with the sum of the electronegativities of the ligands.

Isolated and oxidized Amavadin manifests a ^{51}V resonance at −263 ppm, and $[V(hidpa)_2]^-$ and $[V(hida)_2]^-$ (Sec. 2.4.2) display ^{51}V resonances at a very similar value [35,36].

There is a general expectation that ^{51}V resonances for vanadium(V) bound in a low-symmetry environment to a large macromolecule will display very broad resonances due to quadrupolar relaxation. However, this is not always the case. Thus, for tight vanadium(V) binding to transferrin, the relatively sharp (~400 Hz) resonance is a consequence of the motional characteristics of the protein, which place the transferrin-bound vanadium(V) outside the extreme narrowing limit but within the motional narrowing limit [117]. The observed signal intensity is 19% of an equimolar aqueous vanadate(V) solution, since only the central transition ($+\frac{1}{2} \rightarrow -\frac{1}{2}$) is observed with the other three spin-spin relaxation components broadened beyond detectability. The ^{51}V NMR and other spectroscopic data obtained for vanadium(V) bound to transferrin have been interpreted in terms of a VO_2^+ moiety coordinated to the Tyr 426, Tyr 517, and Asp 392 residues of the protein and the V=O oxygens hydrogen bonded to His 585 and Arg 456 [5,117].

Vanadate(V) is an effective inhibitor of ribonucleases (RNases) since it blocks the substrate (RNA) recognition site. RNase from *Aspergillus oryzae* and vanadate(V) form a relatively weak binary complex but a stronger ternary complex is formed when inosine or uridine is added. ^{51}V NMR spectra have been interpreted in terms of vanadate(V) binding across the 2′ and 3′ positions of ribose with formation of a trigonal-bipyramidal geometry at the metal [118,119].

A ^{51}V NMR signal has been reported for the native BPO from *Ascophyllum nodosum*; the resonance is very broad (~5 kHz) and the chemical shift is about −1250 ppm [120,121]. The position of this resonance has no clear chemical parallel and the relationship of this information to that obtained from the vanadium K-edge EXAFS studies [26] (Sec. 2.4.1) needs further examination.

ACKNOWLEDGMENTS

We thank our many colleagues without whose efforts this chapter could not have existed. We are also grateful to Dr. J. M. Charnock for assistance with diagrams for this review. Dr. D. Collison thanks The Royal Society for financial support.

ABBREVIATIONS

Arg	arginine residue
Asp	aspartate residue
BPO	bromoperoxidase
BVSA	bond valence sum analysis
CD	circular dichroism
CT	charge transfer
Cys	cysteine residue
c.w.	continuous wave
DMF	*N,N*-dimethylformamide
ENDOR	electron nuclear double resonance
ESEEM	electron spin echo envelope modulation
EPR	electron paramagnetic resonance
EXAFS	extended X-ray absorption fine structure
FeMoco	iron-molybdenum cofactor
FeVaco	iron-vanadium cofactor
FTIR	Fourier transform infrared
FTATR IR	Fourier transform attenuated total reflection infrared
$hidaH_3$	2-hydroxyiminodiacetic acid
$hidpaH_3$	2-hydroxyiminodipropionic acid
His	histidine residue
IR	infrared
LMCT	ligand-to-metal charge transfer
Me	methyl
meox	2-methylquinolin-8-olato
mim	1-methylimidazole
MLCT	metal-to-ligand charge transfer
MCD	magnetic circular dichroism
NIR	near-infrared
NMR	nuclear magnetic resonance
ORD	optical rotatory dispersion
Ph	phenyl

tmu	*N,N,N′*,N′-tetramethylurea
trencam	catecholamide derivative of triaminotriethylamine
Tyr	tyrosine residue
uro P	uroporphyrin I
UV	ultraviolet
Vis	visible
XANES	X-ray absorption near-edge structure
XAS	X-ray absorption spectroscopy

REFERENCES

1. N. D. Chasteen, *Struct. Bond.*, *53*, 105 (1983).
2. D. Boyde and K. Kustin, *Adv. Inorg. Biochem.*, *6*, 311 (1985).
3. R. Wever and K. Kustin, *Adv. Inorg. Chem.*, *35*, 81 (1990).
4. D. Rehder, *Angew. Chem. Int. Ed. Engl.*, *30*, 148 (1991).
5. A. Butler and C. J. Carrano, *Coord. Chem. Rev.*, *109*, 61 (1991).
6. D. C. Koningsberger and R. Prins (eds.), *X-Ray Absorption*, Wiley, New York, 1988.
7. E. A. V. Ebsworth, D. W. H. Rankin, and S. Cradock, *Structural Methods in Inorganic Chemistry*, Blackwell, Oxford, 1987.
8. F. E. Mabbs and D. Collison, *Electron Paramagnetic Resonance of d Transition Metal Compounds*, Elsevier, Amsterdam, 1992.
9. J. R. Pilbrow, *Transition Ion Electron Paramagnetic Resonance*, Clarendon Press, Oxford, 1990.
10. D. Rehder, in *Multinuclear NMR Spectroscopy* (J. Mason, ed.), Plenum Press, New York, 1987, Chap. 19.
11. B. M. Kincaid and P. Eisenberg, *Phys. Rev. Lett.*, *34*, 1361 (1975).
12. S. P. Cramer and K. O. Hodgson, *Prog. Inorg. Chem.*, *25*, 1 (1979).
13. C. D. Garner, *Adv. Inorg. Chem.*, *36*, 303 (1991).
14. E. M. Armstrong, S. N. Ertok, J. M. Charnock, D. Collison, and C. D. Garner, to be published.
15. P. A. Lee and J. B. Pendry, *Phys. Rev.*, *B11*, 2795 (1975).
16. C. A. Ashley and S. Doniach, *Phys. Rev.*, *B11*, 1279 (1975).
17. S. J. Gurman, N. Binsted, and I. Ross, *J. Phys.*, *C17*, 143 (1986).

18. D. E. Sayers, E. A. Stern, and F. W. Lytle, *Phys. Rev. Lett.*, *27*, 1204 (1971).
19. R. W. Joyner, K. J. Martin, and P. Meehan, *J. Phys.*, *C20*, 4005 (1987).
20. S. J. Gurman, N. Binsted, and I. Ross, *J. Phys.*, *C19*, 1845 (1986).
21. N. J. Blackburn, R. W. Strange, L. M. McFadden, and S. S. Hasnain, *J. Am. Chem. Soc.*, *109*, 7162 (1987).
22. N. Binsted, R. W. Strange, and S. S. Hasnain, *Biochemistry*, *31*, 12117 (1993).
23. A. M. Flank, M. Weininger, L. E. Mortensen, and S. P. Cramer, *J. Am. Chem. Soc.*, *108*, 1049 (1986).
24. G. N. George, R. C. Prince, and S. P. Cramer, *Science*, *243*, 789 (1989).
25. H. Vilter, this volume in Chapter 10 and references therein.
26. J. M. Arber, E. D. Boer, C. D. Garner, S. S. Hasnain, and R.Wever, *Biochemistry*, *28*, 7968 (1989).
27. J. Wong, F. W. Lytle, R. P. Messmer, and D. H. Maylotte, *Physiol. Rev.*, *B30*, 5596 (1984).
28. I. D. Brown and D. Altermatt, *Acta. Crystallogr.*, *B41*, 244 (1985).
29. E. Pidcock, D. Collison, and C. D.Garner, *J. Inorg. Biochem.*, *51*, 173 (1993).
30. C. J. Carrano, M. Mohan, S. M. Holmes, R. de la Rosa, A. Butler, J. M. Charnock, and C. D. Garner, *Inorg. Chem.*, *33*, 646 (1994).
31. E. Bayer, this volume in Chapter 12 and references therein.
32. H. Kneifel and E. Bayer, *J. Am. Chem. Soc.*, *108*, 3075 (1986).
33. M. A. Nawi and T. L. Riechel, *Inorg. Chim. Acta*, *136*, 33 (1987).
34. M. A. A. F. de C. T. Carrondo, M. T. L. S. Duarte, J. C. Pessoa, J. A. L. Silva, J. J. R. Fraústo da Silva, M. C. T. A. Vaz, and L. F. Vilas-Boas, *J. Chem. Soc. Chem. Commun.*, 1158 (1988).
35. E. M. Armstrong, R. L. Beddoes, L. J. Calviou, J. M. Charnock, D. Collison, N. Ertok, J. H. Naismith, and C. D. Garner, *J. Am. Chem. Soc.*, *115*, 807 (1993).
36. E. M. Armstrong, M. S. Austerberry, D. Collison, S. N. Ertok, C. D. Garner, M. Helliwell, and F. E. Mabbs, unpublished results.
37. T. D. Tullius, W. O. Gillum, R. M. K. Carlson, and K. O. Hodgson, *J. Am. Chem. Soc.*, *102*, 5670 (1980).

38. P. Frank, B. Hedman, R. M. K. Carlson, T. A. Tyson, A. L. Roe, and K. O. Hodgson, *Biochemistry*, *26*, 4975 (1987).
39. P. Frank, R. M. K. Carlson, and K. O. Hodgson, *Inorg. Chem.*, *27*, 118 (1988).
40. G. S. Brand, C. J. Hawkins, and D. L. Parry, *Inorg. Chem.*, *26*, 627 (1987).
41. G. R. Hanson, personal communication.
42. S. P. Cramer, K. O. Hodgson, W. O. Gillum, and L. E. Mortenson, *J. Am. Chem. Soc.*, *100*, 3398 (1978).
43. S. P. Cramer, W. O. Gillum, K. O. Hodgson, L. E. Mortenson, E. I. Stiefel, J. R. Chisnell, W. J. Brill, and V. K. Shah, *J. Am. Chem. Soc.*, *100*, 3814 (1978).
44. S. D. Conradson, B. K. Burgess, W. E. Newton, L. E. Mortenson, and K. O. Hodgson, *J. Am. Chem. Soc.*, *109*, 7507 (1987).
45. H. J. Liu, A. Filipponi, N. Gavini, B. K. Burgess, B. Hedman, A. Di Ciccio, C. R. Natoli, and K. O. Hodgson, *J. Am. Chem. Soc.*, *116*, 2418 (1994).
46. J. Kim and D. C. Rees, *Science*, *257*, 1677 (1992).
47. J. Kim and D. C. Rees, *Science*, *260*, 792 (1993).
48. R. R. Eady, this volume, Chapter 11, and references therein.
49. J. M. Arber, B. R. Dobson, R. R. Eady, P. Stevens, S. S. Hasnain, C. D. Garner, and B. E. Smith, *Nature*, *325*, 372 (1987).
50. J. M. Arber, B. R. Dobson, R. R. Eady, S. S. Hasnain, C. D. Garner, T. Matsushita, M. Nomura, and B. E. Smith, *Biochem. J.*, *258*, 733 (1989).
51. G. N. George, C. L. Coyle, B. J. Hales, and S. P. Cramer, *J. Am. Chem. Soc.*, *110*, 4057 (1988).
52. J. A. Kovacs and R. H. Holm, *J. Am. Chem. Soc.*, *108*, 340 (1986).
53. J. M. Arber, A. C. Flood, C. D. Garner, C. A. Gormal, S. S. Hasnain, and B. E. Smith, *Biochem. J.*, *252*, 421 (1988).
54. I. Harvey, J. M. Arber, R. R. Eady, B. E. Smith, C. D. Garner, and S. S. Hasnain, *Biochem. J.*, *266*, 929 (1990).
55. J. Goulon, A. Retournard, P. Friant, C. Goulon-Ginet, C. Berthe, J.-F. Muller, J.-L. Poncet, R. Guilard, J.-C. Escalier, and B. Neff, *J. Chem. Soc., Dalton Trans.*, 1095 (1984).
56. D. H. Maylotte, J. Wong, R. L. St. Peters, F. W. Lytle, and R. B. Greegor, *Science*, *214*, 554 (1981).

57. Y.-P. Zhang and R. H. Holm, *Inorg. Chem.*, *27*, 3875 (1988).
58. A. Müller, J. Schimanski, and H. Bögge, *Z. Anorg. Allg. Chem.*, *544*, 107 (1987).
59. Y. K. Agrawal and G. D. Mehd, *Int. J. Env. Anal. Chem.*, *10*, 183 (1981).
60. C. Djordjevic, B. C. Puryear, N. Vuletic, C. J. Abelt, and S. J. Sheffield, *Inorg. Chem.*, *27*, 2926 (1988).
61. C. K. Jørgensen, *Acta Chem. Scand.*, *11*, 73 (1957).
62. C. J. Ballhausen and H. B. Gray, *Inorg. Chem.*, *1*, 111 (1962).
63. D. Collison, B. Gahan, C. D. Garner, and F. E. Mabbs, *J. Chem. Soc., Dalton Trans.*, *667* (1980).
64. D. Sutton, *Electronic Spectra of Transition Metal Complexes*, McGraw-Hill, London, 1968.
65. C. J. Ballhausen, *Ligand Field Theory*, McGraw-Hill, New York, 1962.
66. A. R. Bulls, C. G. Pippin, F. E. Hahn, and K. N. Raymond, *J. Am. Chem. Soc.*, *112*, 2627 (1990).
67. U. Auerbach, B. S. P. C. Della Vedova, K. Wieghardt, B. Nuber, and J. Weiss, *J. Chem. Soc., Chem. Commun.*, 1004 (1990).
68. K. F. Purcell and J. C. Kotz, *Inorganic Chemistry*, W. B. Saunders, Philadelphia, 1977.
69. D. M. Dooley and J. H. Dawson, *Coord. Chem. Rev.*, *60*, 1 (1984).
70. D. J. Robbins, M. J. Stillman, and A. J. Thomson, *J. Chem. Soc., Dalton Trans.*, 813 (1974).
71. K. Nakamoto, *Infrared and Raman Spectra of Inorganic and Coordination Compounds*, Wiley, New York, 1986.
72. R. G. Alders, L. D. Sparks, M. R. Ondinas, B. A. Crawford, and J. A. Shelnutt, *J. Phys. Chem.*, *94*, 1440 (1990).
73. A. Bencini and D. Gatteschi, *Trans. Met. Chem.*, *8*, 1 (1982).
74. J. A. Weil, T. Buch, and J. E. Clapp, *Adv. Magn. Reson.*, *183* (1973).
75. R. Wilson and D. Kivelson, *J. Chem. Phys.*, *44*, 154 (1966).
76. P. W. Atkins and D. Kivelson, *J. Chem. Phys.*, *44*, 169 (1966).
77. F. A. Walker, R. L. Carlin, and P. H. Rieger, *J. Chem. Phys.*, *45*, 4181 (1966).
78. M.-K. Ahn and D. E. Ormond, *J. Phys. Chem.*, *82*, 1635 (1978).

79. M. M. Iannuzzi, C. P. Kubiak, and P. H. Rieger, *J. Phys. Chem.*, *80*, 541 (1976).
80. F. E. Mabbs and D. Collison, *Electron Paramagnetic Resonance of d Transition Metal Compounds*, Elsevier, Amsterdam, 1992, Chap. 2.
81. F. E. Mabbs, *Chem. Soc. Rev.*, *22*, 313 (1993).
82. F. E. Mabbs and D. Collison, *Electron Paramagnetic Resonance of d Transition Metal Compounds*, Elsevier, Amsterdam, 1992, Chap. 7.
83. F. E. Mabbs and D. Collison, *Electron Paramagnetic Resonance of d Transition Metal Compounds*, Elsevier, Amsterdam, 1992, Chap. 4.
84. D. L. Liczwek, R. L. Belford, J. R. Pilbrow, and J. S. Hyde, *J. Phys. Chem.*, *87*, 2509 (1983).
85. G. L. Wilson, R. J. Greenwood, J. R. Pilbrow, J. T. Spence, and A. G. Wedd, *J. Am. Chem. Soc.*, *113*, 6803 (1991).
86. E. de Boer, Y. van Kooyk, M. G. M. Tromp, H. Plat, and R.Wever, *Biochim. Biophys. Acta*, *869*, 48 (1983).
87. E. de Boer, M. G. M. Tromp, H. Plat, G. E. Krenn, and R. Wever, *Biochim. Biophys. Acta*, *872*, 104 (1986).
88. E. de Boer, K. Boon, and R. Wever, *Biochemistry*, *27*, 1629 (1988).
89. D. Collison, F. E. Mabbs, and J. Temperley, *Spectrochim. Acta*, *47A*, 691 (1991).
90. (a) C. Gemperle and A. Schweiger, *Chem. Rev.*, *91*, 1481 (1991). (b) M. W. Makinen and D. Mustafi, this volume in Chap. 3.
91. G. Rist and J. Hyde, *J. Chem. Phys.*, *49*, 2449 (1968).
92. G. Rist and J. Hyde, *J. Chem. Phys.*, *52*, 4532 (1969).
93. G. Rist and J. Hyde, *J. Chem. Phys.*, *52*, 4633 (1969).
94. G. C. Hurst, T. A. Henderson, and R. W. Kreilick, *J. Am. Chem. Soc.*, *107*, 7294 (1985).
95. A. Schweiger, *Angew. Chem. Int. Ed. Engl.*, *30*, 238 (1991).
96. T. Wacker, G. A. Sierra, and A. Schweiger, *Isr. J. Chem.*, *32*, 305 (1992).
97. E. J. Hustedt, A. Schweiger, and R. R. Ernst, *J. Chem. Phys.*, *96*, 4954 (1992).

98. A. Schweiger, *Appl. Magn. Reson.*, *5*, 229 (1993).
99. C. F. Mulks, B. Kirste, and H. van Willigen, *J. Am. Chem. Soc.*, *104*, 5906 (1982).
100. P. M. Hanna, N. D. Chasteen, G. A. Rottman, and P. Aisen, *Biochemistry*, *30*, 9210 (1991).
101. S. S. Eaton, J. Dubach, K. M. More, G. R. Eaton, G. Thurman, and D. R. Ambruso, *J. Biol. Chem.*, *264*, 4776 (1989).
102. G. J. Gerfen, P. M. Hanna, N. D. Chasteen, and D. J. Singel, *J. Am. Chem. Soc.*, *113*, 9513 (1991).
103. A. V. Astashkin, S. A. Dikanov, and Yu. D. Tsvetkov, *J. Struct. Chem.*, *26*, 363 (1985).
104. E. J. Reijerse, J. Shane, E. de Boer, P. Höfer, and D. Collison, *Proc. 2nd Int. Workshop Electron Magn. Reson. Disord. Syst.*, *253* (1991).
105. S. A. Dikanov, C. Burgard, and J. Hüttermann, *Chem. Phys. Lett.*, *212*, 493 (1993).
106. D. Collison, B. Gahan, and F. E. Mabbs, *J. Chem. Soc., Dalton Trans.*, 111 (1987).
107. M. Shiro and Q. Fernando, *Anal. Chem.*, *43*, 1222 (1971).
108. L. J. Calviou, J. M. Arber, D. Collison, C. D. Garner, and W. Clegg, *J. Chem. Soc., Chem. Commun.*, 654 (1992).
109. H. van Willigen, *J. Magn. Reson.*, *39*, 37 (1980).
110. P. A. Tipton, J. McCraken, J. B. Cornelius, and J. Peisach, *Biochemistry*, *28*, 5720 (1989).
111. L. Petterson, B. Hedman, A.-M. Nenner, and I. Andersson, *Acta Chem. Scand.*, *A39*, 499 (1985).
112. A. T. Harrison and O. W. Howarth, *J. Chem. Soc., Dalton Trans.*, 1173 (1985).
113. A. T. Harrison and O. W. Howarth, *J. Chem. Soc., Dalton Trans.*, 1405 (1986).
114. A. S. Tracey and J. S. Jaswal, *J. Am. Chem. Soc.*, *114*, 3835 (1992).
115. D. Rehder, C. Weidemann, A. Duch, and W. Priebsch, *Inorg. Chem.*, *27*, 584 (1988).
116. W. Priebsch and D. Rehder, *Inorg. Chem.*, *24*, 3058 (1985).
117. A. Butler and H. Eckert, *J. Am. Chem. Soc.*, *111*, 2802 (1989).

118. C. F. G. C. Geraldes and M. M. C. A. Castro, *J. Inorg. Biochem.*, *35*, 79 (1989).

119. A. S. Tracey, M. J. Gresser, and S. Liu, *J. Am. Chem. Soc.*, *110*, 5869 (1988).

120. H. Vilter and D. Rehder, *Inorg. Chim. Acta*, *136*, L7 (1987).

121. D. Rehder, H. Vilter, A. Duch, W. Priebsch, and C. Weideman, *Recl. Trav. Chim. Pays-Bas*, *106*, 408 (1987).

20

Analytical Procedures for the Determination of Vanadium in Biological Materials

Hans G. Seiler

Institut für Anorganische Chemie, Universität Basel, Spitalstrasse 51, CH-4056 Basel, Switzerland

1. INTRODUCTION

Analytical methods for the determination of vanadium at trace levels in biological and environmental materials have been and still are very controversial with respect to their validity and applicability. There are different reasons for this. The ubiquitous occurrence of vanadium in the earth's crust is comparable to that of zinc, but elevated concentrations are found in fossile carburants, above all those of maritime origin like fuel oil and gasoline [1]. By their combustion, airborne microparticulate compounds of vanadium, mostly V_2O_5, are emitted into the environment. These are responsible for high risks of contamination, particularly in urban and industrial regions and in those with high traffic frequencies [2,3]. Another risk for contamination arises from the use of vanadium as a component of different alloys, especially in stainless steel. Besides this, the somewhat complicated chemistry of vanadium and the similarity of its reactions to other elements sometimes renders its separation and determination very difficult. Sometimes enrichments by complexation and extractions are necessitated by the low sensitivity of a detection method.

Most of the recently published methods involve the determination of vanadium in crude oil and petroleum products [4–6], in tap, sea, river, lake, mineral, and rainwater [7–12]. Earlier publications concern the determination of vanadium in vegetables, fruits, as well as various other foods and condiments [13–15]. Several important publications describe the determination of vanadium in body fluids and tissues [16–24].

In the following sections a short review of different selected methods for the determination of vanadium in biological and environmental materials will be given.

2. SPECIMEN COLLECTION, SAMPLING, TRANSPORTATION, AND STORAGE

2.1. Choice of Implements and Chemicals

As in all methods concerned with the determination of trace elements, the materials and chemicals coming into contact with the specimen or the sample derived from it should be selected with regard to possible contaminations, loss, and interfering reactions. Their exact composition, mode of fabrication, and behavior during employment should be well known. But even the best information does not release the analyst from scrupulous testing of all materials and chemicals with respect to the element to be determined. All chemicals, especially those used in larger quantities, must be of the highest purity available. The selection of chemicals should be made not only with respect to their chemical properties but also with regard to the ease and quality of their purification. Constructional materials for the apparatus, vessels, and tools for specimen collection and sample preparation must not be contaminating [25]. Special attention must be given to the use of metallic implements made from steel, like knives, scissors, grinders, and mixers. The determination of traces of vanadium demands that a special attention be paid to possible contaminations by dusts from the air of the laboratory. These analyses should be made in a clean-air room or within a laminar flow workstation fitted with a high-efficiency particulate air filter.

2.2. Specimen Collection and Sampling

Specimen collection and sampling are the most critical steps in every trace analysis in biological materials with respect to obtaining erroneous results. The specimen and the sample derived from it must exactly reflect the properties of the matrix. Therefore the planning of the analysis and the specimen collection should only be undertaken in close interdisciplinary collaboration between a scientist in the field of speciality and the analyst. The specimen collection, its transport and storage, the possibly necessary homogenization and the aliquotation of the sample must be undertaken in such a manner that neither contaminations with the element of interest nor its loss can occur. The specimen of biological material should not be too small, even when the sensitivity of the method of determination would require sample quantities of several μg or μL because the trace elements are not uniformly distributed within the biological material. Therefore specimens of tissues must be very well homogenized before aliquotation of the sample occurs. However, liquid specimens such as body fluids or plant juices may have become inhomogeneous during transport and storage. In this so-called preanalytical phase there is no possibility of recognizing the exactitude of the operations by statistical quality control. Extensive articles on these problems concerning different kinds of biological materials have been published [25–33].

3. SAMPLE PREPARATIONS

Sample preparation comprises every treatment the biological material must undergo between sampling and definitive determination of the element of interest. This is the first step of the analytical phase, i.e., the interference factors of unknown effectiveness can be excluded. Random and systematic errors can be recognized and overcome by means of statistical quality control [34,35]. The purpose of sample preparation is to bring the sample and the element of interest in a physical and chemical form suitable to the method of determination, eliminate disturbing substances, and in many cases enrich the element of interest. Typical kinds of preparation procedures are dilution, solubilization, acid extraction, solvent extraction after complexation, ion exchange, oxidation, mineralization, and the elimination of interfering substances by extracting, filtration, and centrifugation.

The procedure of choice depends on the kind of the biological matrix, the chemical behavior and concentration of the element to be determined, and the sensitivity and specificity of the determination method. All preparation procedures have a certain potentiality for contaminations and losses. These aspects must determine the selection of a preparation procedure.

3.1. Mineralization

Different methods of determination of vanadium demand complete mineralization of a biological material, above all those materials which require the complete elimination of disturbing substances, a defined oxidation state of the element, and a homogeneous solution. Generally, there are two mineralization methods—dry ashing or incineration, and wet mineralization.

With respect to possible contaminations, dry ashing is a preferable method. It is easy to perform, allows great series at the same time, and the only needed reagent is O_2. The main oxidation product of V, mostly V_2O_5, is not volatile [22]. In order to prevent the formation of melts, the temperature should not exceed 500°C. The preferred material for recipients is pure SiO_2. Mostly the resulting ash can be easily dissolved in mineral acids. At insufficient oxygen concentration during incineration of biological material, different lower oxidation states of V can be formed which may give sparingly soluble products with other constituents of the matrix, e.g., vanadium carbide (VC), which are difficult to dissolve. For the determinations of V by spectrophotometric and voltammetric methods, as well as for different procedures for enrichment and separation such as complexation and ion exchange, the oxidation state V(+V) is indispensable.

Wet mineralization is the most applied preparation method for biological material. Overviews on the different methods, implements, and reagents have been published [25,36]. Sometimes this method is used only to bring the biological material into homogeneous solution without complete elimination of the organic constituents, as for subsequent determinations by atomic absorption spectrometry (AAS), inductively coupled plasma atomic emission spectrometry (ICPAES), inductively coupled plasma mass spectrometry (ICPMS) and for separations and enrichment procedures by complexation and extraction.

Contaminations in wet mineralization are mainly due to the reagents and to the materials of vessels and apparatus, but also to dusts in the air at the laboratory. The preferred construction material for vessels is pure SiO_2, but glasses with a high SiO_2 content, such as Pyrex or Vycore, and high-temperature-resistant polytetrafluoroethylene (PTFE) can be used. In the case of V, losses by volatile compounds and adsorptions to the walls of the vessels are not to be expected in strong acid milieu.

In wet mineralization three groups of reagents are generally used:

1. Oxidants: HNO_3, H_2O_2, $HClO_4$, etc.
2. Solubilizers and neutralizers: H_2SO_4, HCl, NH_3
3. Catalysts: H_2SO_4, Fe^{2+}, Ag^+

The consumption of reagents is as follows: oxidants > solubilizers > catalysts.

In the choice of the reagents the following factors should be considered:

1. Their purity and consumption. The blank values of the reagents establish the determination limit of the analytical method.
2. The possible formation of interfering substances with constituents of the matrix, e.g., materials with high content of Ca should not be mineralized in the presence of H_2SO_4 because vanadium species may be adsorbed to the precipitated sparingly soluble $CaSO_4$. H_2SO_4 should also be avoided for samples to be measured by graphite furnace atomic absorption spectrometry (GFAAS), because in the atomization step residual H_2SO_4 may produce strong fumes which disturb the measurement.
3. The ease with which excess reagent can be removed.
4. If the subsequent determination method or the separation or enrichment procedures demand the +V oxidation state, it should be kept in mind that in acid milieu VO_2^+ is a stronger oxidant ($E^0 = +1.00$ V) than HNO_3 ($E^0 = +0.94$ V). Therefore the addition of H_2O_2 to the oxidant mixture is to be advised.

There are three different systems generally used for wet mineralization:

1. The open system without controlled reflux
2. The open system with forced reflux
3. The closed system with elevated temperature and pressure

In the open system without controlled reflux mostly long-necked Kjeldahl flasks of borosilicate glass are used. The volatile oxidation products of the matrix can be distilled off, and in this way the reaction equilibrium favors the products. The maximum applicable temperature is limited by the lowest boiling point of one of the constituents of the oxidant mixture. The consumption of the reagents is comparatively high and because of the limited temperature a long time is needed for complete mineralization.

The open system with forced reflux consists of a round-bottom flask with attached efficient air cooler. The equilibrium of the reaction also tends to the reaction products, but the maximum applicable temperature is limited by the highest boiling point of one of the components of the reagent mixture; the higher applicable temperature considerably shortens the duration of the mineralization. By forced reflux the consumption of reagents is decreased by more than one order of magnitude compared with the open system without forced reflux.

The closed system consists of a pressure container (stainless steel or PTFE) in which a sealed vessel of an acid-resistant, noncontaminating material, e.g., SiO_2, is inserted. The applicable temperature is considerably elevated compared with the open systems and limited by the pressure resistance of the container. These systems are fitted with security devices to prevent bursting. The volatile reaction products cannot escape, i.e., the oxidation of the organic constituents of the matrix is never complete.

In recent years a modification has become usual. Instead of heating the pressure vessels in an oven, the heat is transferred by microwaves. Thus, not the container but the solution is heated. The heat transfer is much faster and therefore the reaction time is considerably shortened. The pressure vessels must be made from a material that does not absorb microwaves. As metallic materials cannot be used, special polymers (mostly on the basis of PTFE) have been developed.

Liquid specimens which do not contain high concentrations of organic materials like juice of plants or acid extractions of solids can efficiently be mineralized by short-wave UV light. The liquid specimen in an SiO_2 tube is acidified to pH 0.5—preferably with $HClO_4$—and after addition of a suitable quantity of H_2O_2, the mixture is exposed to an intensive UV source. The temperature should be controlled and not exceed 90°C. The reaction time depends on the kind and quantity of organic material (minutes to hours).

3.2. Separation and Enrichment Methods

Even after mineralization, different methods of determination necessitate the separation of the vanadium species from other interfering ions or an enrichment step due to a lack of sensitivity on the part of the methods. These additional procedures can be performed by complexation at controlled pH with subsequent extraction into an adequate with water immiscible solvent or by ion exchange.

Such complexing agents for VO_2^+ are, for example, cupferron (*N*-nitrosophenylhydroxylamine, ordinarily used as its ammonium salt), 8-hydroxyquinoline (oxine), and diethyldithiocarbamates. These complexes can be extracted into organic solvents like CCl_4 and $CHCl_3$ with very good partition coefficients. But most of the ligands are not specific to vanadium; many other metal ions are also extracted. To improve the separation often a back-extraction into an aqueous phase at a different pH, where the vanadium complex is not stable, is necessary. These procedures are generally very time consuming, cumbersome, and give raise to losses and contaminations.

4. METHODS OF DETERMINATION

Comprehensive overviews on the different methods of determinations have recently been published: spectrophotometry [37], AAS [38], voltammetry [39], ICPAES and ICPMS [40], and neutron activation analysis (NAA) and γ-spectrometry [41].

4.1. Spectrophotometry

In the past colorimetric methods have been the only ones to determine V at lower concentrations. All these methods required an extensive sample preparation with separation and enrichment steps. Generally they can only be applied to samples with concentrations of V up from the ppm range. Among the different ligands which have been used, only some will be mentioned here. *N*-*m*-Tolylbenzohydroxamic acid in $CHCl_3$ for the determination of V in fruits, vegetables, tobacco, and rice [13],

N-cinnamoyl-*N*-(2,3-xylyl)hydroxylamine for the analysis of V in foods [14], and *N*-phenylbenzohydroxamic acid in *o*-dichlorbenzene for the extraction and determination of V(+V) in cauliflower, spinach, algae, and fungi [42].

Besides the direct colorimetric determination by measurement of the extinction of the V complexes, the catalytic effect of V on oxidation reactions of different substances has been used for its determination. Gallic acid is oxidized by persulfate in acidic medium to a yellow-red color. In the presence of vanadium this normally very slow reaction is accelerated and proportional to the concentration of V. This method has been applied to the determination of V in urine after wet mineralization with HNO_3 and H_2O_2, extraction of V with 8-hydroxyquinoline in $CHCl_3$, and back-extraction into a basic water phase prior to the addition of gallic acid and persulfate. The measurement at 415 nm must be made after an exactly defined time. The detection limit is 1 μg V/L [16]. Another kinetic-spectrophotometric determination of V is based on the V(+V)-catalyzed oxidation of 3,5-diaminobenzoic acid dihydrochloride by $KBrO_3$. This method allows the determination of V by measuring the rate of change in absorbance with time at 480 nm. It is applicable in the range of 3–50 ng/mL [7].

By their lack of sensitivity, the laborious sample preparation, and the great risk of obtaining erroneous results, spectrophotometric methods have limited suitability for the determination of V in biological materials. They cannot be used for the determination of V in human body fluids of the general population.

4.2. Atomic Absorption Spectrometry

AAS is the most applied method for the determination of metals in biological materials. AAS methods can be automated thus enabling the analyses of greater series. Among the different techniques of AAS, only flame atomic absorption spectrometry (FAAS) and graphite furnace atomic absorption spectrometry (GFAAS) can be used for the determination of vanadium. Normally the sample must be in a homogeneous solution, i.e., solid samples and inhomogeneous liquids must be prepared by dry or wet mineralization, but in most cases a complete mineralization is not necessary and a good solubilization is sufficient.

FAAS is applicable only to samples which contain V at least in the ppm range. As V forms refractory V_2O_5 the temperature of an air-acetylene flame would be insufficient to dissociate this compound into atoms. This can be achieved by the nitrous acetylene flame, which is hotter and has a much higher reducing power. The addition of an ionization buffer to the sample and standards, e.g., KCl, is recommended. V is measured at the 318.5-nm line. If the analysis is made by a calibration curve, the solution of the sample and of the calibration standards must have the same viscosity in order to achieve the aspiration of the same volume in the time.

GFAAS is more sensitive than FAAS by about three orders of magnitude and enables the determination of V in the low ppb range. Because of the high temperature needed for the dissociation of vanadium compounds, pyrolytic graphite tubes should be used. Because the V content in most biological material is rather low, a Zeeman effect background corrector is highly recommended. As compounds of vanadium are not very volatile, ashing of the sample within the graphite tube may occur at temperatures up to 1500°C whereas atomization happens at 2700°C. There are no problems from the different viscosities of the solutions because the small volumes of sample and standard (10–50 μL) are directly injected into the tube. Mostly the calibration occurs by the standard addition method. In spite of the high sensitivity of GFAAS, the very low concentrations of V in serum and urine of the general population cannot be directly determined by this method. This could be achieved after wet mineralization of the body fluids and a subsequent enrichment of V by an extraction with 8-hydroxyquinoline [43].

4.3. Inductively Coupled Plasma Atomic Emission Spectrometry and Inductively Coupled Plasma Mass Spectrometry

ICPAES and ICPMS have the great advantage of permitting the determination of different elements sequentially or simultaneously in the same aliquot of the sample. The Ar plasma normally used by commercial ICP instruments has a gas kinetic temperature of 6000–10 000 K

[44]. Samples introduced into the plasma are rapidly desolvated, vaporized, atomized, and ionized. Although samples can be introduced into the ICP as solids, liquids, or gases, most samples of biological interest are first put into solution. For the determination of V the solid and the liquid samples are mostly digested with mixtures of HNO_3 and H_2O_2. As in FAAS, the transport of the sample solution into the plasma is governed by two critical factors: the analyte mass transport rate and the solvent loading of the plasma. Therefore solutions of the samples and of the standards should have the same characteristics. Special attention must be given to the absence of particles, and the total concentration of salts should not be too high because the capillaries of the nebulizers are very fine and may be clogged. Most analyte atoms are singly ionized in the ICP.

In the ICPAES mode the emission lines of the neutral (M^0) and the singly ionized metal ion (M^+) are seen. The main requirement for this method is a spectrometer with a very high resolution of spectral lines and effective rejection of stray light from other intense emission lines. As the detection limit for V is in the range of 1 μg V/L, the limit of determination is 10 times less sensitive compared with GFAAS. Therefore, the determination of V in the low ppb range needs an enrichment step. For example, by adsorption of the thiocyanate complex on a strongly basic anion exchange resin and elution with a mixture of 2 M $HClO_4$/1 M HCl, subsequent evaporation of the eluates to dryness and dissolution of the residues in 1.2 M HCl [9]. ICPAES has been used for the determination of V in bone [45] and different rat organs [46]. It is not sensitive enough to determine V in serum and urine of the general population.

In the ICPMS mode, the ionized species from the ICP are introduced via a specially designed interface into the vacuum chamber of the mass spectrometer, mostly a quadrupole filter, and the mass-to-charge ratio (*m/z*) is measured. Thus only ionized species are recognized. The sensitivity is comparable to or better than that of GFAAS. The major advantage is that different elements in a single sample aliquot can simultaneously be determined. The main interferences in ICPMS are mass overlaps resulting from coincident isotopic masses of different elements, oxides, and other polyatomic species. The main interference in the determination of V is the presence of $^{35}Cl^{16}O^+$ in the plasma, which has the same *m/z* ratio as $^{51}V^{+}$. This problem has been overcome for the determination of V in urine by using cryogenic desolvation. The urine was diluted with high-purity water and acidified with HNO_3. By the cryogenic desolvation of the nebulized water and the HCl the corresponding densities of oxygen and chloride in the ICP were reduced [24]. The calibration is preferably performed by the standard addition method.

4.4. Neutron Activation Analysis

NAA is a very useful method for the determination of various elements in biological materials. It is unique because it is based on the reaction of neutrons with the atomic nucleus of a particular isotope of the element. This reaction takes place without respect to the physical and chemical state of the element. Each element is characterized by its isotopic composition and its corresponding ability to form neutron-induced radioactive indicators with individual half-lives and decay characteristics. In most cases the detection and determination of an element is performed by measuring the energy and the counting rates of the specific γ emissions by means of an efficient detector with high spectral resolution, e.g., a well-type Ge detector combined with a multichannel analyzer. Vanadium can be measured by the neutron-induced isotope ^{52}V formed by n,γ reaction from ^{51}V (natural abundance 99.76%, n,γ cross-section: 5.1 barns). ^{52}V has a half-life of 3.75 min and a γ emission of 1.433 MeV. Due to the short half-life already after a 10-min irradiation with neutrons more than 80% of the maximum attainable activity for a given neutron flux is reached. There are two kinds of NAA: the nondestructive instrumental kind (INAA) and the NAA with chemical treatment of the sample (RNAA resp. CNAA).

In INAA the irradiated sample is measured without any chemical treatment; thus all risks of contamination by the sample treatment are excluded. This method can be used if there are no spectral interferences from the matrix. In most analyses of biological materials the main interferences arise from K, Na, and Cl, which are also activated to the nuclides ^{42}K, ^{24}Na, and ^{38}Cl. These have longer half-lives than ^{52}V, are mostly present in relatively high concentrations, and emit γ energies disturbing the spectrum of ^{52}V. Therefore INAA can only be used for the determination of V if the concentration is above 15 μg V/kg. At lower concentrations of V, the disturbing elements must be eliminated.

In the CNAA the sample is digested by wet mineralization and the disturbing elements are eliminated by ion exchange chromatography prior to the irradiation with neutrons [17,21,47]. In this mode of sample preparation there is a certain risk for contamination, but the irradiated sample can be measured a short time after irradiation.

In the RNAA the wet mineralization and the chemical separation take place after activation. The separation of ^{52}V from disturbing nuclides is mostly performed by the addition of inactive V, complexation with a suitable ligand, e.g., cupferron, *N*-benzoylphenylhydroxylamine, or oxine, and extraction of the complexes into an organic solvent before the measurement [17,21,22,48,49]. This method has the great advantage that it is not prone to contamination. The disadvantage is that samples of rather high radioactivity must be handled and with chemical preparation some time elapses between the end of irradiation and the measurement.

The great merit of NAA is that by this method erroneous results of earlier methods of determination of V in biological material, above all in body fluids, can be detected. This method allows the determination of V in the pg V/g level. Unfortunately, this method requires the availability of a nuclear reactor fitted with a high-speed transport system for the irradiated samples. Only a few specialized laboratories are equipped to perform this kind of analysis.

4.5. Voltammetric Methods

The voltammetric determinations of vanadium in biological and environmental materials at the trace level demand a complete elimination of the organic constituents of the matrix, mostly by wet mineralization. By this sample treatment not only can interfering substances be eliminated but the needed highest oxidation state of V(+V) is guaranteed. All reagents used for mineralization and voltammetric determination must be of the highest purity.

The determination of vanadium by means of the direct differential pulse polarography (DPP) at the dropping mercury electrode is by its relatively low sensitivity not applicable to biological materials. Therefore a method with preconcentration of V at a stationary mercury electrode, above all the hanging mercury drop electrode (HMDE), must be used. Because V forms neither amalgams nor sparingly soluble compounds with mercury, it cannot be accumulated by a simple electrolytic process at the electrode as it is done in anodic or cathodic stripping voltammetry (ASV resp. CSV). The accumulation of V can be reached by adsorption of a suitable complex of V to the surface of the working electrode during a well-defined time and efficient stirring of the solution at a voltage where no faradaic reaction takes place. After a short time without stirring the voltage of the working electrode is continuously changed to more negative values and the current in function of the voltage is enregistered, mostly in the differential pulse mode. At a specific voltage the V in the complex is reduced. This faradaic current is peak-shaped and the peak height is proportional to the concentration of V in the solution. This method is called adsorptive stripping voltammetry (ADSV). Different ligands for the adsorption of V have been investigated for their suitability, e.g., solochrome violet RS [8] and cupferron [50]. These methods allow the determination of V down to the ppt range in water.

Recently, an ADSV method for the determination of V in urine was developed and tested for reliability and accuracy by different laboratories. The urine was wet-mineralized with HNO_3 and a small quantity of H_2SO_4 in the open system with forced reflux until total discoloration of the remaining solution occurred; then H_2O_2 (30%) was added and evaporated to near dryness. The residue was dissolved in high-purity water, neutralized with conc. NH_3, and again evaporated to dryness. After dissolution of the solid residue with high-purity water, ammoniacal nitrilotriacetate, NH_4Cl/NH_3-buffer (pH 8.5), and a freshly prepared solution of cupferron were added. This mixture was heated to 50°C for 5 min. After cooling the mixture was transferred to the voltammetric cell. The accumulation of the complex to the working electrode occurred at −0.4 V over 10 sec with stirring and 10 sec without stirring. After that the voltage of the work electrode was changed to more negative values with a voltage sweep rate of −10 mV/sec. The observed height of the current peak at −0.75 V is representative of the concentration of V in the solution. The voltages are measured against an AgCl/Ag 3 M KCl reference electrode. The calibration is made by the standard addition method. The sensitivity of the method is 0.1 ng V/mL urine [51].

ABBREVIATIONS

AAS	atomic absorption spectrometry
ADSV	adsorptive stripping voltammetry
ASV	anodic stripping voltammetry
CNAA	chemical neutron activation analysis
CSV	cathodic stripping voltammetry
DPP	differential pulse polarography
FAAS	flame atomic absorption spectrometry
GFAAS	graphite furnace atomic absorption spectrometry
HMDE	hanging mercury drop electrode
ICPAES	inductively coupled plasma atomic emission spectrometry
ICPMS	inductively coupled plasma mass spectrometry
INAA	instrumental neutron activation analysis
NAA	neutron activation analysis
PTFE	polytetrafluoroethylene
RNAA	radiochemical neutron activation analysis

REFERENCES

1. H. T. Evans, Jr. and S. Landergren, in *Handbook of Geochemistry* (K. H. Wedepohl, ed.), Springer-Verlag, New York, 1974.
2. R. Wennig and N. Kirsch, In *Handbook on Toxicity of Inorganic Compounds* (H. G. Seiler, H. Sigel, and A. Sigel, eds.), Marcel Dekker, New York, 1988, p. 749.
3. B. Sansoni and G. V. Iyengar, in Internat. Atomic Energy Agency, Vienna, Technical Report, 197 (*Element. Anal. Biol. Materials*), 57 (1980).
4. R. A. Reimer and A. Miyazaki, *Anal. Sci.*, *9*, 157 (1993).
5. C. C. Gamez, P. V. Gonzales, M. J. Arrabal, V. L. Moreno, and A. S. Misiego, *Fresenius J. Anal. Chem.*, *340*, 169 (1991).
6. T. Aman, A. Rashid, N. Kausar, and N. A. Ginnai, *Microchem. J.*, *47*, 263 (1993).
7. A. Ch. Zotou and C. G. Papadopoulos, *Analyst*, *115*, 323 (1990).
8. P. A. M. Farias and I. Takase, *Electroanalysis*, *4*, 823 (1992).
9. I. Steffan and G. Vujicic, *Microchim. Acta*, *110*, 89 (1993).

10. J. Miura, H. Hoshino, and T. Yotsuyanagi, *Anal. Chim. Acta, 233*, 121 (1990).
11. H. Ohashi, N. Uehara, and Y. Shijo, *J. Chromatogr., 539*, 225 (1991).
12. K. Hirayama, S. Kageyama, and N. Unohara, *Analyst, 117*, 13 (1992).
13. Y. K. Agrawal and K. P. S. Raj, *Microchem. J., 25*, 219 (1980).
14. S. Tsukada and R. Demura, *Eisei Kagaku, 30*, 79 (1984).
15. P. Berjemo-Barrera, A. Berjemo-Barrera, and F. Berjemo-Martinez, *Ann. Bromatol., 33*, 17 (1981).
16. K. H. Schaller and A. Eben, in *Analysen in biologischem Material*, Vol. 2 (D. Henschler, ed.), Verlag Chemie, Weinheim, 1978, p. D1.
17. A. J. Blotcky, W. C. Duckworth, F. G. Hamel, and E. P. Rack, in *Handbook on Metals in Clinical and Analytical Chemistry* (H. G. Seiler, A. Sigel, and H. Sigel, eds.), Marcel Dekker, New York, 1994, p. 651.
18. J. Versieck, in *Critical Reviews in Clinical Laboratory Science*, Vol. 22, Part 2, CRC Press, Boca Raton, 1985, p. 97.
19. A. J. Blotcky, C. Falcone, V. A. Medina, and E. P. Rack, *Anal. Chem., 51*, 178 (1979).
20. A. J. Blotcky, F. G. Hamel, A. Stranik, A. Ebrahim, R. B. Sharma, E. P. Rack, and S. S. Solomon, *J. Radioanal. Nucl. Chem., 131*, 319 (1989).
21. A. J. Blotcky, W. C. Duckworth, A. Ebrahim, F. K. Hamel, E. P. Rack, and R. B. Sharma, *J. Radioanal. Nucl. Chem., 134*, 151 (1989).
22. A. R. Byrne and J. Kučera, *Fresenius J. Anal. Chem., 340*, 48 (1991).
23. B. Patel, G. E. Henderson, S. J. Haswell, and R. Grzeskowiak, *Analyst,, 115*, 1063 (1990).
24. L. C. Alves, L. A. Allen, and R. S. Houk, *Anal. Chem., 65*, 2468 (1993).
25. H. G. Seiler, in *Metal Ions in Biological Systems*, Vol. 20 (H. Sigel and A. Sigel, eds.), Marcel Dekker, New York, 1986, p. 305.
26. J. Versieck, F. Barbier, R. Cornelis, and J. Hoste, *Talanta, 29*, 973 (1982).
27. J. Versieck and R. Cornelis, *Anal. Chim. Acta, 116*, 217 (1980).

28. J. Versieck, *Trends Anal. Chem.*, *2*, 110 (1983).
29. J. Angerer, K. H. Schaller, and H. Seiler, *Trends Anal. Chem.*, *2*, 257 (1983).
30. D. Behne, *J. Clin. Chem. Clin. Biochem.*, *19*, 115 (1981).
31. K. Heydorn, E. Damsgard, N. A. Larsen, and B. Nielsen, in *Nuclear Activation Techniques in the Life Sciences*, International Atomic Energy Agency, Vienna, 1978, p. 129.
32. J. Versieck and L. Vanballenberghe, in *Handbook on Metals in Clinical and Analytical Chemistry* (H. G. Seiler, A. Sigel, and H. Sigel, eds.), Marcel Dekker, New York, 1994, p. 31.
33. H. G. Seiler, in *Handbook on Toxicity of Inorganic Compounds* (H. G. Seiler, H. Sigel, and A. Sigel, eds.), Marcel Dekker, New York, 1988, p. 39.
34. K. H. Schaller, J. Angerer, G. Lehnert, H. Valentin, and D. Weltle, *Arbeitsmed. Socialmed. Praeventivmed.*, *19*, 79 (1984).
35. J. M. Christensen, O. M. Poulsen, and T. Anglov, in *Handbook in Metals on Clinical and Analytical Chemistry* (H. G. Seiler, A. Sigel, and H. Sigel, eds.), Marcel Dekker, New York, 1994, p. 45.
36. H. Seiler and K. H. Schaller, in *Analyses of Hazardous Substances in Biological Materials*, Vol. 2 (J. Angerer and K. H. Schaller, eds.), VCH, Weinheim, 1988, p. 1.
37. G. Gauglitz, in *Handbook on Metals in Clinical and Analytical Chemistry* (H. G. Seiler, A. Sigel, and H. Sigel, eds.), Marcel Dekker, New York, 1994, p. 63.
38. B. Welz, in *Handbook on Metals in Clinical and Analytical Chemistry* (H. G. Seiler, A. Sigel, and H. Sigel, eds.), Marcel Dekker, New York, 1994, p. 85.
39. H. G. Seiler, in *Handbook on Metals in Clinical and Analytical Chemistry* (H. G. Seiler, A. Sigel, and H. Sigel, eds.), Marcel Dekker, New York, 1994, p. 123.
40. D. M. Templeton, in *Handbook on Metals in Clinical and Analytical Chemistry* (H. G. Seiler, A. Sigel, and H. Sigel, eds.), Marcel Dekker, New York, 1994, p. 167.
41. K. Heydorn, in *Handbook on Metals in Clinical and Analytical Chemistry* (H. G. Seiler, A. Sigel, and H. Sigel, eds.), Marcel Dekker, New York, 1994, p. 181.
42. A. K. Shrivastava and S. G. Tandon, *Ind. J. Environ. Health*, *24*, 39 (1982).

43. O. Ishida, K. Kihira, Y. Tsukamoto, and F. Marumo, *Clin. Chem.*, *35*, 127 (1989).

44. K. A. Marshall and G. M. Hieftje, *J. Anal. Atom. Spectrosc.*, *2*, 567 (1987).

45. H. S. Mahanti and R. M. Barnes, *Anal. Chim. Acta*, *151*, 409 (1983).

46. S. S. Q. Hee and J. R. Boyle, *Anal. Chem.*, *60*, 1033 (1988).

47. R. R. Greenberg, H. M. Kingston, R. Zeisler, and J. Woittiez, in *Nuclear Analytical Methods in the Life Sciences* (R. Zeisler and V. P. Guinn, eds.), Humana Press, Clifton, NJ, 1990, p. 17.

48. K. Heydorn, in *Nuclear Analytical Methods in the Life Sciences* (R. Zeisler and V. P. Guinn, eds.), Humana Press, Clifton, NJ, 1990, p. 541.

49. A. R. Byrne and J. Versieck, in *Nuclear Analytical Methods in the Life Sciences* (R. Zeisler and V. P Guinn, eds.), Humana Press, Clifton, NJ, 1990, p. 529.

50. J. Wang, B. Tian, and J. Lu, *Talanta*, *39*, 1273 (1992).

51. H. G. Seiler, in *Analyses of Hazardous Substances in Biological Materials*, Vol. 4 (J. Angerer and K. H. Schaller, eds.), VCH, Weinheim, 1994, p. 241.

Author Index

Numbers in parentheses are reference numbers and indicate that an author's work is referred to although his name may not be cited in the text. Underlined numbers give the page on which the complete reference is listed.

A

C

D

E

G

H

I

J

K

L

M

N

P

S

T

W

X

Y

Z

Subject Index

A

H

N

Q

R

S

U

V

W

X

Y

Z